Artificial Intelligence in Industrial Decision Making, Control and Automation

W0245828

International Series on
MICROPROCESSOR-BASED AND INTELLIGENT SYSTEMS ENGINEERING

VOLUME 14

Editor

Professor S. G. Tzafestas, *National Technical University, Athens, Greece*

Editorial Advisory Board

Professor C. S. Chen, *University of Akron, Ohio, U.S.A.*
Professor T. Fokuda, *Nagoya University, Japan*
Professor F. Harashima, *University of Tokyo, Tokyo, Japan*
Professor G. Schmidt, *Technical University of Munich, Germany*
Professor N. K. Sinha, *McMaster University, Hamilton, Ontario, Canada*
Professor D. Tabak, *George Mason University, Fairfax, Virginia, U.S.A.*
Professor K. Valavanis, *University of Southern Louisiana, Lafayette, U.S.A.*

Artificial Intelligence in Industrial Decision Making, Control and Automation

edited by

SPYROS G. TZAFESTAS
Department of Electrical and Computer Engineering,
National Technical University of Athens,
Athens, Greece

and

HENK B. VERBRUGGEN
Department of Electrical Engineering,
Delft University of Technology,
Delft, The Netherlands

SPRINGER SCIENCE+BUSINESS MEDIA, B.V.

Library of Congress Cataloging-in-Publication Data

Artificial intelligence in industrial decision making, control, and
   automation / edited by Spyros G. Tzafestas and Henk B. Verbruggen.
       p.   cm. -- (International series on microprocessor-based and
   intelligent systems engineering ; v. 14)
     Includes index.
     ISBN 978-94-010-4134-8   ISBN 978-94-011-0305-3 (eBook)
     DOI 10.1007/978-94-011-0305-3
     1. Decision support systems. 2. Intelligent control systems.
   3. Automation. 4. Artificial intelligence.   I. Tzafestas, S. G.,
   1939-   . II. Verbruggen, H. B. III. Series.
   T58.62.A78 1995
   658.4'03--dc20   94-46547

ISBN 978-94-010-4134-8

Printed on acid-free paper

All Rights Reserved
© 1995 Springer Science+Business Media Dordrecht
Originally published by Kluwer Academic Publishers in 1995
Softcover reprint of the hardcover 1st edition 1995
No part of the material protected by this copyright notice may be reproduced or
utilized in any form or by any means, electronic or mechanical,
including photocopying, recording or by any information storage and
retrieval system, without written permission from the copyright owner.

CONTENTS

PART 1
GENERAL ISSUES

CHAPTER 1
ARTIFICIAL INTELLIGENCE IN INDUSTRIAL DECISION MAKING, CONTROL AND AUTOMATION: AN INTRODUCTION
S. Tzafestas and H. Verbruggen

CHAPTER 2
CONCEPTUAL INTEGRATION OF QUALITATIVE AND QUANTITATIVE PROCESS MODELS
E. A. Woods

CHAPTER 3
TIMING PROBLEMS AND THEIR HANDLING AT SYSTEM INTEGRATION
L. Motus

CHAPTER 4
ANALYSIS FOR CORRECT REASONING IN INTERACTIVE MAN ROBOT SYSTEMS: DISJUNCTIVE SYLLOGISM WITH MODUS PONENS AND MODUS TOLLENS
E. C. Koenig

PART 2
INTELLIGENT SYSTEMS

CHAPTER 5
APPLIED INTELLIGENT CONTROL SYSTEMS
R. Shoureshi, M. Wheeler and L. Brackney

CHAPTER 6
INTELLIGENT SIMULATION IN DESIGNING COMPLEX DYNAMIC
CONTROL SYSTEMS
F. Zhao

CHAPTER 7
MULTIRESOLUTIONAL ARCHITECTURES FOR AUTONOMOUS SYSTEMS WITH INCOMPLETE AND INADEQUATE KNOWLEDGE REPRESENTATION

A. Meystel

CHAPTER 8
DISTRIBUTED INTELLIGENT SYSTEMS IN
CELLULAR ROBOTICS
T. Fukuda, T. Ueyama and K. Sekiyama

CHAPTER 9
DISTRIBUTED ARTIFICIAL INTELLIGENCE IN
MANUFACTURING CONTROL
S. Albayrak and H. Krallmann

PART 3
NEURAL NETWORKS IN MODELLING, CONTROL AND SCHEDULING

CHAPTER 10
ARTIFICIAL NEURAL NETWORKS FOR MODELLING
A.J. Krijgsman, H.B. Verbruggen and P.M. Bruijn

CHAPTER 11
NEURAL NETWORKS IN ROBOT CONTROL
S.G. Tzafestas

CHAPTER 12
CONTROL STRATEGY OF ROBOTIC MANIPULATOR BASED ON FLEXIBLE NEURAL NETWORK STRUCTURE
M. Teshnehlab and K. Watanabe

CHAPTER 13
NEURO - FUZZY APPROACHES TO ANTICIPATORY CONTROL
L.H. Tsoukalas, A. Ikonomopoulos and R.E. Uhrig

CHAPTER 14
NEW APPROACHES TO LARGE - SCALE SCHEDULING PROBLEMS:
CONSTRAINT DIRECTED PROGRAMMING AND NEURAL
NETWORKS
Y. Kobayashi and H. Nonaka

PART 4
SYSTEM DIAGNOSTICS

CHAPTER 15
KNOWLEDGE - BASED FAULT DIAGNOSIS OF TECHNOLOGICAL
SYSTEMS
H. Verbruggen, S. Tzafestas and E. Zanni

CHAPTER 16

MODEL - BASED DIAGNOSIS: STATE TRANSITION EVENTS AND CONSTRAINT EQUATIONS

K.-E. Arzen, A. Wallen and T.F. Petti

CHAPTER 17

DIAGNOSIS WITH EXPLICIT MODELS OF GOALS AND FUNCTIONS

J.E. Larsson

PART 5
INDUSTRIAL ROBOTIC, MANUFACTURING AND ORGANIZATIONAL SYSTEMS

CHAPTER 18
MULTI-SENSOR INTEGRATION FOR MOBILE ROBOT NAVIGATION
A.Traca de Almeida, H. Araujo, J. Dias and U. Nunes

CHAPTER 19
**INCREMENTAL DESIGN OF A FLEXIBLE ROBOTIC ASSEMBLY CELL
USING REACTIVE ROBOTS**
E.S. Tzafestas and S.G. Tzafestas

CHAPTER 20
**ON THE COMPARISON OF AI AND DAI BASED PLANNING
TECHNIQUES FOR AUTOMATED MANUFACTURING SYSTEMS**
A.I. Kokkinaki and K.P. Valavanis

CHAPTER 21
KNOWLEDGE-BASED SUPERVISION OF FLEXIBLE MANUFACTURING SYSTEMS
A. K. A. Toguyeni, E. Craye and J.-C. Gentina

CHAPTER 22
A SURVEY OF KNOWLEDGE-BASED INDUSTRIAL SCHEDULING
K. S. Hindi and M. G. Singh

CHAPTER 23
REACTIVE BATCH SCHEDULING
V. J. Terpstra and H. B. Verbruggen

CHAPTER 24
APPLYING GROUPWARE TECHNOLOGIES TO SUPPORT
MANAGEMENT IN ORGANIZATIONS
A. Michailidis, P.-I. Gouma and R. Rada

PREFACE

This book is concerned with Artificial Intelligence (AI) concepts and techniques as applied to industrial decision making, control and automation problems. The field of AI has been expanded enormously during the last years due to that solid theoretical and application results have accumulated. During the first stage of AI development most workers in the field were content with illustrations showing ideas at work on simple problems. Later, as the field matured, emphasis was turned to demonstrations that showed the capability of AI techniques to handle problems of practical value. Now, we arrived at the stage where researchers and practitioners are actually building AI systems that face real-world and industrial problems.

This volume provides a set of twenty four well-selected contributions that deal with the application of AI to such real-life and industrial problems. These contributions are grouped and presented in five parts as follows:

Part 1: General Issues
Part 2: Intelligent Systems
Part 3: Neural Networks in Modelling, Control and Scheduling
Part 4: System Diagnostics
Part 5: Industrial Robotic, Manufacturing and Organizational Systems

Part 1 involves four chapters providing background material and dealing with general issues such as the conceptual integration of qualitative and quantitative models, the treatment of timing problems at system integration, and the investigation of correct reasoning in interactive man-robot systems.

Part 2 presents a number of systems with built-in intelligence. It starts with an introduction to the concept of intelligent control systems and continues with the demonstration of an autonomous control synthesis system (called phase space navigator) for nonlinear control systems. Then, an overview of the hierarchical and behavioural approaches to autonomous robotic systems is provided, and a combined (behavioural plus planning) approach is developed which possesses a multiresolutional hierarchy of behaviours. Then, a study on distributed intelligent systems in robotics is provided, which is based on an intelligent cellular robotic system (CEBOT) that consists of a number of autonomous robotic units called cells. This part finishes with a contribution showing that subtasks of manufacturing control are so complex and interconnected that cannot be modelled by a single agent system. The problem solution can be achieved using only intensive goal oriented cooperation with other experts.

Part 3 is devoted to artificial neural networks (ANN). First, the application of ANNs to systems modelling and identification is examined including some experimental results. Then, the application of ANNs to robot control is reviewed. The basic architectures of neural control are described, and several illustrative robotic exampes are included. Then, a robotic neurocontroller is described which makes use of bipolar neurons to learn the inverse model of the system. The backpropagation algorithm is used to learn the inverse dynamic model, and the feedback-error-learning scheme is employed as a learning method for the feedforward controller. A 2-link robotic example is included. Next, the neuro-fuzzy approach to anticipatory control is considered. Anticipatory systems can utilize fuzzy predictions about the future in regulating their behaviour through "virtual measurement" which is mapped using ANNs. Finally, the class of large-scale scheduling problems is investigated through interactive and automated approaches. The ANN here is used to treat the combinatorial optimization problems resulting from the scheduling problems.

Part 4 contains three contributions on fault diagnosis. The first contribution provides an overview on the knowledge-based approach to the fault diagnosis of technological systems. First-generation and second-generation diagnostic expert systems are discussed, a survey of digital systems diagnostics tools is presented, and two general methodologies for the development of fault diagnostic tools are developed. The second, presents and compares two methods for model-based diagnosis, namely the *diagnostic model processor* (DMP) and the *model integrated diagnosis analysis system* (MIDAS). Finally, the third contribution describes the multilevel flow model (MFM) which belongs to the class of means-end models. The basic ideas of MFM are outlined, and three diagnostic reasoning methods which can be efficiently implemented with the aid of MFM are developed. These methods have been implemented on the G2 programming tool.

Part 5 involves a number of useful applications of AI. The first contribution is concerned with the problem of multisensor integration for mobile robot navigation. In particular, a mobile platform navigating in a 2D environment with unknown obstacles is considered. The second, is concerned with the use of reactive robots for incremental design of flexible robotic assembly cells. A layered reactive architecture for assembly robots that possesses robustness, reactivity and incrementality is proposed, and a series of simulation experiments are described. The next contribution provides a comparative review of AI and DAI (distributed AI) based planning techniques for manufacturing systems. Planning is a central function in all automated systems, and consists in the selection of the sequence of compatible tasks/actions by which the system goals are achieved. This part continues with a contribution on knowledge-based supervision of flexible manufacturing systems (FMS), and a survey of knowledge-based techniques for industrial scheduling. Supervision of FMSs covers different kinds of activities such as the piloting, the management of working models and the monitoring of the failures. Knowledge-based techniques, in contrast to operational research techniques, are suitable for generating on-line dynamic schedules based on the actual system state. Then a contribution on reactive on-line batch scheduling is presented. A design method for a robust on line scheduler is provided that makes a prediction of the effects of the schedule and tries to optimize the global plant performance. This scheduler is composed by a planner, an integer scheduler and a non-integer scheduler, and was implemented on the real-time expert system shell G2. Finally, a contribution is given on technological support which becomes a "must" in modern organizations and extends the area of management. Here the groupware technology is adopted, which can provide the kind of support the manager needs to deal with uncertainty and ambiguity, and a tool is developed that can supervise and coordinate the overall use of the system and mediate the interactions among its users.

Taken together the contributions of the book provide a well balanced and representative picture of the capabilities of current AI techniques to treat important decision-making and control problems in real-scale robotic, manufacturing and other industrial systems. These techniques which are in the center of the *computer revolution* relieve us of a great deal of a mental effort, in the same way that the techniques of the *industrial revolution* relieve us of a great deal of physical labour. The results on the actual applications of AI are widely sparse in the literature, and only a few books of a nature similar to the present book exist on the subject. Thus the editors feel that this book provides an important addition, since it presents in collective form several angles of attack, methodologies and applications. Each chapter is self-contained, and in many cases includes review material and how-to-do issues.

The book is suitable for the researcher and practitioner of the field, as well as for the educator and senior graduate student. The editors are indebted to all contributors for their high quality contributions, and to Kluwer's (Dordrecht) editorial staff members for their particular care throughout the editorial and printing process.

Spyros G. Tzafestas
Henk B. Verbruggen

CONTRIBUTORS

ALBAYARAK S.	T.U. Berlin, Berlin, Germany
ARAIJO H.	Univ. of Coimbra, Coimbra, Portugal
ARZEN K.-E.	Lund Inst. of Technology, Lund, Sweden
BRACKNEY L.	Purdue Univ. West Lafayette, U.S.A.
BRUIJN P. M.	Delft Univ. of Technology, Delft, The Netherlands
CRAYE E.	Ecole Centrale de Lille, Lille, France
DIAS J.	Univ. of Coimbra, Coimbra, Portugal
FUKUDA T.	Nagoya Univ., Nagoya, Japan
GENTINA J.-C.	Ecole Centrale de Lille, Lille, France
GOUMA P.-I.	Univ. of Liverpool, Liverpool, U.K.
HINDI K.S.	Dept. of Computation, UMIST, Manchester, U.K.
IKONOMOPOULOS A.	The Univ. of Tennessee, Knoxville, U.S.A.
KOBAYASHI Y.	Energy Res. Lab., Hitachi Ltd., Ibaraki-ken, Japan
KOENIG E.C.	CS Dept., Univ. of Wisconsin-Madison, U.S.A.
KOKKINAKI A.I.	Univ. of Southwestern Louisiana, Lafayette, U.S.A.
KRALLMANN H.	T.U. Berlin, Berlin, Germany
KRIJGSMAN A.J.	Delft Univ. of Technology, Delft, The Netherlands
LARSSON J.E.	Lund Inst. of Technology, Lund, Sweden
MEYSTEL A.	Drexel Univ., Philadelphia, U.S.A.
MICHAILIDIS A.	Univ. of Liverpool, Liverpool, U.K.
MOTUS L.	Tallinn Techn. Univ., Tallinn, Estonia
NONAKA H.	Energy Res. Lab., Hitachi Ltd., Ibaraki-ken, Japan
NUNES U.	Univ. of Coimbra, Coimbra, Portugal
PETTI T.	Washington Res. Center, Columbia, MD, U.S.A.
RADA R.	Univ. of Liverpool, Liverpool, U.K.
SEKIYAMA K.	Nagoya Univ., Nagoya, Japan
SHOURESHI R.	Purdue Univ., West Lafayette, U.S.A.
SINGH M.G.	Dept. of Computation, UMIST, Manchester, U.K.
TERPSTRA V.J.	Delft Univ. of Technology, Delft, The Netherlands
TESHNEHLAB M.	Saga Univ., Graduate School, Japan
TOGUYENI A.K.A.	Ecole Centrale de Lille, Lille, France
TRACA de ALMEIDA A.	Univ. of Coimbra, Coimbra, Portugal
TSOUKALAS L.	The Univ. of Tennessee, Knoxville, U.S.A.
TZAFESTAS E.S.	Univ. P. et M. Curie, Paris, France
TZAFESTAS S.G.	Natl. Tech. Univ. of Athens, Athens, Greece
UEYAMA T.	Nagoya Univ., Nagoya, Japan
UHRIG R.E.	The Univ. of Tennessee, Knoxville, U.S.A.
VALAVANIS K.P.	Univ. of Southwestern Louisiana, Lafayette, U.S.A.
VERBRUGGEN H.B.	Delft Univ. of Technology, Delft, The Netherlands
WATANABE K.	Saga Univ., Mech. Eng. Dept., Saga, Japan
WALLEN A.	Lund Inst. of Technology, Lund, Sweden
WHEELER M.	Purdue Univ., West Lafayette, U.S.A.
WOODS E.A.	SINTEF Automatic Control, Trondheim, Norway
ZHAO F.	CIS Dept., Ohio State Univ., Ohio, U.S.A.

PART 1
GENERAL ISSUES

1
ARTIFICIAL INTELLIGENCE IN INDUSTRIAL DECISION MAKING, CONTROL, AND AUTOMATION: AN INTRODUCTION

SPYROS TZAFESTAS[(*)] , HENK VERBRUGGEN[(**)]
[(*)] Intelligent Robotics and Control Unit,
National Technical University of Athens
Zographou 15773, Athens, Greece

[(**)] Control Laboratory,
Department of Electrical Engineering,
Technical University of Delft,
Mekelweg 4, 2628 CD Delft, The Netherlands

1. INTRODUCTION

Artificial Intelligence (AI) may be defined as the science of enabling computers and machines to learn, reason and make judgments. According to Elaine Rich, AI is the study of how to make computers do things for which, at the moment, people are better. Why is AI such a hot branch in our days? One might argue that AI has only very recently started building systems that are able to display human-level intelligence in certain domains. The necessity for applying AI to managerial and engineering control problems originates from the growing complexity of current systems, as well as from the traditional expense, time constraints, and limited availability of human expertise. AI technology offers the tools that enable us to:

- capture and retain expertise that was gained over many years of engineeering,

-amplify expertise that is needed to successfully deploy new methods and applications,

- design systems that reason intelligently about necessary actions to take in real time, thus freeing operational staff.

AI tools include the so-called "knowledge-based expert systems" (ES) which are suitable for specialized tasks, and formed the branching point away from the quest for generality that characterized early AI. The key issues of AI expert system building tools are:

i) symbolic representation of the knowledge (predicate logic, semantic networks, frames, objects, production rules, etc.),

ii) symbolic reasoning using rules and methods able to deduce, examine, judge, determine, and so on,

iii) graphic explanations using developer-oriented graphic means (knowledge-based graphs, rule graphs, etc.), and end-user-oriented graphic means to provide developers and users the representation and reasoning.

Decision making, control and manufacturing are three most attractive areas of application of AI techniques. However, considerable effort is required to capture and organize the accumulated knowledge of decision makers, and control or manufacturing

S. G. Tzafestas and H. B. Verbruggen (eds.),
Artificial Intelligence in Industrial Decision Making, Control and Automation, 1–39.
© 1995 *Kluwer Academic Publishers.*

engineers. There is a vast amount of knowledge here, extending over many processes, diverse situations and an infinite array of parts and products. Particular attention should be given to representing this knowledge and adequately representing explicitly the characteristics of plants,machines and processes.

For example to automate manufacturing tasks one can follow two radically different approaches:

i) By duplicating (imitating) human actions (e.g. in spray painting a robot is programmed to move a spray gun in exactly the same patterns as an expert painter).

ii) By studying the task's goal or expected functions and applying a convenient strategy that may bear little or no resemblance to human methods (e.g. automatic dishwashing or automatic serving). Both approaches have relative advantages and disadvantages.

Our purpose in this chapter is to provide an introductory survey of AI and knowledge-based techniques applied to decision making, control and automation processes. For completeness, the chapter starts with a look at the decision making, control and automation systems themselves. Section 3 discusses three basic AI approaches to decision making, control and automation, namely: reasoning under uncertainty, qualitative reasoning, and neural nets reasoning. Section 4 is devoted to the study of AI as applied to decision analysis and decision making processes. Here the multiattribute utility concept is employed and several contributions are discussed on the use of the synergy between the prescriptive problem structuring techniques in decision analysis and the rule based expert system approaches, as well as on the role played by the user's preference in these decisions. Sections 5 and 6 discuss the artificial intelligence application in the control and supervision, and fault diagnosis areas, respectively. Finally, Section 7 is concerned with the work being done on the application of AI to the robotics and manufacturing systems areas. Here, the work on intelligent robotic systems (IRSs) and intelligent machines (IMs) is reviewed, the principal functions of manufacturing systems that provide a challenge for AI techniques are discussed, and a number of working knowledge-based expert systems for manufacturing operations are listed.

2. DECISION MAKING, CONTROL AND AUTOMATION

2.1. Decision Making Theory

According to Castles [1] a decision is a conscious choice between at least two possible courses of action. Formal processes for decision making have been discussed for a very long time and a plythora of sophisticated mathematical techniques for making rational decisions have been developed in economics, statistics, control and operational research [2]. Applications to assist human decision making have been considered in business, medicine, engineering, politics and many other areas [3-11]. One of the clearest statements of what classical decision making theory is was provided by Lindley [2] according to whom there is essentially only one way to reach a decision sensibly. "First, the uncertainties present in the situation must be quantified in terms of values called probabilities. Second, the various consequences of the courses of action must be similalry described in terms of utilities. Third that decision must be taken which is expected - on the basis of the calculated probabilities - to give the greatest utility. The force of "must", used in three places there, is simply that any deviation from the precepts is liable to lead the decision maker into procedures which are demonstrably absurd."

As we observe this is a definition of statistical decision theory, where namely *probability* and *utility*, are used. From them a range of measures for decision can be derived.

Probability: The principal method for computing posterior probabilities is Bayes' rule

$$p(H_i|e_j) = \frac{p(e_j|H_i)p(H_i)}{\sum_i p(e_j|H_i)p(H_i)} \tag{1}$$

where H_i is a hypothesis and e_j an item of evidence. A rational decision rule is to select the hypothesis for which the posterior probability is highest after all items of evidence have been acquired.

In many situations Bayes' rule is combined with infomation theoretic concepts leading to the so called *expected information yield* of a test case over all of the possible outcomes weighted by their likelihood [2]. If the test cases give a yes/no or positive/negative answer, the expected information yield is computed using the entropy before and after all information outcomes, weighted by the likelihood of each hypothesis and each outcome x for the test case. In the case of two outcomes the expected information yield of the test case is

$$\sum_i \{p(H_i|x)\log p(H_i|x)\}p(x) - \sum_i p(H_i)\log p(H_i) \tag{2}$$

where the values of x are "yes" or "no" and

$$p(x) = \sum_i p(x|H_i)p(H_i)$$

The test case with the maximum expected information yield is selected. For example the test cases may be laboratory tests for diagnostic purposes.

Utility: Utility (U) is a quantitative expression of the values of different events. Utilities may be employed for representing costs and benefits of alternative actions in objective or subjective terms. The decision is based on the *subjective expected utility* *(SEU)*, i.e. given a set of alternative actions, the action A is selected which produces the outcome that maximizes *SEU* computed by weighting the utility of each event by its probability: $SEU = p(A_i)U(A_i)$. The concept of utility is actually problematic because costs and benefits are individualistic and subjective. In many cases the utilities of outcomes (e.g. "quality of life" following surgery) may be almost imposible to determine.

Structure and Levels of Decision Making: According to Hill et al [3] the complete decision process has the structure shown in Fig. 1, i.e. the process begins with the problem definition and ends with the implementation of the decision. Iteration of steps occurs when there is insufficient information to complete a given step. The previous step must then be repeated until the necessary data are in hand.

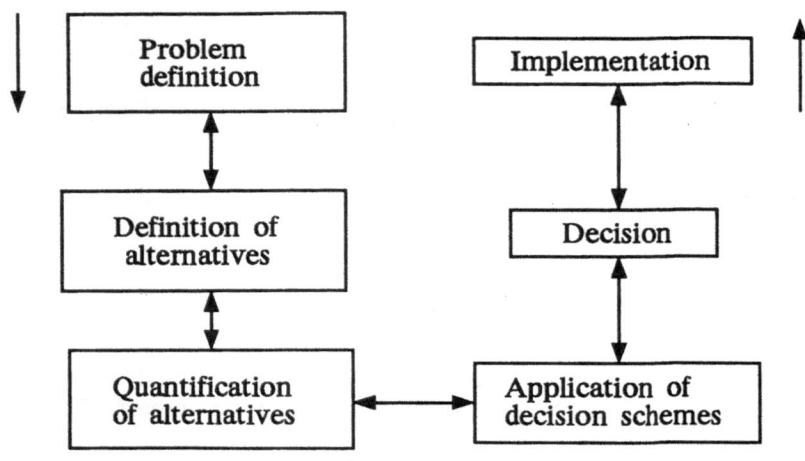

Fig.1. Classical structure of the decision making process

Managerial and organizational decisions can be grouped in several ways, e.g. through a hierarchy of impact, time horizon of interest, data needs, degree of structure etc. [12]. According to Anthony [13] decisions can be classified in the following four levels:

1. Strategic planning decisions
2. Management control decisions
3. Operational control decisions
4. Operational performance decisions

The first level involves decisions that are related to selecting the highest level policies and objectives, the second refers to decisions made for the purpose of assuring effectiveness in the acquisition and use of resources, the third concerns the decisions made for the purpose of assuring effectiveness in the performance of operations, and the fourth level involves the day-to-day decisions made while performing operations. As we go down from level 1 to level 4 the frequency of decisions increase, whereas in the opposite direction it is the importance of decisions that increases.

2.2. Control and Automation

Control is the science and technology of feedback and automation is the discipline of automating the operation of technological processes. Here, a large number of subfields are included such as computer-aided control system design, computer-aided process control operation, automated fault diagnosis, real time control, system alarm monitoring, equipment selection, facility lay out, supervisory control, etc. Automation involves three primary phases namely the design of the product, the design of the process to manufacture the product and finally the design of the physical implementation of the process, the actual industrial (or nonindustrial) plant. The distinction between process and plant is often blurred, however it always exists. The process is the abstract definition of steps to be carried out to produce the product. Its physical existence is on paper or in a computer only. The plant is the hardware built to carry out these steps.

As an example of automation we mention the CIM systems that have the hierarchical architecture of Fig. 2.

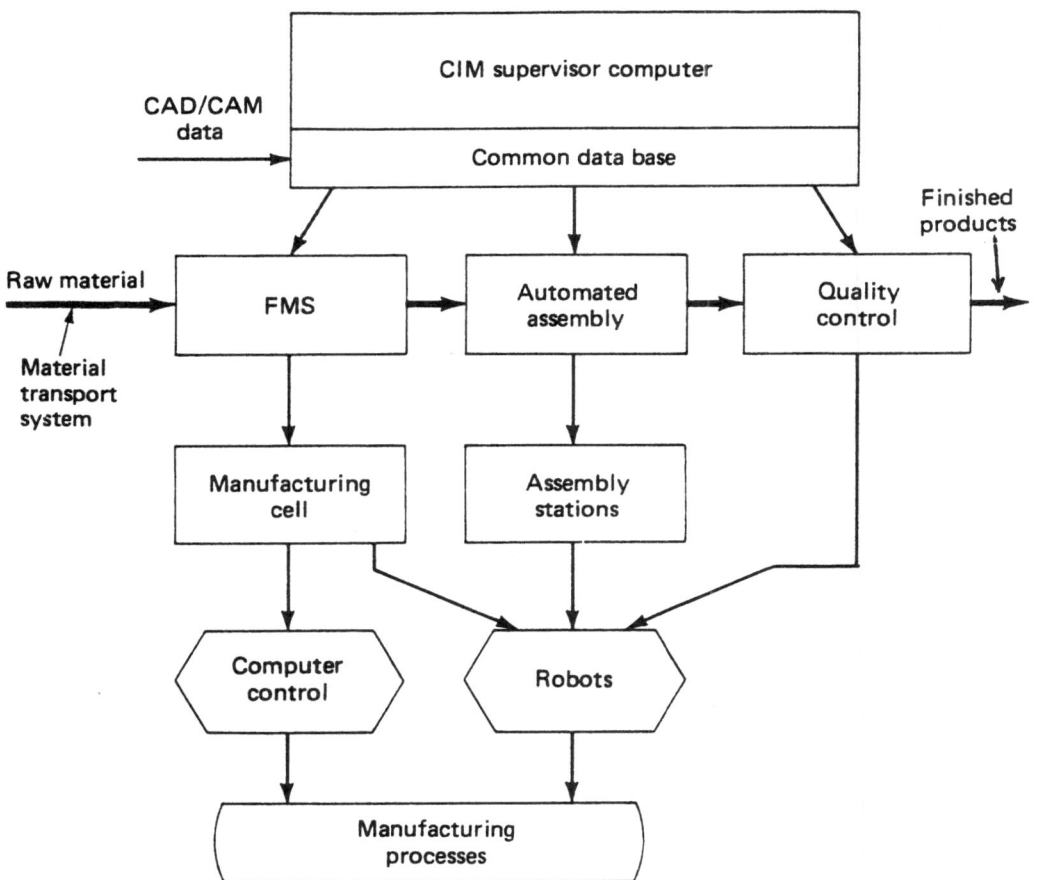

Fig.2. Hierarchical architecture of a CIM system

At the lowest level of this architecture there are stand-alone computer control systems of manufacturing plants and industrial robots (manufacturing cells). The operation of several manufacturing cells is coordinated by the central computer through a materials-handling system. This constitutes the intermediate level of the CIM system, which is known as the flexible manufacturing system (FMS). The products (final or semifinal) of the FMS are tested by appropriate automatic inspection stations. The integration of the design and manufacturing of an FMS can be made with the aid of a CAD/CAM (computer-aided design / computer-aided manufacturing) system. This leads to substantial reductions of the factory's production cycle. Robots and computerized numerical control (CNC) machines can only replace human power and skill. CIM systems can also replace human intelligence and achieve an incomparably higher efficiency.

3. ARTIFICIAL INTELLIGENCE METHODOLOGIES

Three important AI methodologies that are applicable to decision, control and automation systems are:
- Reasoning under uncertainty
- Qualitative reasoning
- Neural nets reasoning

The tools used to implement these methodologies are the "expert systems" (ES) and "knowledge based systems" (KBS). Expert systems are computer programs that embody the knowledge of human expert enriched with some common sense. KBSs are ESs that also embody knowledge obtained from other (than human) knowledge sources. Expert systems are profitable if the following are true:
- At least one human expert exists who can carry out the task very well,
- The expert ability is manily due to special knowledge, experience and judgment
- The task belongs to a well defined application,
- The expert can combine knowledge, judgment and experience, and can explain his path of reasoning.

Some of the benefits of the above AI issues are the following:
- Explicit representation of symbolic structures, performance, and reasoning and easy integration of various representation types,
- Visibility, flexibility, and adaptability obtained from the declarative nature of the representation,
- Reasoning in radically different ways from those tried by non-AI systems,
- Intelligent interaction, based on the representation.

Modern AI/ expert systems attempt to remove the programmers as problem translators and allow the expert to interact directly with the system to explore the problem space.

An expert system consists of the following clearly separated components (Fig. 3) [14]:
- Knowledge base (general knowledge about the problem, i.e. facts and rules)
- Data base (information about the current problem, i.e. input data)
- Inference engine (schemes for controlling the general knowledge of the problem)
- Explanation component (which can inform the user on why and how the conclusions are obtained).
- User interface and knowledge acquisition component
- Work space (i.e. an area of memory for storing a description of the problem constructed from facts supplied by the user or inferred from the knowledge base).

In the following a brief outline of the AI methodologies mentioned above will be provided.

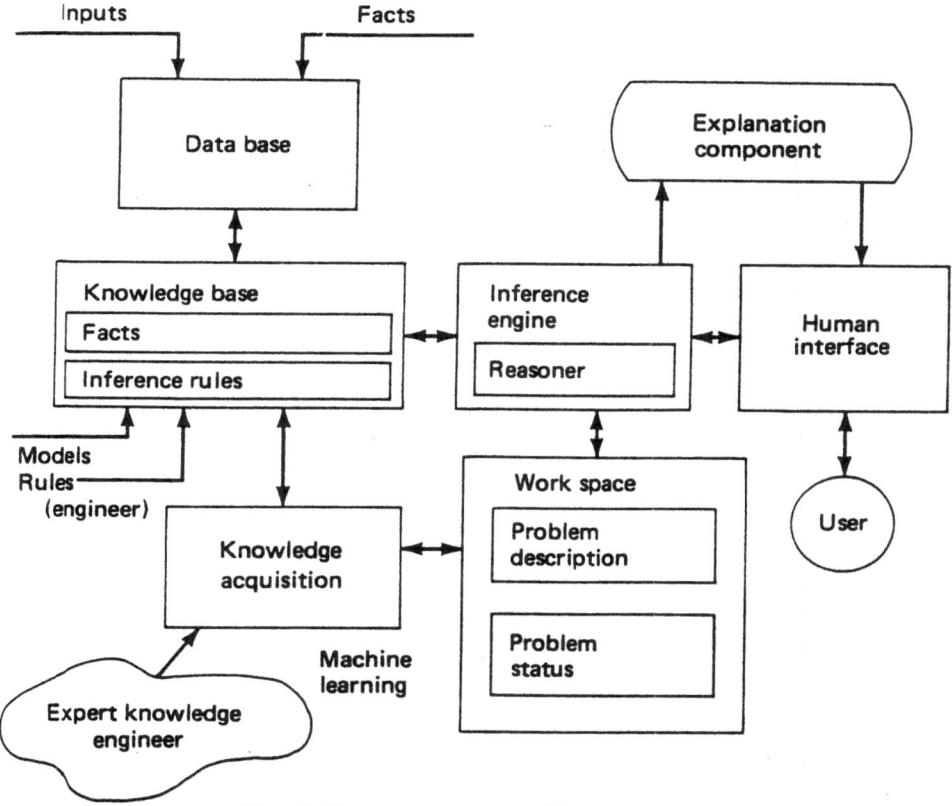

Fig. 3. Expert system architecture

3.1. Reasoning Under Uncertainty

Uncertainty can be originated from several sources and appears to have many different forms. A reasonable list of them is the following [15]:

- random event
- experimental error
- computational error
- uncertainty in judgment
- lack of evidence
- lack of certainty in evidence

Reasoning under uncertainty is the process of drawing conclusions (and taking decisions) in the presence of uncertainty with regard to the pertinent data, and knowledge is used to combat the uncertainty in arriving at the "best" conclusion (decision).

The conclusion-drawing/decision-making process can be defined as a set of applicable rules of the form

$$\text{IF } D \text{ IS } D_x \text{ THEN } A \text{ IS } A_x$$

where the D are data points in some n-dimensional data-space, and the A are actions (conclusions) in some one-dimensional conclusion (action) space. Thus $D_x = \{D_1, D_2, ..., D_n\}$. In many cases the components D_i of D_n involve uncertainty (instrumental erros, conceptual error, etc.) in which case A_x also involves uncertainty. The problem is to assign the uncertainty value to A_x knowing the uncertainty values of D_i ($i = 1,2,...,n$) expressed in some well defined way.

The methods proposed for dealing with uncertainty in computer systems fall into two broad classes; non-numerical methods (such as endorsements, high granularity symbol manipulation and nonmonotonic logic) and various numerical methods. Here the following methods will be reviewed:

    i) probabilistic approach
    ii) certainty factors approach
    iii) fuzzy sets approach
    iv) Shafer-Dempster approach

Probabilistic approach: Probability is associated with randomness - the estimation of the likelihood of a given event occuring out of a possible set of events which can be enumerated. Probability theory rests on three postulates:

    - there is a value which may be assigned in advance to the probability of an event

    - decisions will adhere consistently to a rule stated in advance, irrespective of the accruing experience

    - events themselves are the result of a randomizing process.

The key concept is the concept of "inverse probability" i.e. the likelihood of an event having a given cause, rather than a cause giving rise to a given event. Bayes' rule [see Eq. 1] is used here to compute updated probabilities. Thus, given some degree of uncertainty concerning reality, one makes some hypotheses, and assign to each hypothesis an initial likelihood, called the "prior probability" of the hypotheses. Then some test is performed, and on the basis of the outcome, the probabilities associated with each hypothesis are revised. The revised probabilities are called the posterior probabilities. They can then be used as the prior probabilities for some further test, provided of course that the further test is measuring some parameter truly independent of the measurement taken in the first (or many previous) test. Central to Bayes' approach is the "Principle of Insufficient Reason". This means that in assigning our prior probabilities to alternative hypotheses, one is free to assign equal probabilities to each, in the absence of any "Sufficient Reason" to the contrary [99].

Certainty factors approach: This approach is due to Shortliffe [16] and was developed within the framework of the Stanford Heuristic Programming Project in the early 1970s which produced the well known medical expert system MYCIN.

The basic issue is that there is at any point in time a certainty factor CF associated with any given a priori hypothesis. The values of CF range from -1 representing the statement "believed to be wholly untrue" to +1 representing the statement "believed to be wholly true". The method is not in any sense statistical in its origins, i.e. no assumptions are made that N mutually exclusive hypotheses cover the universe of discourse, nor $\Sigma_i CF(i) = 1$, either initially or later.

Introducing the two measures MB (current measure of belief) and MD (current measure of disbelief), CF is given by:

$$CF(H{:}E) = MB(H{:}E) - MD(H{:}E) \tag{3}$$

for hypothesis H given evidence E. The belief and disbelief measures both range from

0 to 1.

Shortliffe has introduced the following formula for updating the belief in a hypothesis H as evidence is accumulated:

$$MB(H:E_1,E_2) = MB(H:E_1) + MB(H:H_2)\{1 - MB(H:E_1)\} \qquad (4)$$

This is known as Shortliffe formula and states that the belief in hypothesis H after evidence E_2 is increased by the weight of the evidence associated with E_2 proportionaly to the current degree of belief. Clearly Shortliffe's formula is symmetric with respect to the accrual of evidence from different sources and is cumulative asymptotically to certainty (i.e. linear with respect to the logarithm of additional evidence).

The link between certainty factors and Bayesian probabilities is the formula

$$CF(H:E) = \frac{Pr(H:E) - Pr(H)}{Pr(H)} \qquad (5)$$

If the total certainty factor of the premises (assumptions) of a rule is CF_p, and the certainty factor of the rule itself is CF_r, then the certainty factor CF_c of the conclusion drawn from this rule is given by the product

$$CF_c = CF_r \cdot \max\{0, CF_p\} \qquad (6)$$

To compute the certainty factor of the "conjuction" and "disjunction" of two hypotheses H_1 and H_2 the following formulas are used:

$$\begin{aligned}
MB(H_1 \text{ AND } H_2:E) &= \min\{MB(H_1:E), MB(H_2:E)\} \\
MB(H_1 \text{ OR } H_2:E) &= \max\{MB(H_1:E), MB(H_2:E)\} \\
MD(H_1 \text{ AND } H_2:E) &= \max\{MD(H_1:E), MD(H_2:E)\} \\
MD(H_1 \text{ OR } H_2:E) &= \min\{MD(H_1:E), MD(H_2:E)\}
\end{aligned} \qquad (7)$$

A diagnostic expert system designed using the certainty factors model is described in [100].

Fuzzy sets approach: Fuzzy set theory was initiated by Zadeh and permits the treatment of vague, uncertain, imprecise and ill defined knowledge and concepts in an exact mathematical way [17, 18]. Actually, our life and world obey the *principle of compatibility* of Zadeh according to which "the closer one looks at a real world problem, the fuzzier becomes its solution, or in other words, as the complexity of a system increases our ability to make precise and yet significant statements about its behaviour diminishes until a threshold beyond which precision and significance (relevance) become almost exclusive characterisitcs. Let X be a reference superset and A a subset of X. Then A is said to be a fuzzy subset of X if and only if

$$A = \left\{(x, \mu_A(x)) \mid x \in X, \mu_A(x): X \to [0,1]\right\} \qquad (8)$$

where $\mu_A(x)$ is the so called membership function of x. Clearly, in the special case where we have $\{0,1\}$ instead of $[0,1]$ the fuzzy subset A degenerates to the crisp set A. The crisp subset $Supp(A)$ of X is called support of A if and only if

$$Supp(A) = \left\{x \in X, \mu_A(x) \geq 0\right\}, \quad \text{with } Supp(A) \subset X \qquad (9)$$

The crisp subset $L_\alpha A$ of X is called α-cut of A if and only if

$$L_\alpha A = \{x \in X, \mu_A(x) \geq \alpha\} \quad \text{with } L_\alpha A \subset X \tag{10}$$

The quantity $|A| = \Sigma_{x \in X} \mu_A(x)$ is called the cardinality of A and the quantity $|A| = |A| / |X|$ is called the relative cardinality of A.

Given the fuzzy subsets A and B of X, then the *section* and *union* of A, B, are defined as

Section: $C = A \cap B = \{(x, \mu_C(x)) \mid x \in X, \mu_C(x) = \min\{\mu_A(x), \mu_B(x)\}\}$ (11)

Union: $C = A \cup B = \{(x, \mu_C(x)) \mid x \in X, \mu_C(x) = \max\{\mu_A(x), \mu_B(x)\}\}$ (12)

and the complement A^c of A as

$$A^c = \{(x, \mu_{A^c}(x)) \mid x \in X, \mu_{A^c}(x) = 1 - \mu_A(x)\} \tag{13}$$

The algebraic sum $A + B$ and product $A \cdot B$ of the fuzzy subset A and B of X are defined as:

$$A + B = \{(x, \mu_{A+B}(x) \mid x \in X, \ \mu_{A+B}(x) = \mu_A(x) + \mu_B(x) - \mu_A(x)\mu_B(x)\} \tag{14a}$$

$$A \cdot B = \{(x, \mu_{A \cdot B}(x) \mid x \in X, \ \mu_{A \cdot B}(x) = \mu_A(x)\mu_B(x)\} \tag{14b}$$

A fuzzy relation $f(A_1, ..., A_n)$ of the fuzzy sets $A_1 \subset X_1, ..., A_n \subset X_n$ is defined to be the fuzzy set

$$f(A_1, ..., A_n) = \{((x_1, ..., x_n), \mu_{f(A_1, ..., A_n)}(x_1, ..., x_n)) \mid (x_1, ..., x_n) \in X_1 \times ... \times X_n\} \tag{15a}$$

where

$$\mu_{f(A_1, ..., A_n)}(x_1, ... x_n) = f(\mu_A(x_1), ..., \mu_A(x_n)) \tag{15b}$$

The Cartesian product of $A_1, ..., A_n$ is defined to be the fuzzy set

$$g(A_1, ..., A_n) = \{((x_1, ..., x_n), \mu_{g(A_1, ..., A_n)}(x_1, ..., x_n)) \mid (x_1, ..., x_n) \in X_1 \times ... \times X_n\} \tag{16a}$$

where

$$\mu_{g(A_1, ..., A_n)}(x_1, ... x_n) = \min\{\mu_A(x_1), ..., \mu_A(x_n)\} \tag{16b}$$

A concept that plays a dominant role in fuzzy reasoning is the concept of linguistic variable. Linguistic variable is a variable the values of which are not numbers but words or sentences or propositions in a natural or artificial language.

As an example we give the linguistic variable $x =$ "age" with values "very young", "not very young", "pretty young", "a little old", "pretty old", "not very old", and "very old".

Then a meaning of the fuzzy set "old" is $M_{old} = \{(u, \mu_{old}(u)), \ u \in [0, 100]\}$ where

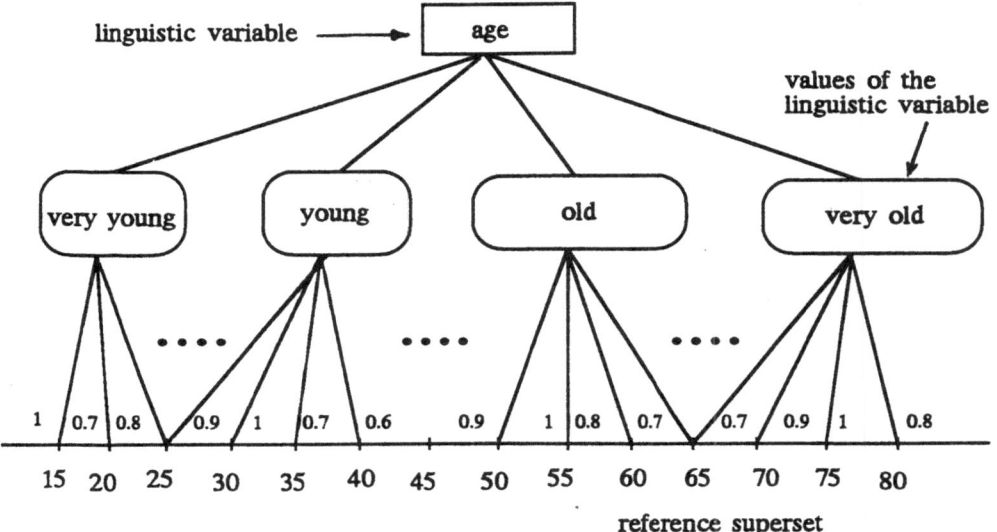

Fig . 4. Linguistic variable "age" and its "linguistic values"

$$\mu_{old} = \begin{cases} 0 & u \in [0,50] \\ \{1+((u-50)/5)^{-2}\}^{-1} & u \in [50,100] \end{cases}$$

The linguistic variable "age" and its values are pictorially depicted in Fig.4.

We close our discussion on fuzzy reasoning with Zadeh's max - min compositional rule which provides a way of reasoning under uncertainty [19].

Let A be a fuzzy set on X, B a fuzzy set on Y, and R a fuzzy relation on $X \times Y$, ie.

$$A = \{(x,\mu_A(x)) \mid x \in X, \mu_A(x): A \rightarrow [0,1]\} \tag{17a}$$

$$B = \{(y,\mu_B(y)) \mid y \in Y, \mu_B(y): B \rightarrow [0,1]\} \tag{17b}$$

$$R = \{((x,y)\mu_R(x,y)) \mid (x,y) \in X \times Y, \mu_R(x,y) = \min\{\mu_A(x),\mu_B(y)\}\} \tag{17c}$$

Then, given A and R, B is found as

$$B = A \circ R, \quad \mu_B(y) = \max_x \{\min\{\mu_A(x),\mu_R(y)\}\} \tag{18}$$

where "$\circ$" is the symbol of the max - min composition operator.

As an example consider the following:
(a) $X = \{1,2,3,4\}$, (b) A = "x small" = $\{(1,1), (2,0.6), (3, 0.2), (4,0)\}$,
(c) R = "x nearly equal to y" with fuzzy matrix as shown below.

R			x		
		1	2	3	4
y	1	1	0.5	0	0
	2	0.5	1	0.5	0
	3	0	0.5	1	0.5
	4	0	0	0.5	1

It is desired to find what happens with the variable y, i.e. to determine the fuzzy set B ="y". To this end, use of the max - min composition rule of inference is made, i.e. $B = A \circ R$ where

$$\mu_B(y) = \max_x \left\{ \min\{\mu_A(x), \mu_R(x,y)\} \right\}$$
$$= \{(1,1), (2,0.6), (3,0.5), (4,0.2)\}$$

Thus, IF "x small" AND "x nearly equal to y", THEN "y nearly small".

Shafer - Dempster approach: The Shafer Dempster theory provides a more rigorous basis for the use of certainty factors and other indicators (measures) of uncertainty [20 -22]. This theory uses the numbers of the interval [0,1] to represent the support of some hypothesis from a piece of evidence. Let X denote the *frame of discerment* (reference superset). The effect of the evidence items upon the subsets of X are expressed via a function m which is defined as

$$m: S \rightarrow [0,1], \quad \sum_{A \subseteq X} m(A) = 1 \text{ with } m(\emptyset) = 0 \qquad (19)$$

where S is the set of all subsets of X, or equivalently:

$$m: P \rightarrow [0,1], \quad \sum_{i=1,n} m(p_i) = 1, \quad p_i \in P, \quad m(0) = 0 \qquad (20)$$

where P is the Boolean lattice $(P, \wedge, \vee, \sim)$.

The function m is called the basic probability assignment and it is a generalization of the classical concept of probability density function which distributes the unit probability among the elements of X (while m distributes 1 among the elements of S). The quantity $m(A)$, $A \subseteq X$ represents the part of the unit probability that is exactly assigned to the set A and cannot be assigned to the subsets of A. In other words $m(A)$ is the belief that is assigned to the set A ignoring all of its subsets.

Any function g from S to [0,1] that possesses the following properties

i. $g(\emptyset) = 0$

ii. $g(X)=1$ (21)
iii. If $A \subseteq B$ then $g(A) \le g(B)$

is called a measure of belief (or a fuzziness measure according to Sugeno).

Shafer has introduced the *credibility function* (measure) Cr which satisfies (21) and is generated from the function m as:

$$\forall A \in S \quad Cr(A) = \sum_{B \subset A} m(B) \qquad (22)$$

By duality one can define the *plausibility* function Pl from Cr as

$$\forall A \in S \quad Pl(A) = 1 - Cr(\sim A) \qquad (23)$$

where $\sim A$ is the complement of A. Using (22) one obtains

$$\forall A \in S \quad Pl(A) = \sum_{B \cap A \ne \emptyset)} m(B) \qquad (24)$$

The interval $[Cr, Pl]$ is called *evidential interval* and expresses the uncertainty that exists in the validity of a hypothesis. Some properties of Cr and Pl are the following:

$$\forall A \in S \quad 0 \le Cr(A), \; Pl(A) \le 1 \qquad (25a)$$

$$\forall A \in S \quad Cr(A \cup B) \ge Cr(A) + Cr(B) - Cr(A \cap B) \qquad (25b)$$

$$\forall A \in S \quad Pl(A \cap B) \le Pl(A) + Pl(B) - Pl(A \cup B) \qquad (25c)$$

Setting $B = \sim A$ in (25a,b) one obtaines the properties

$$\forall A \in S \quad Cr(A) + Cr(\sim A) \le 1, \; Pl(A) + Pl(\sim A) \ge 1 \qquad (25d)$$

If

$$\forall B \in S, \; m(B) > 0 \; \rightarrow \; m(B \cap A) = 0$$

then the following are true

$$Cr(A) = 0, \; Cr(\sim A) = 0 \qquad (25e)$$

$$Pl(A) = 1, \; Pl(\sim A) = 1 \qquad (25f)$$

This means that two contradictive hypotheses may be plausible but not credible at all. From $Cr(A) = 0$ it does not follow that

$$Cr(\sim A) = 1 \; (\leftrightarrow Pl(A) = 0)$$

whereas from $Cr(\sim A) = 1$ it follows that $Cr(A) = 0$. Also from (23) and (24) it follows that

$$\forall A \in S \quad Cr(A) \le Pl(A) \qquad (25g)$$

i.e. the plausibility of a hypothesis is greater than or equal to its credibility Finally, one can easily see that

$$Cr(X) = 1 \tag{25h}$$

The above show that the certainty (fidelity) of a hypothesis A is expressed by the interval $[Cr(A), Pl(A)]$.

In the Shafer-Dempster approach the updating of the belief of a hypothesis $A \subseteq X$ is performed via the combination of the basic probability assignments as they result from the several sources of information, Let m_1 and m_2 two probability assignments. Dempster's formula gives their combination m_{12} as

$$m_{12} = \frac{\sum_{A \cap B = C} m_1(A) m_2(B)}{\sum_{A \cap B \neq \emptyset} m_1(A) m_2(B)} \quad \forall C \subseteq X, \; C \neq \emptyset \tag{26a}$$

with $m_{12}(\emptyset) = 0$.

The denominator of (26a) can be written as $1-k$ where

$$k = \sum_{A \cap B = \emptyset} m_1(A) m_2(B) \tag{26b}$$

Here, the quantity k represents the amount of probability assigned by the section process of the numerator to the empty set.

Dempster formula is *associative* and *commutative* and so the combination of pieces of evidence is independent of the order in which they appear. Thus (26a) can be generalised to more than two probability assignments m. The rule (26a) is a generalization of Bayes' rule, and for $m_1(A) = 1 - m_1(X)$, $m_2 = 1 - m_2(X)$ it gives

$$m(A) = m_1(A) + m_2(A) - m_1(A) m_2(A)$$
$$m(X) = (1 - m_1(A))(-m_2(A))$$

which are relations that correspond to the way $MB(H:E)$ and $MD(H:E)$ are combined in the Shortliffe's model.

The drawback of the rule (26a) is its sensitivity in the case of strongly conflicted pieces of evidence. It can only be applied with complete safety when

$$\forall \, A, B \subset X, \; m_1(A) m_2(B) > 0 \rightarrow A \cap B = \emptyset \tag{27}$$

3.2 Qualitative Reasoning

A discussion of the qualitative reasoning (QR) for the purposes of fault diagnosis is provided by the present authors in another chapter of the book (Ch. 15). Actually, the research in the fields of *qualitative reasoning* and *naive physics* has led to several important approaches to the problem of describing the general behaviour of physical systems. For example the *qualitative simulation* of Kuipers [23,24] produces a state transmition tree of the system starting with some initial conditions, while the envisioning technique of Dekleer and Williams [25] is a state diagram describing all the possible states that the system can take and all possible transitions between these states. Qualitative-based systems utilize *deep knowledge* in contrast to other techniques that employ *shallow knowledge* [26]. The qualitative approach relies on the abstraction of the numerical values of the physical variables, into a limited number of intervals on the real line, and the abstraction of the analytical functions that describe the relationship between the variables into monotonically increasing or decreasing qualitative functions. Actually a purely qualitative process is not sufficient to simulate the behaviour correctly as it may lead to erroneous solutions. A filtering process is needed for separating the correct predictions from the erroneous ones. This filtering process is usually designed using quantitative information about the system at hand.

Our purpose here is to discuss the application of qualitative reasoning for control purposes. The control problem via this approach was studied by Kuipers and applied to a liquid control problem. Another interesting piece of work in this direction has been completed by Francis and Leitch [27,28]. They produced a Prolog-based expert system shell (called ARTIFACT) for the automation of the decision and control process. Work in this direction was also done in [29] by Clocksin and Morgan. A more efficient control rules set for the liquid control problem derived from qualitative considerations was presented by Abdulmajid and Wynne [30]. The controller derived (as all the above qualitative controllers) was applied at plant level and it can be considered as a replacement of the conventional controllers/compensators (say of the PID type).

In QR the system at hand is regarded as a series of sub-systems where the state of each subsystem is represented by its associated "state variable" using a qualitative quantitiy space (say +, 0, -). Thus the state X_i of the subsystem i can at any instant of time take one og the values $(+)$, (0) or $(-)$. The dynamic interaction between adjacent subsystems is represented by either a monotonically increasing or monotonically decreasing function, i.e.

$$X_{i+1} = M-\text{plus}(X_i) \quad \text{or} \quad X_{i+1} = M-\text{minus}(X_i)$$

In [30] a two subsystems plant was considered with an M-plus dynamic relationship (Fig. 5).

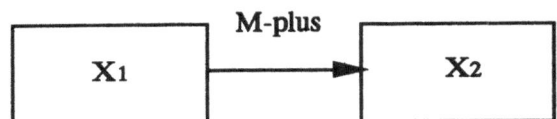

Fig. 5. A 2-sub-system plant with M-plus interaction

The problem is to determine the change in X_1 to cause X_2 to change from $(+)$ to $((+),(0),(-))$ respectively, then from (0) to $((+),(0),(-))$, and finally from $(-)$ to $((+), (0), (-))$. The corresponding nine rules are shown in Table 1.

These rules together with a similar set for a monotonically decreasing relationship provide the basis for determining the control rules for the system. For a review of QR methods and their comparison with other AI methodologies the reader is referred to [33]. The above controller was applied in [30] to a two coupled tanks system which has intrisically non-linear characteristics. The control objective was to regulate the level of the liquid in tank 2.

We close our discussion on QR by mentioning the work on Order of Magnitude (OM) of Mavrovouniotis and Stephanopoulos [38-39]. The key idea is to extend the relational algebra concept of greater-than, equal or less-than $(+, 0, -)$ to the

Table 1: Control Rules for a Monotonically Increasing Relationship			
Rule	X_1	Current X_2	Required X_2
1	0^*	+	+
	+	+	+
2	-	+	0
3	-	+	-
4	+	0	+
5	0	0	0
6	-	0	-
7	+	-	+
8	+	-	0
9	0^*	-	-
	-	-	-

following seven primitives

> much-smaller than
> moderately-smaller than
> slightly-smaller than
> exactly-equal to
> slightly-larger than
> moderately-larger than
> much larger than

The authors discuss the application of the O(M) reasoning to some process engineering and bio-engineering problems. They claim that the O(M) formalism can be used to bridge the gap between qualitative reasoning with signs and full quantitative reasoning with exact numbers. They point out that in actual practice there are always many positive and negative effects on any aggregate result.

3.3 Neural Nets Reasoning

The idea of subsymbolic reasoning using neural networks, which is defined as the counterpart of symbolic (e.g., rule based) reasoning is not new. The idea was originally meant as an attempt to model the biophysiology of the brain (i.e., to understand the working and functioning of the human brain).

A biological neuron forms the basis on which artificial neural networks are based. Each of these neurons is composed of a body, an axon, and a large number of dendrites. These dendrites form a very fine "filamentary brush" surrounding the body of the neuron. The axon can be seen as a very fine long, thin tube which splits into branches terminating in little endbulbs almost touching the dendrites of other cells.

The small gap between such an endbulb and a dendrite of another cell is called a synapse. The axon of a single neuron forms synaptic connections with many other neurons. Impulses propagate down the axon of a neuron and impinge on the synapses, sending signals of various strengths down the dendrites of other neurons. The strength of the signal is determined by the efficiency of the synaptic transmission. A neuron will send an impulse down its axon if sufficient signals from other neurons impinge on its dendrites in a short period of time. A signal acting on a dendrite may be either inhibitory or excitatory. A biological neuron fires (i.e., sends an impulse down its axon) if the excitation exceeds its inhibition by a critical amount, the threshold of the neuron.

A human brain consists of many neurons, called a neural network, consisting of many interconnected neurons. In artificial neural networks (ANN) a layered structure of artificial neurons is used. The artificial neuron is a model of a biological neuron and is used to mimic its basic behaviour (see Figure 6). It consists of inputs, x_i (axons), which are weighted, w_{ij} (synapses), and fed to (dendrites) a summer, z_j (body). An output, y (axon), is produced after passing a threshold, represented by an arbitrary linear or nonlinear function, called a transfer function, $f(z_j)$.

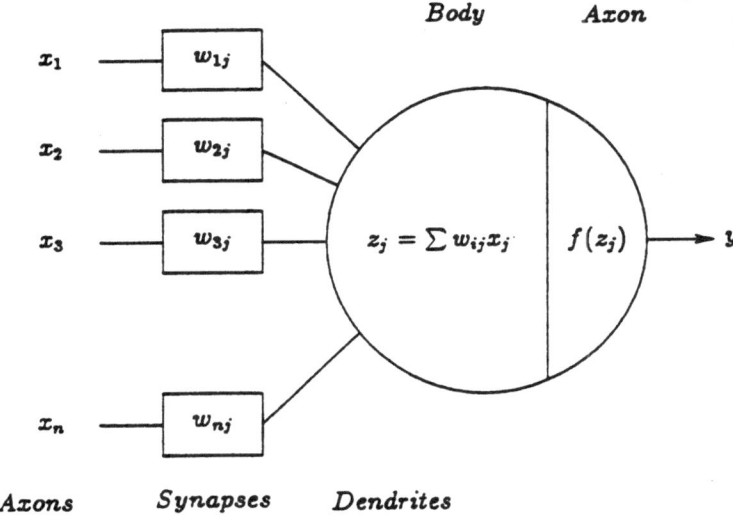

Fig. 6. McCulloch-Pitts neuron

The neuron just described is called the McCulloch-Pitts neuron. When a binary threshold is used as transfer function, this neuron is called a perceptron. In a ANN the artificial neurons are called nodes. They are usually organized into layers (see Figure 7). Data are presented to the network via the input layer, hidden layers are used for storing information, and an output layer is used to present the output of the network. In most cases ANNs have adaptable weights. Learning rules are used to adjust the interconnection strengths, w_{ij}. These rules determine how the network adjusts its

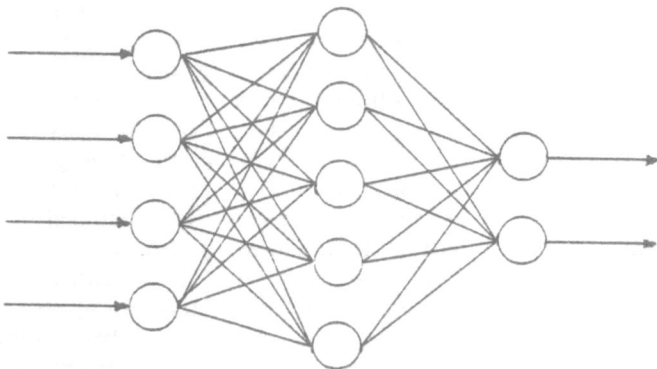

Fig. 7. Typical multilayered network

weights using some error function of any kind of criterion. In a strict forward network, information flows from the input layer, through a number of hidden layers, to the output layer. In the case of feedback connections, information reverberates around the network under control of an algorithm, taking into account a convergence algorithm. In that case the network can be described by a nonlinear dynamical system. The application of a neural network can be divided into two main phases:
1.   *Training phase*. The weights of the network are adjusted by a number of input (and output) signals in order that the network performs as desired.
2.   *Recall phase*. The network computes an output for a given input in the range of signals that has been trained.

In learning there are two essentialy different methods: *Supervised* and *unsupervised*:
1.   *Supervised learning*. The network is supplied with both the input values and the correct output values, and the network adjusts its weights based on the error of the computed output.
2.   *Unsupervised learning*. The network is provided only with the input values, and the network adjusts the interconnection strengths based only on the input values and the current network output.
        The number of algorithms developed and parameters to be chosen is very large, resulting in a large number of names connected to the various networks introduced. Among the parameters to be chosen are:
                - Number of inputs to the network
                - Number of layers (minimal one representing a linear system, called a flat network)
                - Number of nodes (the total number and the number by layer)
                - Type of the transfer function $f(z_j)$: linear, nonlinear, binary
                - Fixed or adaptable weights
        Among the algorithms to be chosen, points to be considered are:
                - Forward or feedback connections

- Supervised or unsupervised learning
- Choice of the learning law

With the introduction of multilayered neural networks more complex learning could be performed using the *back propagation learning rule*. This rule was introduced by Rumelhart et al. [34]. In literature many applications of these networks for pattern recognition and classification have been shown. An extensive overview of all possible neural network configurations is given by Khanna [35].

The main features of ANNs that are of importance for control engineering applications are the following:

1. *Learning capability*. A process model or controller is learned by experiments that correlate input examples and expected outputs. There is no direct need for an analytical description of the process to be controlled.
2. *Generalization*. In the recall phase, the network, when properly learned in the training phase, generalizes relationships for undemonstrated inputs and outputs.
3. *Cause and effect relations*. There is no strict previous structuring of the process model necessary, such as order and amount of nonlinearity.
4. *High-speed computation*. In the recall phase a minimal amount of simple computations produces the output. The calculations are straightforward and simple.
5. *Robustness*. The network can produce reasonable outputs when incomplete data are provided or a part of the network does not function adequately.

These features make it challenging to apply artificial neural networks in control engineering applications. Narendra and Parthasarathy [36] proposed many control schemes. These schemes can be used as a basis for an ANN-based control and identification scheme for practical applications (den Hertog, [37]).

4. ARTIFICIAL INTELLIGENCE IN DECISION MAKING

Many attemps have been made for applying AI concepts and techniques in decision analysis (DA) models and for producing combined AI-AD systems possessing key features of both of them [40-52].

Normative decision making models are founded on axioms from decision and measurement theories, which guarantee that if the axioms are satisfied in a problem domain, then the problem decomposition and the form of the corresponding decomposition equations are necessarily correct. For example, for additive *multiattribute utility* (MAU) models it is known that if the value contibuted by any element in the MAU model is not affected by changes to other parameter values in the model, then the use of the MAU model is normatively correct for a problem domain in which this "value independent" axiom is satisfied. Thus, representing expert knowledge becomes a process of developing decision models that reflect a problem decomposition that satisfies the axioms of a normative decision model.

In [43] the natural synergy between the prescriptive problem structuring techniques in decision analysis and the rule-based program architectures used in AI expert systems is revealed and used for developing combined AI and DA aids. Pure DA modeling procedures are appropriate for problem structuring, whereas AI rule-based architectures are suitable for making the problem structure incrementally modifiable and developing a user interface that uses only terms and references familiar to users. A possible combined AI/DA scheme is shown in Figure 8 [43]. In this scheme the decision model is encoded in the form of a set of separate composition rules that can be individually added, deleted, or modified by a general rule editor. The primary advantage of this scheme is that it supports a model-knowledge base development and enhancement process that not only can start with strict DA models for the first-cut composition trees but also can allow for incremental modification and enhancement of the initial model. This is because the scheme allows modifications to individual models.

Two particular cases of the above scheme are the following:

(i) Rule based control of a decision analytic aid [40]

(ii) Generic indicator monitoring and analytic system (GIMAS) [43]

The first is a decision aid for course-of-action (COA) evaluation that combines an additive MAU model for COA evaluation with rule based procedures for assigning parameter values (scores and weights). Another example of an aid which uses rule-based procedures for assigning (or modifying) parameter values in a decision analytic model is provided in [42].

The second provides two alternative aids thar support intelligence indicator monitoring and analysis problems and both use the GIMAS core modules. GIMAS reflects a general approach for evaluating indicators. It uses decision analytic models to identify significant low-level observable indicators and to define higher level aggregate indicators that should be monitored. The enhancements of the analyst's capability to detect and interpret changes in indicator values is performed using rule-based hypothesis-selection techniques together with other supporting technologies.

In [46] a survey is provided of the works that apply a DA system and an Expert System (ES) to the same problem-solving situation, as well as a discussion of papers that indicate how DA and ES have been and/or can be combined to form the basis of enhanced problem solving of aids with reference to the key role played by user preference. A full discussion on how the user preference concept can be used for conflict resolution in rule-based expert systems can be found in [45]. For conflict set A, suppose that I is the finite set of objectives, presumably conflicting and noncommensurate, to be considered in the selection process. It is assumed that the highest level objective is to maximize utility and that the set I is homeomorphic to a set of a metarules. Then conflict resolution can be viewed as a multi-attribute decision analysis problem. Under the assumption that the objectives are additive independent, the utility of selecting rule $a \in A$ is

$$\sum_i w_i U_i(a) = wU(a)$$

where $U_i(a)$ is the utility of rule "a" relative to objective i, and w is the tradeoff weight associated with objective i. The tradeoff weights are nonnegative and sum to one, and we assume that there is a most (least) preferred rule having utility score equal to 1 (equal to 0) for each objective and that rule a' is at least as preferred to rule a with respect to objective i if and only if $U_i(a') \geq U_i(a)$. Now, given the utility scores and tradeoff weights, the rules in A can be totally observed as follows:

"rule a' is at least as preferred as rule a if and only if $wU(a') \geq wU(a)$".

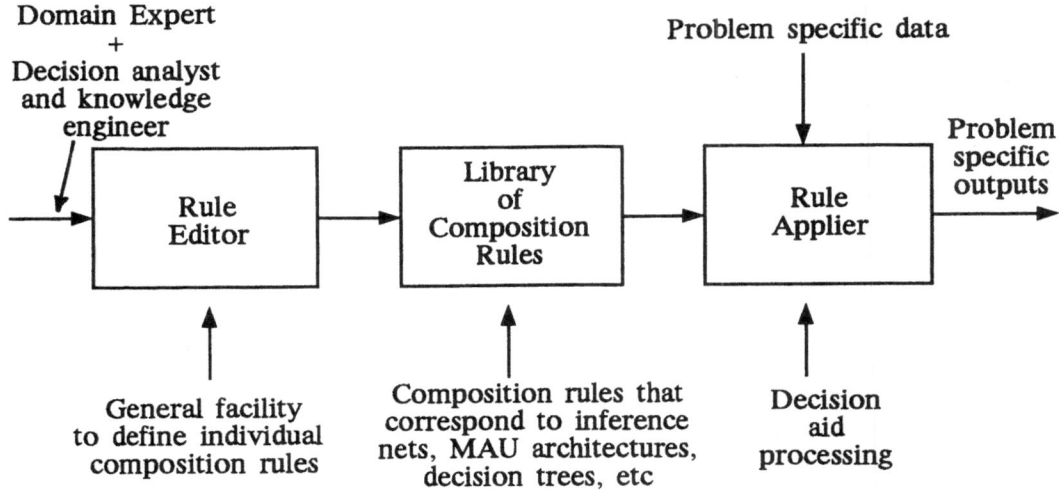

Domain Expert
+
Decision analyst
and knowledge
engineer

Problem specific data

Problem
specific
outputs

Rule
Editor

Library
of
Composition
Rules

Rule
Applier

General facility
to define individual
composition rules

Composition rules that
correspond to inference
nets, MAU architectures,
decision trees, etc

Decision
aid
processing

Fig.8. Hybrid AI/DA Architecture

This means that the determination of the most preferred rule to select for execution is straightforward if paramerer values are known precisely.

Preferences are often encountered in the form of rules about rules (metarules). As an example, the following set of rules concerning *the pollutant spill crisis management situations that permeates* [45,62] is given, where the last three rules are metarules:

R1   If the spill is sulfuric acid, then use an anion-exchanger.
R2   If the spill is sulfuric acid, then use acetic acid.
R3   Use rules that employ cheap materials beore those that employ more expensive materials.
R4   Use less hazardous methods before more hazardous methods.
R5   Use rules entered by an expert before rules entered by a novice.

One can see that rules R3 and R4 are directly related to objectives "minimize cost" and "maximize safety", respectively.

Four basically inconclusive efforts involving the construction of a DA system and an ES applied to the same problem situation were made in [49. 54-56]. In [49] ESs and DA systems are compared in the domain of diagnosis and treatment of root disorders in apple trees. The conclusion is that the knowledge engineering effort needed for the DA system is greater than that required for the ES approach and that there are significant merits in applying problem solving approaches based on normative theory over those based on *ad-hoc* procedures. In [54] a DA model and an ES have been developed for motorcycle engines' diagnosis. The comparison of the two approaches showed that the DA approach needs about 30% less time than the ES to arrive at a diagnosis. In [55] a comparison is provided of the statistical and knowledge-based approach when applied to a clinical decision making problem in gastroenterology. Finally, in [56] an ES is developed together with several mathematically based approaches for assisting managers with investment decisions referred to R&D engineeriing and production contact bidding. The result is that the ES compared favorably to the mathematically-based approaches.

Actually, due to the nature of the problem, there does not exist a standard and generally accepted way for integrating the DA and ES models. A representative but nonexhaustive set of examples of this integration can be found in [46] which are fully presented in [48, 52, 57-60]. Thurston [47] provides a methodology for incorporating the MAU model into databases, expert systems and CAD systems, with particular emphasis on CAD systems for materials selections and design for automotive structural and body skin systems. It is useful to note that the obvious, but not necessarily strong analogy between *design* and *search* has motivated many authors to integrate multicriterion optimization and heuristic search [61].

In [12] a thorough discussion is provided on how the AI and DA approaches can be usefully merged to tackle the class of tactical decision-making problems. This discussion is based on a number of combined or hybrid knowledge-based systems that have been developed by the author to help companies tackle key problems in different industries. These examples are:

Price-Strat: This is a system that can help companies price their products in competitive markets [50].

BS-opt: This is a system for banks and building societies to determine interest-rate mixes in the face of competitors [51,63-64].

Resource-opt: A system for retail space allocation or sales force allocation [52].

TAPS: A system for marketing budget allocation and sizing.

In all these examples, a three phase procedure is used for constructing the hybrid knowledge-based system, namely:

phase 1: Construction of the market knowledge base

phase 2: Determination of the decision and its sensitivity

phase 3: Enrichment of the knowledge base

A number of systems have also been developed by Singh and co-workers for strategic decision making problems [53].

5. ARTIFICIAL INTELLIGENCE IN CONTROL AND SUPERVISION

Artificial inntelligence techniques can be used at both the control system design and process supervisory control levels [65-82]. The control system design is a cut-and-try process since the control engineer develops tentative plans and then tests them by analyzing the results and executes one or more steps. Control system design needs reasoning at two levels. The higher level reasoning produces a general plan for the approach to the design and is based on basic rules about the control system design. The lower level reasoning employs backward chaining and concerns the mathematical algorithm used for the control design. However forward chaining is more appropriate when the designer wants to have the ability to replan (redesign) the system following failure of a tentative approach. The best results are obtained by a suitable combination of backward and forward reasoning.

In a conventional control system the controller design is based on fundamental knowledge described by mathematical equations (state equations, transfer functions, etc.), which are deduced from physical laws and experimental data. In *direct expert control* (d.e.c.) a knowledge-based system replaces the controller based on fundamental knowledge. It is based on the experience of the operator and the control engineer and on the observations of the process and control variables. Relationships among variables may be known or assessed in qualitative terms. Usually the knowledge based system contains a mixture of qualitative and fundamental knowledge. The d.e.c. approach can be applied successfully in those cases where the process is highly non-linear or is hard to describe because existing theories do not cover the analysis and design methods required for those systems. Because of their very nature d.e.c. systems lack the confirmation of conventional characteristics, such as the guaranteed stability of the control loop, consistency and desired prescribed performance. However, with careful supervision, also based on an expert system, acceptable control behaviour and certain amount of learning capability can be realized. It is obvious that in d.e.c. guaranteed response times are crucial and high demands are put on the processing speed of the system. An attractive proposition for real-time expert systems is the application of the progressive reasoning concept. The knowledge about the system is divided into separate knowledge bases each with its own kind of knowledge and therefore its own rule base, and the knowledge bases are hierarchically ordered. The inference engine starts with the lowest layer. When a conclusion in obtained in this layer, it is stored and the reasoning continues in the next upper layer. This reasoning/conclusion cycle continues upwards to the next layers. As soon as the system is interrupted by the process, the best conclusion up to then is taken and the related control action executed. In [83-84] a d.e.c. system based on the progressive reasoning concept has been realized. The controller consists of five hierarchically ordered layers. The first three layers classify the process in an area of the phase plane, spanned by the error signal and its difference, and calculates a proper control action. The conclusion about the conrtol action is overruled as soon the next layer comes to a conclusion. The control actions are related to proportional + reset action in the neighborhood of the origin of the phase plane. When the distance to the origin is large the controller action is similar to bang-bang control. To achieve a smooth response, a fourth layer has been implemented. The actions to be undertaken by this layer are based on forcing the response to the response of a first-order reference model. Finally, a fifth layer has been implemented to supervise the performance of the lower layers. By observing the responses of the system to the actions initiated by the lower layers, the supervisory layer decides whether the system is slow or fast, whether it contains time delay and recognizes the vicinity of the steady state. Also the adaptation of the boundaries used to classify the system is possible. This layer does not provide a control signal but it enables the rule-based system to make a choice between different strategies, and it is able to set parameters in the lower-level layers of the d.e.c.

We now come at the Supervision code which is the major part of the coding of a controller outnumbering the code for the actual control algorithm several times. Supervision code is easily expressed in logic and is a natural candidate for use in an expert system. By the merging of PLC and DDC, the combination of continuous time control and discrete logic and sequencing, expert systems can be used as an alternative or an extension to logic and sequencing and can be intermingled with logic surrounding the DDC algorithms. A number of possible applications of expert systems in this field can be distinguished: automatic tunning of controllers, supervision of adaptive and sophisticated controllers and strategy switching. The automatic tunning of PID controllers in most industrial control has been solved successfully by various approaches. One approach based on pattern recognition was developed at Foxboro (see Kraus [85]). The tunning approach mimics the procedure used by an experienced process engineer. It can be implemented in a rule-based expert system. This expert system is composed of a transient analyzer that determines the damping and the frequency of the closed loop system, followed by a design algorithm which is a collection of empirical tunning rules to tune the PID controller. Returning of the parameters can be initiated automatically or be performed on the initiative of the operator. Another approach is based on relay feedback [86]. This auto-tuner is based on the idea that the ultimate frequency, i.e. the frequency where phase lag of the open loop is 180 degrees, is the crucial information for tuning a PID-controller. The ultimate frequency can be determined from an experiment with relay feedback. A limit cycle oscillation is obtained and the controller settings are calculated from the amplitude and the period of the limit cycle. Next, the controller is automatically switched to PID control. A number of measures are taken to keep the limit cycle oscillation within specified limits.

Many times control engineers do not realize that a sophisticated control algorithm does not work properly without extensive supervision logic. Especially for complex adaptive control algorithms it is rather difficult to know all the effects of a parameter change, unexpected change in noise level and colouring, coupling of interacting loops, etc. Many parameters of the algorithm should be chosen during the design phase, probably supported by an advisory expert system, and a number of parameters, related to the given and desired system behaviour, should be adapted during operation. A rule base can be set up to manage this task based on the performance of the separate parts the adaptive controller consists of. Rules can be added relatively easy when more experience about the controller actions and process behaviour is built up. A sophisticated controller consisting of several algorithms for control, estimation, controller settings calculation and performance calculation, can be orchestrated by an expert system [87]. The expert system architecture separates the control algorithms from the logic and it supplies a convenient way to interact with the system.

6. ARTIFICIAL INTELLIGENCE IN ENGINEERING FAULT DIAGNOSIS

Actually, there is a lot of interest in the fault diagnosis field both from the process industry and from the vendors of the process control systems. The following items can be distinguished: alarm reduction, alarm diagnosis, alarm prediction, and alarm handling. *Alarm reduction* can be defined as the suppression of certain alarm messages out of a number of alarms generated in a short period. An avalanche of alarm messages and conflicting indications confuses operator and delay corrective actions. An expert system can investigate patterns of alarm messages, including their sequence in time, and can take into account the dynamics of cause-and-effect relations and predictions of some variables to advise the operator in these most stressing situations. *Fault diagnosis* is the next step to be taken. The experience built up in previous situations, the expertise of experienced operators, and all other available process knowledge (e.g. reconciliation of data, information from adjacent automation levels, etc.) can be applied in an expert system. It leads to a message describing the possible fault and recommending remedial action. In case of uncertainty, a number of alternatives with descending probability are presented to the operator, allowing him or her to select the most adequate solution for the time being, eventually supported by additional information not taken into account by the expert system. It is often possible to detect in an early stage possible malfunctioning of (a part) of the system. Measures can be taken just in time to prevent alarming conditions by using predictive models. Also, a well-scheduled maintenance scheme, eventually based on an expert system, can reduce considerably the number of emergency situations. In most cases the operator will perform the remedial action taking into account the presented fault diagnosis. However, some of these actions can also be performed automatically in an efficient way without unnecessarily violating other plant conditions. The measures to be taken could be adapted when time is passing and new information becomes available. A number of alarm monitoring and fault diagnosis expert sytems has been developed by software houses, process control systems vendors, and large process industries [88-91].

A problem related to fault diagnosis and alarm handling is real-time trajectory planning, in which an optimal path has to be followed from certain process state to a desired state, given a number of constraints related to the physical process and operations taking place. Only in very severe circumstances is this procedure done automatically by a so-called trip system. In all other cases the operator plans the measures to be taken. A real-time planning system should be able to perform this task given a rough model of the process which is based on an appropriate plant representation.

Process fault diagnosis has gained much attention in process control during the last few years [92,93]. Early detection and localization of faults is necessary to obtain a high level of performance for complex systems. A possible approach is to use very accurate (mathematical) models in which not only must the parameters be accurately estimated but also a transformation is necessary to the real physical quantities involved; that is, when a time constant is estimated, this time constant consists of at least two quantities, of which one or more can cause a fault that has to be detected.

Also, global models can be used. Techniques are being developed to build such global models of the process by means of qualitative reasoning methods [23,24,94]. It is claimed that even with incomplete knowledge, there is enough information in a qualitative description to predict the possible behaviour of an incompletely described system. Most promising are methods that combine the analytical problem solution, the qualitative reasoning approach, and heuristic knowledge in the form of fault trees and fault statistics.

A major problem is how to integrate information-processing tasks and knowledge representation in an object-oriented way. Looking at the information processing flow, four entities can be recognized:

1. *Reality or outside world*: represented by the real plant
2. *Plant representation*: portrayal of the reality
3. *Analysis task*: updating the plant representation by observing the reality
4. *Synthesis*: planning and control actions to influence the reality so that it behaves as desired.

Artificial neural networks and heuristics can be used to represent (part of) the plant. Virtual sensors can be introduced to describe numerical routines such as filtering and estimation, methods to estimate nonmeasurable variables, and so on. Next this information is used to update the plant representation. In the synthesis part the behaviour of the plant is observed and it is decided to choose a target state. Control actions are planned to reach this target state in a well-defined way [95]. Stephanopoulos [96] and Terpstra et al. [97] proposed a plant representation in which regions, fault modes, and views or scopes are recognized. A region comprises a set of constraints defining the operating state in which the model is valid. The plant can be represented in different regions. Each fault mode represents one fault, defined by a deviation of the model relative to a normal model. The same system can play a role in various contexts. Each context highlights different aspects of the same component. Views offer a "filter" that selects the aspects which are relevant to that view. The plant is divided into sections, units, subunits, components, an so on. Each is composed of objects that have the responsibility of fulfilling a function. The multilevel flow modeling (MFM) method deals with goals and functions and can be used to describe the process for fault detection and diagnosis [98].

7. ARTIFICIAL INTELLIGENCE IN ROBOTIC AND MANUFACTURING SYSTEMS

Two areas where AI/expert systems technology has found wide application are robotics and manufacturing systems. The evolution of robotic systems was done in three generations, from systems with no computing and sensory capabilities, to systems with limited computational and feedback capabilities, and finally to systems with multisensory and decision making capabilities. Third generation robotic systems, which are called intelligent robotic systems (IRS), are equipped with a diverse set of visual and non-visual sensors and are being designed to adapt to changes within their workspace environment [101]. IRS are a subclass of *intelligent machines* (IM) [102] which require utilization and implementation of concepts, ideas and techniques from AI/ES, Operational Research and Control Systems Theory.

A representative but not exhaustive classification of the areas of IRS is the following [103]:

1. Design of Intelligent Systems with learning capabilities from the control system theory perspective [104-106],

2. Design of intelligent machines with Petri nets [107-109],

3. Task planning for robotic assemblies using different AI approaches [110-115],

4. Moddelling of IMs and IRSs on the basis of Saridis' principle of increasing intelligence and decreasing precision [116-122].

From a systemic theoretic point of view, a major result in this area has been the new theory, mainly due to Saridis, defined as hierarchically Intelligent Control, directly applicable to IMs and IRSs, to organize, coordinate, and execute in a tree-like structure, anthropomorphic tasks with minimum supervision and interaction with a human operator [117]. The system hierarchical structure obeys the Principle of Increasing Intelligence and Decreasing Precision, mathematically proven by Saridis [121]. Meystel [28,29] introduced the so-called "Nested Hierarchical Information

Structure" which, although not tree-like is equally beneficial and applicable to problems formulated within the theory of intelligent controls.

A major area of development from the pure AI point of view has been the plan generation for intelligent and robotic systems. Of central importance in plan generation are the state representation, the system architecture and the planning strategies. Planning strategies are dominated by heuristic methodologies like depth-first or breadth-first search, best first search, A* algorithm and means ends analysis [122-124]. AI planning systems employ explicit symbolic, logical or temporal logic models of discrete operations to construct sequences of tasks to achieve specific task goals either at the motion and path planning level or at the task planning level [125-128]. In AI planning systems, the problem domain is considered to be the knowledge base and the planner the inference engine [128,129]. Searching for the most efficient plan requires criteria to decide whether one plan is better than another. The information necessary to make this decision is usually available during the actual execution of the plan. Thus, there is a choice between off-line and on-line (real time) planning. AI planning approaches have dominated most of the research in robot assembly task planning using domain independent methods [130]. The central idea of these methods is to have one general purpose inference engine that may be used for any domain by describing initial state, goal and the operators in a logic formalism. The major problem with linear planners is the goal interaction. That is in many situations, goal interactions cannot be removed by simply reordering the operator sequence by which these goals are achieved. Rather it requires that operations be intermixed. This observation has led to the so called class of nonlinear planners. Although nonlinear planners were shown to be more efficient than linear ones, however it can be proved that they have by no means solved general planning problems. Task planning for RAs is critically dependent on the task representation. Systematic procedures are needed to guarantee that every assembly sequence is generated. Most assembly type planners are hierarchical planners that use data structures such as graphs or trees to represent objects. Among the representation schemes that are often used for storing all posible assembly plans for a given product we mention here the following [111,112,130]; (i) ordered lists of actions, (ii) triangle tables, (iii) precedence diagrams, and (iv) AND/OR graphs. Task planning for IRSs involves similar issues with task planning for RAs. When an IRS is designed on the basis of Saridis' hierarchical structure composed of three interactive levels (organization, coordination, execution) all planning and decision making operations are taking place within the highest level, i.e. the organizer. In general, performance of such systems is improved through self-modification with learning algorithms and schemes interpreted as iterative procedures for the determination of the best possible action given a specific user command [116-122]. Learning in IRSs is considered as finding an optimal action out of a set of allowable actions. Similar ideas are applicable even when the IRS is not designed on the basis of the three-level hierarchical model [104,105,131]. In [120] a hardware and sotware methodology and an expert system was introduced for IRSs, where the ES works at the organization level and is being used to develop off-line plan scenarios to execute a user requested job. The knowledge base and the inference engine have been designed so as, although they are separate entities, they operate closely together as a whole.

We are now turning our discussion on the computer-aided manufacturing systems [132-144]. The principal functions of such systems that provide a challenge for

AI and knowledge-based techniques are:
- Product design and design for assembly (DFA)
- CAD/CAM in terms of features
- Process planning, scheduling, and control
- Dynamic simulation of flexible manufacturing systems (FMS)
- Equipment selection
- Fault diagnosis
- Facility layout
- Quality assurance

Three important AI areas that are mostly used in computer-integrated manufacturing (CIM) systems are automatic planning, automatic learning and qualitative simulation. The first and third have already been discussed in the present chapter. Two planning systems are reviewed in [133]. These are the systems STRIPS and NOAH (see also [130],[134]). STRIPS can also be found in [123]. Other examples of AI-based planners are GPS (General Problem Solver) [135], WARPLAN [136], AL3 [137], and SIPE [138]. STRIPS is actually a Lisp implementation of GPS [112], and WARPLAN is an improved version of STRIPS in Prolog.

STRIPS uses formal logic for representing domain knowledge, and so the current state is represented by a logical statement involving terms of relations among objects. Permissible operations actually remove some of the relations from the current state and introduce new relations to it. STRIPS is appropriate for robot planning systems. WARPLAN allows nonlinear construction of plans and is somewhat superior to STRIPS, but it possesses the drawbacks of the depth-first search strategy it employs. NOAH is a planner that can examine the nature of incomplete ordering for actions in plans. SIPE has the capability of replanning whenever some plan fails to achieve the desired goal. During the execution of a plan by a robot, some action in the plan may not lead to the desired goal. This can be detected by the robot sensors (e.g. visual or tactile), and the robot has to make a new plan. SIPE modifies the original plan so as to work well with the new, unforeseen conditions. Planning in CIM is needed not only for generating robot actions on the basis of sensory feedback, but also for planning the procedures of collecting information.

Automatic Learning: Automatic learning can be accomplished using the following approaches:
- Learning from instruction
- Learning by analogy
- Learning from examples (inductive learning)
- Learning by discovery

Of course, there are other styles of learning, such as learning through visual imagery or through tactile sensing. So far, machine learning has mainly been connected with symbolic forms of learning.

Generally the learning problem is the following:

Given a set of objects specified in some description language and the classes to which the objects belong,

Find a general calssification rule that "explains' the learning set of objects and can also be used for the classification of new objects.

Explaining a set of objects means classifying them into classes as specified in the initial specification. Learning algorithms differ in their use of particular generalizing and specializing rules, rule selection criteria, whether negative instances are included in the training set, and whether a bottom-up or top-down learning strategy is used.

A useful collection of chapters on various approaches of learning is provided in Michalski et al. [139]. A discussion on the relation of machine learning techniques and expert systems is presented in Bundy [140]. Several examples of learning systems are given in Quinlan [141], Shepherd [142], Dechter and Michie [143], and Dufay and Latcombe [144].

A brief list of a number of knowledge-based expert systems for manufacturing systems that are available in the open literature is the following

i) **PROPLAN (PROcessPLANning) [145,146]:**
This is a knowledge-based environment that integrates the design and planning phases for mechanical parts. Its main advantage is the reduction of human intervention between design and planning.

ii) **AIFIX [147]:**
This is an expert system suitable for designing fixtures for parts being produced on milling machines. It is an if-then rule based system and involves FORTRAN routines that are called by LISP functions.

iii) **DOMINIC I [148]:**
This is a rather domain-independent expert system (coded in Common Lisp) for iterative design of mechanical parts. Design and performance data are introduced through a knowledge acquisition module. The system asks the user to provide an initial design or generates such a design automatically.

iv) **GARI [149]:**
This is an if-then rule based expert system that employs a general problem solver (GPS) and appropriate knowledge for planning the sequence of machining cuts for mechanical parts. The approach of iterative refinement is followed, and the part geometry is represented in terms of features.

v) **TOM (TechnOstructure Machining) [150]:**
This is written in Pascal and generates a plan for a given finished geometry in the hole-making domain. It is again an if-then rule-based system with backward inference (alpha-beta) strategy.

vi) **HI-MAPP (Hierarchical Intelligent Manufacturing Automated Process Planner) [151]:**
This is similar to GARI with regard to part presentation but differs in that it produces hierarchical plans and produces initially an abstract of a correct plan, whereas GARI generates initially a loosely constrained plan and uses a time-consuming refinement process.

vii) **EXCAPP (EXpert Computer-Aided Process Planning) [152]:**
It is written in Pascal and provides process plans for rotational parts only.

viii) **OPEX (OPEration Sequence planning) [153]:**
This is an expert system appropriate for operation sequence planning and can be used either as a stand-alone shell or as a numerical control (NC) part programming tool.

ix) **SIPP [154]:**

This is a frame-based expert system for planning parts requiring metal removal processes.

x) **SAPT (System for Automatic Programming Technology) [155]:**

This is a part designer expert system consisting of a product designer segment, a manufacture designer segment, and a production planner controller segment. These three segments employ the same engineering knowledge base but are mutually independent with respect to modularity.

xi) **CEMAS (CEll MAnagement System) [156]:**

This is designed to supervise a manufacturing cell. It contains a data-base, a knowledge-base, an inference engine, a knowledge acquisition block, and an explanation block. It is realized in Lisp supported by a data structure array. CEMAS has three activity levels: (1) the execution level, which controls the moves of the machine tools, (2) the information level, which records the actual state, and (3) the decision level, which schedules work to the workstations.

We close our discussion on computer aided manufacturing systems by mentioning some other important works that deal with several aspects. Fox and Smith [157] describe the system ISIS which is suitable for process scheduling in a job shop facility. Bruno and Marchetto [158] propose a production scheduling system that combines the expert system methodology with queueing network analysis for fast performance evaluation. Mill and Spraggett [159] present a rule-based expert scheduling system that includes the design of a database of part geometry. An integrated environment for intelligent manufacturing automation is described in [160]. This environment contains the following components: interface to external environment, meta-knowledge base, data base, inference mechanism, static blackboard, and interface to other subsystems. This system has been applied for the product design in a manufacturing process. A knowledge-based approach for manufacturing systems is proposed in [161] and its advantages over the optimization-based approach are discussed. An approach to building up a distributed expert system for planning, scheduling and fault diagnosis is presented in [162]. The system is integrated with the conventional FMS. The overall system is composed of an intelligent control, planning, diagnosis, supervisor and expert system. A manufacturing environment for the production of step controllers is used as a case study. Finally, a number of important contributions in the area of AI application in manufacturing systems can be found in [93,101,163-165].

CONCLUSIONS

This chapter was designed to be an introduction to the present book on AI in industrial decision making, control and automation. We have included a representative set of issues and concepts, sufficient for this purpose but by no means exhaustive. The chapter includes a discussion of three fundamental AI methodologies, and a rich set of references where the reader can find detailed derivations and discussions on a large variety of AI systems as applied to decision making, control and automation. Further issues can be found in the recent books [166-170]. Fault diagnosis is a primary function of automated systems and was discussed separately. The discussion on neural nets is very general. More detailed studies can be found in other chapters of the book.

REFERENCES

1. F.G. Castles, D.J. Murray and D.C. Potter, Decisions, Organisations and Society, *Penguin*, Harmondsworth, U.K., 1971.
2. D.V. Lindley, Making Decisions, *John Wiley*, New York, 1985.
3. P.H. Hill et. al. (eds), Making Decisions: A Multidisciplinary Introduction, *Addison-Wesley*, Reading, Mass., 1979.
4. S. Schwartz and T. Griffin, Medical Thinking, *Springer*, Berlin, 1986.
5. D. Kahneman et al. (eds), Judgement and Uncertainty: Heuristics and Biases, *Cambridge Univ. Press*, Cambridge, U.K., 1982.
6. B. Fischoff et al., Subjective Expected Utility: A Theory of Decision Making, In: *Decision Making Under Uncertainty* (R. Scholz, ed.), *Elsevier/North Holland*, Amsterdam, pp.160-175, 1983.
7. R. Hogarth, Judgement and Choice: The Psychology of Decision, *John Wiley*, New York, 1980.
8. D.E. Broadbent, Decision and Stress, *Academic Press*, New York, 1972.
9. S.G. Tzafestas, Systems, Management, Operational Research and Control: A Unified Look, *EURO III: Europ. Symp. Op. Res.*, Amsterdam, April 1979.
10. S.G. Tzafestas (ed.), Optimisation and Control of Dynamic Operational Research Models, *North-Holland*, Amsterdam, 1982.
11. J. Fox, Symbolic Decision Procedures for Knowledge-Based Systems, In: *Knowledge Engineering* (H. Adeli, ed.), *McGraw-Hill*, New York, pp.26-55,1990.
12. M. Singh, Artificial Intelligence and Decision Support Systems for Managerial Decision Making, In: *Applied Control: Current Trends and Modern Methodologies* (S.G. Tzafestas, ed.), *Marcel Dekker*, New York, pp.913-935, 1993.
13. R.N. Anthony, Planning and Control Systems: A Framework for Analysis, *Harvard University Press*, Cambridge, Mass., 1965.
14. R. Forsyth, Expert Systems: Principles and Case Studies, *Chapman and Hall*, New York, 1984.
15. I. Graham and P.L.Jones, Expert Systems: Knowledge, Uncertainty and Decision, *Chapman and Hall*, London, 1988.
16. E.H. Shortliffe, Computer Based Medical Consultations: MYCIN, *American Elsevier*, 1976.
17. L.A. Zadeh, Fuzzy Sets, *Information and Control*, Vol.8, pp.338-353,1965.
18. L.A. Zadeh, Fuzzy Algorithms, *Information and Control*, Vol.11, pp.323-339, 1969.
19. L.A. Zadeh, Fuzzy Logic, *IEEE Computer*, pp.83-93, April 1988.
20. G. Shafer, A Mathematical Theory of Evidence, *Princeton Univ. Press*, 1976.
21. G. Shafer, Belief Functions and Possibility Measures, In: *The Analysis of Fuzzy Information* (J.C. Bezdek, ed.), Vol.1, CRC Press, Boca Raton, Fl., pp.51-84, 1987.
22. A.P. Dempster, A Generalization of Bayesian Inference, *J. Royal Stast. Soc.*, Series B30, pp.205-247, 1968.

23. B. Kuipers and C Chiu, Timing Intractable Branching in Qualitative Simulation, *Proc. 10th IJCAI*, pp.1079-1085, 1987.
24. B. Kuipers and D. Berleant, Using Incomplete Quantitative Knowledge in Qualitative Reasoning, *Proc. AAAI'88*, pp.324-329, 1988.
25. J. De Kleer and B.C. Williams, Diagnosis with Behavioural Models, *Proc. 11th IJCAI*, 1989.
26. C. Price and M. Lee, Why do Expert Systems Need Deep Knowledge? , *Proc. 12th IMACS World Congress*, pp.264-266, 1988.
27. J. Francis and R.R. Leith, Intelligent Knowledge-Based Process Control, *Proc. IEE Conf. on Control*, Cambridge, 1985.
28. J. Francis, Qualitative System Theory, *Ph.D. Thesis*, Electr. & Electronic Eng. Dept., Heriot Watt Univ., 1986.
29. W.F. Clocksin and A.J. Morgan, Qualitative Control, P*roc. 7th Europ. Conf. on AI*, pp.350-356, 1986.
30. B.A. Abdulmajid and R.J. Wynne, The Application of Qualitative Reasoning to Real Time Control, *CCAI J.*, Vol.9, No.2-3, pp.175-193, 1992.
31. B.A. Abdulmajid and R.J. Wynne, Initial Experience in Applying Qualitative Reasoning to Process Control, *Proc. 28th IEEE Conf. on Decision and Control*, pp.783-784, 1989.
32. B.A. Abdulmajid and R.J. Wynne, An Improved Qualitative Controller, *Proc. 1990 ACC*, pp.1455-1460, 1990.
33. B.A. Abdulmajid, *Ph.D. Thesis*, Univ. of Manchester, Dept. of Engrg., 1992.
34. D.E. Rumelhart, G.E. Hinton and R.J. Williams, Learning Internal Representations by Error Propagation, In: *Parallel Distributed Processing* (D.E. Rummelhart et. al. eds.), *MIT Press*, Cambridge, MA, pp.318-362, 1986.
35. T. Khanna, Foundations of Neural Networks, Addison Wesley, *Reading*, USA, 1990.
36. K.S. Narendra and K. Parthasarathy, Identification and Control of Dynamical Systems Using Neural Networks, *IEEE Trans. on Neural Networks*, Vol.1, pp.4-27, 1990.
37. P.A. den Hertog, Neural Networks in Control, *Report A91.003(552)*, Control Lab., TUDelft, 1991.
38. M.L. Mavrovouniotis and G. Stephanopoulos, Reasoning with Order of Magnitude and Approximate Relations, *Proc. AAAI-87*, 1987.
39. M.L. Mavrovouniotis and G. Stephanopoulos, Formal Order of Magnitude Reasoning in Process Engineering, *Computers and Chemical Engineering*, Vol.12, Nos.9-10, pp.867-880, Pergamon Press, 1988.
40. P.E. Lehner, Combining Decision Analysis and Artificial Intelligence Techniques in a Decision Aid, *Tech. Report no.84-131*, PAR Technology Corp., McLean, VA 1985.
41. R. Steeb and S.C. Johnson, A Computer-Based Interactive System for Group Decision Making, *IEEE Trans. Syst. Man and Cybern.*, vol.SMC-8, No.8, pp.544-552, 1981.

42. J.J. Weiss and K.A. Waslov, CTA-A C^3CM Planning Aid: Expert Knowledge Base Description, *DDI/TR 83-2-183*, Decision and Designs Inc., McLean, VA, April, 1983

43. P.E. Lehner, M.A. Probus and M.L. Donnell, Building Decision Aids: Exploiting the Synergy Between Decision Analysis and Artificial Intelligence, *IEEE Trans. Syst., Man, and Cybern.*, Vol SMC-15, No 4, pp 469-474, 1985.

44. A. Ligeza, Expert Systems Approach to Decision Support, *Europ. J. Op. Res.*, Vol. 37, pp. 100-110, 1988

45. C.C. White, III and E.A. Sykes, A User Preference Guided Approach to Conflict Resolution in Rule-Based Expert Systems, *IEEE Trans. Syst., Man, and Cybern.*, Vol. SMC-16, No 2, pp. 276-278, 1986

46. C.C. White, III, A Survey of the Integration of Decision Analysis and Expert Systems for Decision Support, *IEEE Trans. Syst., Man, and Cybern.*, Vol. 20, No 2, pp. 358-364, 1990

47. D.L. Thurston, Towards Integration of Multiattribute Utility Analysis and Expert Systems at the Rule Level for Optimal Design, *Working Paper*, Dept. Gen. Eng., Univ. of Illinois, 1988

48. V.S. Jacob, J.C. Moore and A.B. Whinston, Rational Choice and Artificial Intelligence, *Tech. Report*, Krannert Graduate School of Management and Dept. of Computer Science, Purdue Univ., West Lafayette, IN, May 1987

49. M. Henrion and D.R. Cooley, An Experimental Comparison of Knowledge Engineering for Expert Systems and for Decision Analysis, *Proc. AAAI-87 6th Natl. Conf. on AI*, Seattle, WA, pp 471-476, July 1987.

50. M.G. Singh, The Price-Strat Approach for Automobile Pricing, in *Managerial Decision Support Systems and Knowledge-Based Systems* (M.G. Singh, K. Hindi and D. Salassa, eds), North-Holland, Amsterdam, 1988

51. M.G. Singh and R. Cook, Price Strat: A Decision Support System for Determining Bank and Building Society Interest Rate Mixes, *Intl. J. of Bank Marketing*, Vol. 5, No 3, 1986.

52. M.G. Singh, R. Cook and M. Corstjens, A Hybrid Knowledge-Based System for Retail Space and Other Allocation Problems, *Interfaces*, May, 1988.

53. M.G. Singh, J. Bhondi and M. Corstjens, Mark-opt: A Negotiating Tool for Manufacturers and Retailers, *IEEE Trans. Syst., Man, and Cybern.*, Vol. 15, No. 14, pp 483-495, 1985.

54. J. Kaladgnamam and M. Henrion, A Comparison of Decision Analysis and Expert Rules for Sequential Diagnosis, *Proc. 4th AAAI Workshop on Uncertainty and AI*, 1988

55. D.J. Spiegelhalter and R.P. Knill-Jones, Statistical and Knowledge-Based Approaches to Clinical Decision Support Systems with an Application to Gastroenterology, *J. R. Stat. Soc.*, Vol. 147, pp 35-77, 1984.

56. G.R. Madey, M.H. Wolfe and J. Potter, Development of an Expert Investment Srategy System for Aerospace RD & E and Production Contract Bidding, *Tech. Report*, Goodyear Aerospace Corp., Akron, GH, 1987

57. P.H. Farquhar, Applications of Utility Theory in Artificial Intelligence Research, in: *Toward Interactive and Intelligent Decision Support Systems* (Y. Sawaragi et. al. eds.), Springer-Verlag, New York, pp. 155-161, 1987
58. R.L. Keeney, Value-Driven Expert Systems for Decision Support, *Decision Support Systems*, Vol. 4, pp. 405-412, 1988
59. E.J. Horvitz, J.S. Breese and M. Henrion, Decision Theory in Expert Systems and Artificial Inyelligence, *Int. J. Approximate Reasoning*, 1988
60. J. Ch. Pomerol, MCDM and AI: Heuristics, Expert and Multi-Expert Systems, *Working Paper*, Lab. Informatique et Decision, Univ. P. et M. Curie, Paris, 1988.
61. B.S. Stewart, Heuristic Search with a General Order Relation, *Ph. D. Dissertation*, Dept. Syst. Eng., Univ. of Virginia, 1988.
62. F. Hayes-Roth et. al., Building Expert Systems, Addison-Wesley, Reading, MA, 1983
63. M.G. Singh and R. Cook, Improving the Profitability of Bank Lending in a Competitive Environment, *Int. J. on Bank Marketing*, Vol. 5, No. 3, 1987
64. M.G. Singh and R. Cook, Decision Support Systems for Assets and Liability Management in Retail Banking, *Quart. J. Girocentralle Bank*, Austria, 1988.
65. S.G. Tzafestas, S. Abu El Ata-Doss and G. Papakonstantinou, Expert System Methodology in Process Supervision and Control, in: *Knowledge-Based System Diagnosis, Supervision, and Control* (S.G. Tzafestas, ed.), Plinum, New York, pp. 181-215, 1989
66. S. Abu El Ata-Doss and P. Ponty, Supervision of Controlled Processes in Nonstationary Conditions, *Proc. 7th IFAC Symp. on Identification and Syst. Param. Estimation*, York, England (July, 1985).
67. S. Abu El Ata-Doss and J. Brunet, On-line Expert Supervision for Process Control, *Proc. 25th CDC*, Athens, Greece, Dec. 1986.
68. S. Abu El Ata-Doss and J. Brunet, Conception of Real-Time Expert Supervision Based on Quantitative Model Simulation of the Process, *Proc. 1st ICIAM*, Paris, France, June, 1987.
69. K.J. Astrom, J.J. Anton and R.E. Arzen, Expert Control, *Automatica*, Vol. 22, No. 3, 1986
70. R.L. Moore et. al., A Real Time Expert System in On-line Process Control, *Proc. 1st Conf. on AI Applications*, Denver, Dec., 1984
71. R.L Moore and M.A. Kramer, Expert Systems in On-line Process Control, *Proc. ASILOMAR*, U.S.A., Jan., 1986
72. S.P. Sanoff and P.E. Wellstead, Expert Identification and Control, *Proc. 7th IFAC Symp. on Identification and Parameter Estimation*, York, England, July, 1985.
73. S.G. Tzafestas, Artificial Intelligence and Expert Systems in Control: An Overview, *Syst. Anal. Modelling Simul.* (SAMS), Vol 7, No. 3, pp. 171-190, 1990
74. S.G. Tzafestas and A. Ligeza, A Framework for Knowledge-Based Control, *J. Intell & Robotic Syst.*, Vol. 2, pp 407-425, 1989.
75. H.B. Verbruggen et. al., Artificial Intelligence in Real-Time Control, In: *Applied Control: Current Trends and Modern Methodologies* (S.G. Tzafestas, ed.), pp. 785-824, 1993

76. B.P. Butz, and N.F. Palumbo, Expert Systems for Control Systems Design, *ibid*, pp. 825-850, 1993
77. K.E. Årzen, Expert Control: Intelligent Tuning of PID Controllers, *ibid*, pp. 851-874, 1993
78. J. Aguilar-Martin, Knowledge-Based Real-Time Supervision of Dynamic Processes: Basic Principles and Methods, *ibid*, pp. 875-910.
79. J.R. James, D.K. Frederick and J.H. Taylor, The Use of Expert Systems Programming Techniques for the Design of Lead-Lag Compensators, *Proc. Control '85*, Cambridge, England, pp. 1-6, 1985
80. T.L. Trankle, P. Sheu and U.H. Rabin, Expert System Architecture for Control System Design, *Proc. ACC*, Paper TP5, pp 1163-1169, 1986.
81. G.K.H. Pang, J.M. Boyle and A.G.J. MacFarlane, An Expert System for Computer-Aided Linear Multivariable Control System Design, *Proc. IEEE Symp. on CACSD*, pp 1-6, Sept. 1986
82. K.K. Gidwani, The Role of AI in Process Control, *Proc. ACC*, pp. 881-884, 1986.
83. H.M.T. Broeders, P.M. Bruijn and H.B. Verbruggen, Real Time Direct Expert Control, *Engineeiring Applications of Artificial Intelligence*, Vol.2, No. 2, pp. 109-119, 1989
84. A.J. Krijgsman, P.M. Bruijn and H.B. Verbruggen, Knowledge-Based Control, *Proc. 27th IEEE Conf. on Decision and Control*, Austin, Texas, U.S.A., 1988.
85. T.W. Kraus and T.J. Myron, Self-Tuning PID Controller Uses Pattern Recognition Approach, Control Engrg. Magaz., June, 1984
86. K.J. Åstrom and T. Hägglund, A New Auto-Tuning Design, *Proc. IFAC Int Symp. on Adaptive Control of Chemical Processes, Copenhangen,* Denmark, 1988.
87. K.E. Årzen, Realization of Expert-System Based Feedback Control, *Ph. D. Thesis* (Code: LUTFD2/TFRT-1029), Dept. of Automatic Control, Lund Inst. of Technology, Lund, Sweden, 1987
88. A.I. Kokkinaki, K.P. Valavanis and S.G. Tzafestas, A Survey of Expert Systems Tools and Engineering-Based Expert-Systems, In:*Expert Systems in Engineering Applications* (S.G. Tzafestas, ed.) Springer-Verlag, Berlin, pp. 367-378, 1993
89. T. Walker and R.K. Miller, Expert Systems: An Assessment of Technology and Applications, *Madison GA: SEAI Technical Publications*, 1986.
90. S.G. Tzafestas, System Fault-Diagnosis Using the Knowledge-Based Methodology, In: *Fault Diagnosis in Dynamic Systems* (R. Patton, P. Frank and R. Clark, eds.), Prentice Hall, pp. 509-572, 1989.
91. W.T. Scherer and C.C. White, III, A Survey of Expert Systems for Equipment Maintenance and Diagnostics, in: *Knowledge-Based System Diagnosis, Supervision, and Control* (S.G. Tzafestas, ed.), Plenum, London, pp. 285-300, 1989
92. R. Patton, P. Frank and R. Clark, Fault Diagnosis in Dynamic Systems, *Prentice Hall*, New York, 1989.
93. S.G. Tzafestas, Knowledge-Based System Diagnosis, Supervision and Control, *Plenum*, New York-London, 1989
94. B. Kuipers, Qualitative Reasoning: Modelling and Simulation with Incomplete Knowledge, *Automatica*, Vol. 25, No. 4, pp. 571-585, 1989

95. R.H. Fusillo and G.J. Powvers, Operating Procedure Synthesis Using Local Models and Distributed Goals, *Computers and Chemical Enginnering*, Vol.12, No. 9/10, pp. 1023-1034, 1988.

96. G. Stephanopoulos, Artificial Intelligence... What Will its Contributions be to Process Control? In: *Proc. of The Second Shell Process Control Workshop: Solutions to the Shell Standard Control Problem* (D.M. Prett et. al., eds), Butterworths, pp. 591-646, 1990.

97. V.J. Terpstra, H.B. Verbruggen and P.M. Bruijn, Integrating Information Processing and Knowledge Representation in an Object-Oriented Way, *Proc. IFAC Workshop on Comp. Software Structures Integrating AI/KBS Systems in Process Control*, Bergen, Norway, May, 1991.

98. M. Lind, Representing Goals and Functions of Complex Systems: An Introduction to Multilevel Flow Modelling, *Tech. Report*, Technical Univ. of Denmark, Inst. of Automatic Control Systems, 1990.

99. S.G. Tzafestas and L.Palios, Improved Probabilistic Inference Engine for System Fault-Diagnosis, *Systems Science*, Vol.19, No.4, pp. 5-15, 1993.

100. S.G. Tzafestas, L. Palios and F. Cholin, Diagnostic Expert System Inference Engine Based on the Certainty Factors Model, *Knowledge-Based Systems*, Vol.7, No.1, pp. 17-26, 1994.

101. S.G. Tzafestas, Intelligent Robotic Systems, *Marcel-Dekker*, New York, 1991.

102. K.P. Valavanis and G.N. Saridis, Intelligent Robotic Systems: Theory, Design and Applications, Kluwer, Boston, 1992

103. K.P. Valavanis, A Review of Intelligent Control Based Methodologies for the Modeling and Analysis of Hierarchically Intelligent Systems, *Tech. Report*, Dept. ECE, Northeastern Univ., Boston, U.S.A., 1989.

104. A. Meystel, Intelligent Control in Robotics, *J. Robotic Syst.*, Vol.5, No4, August, 1988

105. A. Meystel, Intelligent Motion Control in Anthropomorphic Machines, In: *Applied Artificial Intelligence*, (S. Andriole, ed.) *Pentrocellis Books*, Princeton, N.J., 1985.

106. A A.Meystel (ed.), Special Issue on Intelligent Control, *J. Intell. & Robotic Systems*, Vol.2, Nos2,3, 1989

107. K.M.Passino and P.J. Antsaklis, Artificial Intelligence Planning Problems in a Petri Net Framework, *Proc. 1988 ACC*, Atlanta, Georgia, 1988

108. K.M. Passino, and P.J. Antsaklis, Planning Via Heurisitc Search in a Petri Net Framework, *Proc. 3rd Int. Symp. on Intelligent Control*, Arlington, VA, 1988.

109. K. Jensen, Coloured Petri Nets, In: Petri Nets: Central Models and Their Properties (B. Brauer et. al., eds.), *Springer-Verlag* (Lecture Notes in C.S.), New York, 1987

110. L. De Mello, and A.C. Anderson, AND/OR Representation of Assembly Plans, *Proc. AIII-86*, Philadelphia, PA, August, 1986.

111. B.R.Fox and K.G. Kempf, Opportunistic Scheduling for Robotic Assembly, *Proc. 1985 IEEE Int. Conf. on Robotics and Automation, IEEE Computer Soc. Press*, pp 880-889, 1985

112. R.E. Fikes et. al., Learning and Executing Generalized Robot Plans, *Artificial Intelligence*, Vol.3, pp. 251-288, 1972
113. H. Zhang and A.C. Sanderson, Generation of Precedence Relations for Mechanical Assemblies, CIRSSE Tech. Report No. 34, RPI, 1989
114. E.D. Sacerdoti, Planning in Hierarchy of Abstraction Spaces, *Artificial Intelligence*, Vol.5, N0.2, 1974
115. M. Stefik, Planning with Constraints (MOLGEN: Part1), *Artificial Intelligence*, Vol.15, No.2, 1981
116. G.N. Saridis, Toward the Realization of Intelligent Controls, *Proc. IEEE*, Vol.67, No.8, August, 1979.
117. G.N. Saridis, Intelligent Robotic Control, *IEEE Trans. on Auto. Control*, Vol.AC-28, No.5, May, 1983
118. G.N. Saridis and K.P. Valavanis, Analytical Design of Intelligent Machines, *Automatica*, Vol.24, No.2, 1988.
119. K.P. Valavanis and G.N. Saridis, Information Theoretic Modelling of Intelligent Robotic Systems, *IEEE Trans on Syst., Man, and Cybern.*, Vol.SMC-18, No.6,1988
120. K.P. Valavanis and P.H. Yuan, Hardware and Software for Intelligent Robotic Systems, *J. Intell. & Robotic Syst*, Vol.1, No.4, 1989.
121. G.N. Saridis, Analytical Formulation of the Principle of Increasing Precision with Decreasing Intelligence for Intelligent Machines, *Automatica*, 1989.
122. F. Wang and G.N. Saridis, Structural Formulation of Plan Generation for Intelligent Machines, *CIRSSE Tech. Report No.40*, RPI, 1984
123. R.E. Fikes and N.J. Nilsson, STRIPS: A New Approach to the Application of Theorem Proving to Problem Solving, *Artificial Intelligence*, Vol.2, 1971.
124. J. Pearl, Heuristics, Addison-Wesley, 1984
125. N.J. Nilsson, Problem Solving Methods in Artificial Intelligence, McGraw-Hill, 1971
126. D. McDermott, Reasoning About Plans, In: *Formal Theories of the Common Sence World* (E. Hobbs and Moore, eds), *Ablex Publ.*, NJ., 1985.
127. D.McDermott, A Temporal Logic for Reasoning about Processes and Plans, *Cognitive Science*, Vol.6, 1982.
128. E. Charniak and D. McDermott, Introduction to Artificial Intelligence, *Addison-Wesley*, 1985.
129. K.M. Passino and P.J. Antsaklis, A System and Control Theoretic Perspective on Artificial Intelligence Planning Systems, *Applied Artificial Intelligence*, Vol.3, 1989.
130. P.C.-Y. Sheu and Q. Xue, Intelligent Robotic Planning Systems, *World Scientific*, Singapore-New York, 1993
131. K.P. Valavanis and S.G. Tzafestas, Knowledge Based (Expert) Systems for Intelligent Control Applications, In: *Expert Systems in Engineering Applications* (S.G. Tzafestas ed.), Springer-Verlag, Berlin, pp. 259-268. 1993.
132. S.G. Tzafestas, AI Techniques in Computer-Aided Manufacturing Systems, In: *Knowledge Engineering* (H. Adeli, ed.), McGraw-Hill, New York, pp. 161-212, 1990.

133. L. Daniel, Artificial Intelligence: Tools, Techniques and Applications, Harper & Row, N.Y., 1984

134. E.D. Sacerdoti, A Structure for Plans and Behaviour, *Elsevier*, N.Y., 1977

135. G.W. Ernst and A. Newell, GPS: A Case Study in Generality and Problem Solving, *Academic Press*, N.Y., 1969.

136. D.H.D. Warren, WARPLAN: A System for Generating Plans, *DCL Memo 76*, Dept. of AI, Edinburgh Univ., 1974.

137. I. Bratko, Knowledge-Based Problem in AL3, In: *Machine Intelligence* (J. Hayes, D. Michie and J.H. Pao, eds.), Harwood, Chichester, U.K., pp 73-100, 1982.

138. D.E Wilkins, Recovering from Execution Error in SIPE, *Comput. Intell. J.*, Vol.1, 1985.

139. R.S. Michalski, J.G. Carbonell and T. Mitchell, *Machine Learning*, Tioga, Palo Alto, Calif., 1983

140. A. Bundy, What Has Learning Got To Do with Expert Systems? *Paper No.214*, Dept of AI, Univ. of Edinburgh, 1984.

141. J.R. Quinlan, Discovering Rules by Induction from Collections of Examples, In: *Expert Systems in the Microelectronic Age* (D. Michie, ed.), Edinburgh Univ. Press, Edinburgh, pp. 168-202, 1981

142. B.A. Sepherd, An Appraisal of a Decision Tree Approach to Image Classification, *Proc. 8th Int. Joint Conf. on Artificial Intelligence* (IJC AI '83), Vol.1, pp. 473-475, 1983

143. R. Dechter and D. Michie, Structural Induction of Plans and Programs, IBM, Los Angeles, Calif., 1984

144. B. Dufay and J.-C. Latcombe, An Approach to Automatic Robot Programming Based on Inductive Learning, *Int. J. Robotics*, Vol.3, No.3, p.20, 1987.

145. C.B. Mouleeswaran, PROPLAN: A Knowledge-Based Expert System for Process Planning, M.S. Thesis, Univ. of Illinois, Chicago III, 1984

146. C.B. Mouleeswaran and H.G. Fisher, A Knowledge-Based Environment for Process Planning, In: Applications of Artificial Intelligence in Engineering Problems, Vol.2 (D. Sriram and R. Adey, eds.), Springer-Verlag, N.Y., pp. 1013-1027, 1986.

147. P.M. Ferreira, B. Kochar, C.R. Liu and V. Chandru, AIFIX: An Expert System Approach for Fixture Design: In: *Computer-Aided/Intelligent Process Planning* (C.R. Liu, T.C. Chang and R. Komanduri, eds.), ASME, N.Y., pp. 73-82, 1985.

148. J.R. Dixon, A. Howe, P.R. Cohen and M.K. Simmons, DOMINIC I: Progress Towards Domain Independence in Design by Iterative Redesign, *Eng. Comput.*, Vol.2, pp. 137-145, 1987.

149. Y. Descotte and J.C. Latombe, GARI: An Expert System for Process Planning, *Solid Modelling by Computers: From Theory to Applications*, N.Y., 1984

150. K. Matsusima, N. Okada and T. Sata, The Integration of CAD and CIM by Application of Artificial Intelligence Techniques, In: *Manufacturing Technology*, Techn. Rundschan, Berne, Switzerland, 1982.

151. H.R. Berenji and B. Khoshnevis, Use of Artificial Intelligence in Automated Process Planning, *Comput. Mech. Eng.*, pp. 47-55, 1986.

152. I. Darbyshire and E.J. Davies, EXCAP-An Expert System Approach to Recursive Process Planning, *Proc. 16th CIRP Int. Seminar on Manufacturing Systems*, Tokyo, 1984.

153. A. Sluga, P. Butala, N. Lavrac and M. Gams, An Attempt to Implement Expert Systems in CAPP, *Robotics Integrated Manuf.*, Vol.4, No.1/2, pp. 77-82, 1988.

154. D.S. Nau and T.C. Chang, A Knowledge-Based Approach to Generative Process Planning, In: Computer-Aided/Intelligent Process Planning (C.R. Liu, T.C. Chang and R. Komanduri, eds.), *ASME*, N.Y., pp. 65-71, 1985

155. V.R. Milacic and M. Urosevic, SAPT-Knowledge-Based CAPP System, *Robotics Computer Integrated Manufacturing*, Vol.4, No.1/2, pp. 69-76, 1988.

156. F. Gliviak, J. Kubis, A. Mikovsky and E. Karabinosova, A Manufacturing Cell Management System: CEMAS, In: *Artificial Intelligence and Information: Control Systems of Robots* (I. Plander, ed.), *North-Holland*, Amsterdam, pp. 153-156, 1984.

157. M.S. Fox and S.F. Smith, A Knowledge-Based System for Factory Scheduling, *Expert Syst.*, Vol.1, pp. 25-49, 1984.

158. G. Bruno and G. Marchetto, Process-Translatable Petri Nets for the Rapid Prototyping of Process Control Systems, *IEEE Trans. Software Eng.*, SE-12, No.2, 1986

159. F.G. Mill and S. Spraggett, An Artificial Intelligence Approach to Process Planning and Scheduling for Flexible Manufacturing Systems, *Proc. Int. Conf. Computer-Aided* Engineering, IEE, London, 1984.

160. J. Cha, M. Rao, Z. Zhou and W. Guo, New Progress on Integrated Environment for Intelligent Manufacturing Automation, *Proc. IEEE Int. Symp. on Intelligent Control*, Arlington, VA, 1991

161 A. Kusiak, Manufacturing Systems: A Knowledge and Optimization-Based Approach, *J. Intell. & Robotic Systems*, Vol.3, No.1, 1990

162. L. Mikhailov, R. Schockenhoff, F. Pautzke and, H.A. NourEldin, Towards Building an Intelligent Flexible Manufacturing Control System, *Proc. IFAC Int. Symp. on Distributed Intelligent Systems*, Arlington, VA, 1991

163. S.G.Tzafestas, Expert Systems in Engineering Applications, *Springer-Verlag*, Berlin, 1993.

164. S.G. Tzafestas, Applied Control: Current Trends and Modern Methodologies, *Marcel-Dekker*, N.Y., 1993.

165. O. Kaynak, G. Honderd and E. Grant, Intelligent Systems: Safety, Reliability and Maintanability Issues, *Springer-Verlag*, Berlin, 1993.

166. C.J. Harris, C.G. Moore and M. Brown, Intelligent Control: Aspects of Fuzzy Logic and Neural Nets, *World Scientific*, Singapore, 1993.

167. S.G. Tzafestas, Engineering Systems with Intelligence: Concepts, Tools and Applications, *Kluwer*, Dordrecht/Boston, 1991.

168. N. Nayak and A. Ray, Intelligent Seam Tracking For Robotic Welding, *Springer-Verlag*, Berlin, 1993.

169. Y. Ito, Human-Intelligence-Based Manufacturing, *Springer-Verlag*, Berlin, 1993.

170. R. Bernhardt, R. Dillman, K. Hörmann and K. Tierney, Integration of Robots into CIM, *Chapman & Hall*, London/New York, 1992.

2
CONCEPTUAL INTEGRATION OF QUALITATIVE AND QUANTITATIVE PROCESS MODELS

ERLING A. WOODS
SINTEF Automatic Control
N-7034 Trondheim
Norway

1 INTRODUCTION

Different types of models reveal different subsets of the characteristic properties of a given system. No single type of model is inherently better than any other and the goodness of a model can only be assessed by taking its intended usage into account. In applications such as process monitoring and diagnosis, it will often be beneficial to model disparate properties of the process. For this reason, a multitude of types of models are currently being used to describe industrial process systems.

Traditionally, the most widely used models in process engineering and control belong to the class of quantitative mathematical model formulations. Such models will typically produce accurate results and predictions, unless some of the underlying assumptions are violated. Furthermore, this type of models places heavy demands on the accuracy with which the values of the model parameters must be known. The modeling techniques developed in the Qualitative reasoning subfield of AI typically take account of this by allowing parameters to fluctuate within more or less well defined limits. But as a consequence of this, the qualitative models are unable to make unique predictions and instead generate all possible solutions sanctioned by the qualitative model. The rule based systems, frequently denoted Expert systems or Knowledge Based Systems, have become quite popular among industrial practitioners. The rule base may be viewed as a model, partly capturing a description of the domain and partly capturing a formal description of some kind of human expertise related to that domain. But this approach normally involves extensive use of heuristics capturing only what is typically true. Rule based systems are thus susceptible to violation of underlying assumptions as well.

A joint IFAC and IEEE project have recently addressed several issues surfacing when industrial implementations of control theory is attempted. Some of the findings have been summarized in [Benveniste & Åström, 1993]. One of the recognized emerging challenges is the development of formal modeling frameworks capable of expressing a wide variety of the properties of an industrial process system, notably both discrete and continuous aspects. The need for such modeling frameworks is also reflected by the development of several architectures supporting simultaneous use of different types of models, see [Isermann, 1993] for an example.

The usefulness of this type of architecture is not disputed, but this approach does in some sense obscure the more fundamental issues. Consider a system

S. G. Tzafestas and H. B. Verbruggen (eds.),
Artificial Intelligence in Industrial Decision Making, Control and Automation, 41–65.
© 1995 *Kluwer Academic Publishers.*

where several models of different types are used in parallel to monitor a given physical process. At any given time, the state of a given model provides a description of some aspects of the state in the process. The different models describe the same system, in the same state, but in alternative perspectives. Hence it is reasonable to expect that dependencies between the state and structure of the different models exists. If these dependencies between the different models could be formalized, the reasoning or computation taking place with one model could be propagated to the other models. This could be used to ensure that the different models would always provide a consistent description of the state in the process.

The term *conceptual integration* of models is used to denote a set of models which have been integrated to the extent that the basic conceptual entities of one model component occur as integral parts of the other model formulations. Take a quantitative state space model, SSM, of a process system and a topology model using objects to describe the physical parts of the same system as an example. Two possible basic concepts for the SSM are *state-variable* and *equation.* Both states and equations could be associated with the corresponding objects in the topology model. The complete SSM could be derived from the topology model. If one were to replace an object in the topology model with another slightly different one, the set of equations constituting the SSM would automatically change as well. Both states and equations would in this case be distributed among entities forming the topological model.

This chapter describes the Hybrid Phenomena Theory, HPT, which utilizes other concepts than just variables and equations to integrate SSMs with qualitative models. The HPT claims that the view provided by mathematical state space models based on physical principles may be integrated with two other views describing topology and phenomena respectively. Which physical phenomena may appear in a given process system depend on the topology of the process, e.g. no liquid-flow will occur unless there exists some kind of reservoir and a flow-path. Which phenomena will actually take place also depend on the state of the system, e.g. no fluid flow will occur if the liquid in the reservoir is frozen. The HPT builds on the Qualitative Process Theory, QPT, [Forbus, 1984]. The HPT has been implemented in the Common Lisp Object System CLOS. Since the quantitative type of models are well known in the industrial community, whereas the model formulations developed in Qualitative reasoning is less familiar, a description of some aspects of the qualitative models will be provided before the HPT is introduced.

2   QUALITATIVE REASONING

Qualitative reasoning was initiated in the late seventies. Some early ideas on how a qualitative understanding of classical mechanics could be integrated with a quantitative description to solve problems in the kinetics domain was described by Johan de Kleer [1979]. At about the same time, Patrick Hayes advocated that AI research should move away from simple toy problems, known as "micro worlds", and attack a real-world problem like formalizing the commonsense understanding of physical systems, [Hayes, 1978; Hayes, 1985]. But the non-implementation, first-order-logic approach advocated by Hayes failed to inspire many researchers to work on the idea.

Research in qualitative reasoning was originally motivated by the fact that human beings function well in their physical surroundings, without resorting to quantitative computations. Researchers in qualitative reasoning therefore argued that it should be possible to formalize the knowledge providing humans with this capability

Table 1: A quantity space for the temperature of water

qualitative value	corresponding region on the real line.				
F			T	$<$	0	°C
M			T	$=$	0	°C
N	0	°C $<$	T	$<$	100	°C
B			T	$=$	100	°C
V	100	°C $<$	T			

by means of pure symbolic methods. Lately there has been a shift in attitude, as many researchers now address the issue of describing the behavior of physical systems without making claims about similarities with the way humans function in the real world. Qualitative reasoning is the first field of AI where attempts have been made to describe and reason about dynamic, continuous, physical systems.

2.1 Common concepts

The state of a physical process system may be characterized by the values of a set of variables. In a qualitative approach, the values assigned to the variables should reflect some kind of *qualitative difference*. The set of values which a given variable may attain is specified as a set of symbols, the meaning of these symbols is defined in a *quantity space*. Take the temperature, T, of an amount of water as an example. A viable quantity space for this variable is shown in Table 1.

The set of symbols introduced in Table 1 captures some of the characteristic properties of water as a function of temperature, *absolute* values like these typically reflect material properties. *Relative* values, expressed as the order of the values of the variables, may provide additional information. E.g., knowing that the temperatures of both fluids in a heat exchanger equal "N" is not very helpful, but knowing that the temperature of the first fluid is greater than that of the second may be used to determine the direction of the heat transfer.

Some approaches require a complete ordering among the symbols in the quantity space. For those which do not, the order will sometimes change during the analysis. Consider the system in Figure 1. Assume that the initial situation is as shown in the figure. The initial quantity space might then look like the following:

$$\text{BOTTOM-A} = \text{BOTTOM-B} < \text{INIT-LEVEL-B} \begin{array}{l} < \text{INIT-LEVEL-A} < \text{TOP-A} \\ < \text{TOP-B} \end{array}$$

This partial ordering expresses that it is known that INIT-LEVEL-B is less than both INIT-LEVEL-A and TOP-B, but initially it is not known whether INIT-LEVEL-A is less than or greater than TOP-B. Simulating the system above, a behavior of decreasing oscillations would be expected. This will show up as repeated changes in the ordering of *level-a* and *level-b*.

The values of the derivatives of the variables play an important role in the qualitative analysis. All approaches use the same set of values for the derivatives; $\{+,0,-\}$. Consequently, it is possible to express whether a variable is increasing, steady or decreasing. But it is generally not possible to express anything about the magnitude of the rate of change.

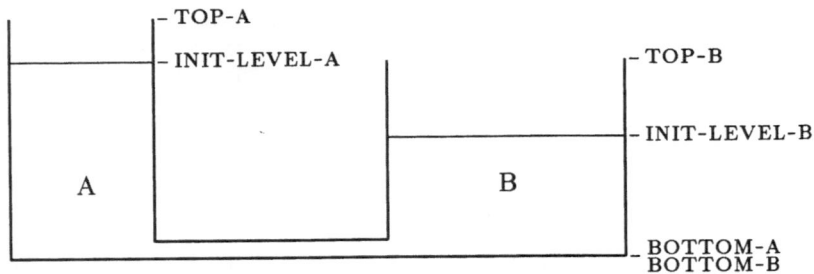

Figure 1: Connected liquid containers

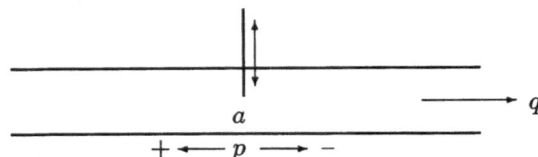

Figure 2: a simple valve mechanism

2.2 Qualitative mathematics

Where the quantitative approaches make use of algebraic and differential equations to express relationships between variables, the qualitative approaches utilize qualitative constraints. Consider the valve in Figure 2. This works by sliding the vertical plate up or down to change the flow area. There will be relationships between the values of the area, a, the drop in pressure over the restriction, p, and the flow, q. Denoting the qualitative derivatives of a, p and q with da, dp and dq respectively, the constraint 1 relates the values of the derivatives.

$$dp \oplus da = dq \qquad (1)$$

Assuming that da equals zero, dp and dq must always attain identical values. If the pressure increases, the flow must increase as well. A qualitative model consists of a collection of such constraints. A set of operations to combine qualitative values is needed to identify the combinations of values for the involved variables sanctioned by the collection of constraints. The two most interesting qualitative operations are addition and multiplication, defined in Table 2.

When the values of all but one variable are known, a quantitative equation unambiguously determines the value of the last variable. This is not necessarily the case with qualitative constraints. A qualitative constraint may allow certain possibilities to be ruled out, but it will not in general provide a unique result. Table 2

Table 2: definition of qualitative addition and multiplication

$\oplus$	$-$	0	$+$
$-$	$-$	$-$	?
0	$-$	0	$+$
$+$	?	$+$	$+$

$\otimes$	$-$	0	$+$
$-$	$+$	0	$-$
0	0	0	0
$+$	$-$	0	$+$

illustrates this point. The qualitative sum of two variables of opposite sign is unde-fined. Again consider constraint 1 and assume that dp equals [–] while da equals [+]. In this case, the constraint does not help to identify the value of dq. If the drop in pressure is decreasing while the area available for flow is increasing, the change in flow depends on the *magnitudes* of the derivatives of pressure and area. Since these are unknown, it is not possible to determine the value for dq.

2.3 The notion of state

The three variables p, a and q describe the system in Figure 2. Each variable is associated with a quantity space with three values; $\{+,0,-\}$, i.e. three different states. This gives 27 possible qualitative states for the system. But all 27 states will not satisfy the constraints describing the system. Together, the constraints 1 previously given and 2 given below constitute a viable description of the system in Figure 2.

$$p \oplus a = q \qquad (2)$$

Substituting values: $p = [+]$, $a = [+]$ and $q = [-]$ in constraint 2 con-tradicts the definition of $\oplus$ in Table 2. This state is consequently not permitted. Physically, this expresses that the direction of flow cannot oppose the direction of the drop in pressure. The identification of the states which are consistent with the collection of constraints is an important part of the qualitative analysis.

2.4 Describing behavior

In qualitative reasoning, behavior is described as a sequence of qualitative states. This sequence is derived by means of qualitative simulation. The basic strategy is to start with a known qualitative state. By applying the constraints which describe the relations between values and derivatives, the possible values of the derivatives are identified. By extrapolating from the current state in the direction given by the derivatives, the potential successor states for the system are derived. This procedure is iterated until either a state where all the derivatives are zero is reached, or a previously explored state is revisited. The first case corresponds to quiescence, the second to an oscillation.

The behavior of the system can then be summarized graphically as a set of permitted combinations of values for variables and derivatives, represented as boxes, and a set of permitted transitions, represented as directed lines between the boxes. Assuming that container B will not overflow, Figure 3 provides a simplified graphical illustration of the behavior of the system in Figure 1. The inherent ambi-guity of a qualitative analysis manifests itself in Figure 3, two boxes have multiple exits corresponding to alternative sequences of events.

2.5 Components of qualitative reasoning

Figure 4 illustrates that qualitative reasoning is more than just qualitative simula-tion. Each of the components in this figure is responsible for a specific task, but a given approach need not implement all three components. One of the best known approaches, QSIM, [Kuipers, 1986], only includes component 3. This means that the user will have to establish the model in terms of a suitable set of constraints manually.

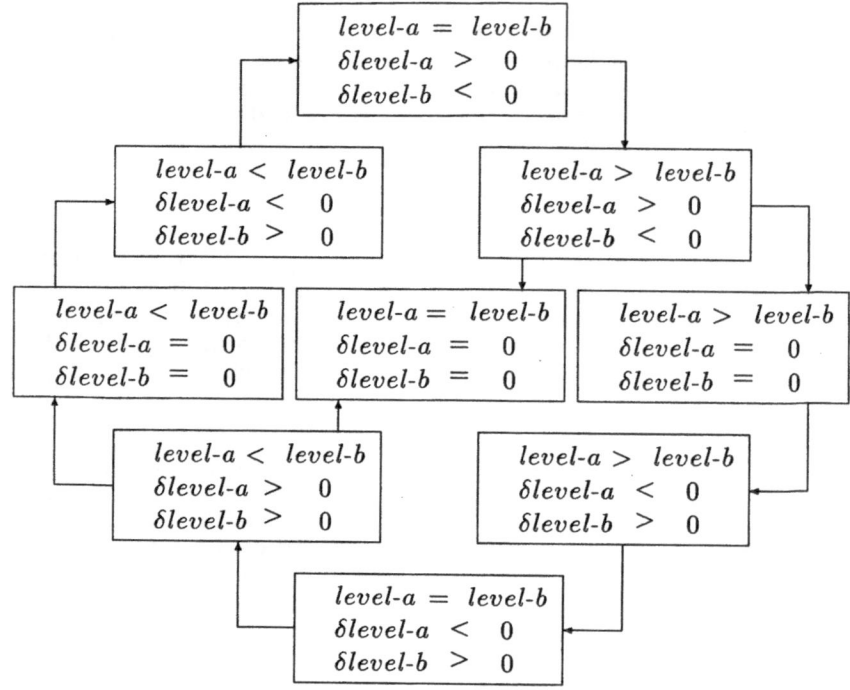

Figure 3: A qualitative behavioral description

Two classes of approaches for implementing components 1 and 2 in Figure 4 exists, these may be denoted *component* based and *process* based respectively. An example of the former is Envision, [de Kleer & Brown, 1984], whereas the QPT, [Forbus, 1984], is the original process based approach. In a component based approach, a set of generic component models is developed. Tanks, valves, pipes with flow-restrictions would be typical examples of components. A component model would then comprise a set of constraints offering a generic description of the component in question. A qualitative model for a complete system may then be assembled by combining the set of constraints obtained by applying the generic model to the actual components found in the system.

Component-1 of Figure 4 provides the framework, or vocabulary, used to formalize knowledge about physical interactions. In Envision this is covered by a set of generic component models. In case of the QPT, a formalism for expressing *view* and *process* definitions forms the first component. View and process definitions are used to describe generic abstractions of physical interactions. By applying these abstractions to a topological description of the system to be modeled, it is possible to derive which interactions will actually take place. The set of constraints constituting the qualitative model is established from this set of interactions. In addition to the framework as such, Component-1 typically also includes a library of definitions which may be applied to the analysis of any system belonging to a specific class. In Envision, this is a library of generic component models. In the QPT, this is a set of view and

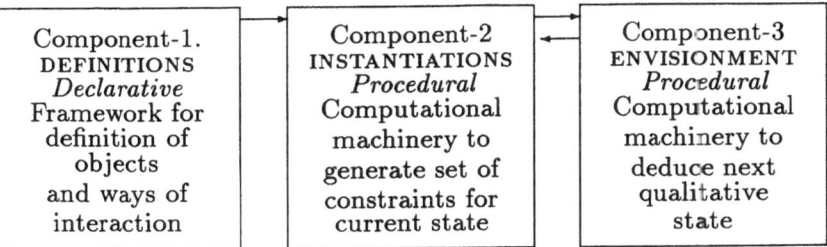

Figure 4: Components of Qualitative reasoning

process definitions.

Component-2 consists of a procedure. The inputs to this procedure are a description of the system to be analyzed, the initial state in the system and the contents of a library expressed in the formalism from Component-1. The procedure produces a set of constraints describing the system in the actual state. These constraints are used to perform qualitative integration by extrapolating from the current value in the direction given by the values of the derivatives.

In general, most approaches to qualitative simulation seems to assume continuity in values and derivatives. The most important distinctions are found in the kind of quantity space permitted. The QPT and QSIM approaches are more flexible in this respect, but at the cost of a more complex analysis. Peter Struss performed a rigorous mathematical analysis, relating qualitative mathematics to constraint operations on intervals of real numbers [Struss, 1988; Struss, 1990]. Amongst other things, he showed that only by restricting the quantity space to the {+,0,-} type, could associativity be preserved for addition. In general, all qualitative simulation approaches suffer from some inherent weaknesses, notably ambiguity leading to branching in the analysis and a tendency to produce spurious solutions.

The QPT makes the notion of processes or phenomena explicit, thereby relating behavior to abstractions of physical phenomena. Forbus claims that the notion of causality is inextricably linked to the notions of physical phenomena. E.g. it is a boiling process which causes the generation of steam, not the fact that the temperature of water exceeds a certain temperature. Neither QSIM nor Envision embodies a counterpart to the notion of a phenomenon. The QPT therefore provides a stronger conceptual framework for formalizing available knowledge on physical systems. This conceptual framework constitute the basis for the HPT.

2.6 Towards more Quantitative Models

Order of Magnitude Reasoning, OMR, first introduced by the FOG system, [Raiman, 1988], attempts to circumvent the problems by incorporating a more fine grained description of the relations between the values of the variables describing a system. But this approach does not solve all problems. One problem is that as new values are computed from old, the inherent uncertainty tends to cause a smearing out of the results. Even if it is known that x_1 is much larger than x_2 and that x_3 is much larger than x_4, it is still not possible to say anything about the relationship between the two fractions x_1/x_3 and x_2/x_4.

A new school of thought has recently started to gain ground within qualitative reasoning. Rather than making attempts to express the relations which exists between variables in a qualitative way, the idea of this new school is to start from traditional mathematical models. By building programs combining different kinds of numerical analysis, one attempts to derive qualitative descriptions of the overall behavior of the physical systems modeled by the equations. Typically, these approaches rely on phase space analysis and use numeric techniques to identify characteristic points and curves in the phase-space. Both fixed point solutions and limit cycles are identified and attractors, repellors and saddle points detected. Two examples of this approach are POINCARE, [Sacks, 1991] and MAPS [Zhao, 1991]. An overview of the major ideas and objectives may be found in [Abelson & al., 1989].

3 Formal concepts and relations in the HPT

This section introduces the basic concepts of the HPT. These concepts are formalized as class definitions and a given concept may thus inherit the properties of other more basic concepts. Some concepts are purely abstract, the corresponding classes are never instantiated but are only used to formalize common properties of other more specialized classes.

3.1 Quantities

The concept of a *quantity* forms the common basis for all quantitative magnitudes encountered in the HPT. The following properties are defined for **quantity**: *value*, *symbolic name*, *physical unit*, and the *identifier* of the physical object which is characterized by the quantity. The class **constant** has no additional properties but only inherits the ones defined for quantity. The class definition establishes the concept of a constant and allows methods which are applicable to constants only to be defined.

The class **parameter** also inherits quantity and includes a slot[1] which is intended to store a function as the value of the slot. This enables a model builder to formalize how to compute the value of the parameter from the values of other quantities. Similarly, the **variable** class inherits quantity and includes additional properties which will enable the program to keep track of which equations the various instances of variable are involved with at any given time. The class of **state-variable** specializes variable. For any instance i of either **variable** or **state-variable**, it may be stated that i *is a* quantity with the same right as it is stated that i *is* a variable.

3.2 Physical objects, Process Equipment, Materials and Substances

The *types* of objects used to model the topology of the process are also defined in terms of a class hierarchy expressed as CLOS definitions. The term *physical object* is used as a common name for all such objects. Physical objects may be categorized in terms of process equipment, materials and substances.

In the HPT, all objects representing topological components or physical interactions originate from the base class **object**. Amongst other things, this class defines the properties for *symbolic name* of the physical object described by the

---

[1]A slot may both be considered as a variable within an object, and as something which defines a property of the object containing the slot.

instances of this class, as well as properties enabling sets of *variables, parameters* and *constants* to be associated with the physical object.

The class **object-with-heat-capacity** inherits **object** and provides an example of how abstract properties may be described in terms of class definitions. This class incorporates two default quantities, the parameter *heat-capacity* and the variable *temperature*. All objects instantiated from a class which includes **object-with-heat-capacity** as a superclass will thus inherit these quantities. The generic aspects of process equipment is defined by the class **device**. As an example, the class **electric-device** includes the additional common characteristics of devices which consume or produce electricity.

In the class hierarchy currently used, all types of materials descend from **stuff**. The present description is clearly influenced by a "naive physics" approach. Here, the substances are first classified according to the phase they appear in, then split depending on the chemical contents of the substance. This approach suffices for the example. In general, when a class hierarchy for physical objects is being established, attempts should be made to attribute sets of characteristic properties to abstract class definitions forming building blocks for more complex objects. At the most abstract level, attempts should be made to isolate the aspects which may conceivably appear independently of any other characteristic properties of any object. At the next level, aspects which tend to appear together should be grouped by mixing the appropriate classes from the level above. It is important to maintain a balance between the desire to keep the number of class definitions manageable, and the desire to provide ready-made definitions for specific types of equipment. The latter requirement becomes less important if the end users are provided with the means to define their own types of equipment simply by mixing the appropriate combination of predefined blocks.

3.3 The input file

The input file to the current HPT implementation comprises two parts. The first defines a set of physical objects while the second part defines the relations between the objects. An example of an object specification is shown below;

```
(defobject pan 'container-with-open-top
   (construct-material steel)
   (bottom-area 0.018)
   (capacity 0.003)
   (has-heat-leading-bottom) )
```

This statement defines an object PAN to be represented as an instance of the class container-with-open-top. This implies that PAN will inherit all the properties defined for this class and all of its superclasses. In this manner, it also acquires all the default values specified for each property including the default values for the quantities associated with this class. The example shows how the default values of the properties inherited from the class definitions may be overridden by including the relevant statements in the defobject statement. The last line in the defobject statement above specifies a property to be associated with the specific object PAN.

3.4 Activity Conditions

An *activity condition* in the HPT may be considered as a *daemon*, an entity monitoring the values of certain data constructs and taking action when specific conditions are

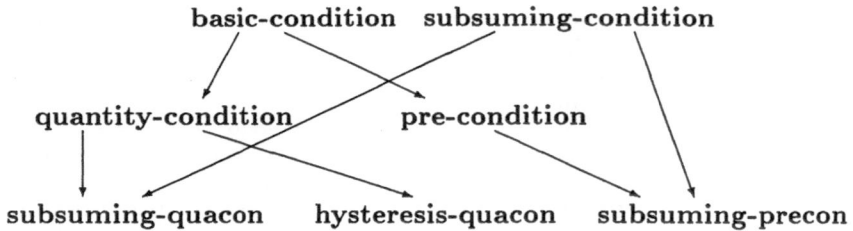

Figure 5: Hierarchies of activity conditions.

met. Activity conditions are used to monitor whether a specific relationship, affecting one or more instances of physical phenomena, is satisfied or not.

Figure 5 shows some types of activity conditions employed in the current HPT implementation. The most commonly used activity conditions are those of **quantity-condition** and **pre-condition**. The three classes at the lower row implement special cases which, although conceptually important, are not frequently used. Instances of **quantity-condition** implement conditions on the magnitudes of a set of quantities. It is thus typically used to formalize how the activity of a phenomenon depends on the state in the system. An example is that no boiling will occur unless the temperature in the liquid exceeds the boiling temperature for that liquid at the current pressure.

Instances of **pre-condition** implement logical conditions. There are two major forms of usage. The first is to formalize the effects of outside interaction, e.g. operator interference such as the closure of a manual valve. The second is to formalize the effect of making, or not making, an assumption. In practice, this means that the modeler may choose whether to include the effects of a specified phenomenon, or group of phenomena, in the model.

Instances of **hysteresis-quacon** is similar to instances of the ordinary quantity condition, except for the fact that the former incorporates two different conditions. One of these is used when the state of the condition is false, the other is applied when the state of the condition is true. This allows the implementation of hysteresis effects when switching phenomena on and off. Instances of a class including **subsuming-condition** as a superclass implement conditions which are special in the sense that when these conditions changes state, the implications are somewhat different from the normal type of conditions.

3.5 Numerical functions and Influences

The HPT employs the Lisp notation[2] to express numerical functions. In this notation, the mathematical operators precede their arguments. Many operators, which in other notations can only accept two arguments, will in Lisp accept a sequence of arguments. E.g. (+ 1 2 3 4) is a valid construct whose result is 10. Both system-defined and user-defined operators are allowed. For a further elaboration on the Lisp syntax, see[Steele, 1990]. In the HPT, numerical functions constitute essential building blocks in both equations and quantity-conditions. Each of the arguments to a function can

---

[2]This is a prefix notation sometimes denoted "Polish" notation. It was developed by the Polish mathematician Jan Lukasiewicz who was the unfortunate bearer of a name to complex for his invention to become known as Lukasiewicz notation.

be of three different types; a number, a reference to a quantity, or another numerical function. The latter alternative implies that the building blocks for both equations and quantity conditions can extend to an arbitrary depth; there are no limits to the complexity of the functions.

In the HPT, equations merely emerge from the manipulation of more basic structures, the *influences*. Each influence instance identifies a set of quantities and a numerical function relating these quantities. The HPT employs two types of influences, dynamic influences and algebraic influences. The syntax for a dynamic influence is as follows;

```
(dyn-infl <influenced variable>
          <list of influencing variables>
          <numerical function>)
```

The term **numerical function** was explained in the previous section. An example of a dynamic influence is given below;

$$(\text{dyn-infl } x_1 \ (x_2 \ x_3) \ (\text{sqrt } (* \ x_2 \ x_3)))$$

This influence is said to be *affecting* x_1 since x_1 is the influenced variable. If this is the only influence affecting x_1, the derivative of x_1 will be computed from the following expression; $\dot{x}_1 = \sqrt{x_2 x_3}$

The complete semantic interpretation for dynamic influences is that; *The derivative of a dynamically influenced variable equals the sum of the numerical functions specified in the dynamic influences affecting the variable.*

Each dynamic influence may thus be interpreted as the specification of a term in the equation defining the derivative of the dynamically influenced variable. There is a one-to-one correspondence between the dynamically influenced variables and the state variables. Also note that the term *numerical function* means a function which will return a number, but the numerical functions are specified as *symbolic constructs*. Consequently, symbolic equations may be extracted.

Apart from a different name, the syntax for an algebraic influence is similar to the syntax for a dynamic influence;

```
(alg-infl <influenced variable>
          <list of influencing variables>
          <numerical function>)
```

For algebraic influences, the semantic interpretation is that; *An algebraically influenced variable equals the sum of the numerical functions specified in the algebraic influences affecting that variable.*

Each algebraic influence may thus be interpreted as the specification of a term in the equation defining the value of the algebraically influenced variable. There is a one-to-one correspondence between the algebraically influenced variables and the dependent variables.

To illustrate how influences are combined, consider a closed container which is partially filled with liquid and with a gas taking up the remaining volume. The liquid has a level l, the gas a pressure p_1. The pressure at the bottom of the container, p_2, is affected by two different influences as specified below.

$$(\text{alg-infl } p_2 \ (p_1) \ (+ \ p_1))$$
$$(\text{alg-infl } p_2 \ (l) \ (+ \ \rho g l))$$

If no other influence specifies p_2 as the influenced variable, this implies that p_2 is given by the following expression; $p_2 = p_1 + \rho g l$

3.6 Logical relations and rules

The HPT makes use of logical relationships to specify both attributes of physical objects and the geometry of a process system. The input description of the system will include an initial set of logical relationships, others will later be added by the HPT reasoning mechanisms. This is partly a consequence of the application of a rule-based system, partly a consequence of the instantiation of views and phenomena which explicitly specifies that certain logical relationships may be created when the definition is instantiated. An example of a rule is given below:

```
(heat-connected (a b)      (rests-on (a b))
                           (rests-on (b a)))
```

Since `rests-on` (PAN BOX) may be matched with both of the antecedents in the rule specified above, both `heat-connected` (PAN BOX) and `heat-connected` (BOX PAN) may be derived.

A simple goal-driven inference engine with a global working memory is employed. In case the inference engine is forced into a recursive search, thus establishing some intermediate conclusions in order to prove the requested one, all the intermediate conclusions are asserted as valid relationships in the global working memory in addition to the relationship which was actually requested.

4 Defining views and phenomena

This section describes the formal framework used to describe *views* and *phenomena*. Views and phenomena definitions capture the generic aspects of physical interactions. The basic concepts and relations described in the two previous chapters are used as building blocks in the definitions. Views describe physical interactions which do *not* incorporate any dynamic aspects. Phenomena describe physical interactions which *do* incorporate dynamic aspects. Both types of definitions share four common parts; *Individuals, Quantityconditions, Preconditions* and *Relations*. Definitions of phenomena incorporate a fifth part; *Dynamics*. A view may thus be considered as a phenomenon without dynamics.[3] Examples of phenomena definitions are found in Figure 6.

The distinction between the *definition* of a phenomenon and the *instantiations* of that definition is emphasized. A definition is *generic* and does not relate to any specific physical interaction. Instantiations are created by applying the definition to a particular set of physical objects which satisfies the requirements of the definition. An instantiation thus describes a particular appearance of a phenomenon.

The simple system in Figure 7 will be used to illustrate how the HPT mechanisms works. Three objects, the HOTPLATE, the PAN and the WATER are defined in the input description. In addition, two relations are defined. These specify that the WATER is positioned inside the PAN and that the PAN is resting on the HOTPLATE.

4.1 Individuals and individual conditions

A list of individuals appears immediately after the name of the phenomenon in a definition. The two individuals in `heat-bridge`, see Figure 6, are named `obj1` and

---

[3]For reasons of brevity, the term phenomenon will be used as a common term for both views and phenomena whenever common aspects for both concepts are being discussed.

```
(defview heat-bridge (obj1 obj2)              (defphenomenon heat-flow (src dst hbr)
  (individuals                                  (individuals
    (obj1 (is-a object-with-heat-capacity))       (hbr (instance-of heat-bridge(src dst)))
    (obj2 (is-a object-with-heat-capacity))       (src (is-a object-with-heat-capacity))
    (heat-connected obj1 obj2))                    (dst (is-a object-with-heat-capacity)))
  (relations                                    (quantityconditions(> temp(src) temp(dst)))
    (define-parameter \alpha                    (relations
      (:value 1200 :unit "J s-1 K-1 m-2"))        (define-variable hstr
    (define-parameter \kappa                         (:unit "J s-1"))
      (:unit "J s-1 K-1"                          (alg-infl hstr (temp(src) temp(dst))
       :compfunc                                     (* \kappa(hbr) (- temp(src) temp(dst)))))
        (* \alpha (MIN area(obj1)               (dynamics
                      area(obj2)))))))             (dyn-infl temp(dst) (hstr) (/ hstr c(dst)))
                                                   (dyn-infl temp(src) (hstr) (/ hstr c(src)))))
(defview container-with-liquid (c l)
  (instance-name
    view-inst-with-heat-capacity)             (defphenomenon boiling (lc hf)
  (individuals                                   (individuals
    (c (is-a container-with-open-top)             (lc (instance-of
       (can-contain-liquid))                          container-with-liquid(nil nil)))
    (l (is-a liquid))                             (hf (instance-of heat-flow(nil lc nil))))
    (placed-in l c))                            (quantityconditions
  (preconditions                                  (> temp(lc) btem(lc(l))))
    (subsuming                                  (dynamics
      (assume-one-object-liquid-container         (dyn-infl mass(lc(l)) (hstr(hf))
        (c l))))                                    (- (/ hstr(hf) h(lc(l)))))
  (quantityconditions                            (dyn-infl temp(lc) (hstr(hf))
    (subsuming (> mass(l) 0)))                      (- (/ hstr(hf) c(lc)))))
  (relations
    (define-variable h (:value 0 :unit "meter" :what? "liquidlevel in container"))
    (define-variable p (:value 0 :unit "Pa" :what? "bottompressure in container"))
    (define-variable temp (:value 0 :what? "temperature" :unit "K"))
    (define-variable m (:what? "mass" :unit "kg"))
    (define-parameter c (:what? "heat capacity" :compfunc (+ c(c) (* sc(l) mass(l)))))
    (connect-quantity area area (c))
    (if (rests-on(c x)) then (rests-on(self x)))
    (if (has-heat-leading-bottom(c)) then (has-heat-leading-bottom(self)))
    (alg-infl m (mass(l)) (+ mass(l) mass(c)))
    (alg-infl h (mass(l)) (/ mass(l) (* density(l) area(c))))
    (alg-infl p (h) (* h density(l) g(global)))))
```

Figure 6: Definitions of physical interactions.

obj2. An individual may be considered as a variable. In the following, an object
is said to be *assigned* to an individual when a collection of objects are being tested
against the conditions in a given definition. Any HPT-object may be assigned to
any individual, but no HPT-object may be simultaneously assigned to more than one
individual.[4] The individual conditions governs which combinations of HPT objects will
give rise to the creation of an instantiation of the definition. This will happen for
every combination of HPT objects which:

1. comprises as many objects as there are individuals in the list of individuals
   following the name of the phenomenon, and;

2. allows a one-to-one assignment between the set of individuals and the set of
   objects to be established such that;

3. all of the individual conditions are satisfied when the objects replace the individuals they have been assigned to.

---

[4]This restriction ensures that no single object may simultaneously play several roles intended to
be played by different objects.

When the test procedure succeeds in identifying a set of objects complying with the conditions, the definition in question is instantiated. For every resulting instantiation, each object is said to be *bound* to the individual it was assigned to when the test procedure succeeded. Objects are thus bound to individuals in *instantiations* of a phenomenon, never to individuals in the phenomena definitions as such. Each individual in a given instantiation binds a single object. A given object may be bound to numerous individuals in different instantiations arising from the same or several phenomena definitions.

The ordered list of individuals following the name of the definition allows the identification of a specific instantiation. An instantiation is uniquely identified by the name of its definition followed by the names of the objects listed in the same sequence as the individuals they are bound to were listed in the definition. E.g. HEAT-BRIDGE (PAN WATER) describes the particular instantiation of heat-bridge where PAN is bound to obj1 and WATER to obj2.

The individual conditions follow the individuals keyword. The first two conditions in heat-bridge expresses that in order to be bound to either of these individuals, an object must satisfy is-a object-with-heat-capacity. This translates into a requirement that only objects instantiated from a class having object-with-heat-capacity as a superclass may be bound to either of these individuals. In general, an is-a type of condition always expresses a constraint on the location in a class hierarchy for the object to be bound to the individual in question. The final condition requires that there must exist a logical relationship, heat-connected, between the two objects assigned to the obj1 and obj2 individuals.

Applying the definition of heat-bridge to the input description yields four combinations of objects complying with the conditions. This gives rise to the following instantiations; HEAT-BRIDGE (PAN WATER), HEAT-BRIDGE (WATER PAN), HEAT-BRIDGE (HOTPLATE PAN) and HEAT-BRIDGE (PAN HOTPLATE). Note that the rule-based system has to be invoked to derive the required relationships for the heat-connected condition.

4.2 Quantity conditions and Preconditions

A quantity condition defines a constraint on how the values of certain quantities must relate for a phenomenon instantiation to be applicable. The quantities in the conditions either characterize the instantiation as such, or they characterize objects that are in some sense related to that instantiation.

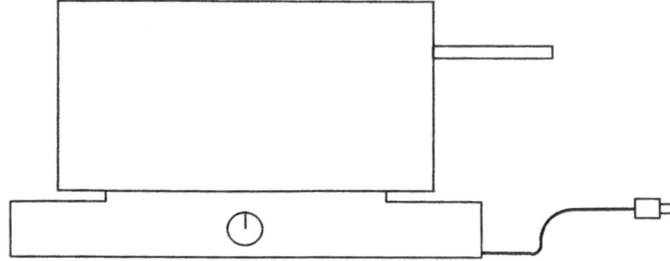

Figure 7: PAN with WATER on HOTPLATE example system.

The condition (> temp(src) temp(dst)) is included in the heat-flow definition. In any instantiation of heat-flow, the object bound to src represents the source of the heat flow while the object bound to dst represents the destination. This quantity condition thus specifies that for any particular instantiation of heat-flow to be applicable, the temperature of the source must be greater than that of the destination. The expression temp(src) in effect specifies a path where a quantity will be found. To gain access to the quantity, the program should get hold of the quantity temp associated with the object bound to src.

Any quantity in a HPT model is associated with the HPT-object which the quantity was created to characterize. The temperatures above were created to describe the objects which have subsequently become bound to either src or dst in a particular instantiation of heat-flow. Each quantity is in this sense "owned" by an HPT-object. The variable temp describes the temperature of the WATER and is thus said to be owned by the WATER object. But the owning objects do not enjoy any special privileges when it comes to accessing or updating a given quantity. Quantities, whether constant, parameter or variable, are entities in their own right and are directly accessible from any entity which knows about the existence of the quantity. The owning object will provide a link to any of its quantities to every other object which requests such a link. Thus, if WATER is bound to src, a link to the object representing the temperature of WATER may be found by querying the object bound to src for a link to the quantity it knows as temp. Once this link is established, the quantity condition accesses this variable directly.

Only those quantities which are inside the scope of a given instantiation will be accessible from that instantiation. A quantity Q_i is said to be within the scope of an object O_j if and only if; Q_i is owned by O_j, or, Q_i is within the scope of an object bound to an individual in O_j, or, Q_i is owned by the special object GLOBAL. Note that this is a recursive definition. The special HPT-object denoted GLOBAL is accessible from any instantiation. All global constants will be owned by GLOBAL, thus ensuring that any object may access fundamental constants such as the Avogadro and Boltzmann constants.

Every instance of some type of quantity condition is always in one of two states; *true* or *false*. The state is determined by the truth value of the condition and thus depends on the actual values of the quantities involved. The state of a quantity condition may therefore change when the value of any given quantity involved with the condition changes its value. The state of an instantiation is a function of the state of its quantity conditions.

Preconditions are similar to quantity conditions in that they affect the state of the instantiations they are linked to. But while the truth value of a quantity condition is determined from the numerical values of quantities characterizing a HPT-object, the truth values of the preconditions are determined by factors which cannot normally be derived by reasoning with the HPT-model. Examples of such factors are the opening or closing of a by-pass valve by a human operator, or a change in an assumption affecting the model. The definition of electric-heat-hotplate provides an example of a standard precondition: (power-on(elh)). This condition merely expresses that the power must be turned on for any heat to be generated in the object bound to elh. The truth value of this condition could, depending on application and available instrumentation, either be specified by an operator or be tested.

4.3   Relations

The relations part of a definition may include four distinct types of specifications creating quantities, relationships, additional physical objects and algebraic influences respectively. In addition, a specification may be made conditional.

The definition of heat-flow includes a definition of a variable hstr. Each instantiation of heat-flow will thus have a variable hstr describing the amount of heat actually flowing in that specific instantiation. Each of these variables has an associated text string explaining its purpose; note that src and dst will be replaced by the print names of the objects actually bound to these individuals in the respective instantiations. The resulting text strings are used to document the purpose of the variables. Each variable also has a physical unit associated with it, this will be common for all resulting hstr variables.

The connect-quantity command makes it possible to specify that a quantity which is within the scope of the instantiation being created will be accessed as a local quantity. An illustration of this is found in the definition of container-with-liquid.

Sometimes, it may be desirable to create a new object to capture the consequences of a newly instantiated phenomenon. Examples are chemical reactions leading to new substances or phase-changes giving rise to new phases; these are sometimes best described by new objects. A boiling process in a closed container might be specified to give rise to a steam object situated above the liquid phase. Such objects may be created as a side effect of the instantiation of a phenomenon definition by including the defobject construct in the definition. This construct was previously discussed in Subsection 3.3.

Several definitions in Figure 6 provides examples showing how algebraic influences are defined. The quantities are here addressed in the same way as was the case with the quantity conditions. The scoping rules are identical as well. A given quantity may thus be addressed by any number of instantiations.

All specifications in the relations part of a definition may be made conditional. The container-with-liquid definition includes two examples of conditional definitions of relationships. In the definition of rests-on, c is an individual, it will therefore always be replaced by the object bound to it in the actual instantiation about to be created. Another type of variable used in this context is the special symbol self. This variable will always be bound to the actual instantiation being created. The final type of variable used in this context may be bound to any existing HPT-object satisfying the relationship in which the variable is specified. This is exemplified by the variable x in the same conditional definition.

4.4   Dynamic influences

The part of the definition denoted *dynamics* is reserved for true phenomena definitions and is *never* found in the definition of a view. The dynamics part comprises the specification of dynamic influences. The scoping rules for accessing quantities are identical to the rules previously described for quantity conditions and algebraic influences. The vital difference is in the semantics, as was explained in Section 3.5.

4.5 Instantiating a definition

The instantiation procedure combines the information provided in the phenomenon definition with information retrieved from the objects about to be bound to the individuals in the new instantiation. The procedure creates an object describing the new instantiation. Other objects representing preconditions, quantity conditions, influences, compfuncs, relationships, quantities referenced by the instantiation and additional basic objects are created and the required cross references are inserted.

Roughly speaking, the set of phenomena instantiations existing when all possible instantiations have been created, their associated preconditions and quantity conditions as well as the logical relationships involving these objects, constitute the *phenomenological* model component in the HPT. The topological model includes all objects specified in the input description, any additional basic objects resulting from the application of defobject statements during instantiation of phenomena, as well as all logical relationships involving these objects only. The program has now established a set of "Conceivable Phenomena", see Figure 8. The next step is to determine which of these phenomena are applicable, or *active*, given the state of the system and the set of assumptions which are in effect.

4.6 Activity levels

At any given time, an instantiation is in one of the following states; subsumed, passive, subsumactive, active.

Subsumed objects may be thought of as non-existing entities in the current context. Think of subsumption as a mechanism which sweeps away any instantiation which has become obsolete because it turns out that there exists a competing instantiation of the same phenomenon definition that provides a more suitable description in the current context. An object may enter state *subsumed* for two different reasons. The first is if the object is identified by the algorithm applying the criteria for when an instantiation should be subsumed; this is denoted *direct subsumption*. The second is if one of the objects bound to an individual in the instantiation enters state subsumed; this is denoted *indirect subsumption*. Lack of space prohibits a complete description of the subsumption mechanism. The subsumption mechanism takes precedence over all other criteria determining the state of an instantiation. The rest of the description in this section thus presumes that the instantiations are not subsumed.

All instantiations are passive upon creation. An instantiation remains passive if and only if EITHER;

1. None of the objects bound to the individuals in the instantiation is subsumed and at least one of them is in state passive, **or**

1. The instantiation comprises at least one subsuming activity condition, and
2. One or more of these subsuming activity conditions are in state false, and
3. None of the objects bound to the individuals in the instantiation is subsumed.

A passive HPT-object may be bound to individuals in other phenomena instantiations and may thus cause these instantiations to remain passive as well. But apart from this, such objects have no impact on the reasoning in the HPT. Specifically, passive objects have no impact on the structure of the state space model.

For an HPT-object to leave state passive and enter state *subsumactive*, none of the objects bound to the individuals in the instantiation must be in state

58

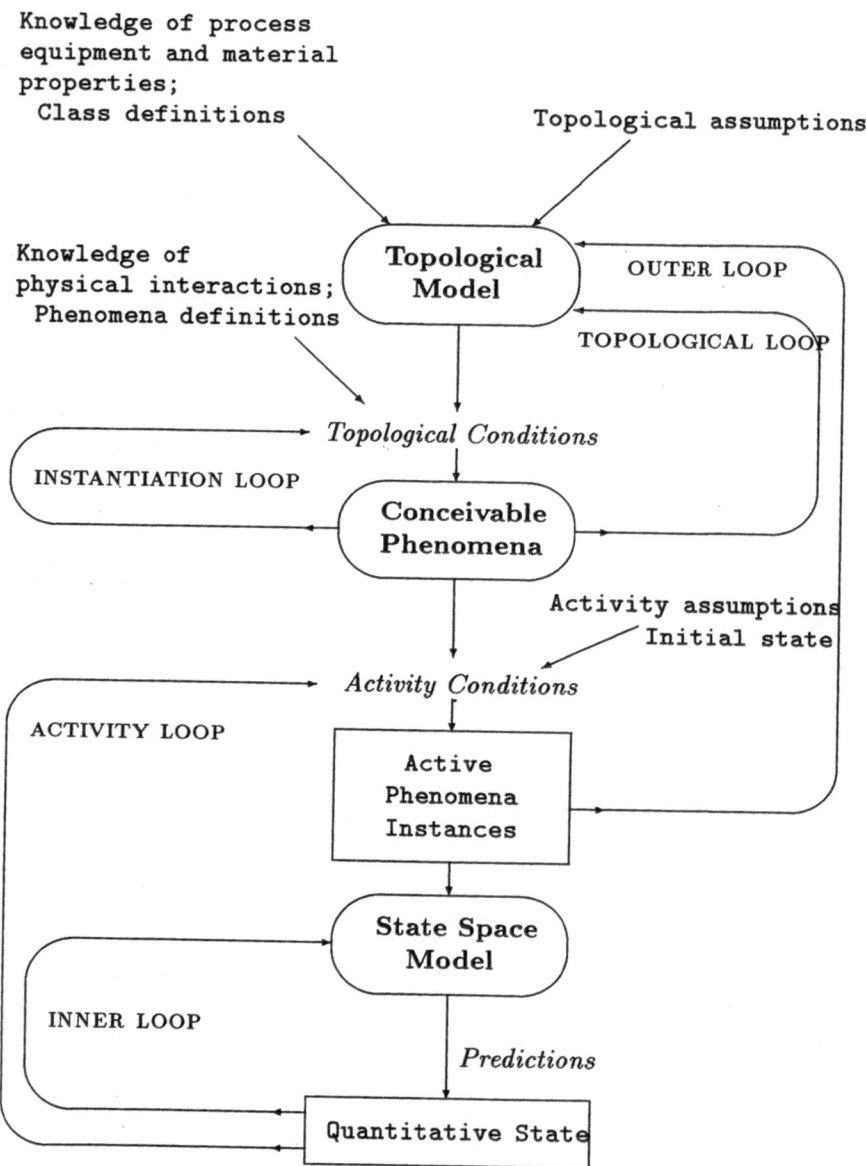

Figure 8: Reasoning with the model components

passive or subsumed, and all of the definition's subsuming activity conditions must be in state true. Assuming that a definition does not include any subsuming activity conditions, all instantiations of that definition will be in state subsumactive provided that none of the objects bound to an individual remains in state passive or is subsumed. Subsumactive HPT-objects are like passive objects in the sense that such instantiations will not affect the state space model derived by the program. But, unlike passive objects, subsumactive objects may cause other instantiations to be subsumed.

For an HPT-object to be in state *active*, all of the objects bound to the individuals in the instantiation must be in state active, and all of the object's activity conditions must be in state true. To be in state active, an HPT-object must satisfy all conditions required for it to be in state subsumactive. Objects which are active are thus also considered to be subsumactive, active objects may thus subsume other objects. In addition, all influences in the active objects are considered applicable and will thus be accounted for in the resulting state space model. The active instantiations describe those phenomena which are actually expected to be occurring.

5 Deriving and reasoning with an HPT Model.

This section will illustrate the HPT mechanisms using the system in Figure 7. All models shown in this chapter have been produced by the HPT implementation. The indexes and print names shown for all phenomena instantiations and quantities are those selected by the program.[5] Some of the explanatory text strings have been edited to provide a more concise description, but, apart from this, all temptations to improve on the resulting models have been resisted. The reasoning process is outlined in Figure 8. The following subsection describes how the initial topological model is extended by the HPT program.

5.1 Extending the topological model

The topological model consists of two sets of objects. The first set represents the physical entities in the system to be modeled. In the example, this is the HOTPLATE, the PAN and the WATER. The second set represents properties of the individual physical entities and relationships among the two or more of them. These objects describe the topological organization of the physical entities. In the input description, two properties are specified; that the HOTPLATE has a has-heat-leading-top and that the PAN has a has-heat-leading-bottom. In addition, there are two relationships; placed-in (WATER PAN) and rests-on (PAN HOTPLATE).

None of the phenomena definitions employed in this chapter specify that additional basic objects shall be created upon instantiation of the phenomenon definition. Hence, no additional basic objects are produced by the program. Several additional properties and relationships will be created, thus extending the topological model from what was specified by the user. The complete set of logical relationships created for the example is listed below.

---

[5]The algorithm designates variables that influence one or more variables and which are not themselves influenced by another variable as control variables. Variables which are at first identified as control variables may in revised versions of the model be influenced by other variables. The variable will then no longer qualify as a control variable. For this reason, the same variable may have several different print representations. As an example, see the mass of WATER in Figure 9.

```
has-heat-leading-bottom (PAN)
has-heat-leading-top (HOTPLATE)
can-contain-liquid (PAN)
placed-in (WATER PAN)
rests-on (PAN HOTPLATE)
has-heat-leading-bottom (CONTAINER-WITH-LIQUID-1)
rests-on (CONTAINER-WITH-LIQUID-1 HOTPLATE)
heat-connected (PAN WATER)
heat-connected (WATER PAN)
heat-connected (HOTPLATE PAN)
heat-connected (PAN HOTPLATE)
heat-connected (HOTPLATE CONTAINER-WITH-LIQUID-1)
heat-connected (CONTAINER-WITH-LIQUID-1 HOTPLATE)
```

5.2 Deriving the phenomenological model

The phenomenological model includes all instantiations of phenomena created by the program. Assuming all relevant phenomena are included in the knowledge base, the instantiations describe all phenomena instantiations that can conceivably affect the system being modeled, given the specified input description. This model also includes a number of objects representing how the activity level of the instantiations depend on logical facts and the quantitative state of the system. The combined truth values of these relationships determine which instantiations are active. This set of active instantiations is denoted the *qualitative state*. All instantiations in the current example are listed below.

```
HEAT-BRIDGE-1 (PAN WATER)
HEAT-BRIDGE-2 (WATER PAN)
HEAT-BRIDGE-3 (HOTPLATE PAN)
HEAT-BRIDGE-4 (PAN HOTPLATE)
ELECTRIC-HEAT-HOTPLATE-1 (HOTPLATE)
CONTAINER-WITH-LIQUID-1 (PAN WATER)
HEAT-FLOW-1 (PAN HOTPLATE HEAT-BRIDGE-4)
HEAT-FLOW-2 (HOTPLATE PAN HEAT-BRIDGE-3)
HEAT-FLOW-3 (WATER PAN HEAT-BRIDGE-2)
HEAT-FLOW-4 (PAN WATER HEAT-BRIDGE-1)
HEAT-BRIDGE-5 (HOTPLATE CONTAINER-WITH-LIQUID-1)
HEAT-BRIDGE-6 (CONTAINER-WITH-LIQUID-1 HOTPLATE)
HEAT-FLOW-5 (CONTAINER-WITH-LIQUID-1 HOTPLATE HEAT-BRIDGE-6)
HEAT-FLOW-6 (HOTPLATE CONTAINER-WITH-LIQUID-1 HEAT-BRIDGE-5)
BOILING-1 (CONTAINER-WITH-LIQUID-1 HEAT-FLOW-6)
```

The first four instantiations of **heat-bridge** characterize the heat paths between the three basic objects; there are two instantiations for each pair of objects as the descriptions carries a notion of direction. For each of these instantiations of **heat-bridge**, there is an instantiation of **heat-flow** which binds the corresponding **heat-bridge** instantiation.

The next instantiation describes the heat generation in the HOTPLATE. Then follows the object implementing the assumption that the PAN and WATER objects may be considered as one object with respect to any heat-flow. The latter instantiation breeds two new instantiations of **heat-bridge**; these subsequently give rise to two additional instantiations of **heat-flow**. Finally, there is an instantiation describing the boiling phenomenon.

5.3 Activity and state space models

In order to determine which instantiations are active, the numerical values for the variables and parameters involved in the relevant activity conditions must be specified. For this purpose, the program presents the user with a menu listing all identified quantities together with any currently existing values. The user may change the value of any number of quantities. For the example, all variables describing the mass of an object are initialized to 2, all temperatures are set to 0, and the power consumption of the hotplate is set to 1000.

Once the initial values for the relevant variables have been specified, the program tests the quantity conditions. Next, the user is presented with a list of preconditions. Each precondition must be set to either true or false. At first it is specified that the hotplate is not turned on, but the assumption that WATER and PAN be considered as one object is set to true. This means that the first four instantiations of `heat-bridge` and the corresponding instantiations of `heat-flow` are all inhibited by the subsumption mechanism. The only active instantiations are CONTAINER-WITH-LIQUID-1, HEAT-BRIDGE-5 and HEAT-BRIDGE-6. The resulting state space model comprises no state variables, only some algebraic relationships. It is reproduced below.

$$z_1 = z_2 \cdot \rho_1 \cdot g \tag{3}$$

$$z_2 = \frac{u_1}{(\rho_1 \cdot a_1)} \tag{4}$$

$$z_3 = u_1 + m_2 \tag{5}$$

All of the equations originate from the CONTAINER-WITH-LIQUID-1 instantiation and are describing the effects of the mass of liquid on the level and bottom pressure in the container.[6] Figure 9 defines the variables and parameters used in all models in this chapter. Next, the power in the HOTPLATE is turned on. The instantiation ELECTRIC-HEAT-HOTPLATE-1 changes state to active, but no other instantiations are affected. The dynamic influence of this instantiation will now give rise to a dynamic equation describing how the temperature in the HOTPLATE is affected by the heat generated. The corresponding model is reproduced below.

$$\dot{x}_1 = \frac{u_2}{c_3} \tag{6}$$

Next, the value of the temperature in the HOTPLATE is modified to 0.1 to anticipate the consequences of the now active heat generation in the HOTPLATE.[7] This causes the HEAT-FLOW-6 phenomenon to change to active since there is now a

---

[6]These equations will be included in all models as long as this view is active, to save space they are not shown in the subsequent models.

[7]No simulation capabilities have currently been implemented. Gradually changing the values of variables illustrates how the state space model would have evolved automatically during a simulation with a complete HPT model.

x_1 - temperature associated with HOTPLATE.
x_2 - temperature associated with CONTAINER-WITH-LIQUID-1.
x_3 - mass associated with WATER.
x_4 - temperature associated with PAN.
u_1 - mass associated with WATER.
u_2 - power consumption associated with HOTPLATE.
z_1 - bottom pressure associated with CONTAINER-WITH-LIQUID-1.
z_2 - liquid level associated with CONTAINER-WITH-LIQUID-1.
z_3 - mass associated with CONTAINER-WITH-LIQUID-1.
z_4 - heatflow from src to dst associated with HEAT-FLOW-6.
z_5 - heatflow from src to dst associated with HEAT-FLOW-2.
g - gravity constant associated with GLOBAL.
m_2 - mass associated with PAN.
a_1 - bottom-area associated with PAN.
c_3 - heat-capacity associated with HOTPLATE.
c_4 - heat capacity associated with CONTAINER-WITH-LIQUID-1.
c_6 - heat-capacity associated with PAN.
h_2 - vaporization-heat associated with WATER.
κ_5 - heat transfer pr degree Kelvin associated with HEAT-BRIDGE-5.
κ_7 - heat transfer pr degree Kelvin associated with HEAT-BRIDGE-3.
ρ_1 - density associated with WATER.

Figure 9: Parameters and variables used in the present chapter.

difference in temperature between the HOTPLATE and the CONTAINER-WITH-LIQUID-1. The resulting model is reproduced below. The heat-flow specifies two dynamic influences. The first describes the effect on the temperature of the object receiving the heat, CONTAINER-WITH-LIQUID-1, and gives rise to an additional dynamic equation. The second describes the negative impact on the temperature of the object giving up the heat, HOTPLATE. This gives rise to an additional term in the dynamic equation from the previous model.

$$\dot{x}_1 = \frac{u_2}{c_3} - \frac{z_4}{c_3} \tag{7}$$

$$\dot{x}_2 = \frac{z_4}{c_4} \tag{8}$$

$$z_4 = \kappa_5 \cdot (x_1 - x_2) \tag{9}$$

The temperatures in both HOTPLATE and CONTAINER-WITH-LIQUID-1 may now be increased gradually. However, there will be no change in the qualitative state until the temperature of the liquid reaches the point where boiling occurs. The boiling temperature for the liquid was specified to 100. Changing the temperature in the WATER to 100.1 thus causes BOILING-1 to change to active. The resulting model is given below.

$$\dot{x}_1 = \frac{u_2}{c_3} - \frac{z_4}{c_3} \tag{10}$$

$$\dot{x}_2 = \frac{z_4}{c_4} - \frac{z_4}{c_4} \tag{11}$$

$$\dot{x}_3 = -\frac{z_4}{h_2} \tag{12}$$

$$z_4 = \kappa_5 \cdot (x_1 - x_2) \tag{13}$$

The first dynamic influence of the boiling instantiation specifies how the boiling phenomenon will cause the liquid to evaporate. This causes the mass of WATER to be converted to a state variable and gives rise to an additional dynamic equation. The second dynamic influence describes how the boiling phenomenon consumes heat. This gives rise to an additional term in the equation describing x_2, the temperature in CONTAINER-WITH-LIQUID-1.[8]

If this situation persists long enough, the water will eventually evaporate. The value of the mass of WATER may gradually be reduced until it reaches zero. At this point, the quantity condition in CONTAINER-WITH-LIQUID-1 shifts to false. Since this is a subsuming quantity condition, CONTAINER-WITH-LIQUID-1 will loose its capability to subsume other objects. For all practical purposes the WATER has ceased to exist. All objects binding the water, including the CONTAINER-WITH-LIQUID-1 , must thus be inhibited. All of the previously active instantiations except for ELECTRIC-HEAT-HOTPLATE-1 are now deactivated. The only instantiation containing any influences that switches to active in the new situation is HEAT-FLOW-2. This instantiation describes the heat flowing from the HOTPLATE to the PAN. The state space model corresponding to this qualitative state is reproduced below.

$$\dot{x}_1 = \left(-\frac{z_5}{c_3}\right) + \frac{u_2}{c_3} \tag{14}$$

$$\dot{x}_4 = \frac{z_5}{c_6} \tag{15}$$

$$z_5 = \kappa_7 \cdot (x_1 - x_4) \tag{16}$$

6 DISCUSSION AND CONCLUSION

The HPT supports the three components introduced in Figure 4. But the third component is no longer qualitative, it is realized through simulation with a quantitative SSM. Further, the first and second components have been extended to support the derivation and on-line modification of the SSM.

The three component models of the HPT work together and the complete HPT model may thus support applications which have hitherto not been feasible in practice. The example in this chapter illustrates the principles of two such novel applications. The first is the automatic reformulation of a state space model in the face of changing assumptions. The second is the simulation of a system moving through different operating regimes. Essentially, perceived changes in the state of one component model affecting the structure of another component model propagate through, leading to the automatic updating of that model. Figure 10 depicts the conceptual dependencies which directs this updating of the models.

---

[8]Note that the two terms in the equation describing x_2 are identical except for opposing signs. These terms thus cancel out and leave the temperature in CONTAINER-WITH-LIQUID-1 constant. A future version of the implementation may take advantage of this and cancel the equation altogether.

64

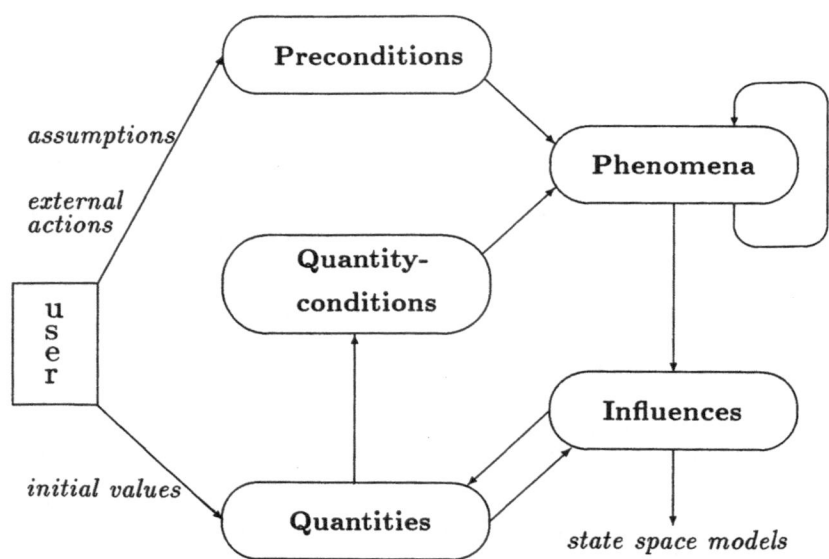

Figure 10: Conceptual dependencies in HPT models.

This figure describes the situation after the instantiation procedure has been completed, it shows how instances of five central HPT concepts may affect each other. Each oval represents all instances of the concept it names. Note that the role of the user could be filled by other application programs reasoning with the HPT model. In applications such as diagnosis, the HPT could be used to adapt the structure of the mathematical models to fit a possible error. E.g. if it is assumed that a pipe is clogged or a valve is stuck, this could be expressed in terms of a set of changed assumptions. The HPT could then return a SSM which could be use to establish or reject this hypothesis.

This chapter has focused on the use of the HPT to maintain a consistent state space model in the face of changing operating regimes and assumptions. Whether a given influence is relevant in the current state depends on the state of the phenomenon from which it originates. The value of an influence depends on the variables and parameters which it involves; the value of state and dependent variables depend on the influences affecting these variables. A change in the value of a quantity may cause a change of state of a quantity condition, and thus indirectly the state of a instantiation. The HPT framework as such does not incorporate any explicit notion of a state space model or even an equation. The state space model is implicit in the HPT description. But it is easy to establish a printed representation of the valid SSM.

References

[Abelson & al., 1989] H. Abelson, M. Eisenberg, M. Halfant, J. Katzenelson, E. Sacks, G. J. Sussman, J. Wisdom and K. Yip. Intelligence in Scientific Computing. *Communications of the ACM,* 32, may 1989.

[Benveniste & Åström, 1993] A. Benveniste & K.J. Åström Meeting the Challenge of Computer Science in the Industrial Applications of Control: An introduction to the Special Issue. In: *IEEE Transactions on Automatic Control*, Vol 38, No7, July 1993.

[Forbus, 1984] K. D. Forbus. A Qualitative Process Theory. *Artificial Intelligence* 24, 1984.

[Hayes, 1978] P. Hayes. The naive physics manifesto. in: Michie D. (Ed) *Expert systems in the micro electronic age.* Edinburgh Scotland: Edinburgh University Press, 1978.

[Hayes, 1985] P. Hayes. The second naive physics manifesto. In J. R. Hobbs and R. C. Moore, (Eds). *Formal Theories of the Commonsense World.* Ablex Publishing Corporation, 1985.

[Isermann, 1993] R. Isermann. Fault Diagnosis of Machines via Parameter Estimation and Knowledge Processing - Tutorial paper. *Automatica* 29, 1993.

[de Kleer, 1979] J. de Kleer. Qualitative and quantitative reasoning in classical mechanics, in: P. H. Winston, and R. H. Brown. *Artificial Intelligence: An MIT perspective, Volume I.* The MIT press series in artificial intelligence, 1979.

[de Kleer & Brown, 1984] J. de Kleer and J. S. Brown. A Qualitative Physics based on Confluences. *Artificial Intelligence* 24, 1984.

[Kuipers, 1986] B. Kuipers. Qualitative Simulation. *Artificial Intelligence* 29, 1986.

[Raiman, 1988] O. Raiman. Order of Magnitude Reasoning. In: *Proceedings of AAAI-88*

[Sacks, 1991] E. P. Sacks. Automatic Analysis of one-parameter planar ordinary differential equations by intelligent numeric simulations. *Artificial Intelligence* 48, 1991.

[Steele, 1990] G. L. Steele jr. *Common Lisp, The Language* Second Edition. Digital Press, 1990.

[Struss, 1988] P. Struss. Mathematical aspects of qualitative reasoning. *Artificial Intelligence in Engineering* 3.3 1988.

[Struss, 1990] P. Struss. Problems of Intervall-Based Qualitative Reasoning. In *[Weld and de Kleer, 1990]*

[Weld and de Kleer, 1990] D. S. Weld and J. de Kleer, (Eds.), *Readings in Qualitative Reasoning about Physical Systems.* Morgan Kaufman, 1990.

[Zhao, 1991] F. Zhao. Extracting and Representing Qualitative Behaviors of Complex Systems in Phase Spaces. In: *Proceedings of the 12th IJCAI*, Sydney, Morgan Kaufmann 1991.

3
TIMING PROBLEMS AND THEIR HANDLING AT SYSTEM INTEGRATION

LEO MOTUS
Chair of Real-time systems
Tallinn Technical University, ee-0026,
Tallinn, Estonia

1. Introduction

The number of artificial intelligence methods applied to actual industrial process control and/or to plant-wide control has been rapidly increasing. For example, the application of fuzzy controllers in continuous process control and manufacturing, expert systems for automatic and/or automated decision making in production control and in supporting the managerial activities, have proven their effectiveness.

Additional AI application examples can be drawn from the flexible manufacturing (car production, food processing, etc) where they can be, and are being, used for controlling separate elements of a production line, separate cells in the line, as well as for planning the whole production process.

Strictly speaking, majority of artificial intelligence applications in industrial decision-making, control, and automation should obey certain timing restrictions imposed by the other parts of the application which interact with the AI based components. Sometimes the timing restrictions have a very relaxed form and breaking the restrictions will not necessarily cause serious losses. Still, a realistically minded engineer is aware that such relaxed restrictions are more of an exception than a rule. Consequently, in the general case, the AI based component in a plant-wide control system is supposed to obey given/imposed timing restrictions.

This leads the system developers to the following non-trivial problems. First, it is recommendable to have (or to develop) artificial intelligence methods, algorithms and their implementations which have time-deterministic behaviour and well-defined, easily applicable performance evaluation tools [1].

The question of obtaining time-deterministic behaviour for AI based methods is extremely intriguing since it triggers many fundamental issues. For example, formalization of general problem solving, the essence of reasoning (especially reasoning with incomplete information and/or with imposed strict deadlines), the essence of reflexes (i.e. dynamic reorganization of a deduction mechanism, based on the obtained feedback experience and with a general goal of improving effectiveness of one's behaviour). Unfortunately, the author of this chapter is not yet ready to tackle such fundamental issues. Much more experimental study and theoretical research is needed before, for example, the approximate reasoning methods could be applied with clear conscience in hard real-time systems -- in other words, in systems which require that decisions with precisely measured quality be reached at strictly fixed deadlines.

S. G. Tzafestas and H. B. Verbruggen (eds.),
Artificial Intelligence in Industrial Decision Making, Control and Automation, 67–87.
© 1995 *Kluwer Academic Publishers.*

Second problem is concerned with ensuring appropriate/expected interaction between system's components and thus the planned behaviour of the overall system. Industrial automation systems are built and updated during an extended time period. During the system's design, and each time a new component is introduced to the existing system, raises the problem of matching the behaviour of a new component with the rest of the components in the system. Pretty good methods are available for guaranteeing the logical match, i.e. the match between interfaces of the new component and those of the other components in the system. One could use the widespread system design methodologies as described in [2] -- e.g. SSADM, DSSD, HOOD, Z, VDM -- majority of which are based on data flow diagrams and/or physical flow diagrams, and rely heavily on structured design ideas. It is, however, a common knowledge that, although practically very useful, the abovementioned system design methodologies are not too helpful in timewise matching the behaviour of system's components and for proving time correctness of the system as a whole. In fact, majority of those methodologies and corresponding tools are quite helpless in system's timing analysis.

It is clear that solution of the second problem, i.e. determining the required time constraints for a new component in a system, is one of the major preconditions to the successful solution of the first problem. No reasoning system can provide results with the required quality for any given deadline -- if a well-founded deadline (and/or other timing requirements) would exist, the task of finding an appropriate reasoning method would become better motivated.

This chapter suggests a solution to the second problem and discusses also why such an obvious solution has never been suggested before. The basic line of thoughts followed in this chapter is:

- considering the truly complex nature of industrial decision and control systems, and the extremely wide variety of applications, one has to fix essential properties that influence system's time correctness and abstract away the details which are specific to a particular application; examples of such essential properties for each component could be, frequency of execution, single execution time, input/output data and time-constraints imposed on that data, timing requirements to intercomponent communication (i.e. exhange of information, or of physical substance)

- a mathematical model is developed/selected which focuses on the abovelisted properties and enables comparatively easily to prove time correctness of the system; of course, before one can start proving system's time correctness it is necessary to reach an agreement on what is meant by time-correct behaviour of a system.

The chapter is partitioned as follows. Section 2 fixes the essential features which influence control system's behaviour. Section 3 states what is meant by time correctness in this chapter. In section 4, the background and some paradigms for developing a mathematical model are discussed, and the mathematical model itself is presented. Section 5 is devoted to analytical study of timing properties, section 6 illustrates the presentation by an application example (a cascade controller). Section 7 concludes the chapter by discussing the automation of timing analysis and some open problems.

2. Essential features of control systems

A contemporary control system can be partitioned into three large parts:
- an object which is to be controlled,
- a computer based system (often geographically distributed) which monitors the status of the object, makes control decisions (considering also the goals set by its developers and/or users), and executes the decisions in order to influence on the behaviour of the control object,

- an interface between the object and the computer based system (typically consisting of sensors, actuators and communication media)

In the context of a control system's correct behaviour, one usually assumes that the co-operation of all three parts of the system results in a behaviour which satisfies the requirements and constraints imposed by the control object, the user, and the interface.

A system designer usually accepts the control object as given -- i.e. its properties and requirements to the other parts of the system have the highest priority and can be modified only slightly, if at all. At the same time the computer system and the interface are to be designed so as to satisfy the requirements and constraints imposed by the object.

Such systems are being studied under various names. For example, real-time systems is used by control engineering and computer science communities [3,4], reactive systems is used by computer science community [5], hybrid systems is used by computer science and control engineering communities [6,7]. It has also been suggested that real-time systems and embedded systems are equivalent terms [8,9]. Throughout this chapter the term "real-time system" is used as an extension of the term "industrial decision and control system".

Real-time systems form a typical border-area where many mature research domains intersect: applied mathematics, artificial intelligence, communication, computer engineering, control engineering, diagnostics, fault-tolerance, instrumentation, software engineering, and system engineering are just some alphabetically ordered examples of the involved research areas. With such a variety of involved aspects it is quite difficult to agree upon the essential features.

The existing theories which have been applied to real-time systems -- such as temporal logic [5], Petri net extensions [10], calculus of duration [11], hybrid systems [6] -- try to cover many aspects of a system at the same time. Please note that the references given are not necessarily representative for the corresponding theory, they only give an idea of the theory. Also, the list itself is not complete, its aim is to give just examples of the suggested theories.

All the abovelisted theories are useful and have remarkably contributed to understanding the nature of problems in real-time systems. Some of the theories are even focused on timing problems. Nevertheless, one should admit that they share two disadvantages, at least as far as computer control practice is kept in mind. First, it seems that all the abovelisted theories have had a common pardigm -- a system with completely known causal relations, which is unfortunately quite rear in control practice. Handling of incompletely known causal relations, or of the cases where unknown causal relations have been approximated by quantitative time constraints, tend to be too sophisticated in the abovelisted theories.

The other disadvantage stems from the incomplete coverage of practically important timing properties. This is a more serious drawback since it indicates that the requirements to time correctness of systems have not been systematically studied (for further comments see section 3 of this chapter). This problem was pointed out already in 1988 by Stankovic [3]. Nevertheless, many of the theories are still interested in (and able to analyse) performance bound timing properties only. In the majority of cases such simplistic approach to time correctness is caused by insufficient understanding of the role of time in a system and in a system's model and by the consequent trivial handling of time. To a certain extent such sad outcome has also been fostered by using the not quite appropriate paradigm (i.e. a real-time system is a single non-terminating program).

The presented search for essential features of real-time systems has been influenced and supported by a series of papers [3,12,13], which have provided a quasi-philosophical view on critical properties of those systems and recommendations about the features to be addressed by new formal methods. Some inspiration has also been recieved from the critical analysis of widespread conventional approaches (for example, [14]).

In spite of the critique made about the widespread conventional approaches, one should give full appreciation to the work done in this area. It should be admitted that information system's design and data processing aspects in real-time systems have been well covered. Therefore, further in this section the attention is paid to presenting features which differ from those of conventional information processing systems and which are characteristic to real-time systems.

Taking an impartial look at the existing, operational, real-time systems and at the expected future systems, one can distinguish the following three characteristic features, potentially present in any real-time system. None of those three features have been (or can be) sufficiently studied in the framework of the wide-spread, existing methods.

2.1. *Essential (forced) concurrency*, which can be present in real-time systems, in addition to conventional concurrency. This phenomenon is known in circuit theory and is also inevitable in real-time systems. Parallel (concurrent) programs as considered normally in computing science are designed/implemented by applying only the "good will" of the designer -- e.g. because one has a multiprocessor at hand, the algorithm enables parallel/concurrent processing and one needs the answer faster than sequential processing can provide.

In real-time systems, quite often, the computer has to control more than one physical process which all are simultaneously carried on in the environment. If the characteristics of the physical processes (and the computer) do not allow pseudo-parallel mode, one faces the essential (forced) concurrency. Essential concurrency is more demanding than the conventional one in a sense that:

- one has to obey (often strict) timing requirements imposed by the environment, consequently the free choice of the designer is severely limited

- timing (or just synchronization) requirements and constraints imposed upon the interaction of concurrent processes in the computer are more sophisticated than one would expect in traditional concurrent programming.

The essential (forced) concurrency is unavoidable during the specification stage when one normally is concerned with understanding the functioning of the control object (and assumes resource adequacy). In some cases the essential concurrency can be reduced to pseudo-parallel execution (and/or to conventional concurrency) during the physical design and implementation stages.

2.2. *Truly asynchronous mode of execution of interacting processes* is a consequence of the essential concurrency -- causal reasons which normally control the concurrent execution, are often invoked outside of the computer and are out of the designer's control (or even unknown to her/him). Therefore it is natural to assume that the activation instants of some of the concurrently executed processes may in certain cases be independent of each other, or the actual causal reasons are not known, or they are too complicated to be considered in a computer system. In such cases one can just state that these processes execute truly asynchronously.

Two modes of process execution are typically considered in computing science:

- *synchronous*, when the activation instants of a process are synchronized with some, usually periodic, event (not necessarily occurring in the same system), e.g. a pulse generator

- *asynchronous*, when the activation instants of a process are synchronized with (usually non-periodic) events occurring in the same system; as a rule, the synchronization requirements are looser than in the previous case.

As one could expect, these two modes are not disparate. This has been demonstrated in [15,16].

The truly asynchronous mode has always been present in real-time systems -- a trivial example is periodic data acquisition and an operator's question about the status of the control object, or an alarm condition and the data integrity problem of the correspondingly displayed information.

The truly asynchronous mode has never been seriously accepted by the computing science community, thus the abuse of term "asyncronous" -- there is no sense in talking about an algorithm with non-existent (or not known) causal reasons (remember the computer science paradigm of a single non-terminating program for a real-time system). The practical existence of the truly asynchronous mode has forced us to look for a new paradigm for real-time system -- a real-time system is considered here to be a collection of loosely coupled, repeatedly activated, terminating programs (for details see section 4 of this chapter)

The interacting processes, executed in truly asynchronous mode may cause the most sophisticated timing errors. Those errors may appear very seldom and at seemingly random time instants, a more detailed handling of such errors is given in section 5 of this chapter.

2.3. Time-selective interprocess communication. Such communication has always implicitly been present in real-time systems. The necessity of explicit handling of time-selective interprocess communication has been caused by the new paradigm for real-time systems -- a collection of loosely coupled, repeatedly activated, terminating programs.

The basic idea of time-selective communication is, in a nutshell, that data received by the consumer must be of exactly the right age (not too fresh, not too old). This can be achieved by applying the "producer-consumer" communication schema, whereas the consumer process has a possibility to subscribe to the data of required age. This idea does not quite conform with the conventional computing science data exchange disciplines (stacks (LIFO), queues (FIFO), etc). Also the definition of fairness, as usually applied to a data communication subsystem -- " eventually each sent message reaches its destination" -- is not valid any more, it turns out to be normal that some messages are consumed several times, some will never be consumed.

Only a few of the existing computational models enable to describe and analyse time-selective interprocess communication. For the examples, see the history transformers in [17], and the Q-model which is based on the work of Quirk and Gilbert [18], developed further in [19] and also discussed in this chapter (section 4).

The above emphasised three features, in spite of their harmless appearence, pose serious new requirements to the system models, and also influence the spectrum of properties which must be checked to guarantee correct functioning of a system.

3. Concerning time-correct functioning of systems

As the number of computer based decision and control applications grows, increases also the number of malfunctionings caused by timing errors. Correspondingly increases the number of serious research papers published on various aspects of timing problems. Some characteristic examples are referred in the following: He, [20], attempts to introduce time into CSP; [10,21] suggest methods for proving performance-type properties (i.e. upper and lower bounds on the execution time). An interesting effort has been made in [22] by studying timing characteristics of a process in the context of pre-run-time scheduling.

Surprisingly little attention, however, has been devoted to analysing the essence of timing correctness. The problem was mentioned in [13], some further comments were added in [23]. In this chapter, the classification suggested in [13,23] is adopted. According to this classification the timing properties and the criteria which define timing correctness, are classified into three -- the performance-bound properties, those characterizing the timewise correctness of events and data, and the properties important for timewise correctness of interprocess communication.

3.1. *Performance-bound properties* are the most thoroughly studied group of timing properties. Earlier performance-bound properties were considered as the only important timing feature of a real-time system. Performance characteristics were even used as definitive for real-time systems. This misconception was pointed out in [3]. The group of performance-bound properties is addressed by a large number of researchers. Some examples of the published research results are given in [20,21,22,24,25,26]. Typical attributes studied by the researchers of this group are execution time of a program (or a sequence of programs), response time, and deadlines. All the basic classes of formal methods -- temporal logic, algebraic methods, Petri nets -- support, at least in principle, the study of performance-bound properties. A nasty and cynical person could argue that no principally new results have been published since [26] in 1979 -- majority of authors just apply the well-known, traditional, ideas to a specific formalism, or concentrate on a slightly different aspect of the problem. Actually it has been a steady step-by-step progress which have resulted in good understanding of the essence of, and the methods for, evaluating performance-bound properties. A larger step, perhaps, has been taken in [22], where the study of performance-bound parameters has been shifted into the specification/design stage.

3.2. *Timewise correctness of events and data.* A closer look at this group of timing properties gives some additional evidence to LeLann's claim, [12], that it does not suffice to provide system's parts with access to some global centralized clock in order to establish consistent time reference. In addition to the centralized clock one has to provide mappings from the autonomous/independent time measuring mechanisms of various system parts onto the centralized time counting/measuring mechanism. Even then one would face some problems with data integrity unless validity times of data and events are not given. Typical parameters required for demonstration of timewise correctness of events and data are:

- validity time interval for data and/or event [27]
- tolerance, equivalence and simultaneity intervals for matching time instants and events in different time counting/measuring systems [8]
- time parameters (like activation period, activation instants) required to state and solve "specification-time (pre-run-time) scheduling" type problems , see [8,22].

The idea of attaching a validity time interval to each data item (or, at least to majority of data items) that is moved around in a computer system is not very old, it was mentioned in 1979 in [28].

Tolerance, equivalence and simultaneity intervals were first mentioned in [29]. This set of intervals gives a possibility to check the pairwise match between the parameters belonging to two different time measuring systems. These intervals form well-founded requirements for selecting the proper granularity of global system time, and the required accuracy of synchronization between various parts of a system. The same set of intervals has a decisive role in matching the dynamic behaviour of a computer system with that of its environment. Such a match is of primary importance in order to achieve the goals of a real-time system.

Activation periods in connection with specifications is also a comparatively recent innovation [18]. "Specification-time scheduling" is important for discovering major inconsistencies in the specified time requirements. When speaking about timing correctness of system functioning in this chapter, a strong similarity with the ideas presented in Xu [22] must be admitted. The major difference, however, is that here not everything is reduced to the question of satifisfying deadlines. Often it is just the contrary, in many cases timing errors occur due to quite tricky combinations of various timing parameters and can not be easily related to deadlines. This is partly caused by differences in handling sporadic/aperiodic tasks. In this chapter sporadic/aperiodic tasks are approximated as periodic tasks with given intervals of uncertainty. The corresponding method was first suggested in [18], and provides rather flexible environment for modelling aperiodic task execution as regular activities.

Timing analysis at the specification stage results in a sort of "discrimination" as compared to true computing science approach where all the possible sequencies of process execution are considered as equally interesting and useful. As the result of timing analysis, certain potentially possible sequencies are eliminated by imposing timing constraints on the computational model, and consequently on the future system.

Note that such "discrimination" is not necessary when one is interested only in performance type characteristics (as is the case in computing science where all the other constraints are determined by the causal reasons). However, the elimination of some execution sequencies becomes unavoidable as soon as timewise correctness of events and data becomes important, and even more so when time-selective interprocess communication is considered.

3.3 *Time correctness of interprocess communication* is critical for hard real-time systems. Still, in any real-time system this feature is a good indicator of whether or not time parameters and constraints -- usually imposed upon the interacting parties independently of each other, are quite often obtained from different sources and specified by different persons -- are consistent and non-contradicting. Surprisingly little attention has been paid to studying this excellent integral indicator of correct functioning of a specification, design and implementation of real-time systems.

Many researchers have addressed access and transport delays caused by communication protocols and media. A few researchers are interested in fixing the starting instant of the interaction (see, for example, [30]). The actual age of the received message has not yet been of interest to the majority of specifiers. As an exception, [17] and a series of publications invoked by Quirk and Gilbert (e.g. [8,17-19]) could be pointed out.

It is interesting to note that in order to be able to analyse time correctness of interprocess communication (including time-selective communication) one needs a sophisticated handling of time . The traditional computing science understanding of time as just an ordering of activities in a program, is not sufficient. It has been demonstrated in [13] that simultaneous use of three different aspects of time is necessary to provide full support to complete analysis of timing properties in a real-time system. The required aspects are covered by the following philosophical concepts of time:
   - time as used in theoretical physics (fully reversible time), which is typically present in transformational programs,
   - time as used in thermodynamics, needed for describing physical processes with which the computer system interacts during its functioning and for describing the evolution of a real-time system itself; also needed for matching dynamical behaviour of the computer system and its environment;
   - time as used in our conscious awareness, needed for describing and analysing time-selective interaction; the origin of this time is always at the instant when the consumer process requests a message; the age of the requested data is specified with respect to the origin of this time (further on, in this chapter, such time will be called relative time)

Now that the essential features of real-time systems, and the requirements for timing correctness of such systems have been fixed, remains the question of selecting/developing a suitable mathematical model which desribes adequately the features and enables to check the consistency and non-contradiction of the requirements.

4. A mathematical model for quantitative timing analysis (Q-model)

It is necessary that the model includes all the time parameters needed for scheduling -- timing analysis is closely related to "pre-run-time-scheduling". The ultimate goal of such a model is to provide a specification (and a design) which will lead to an applicable implementation -- i.e. it becomes possible to exclude obvious timing controversies by analytical study of the specification (design) before the implementation starts. The parameters should characterize the individual structural elements (processes) of a real-time system, and all the timing constraints imposed upon the interaction of processes and upon the overall functioning of the system. As it was stated earlier, the existence of the parameter values alone does not suffice to enable full-scale analytical study of timing properties. A sufficient condition is that the model captures the required multitude of time concepts, because only this supports the analytical study of timing properties and leaves to testing and simulation an auxiliary role.

The model proposed in this chapter (the Q-model) is based on the thermodynamical time concept (also used as a basis in Petri Nets, temporal logic, etc). In addition, the Q-model relies heavily on the use of many relative times (one for each process and one for each producer-consumer pair of interacting processes). For each process the thermodynamical time advances in grains defined by the execution time interval and/or by the repeated (periodic) activation interval. The thermodynamical time may be reversible inside one grain -- inside each grain time is used as in theoretical physics. A detailed presentation of time handling in the Q-model is given in [13].

The basic framework of the model was originally described in [18], and developed further in [19]. The most complete presentation of the model, so far, is given in [8].

4.1 Paradigms used. It was mentioned earlier in this chapter that the traditional computing science paradigm for real-time (reactive) systems has been a single, non-terminating program. The Q-model is based on another paradigm since the conventially used one:

- is too abstract, and difficult to understand for engineers who have to map the reality into a specification or a program,
- is intrinsically based on a single, thermodynamical, time concept and thus hinders the handling of many essential features, e.g. truly asynchronous execution of processes, and time-selective communication,
- can not support the complete analysis of timing properties (see section 3, this chapter).

The Q-model is based on a paradigm that **a real-time system is a collection of interacting, loosely coupled, repeatedly activated and terminating programs.**

This paradigm enables better to focus on a fundamental theoretical problem in connection with real-time systems -- under which conditions, if at all, can a Turing machine (an algorithm) give reasonable results in a dynamically changing environment --this problem was posed by Maler in [7]. In the other words the same question can be reformulated as a statement that the fundamental problem in real-time systems is: how to match the dynamics of the computer system to that of the environment? [8].

The newly introduced paradigm also facilitates mapping of the actual industrial environment onto a specification/design of a computer system. Practically any technical device can be seen as a cyclic/sporadic execution of the same routine job which terminates each time with a positive or negative result. Emphasising the repeated, terminating execution of processes enables the specifier/designer to equip all the data items in the system with time labels, which is important for timing analysis. Also, this exposes explicitly the true nature of time-selective interprocess communication -- it is actually time-selective, interprocess and intercycle communication, i.e. communication between different execution cycles of interacting processes.

4.2 The Q-model. A detailed description of the Q-model is given in [8]. Here is presented a minimal description, required for following the timing analysis section.

The Q-model consists of *processes* (which transform data and physical matter in a system) and of *channels* (which implement process interaction).

A process is a repeatedly executed mapping

$$p_i : T(p_i) \ x \ dom \ p_i \rightarrow val \ p_i \, ,$$

where $T(p_i)$ is a well-defined timeset with elements determining the activation instants of the mapping, or its activation period; $dom \ p_i$ is domain of definition, and $val \ p_i$ is value range of the mapping.

Please note that cyclic/repeated execution is practically not possible unless each process has a finite execution time which in the Q-model is given as an interval extimate:

$$\zeta(p_i,t) \in [\alpha(p_i), \beta(p_i)].$$

Sporadically activated processes are considered as regularly cyclic with an indeterminacy in their period. Therefore two consecutive elements ($t,t' \in T(p_i)$), of the process timeset should satisfy the constraint:

$$t_{min}(p_i) \le t' - t \le t_{max}(p_i).$$

A channel, as means of interprocess communication, is a mapping from the producer process value range (val p_i) onto the consumer process domain of definition (dom p_j):

$$\sigma_{ij}: T(p_i) \ x \ T(p_j) \ x \ val \ p_i \rightarrow proj \ _{val \ pi} \ dom \ p_j.$$

The channel throughput is limited by a channel function so as to enable time-selective interaction. The channel function is mathematically a subset of the producer's timeset

$$K(\sigma_{ij}, \ t) \subset T(p_i), \qquad\qquad t \in T(p_j).$$

In practice, however, the channel function is given as an interval in relative backward time which is defined for this particular interaction, i.e. $K(\sigma_{ij}, \ t) = [\mu,\nu]$. The origin of this time is the instant when the consumer requests data from the channel. The origin of time is then projected to the producer's timeset and each data generation produced by the corresponding activation is labelled by a timetag. The timetag's value is determined by counting, in the producer's well-ordered timeset, the activation instants backwards from the projected origin of the relative time.

For practical purposes it is good to distinguish between different types of channels, depending on the relationship of the interacting partners' timesets. Everybody may, in principle, introduce new channel types -- the set of channel types is open. Nevertheless, a vast majority of applications can be described by using just four types of channels:
- synchronous channel, if the corresponding timesets coincide,
- semisynchronous channel, if the producer's timeset generates the consumer's timeset; this means that each termination of the producer causes activation of the consumer process,
- asynchronous channel, if the corresponding timesets are proven, or believed, to be independent,
- null channel which synchronizes only the activation instants of the partners and carries no application oriented messages.

Note the shift in terminology, semisynchronous channel implements the asynchronous mode, as used conventionally, whereas asynchronous channel implements the truly asynchronous mode (see section 2, this chapter).

In the Q-model a process need not have its input data at its activation. The request for data may be sent later, during process execution. This allowable delay with respect to the activation instant is given in the Q-model as an interval estimate:

$$\eta(\sigma_{ij},t) \in [\ \gamma(\sigma_{ij}),\ \delta(\sigma_{ij})\ ].$$

The abovedescribed set of parameters together with the underlying time concepts is sufficient for full scale timing analysis as demonstrated in the next section.

The Q-model can easily be used to extend the description obtained by one of the wide-spread data-flow-diagram based models (see, for example [31]) and in this way add timing analysis facilities to that class of models.

It has been demonstrated that the Q-model is a superset of ordinary Petri Nets, and that the Q-model can be mapped into a weak second order predicate calculus (see, for example, [32-34).

5. The Q-model based analytical study of system properties

Due to the sophisticated handling of time, the Q-model properties enable full-scale timing analysis. The timing analysis of a specification/design described in the Q-model notations leads, in the most cases, to straightforward analytical procedures. Given a system specification, the analysis proceeds in three stages [8].

5.1 Separate elements of a specification. In many cases this analysis has been reduced to trivial checks on the specified values for execution time and for allowable delay of input data consumption, on definitions of the timeset, domain of definition and value range of a process, and on the specified channel function. These checks detect obvious specification errors and strong deviations of given values from reasonable limits (if known).

Specific problems at this stage are the termination order of process copies and data consumption by process copies which become explicit in a system comprising of cyclically executed terminating processes. Simultaneously executed copies of a process appear as soon as the period of process activation is smaller than its execution time. In the case of the Q-model, all the parameter values are given as interval estimates, consequently sufficient condition for occurrence of simultaneous copies for a process p_i is

$$\beta(p_i)\ /\ t_{min}(p_i) > 1.$$

This ratio also gives an estimate of the maximum number of simultaneously executed copies - - a valuable hint for assessing the actual resource requirements.

5.1.1 Termination order of copies could be a source of timing errors. As soon as the later activated copy of a process terminates before the earlier activated copy, the system has a timing error. In a real system, where the activation instants fluctuate, even if the process was specified as strictly periodic, it is difficult to detect such an error by testing or simulation. It has been proven in [18] that the correct termination order is guaranteed iff the inequality

$$\beta(p_i) - \alpha(p_i) < t_{min}(p_i)$$

holds.

5.1.2 The channel access order by copies of a process is equally nasty to detect, since the number of process copies does not change the interaction structure of the program and the same channel is used by all the copies. Therefore it is important that simultaneously executed copies of a process request data from the channel in the order of their activation. It has been proven in [18] that this order is maintained iff the inequality

$$\delta(\sigma_{ij}) - \gamma(\sigma_{ij}) < t_{min}(p_i)$$

holds.

5.2 Pairs of interacting processes. Analysis of pairwise interaction is carried out separately for each channel type. At the specification and design stage the main concern is that the requirements for time-selective communication are not contradicting to the specified time parameters of the involved partners.

5.2.1 Interaction via synchronous channel is a widely used and thoroughly studied mode of communication. Main problem here is that the requested data may not be ready and it is difficult to say how long the consumer can wait for the data without any harm to the system. A proposition has been proven in [8] that the specified time parameters of the system will not be violated because of the interaction iff

$$\beta(p_i) < \gamma(\sigma_{ij}) + \nu \, t_{min}(p_i),$$

and the parameters of the consumer process are violated slightly, still additional simultaneous copies are not created because of those violations, iff

$$\beta(p_i) < \gamma(\sigma_{ij}) + \nu \, t_{min}(p_i) + [t_{min}(p_j) - \beta(p_j)],$$

where p_i is the producer process, p_j is the consumer process, and $\nu \geq 0$.

The above inequalities give the worst case estimates which can be loosened if the designer agrees to take a reasonable risk. The following case where the producer subscribes to the future data, i.e. to the data which will be produced after the consumer has been activated, gives an idea of a looser inequality. The specified parameters will not be violated because of the interaction (the case with $\nu < 0$), iff

$$\delta(p_i) > \beta(p_i) + (|\nu| - 1)(t_{max}(p_i) - t_{min}(p_i))/2 + t_{min}(p_i),$$

where $|...|$ denotes the absolute value of the variable or expression. The inequality has been loosened substituting minimum activation period by average activation period. At the moment, however, there are no ready results for evaluating the involved risk quantitatively since the estimate depends on too many factors.

5.2.2 Interaction via semisynchronous channel combines two functions -- synchronization of consumer activation with the producer termination, and transfer of a message similarly to the case of synchronous channel. Qualitatively new potential error type is that the activation attempted by the producer could be neglected by the consumer -- for example, because of its specified time parameters, especially the equivalence interval (see section 3, this chapter), and because of structural peculiarities of the model/system (e.g. the consumer has many incoming semisynchronous channels).

The potential loss of an activation attempt can be detected by comparing the distance between two successive activation attempts with the length of equivalence interval specified for this process.

Waiting for data is not a problem in the case of $\nu \geq 0$. For the cases with $\nu < 0$, i.e. for the cases when the consumer has subscribed to the future data, inequalities which are similar to those described in the previous section, have been developed.

5.2.3 Interaction via asynchronous channel is the most liberal form of communication since the timesets of interacting partners are independent and the interaction does not need any synchronization. Therefore the attention is focused on the age of transmitted data only.

Any communication causes a transport delay which characterizes time spent in the communication media -- components of the transport delay are caused, for example, by physical propagation speed of signals, by competition for communication line, by pre- and post-processing of the message. Communication via asynchronous channel adds to this a non-transport delay. Non-transport delay is caused by the fact that interacting processes are activated at independent, from each other, time instants. The non-transport delay behaves like a saw-tooth function [8], which means that the actual value of the delay changes in time. Accordingly the delay may exceed the maximum specified allowable value at seemingly random (for the external observer) time. This is the type of timing errors which emerge once in a long period of successful operation of a system, and are extremely hard to detect and eliminate on the basis of testing or simulational runs.

The Q-model properties allow to obtain an upper bound for non-transport delay as a function of time parameters of the interacting partners. The presently existing upper bound estimate is quite pessimistic, still it is better than no estimate at all. It has been proven in [8] that an upper bound for the non-transport delay (φ_{max}) in an asynchronous channel between the producer p_i and the consumer p_j is

$$\varphi_{max} \leq t_{max}(p_i) + \beta(p_i) - \gamma(\sigma_{ij}) - 1.$$

5.3 Group of interacting processes is analysed for the absence of informational deadlocks, for analytical evaluation of performance-type characteristics, and for detecting specific malfunctioning caused by a mixture of structural and timing factors.

5.3.1 Informational deadlock is a circular wait condition where all the partners connected via synchronous channels need input message of age 0 in backward relative time so as to produce output message of the same age. Informational deadlock may occur if a synchronous cluster of processes contains a synchronous loop (i.e. a closed contour of processes and channels where all the channels are synchronous with channel function parameters $\nu = 0$ and $\mu \geq 0$ is optional) [8].

5.3.2 Analytical evaluation of performance type characteristics is an important innovation since the majority of existing methodologies in system and software engineering are based on simulational study of such characteristics. The unique properties of the Q-model enable to obtain easily computable lower and upper bounds of the time required to pass a message (or a physical matter) through a network of processes.

The most labor consuming part here is determination of paths which are of interest for this particular application in a, potentially cyclic, graph that corresponds to the Q-model description of the system. Once the set of interesting paths has been determined, formulae for analytical evaluation of the passing time are applied.

The key idea is that a path is partitioned into legs, each of which contains only one type of channels -- i.e. into a set of synchronous legs, semisynchronous legs, and asynchronous legs. For each leg formulae for lower and upper bounds of the passing time have been developed (see, for example, [8]). As the result of applying this procedure, each leg can be substituted by a single process with the execution time given as an interval estimate, the endpoints of the interval are determined by the mentioned formulae. The same procedure is repeated until the whole path is aggregated into one process with given interval estimate for its execution time.

5.3.3 Specific malfunctioning caused by a mixture of structural and timing properties comprises a variety of problems related to incorrect activation of processes, ordering distortion of messages in a path, and others. Usually such disorders have been connected only to data-dependent conditional control transfer and/or to incorrect structure of the system.

The Q-model study indicates that such disorder may appear, in many cases, due to inappropriately specified time parameter values. This chapter gives only a couple of examples of the problem. The subject is still under study and the results will be extensively described in a future publication in the context of a software engineering environment CONRAD being developed on the Q-model basis.

The first example is related to permanent or occasional neglecting of some input messages and activation attempts -- this may happen in the case of multiple semisynchronous channels connected to an input port of a process due to the specified value of the equivalence interval. It is true that this situation is not necessarily erroneous. However, each particular case needs special study. A simulational study of the system will often not reveal such peculiarities, or various simulation runs may give contradicting results depending on the selected simulation scenario.

The second example illustrates a retarder process in a sequence of processes -- i.e. a process whose response is much slower than that of its predecessors in the sequence and therefore is not as productive as the others are. A retarder process typically appears in a sequence of processes interacting via semisyncronous channels and may cause anomalies in the order of messages moving through the sequence of processes. The existence of such a retarder depends on the activation frequency of the sequence, on the specified execution times of individual processes in the sequence and on the specified value of equivalence intervals. It is interesting to note that the same situation may happen in any pipeline or array processor unless special synchronization measures have been taken.

6. An example of the Q-model application

There is no point in illustrating how the formulae are used for computation. Therefore this example concentrates on demonstration of the effectiveness of the Q-model in fostering the understanding, even without sophisticated mathematical analysis, of the essential problems in the system development and on pointing out some important differences from the corresponding Petri model.

The following example describes a direct digital controller which, on the basis of a given algorithm, computes a required position of a valve. The valve position is allowed to be modified in small steps, therefore the actuator takes the new required valve position as a set-point and implements the required change in accordance with the restrictions prescribed by the plant's dynamics. The combination of the controller's algorithm and the intelligent actuator form, in fact, a cascade controller where one of the controllers (A) computes set-point values and the other (B_1, B_2) realizes the set-point values by controlling the actuator. Graphically a cascade controller and its corresponding Q-model description are presented in Fig. 1.

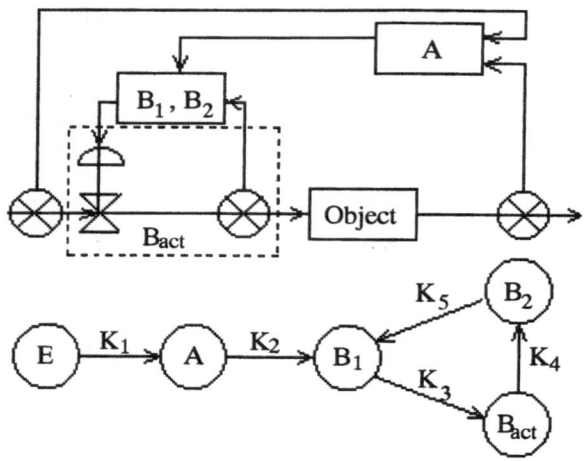

Fig. 1 A cascade controller and its Q-model description

The Q-model processes in Fig.1 have the following meanings (more details can be found in Table 1):

- process **E** describes the object as required by the control algorithm
- process **A** implements the control algorithm and computes new values for the control variable
- process **B₁** transmits the admissible changes in the control variable value to the actuator
- process **B₂** keeps track of the current actual value of the control variable in the actuator
- process **B_act** simulates the actuator (so as to include interface to the environment into the Q-model, but also to enable autonomous testing and simulational study of the controller).

It is possible to achieve several different behavioural patterns of the controller by changing the interaction characteristics in the same structure. In Table 2 three versions of channel types have been described. The versions given in Table 2 are superficially analysed in this chapter, the resulting time diagrams are presented in Fig.2. In order to build a time diagram of controller's behaviour, one only needs to know time parameters of processes and the basic properties of the corresponding channels Three more combinations of channel types which could be used for building the controller are given in Table 3 and not analysed in this chapter.

Table 1

Process name	Execution time	Input channels	Output channels	Timeset	Comments
E	not important	-	K_1	$T(E)$	The process and its timeset are completely determined by the properties of the object
A	4	K_1	K_2	-	This process is activated as soon as new measurements come from E
B_1	2	K_2, K_5	K_3	$T(B_1)$	$T(B_1)$ may be generated by $T(E)$ via K_1 and K_2, or may be determined independently (if required by the actuator and the object
B_2	1	K_4	K_5	-	May be activated either by B_{act} (via K_4) or by B_1 (via K_5)
B_{act}	3	K_3	K_4	-	Is activated by B_1

Table 2

CHANNEL NAME	1. VERSION		2. VERSION		3. VERSION	
	TYPE	FUNCTION	TYPE	FUNCTION	TYPE	FUNCTION
K_1	ss	[0,0]	s	[1,1]	s	[0,0]
K_2	ss	[0,0]	s	[1,1]	s	[0,0]
K_3	ss	[0,0]	s	[1,1]	s	[0,0]
K_4	ss	[0,0]	s	[1,1]	s	[0,0]
K_5	a	[0,0]	s	[1,1]	s	[1,1]

The project based mostly on semisynchronous channels (version 1, Table 2) has the minimum feasible activation period (input of the measured data activates the controller in this case) of 6 time units and the total length of a complete execution cycle is 10 time units. These results are obtainable by considering the process and channel parameters given in Table 1 and Table 2.

The version 1 (Table 2) is possible to implement on one processor. If one tries to shorten the activation period of the controller below the abovegiven limit of 6 time units, the feedback signal which informs the control algorithm about the actual position of the actuator, will be delayed more than the specified channel function allows. If the delayed feedback signal is acceptable, it is possible to soften the requirement to the channel K_5. It is interesting to note that even if the number of processors will be increased, the presented requirements can not be satisfied when the minimal activation period is shorter than the abovegiven limit of 6 time units.

When buying/building a controller based on the abovedescribed project (i.e. the first version in Table 2) one must be prepared that the controller is sensitive to the actuator characteristics. If the controller functions with minimal activation period and the actuator has to be substituted by a slower one, the danger of timing errors increases remarkably -- the feedback signal which gives the actual position of the actuator may occasionally be delayed.

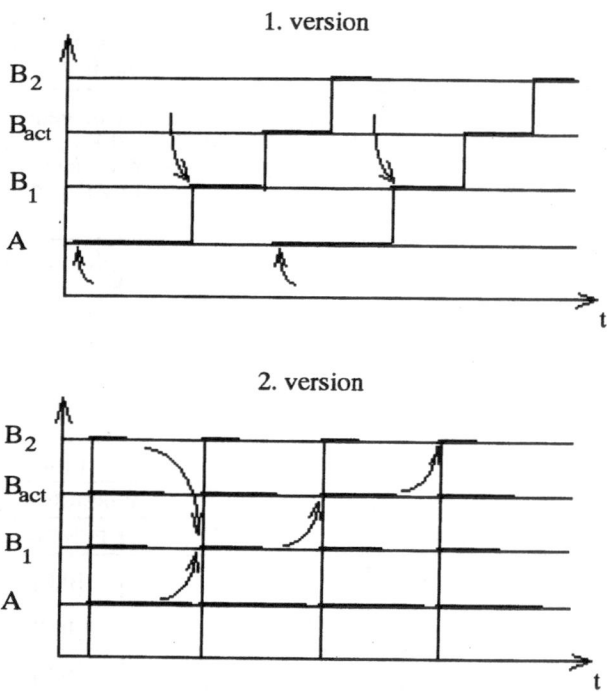

Fig. 2 Time diagrams of the Q-model controller

The project based on synchronous channels, with channel functions [1,1] (version 2, Table 2) has the minimal activation period of $4+\varepsilon$ time units and the total length of an execution cycle of $16+\varepsilon$ time units. Here ε denotes a technically inevitable vacant time between two consecutive executions of a program on a processor, ε is small as compared to the time unit of the specification. This project has the following peculiarities:

- the feedback signal (the actual position of the actuator) is delayed too much; it would be possible to decrease the delay by changing the given structure, e.g. processes B_1 and B_2 should be aggregated into one which eliminates channel K_5, then the resulting delay would be the same as in the first version of Fig. 2.

- a set of synchronous processes is easier to implement on a multiprocessor than on a network, and is impossible to implement (with the specified time constraints) on a single processor.

The project, based on synchronous channels with channel function [0,0] -- the third version in Table 2 -- is not implementable with the given execution times of processes. If we were allowed to modify freely the completion times, this project would end up with the characteristics similar to those of the first project. Still, a substantial difference for the user is that the minimal activation period of the corresponding controller can not be less than 10 time units. Note, that if the channel function for K_5 is [0,0], the processes B_1, B_2 and B_{act} form a synchronous loop and consequently the controller's description contains an informational deadlock.

Slightly more sophisticated time diagrams can be derived if one decides to determine $T(B_1)$ independently of $T(E)$, this possibility follows from Table 1. This, however, will need modification of the corresponding channel types as illustrated in Table 3.

Table 3

Channel name	4. version	5. version	6. version
K_1	s	ss	s
K_2	a	a	a
K_3	ss	ss	ss
K_4	ss	a	ss
K_5	a	s	s

The analysis of controller's versions from Table 3 is left to the reader. Instead, due to the popularity of Petri nets it would be natural that any new modelling methodology is compared with Petri nets. The controller, discussed above, can be described in Petri nets as a modification of a conventional producer-buffer-consumer problem (see Fig.3). The control algorithm is a producer, the consumer is a complex consisting of the actuator and the algorithms which compute changes to the actuator's position and organize access to feedback information about the actuator's real position.

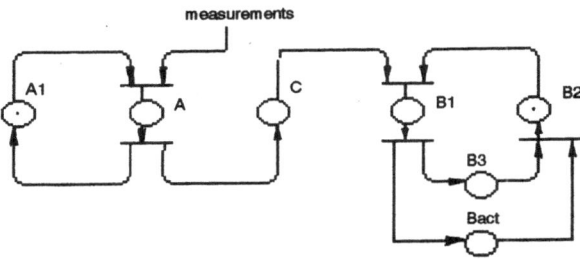

Fig.3 The direct digital controller as a Petri net

84

The activities (processes) are described as places and data communication as transitions in order to facilitate comparison with Fig.1. In this context it is assumed that transitions fire instantly whilst places require time (as specified in Table 1). The notations used for places coincide basically with those used in Fig.1. However, the Petri net functioning and synchronization rules force the introduction of three additional places (processes):

- as usually done in Petri nets, **A1** synchronizes dynamically the input data
- **C** describes data transmission
- since the actuator is really a device which lies outside of the computer system, **B3** is used as a "watchdog" for the actuator.

The execution time of the additional processes (places) may be neglected. The resulting firing diagram of the Petri net is presented in Fig.4.

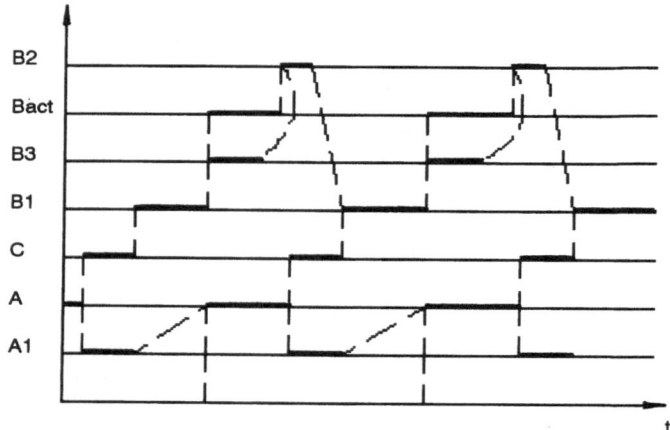

Fig.4 A firing diagram for a Petri net presenting the direct digital controller

A similar Petri net has been thoroughly analysed by Sifakis [26], it was demonstrated that such a Petri net cannot function faster than its "natural" rate is. In this case the natural rate is closely related to the minimal allowable activation period of the controller -- which in this particular case is 6 time units, whereas the total length of a complete execution cycle is $10+\varepsilon$ time units. Note that those integral characteristics for the Petri net version of the controller correspond closely to the version 1 (Table 2) of the Q-model description.

Leaving aside the analysing power and other technical advantages and disadvantages of the involved models, this chapter further concentrates on their effectiveness in specific application oriented situations which occur during the design of the controller. For serious assessment of the two models one should not forget that they have been developed for different goals, and therefore are comparable only to a certain extent. More details about the comparison of Petri nets and the Q-model can be found in [32-34].

Situation A. The case when different parts of a system operate with different frequencies. As it happens, the control algorithm executes more seldom than the algorithm controlling changes in the actual output of the actuator. Nevertheless, the two algorithms should coordinate their activities quite strictly, in many cases quantitative timing restrictions should be satisfied. The Q-model can take care of this situation quite easily by using one of its channel types with the appropriate channel function.The case where two sub-nets function with independent frequencies and exchange information is possible to describe in a Petri net -- unfortunately it is extremely difficult to control whether or not the sub-nets' interaction satisfies time constraints imposed by the application.

Situation B. The case of comparing various descriptions and/or design decisions. Consider, as an example, specification/design of a direct digital controller . It would be nice to have a quick, not too comprehensive, overview of various alternatives before serious efforts will be applied to the most promising one. With Petri nets the different versions, given in Tables 2 and 3, lead to different structures of the corresponding nets; construction of different Petri net versions is obviously more labour consuming than comparing the Q-model versions. In addition, the Q-model presentation covers both data and control flow explicitly which is normally not the case with Petri nets.

7. Conclusions

The results presented in this chapter form a small part of necessary preconditions for developing well-founded, time-deterministic, and safe applications based on artificial intelligence methods. These applications should satisfy strict restrictions imposed by real-time environments, such as industrial control and automatic industrial decision making systems.

Although auxiliary in the context of this book, the results are based on substantial innovations which have gradually been introduced by many researchers and have, mostly, emerged in the last decade. A new paradigm for describing real-time systems, and understanding the necessity of more sophisticated time handling are just two examples of the innovations. The author is convinced that systematic study for developing artificial intelligence based methods which should have time-deterministic behaviour, becomes reasonable and will yield results only after such auxiliary topics have been fully understood.

In order to become practically significant, the abovedescribed results should be supported by a computer-based tool. The author of this chapter has been involved in developing such a tool for several years. The tool supports control systems description in terms of the Q-model, its formal analysis (as discussed in this chapter), and informal analysis of the formally proven description based on animation/simulation study of "what-if" scenarios. Since the ultimate goal of the tool -- called CONRAD (CONtrol systems Requirements, Analysis, and Design) -- is to obtain timewise correct system, it is planned as a post-processor for existing data-flow-diagram based methodologies (such as [31], for example). In principle, CONRAD can also work with object-oriented methodologies, provided that some restrictions on dynamic creation and dynamic migration of objects are introduced.

References

1. L.Motus "Artificial Intelligence in Hard Real-time -- a new paradigm needed?" 3rd IFAC Workshop on Artificial Intelligence in Real-time Control, 1991, California, U.S.A.

2. ButlerBloor report "CASE products: comparison and evaluation", 1992, ButlerBloor Ltd

3. J.A.Stankovic "Misconceptions about Real-time Computing: a Serious Problem for Next Generation Systems" - IEEE Computer, 1988, vol.21, no.10

4. H.W.Lawson "Engineering Predictable Real-time Systems" - Lecture Notes for the NATO Advanced Study Institute on Real-time Computing, Oct. 1992

5. Z.Manna, A.Pnueli " The temporal Logic of Reactive and Concurrent Systems: Specifications" - Springer Verlag, New-York, 1991

6. Z.Manna, A.Pnueli " Verifying Hybrid Systems" - Workshop on Theory of Hybrid Systems, Technical Univ. of Denmark, Oct. 1992

7. O.Maler "Hybrid Systems and Real World Computations" - Workshop on Theory of Hybrid Systems, Technical University of Denmark, Oct. 1992

8. L.Motus "Dynamics of Embedded Software" - Valgus, Tallinn, 1990 (in Russian)

9. A.T.Kuendig "A note on the meaning of "Embedded Systems"" in Embedded Systems, Eds. A.T.Kuendig, R.E.Buehner, J.Daehler, Lecture Notes in Computer Science, 1987, no.284, Springer Verlag, 207 pp

10. C.Ghezzi, D.Mandrioli, S.Morasca, M.Pezzè " A Unified High-level Petri Net Formalism for Time-critical Systems" - IEEE Trans. on Software Engineering, 1991, vol. SE-17, no. 2

11. Zhou Chaochen, C.A.R. Hoare, A.P.Ravn " A Calculus of Durations" - Information Processing Letters, 1991, vol.40, no.5

12. G.LeLann "Critical Issues for the Development of Distributed Real-time Computing Systems" - Proc. IEEE Workshop on Future Trends of Distributed Computing Systems, Cairo, Egypt, 1990.

13. L.Motus "Time Concepts in Real-time Software" - Control Engineering Practice, 1993, vol.1, no.1

14. E.J.Cameron, Y.-J.Lin " A Real-time Transition model for Analyzing Behavioral Compatibility of Telecommunications Services" - Software Engineering Notes, 1991, vol.16, no.5

15. R.Milner " Calculi for Synchrony and Asynchrony" - Theoretical Computer Science, 1983 vol.25, no.3.

16. V.Varshavskii (ed.) " Control Automata for Asynchronous Processes in Computers and Discrete Systems" - Nauka, Moscow, 1986 (in Russian)

17. P.Caspi, N.Halbwachs " A Functional Model for Describing and Reasoning about Time Behaviour of Computing Systems" - Acta Informatica, 1986, vol.22

18. W.J.Quirk, R.Gilbert " The Formal Specification of the Requirements of Complex Real-time Systems" - AERE, Harwell, 1977, no. 8602

19. L.Motus, K.Kääramees " A Model Based Design of Distributed Computer Control System Software" - Proc. 4th IFAC Workshop on DCCS, Pergamon Press, 1983

20. He Jifeng "A Dual-time Model for Communicating Sequential Processes" - Workshop on Theory of Hybrid Systems, Technical University of Denmark, Oct. 1992

21. T.A.Henzinger, Z.Manna, A.Pnueli "Temporal Proof Methodologies for Real-time Systems" - School on Formal Techniques in Real-time and Fault-tolerant Systems, University of Nijmegen, The Netherlands, Jan. 1992

22. J.Xu, D.L.Parnas "On Satisfying Timing Constraints in Hard-Real-time Systems" - Software Engineering Notes, 1991, vol.16, no.5

23. L.Motus "Analytical Study of Quantitative Timing Properties of Software" 5th Euromicro Workshop on Real-time Systems, 1993, IEEE Computer Society Press, 218-223

24. N.A.Lynch, H.Altiga "Using Mappings to Prove Timing Properties" - Distributed Computing, 1992, vol.6, no.2

25. J.Ostroff "Temporal Logic for Real-time Systems" - Research Studies Press; John Wiley and Sons Inc., 1989

26. J.Sifakis "Use of Petri Nets for Performance Evaluation" - Acta Cybernetica, vol. 4, no.2, 1979

27. I.M.MacLeod " A Study of Issues Relating to Real-time in Distributed Computer Control Systems" - Ph.D. Thesis, University of Witwatersand, Johannesburg, 1983

28. L.J.Sloan " Limiting the Lifetime of Packets in Computer Networks" - Computer Networks, 1979, no.3

29. L.Motus "Semantics and Implementation Problems of Interprocess Communication in a DCCS Specification" - Proc. IFAC Workshop on DCCS, Pergamon Press, 1986

30. J.Hooman "Compositional Verification of Distributed Real-time Systems" - School on Formal Techniques in Real-time and Fault-tolerant Systems, University of Nijmegen, The Netherlands, Jan. 1992

31. P.T.Ward and S.J.Mellor "Structured Development for Real-time Systems", Yourdan Press Computing Series, 1985, 156 pp.

32. J.Vain "Comparison of the modelling power of the Q-model and Petri nets" Proc. Estonian Academy of Sciences, vol 36, no 3, 324 - 333 (in russian)

33. J.Tekko "Comparison of the Q-model and Petri nets based on the Q-model language" Proc. Estonian Academy of Sciences, vol 37, no.1, 18 - 25 (in russian)

34. J.Tekko "A formal model for LSD language and CSD calculus" Proc. Estonian Academy of Sciences, vol. 41, no.4, 266 -278 (in russian)

4

ANALYSIS FOR CORRECT REASONING IN INTERACTIVE MAN-ROBOT SYSTEMS: DISJUNCTIVE SYLLOGISM WITH MODUS PONENS AND MODUS TOLLENS

ELDO C. KOENIG
Computer Sciences Dept.
University of Wisconsin-Madison
35005 West Fairview Road
Oconomowoc, Wisconsin 53066, U.S.A.

1 INTRODUCTION

In this paper, reasoning by a robot relates to commands it receives in the natural language. The statements supplied by the master are considered to be incompletely stated arguments, i.e., enthymemes. An attempt is made by the robot to provide missing premises or conclusions to produce valid arguments. This is done on the basis of the inference rule, disjunctive syllogism. The analysis utilizes, in part, the results of work covering correct reasoning by robots based on the two inference rules of modus ponens and modus tollens [1]. The robot seeks out missing premises or conclusions from its knowledge structures for obeying commands and from the environment through its sensors. In the final analysis, the paper establishes composite plausible commands as enthymemes that can be supplied by the master and corresponding missing premises or conclusions that the robot must seek out in an attempt to achieve a primary goal.

Much logical work has been done on knowledge systems but little has been done on robot knowledge systems that treat commands as incompletely stated arguments. Examples of efforts on other knowledge systems are those of Schott & Whalen [2], Trillas & Valverde [3], and Koenig [4, 5]. In the more recent work of Schott & Whalen, an experimental inference engine was established which performs goal-directed backward chaining and forward inferencing by scanning the rule base for opportunities to use both modus ponens and modus tollens. A previous work by Koenig [4] used rules adapted from Gentzen.

It is believed that the analysis previously presented [1, 6] and the analysis presented here is necessary in establishing intelligent interaction between master and robot. And when the robot also expresses its intended responses in the natural language, the interaction is further enhanced with an intelligent conversation. The results of this paper makes this very evident.

89

S. G. Tzafestas and H. B. Verbruggen (eds.),
Artificial Intelligence in Industrial Decision Making, Control and Automation, 89–97.
© 1995 Kluwer Academic Publishers.

2 VALID COMMAND ARGUMENTS

A symbolized argument

$$X \lor Y, \neg X \therefore Y$$

may be established as valid on the basis of disjunctive syllogism. In this paper, the argument unites with other arguments to convey commands. Component sentences are symbolized for the analysis. Their meaning assignments and example sentences that they symbolize are:

A – a primary goal
  "Dust the table."
A2 – a second primary goal
  "Dust the car."
B – an alternate goal
  "Move from the closet to home base."
C – condition for achieving a primary goal
  "A cloth is in the closet."
¬Z – read as "not Z", Z = A, A2, B, C.

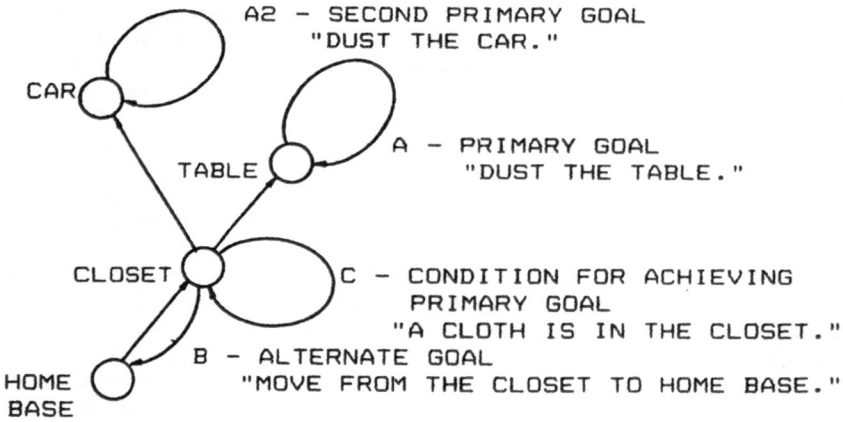

Figure 1. Knowledge structure describing the physical environment in which a robot operates.

The knowledge structure of Figure 1 in the form of an environment graph describes the physical environment in which the robot operates [7, 8]. In application, the structure could contain many more points and lines. The meaning of a symbolized sentence is represented by either two points and an adjacent line or a single point and a loop when the two points are identical. For the analysis, for C true, either A, ¬A2, ¬B are true or ¬A, A2, ¬B are true, and for ¬C true, ¬A, ¬A2, B are true. That is, from the above meaning assignments when the condition C exists, either the primary goal A is accomplished, and the second primary goal A2 and the alternate goal B are not accomplished, or the primary goal A and the alternate goal B are not accomplished, and the second primary goal A2 is accomplished. And when the condition C does not exist, the primary goal A and the second primary goal A2 are not accomplished, and the alternate goal B is accomplished.

In the previous analysis [1] when only modus ponens (MP) and modus tollens (MT) were considered, A2, ¬A2 were not included in the above set of symbols. Since the analysis here utilize, in part, the results of [1], the following definitions and theorems are repeated for later reference. To distinguish the definitions and theorems from those originating in this paper, the identifying numbers are given a prime mark.

Definition 1′. Valid command arguments (determined valid by either modus ponens or modus tollens) are those valid arguments that contain only symbols from the set N = {A, B, C, ¬A, ¬B, ¬C} and that are valid for either A, C, ¬B assumed to be true or for ¬A, ¬C, B assumed to be true.

Definition 2′. A plausible command is a command that is seemingly executable.

Definition 3′. A plausible command argument is a valid command argument that is required in conveying a plausible command; the primary goal must be expressed in one of the plausible command arguments conveying a plausible command.

Definition 4′. A sound command argument is a plausible command argument whose premises and conclusion are in fact true.

Definition 5′. Sound commands are conveyed by sound command arguments.

Theorem 2′. For a plausible command argument (determined to be valid by either modus ponens or modus tollens) either A, ¬B, B, ¬A is the conclusion.

Theorem 3′. The plausible command arguments determined valid by modus ponens are: C ⊃ A, C ∴ A; C ⊃ ¬B, C ∴ ¬B; ¬C ⊃ B, ¬C ∴ B; ¬C ⊃ ¬A, ¬C ∴ ¬A.

Theorem 6′. The plausible command arguments determined to be valid by modus tollens are: ¬A ⊃ ¬C, C ∴ A; B ⊃ ¬C, C ∴ ¬B; A ⊃ C, ¬C ∴ ¬A; ¬B ⊃ C, ¬C ∴ B.

3 CORRECT REASONING: DISJUNCTIVE SYLLOGISM

Consider a plausible command argument X ∨ Y, ¬X ∴ Y to be incompletely stated by the master. The robot must seek out the missing premises or the conclusion from its stored knowledge and from the environments through its sensors. To say X ∨ Y, ¬X ∴ Y is correct reasoning by the robot means that whenever X ∨ Y and ¬X are both true, then Y is also true. This is the case whether or not the premises X ∨ Y and ¬X are in fact both true.

Definition 1. Valid command arguments determined valid by disjunctive syllogism (DS) are those that contain only symbols from the set {A, A2, ¬A, ¬A2} and that are valid for either A, ¬A2 assumed to be true or for ¬A, A2 assumed to be true.

Theorem 1. The number of valid command arguments, determined to be valid by disjunctive syllogism for either A, ¬A2 assumed to be true or for ¬A, A2 assumed to be true, is two.

Proof. There are two different symbols appearing in an argument, and these symbols appear as a pair in one of the premises. Then the number of possible arguments is two for either A, ¬A2 assumed true or for ¬A, A2 assumed true.

Plausible Composite Command Arguments 3.1

Plausible composite command arguments are to be established by uniting select valid command arguments determined to be valid by DS with select plausible command arguments of Theorems 3′ and 6′ determined to be valid by MP and MT.

Theorem 2. The plausible composite command arguments determined to be valid by modus ponens followed by disjunctive syllogism are

$$C \supset (A2 \lor A), C, \neg A2 \therefore A$$
$$C \supset (A \lor A2), C, \neg A \therefore A2,$$

and those determined to be valid by modus tollens followed by disjunctive syllogism are

$$(\neg A2 \lor \neg A) \supset \neg C, C, \neg A2 \therefore A$$
$$(\neg A \lor \neg A2) \supset \neg C, C, \neg A \therefore A2.$$

Proof. Consider the following two arguments determined valid by DS and that satisfy Definition 1:

$$A2 \lor A, \neg A2 \therefore A; \quad A \lor A2, \neg A \therefore A2$$

The premises with the connectives are the conclusions of the following command arguments determined valid by MP and MT:

From Theorem 3′ $C \supset (A2 \lor A), C \therefore A2 \lor A$
and $C \supset (A \lor A2), C \therefore A \lor A2$
From Theorem 6′ $(\neg A2 \lor \neg A) \supset \neg C, C \therefore A2 \lor A$
and $(\neg A \lor \neg A2) \supset \neg C, C \therefore A \lor A2$

The above arguments unite to give the valid composite command arguments of this Theorem 2, and these are the only valid command arguments of this type. The arguments of Theorems 3′ and 6′ containing B or ¬B will not unite with those that are valid by DS.

Plausible Composite Commands 3.2

Theorem 3. The plausible composite command arguments of Theorem 2 convey plausible composite commands and may be combined with plausible command arguments of Theorems 3′ and 6′ for conveying plausible composite commands that are more complete:

i. $C \supset (A2 \lor A), C, \neg A2 \therefore A$
ii. $C \supset (A2 \lor A), C, \neg A2 \therefore A$ and $\neg C \supset B, \neg C \therefore B$
iii. $(\neg A2 \lor \neg A) \supset \neg C, C, \neg A2 \therefore A$
iv. $(\neg A2 \lor \neg A) \supset \neg C, C, \neg A2 \therefore A$ and $\neg B \supset C, \neg C \therefore B$

The interchanges of A and A2 in i, ii, iii, iv above give v, vi, vii, viii.

Proof. The primary goal is expressed in each of the plausible command arguments conveying a plausible command as required by Definition 3'. A plausible command can contain premises or a conclusion from either of the single plausible command arguments, i, iii, v, vii, but after a command is received by the robot, the robot must inquire of information on an alternate goal should the premise C be false. For one of the arguments of the dual arguments, C is a premise and for the other argument, ¬C is a premise so that a command conveyed by a dual argument can contain reasoning information on the primary goal and the alternate goal. For the first argument of the dual arguments ii and iv to be sound, C, ¬A2, A are required to be in fact true. The second argument of these two dual arguments comes from Theorems 3' and 6', and in order to be sound, ¬C and B are required to be in fact true. For the first argument of the dual arguments vi and viii to be sound, C, A2, ¬A are required to be in fact true, and in order for the second argument of these two dual arguments (which also come from Theorems 3' and 6') to be sound, ¬C and B are required to be in fact true.

The premises and the conclusion of a plausible composite argument of Theorem 3 are all involved in commands as enthymemes in various ways. Some of the commands are included in Theorems 5' and 8' involving MP and MT and become enthymemes included there. They are not duplicated here. The remaining premises become involved as components of a composite command. That is, a response is made by the robot to a first component of the composite command before a second component of the composite command is given. This first response by the robot includes a conclusion that preceeds the final conclusion of the second response. For the example argument C ⊃ (A2 V A), C, ¬A2 ∴ A, the commands symbolized by C and A are included in Theorem 5' involving MP and become the enthymemes included there. The remaining premises, C ⊃ (A2 V A) and ¬A2, become involved as components of a composite command presented here.

The detailed analysis for each of the plausible composite command arguments of Theorem 3 follows.

   i. Plausible Composite Command Argument:
      C ⊃ (A2 V A), C, ¬A2 ∴ A
     1. Master: C ⊃ (A2 V A)
         "Robot, if a cloth is in the closet, dust the car or dust the table."
       Robot: C, A2 V A
         (Robot moves to the closet.)
         "A cloth is in the closet.
         Therefore, I will dust the car or the table."
        Master: ¬A2
          "Robot, do not dust the car."
        Robot: A
         "I will dust the table."
      or (when argument not sound),
       Robot: ¬C
        "A cloth is not in the closet.

94

I await further instructions."

ii. Plausible Composite Command Argument of i Combined with a Plausible Command Argument of Theorem 3′:
C ⊃ (A2 V A), C, ¬A2 ∴ A and ¬C ⊃ B, ¬C ∴ B
1. Master: C ⊃ (A2 V A), ¬C ⊃ B
"Robot, if a cloth is in the closet, dust the car or the table.
If a cloth is not in the closet, move from the closet to home base."
Robot: Either ¬C, B or C, A2 V A
(Robot moves to the closet.)
"A cloth is not in the closet.
Therefore, I will move from the closet to home base."
(End of conversation.)
or (when first argument is sound),
"A cloth is in the closet.
Therefore, I will dust the car or the table."
Master: ¬A2
"Robot, do not dust the car."
Robot: A
"I will dust the table."

iii. Plausible Composite Command Argument:
(¬A2 V ¬A) ⊃ ¬C, C, ¬A2 ∴ A
1. Master: (¬A2 V ¬A) ⊃ ¬C
"Robot, do not dust the car or the table
only if there is no cloth in the closet."
Robot: C, (A2 V A)
(Robot moves to the closet.)
"But there is a cloth in the closet.
Therefore, I will dust the car or the table."
Master: ¬A2
"Robot, do not dust the car."
Robot: A
"I will dust the table."
or (when argument not sound),
Robot: ¬C
"A cloth is not in the closet.
I await further instructions."

iv. Plausible Composite Command Argument of iii Combined with a Plausible Command Argument of Theorem 6′:
(¬A2 V ¬A) ⊃ ¬C, C, ¬A2 ∴ A and ¬B ⊃ C, ¬C ∴ B
1. Master: (¬A2 V ¬A) ⊃ ¬C, ¬B ⊃ C
"Robot, do not dust the car or the table
only if there is no cloth in the closet.
Do not move from the closet to home base only
if there is a cloth in the closet."
Robot: Either ¬C, B or C, (A2 V A)
(Robot moves to the closet.)

"A cloth is not in the closet.
Therefore, I will move from the closet to home base."
(End of conversation.)
or (when first argument sound),
"But there is a cloth in the closet.
Therefore, I will dust the car or the table."
Master: ¬A2
    "Robot, do not dust the car."
Robot: A
    "I will dust the table."

v, vi, vii, viii. The interchanges of A and A2 in the detailed analysis of i, ii, iii, iv above give the detailed analysis for v, vi, vii, viii.
    The above discussion establishes the following Theorem:
    Theorem 4. For the plausible composite command arguments of Theorem 3, determined to be valid by modus ponens followed by disjunctive syllogism and by modus tollens followed by disjunctive syllogism, eight dual plausible commands as enthymemes can be supplied to a robot by the master, and the robot seeks out corresponding missing premises or conclusions in an attempt to achieve a primary goal:

i. $C \supset (A2 \vee A)$, C, ¬A2 $\therefore$ A
    Master: $C \supset (A2 \vee A)$;
    either,
    Robot: C, A2 ∨ A; Master: ¬A2; Robot: A;
    or (when argument not sound),
    Robot: ¬C, and requests further instructions.

ii. $C \supset (A2 \vee A)$, C, ¬A2 $\therefore$ A and ¬C $\supset$ B, ¬C $\therefore$ B
    Master: $C \supset (A2 \vee A)$, ¬C $\supset$ B;
    either,
    Robot: ¬C, B;
    or (when first argument sound),
    Robot: C, A2 ∨ A; Master: ¬A2; Robot: A.

iii. (¬A2 ∨ ¬A) $\supset$ ¬C, C, ¬A2 $\therefore$ A
    Master: (¬A2 ∨ ¬A) $\supset$ ¬C;
    either,
    Robot: C, (A2 ∨ A); Master: ¬A2; Robot: A;
    or (when argument not sound),
    Robot: ¬C and requests further instructions.

iv. (¬A2 ∨ ¬A) $\supset$ ¬C, C, ¬A2 $\therefore$ A and ¬B $\supset$ C, ¬C $\therefore$ B
    Master: (¬A2 ∨ ¬A) $\supset$ ¬C, ¬B $\supset$ C;
    either,
    Robot: ¬C, B;
    or (when first argument sound),
    Robot: C, (A2 ∨ A); Master: ¬A2; Robot: A.

v, vi, vii, viii. The interchanges of A and A2 in i, ii, iii, iv above give v, vi, vii, viii.

4 CONCLUSIONS

Component sentences of command arguments state a primary goal A, a second primary goal A2, an alternate goal B, and a condition C for achieving a primary goal. Command arguments are valid for either A, not A2, not B, C assumed to be true or for not A, A2, not B, C assumed to be true or for not A, not A2, B, not C assumed to be true.

The number of command arguments, that are valid by reason of disjunctive syllogism for either A, not A2 assumed true or for not A, A2 assumed true, is two. The number of plausible composite command arguments determined valid by either modus ponens followed by disjunctive syllogism or by modus tollens followed by disjunctive syllogism is two. These plausible composite command arguments convey plausible composite commands and may be combined with plausible command arguments determined valid by either modus ponens or modus tollens. The number of single arguments is four, and the number of combined arguments is also four. Eight different dual plausible commands as enthymemes can be supplied by the master. Corresponding missing premises or conclusions that the robot seeks out in an attempt to achieve a primary goal are also identified.

REFERENCES

[1]   E. C. Koenig, "Analysis for correct reasoning by robots: modus ponens, modus tollens," Proceedings of the 1989 IEEE International Phoenix Conference on Computers and Communications, U.S.A., pp. 584-589, 1989.

[2]   B. Schott and T. Whalen, "Modus ponens, modus tollens, and fuzzy relations in goal directed inferences," Proceedings of the 1987 IEEE International Conference on SMC, Vol. 1, pp. 173-176, 1987.

[3]   E. Trillas and L. Valverde, "On mode and implication in approximate reasoning," Approximate Reasoning in Expert Systems, edited by M. M. Gupta, A. Kandel, W. Bandler, and J. Kiszka. New York: North Holland, pp. 157-166, 1985.

[4]   E. C. Koenig, "Intelligent systems: knowledge association and related deductive processes," Kybernetes, Vol. 7, pp. 99-106, 1978.

[5]   E. C. Koenig, "Generating inferences from knowledge structures based on general automata," Cybernetica, Vol. XXVI, No. 2, pp. 80-97, 1983.

[6]   E. C. Koenig, "Analysis for correct reasoning by robots: hypothetical syllogism with modus ponens, modus tollens," Proceedings of the 8th International Congress of Cybernetics and Systems, Hunter College, City University of New York, New York City, N.Y., June 11-15, 1990.

[7]   E. C. Koenig, "Some principles for robotics based on general automata," Robotica, Vol. 4, pp. 43-46, 1986.

[8]   E. C. Koenig, "A model knowledge structure for robots," <u>Proceedings</u>
       <u>of the 1987 IEEE International Conference on SMC</u>, U.S.A., Vol. 3,
       pp. 1154-1159, 1987.

PART 2
INTELLIGENT SYSTEMS

5

APPLIED INTELLIGENT CONTROL SYSTEMS

R. SHOURESHI, M. WHEELER, L. BRACKNEY
School of Mechanical Engineering
Purdue University
1288 Mechanical Engineering Building
West Lafayette, IN 47907-1288,U.S.A.

1. INTRODUCTION

Complexity of dynamic systems has manifested the need for design of new controllers that have abilities beyond those provided by the existing conventional control theory. The need for qualitative methods to model and analyze complex, uncertain, nonlinear, and high dimensional systems presents the first challenge in forming an intelligent control system (ICS).

In terms of intelligence, it is expected that an ICS would replicate human abilities of creation, sensation, perception, adaptation, inferencing, assessment, self-modification (healing) and development of experiences. Thus it is expected that an ICS would have the objective to: respond autonomously to discrete symbolic commands expressed at a high level of abstraction; identify changes in the system structure or configuration; provide more human centered responsiveness; apply sensory data to infer faults and/or operator errors; facilitate the use of heterogeneous (qualitative, symbolic) knowledge sources; reconfigure feedback control structures to respond to significant variations in the system and/or environment.

Given the current state-of-the-art in the area of machine intelligence and A.I., it does not seem realistic to expect an ICS with the above attributes to be developed in the near future. However, systems with a subset of those abilities defined for an ICS can be configured and apply to class of problems. In the following sections a general structure for an ICS is proposed and applied to two specific problems, namely, intelligent automatic generation control in an electric power plant and human comfort control in large buildings.

All future correspondence should be addressed to Prof. Shoureshi

101

S. G. Tzafestas and H. B. Verbruggen (eds.),
Artificial Intelligence in Industrial Decision Making, Control and Automation, 101–126.
© 1995 *Kluwer Academic Publishers.*

2. A PROPOSED STRUCTURE FOR ICS

Although recent literature provides a variety of configurations for ICS, we consider the following structure as one that generalizes the approach and demonstrates the relationship between ICS and conventional feedback control systems.

Consider the closed loop control system of Fig. 1, where in addition to the conventional feedback control, a feedforward and an open loop optimal predictive controllers are present. For simplicity and without the loss of generality, a tracking problem is assumed where the plant output deviates from a desired output, caused by disturbances or uncertainties associated with the system and/or environment. Superposition of the three controllers shown on the block diagram provides the following input-output relationship for this system.

$$\Delta Y = P\frac{(1 + CF)}{1 + CSP}D + \frac{PC}{1 + CSP}Y \equiv PGD + PHU$$

$$G \equiv \frac{1 + CF}{1 + CSP} \qquad\qquad H \equiv \frac{C}{1 + CSP}$$

Where ΔY is deviations from steady state value of fy. In order to have a complete elimination of the output deviations, the foregoing control structure requires

$$F \to -C^{-1}$$

and/or
$$u \to -(C^{-1} + F)D$$

Feedback + Feedforward + Open-Loop Predictive Control

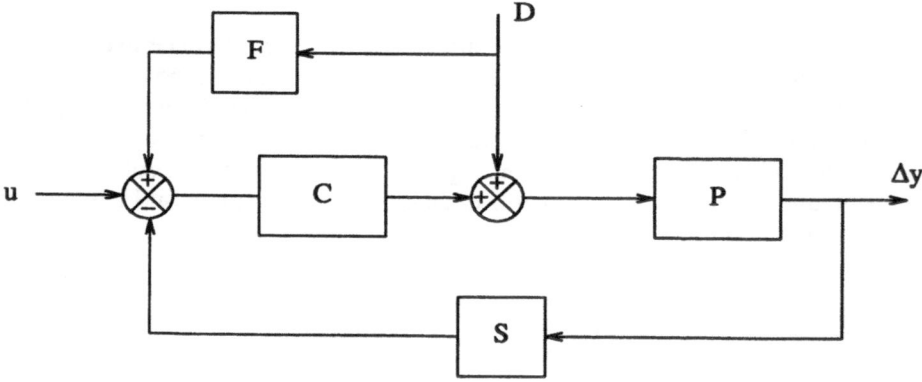

Fig. 1 Proposed Architecture

Namely, while a simple feedback requires infinite control effort (C → ∞) for complete disturbance rejection, the above structure provides a more practical and flexible approach to desired performance. This high performance can be achieved provided an exact feedforward control, and/or" an optimal open loop predictive control are available. Thus it requires disturbance estimation for D and on-line learning and identification for system dynamics and optimal open-loop control. If a direct measurement of disturbance was available, and if full knowledge of system dynamics (no uncertainty) was given, then an ideal control system would have been designed using conventional techniques. Neither of these two requirements can always be satisfied, and in fact, these are shortcomings of conventional control systems. In order to overcome these difficulties, neural networks and fuzzy logic have been used to predict or synthesize disturbances, identify system dynamics, and to describe uncertainties by qualitative measures.

Considering the above structure for ICS, the following points can be deduced.

- The optimal open-loop predictive controller may be derived by performing a constraint, dynamic, nonlinear optimization. This optimization should be able to integrate both qualitative and quantitative information. Thus, it may require new mathematical foundations.

- Constraints on optimization of u may be in the form of rules which may have associated degrees of uncertainty.

- A predictive controller may use neural network-based identification to characterize time varying and nonlinear elements of the system and/or environment.

- The resulting predictive controller may be in the form of an expert control system, expressed in terms of rules.

- When the actual state of the plant and environment matches those anticipated and predicted in the derivation of u, then no feedforward or feedback control would get activated.

- The feedforward controller requires a fast, on-line, disturbance predictor. This may be achieved by application of neural networks, pattern recognition techniques, and rule-based monitoring systems.

- The feedback controller can benefit from advances in the areas of robust control, QFT, and fuzzy control. It would compensate, on-line, for events not anticipated by the predictive controller.

- A hierarchical structure may be used to implement the proposed ICS structure.

- Learning control schemes may be incorporated into the three levels of control for integrating experience into enhancement of the future actions.

Thus an intelligent controller can be composed of a neural network that learns and identifies system dynamics. This information can be used by a qualitative (fuzzy) control and/or quantitative (conventional) controller to develop appropriate input signals for the system.

Example: Develop an intelligent control structure for a robotic manipulator.

Let us consider two cases: i) a neurocontrol is used to replace a human controller, ii) a combined model reference adaptive control with a neurocontroller is used.

Figure 2 shows how a combined human expert and neural network can be used. During a training period the output torque from the neurocontroller is compared with those recommended by the human expert. The difference (torque error) is used to train the neural network. When training is completed, then the human expert can be removed from the loop.

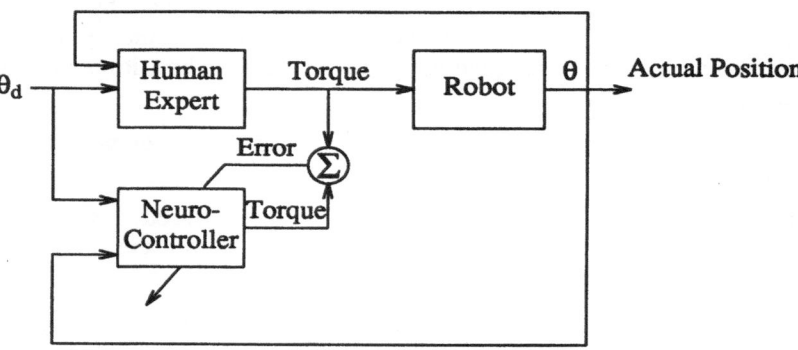

Fig. 2: Human-Neurocontrol for Robot

Figure 3 shows a model reference adaptive control approach combined with a neural network. In this scheme, the objective is to make the combined robot and neural network (the system shown by dashed-lines) to behave like a reference model. For example, although manipulators are nonlinear dynamic systems, but the combined robot and neural network could behave like a linear system. Thus all linear control theory can be applied. In this case the aim of neural network is to learn and cancel all nonlinear effects of the robot.

In the following sections two applications of intelligent control systems are described and implementation results are presented.

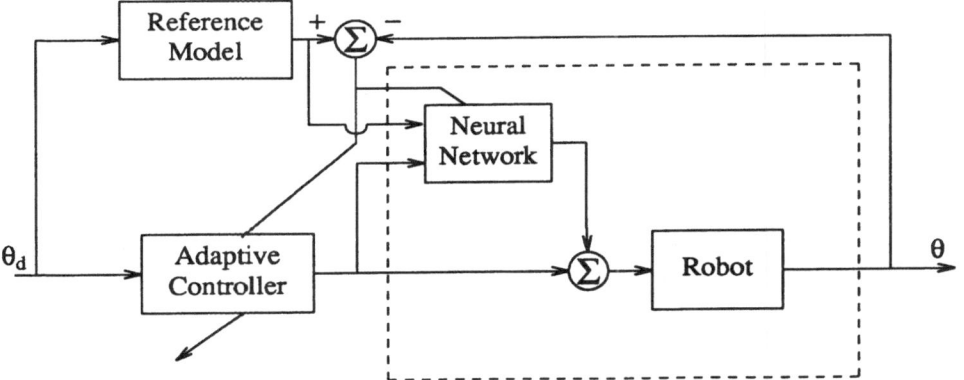

Fig. 3 Neural-Based Model Reference Adaptive Control

3. INTELLIGENT AUTOMATIC GENERATION CONTROL (IAGC)

Electric power generation, transmission and distribution on a national level is composed of different generation areas that are interconnected, as shown in Fig. 4. This area control interconnection enhances power reliability, security, and efficiency. However, in order to have a reasonable match between the power demand and generation, each control area attempts to drive its net power flow to zero.

Present automatic generation control (AGC) systems for different areas operate based on a classical proportional plus integral control approach, as shown in Fig. 5. This form of AGC has a slow reaction which is magnified by the generating unit constraints. Further the PI based AGC always responds to the variations in the interchange power and frequency errors, as measured or observed in the past. Thus the control response always lags the system response. This phasing of the control input and the generation level would not have a significant effect so long as the load on the system is repeatable and slowly time varying.

Recent challenges facing the electric utilities in terms of the highly varying loads, growing applications of digital control systems for manufacturing equipment, and availability of information transfer/data acquisition systems, demand a completely new and more intelligent automatic generational control system. An AGC, regardless of its design structure, has to perform the following four tasks:

- Forecasting and estimation of area load,
- Economic power dispatch targeting,
- Control of power generating units,
- Enhancement of power quality in the transmission lines.

106

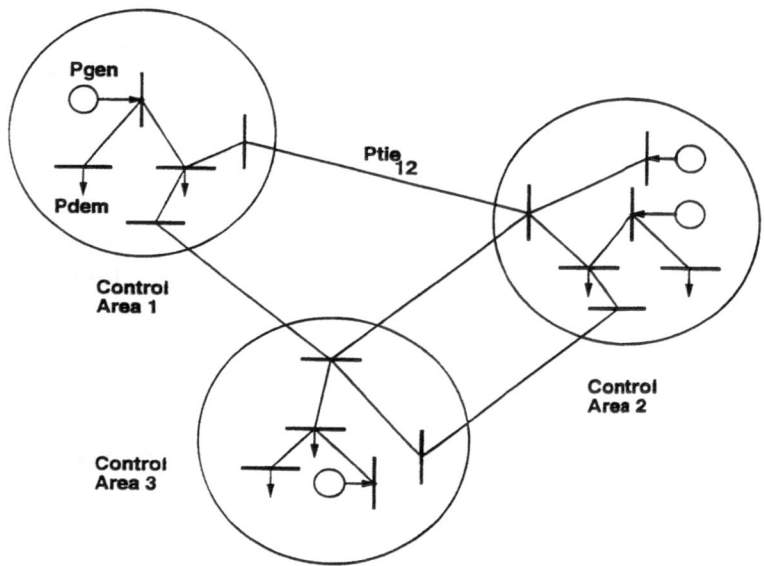

Fig. 4   Interconnected Area Power Control

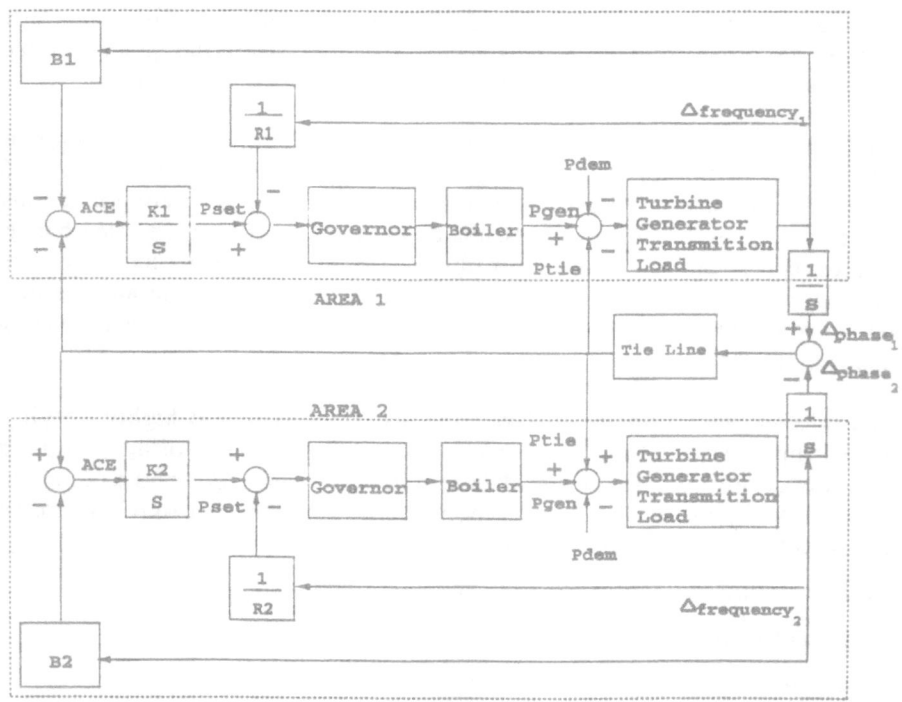

Fig. 5   Conventional Tie-Line Bias Control for Two Area System

Present feedback only AGC systems are unable to handle transient loads with a wide frequency band and unknown magnitudes. The highly varying loads require an AGC system capable of anticipating upcoming demands, while satisfying economic dispatch, generation constraints, and transmission network topology and dynamics. Due to the uncertainties associated with highly varying loads, an AGC should be able to operate with incomplete and qualitative forms of information.

Based on the structure of an ICS, as shown in Fig. 1, an intelligent AGC system structure is proposed in Fig. 6. In order to derive such control system a fuzzy load predictor is developed that generates load patterns that are used for anticipating control actions. Further, an optimal model predictive control system, using partial load measurements, is used to perform an optimal dispatching. Figure 7 shows typical highly varying loads introduced by an electric arc furnace and a rolling mill.

Assume that the transient power demand signals, as shown in Fig. 7, are sequences of discrete events. If the events in the process are labeled, then the sequence might be

$$S = [c,b,a,d, |e,b,c,a,e,b|, a,d,a,b,c,...]$$

a sliding window is used to isolate subsequence of events. As the window is moved across the event sequence, then instance of patterns are recorded, as shown in Fig. 8. Each possible pattern corresponds to a bin in the data structure P, then votes are cast in bins to record the instances. Of course, more predictable signals have a more sparse data structure.

Based on the data structure P, partial patterns can be reconstructed. Namely if the partial event sequence is given as

$$S_i = [e, ?, c, a, ?, b]$$

then the most likely match is found by maximizing P*.

$$P* = \max_{(x,y)} P(e, x, c, a, y, b)$$

The same technique can be used for prediction. Therefore, based on the trailing edge of a highly varying load, a window of raw power demand signal is taken and sent through a power level histogram. A set of membership functions are generated based on this histogram. Then signal segments at each time step are assigned a degree of membership in the overall fuzzy sets. By ranking the patterns in the database in terms of how closely they match the fuzzy representation of the signal, the optimal solution is achieved. Figure 9 shows the results of implementing this fuzzy load identification scheme applied to an electric arc furnace as a highly variable load.

In order to develop an anticipating model predictive control, two steps of path planning and control optimization are taken. A discrete MIMO model is assumed for the system as following:

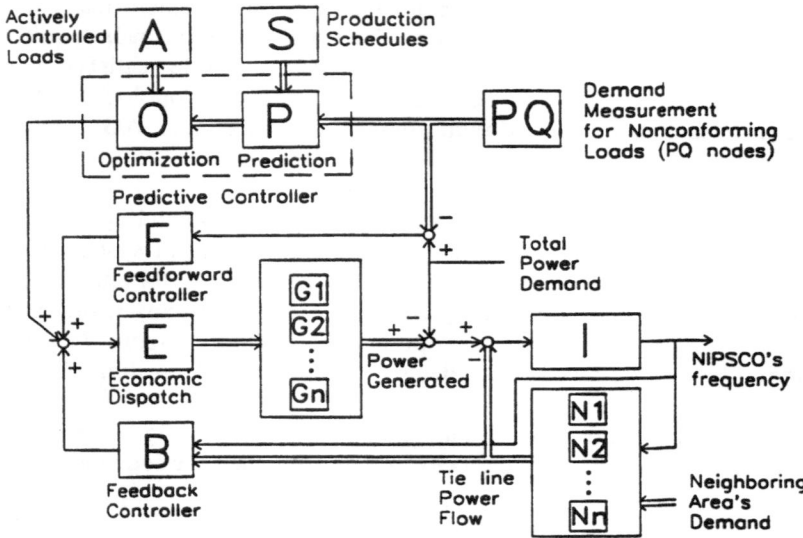

Fig. 6   Proposed Intelligent AGC System

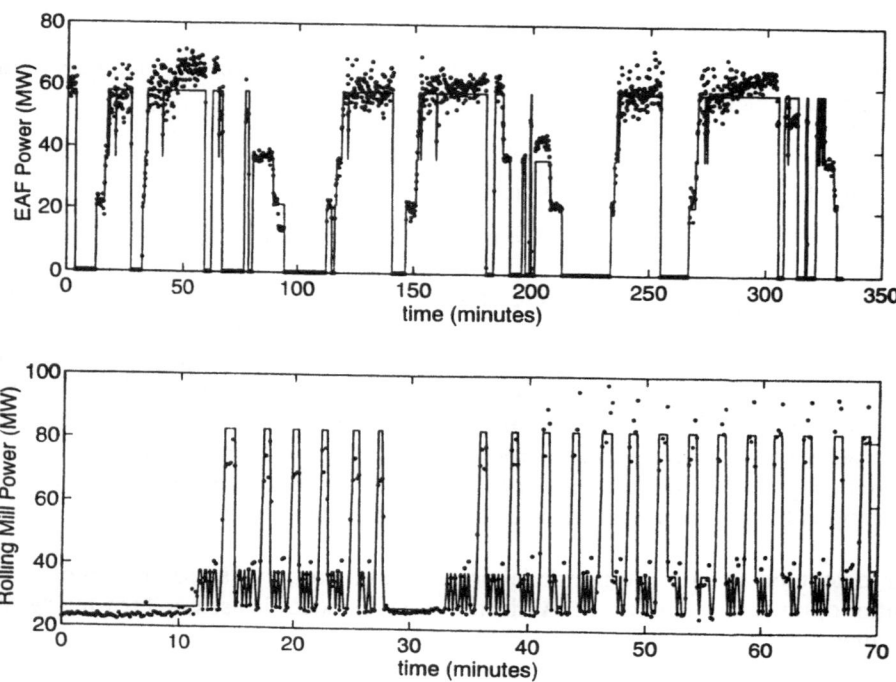

Fig. 7   Sample of Industrial Highly Varying Loads

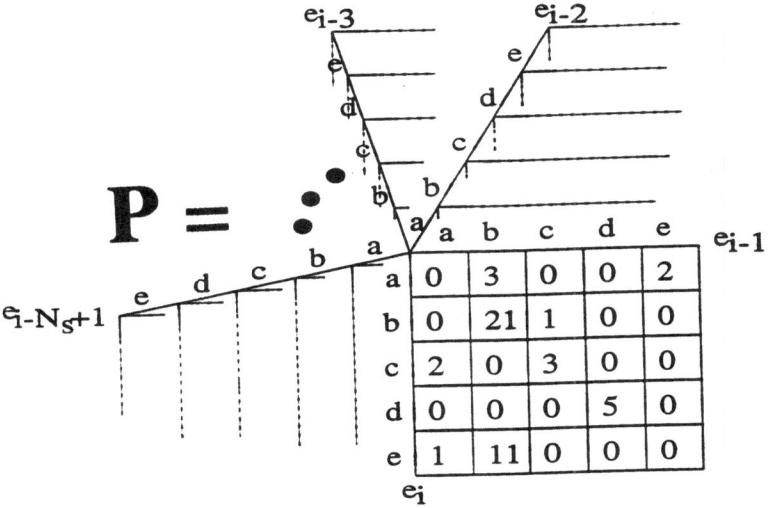

Fig. 8   Database for Recorded Patterns

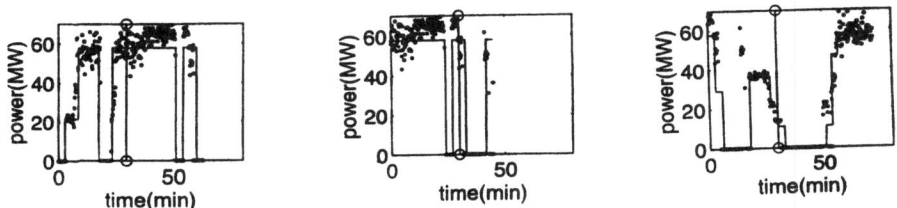

Fig. 9   Prediction of EAF Signal

$$X_{k+1} = AX_k + BU_k + B(V_k + Z_k)$$

$$Z_{k+1} = Z_k$$

$$y_k = CX_k$$

where X is the state vector, U is the control vector, y is the measurement vector, Z is the unmeasurable disturbance vector, and V is the measurable portion (if any) of the system disturbances. To find the optimal comfort effort, U, a quadratic performance index needs to be minimized subject to the following constraint equations.

$$\psi_k([X_k, U_k]) = \begin{bmatrix} y_k^* - CX_k \\ U_k - U \end{bmatrix}^{-1} \begin{bmatrix} Q^* & 0 \\ 0 & R^* \end{bmatrix} \begin{bmatrix} y^* - CX_k \\ U_k - U \end{bmatrix}$$

Subject to

$$[I - A \ -B] \begin{bmatrix} X_k \\ U_k \end{bmatrix} = [B(V_k + Z_k)]$$

$$\begin{bmatrix} 0 & I \\ 0 & -I \end{bmatrix} \begin{bmatrix} X_k \\ U_k \end{bmatrix} < \begin{bmatrix} U_{max} \\ U_{min} \end{bmatrix}$$

Thus the problem can be formulated in terms of a quadratic optimization with constraints equations. Figure 10 shows the results of this intelligent control system applied to the AGC problem of an electric utility, once it has gone through a series of generation set point variations.

4. INTELLIGENT COMFORT CONTROL SYSTEM

The objective of an office building or residential high rises is to provide maximum occupant comfort to attain the highest level of productivity. The heating, ventilation, air-conditioning (HVAC) system maintains a comfortable work environment in an office building. Since buildings are designed as people productivity centers, an increase in comfort creates increase in productivity. The objective of a building is to maximize comfort while minimizing the energy cost.

Each building and its HVAC system are unique. Buildings encompass a variety of sizes and shapes. Even similar buildings built from the same set of plans will be different because of orientation, occupant preferences, or variations in the quality of construction. The result is a wide variety of HVAC systems to match the comfort requirements in a variety of building stock. A fundamental problem in building systems is a controller that is designed for a building becomes outdated quickly since the building is changing with time. For example, heat exchangers become fouled, or a fan system fails and is replaced with a different model. Thus, buildings are diverse with thermal characteristics changing with time. There is a need to develop a control system that can learn the characteristics of the building and determine the best control strategy on its own and without any interventions.

This section of the paper presents formulation and implementation of an intelligent building system. In this context, intelligent refers to a learning system that can develop new strategies to optimize the energy costs while maintaining comfort. This process occurs even if there are changes in weather, occupancy, or building layout and structure without manual or operator reprogramming.

5. CONTROL SYSTEM DEVELOPMENT

The main issue with the control of buildings is that they include system dynamics that have very disperse eigenvalues. Furthermore, it is costly to engineer the control algorithms for each building in an attempt to reach optimal control. Self learning is an important requirement for a building control system. For example, if the computer and its control system were moved to another building, in time it would be able to adapt to the new building. The control system is constantly changing to account for the time varying nature of the building system; and furthermore, the computer system always is searching for the optimal control strategy. Ultimately, a generic control system could be installed in any building and with time would identify the building and establish a method of control to maximize comfort while minimizing cost. This idealistic approach is not presently feasible and this research takes the first step towards formulating an analysis of a true intelligent building.

Analysis of the building dynamics, the types of inputs, and the interaction of the HVAC equipment with local controls dictates a three-level hierarchical control approach to achieve minimum cost and maximum comfort. The first two levels can be isolated based on a distinction in the disturbances. These two logical divisions are the supervisor and the coordinator. These two controllers replace the traditional supervisory level. The third level of control in the intelligent building system (IBS) is the local controller. Figure 11 shows the proposed three-level control system.

The supervisory level makes decisions based on deterministic information. This level is an open-loop control system since it does not make decisions on current information. The decisions are only based on past data. Thus, one objective of the supervisor is to predict disturbances using modeling and forecasting. The coordinator can modify the information from the supervisor if a difference between the predicted information and the actual information is found in real time. The coordinator responds to a set point error and provides an update on the supervisor's commands based on the current situation. The local controller regulates each system component to maintain a specified set point. This instructs the building HVAC system to respond. This regulation affects the building dynamics that are analyzed by each level of the hierarchy.

As stated previously, the coordinator is the middle level of the hierarchy. The first responsibility of the coordinator is to adjust set points and control variables from the supervisor in the case of unexpected events. The second is an extension to the first responsibility. Since the coordinator monitors operation of equipment, it can also be used for failure diagnostics. The coordinator uses a set of rules, or rule bases, to adjust set points of the HVAC equipment. The coordinator is always searching for the least energy method to correct current deviations from the set points. Furthermore,

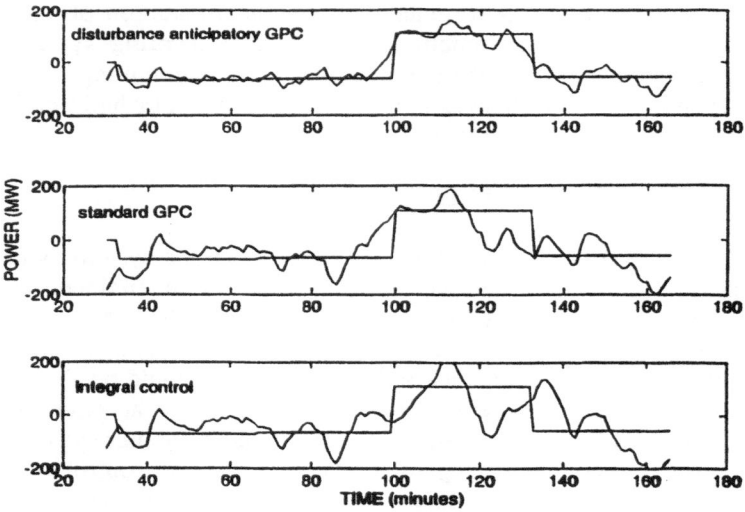

Fig. 10   Results of Implementing Intelligent AGC

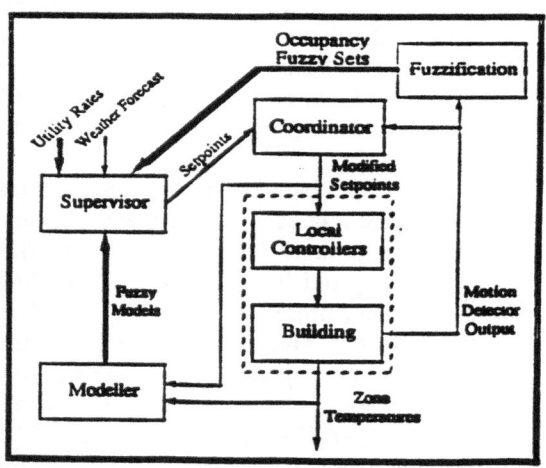

Fig. 11   Proposed Three-Level Comfort Control System

deviations that cannot be corrected may be the result of hardware failure. When a failure occurs, the rule base can search for the next best solution that maintains occupant comfort. For example, a chiller fails. The coordinator detects a problem and responds by initiating startup of another chiller, selecting the chiller that can meet the load deficit quickly with minimum cost. Most decisions are not this clear. A person is "uncomfortable" in a zone. The coordinator modifies the set point on the VAV box "a little." This decision making process is "fuzzy" in nature. The coordinator makes decisions based on fuzzy logic. If the supervisor is always perfect in predicting the building behavior, and if the equipment does not fail, the coordinator is inactive.

Consider the dynamic system represented in a state space form. If the system was deterministic, then the optimal control input would have been resulted from the following equations:

$$U = -KX$$

$$-KABR^{-T} - R^{-1}B^T A^T K^T + KBR^{-1}B^T K^T - R^{-1}B^T QBR^{-T} = 0$$

when the system is deterministic with crisp parameters, then the above equations provide an optimal control input.

Due to the existence of disturbances, the actual system matrices would be different from those of the nominal system. Thus, the resulting feedback gains would not be optimal unless full knowledge of the new A and B matrices is provided. This study assumes uncertainties and disturbances affect the A and B matrices and can be represented in terms of fuzzy variables, thus making the A and B fuzzy. Fuzzy A and B causes the resulting feedback matrix, K, to be fuzzy. Thus the issue of optimization for the resulting fuzzy controller is to find a K matrix that has the highest degree of belongingness (maximum membership value) to the set of optimal gain matrix resulting from the above, based on crisp and actual values of the A and B matrices.

In order to derive a fuzzy formulation for the problem, a fuzzy set in terms of disturbances is defined and is assumed that the lower and upper bounds on variations of disturbances are known. Therefore, two fuzzy sets for variations of the A and B matrices can be obtained. Because the actual A and B matrices are different from their nominal values, the above equation would transfer into a fuzzy inequality. Therefore, this problem can be converted into a fuzzy representation by the fuzzy equality constraint concept that was proposed by Bellman and Zadeh. Let F, F*, D, and D* matrices be defined as

$$F = -PA - A^T P + PBR^{-1}B^T P - Q = \begin{bmatrix} f11 & \cdot & f1n \\ & \cdot & \\ fn1 & \cdot & fnn \end{bmatrix},$$

$$F^* = R^{-1} B^T F B R^T = \begin{bmatrix} f11^* & \cdot & f1n^* \\ & \cdot & \\ fn1^* & \cdot & fnn^* \end{bmatrix},$$

$$E \overset{\Delta}{=} 0 = \begin{bmatrix} e11 & \cdot & e1n \\ & \cdot & \\ en1 & \cdot & enn \end{bmatrix},$$

$$D^* \overset{\Delta}{=} F(\text{extreme}) - F(\text{nominal}),$$

$$D \overset{\Delta}{=} R^{-1} B^T D^* B R^{-T} = \begin{bmatrix} d11 & \cdot & d1n \\ & \cdot & \\ dn1 & \cdot & dnn \end{bmatrix},$$

where F* is calculated based on actual values of A and B matrices. Since the knowledge of extreme values of disturbances is assumed to be known, then A (extreme) and B (extreme) are known. Thus, F (extreme) is known, and D* and D can be calculated based on the nominal system parameters. The following linear membership functions are defined for the gains obtained.

$$\mu(cij) = 1 - \frac{1}{|dij|}(f^*ij - eij) \quad \text{when } eij \leq f^*ij \leq dij{+}eij \text{ and } dij > 0$$

$$\mu(cij) = 1 - \frac{1}{|dij|}(eij - f^*ij) \quad \text{when } dij{+}eij \leq f^*ij \leq eij \text{ and } dij < 0$$

$$\mu(cij) = 0 \quad \text{otherwise}$$

In order to satisfy maximization of membership functions, the optimal value of F* has to be obtained. This results in the following equation.

$$\frac{\partial F^*}{\partial B} = 2R^{-1}BFR^{-T} + R^{-1}B^T \frac{\partial F}{\partial B} BR^{-T} = 0$$

In order to find $\partial F/\partial B$, state covariance matrix along with performance index, represented in the trace form, is used. The state covariance matrix, x_{co} has to satisfy the following equation

$$x_{co}(A - BK)^T + (A - BK)x_{co} + x_{co} = 0$$

Thus to calculate $\frac{\partial F}{\partial B}$, the following can be obtained.

$$F = \frac{\partial(PI)}{\partial x_{co}} = -PA - A^T P + PBR^{-1}B^T P - Q$$

$$\frac{\partial F}{\partial B} = \frac{\partial^2 (PI)}{\partial x_{co} \partial B}$$

Based on these equations F* and thus the fuzzy optimal control input for the fuzzy system would be obtained. This control system would be represented by a set of rules. Based on the system sensors, certain rules would be fired. It should be noted that the results of fuzzy controller rules has to be defuzzified, using centroid of area approach, in order to come up with set points, e.g. reference voltages, for the actuators.

Performance index is the important part of the optimization problem. It trades off the energy cost and the comfort of the occupants. The inputs for the performance index are the room occupancy schedules and the energy rate structures, usually obtained from the fuel or utility companies. Following is a mathematical derivation for the performance index (PI).

$$\text{Energy Cost} = \sum_{j=1}^{n} \left[C_j(t) \sum_{i=1}^{m} \int_{o}^{t_f} P_{ij}(t)dt \right]$$

Where P_{ij} is the instantaneous power for the ith piece of equipment using the jth energy source. The cost per unit energy is expressed as C(t), and t_f represents the supervisor's time horizon for prediction.

Demand charges are typically fixed unless the demand level has exceeded a prescribed value. In order to mathematically represent demand charges, a sigmoid function is used.

$$\text{Demand Charges} = P_o S \left[\frac{1}{1 + e^{-\phi(P - P_o)}} \right]$$

$$P_o = \max \left\{ \frac{1}{t_f} \int_{o}^{t_f} P(t)dt \right\}$$

where P_o is the maximum-to-date demand and P(t) is the total power consumption. The parameter ϕ is a constant representing the rate of penalty amount as the power approaches P_o.

The comfort of a building occupant in a particular space determines the work performance or productivity level. The major comfort criteria are temperature and humidity. In order to mathematically represent the comfort expense, a penalty function is expressed as a deviation from a given set point. Instead of a temperature, there is range of temperatures in which a person is comfortable. Outside of this temperature dead band, a person is uncomfortable, resulting in a penalty to the building's operating cost. This is mathematically expressed as

$$\text{Comfort Expense} \int_{o}^{t_f} \sum_{i=1}^{k} L_i \left[\frac{1}{1 + e^{-\gamma T^*}} \right] dt$$

$$T^* = (T_i - T_{sp,j})^2 - (T_{db,i})^2$$

In this formulation L_i is the cost associated with a loss of productivity as a result of a decrease in comfort, and γ determines the cost associated with approaching the temperature dead band.

Mathematical formulation for prediction of occupancy level is derived that operates based on motion sensors indicating presence or absence of people in a zone. Based on the formulated performance index and building system model, a nonlinear optimization problem is resulted. Derivation of this optimization scheme and incorporation of occupancy predictions are beyond the scope of this paper and can be found.

6. EXPERIMENTAL RESULTS

In order to evaluate performance of the preposed intelligent control system, a building on the campus is instrumented with motion detectors and direct digital control system, as shown in Fig. 12. The three-level intelligent control structure is implemented on a microprocessor system, presented in Fig. 13. A fuzzy occupant pattern generation scheme is developed that generates potential zone occupancy based on the past experience. Thus the resulting heating, cooling, and light profiles during the day depends on the predicted occupancy patterns. Inclusion of the load (occupancy) anticipation has potential for introducing major energy savings in the comfort conditioning of large buildings. Figure 14 demonstrates potential energy savings due to derivation of optimal zone temperature profile in the building where our control scheme is being implemented. Energy savings of over 30% can be realized in building with major occupancy variations.

7. CONCLUSIONS

A structure for formulation of the intelligent control system was proposed. This structure introduces unique, human like, features of the ICS, while providing a suitable bridge to the conventional control systems. Resulting structure was applied to two specific real-life problems with significant results demonstrated. Although the field of intelligent control is in its infancy, its potential for solving more complex problems in a more efficient manner is very significant. The field of ICS will be spreading very fast, as has been the case of neural networks and fuzzy logic. More fundamental and mathematically theorems supporting the behavior of ICS are required.

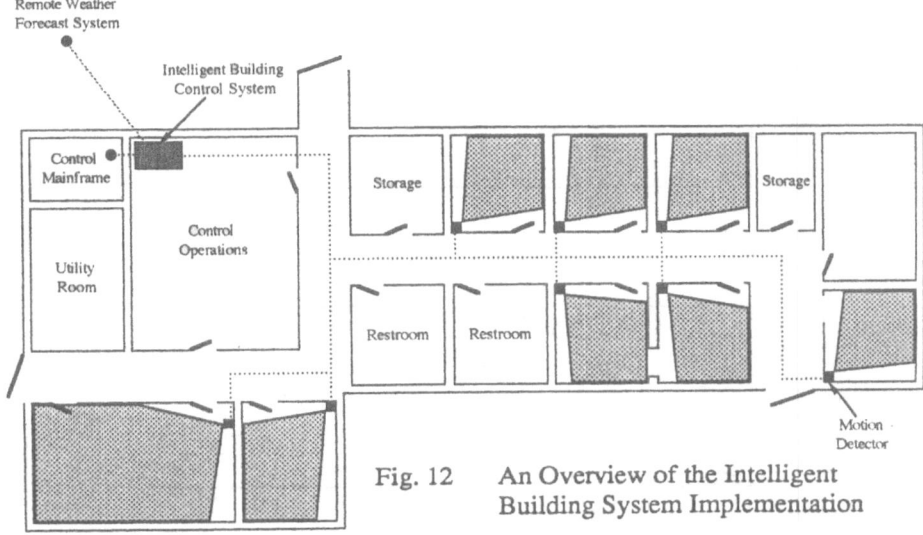

Fig. 12   An Overview of the Intelligent
Building System Implementation

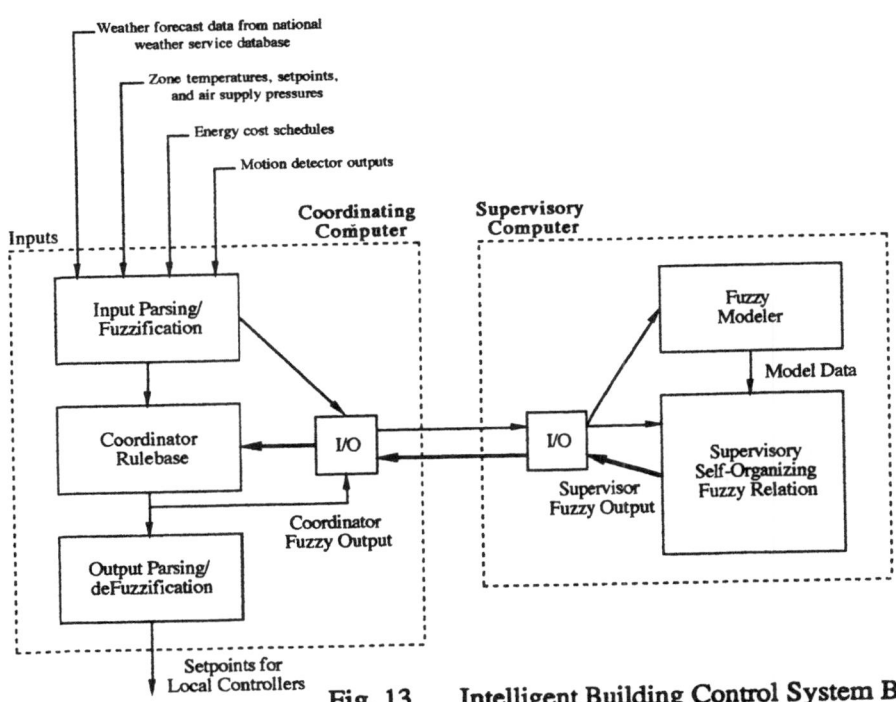

Fig. 13   Intelligent Building Control System Block
Diagram Depicting Processing Allocation

118

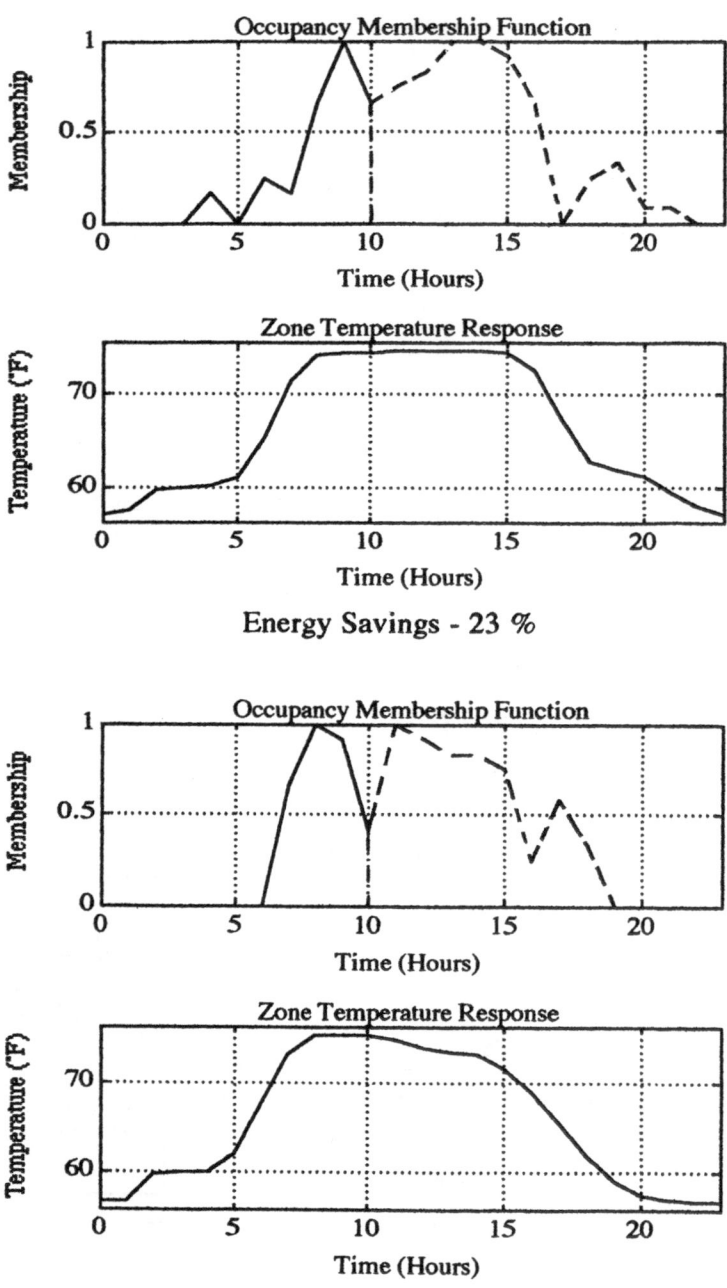

Fig. 14   Potential Energy Savings from Application
of Intelligent Control System

119

8. APPENDIX

Readers interested in more details of ICS are referred to the following literature.

Aarts, E., and Korst, J., 1989, *Simulated Annealing and Boltzmann Machines, A Stochastic Approach to Combinatorial Optimization and Neural Computing*, John Wiley & Sons Ltd., Tiptree Essex GB, 1989.

Ackley, D. H., Hinton, G. E., and Sejnowksi, T. J., 1985, "A Learning Algorithm for Boltzmann Machines," *Cognitive Science,* 9: 147-169.

Albus, J. et al., 1981, "Theory and practice of intelligent control," in Proc. 23rd IEEE COMPCON, pp. 19-39.

Amit, D. J., 1986, "Neural networks--achievements, prospects and difficulties", *International Symposium on the Physics of Structure Formation*, Tubingen, October.

Antsaklis, P. J., Passino, K. M., and Wang, S. J., 1989, "Towards intelligent autonomous control systems: Architecture and fundamental issues," *J. Intelligent Robotic Syst.*, Vol. 1, pp. 315-342.

Antsaklis, P. J., Passino, K. M., and Wang, S. J., 1990, "An Introduction to autonomous control systems," in *Proc. IEEE Int. Symp. Intelligent Control*, Philadelphia, PA, September, pp. 21-26.

Arzen, K. E., 1987, "An Architecture for Expert System Based Feedback Control," *Automatica*, Vol. 25, pp. 813-827. Applications, 1987, pp. 173-178.

Astrom, K. J., Hang, C. C., Persson, P., and Ho, W.K., 1992, "Towards Intelligent PID Control," Automatica, Vol. 28, No. 1, pp. 1-9, January.

Astrom, K. J., et al., 1986, "Expert control," *Automatica*, Vol. 22, pp. 277-286.

Astrom, K.J. and McAvoy, T.J., 1992, "Intelligent Control: An Overview and Evaluation," Handbook of Intelligent Control, Van Nostrand Reinhold, New York.

Astrom, K. J., and Wittenmark, B., 1989, *Adaptive Control*, Addison-Wesley, Reading, MA.

Bellman, R.E., and Zadeh, L.A., 1970, "Decision Making in a Fuzzy Environment," Management Science, 17B (4), pp. 141-164.

Bernard, J.A., 1988, "Use of Rule-Based System for Process Control," *IEEE Contr. Syst. Mag., Vol. 8, No. 5, pp. 3-13.*

Bezdek, J. and Pal, S.K., 1992, "Fuzzy Models for Pattern Recognition," IEEE Press.

120

Borys, B. B., Johansen, G., Hansel, H. G., and Schmidt, J. Jr., 1987, "Task and Knowledge Analysis in Coal-Fired Power Plants," *IEEE Contr. Syst. Mag.*, pp. 26-30, June.

Bredeweg, B., and Wielinga, B., 1988, "Integrating Qualitative Reasoning Approaches," *Proc. Eur. Conf. Artificial Intelligence-88*, London, Pitman.

Breuker, J., and Wielinga, B., 1989, "Models of Expertise in Knowledge Acquisition" *Topics in Experiment System Design*, G. Guida and C. Tasso, Eds. Amsterdam: North Holland.

Carpenter, G. A., and Grossberg, S., 1988, "The ART of Adaptive Pattern Recognition by a Self-Organizing Neural Network," *Computer*, pp. 77-88, March.

Carpenter, G. A., and Grossberg, S., 1987, "A Massively Parallel Architecture for Self-Organizing Neural Pattern Recognition Machine," *Computer Vision, Graphics, and Image Processing* 37, pp. 54-115.

Cybenko, G., 1989, "Approximation by superposition of a sigmoidal function" *Mathematics of Control, Signals, and Systems*, Vol. 2, pp 303 - 314, 1989.

Fahlmann, S. E., 1988, "An Empirical Study of Learning Speed in Back-Propagation Networks," CMU Technical Report, CMU-CS-88-162, June.

Findeisen, W., et al., 1980, *Control and Coordination in Hierarchical Systems*, New York, NY: Wiley.

Firschein, W., et al., 1986, Artificial Intelligence for Space Admission Station, *Automation*, Park Ridge, NJ: Noyes.

Francis, B. A., Helton, J.W., and Zames, G., 1984, "H∞ Optimal Feedback controllers for Linear Multivariable Systems," *IEEE Trans. on Automatic Control*, AC-29.

Freeman, W. J., and van Dijk, B. W., 1987, "Spatial patterns of visual fast EEG during conditioned reflex in a rhesus monkey", *Brain Research*, 422, 267-276.

Freeman, W. J., Yao, Y., and Burke, B., 1988, "Central pattern generating and recognizing in old factory bulb: A correlation learning rule", *Neurla Networks*, 1, 277-278.

Fu, K. S., 1968, *"Sequential Methods in Pattern Recognition and Machine Learning,"* Academic Press.

Fukushima, K., 1988, "Neocognition: A Hierarchical Neural Network Capable of Visual Pattern Recognition," *Neural Networks*, Vol. 1, pp. 119-130.

Funahashi, K., 1989, "On the Realization of Continuous Mappings with Neural Networks," *International Journal of Neural Networks*, Vol. 2, No. 3, pp. 183-192.

Glover, G. E., 1988, "A Hybrid Optical Fourier/Electronic Neurocomputer Machine Vision Inspection System," *Proc. Vision '88 Conference, sponsored by SME/MVA*.

Glover, K., Doyle, J.C., 1988, "State Space for Formulae for all Stabilizing Controllers," *Systems and Control Letters*, No. 11.

Gorman, R. P., and Sejnowski, T. J., 1987, "Analysis of Hidden Units in a Layered Network Trained to Classify Sonar Targets," *Neural Networks*, 1, pp. 75-89.

Grossberg, S., 1969, "Embedding Fields: A Theory of Learning with Physiological Implications," *Journal of Mathematical Psychology*, Vol. 6, pp. 209-239.

Grossberg, S., 1970, "Some Networks that can Learn, Remember, and Reproduce any Number of Complicated Space-Time Patterns, II.," Studies in Applied Mathematics, Volume 49, pp. 135-166.

Grossberg, S., 1971, "Embedding Fields: Underlying Philosophy, Mathematics, and Applications to Psychology, Physiology, and Anatomy," *Journal of Cybernetics*, Vol. 1, pp. 28-50.

Grossberg, S., 1976, "Adaptive Pattern Classification and Universal Recoding: I. Parallel Development and Coding of Neural Feature Detectors," *Biological Cybernetics*, Vol. 23, pp. 121-134.

Haber, R., and Unbehauen, H., 1990, "Structure identification of nonlinear dynamics systems - A survey on input/output approaches", *Automatica*, Vol. 26, No. 4, pp. 651 - 677.

Hashimoto, H., Kubota, T., Kudou, M., and Harashima, F., 1992, "Self-Organizing Visual Servo System Based on Neural Networks," *IEEE Control Systems Magazine*, Vol. 12, No. 2, April.

Hecht-Nielsen, R., 1986, "Nearest Matched Filter Classification of Spatio-temporal Patterns," special report published by Hecht-Nielsen Neuro-Computer Corporation, San Diego, California.

Heer, E., and Lum, H., Eds., 1988, *Machine Intelligence and Autonomy for Aerospace Systems*, Washington, DC, AIAA publication.

Hopfield, J. J., 1984, "Neurons with Graded Response Have Collective Computational Abilities Like Those of Two-State Neurons," *Proc. Natl. Acad. Sci.*, Vol 81, pp. 3088-3092.

Hopfield, J. J., and Tank, D. W., 1985, "Neural Computation of Decisions in Optimization Problems," Biol. Cybernetics, Vol. 52, pp. 141-152.

Horowitz, I. 1992, *Quantitative Feedback Theory, Vol. 1,* QFT Publications, Boulder Colorado.

Kanel, A., 1991, "Fuzzy Expert System," CRC Press.

Kasai, Y. and Morimoto, Y., 1988, "Electronically Control Continuously Variable Transmission," in *Proc. Int. Congress on Transportation Electronics,* Dearborn, MI.

King, P. J., and Mamdani, E.H., 1976, "The Application of Fuzzy Control Systems to Industrial Processes," *Automatica,* Vol. 13, pp. 235-242.

Kinoshita, M., Fukuzaki, T. Satoh, T. and Miyake, M., 1988, "An Automatic Operation Method for Control Rods in BWR Plants," in *Proc. Specialists' Meeting on In-Core Instrumentation and Reactor Core Assessment,* Cadarache, France.

Kosko, B., 1992, *"Neural Networks and Fuzzy Systems,"* Prentice Hall.

Kusiak, A., 1990, *Intelligent Manufacturing Systems.* Englewood Cliffs, NJ: Prentice Hall.

Langari, G., and Tomizuka, M., 1990, "Analysis and Synthesis of Fuzzy Linguistic Control Systems," ASME pub. No. G00544.

Lapedes, A., and Farber, R., "Non-Linear Signal Processing using Neural Networks: Prediction and System Modeling," Los Alamos National Laboratory Report LA-UR-87-2662.

Leitch, R. R., and Stefanini, A., 1988, "QUIC: A Development Environment for Knowledge Based Systems in Industrial Automation," *Proc. 3rd Esprit Technical Conf.,* pp. 674-696.

Lin, C. T., and Lee, C. S. G., 1991, "Neural-Network-Based Fuzzy Logic Control and Decision System," *IEEE Trans. on Computers,* Vol. 40, No. 12.

Lippmann, R. P., 1987, "An Introduction to Computating with Neural Nets," *IEEE ASSP Magazine,* 4-22, April.

Luger, G.F. and Stubblefield, W.A., 1989, "Artificial Intelligence and Design of Expert Systems," Benjamin/Cummings Publishing Co.

Mamdani, E. H., 1974, "Application of Fuzzy Algorithms for Control of Simple Dynamic Plant," *Proc. IEEE,* Vol 121, pp. 1585-1588.

Mamdani, E. H., Assilian, S., 1975, "Fuzzy Logic Controller for a Dynamic Plant," *Int.*

J. Man-Machine Stud. Vol. 7, pp. 1-13.

Mesarovic, M., Macko, D., and Takahara, Y., 1970, *Theory of Hierarchical, Multilevel, Systems*, Orlando, FL: Academic.

Meystel, A., 1985, "Intelligent Control: Issues and Perspectives," *Proc. IEEE Workshop on Intelligent Control*, pp. 1-15.

Minsky, M., 1968, "Semantic Information Processing," MIT Press.

Nabet, B. and Pinter, R.B., 1991, "Sensory Neural Networks: Lateral Inhibition," CRC Press.

Narendra, K. S., and Annaswamy, A. M., 1989, *Stable Adaptive Systems*, Prentice-Hall, Englewood Cliffs, NJ.

Narendra, K. S., and Mukhopadhyay, S., 1991, "Intelligent Control Using Neural Networks," *Proceedings of the 1991 American Control Conference*, June.

Narendra, K., and Parthasarathy, K., 1990, "Identification and Control of Dynamical Systems Using Neural Networks," *IEEE Trans. Neural Networks*, Vol. 1, No. 1.

Nwokah, O. D. I., Jayasuriya, S., and Chait, Y., 1991, "Parametric Robust Control by Quantitative Feedback Theory," *AIAA J. of Guidance, Control and Dynamics*.

Powell, M.J.D., 1987, "Radial Basis Functions for Multivariable Interpolation: A Review," in J.C. Mason and M.G. Cox, editors, Algorithms for Approximation, Clarendon Press, Oxford.

Rahmani, R., 1992, "Development and Implementation of a Fuzzy Optimal Expert Controller for Intelligent Buildings," Ph.D. Thesis, School of Mechanical Engineering, Purdue University.

Reighgelt, H. and van F. Harmelen, 1985, "Relevant Criteria for Choosing an Inference Engine in Expert Systems," *Proc. Expert Systems Conf.*, British Computer Society.

Rosenblatt, F., 1959, *Principles of Neurodynamics*, Spartan Books, New York.

Rumelhart, D. E., and McClelland, J. L., and the PDP Research Group, 1985, Parallel Distributed Processing, Volume 1: Foundations Chapter 7 "Learning and Relearning in Boltzmann Machines" by G.E. Hinton and T.J. Sejnowski, M.I.T. Press, Cambridge, MA, 1985.

Rumelhart, D. E., Hinton, G. E., and Williams, R. J., 1986, "Learning Internal Representations by Error Propagation," in Rumelhart, McClelland, and PDP Research Group, Parallel Distributed Processing: Explorations in the Microstructure of Cognition, Vols. I: Foundations, pp. 318-362, MIT Press, Cambridge, MA.

Saridis, G. N., 1979, "Toward the Realization of Intelligent Controls," *Proceedings IEEE*, Vol. 67, pp. 1115-1133.

Shoureshi, R., 1990, "The Mystique of Intelligent Control," special session at the 1990 American Control Conference, also IEEE Control System Magazine, Vol. 11, No., 1, pg. 33.

Shoureshi, R., and Chu, R. S., 1992, "Convergence Conditions on Hopfield-Based Adaptive Observers," submitted to *IEEE Trans. on Neural Networks*, 1992.

Shoureshi, R., and Chu, R. S., 1992, "Neural-Based Adaptive Nonlinear System Identification," DSC-Vol. 45, Intelligent Control Systems, 1992 ASME-WAM, Annaheim, CA, pp. 55-62.

Shoureshi, R., Chu, R. S., and Tenorio, M. F., 1990, "Neural Networks for System Identification," *IEEE Control Systems Magazine*, Vol. 10, No. 4.

Shoureshi, R., and Rahmani, K., 1992, "Derivation and Application of an Expert Fuzzy Optimal Control System," *Journal of Fuzzy Sets and Systems*, No. 49.

Shoureshi, R., and Wormley, D., 1992, "NSF-EPRI Workshop on Intelligent Control Systems," NSF Final Workshop Report.

Slotine, J-J., and Li, W., 1991, *Applied Nonlinear Control*, Prentice-Hall, Englewood Cliffs, NJ.

Soucek, B., 1991, "Neural and Intelligent Systems Integration," John Wiley.

Sutton, R. S., Barto, A. G., and Williams,R. J., 1992. "Reinforcement Learning is Direct Adaptive Optimal Control," *IEEE Control System Magazine*, Vol. 12, No. 2, pp. 19-22.

Tsuda, I., 1992, "Dynamic link of memory -- Chaotic memory map in nonequilibrium neural networks", *Neural Networks*, 5(2).

Vemuri, V.R., 1992, "Artificial Neural Networks, Concepts and Control Applications," IEEE Computer Society Press.

Wassermann, P. D., 1988, "Combined Back-propagation / Cauchy Machine, Neural Networks" Abstracts of the First INNS Meeting, Boston, Vol. 1, p. 556, Elmsford, NY:Pergammon Press.

Wassermann, P. D., 1989, *Neural Computing, Theory and Practice*, Van Nostrand Reinhold, NY, NY.

Webster's New World, 1988, *"Dictionary of Computer Terms*, 3rd Edition, Prentice Hall.

Werbos, P. J., 1974, Beyond Regression: New Tools for Prediction and Analysis in the Behavior Sciences, Ph.D. thesis, Harvard University, Cambridge, MA.

Werbos, P. J., 1989, "Backpropagation and Neural Control: A Review and Prospectus," *Proceedings of International Joint Conference on Neural Networks (IJCNN)*, New York, June.

Widrow, B., and Hoff, M. E., Jr., 1960 , "Adaptive Switching Circuits," IRE WESCON Conv. Rec., pt. 4, pp. 96-104.

Widrow, B., and Stearns S. D., 1985, *"Adaptive Signal Processing,"* Prentice-Hall.

Wiegand, M. E., and Leitch, R. R., 1989, "A Predictive Engine for the Qualitative Simulation of Dynamic Systems," in *Proc. AI Eng. Conf.*, Cambridge, UK, pp. 141-150.

Wos, L., 1988, *Automated Reasoning: 33 Basic Research Problems.* Englewood Cliffs, NJ, Prentice Hall.

Yagishita, O., Itoh, O. and Sugeno, M., 1985, "Application of Fuzzy Reasoning to the Water Purification Process," in *Industrial Applications of Fuzzy Control*, M. Sugeno. Ed. Amsterdam: North-Holland, pp. 19-40.

Yamakawa, T., 1986, "High Speed Fuzzy Controller Hardware System," in *Proc. 2nd Fuzzy System Symposium*, Japan, pp. 122-130.

Yamakawa, T., 1987, "Fuzzy Controller Hardware System," in *Proc. 2nd IFSA Congress, Tokyo, Japan.*

Yamakawa, T., 1988, "Fuzzy Microprocessors--Rule Chip and Defuzzifier Chip," in Int. Workshop on Fuzzy System Applications, Iizuka, Japan, pp. 51-52.

Yasunobu, S., Miyamoto, S. and Ihara, H. 1983, "Fuzzy Control for Automatic Train Operation System," in *Proc. 4th IFAC/IFIP/IFORS Int. Congress on Control in Transporation Systems*, Baden-Baden, April.

Yasunobu, S. and Miyamoto, S., 1985, "Automatic Train Operation by Predictive Fuzzy Control," in *Industrial Application of Fuzzy Control*, M. Sugeno, Ed. Amsterdam: North-Holland, pp. 1-18.

Yasunobu, S., Sekino, S. and Hasegawa, T., 1987, "Automatic Train Operation and Automatic Crane Operation Systems Based on Predictive Fuzzy Control," in *Proc. 2nd IFSA Congress, Tokyo, Japan, pp. 835-838.*

Zadeh, L. A., 1973, "Outline of New Approach to the Analysis of Complex Structures and Decision Processes," IEEE Trans. on Systems, Man and Cybernetics, SMC-3.

Zadeh, L. A., 1988, "Fuzzy logic," *Computer*, pp. 83-93, April.

Zeigler, B. P., 1989, "DEVS representation of dynamical systems: Event based intelligent control," *Proc. IEEE*, Vol. 77, pp. 72-80.

Zeigler, B. P., and Chi, S. D., 1990, "Model-based concepts for autonomous systems," *Proc. IEEE Int. Symp. on Intelligent Control*, Philadelphia, PA, September, pp. 27-32.

Zimmerman, H. J., 1985, "*Fuzzy Set Theory and Its Applications*," Kluwer-Nijhoff Publishing.

6
INTELLIGENT SIMULATION IN DESIGNING COMPLEX DYNAMIC CONTROL SYSTEMS

FENG ZHAO
Department of Computer and Information Science and
Laboratory for Artificial Intelligence Research
The Ohio State University
Columbus, OH 43210, U.S.A.

1 Introduction

We develop Phase Space Navigator, an autonomous system for control synthesis of nonlinear dynamical systems. The Phase Space Navigator automatically designs a controller for a nonlinear system in phase space. It generates global control laws by synthesizing the desired phase-space flow "shapes" for the system and intelligently planning and navigating the system along desired control trajectories in phase space. It is particularly suitable for synthesizing high-performance control systems that do not lend themselves to traditional design and analysis techniques. It can also assist control engineers in exploring much larger design spaces than otherwise possible.

Computational tools that employ powerful symbolic and numerical techniques and actively exploit nonlinear dynamics play increasingly important roles in the synthesis and analysis of high-performance nonlinear control systems [1]. Our method relies on the phase-space knowledge obtained from an autonomous phase-space analysis and modeling program called MAPS that extracts and represents qualitative phase-space structures characterizing the qualitative aspects of the dynamics [20]. It is made possible by the geometric and dynamical modeling of the phase-space behaviors in so-called *flow pipes* [24]. MAPS generates a high-level description about a dynamical system describing equilibrium points and limit cycles, geometries of stability regions and trajectory flows, and their spatial arrangements and interactions in phase space in a relational graph. The high-level description is sensible to humans and manipulable

*The research was supported in part by the Advanced Research Projects Agency of the Department of Defense under Office of Naval Research contract N00014-89-J-3202, and in part by the National Science Foundation grants CCR-9308639 and MIP-9001651.

S. G. Tzafestas and H. B. Verbruggen (eds.),
Artificial Intelligence in Industrial Decision Making, Control and Automation, 127–158.
© 1995 *Kluwer Academic Publishers.*

by other programs. The control synthesis utilizes flow pipes to model phase spaces and to search for global control paths. We will present the novel idea of phase-space navigation as a paradigm for high-performance nonlinear control design. Algorithms for control trajectory planning and tracking will be described. The synthesis method will be illustrated with an example of synthesizing anti-buckling control laws for a steel column. An application to a maglev control design will conclude the paper.

Because of its autonomous nature and ability to explore rich nonlinear dynamics in phase space, the Phase Space Navigator will be able to synthesize a nonlinear control system whose phase space is difficult to visualize by humans, or on which the desired control properties are impossible to obtain with traditional design techniques. The topological and dynamical modeling of the phase space stability regions and trajectory flows forms the basis for our method.

2 The Control Engineer's Workbench

We have constructed a computational environment, *the Control Engineer's Workbench*, integrating a suite of programs that automatically analyze and design high-performance, global controllers for a large class of nonlinear systems [23]. These programs combine powerful techniques from numerical and symbolic computations with novel representation and reasoning mechanisms of artificial intelligence. The two major components in the Workbench—the Phase Space Navigator and MAPS—work together to visualize and model the phase-space geometry and topology of a given system. They reason about and manipulate the phase-space geometry and topology and search for optimal control paths connecting initial state and the desired state for the system. Given a model of a physical system and a control objective, the Workbench analyzes the system and designs a control law achieving the control objective. The Workbench represents the result of design and analysis in a symbolic form; it produces a high-level summary meaningful to professional engineers.

The design method employed by the Workbench computationally exploits dynamical systems' nonlinearities in terms of phase-space geometries and topology. It uses a technique of flow pipes to group infinite numbers of distinct trajectories into a manageable discrete set that becomes the basis for establishing reference trajectories, and navigates the system along the planned reference trajectories. The phase-space design approach requires powerful computational tools that are able to identify, extract, represent, and manipulate qualitative features of phase space, and is embodied in programs comprised in the Workbench.

3 Automatic Control Synthesis in Phase Space

The Phase Space Navigator synthesizes a control system from a geometric point of view in phase space. The control of a dynamical system is interpreted as the "steering" of the system trajectory, emanating from some initial state, to the desired state by a control signal.

3.1 Overview of the Phase Space Navigator

The Phase Space Navigator consists of a global control path planner, a local trajectory generator, and a reference trajectory follower. The global path planner finds optimal paths from an initial state to the goal state in phase space, consisting of a sequence of path segments connected at intermediate points where the control parameter changes. A brute-force, fine-grain search in high-dimensional phase spaces would be prohibitively expensive. High-level descriptions of the phase space and trajectory flows provide a way to efficiently reason about phase-space structures and search for global control paths. The local trajectory generator uses the flow information about the phase-space trajectories to produce smoothed trajectories. The trajectory follower tracks the planned reference trajectory, reactively corrects deviations, and resynthesizes the reference trajectory if the dynamics of the system changes significantly.

3.2 Intelligent navigation in phase space

The control objective for a stabilization problem is to synthesize a control path, along which the physical system can be brought to the goal state and made to stay there afterwards under control. We are particularly interested in cases in which physical plants to be controlled operate in large regions where nonlinearities of the plants cannot be ignored. When no global stabilization control laws can be found for the desired state with traditional control techniques, composite control paths have to be synthesized. Geometrically, this is the case where initial states of the systems are far away from the desired state. The Phase Space Navigator takes advantage of the underlying dynamics of the phase-space flows to plan the trajectory both locally and globally and switches control at carefully planned time instances and places in phase space[1]. This type of control via phase-space path planning requires relatively smaller control authority and achieves the desired control properties otherwise impossible to obtain or difficult to manually synthesize. It is a small, opportunistic dynamical alteration based on the global knowledge of phase-space structures. As soon as the system enters the neighborhood of the desired state, a local linear controller stabilizes the system at the desired location.

---

[1]Variable-structure control [10] also composes phase spaces along switching surfaces. However, our method differs fundamentally from the variable-structure control in how the phase-space trajectory flows are modeled and utilized in the search for global control paths.

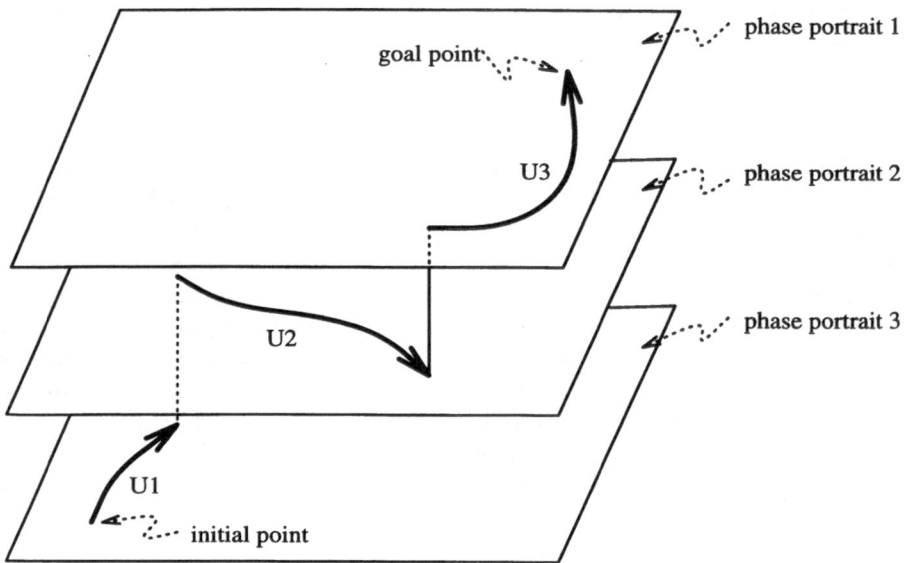

Figure 1: Search for a control path from an initial point to a goal point in a stack of phase portraits.

3.3 Planning control paths with flow pipes

Many nonlinear control systems do not have closed, analytic form solutions. Computational exploration is often the most common means for designing and analyzing global nonlinear controllers. Since a phase space is a union of an infinite number of trajectories, an exhaustive search for plausible control paths in the space of all the trajectories is computationally too expensive and numerically too sensitive to uncertainties. We introduce the technique of flow pipes to model trajectory flows in a phase space. A flow pipe groups a collection of trajectories that exhibit the same qualitative feature into an equivalence class. Reference [24] describes an algorithm for constructing flow pipes.

The geometric modeling of a phase space with flow pipes makes the phase-space control planning and navigation feasible. Given a discrete set of possible control actions, the search for a control path from an initial state to a destination is a reachability problem, i.e., the problem of finding a sequence of connected path segments each of which is under a single control action, as schematically illustrated in Figure 1. This point-to-point planning can be naturally executed in the flow-pipe representation of phase spaces: the system can travel along one flow pipe for a while, switch to a new control action, jump onto another flow pipe, and eventually arrive at the goal.

To make this approach computationally feasible, the phase portraits of the dynamical system indexed by different control actions are first parsed into a discrete set of trajectory flow pipes. These flow pipes are then aggregated to intersect each other and pasted together to form a graph, the *flow-pipe graph*. The flow-pipe graph is a directed graph where nodes are intersections of flow pipes and edges are segments of flow pipes. The graph may contain cycles. The initial state and the goal state are nodes in the graph. Each edge, a segment of a flow pipe, of the graph is weighed according to traveling time, smoothness, *etc.* The weight can be a single value or a range of values. With this representation, the search for optimal paths is formulated as a search for shortest paths in the directed graph.

Since a flow pipe models a bundle of similar trajectories, it provides room for a small deformation of the reference trajectory within the flow pipe at runtime. This is useful as the synthesized reference trajectory often has to be modified to accommodate uncertainties and noise.

4 The Phase Space Navigator

We present a general, two-stage architecture for the autonomous synthesis of nonlinear controllers in phase space [21]. The Phase Space Navigator serves as the prototype for the architecture. It has two main modules: the planning module and the tracking module. The planning module synthesizes a desired trajectory, the reference trajectory, in phase space based on the dynamics of the nominal model. The tracking module follows the reference trajectory and reactively corrects local deviations. The parameters of the nominal model are estimated at runtime and the model is updated. When the dynamics changes significantly, the reference trajectory is resynthesized. Figure 2 shows the interplay between the two modules in the context of controlling a real physical plant.

4.1 Reference trajectory generation

The planning module consists of a global path planner and a local trajectory generator. The global path planner finds coarse global paths from the initial state to the goal state, utilizing the flow pipes of trajectories. The local trajectory generator fits smooth trajectory segments into the global path by slicing out trajectory segments from flow pipes or through local trajectory deformation. Figure 3 illustrates the data flows among components of the planning module.

4.1.1 Global path planner

The global path planner searches for optimal paths from an initial state to a goal state in the phase space, subject to design constraints such as fast convergence and

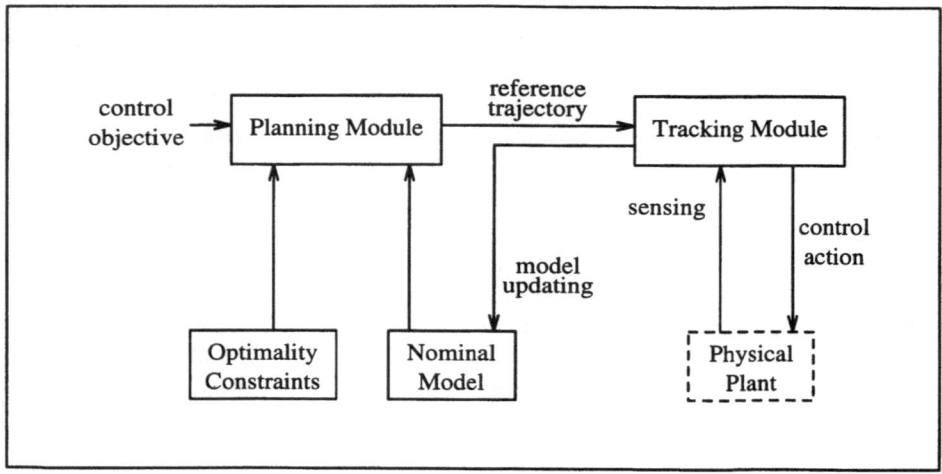

Figure 2: The Phase Space Navigator: a system for autonomous synthesis of global, nonlinear controllers.

little overshooting. The high-level qualitative description of the phase space guides the search.

To synthesize a global path, MAPS first explores in a set of phase portraits indexed by the selected values of the control signal. It generates a sequence of phase-space descriptions summarized in a geometric vocabulary. Flow pipes are then extracted from the descriptions. The global planner searches for a pipe path — a sequence of flow pipes interconnected at intermediate points — from the initial state to the goal state in this collection of flow pipes.

For stabilization problems, we need to synthesize an attractor at the goal state. When the system enters the neighborhood of the goal state, a linear feedback controller is switched in to stabilize the system at the goal state.

4.1.2 Local trajectory generator

The local trajectory generator produces smoothed trajectories connecting intermediate points. More specifically, trajectory segments are extracted from the flow pipes forming the global path. A trajectory segment is extracted from a flow pipe and is continuously deformed to connect intermediate points in the path. The local trajectory generator also smooths out sharp interconnections of trajectory segments to reduce undesired transients, with local linear controllers or local sliding surface insertion [16]. The noise and certain type of parametric uncertainties of the synthesized reference trajectory are modeled with a *thickened trajectory*, of which the thickness

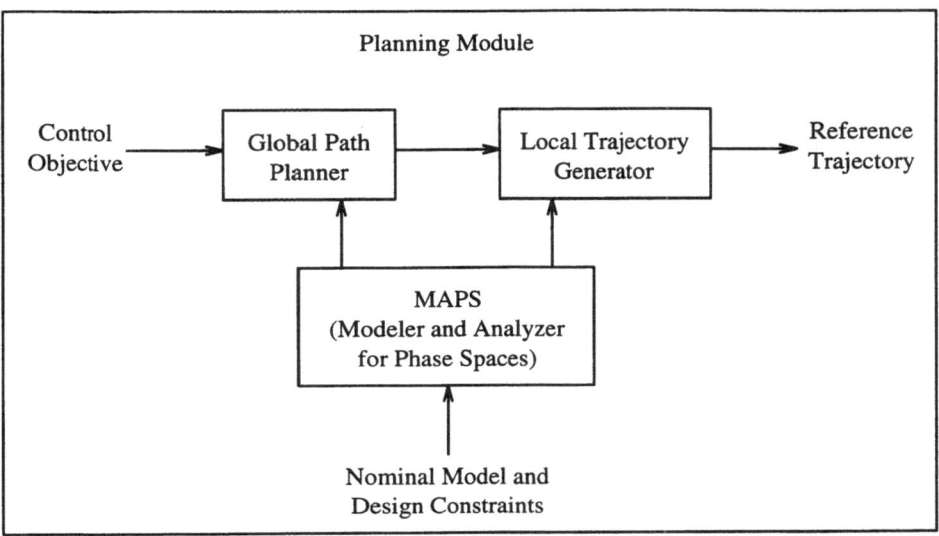

Figure 3: Reference trajectory generation

measures the effect of perturbations. To guarantee the robustness the thickened trajectory segment is restricted within the flow pipe that contains the original reference trajectory segment. However, the method discussed here assumes that the model has no parametric uncertainties or structural uncertainties that change the topological structures of phase space; these kinds of uncertainties include model order uncertainties or unmodeled high-frequency dynamics.

The planning module generates a plan, *i.e.*, the reference trajectory along with the control action for each segment of the reference trajectory.

4.2   Reference trajectory tracking

The tracking module follows the planned trajectory and reactively corrects the deviation due to uncertainties in the modeling of dynamics, disturbances, etc. The tracking is usually local to the region of the phase space containing the trajectory segment; see Figure 4. The planned reference trajectory provides the global feedforward reference term. The sensor and estimator measure the actual state, whose difference from the reference term is the local feedback correction term.

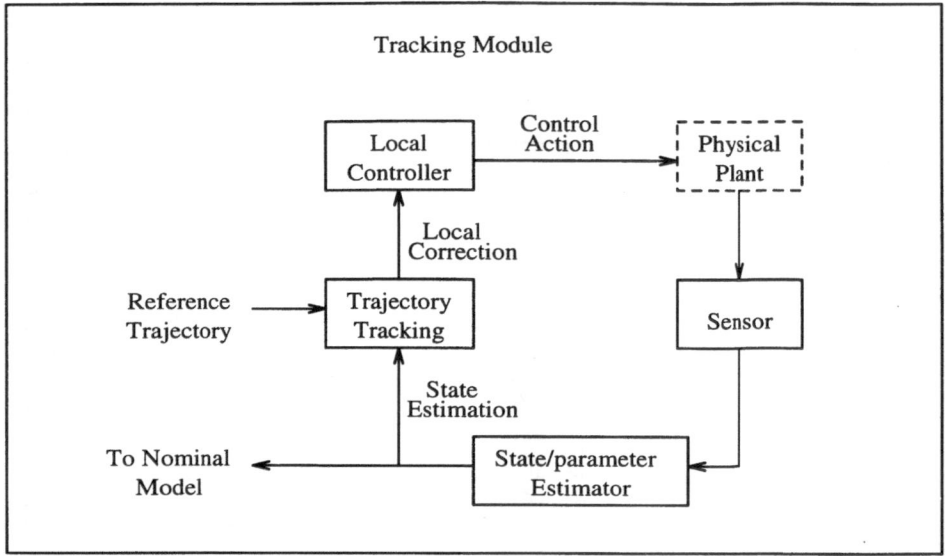

Figure 4: Reference trajectory tracking

4.2.1 Trajectory tracking and reactive correction

The Phase Space Navigator tracks a synthesized reference trajectory by sensing the state of the system, comparing the state with the reference trajectory, and computing the current control action. This tracking requires that the full state of the system be observable. If the system stays within the tolerance of the reference trajectory, the controller simply looks up a control action in the planned trajectory. If the system deviates from the desired trajectory, a local linear controller is switched in to force the system back on track. The trajectory correction is local and reactive. It is designed to ensure the robustness of the system in the presence of noise in measurements and small changes in system parameters that do not lead to bifurcations.

4.2.2 Model updating and reference trajectory resynthesis

The tracking module estimates the model parameters through the observables and updates the nominal model. When the dynamics of the system changes substantially such that the planned global path is no longer feasible, the Phase Space Navigator calls the planning module to synthesize a new reference trajectory.

4.3   The autonomous control synthesis algorithms

We consider a control system $x' = f(x, u)$, where $x \in R^n$ is the state variable and $u \in R^m$ is the control parameter. The initial state of the system is x_s and the goal state for the system to reach is x_g. The set of admissible values for the control parameter u is $\{u_i\}$. The planning algorithm searches for paths from x_s to x_g, subject to certain design constraints, in the phase spaces indexed by control parameter values $\{u_i\}$. The tracking algorithm follows the synthesized reference path.

Algorithm 4.1 *The trajectory planning algorithm:*

1. **Generating phase portraits and parsing them into flow pipes:**

   Generate a collection of phase portraits indexed by $\{u_i\}$. Parse each of the phase portraits into flow pipes.

2. **Constructing flow-pipe graph:**

   Aggregate flow pipes to form flow-pipe graph. Make x_s and x_g to be nodes in the graph. Weigh each edge of the graph with the cost of traversing the edge.

3. **Testing reachability of goal x_g:**

   Cluster the flow-pipe graph into flow-pipe path components. If x_s and x_g are in the same flow-pipe path component, go to Step 6. Otherwise, the goal is unreachable with the given set of parameter values $\{u_i\}$; continue.

4. **Searching for partial paths from x_s to x_g:**

   Compute the reachable set R_{x_s} from x_s and the reverse reachable set R_{x_g} from x_g in the flow-pipe graph. Denote the gap between R_{x_s} and R_{x_g} to be $G = \|R_{x_s} - R_{x_g}\|$.

5. **Tuning control parameter:**

   Search for a value of the control parameter u outside $\{u_i\}$ such that the corresponding phase portrait has flow pipes that, when added to the flow-pipe graph, reduce the value of G. If G is zero, i.e., the gap between R_{x_s} and R_{x_g} is bridged, collect the flow-pipe paths from x_s to x_g and go to Step 7. Otherwise, repeat Step 5 until there are no more parameter values to search, in which case return the collection of partial flow-pipe paths from x_s to x_g.

6. **Finding flow-pipe paths from x_s to x_g:**

   If looking for the shortest paths from x_s to x_g, run a standard shortest path algorithm on the graph. If searching for all paths p from x_s to x_g subject to the constraint $C(p) \leq C$, use the following algorithm[2].

   (a) let $P_{global-paths} = \emptyset$; $P_{partial-paths} = \{(x_s, x_s)\}$.

---

[2]Note that in the discussion $C(p_i)$ is the cost of a path p_i, $W(x_i, x_j)$ is the weight of the edge (x_i, x_j), and "$*$" is the path composition operator.

(b) for each path $p_i \in P_{partial-paths}$, ending at x_i, do
$$P_{p_i} = \{p_i * (x_i, x) | x \text{ one edge reachable from } x_i, x \neq x_g,$$
$$C(p_i) + W(x_i, x) \leq C\};$$
$$P'_{p_i} = \{p_i * (x_i, x) | x \text{ one edge reachable from } x_i, x = x_g,$$
$$C(p_i) + W(x_i, x) \leq C\};$$
$$P_{partial-paths} = (P_{partial-paths} - \{p_i\}) \cup P_{p_i};$$
$$P_{global-paths} = P_{global-paths} \cup P'_{p_i}.$$

(c) if $P_{partial-paths} \neq \emptyset$, goto (6b); otherwise, continue.

(d) if $P_{global-paths} = \emptyset$, goto Step 4; otherwise, continue.

(e) order paths in $P_{global-paths}$ according to their costs and return.

7. **Generating trajectory segments:**

Select a representative trajectory segment from each flow pipe forming the flow-pipe path from x_s to x_g and paste them together at intersections of flow pipes.

8. **Smoothing trajectories:**

Smooth each intersection of trajectory segments through trajectory homotopy deformation. Order the smoothed trajectories with costs and return.

Algorithm 4.2 *The trajectory tracking algorithm:*

1. **Sensing and estimating the physical system:**

Sense the state x of the physical system, estimate the model parameters, and update the nominal model of the system. If x is in the neighborhood of the goal x_g, go to Step 5. If the nominal model changes substantially, signal and go to Step 4. Otherwise, continue.

2. **Computing control action:**

Based on the synthesized reference trajectory and the observation from Step 1, compute control tracking term u_{track} and correction term $u_{correct}$. If $u_{correct}$ is unobtainable with a local linear correction, check alternative paths to the goal from the collection of suboptimal paths. If the goal is unreachable, go to Step 4. Otherwise, the control action is $u = u_{track} + u_{correct}$; continue.

3. **Generating control action:**

Tune the control parameter to u_{track} and generate $u_{correct}$ with a local linear feedback controller. Drive actuators and go to Step 1.

4. **Resynthesizing reference trajectory:**

Call the planning module to resynthesize a reference trajectory and go to Step 2.

5. **Goal stabilizer:**

Stabilize the system at x_g with a local linear controller, subject to uncertainties and noise.

Figure 5 shows the flow chart of the planning algorithm. The algorithm takes as input the system model, allowable control parameter values, design constraints, and desired control objectives. It outputs a synthesized reference trajectory in the form of a list of tuples: (time, switching point, control parameter value). The planning module has been implemented in Scheme, a dialect of LISP [9]. The tracking algorithm described above provides a conceptual framework, whose implementation is an immediate goal of future research.

4.4 Discussion of the synthesis algorithms

4.4.1 On-line and off-line synthesis

The planning module generates smoothed reference trajectories in the phase space. The constraints on response time and the availability of computational resources dictate whether the computation of planning is done on-line or off-line. If the planning is done off-line, the synthesized plan is compiled into a table. At the runtime the controller performs a table lookup for a control action. On-line reference trajectory synthesis and tracking require substantial computational power. The Supercomputer Toolkit [3] provides an ideal platform for experimenting with the above control synthesis algorithm on real-time control applications.

In Step 1 of Algorithm 4.1 for planning, the phase-space structure and flow descriptions can be either computed once for all the control parameter values or generated on demand in the search for optimal paths. The latter case will be more suitable for on-line synthesis.

4.4.2 Generalizations of the point-to-point planning

We present the planning algorithm for a control design problem with one initial state and one goal state. The algorithm also applies to control systems of "one initial state/many goal states", "many initial states/one goal state", or "many initial states/many goal states".

For a "one initial state/one goal state" or "one initial state/many goal states" problem, the Dijkstra's algorithm for the single source shortest paths in a directed graph runs in $O(V^2)$ for a graph with V vertices and E edges [7].

A "many initial states/one goal state" problem can be converted to the "one initial state/many goal states" problem by reversing the directions of all edges; and a "many initial states/many goal states" problem is solved by the Floyd-Warshall algorithm in $O(V^3)$ [7].

4.4.3 Discrete vs. continuous control parameter spaces

A version of the synthesis algorithm is presented for the control parameter initially taking values from a finite discrete set. An example of such finite-valued control

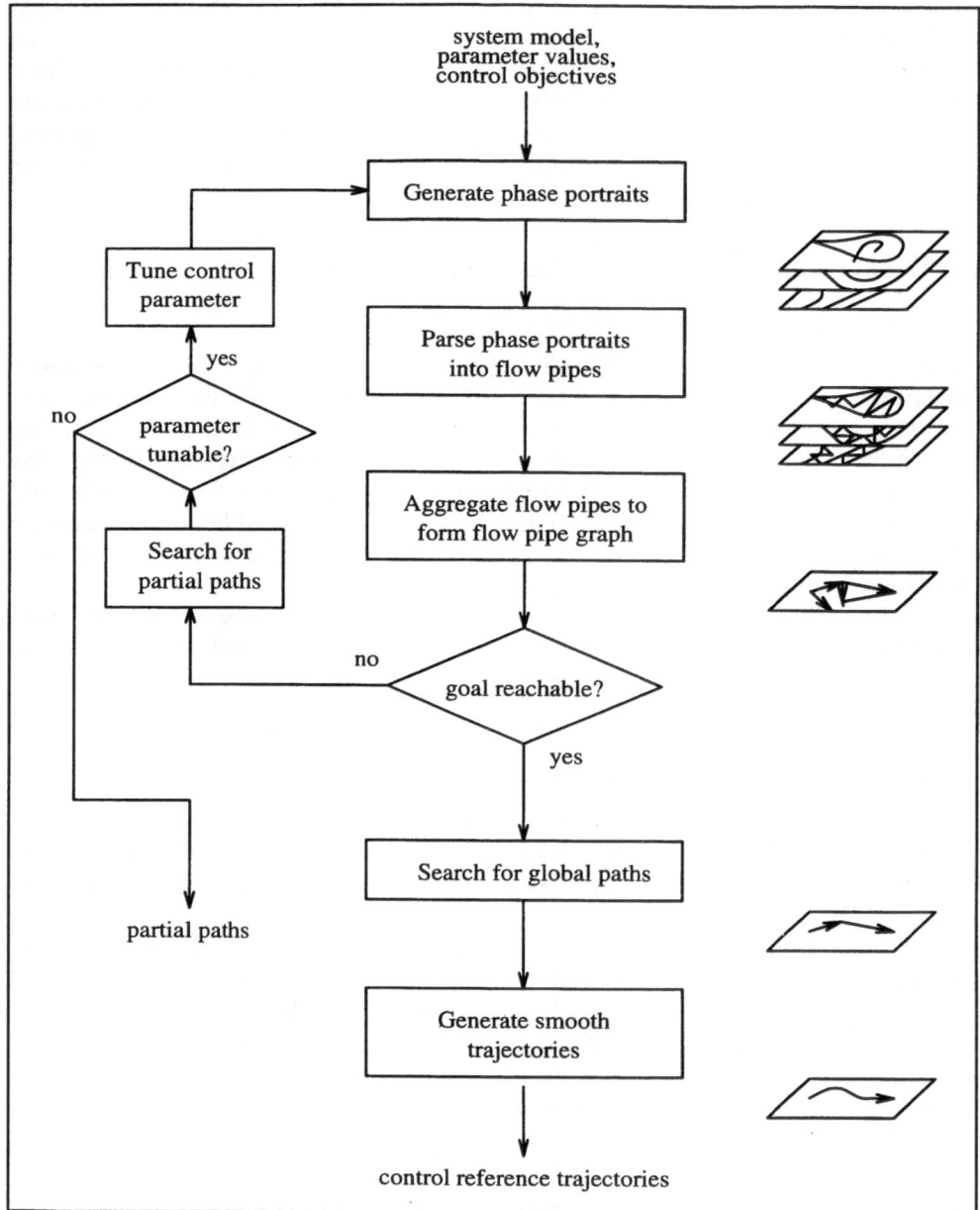

Figure 5: The flow chart of the trajectory planning algorithm of the Phase Space Navigator.

systems is the satellite attitude controller, which stabilizes the antenna direction of the satellite subject to disturbances by turning on or off high thrust jets. Other such examples are switching power regulators.

The method also applies to control systems with continuous, multiple control parameter spaces. The program uniformly samples the continuous parameter space at discrete points and then applies the algorithm on these control points. A much better approach, however, would search for "land-mark points" that delimit distinct behaviors and then partition the parameter space into equivalent subspaces. For example, the Bifurcation Interpreter [2] can be employed to search for the bifurcation points defining qualitative changes in dynamics and decompose the parameter space into topologically equivalent subparts.

4.4.4 Suboptimal control paths

The algorithm finds optimal paths in the flow-pipe graph. However, the suboptimal paths can be useful when the optimal paths are no longer judged feasible. They can be stored in the table in addition to the optimal ones. A controller can opportunistically choose among available trajectory paths according to the desired properties at a control switching point.

4.4.5 Comparison with dynamic programming

To synthesize an optimal control path, dynamic programming discretizes a state space and conducts a fine-grain search in the discretization. The cost of the exhaustive search could be prohibitive for large regions or in higher dimensions. In contrast, the Phase Space Navigator searches for control paths in a manageable set of flow pipes. It is possible to use dynamic programming within a flow pipe, once the global path has been established.

5   An Illustration: Stabilizing a Buckling Column

We illustrate the mechanism of the Phase Space Navigator with a control synthesis for stabilizing a buckling steel column. The buckling motion of the column has been extensively studied in nonlinear dynamics from a theoretical point of view and in structural engineering by practicing engineers. The columns are commonly used as strengthening elements in structures; for example, flexible space structures use columns in large operating regions. Study in nonlinear dynamics shows that a slender steel column can exhibit very complicated dynamical patterns under various operating conditions. Therefore, it is important to understand the dynamical behaviors of the columns and to devise ways of preventing them from failure.

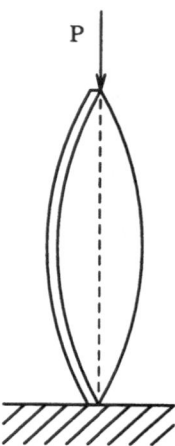

Figure 6: Buckling of a thin elastic steel column due to axial end loads.

5.1 The column model

A thin elastic column, subject to axial end compressive loads, buckles around the
principal axis, as shown in Figure 6. The nonlinearity is introduced by the nonlinear
geometric stiffness of the column [13]. Stoker [17] gave a simplified model for the
column subject to axial compressive force and viscous damping

$$mx'' + cx' + a_1 x + a_3 x^3 = 0, \tag{1}$$

where $a_1 = A + C - 2P/l$ and $a_3 = B + D - P/l^3$. The state x is the characteristic
measure of the column deflection from the principal axis and x' is the velocity. The
column has mass m and length $2l$. The axial load is P, and the coefficient of viscous
damping is c. The bending stiffness is modeled by a primary hard spring with restoring
force $Ax + Bx^3$ and a secondary hard spring with restoring force $Cx + Dx^3$. We rewrite
equation (1) as a system of first-order equations

$$\begin{cases} x_1' = x_2 \\ x_2' = \frac{1}{m}(-a_1 x_1 - a_3 x_1^3 - cx_2), \end{cases} \tag{2}$$

where x_1 represents the deflection and x_2 represents the velocity.

The system (2) describes the buckling motion of the column and represents only
a single mode of vibration. For a long and slender column, vibrations are observed
to occur primarily in the first mode [14].

When the axial load P is less than the critical load $P_{critical}$, the column oscillates
around the principal axis and returns to the vertical state. It has only one stable

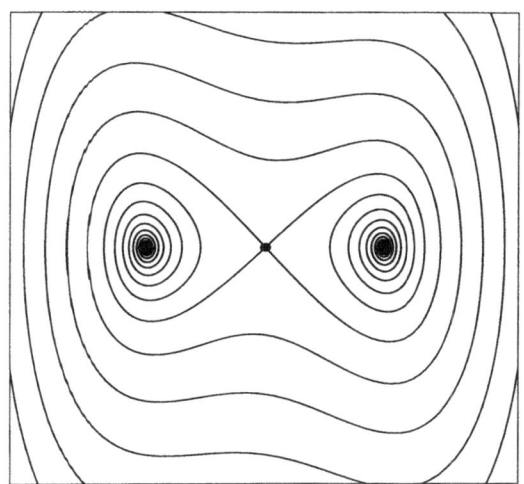

Figure 7: The buckling column: the phase space of a buckling column showing the stability boundaries and connecting trajectories. The horizontal axis x is the characteristic measure of the column displacement from its principal axis and the vertical axis x' is the velocity.

state corresponding to the attractor at the origin of the phase space. Under a heavier load, *i.e.*, $P > P_{critical}$, the column buckles to one side or the other. The phase space has a saddle at the origin and two attractors symmetrically arranged about the saddle; see Figure 7. The positions of the two buckled states depend on the external load P, the stiffness of the column, and the length l. The larger the load is, the farther away the buckled states are from the principal axis. The column breaks when the buckling exceeds certain limit. As P surpasses $P_{critical}$, the column undergoes a pitchfork bifurcation of equilibria [8]: the attractor at the origin gives birth to a saddle at the origin and two attractors on two sides.

5.2 Extracting and representing qualitative phase-space structure of the buckling column

MAPS automatically analyzes the column model in the phase space and extracts and represents the qualitative phase-space structure [20]. For parameter values $a_1/m = -2.0$, $a_3/m = 1.0$, and $c/m = 0.2$ and phase-space region $-3.0 \leq x_1 \leq 3.0$ and $-4.0 \leq x_2 \leq 4.0$, the program reports the following findings and represents them internally in a relational graph:

```
<equilibrium-points:
 equilibrium 1. (attractor at (1.41 0.))
 equilibrium 2. (saddle at (0. 0.))
 equilibrium 3. (attractor at (-1.41 0.))>

<trajectories:
 <boundary-trajectories:
  trajectory 1. (from *infinity* to (0. 0.))
  trajectory 2. (from *infinity* to (0. 0.))>
 <connecting-trajectories:
  trajectory 3. (from (0. 0.) to (-1.41 0.))
  trajectory 4. (from (0. 0.) to (1.41 0.))>>

<stability-regions:
 stability-region 1.
   attractor at *infinity*
   stability-boundary: ()
   connecting-trajectories: ()
 stability-region 2.
   attractor at (1.41 0.)
   stability-boundary: (trajectory 2. trajectory 1.)
   connecting-trajectories: (trajectory 4)
 stability-region 3.
   attractor at (-1.41 0.)
   stability-boundary: (trajectory 2. trajectory 1.)
   connecting-trajectories: (trajectory 3)>
```

The program finds two attractors at $(1.41, 0.0)$ and $(-1.41, 0.0)$ and a saddle at the origin. It generates a high-level description of the phase-space geometry: two banded stability regions associated with the two attractors, separated by the stable trajectories of the saddle at the origin. Figure 7 shows stability boundaries and connecting trajectories of the two stability regions. Based on a triangulation of the stability regions, the phase space is further modeled with two flow pipes formed by aggregating geometric pieces, using a flow-pipe construction algorithm described in [24]. The pipe boundaries consist of the separatrices of the two stability regions that approach the saddle and of the trajectories that connect equilibria, as described by MAPS in the following:

```
<flow-pipes:
 flow-pipe 1. (from *infinity* to (-1.41 0.))
   boundary: (trajectory 2. trajectory 1. trajectory 3.)
 flow-pipe 2. (from *infinity* to (1.41 0.))
   boundary: (trajectory 2. trajectory 1. trajectory 4.)>
```

Figure 8 shows the flow pipe that ends at the left-hand attractor.

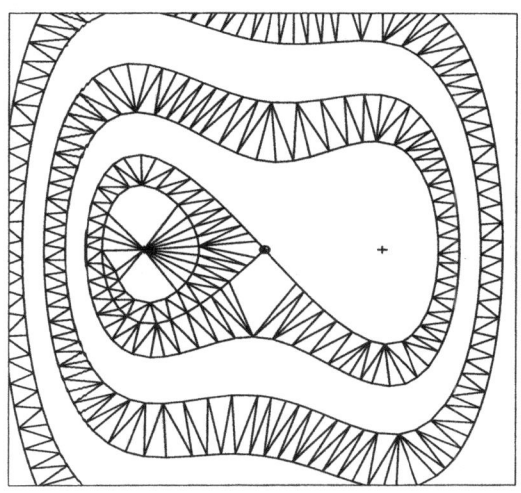

Figure 8: The buckling column: the flow pipe leading to the attractor on the left.

5.3 Synthesizing control laws for stabilizing the column

We want to stabilize the column at its vertical state to support the axial end loads and to prevent it from breaking. Under sufficiently heavy load, the buckling motion of the column leads the column to one of the buckled states and, when the states are far away from the principal axis, induces the failure of the column.

We shall focus on global navigation in phase space that synthesizes global reference trajectories leading to the desired goal state. Since the design of local linear controllers is relatively well understood, we shall not discuss the trade-offs among different designs for linear controllers and shall choose, for the purpose of demonstration, a simple linear feedback design. Local controllable regions of such linear controllers are quantified, given available control strength, and are used in constraining the design of global control paths.

The goal state of the control is the unbuckled state—the saddle at the origin of the phase space that does not have a stability region. We want to synthesize a non-zero stability region for the goal and maximize the region. The controlled column is of the form

$$\begin{cases} x_1' = x_2 \\ x_2' = \frac{1}{m}(-a_1 x_1 - a_3 x_1^3 - c x_2) + u, \end{cases} \tag{3}$$

where u is the control. In the model, mu has the same dimension as the force.

The Phase Space Navigator automatically synthesizes a global trajectory from an

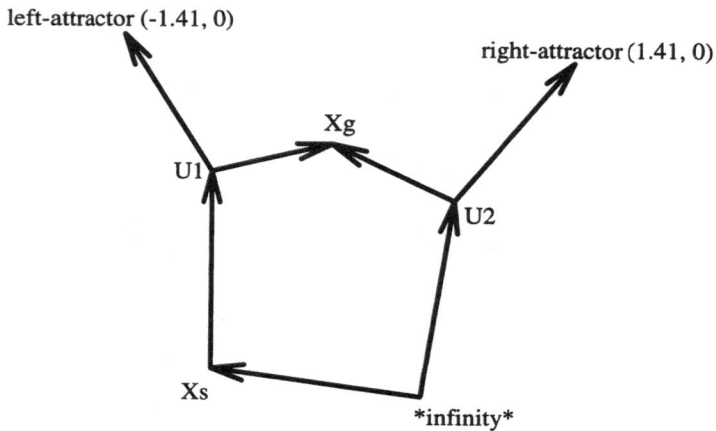

Figure 9: The flow-pipe graph for the buckling column.

initial state x_s to the goal state x_g. We consider two cases for the initial state x_s.[3]

5.3.1 Control design I: stabilizing the buckling motion

The column is initially buckling with sufficient velocity. The initial state x_s is far away from the saddle x_g in phase space in this case. The control parameter u takes its values from a small range around zero.

The Phase Space Navigator constructs a flow-pipe graph from phase spaces of $u = 0$ and $u =*\texttt{local-control}*$, as shown in Figure 9. The graph shows the case when the initial state x_s is in the flow pipe ending at the left attractor as shown in Figure 8. Denote by U_{ctl} the local controllable region at the goal x_g. The two flow pipes of $u = 0$ intersect U_{ctl} at u_1 and u_2 respectively. The edges $\overline{u_1 x_g}$ and $\overline{u_2 x_g}$ represent flows within U_{ctl} produced by the local controller. The paths ($*infinity* \to x_s \to u_1 \to \textit{left-attractor}$) and ($*infinity* \to u_2 \to \textit{right-attractor}$) represent the two flow pipes of the phase space of $u = 0$.

The Phase Space Navigator finds a simple path from x_s to x_g in the graph, consisting of edges $\overline{x_s u_1}$ and $\overline{u_1 x_g}$. Then the program synthesizes an individual trajectory connecting x_s and x_g from the flow-pipe path ($\overline{x_s u_1}$, $\overline{u_1 x_g}$). To construct this desired trajectory, the program deforms an uncontrolled trajectory, emanating from x_s, of the flow pipe represented by the edge $\overline{x_s u_1}$ so that the deformed trajectory enters U_{ctl}.

---

[3]We use this example for the purpose of illustrating the Phase Space Navigator. The system (3) is actually feedback linearizable. A feedback linearization would cancel the nonlinearity of the vector field in x_2 direction. In contrast, our design restrains the control to be less than 10% of the vector-field strength in the first case and less than half of the vector-field strength in the second case.

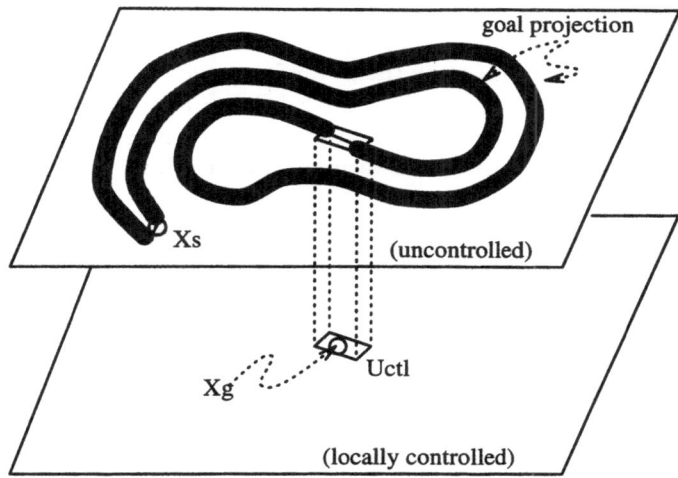

Figure 10: The goal projection and the deformation of the trajectory.

The region U_{ctl} is first projected backwards through the flow pipes to form a goal projection. Figure 10 shows the goal projection from U_{ctl}—two thin pipes illustrated in thick solid lines. Then the uncontrolled trajectory is deformed towards the goal projection. The programmed deformation is designed to push the trajectory towards the nearest goal projection in the controllable direction x_2. Since the control is more effective when the direction of the vector field is relatively orthogonal to the controllable direction x_2, switching points are inserted to turn off the controller when the angle between the two directions are below some threshold.

Consider for example the case when the column is initially at the state $(-1, -3)$. The control design is specified as:

```
control_type:     point_to_point
goal_state:       (0.0, 0.0)
initial_state:    (-1.0, -3.0)
range_of_control: u ∈ [-0.2,0.2]
```

The Phase Space Navigator designed a control law that brings the column back to the unbent state. The control law is represented as a list of tuples, each of which specifies the time, state, and control value for each switching of control:

```
((time 0.) (switching-state #(-1 -3)) (control .2))
((time .284) (switching-state #(-1.82 -2.71)) (control 0.))
((time 1.06) (switching-state #(-1.86 2.49)) (control -.2))
((time 2.71) (switching-state #(1.35 1.82)) (control 0.))
((time 6.76) (switching-state #(-.0023 -.0692)) (control *local-control*))
```

Figure 11(a) shows the synthesized reference trajectory originating at $(-1, -3)$. The circles indicate places where the control of deformation switches. Each segment of the reference trajectory delimited by the switching points is under interval-constant control, as specified by $U1$, $U2$, $U3$, $U4$, or a local linear control law as the trajectory is in the vicinity of the goal. The global portion of the reference trajectory is pushed towards the goal projection with small deformation that is less than 10% of the vector field strength at any state or 0.2 in the normalized unit, whichever is smaller. The position x and velocity $v = x'$ of the controlled column are plotted against the time t in Figures 11(b) and (c), respectively. The control u is shown in Figure 11(d). The synthesized trajectories could be further optimized with variational techniques on the collection of trajectories within the flow-pipe segments [4].

5.3.2 Control design II: restoring from the buckled state

The column is initially near the buckled state $(-1.41, 0.0)$. The control objective here is to pull the column out of the buckled state and to bring it close to the unbuckled state. Since the flow pipe containing the initial state does not intersect U_{ctl} in the down stream, a different control strategy must be employed. The program uses the following strategy to synthesize a control path.

Assume the initial state of the column is $(-1.5, 0.0)$. The control objective is:

control_type:	point_to_point
goal_state:	(0.0, 0.0)
initial_state:	(-1.5, 0.0)
range_of_control:	$u \in [-1.0, 1.0]$

The uncontrolled trajectory from the initial state would spirally approach the attractor in the clockwise fashion. The control is exerted in such a way as to swing the trajectory away from the buckled state to approach the goal projection of the saddle. When the trajectory intersects the goal projection, the control is switched off so that the system slides along the uncontrolled trajectory. As soon as the system enters U_{ctl}, the linear controller is switched in. The control strength is less than half of the vector field strength or 1.0 in the normalized unit, whichever is smaller. The synthesized control law is specified as:

```
((time 0.) (switching-state #(-1.5 0)) (control -.187))
((time .001) (switching-state #(-1.5 .000187)) (control .187))
((time 1.65) (switching-state #(-1.24 -.0011)) (control -.289))
((time 3.19) (switching-state #(-1.66 .0153)) (control .638))
((time 5.92) (switching-state #(-.527 -.00186)) (control -.454))
((time 7.74) (switching-state #(-2.02 .0473)) (control 1.))
((time 8.08) (switching-state #(-1.75 1.47)) (control 0.))
((time 11.1) (switching-state #(-.049 .00163)) (control *local-control*))
```

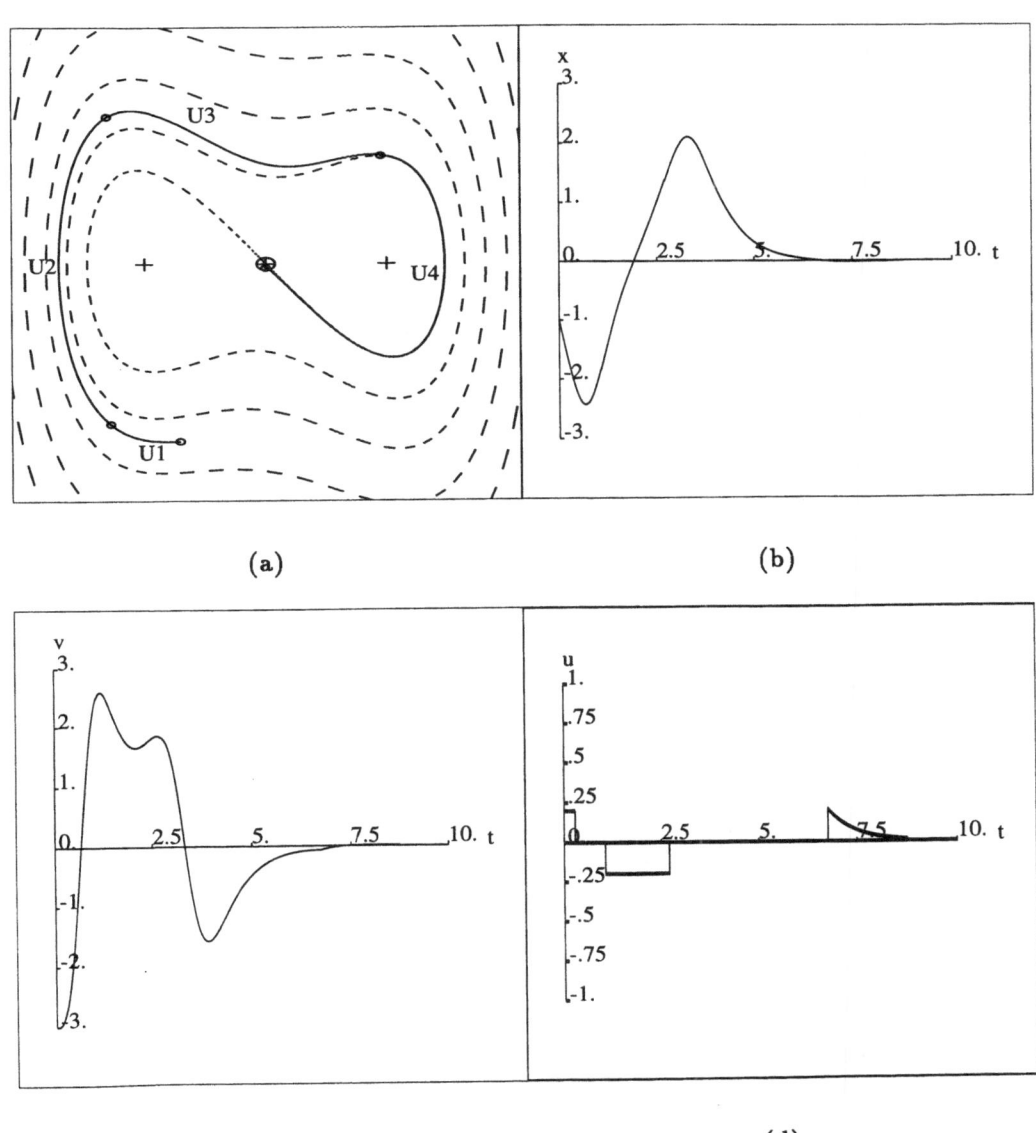

(a)

(b)

(c)

(d)

Figure 11: The synthesized control law that stabilizes the buckling column: (a) the reference trajectory that leads to the unbuckled state corresponding to the saddle at the origin. The column is initially buckling with sufficient velocity. The position x of the column, the velocity $v(= x')$, and the control signal u for stabilizing the column are plotted against time t in (b), (c), (d), respectively.

The corresponding reference trajectory is shown in Figure 12(a). The control parameter changes at places marked by circles. The reference trajectory consists of eight trajectory segments labeled by interval-constant control $U1$, $U2$, ..., $U7$, and a local control law. The first segment ($U1$) and the last segment (local control) are very short in length and thus are invisible in the figure. The position x and velocity $v = x'$ of the controlled column are plotted against the time t in Figures 12(b) and (c), respectively. The control u is shown in Figure 12(d).

5.4 The phase-space modeling makes the global navigation possible

The qualitative description of the phase-space structure and the geometric modeling of the trajectory flows provide a "map" for navigating system trajectory to the goal in phase space.

The directed graph constructed from the flow pipes is used in searching for global paths and in determining whether the goal is reachable. The flow pipes also make it possible to characterize the more microscopic spatial and temporal relations between the current state and the goal state. The deformation of global path segments is constrained by reasoning about the spatial relation with flow pipes.

6 An Application: Maglev Controller Design

The Control Engineer's Workbench has helped synthesize a global, nonlinear controller for a nominal model of a German maglev system [22]. We describe the systematic phase-space design method for determining the global switching points of the controller. The synthesized control system can stabilize the maglev vehicle with large initial displacements from an equilibrium. The simulation shows that our nonlinear controller possesses an operating region more than twenty times larger than that of a classical linear feedback design for the same system.

6.1 The maglev model

Maglev transportation uses magnetic levitation and electromagnetic propulsion to provide contactless vehicle movement. There are two basic types of magnetic levitation: electromagnetic suspension (EMS) and electrodynamic suspension (EDS). In EMS, the guideway attracts the electromagnets of the vehicle that wraps around the guideway. The attracting force suspends the vehicle about one centimeter above the guideway. An attractive system such as the EMS system is inherently unstable. We consider the control design for stabilizing an EMS-mode train traveling on a guideway—a simplified model for the German Transrapid experimental system.

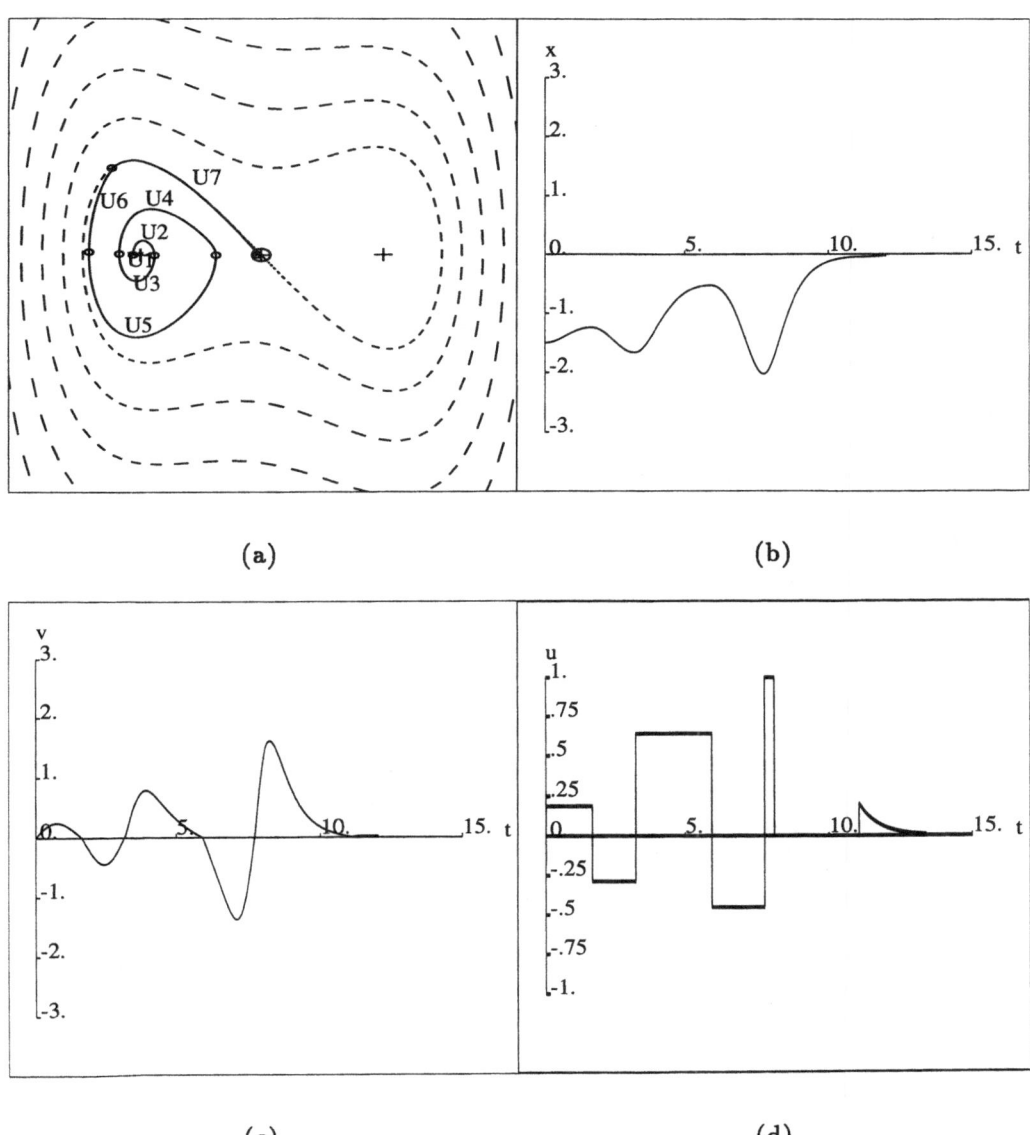

(a)

(b)

(c)

(d)

Figure 12: The synthesized control law for restoring the column from the buckled state: (a) the reference trajectory that swings the column out of the buckled state. The position x of the column, the velocity $v(=x')$, and the control signal u for controlling the column are plotted against time t in (b), (c), (d), respectively.

The state equations for the magnetically levitated vehicle and the guideway are described by

$$\begin{cases} \frac{dx}{dt} = \frac{z(V_i - Rx)}{L_0 z_0} + \frac{xy}{z} \\ \frac{dy}{dt} = g - \frac{L_0 z_0 x^2}{2mz^2} \\ \frac{dz}{dt} = y \end{cases} \tag{4}$$

where the state variables x, y, and z represent coil current in the magnet, vertical velocity of the vehicle, and vertical gap between the guideway and the vehicle, respectively. The control parameter is the coil input voltage V_i. The other parameters are the mass of the vehicle m, the coil resistance R, the coil inductance L_0 and the vertical gap z_0 at the equilibrium, and the gravitational acceleration g. The nonlinearities of the system come from the nonlinear inductance due to the geometry of the magnet and the inverse square magnetic force law.

The system has one equilibrium state at which the magnetic force exactly counterbalances the force due to gravity and the vehicle has no vertical velocity and acceleration. However, the equilibrium is a saddle node which is not stable. The control objective, therefore, is to stabilize the vehicle traveling down the guideway and maintain a constant distance between the vehicle and the guideway despite any roughness in the guideway. The available control input is the coil input voltage V_i in the model (4). We further assume that V_i is produced by a buck converter capable of delivering any voltage from 0 to 300 volts.

A linear control design for the maglev model (4) described in [18] uses the pole-placement method. The system is first linearized around the equilibrium. The linearized system has unstable poles, i.e., the poles in the right-half of s-plane. A linear feedback is introduced to move the poles to the desired locations in the left-half of the s-plane. Such a control design can bring the system back to the equilibrium with an initial displacement of up to 0.2mm from the equilibrium. The linear controller saturates at the beginning for larger initial displacements. This is because the linearized model no longer approximates the original system well in regions far away from the equilibrium. A global, nonlinear control law such as a bang-bang control that respects the nonlinearity of the system must therefore precede the linear feedback control. However, the real challenge for the nonlinear design is to determine the global control law specifying, for instance, the switching points.

6.2   Phase-space control trajectory design

We describe a nonlinear control design—a switching-mode control—in phase space for the maglev system with large initial displacements from the equilibrium. We will show that this controller can be automatically designed by the Workbench comprising MAPS and the Phase Space Navigator. The nonlinear controller brings the system to the vicinity of the equilibrium and then switches to the linear controller.

For the purpose of demonstration, we assume that the vehicle is displaced from the equilibrium in the direction further away from the guideway. We will concentrate on the global design of the control reference trajectories and assume that a linear feedback design is available as soon as the system enters the capture region of the linear controller.

6.2.1 Modeling phase-space geometry

The global control law is designed by analyzing and modeling the phase-space geometry of the system. The Workbench explores the phase space of the system and characterizes the phase space with stability regions and trajectory flow pipes. It composes the phase spaces for different control parameter values and uses flow pipes to synthesize a composite phase space.

The state variables x, y, and z in the model are scaled by 1, 10^3, and 2×10^4, respectively. The parameters of model are assumed to be: $L_0 = 0.1h$, $z_0 = 0.01m$, $R = 1\Omega$, $m = 10000kg$, and $g = 9.8m/sec^2$, typical of a large vehicle lift magnet. Assume the power supply delivers 140 volts, i.e., $V_i = 140$, at the equilibrium. The Workbench explores the phase space in a region bounded by the box $\{(x,y,z)|x \in [0,400], y \in [-300,350], z \in [0,600]\}$ and finds the following equilibrium point:

```
saddle:       #(140. 0. 200.)
eigenvalues:  -17.004+22.963i
              -17.004-22.963i
              24.007
eigenvectors: #(.23604 .97174 0)
              #(.51331 -.55588 .65384)
              #(.30157 .73255 .61027)
```

With the information about the stable eigenvectors of the saddle, the Workbench computes the stable manifold of the saddle, a two-dimensional surface. The Workbench generates a set of trajectories evenly populating the stable manifold to approximate the surface. The trajectories are obtained by backward integrations from initial points in a small neighborhood of the saddle. This neighborhood lies within the plane spanned by the stable eigenvectors of the saddle.

Figure 13 shows the trajectories on the stable and unstable manifolds of the saddle in the yz-projection of the phase space. The stable manifold is two-dimensional and the unstable one is one-dimensional. The stable manifold separates the phase space into two halves: trajectories in the upper-half approach $z \to \infty$ along one of the unstable trajectories, corresponding to the case in which the vehicle falls off the rail; and trajectories in the lower-half approach $z = 0$ plane along the other unstable trajectory, corresponding to the case in which the train collides with the rail.

6.2.2 Synthesizing a global stabilization law

For an initial displacement above or below the equilibrium, the uncontrolled system will follow either a trajectory traveling upwards with increasing z and leaving the bounding box or a trajectory traveling downwards and hitting the $z = 0$ plane. To

152

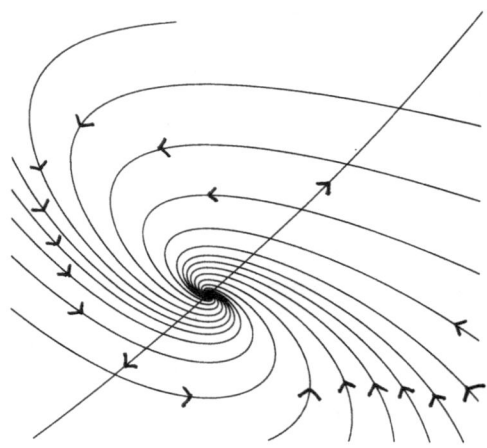

Figure 13: Stable and unstable manifolds of the saddle for $V_i = 140$ (yz-projection).

stabilize the system at the equilibrium, it is necessary to synthesize a new vector field on both sides of the stable manifold so that trajectories travel towards the stable manifold of the saddle in the new vector field. We consider only the top-half here.

The Workbench first explores the phase space of the model for different values of V_i and concludes that the larger the V_i is, the further away the stable manifold is from the $z = 0$ plane. For $V_i = v > 140$, the region sandwiched by the stable manifold of $V_i = 140$ and that of $V_i = v$ has the desired property—the vector field of $V_i = v$ in this region is pointed towards the stable manifold of $V_i = 140$. When $v = 300$, the region is maximized. The model with $V_i = 300$ has a saddle node at $(300., 0., 428.57)$; the stable manifold of the saddle has a similar structure as that of $V_i = 140$ case, but with larger z coordinate.

The Workbench searches for control trajectories in the set of flow pipes indexed by different V_i values and finds a sequence of flow pipes that lead to the desired goal: the composite of a trajectory flow with $V_i = 300$ and a flow with $V_i = 140$, glued together at the stable manifold of $V_i = 140$. As a result, all the trajectories of the flow with $V_i = 300$ can be brought to the equilibrium by switching to $V_i = 140$ as soon as the trajectories hit the bottom boundary. The region comprising these trajectories is therefore the controllable region for the system, as shown in Figure 14. As the trajectories enter a small neighborhood of the equilibrium, a linear feedback controller is used to stabilize the system at the equilibrium. Figure 14 also shows the synthesized control reference trajectories originating at different initial displacements from the equilibrium: 1mm, 4mm, 4.5mm, and 5mm. The controller is able to recover from the first three initial points that are within the region. The last point is outside

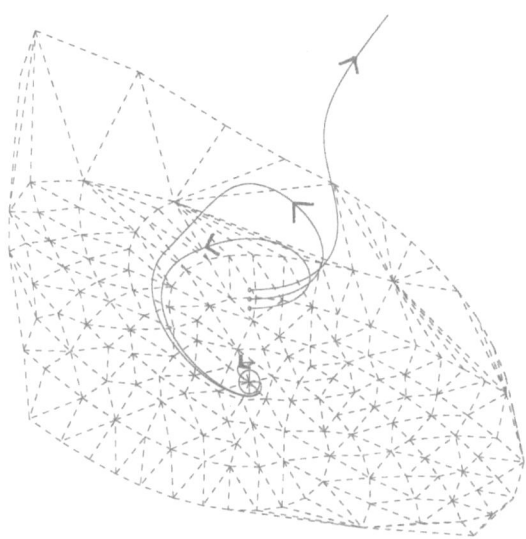

Figure 14: The synthesized control reference trajectories originating from four different initial states, together with the controllable region in yz-projection.

the region and thus uncontrollable; the current in the magnet can not build up fast enough to keep up with the ever-increasing airgap.

6.2.3 Evaluating the control design

The synthesized global control law is a switching-mode one that changes the control parameter at the switching surface—the stable manifold of $V_i = 140$. It is able to bring trajectories originating from any states within the controllable region to a local neighborhood of the saddle.

The responses of the controller with respect to the four different initial displacements are shown in Figure 15. The vertical axis of each graph represents state variables x, y, and z as in the maglev model (4), one for each curve, and the horizontal one represents the time. For all the controllable initial displacements, the controller is able to bring the system back to the equilibrium with errors less than 0.2mm in displacement—a distance within the capture range of the linear feedback controller.

By exploring the phase-space geometries of the maglev system, the Workbench is able to automatically determine the switching points for the global controller. The linear feedback controller can recover from only displacements of less than 0.2mm. The global controller has significantly enlarged the operating region of the linear controller. With the geometric representation of the controllable region in phase space, the Workbench precisely determined that the maximum recoverable displacement is

154

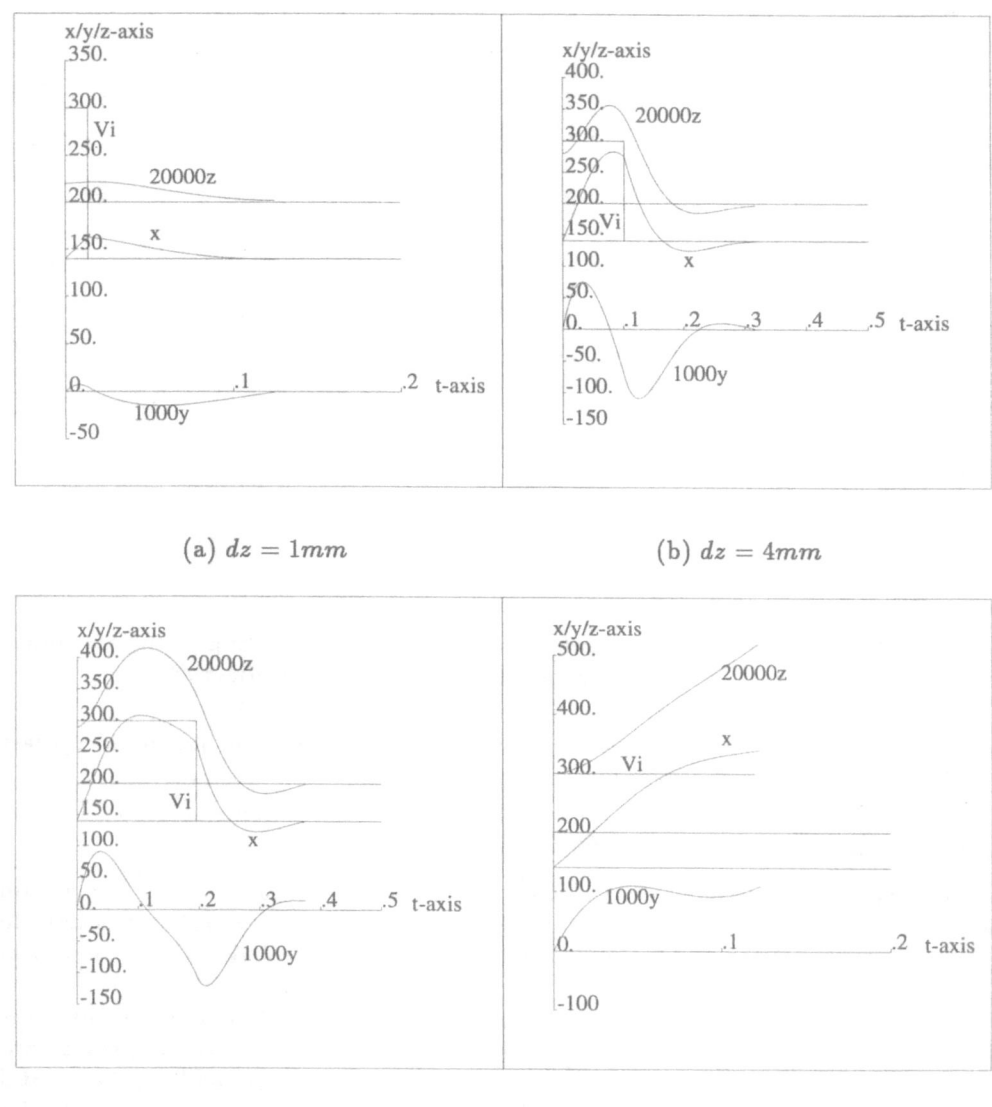

(a) $dz = 1mm$ (b) $dz = 4mm$

(c) $dz = 4.5mm$ (d) $dz = 5mm$

Figure 15: Simulation of the nonlinear control design for different initial displacements.

4.55mm. Our simulation geometrically explained the observation in [18] that the vehicle would fall off the rails with a displacement of 5mm or larger.

Many issues remain to be addressed in order to make the control design practical. Since our control law is designed with the nominal model of the maglev system, the effect of uncertainties in the model and noise in the environment on the design needs to be studied in future research. The design can also be optimized with respect to response time.

The control design described above has been computationally simulated only. How will this design be implemented on the real system and used in real time? The control law specifying the switching surfaces in phase space can be compiled into a table. The control execution will be a table lookup and a geometric inequality test. At each step of the execution, the state of the system is sensed and checked against the switching surface. If the state is on the switching surface, the corresponding control value for the next time interval is read from the table and applied to the physical system. The implementation will be similar to that of dynamic programming.

7   Discussions

The control synthesis method developed here is not meant to replace other nonlinear control design techniques. To the contrary, the method complements the other techniques. The potential of the method in solving real problems can be fully realized when the method works together with other techniques, such as feedback linearization, gain scheduling, and dynamic programming[4].

The area of intelligent control, sometimes called knowledge-based control, has been very active recently. A typical intelligent control system has a hierarchical structure. The hierarchy ranks from high-level decision making to low-level PID controls. Phase Space Navigator can be regarded as one kind of intelligent control, if one wishes. It encodes deep domain knowledge of phase space and dynamics in the form of constraints and procedures. However, unlike the hierarchical structure in a typical intelligent control system, the Phase Space Navigator consists of a global path planner and a local trajectory tracker; the two modules are integrated in a tight loop that allows control decisions to be revised more naturally.

8   Conclusions

We have developed and demonstrated an autonomous control synthesis method, the Phase Space Navigator, for synthesizing controllers for nonlinear systems in phase space. The technique applies to both finite-valued control systems and those with continuous parameter spaces. We have demonstrated our method on the automatic

---

[4]Our method differs from gain scheduling and dynamic programming techniques in the way the phase space is decomposed.

design of a high-quality controller for stabilizing a maglev vehicle. The simulation showed that our design tolerates much larger disturbances in the airgap than a classical linear design does. Other potential engineering applications include large flexible space structures, robot manipulator planning, satellite attitude control, and switching power regulators.

The Phase Space Navigator synthesizes global reference trajectories using knowledge of phase-space structures provided by the modeling and analysis program. The flow-pipe modeling of phase spaces provides a way to efficiently search for and reason about global structures of the phase spaces. The global nonlinear control synthesis becomes a graph search problem with this representation of the phase space. Using the technique of parsing a continuous phase space of infinite numbers of trajectories into a manageable discrete collection of flow pipes, the difficult task of synthesizing a nonlinear controller is formulated as a computational problem that requires a combination of substantial numerical, symbolic, and combinatorial computations and spatial reasoning techniques. Modeling fractal geometries and improving computational efficiency for high-dimensional systems remain as the topics for future research.

Automatic synthesis of control systems that employs latest computational technology to actively exploit nonlinear dynamics has demonstrated great potentials. We have primarily focused on the automatic synthesis of global control laws and have not addressed the important issues of modeling, sensing, estimation, and implementations. Intelligent systems armed with methods tackling all these issues can revolutionize the synthesis of high-performance nonlinear control systems.

9   Acknowledgment

The author is grateful to H. Abelson, A. Berlin, B. Chandrasekaran, M. Raibert, G.J. Sussman, G. Verghese, and D. Waltz for helpful discussions.

References

[1] H. Abelson, M. Eisenberg, M. Halfant, J. Katzenelson, E. Sacks, G.J. Sussman, J. Wisdom, and K. Yip, "Intelligence in Scientific Computing." *CACM*, 32(5), May 1989.

[2] H. Abelson, "Bifurcation Interpreter: A step towards the automatic analysis of dynamical systems." *Int. J. of Computers and Mathematics with Applications*, 20(8), June 1990.

[3] H. Abelson, A. Berlin, J. Katzenelson, W. McAllister, G. Rozas, and G.J. Sussman, "The Supercomputer Toolkit and its Applications," *Proc. of the Fifth Jerusalem Conference on Information Technology*, Oct. 1990.

[4] R. E. Bellman and S. E. Dreyfus, *Applied Dynamic Programming*. Princeton University Press, Princeton, New Jersey, 1962.

[5] E. Bradley and F. Zhao, "Phase-Space Control System Design." *IEEE Control Systems*, 13(2), 1993.

[6] E. Bradley, "Taming Chaotic Circuits." *Technical Report AI-TR-1388*, MIT Artificial Intelligence Lab, 1992.

[7] T. H. Cormen, C. E. Leiserson, and R. L. Rivest, *Introduction to Algorithms*. The MIT Press, Cambridge, MA, 1990.

[8] J. Guckenheimer and P. Holmes, *Nonlinear Oscillations, Dynamical Systems, and Bifurcations of Vector Fields*. Springer-Verlag, New York, 1983.

[9] C. Hanson, "MIT Scheme Reference Manual." *AI-TR-1281*, MIT Artificial Intelligence Lab, 1991.

[10] U. Itkis, *Control Systems of Variable Structure*. John Wiley, New York, 1976.

[11] A. Isidori, *Nonlinear Control: An Introduction*. Springer-Verlag, New York, 1985.

[12] R. E. Kalman, "Phase-plane Analysis of Automatic Control Systems with Nonlinear Gain Elements." *Trans. of AIEE*, 73(2):383-390, Jan. 1954.

[13] F. C. Moon, *Chaotic Vibrations*. John Wiley & Sons, New York, 1987.

[14] F. C. Moon and P. J. Holmes, "A Magnetoelastic Strange Attractor." *J. Sound Vib.*, 65(2), 1979.

[15] E. Sacks, "Automatic Analysis of One-Parameter Planar Ordinary Differential Equations by Intelligent Numerical Simulation." *Artificial Intelligence*, pp. 27-56, 1991.

[16] J. E. Slotine and W. Li, *Applied Nonlinear Control*. Prentice-Hall, Englewood Cliffs, New Jersey, 1991.

[17] J. J. Stoker, *Nonlinear Vibrations*. Interscience, New York, 1950.

[18] R. Thornton, *Electronic Circuits*. MIT course notes, 1991.

[19] K. M. Yip, *KAM: A System for Intelligently Guiding Numerical Experimentation by Computer*. MIT Press, 1991.

[20] F. Zhao, "Extracting and Representing Qualitative Behaviors of Complex Systems in Phase Spaces." *Proc. of the 12th Int'l Joint Conf. on Artificial Intelligence*, Morgan Kaufmann, 1991.

158

[21] F. Zhao, "Phase Space Navigator: Towards Automating Control Synthesis in Phase Spaces for Nonlinear Control Systems." *Proc. of the 3rd IFAC Int'l Workshop on Artificial Intelligence in Real Time Control*, Pergamon Press, 1991.

[22] F. Zhao and R. Thornton, "Automatic Design of a Maglev Controller in State Space." *Proc. of the 31st IEEE Conf. on Decision and Control*, Tucson, Arizona, December 1992.

[23] F. Zhao, "Automatic Analysis and Synthesis of Controllers for Dynamical Systems Based on Phase Space Knowledge." *Technical Report AI-TR-1385*, MIT Artificial Intelligence Lab, 1992.

[24] F. Zhao, "Computational Dynamics: Modeling and Visualizing Trajectory Flows in Phase Space." *Annals of Mathematics and Artificial Intelligence*, 8(3-4), 1993.

[25] F. Zhao, "Intelligent Computing About Complex Dynamical Systems." *Proc. 3rd Int'l Conf. on Expert Systems for Numerical Computing*, Purdue University, Indiana, May 1993.

7

MULTIRESOLUTIONAL ARCHITECTURES FOR AUTONOMOUS SYSTEMS WITH INCOMPLETE AND INADEQUATE KNOWLEDGE REPRESENTATION

A. MEYSTEL
Department of Electrical and Computer Engineering
Drexel University, Philadelphia, PA 19104, USA

1. Introduction

An overview of the area of Intelligent Control Architectures for robots can be a guide in existing achievements in the area of intelligent control since most of the efforts recorded can be identified with our desire to reproduce cognitive activities of human brain. This is why this area is sensitive and any statement in this area can have repercussions for a variety of scientific domains adjacent to cognition. One can see it in the polar stands on the issue of "behavioral vs hierarchical" architectures. We will show later that this issue is actually a non-existent one: it is actually an issue of the degree of distribution in a control structure ("centralized vs distributed" control could be a better formulation for the problem).

The decades of "behaviorism" in psychology led to a desire of definite scientific groups to interpret intelligence in a "stimulus-response" manner without visualizing an entity behind numerous experimental and theoretical results, an entity testifying for existence of a mechanism larger than just collections of look-up tables. This mechanism is a multiresolutional system of knowledge representation and processing which emerges as a result of interrelated processes of generalization, focusing attention, and combinatorial search. The advantages of multiresolutional architectures are broadly demonstrated in the literature for the existing scenarios of application.

We will show that the number of levels in a multiresolutional architectures can be selected in such a way as to minimize the complexity of the planning/control system. In this paper, the theoretical foundations are outlined of decision making in the class of control systems which allows for using nested multiresolutional representation, and nested multiresolutional algorithms of control processes. As a result, nested hierarchies of multiresolutional (multiscale, multigranular) control structures are generated. The core of the theory of nested multiresolutional control is based upon a concept of nested multiresolutional knowledge organization which enables efficient practice of design and control using nested search in the state space. So, existence of knowledge representation is becoming a major issue for the intelligent system. There are researchers that are trying to deny the need in knowledge representation, however, this knowledge turns out to be hidden in a different form and distributed all over the system.

At the present time, two major principles of building the control architectures are applied in a multiplicity of robot control systems: the "hierarchical", and the "behavioral" ones. The "hierarchical" are using "goals" and "tasks" decomposition, and the reactive principle is utilized in a form of feedback control loops compensating for imperfectness of our knowledge stored in the model of the World. "Behavioral" architectures rely on a society of free "autonomous agents" that are specialized each in a particular operation and provide their own self-control using just a reactive principle. However, together these agents are believed to be able to perform the whole job no matter what was the goal. There are systems that are very simple so that they do not require any hierarchy, need just one agent, and reactive

159

S. G. Tzafestas and H. B. Verbruggen (eds.),
Artificial Intelligence in Industrial Decision Making, Control and Automation, 159–223.
© 1995 *Kluwer Academic Publishers.*

control is the only proper solution for them. We will consider here a class of complex robotic systems that are expected to demonstrate a definite level of autonomy: autonomous service robot for flexible manufacturing systems, autonomous mobile robot for functioning in dangerous environments while performing maintenance and particular missions (rescue operations, supply, delivery, repair, etc.)

Hierarchical systems are actually based upon "heterarchical" world representation and control since they allude to a heterarchical information: structures are using more than one principle of classification, trees belonging to different principles of classification are "talking" to each other, etc. The hierarchical principle presumes dealing with a world representation which is constructed because of complexity reducing tools such as generalization, focusing attention, and combinatorial search (or GFACS). The hierarchical principle employs a "planning/control continuum" which presumes performing planning of the motion trajectory consecutively top down with gradual increase of resolution and decrease of the scope of decision making. At the very bottom of this multiplicity of planning we have a planning for one discrete of time ahead, i.e. we have an on-line execution control. The frequency of decisions increases top-down and at a particular level, the on-line feedback information is being introduced and the real-time feedback controllers operate at the bottom of planning-control hierarchy.

The behavioral principle is based upon the multi-agent scheme of operation when each of the agents performs a particular permanently required reactive (reflexive) mission such as "obstacle-avoidance", "edge-of-the-road-following", "striving-to-the-goal", etc. Multiagent behavioral principle is associated with the idea of "subsumption". The researchers in the area of behavioral robot control schemes claim that the multiagent subsumption based robots do not require any world representation, and the robotic operation can be based pure upon the sum of the multiagent reactive behaviors. In Section 9 we will find that some of the behavioral researchers admit that the reactive principle by itself does not work - no matter how many agents are used - unless there is no upper level planning performed on the top of the behavioral system. This merger of behavioral concept with the hierarchical ones is symptomatic: in fact we can show that at each level of resolution the process of feedback control can be arranged in a multiagent reactive manner.

This paper gives an overview of the results in both areas and compares the advantages and shortcomings of the two principles when they are applied independently. It is demonstrated that because of reliance solely on a top-down decision making, one can arrive with extremely computationally intensive systems. On the other hand, the desire to follow purely subsumption based controllers does not allow for any substantial increase in the complexity of tasks performed by the robot. On the other hand, it is very convenient to represent the problem of planning/control and the rule based controller in their feedback part as a multiagent reactive controller.

A combined architecture is described which takes advantage both of 1) the multiresolutional dealing with the world representation, and the independent pro-active decision making schemes accepted in the top-down hierarchical architectures,and 2) the computationally efficient behavioral approach employed in the existing multi-agent solutions. This architecture which we call "behavioral+planning" has a multiresolutional hierarchy of behaviors. It turned out that in the same way as we can build a hierarchy of goal-oriented top-down decision making processes, we also can build a hierarchy of behaviors. Unlike the task-hierarchy, the behavioral hierarchy is a bottom-up hierarchy, and it complements the pro-active top-down hierarchy by the reactive bottom-up functioning. It turns out that behaviors are formed at different levels of resolution that are determined by the generalization of sensor information based upon the experience of operation.

This architecture has been developed at Drexel University for a variety of applications. It has been industrially applied for a practical operation in robot control of spray-casting machine where three levels (both goal-directed top-down, and reactive behavioral bottom-up) are working together. Behavioral+planning architecture was theoretically analyzed in simulation of the autonomous vehicle which allowed for a number of interesting observations on the joint use of information at hand. Finally, we are developing an architecture for controlling a 6-legged robot which is expected to learn the hierarchy of reactive behaviors at 5 levels and also perform goal directed activities.

2. Architectures for Intelligent Control Systems: Terminology, Issues, and a Conceptual Framework

2.1 Definitions

Architectures for Intelligent Control Systems are becoming a focus of attention of a diversified scientific community as the interface between several domains of science: control theory with its subdomain "intelligent control", robotics, cognitive science, computer architectures, neural networks, and artificial intelligence. The variety of views, approaches, and terminology is so dramatic that we shall introduce a number of key definitions illuminating further materials of this paper.

Definition 1. Control.
Control is directing of a system to a preassigned goal.

This directing can be done both in an open-loop as well as in a closed-loop fashion. Open loop control presumes existence of a model of the system. The open-loop control assignment is called "plan". The process of finding this assignment is called "planning". Since the model is usually incomplete and/or inadequate, the closed loop controller is required for error compensation which uses a feedback. Thus, Definition 1 presumes existence of a goal, a model, a plan (open-loop control), and a feedback control law - all determined for a particular resolution of the control level.

Definition 1.1 . Open-loop Control (Feedforward Control, or Active Control). An output motion trajectory which satisfies the specifications, and the input variables trajectory which is supposed to produce this output trajectory if our knowledge of the system is adequate.

Definition 1.2 . Closed-loop Control (Feedback Control, or Reactive Control). A law of transforming the measured error (the deviation of the actual output from the desirable one) into an input variable which is supposed to cancel-out (eliminate, or reduce) the measured error.

Examples: Positioning ("pick-and-place" operation) has a final position preassigned. Knowing a model of the system we can contemplate what is the desirable motion trajectory, what is the open-loop control sequence that will drive the system along with this trajectory. It is presumed that the customer is interested to make the motion either minimum-time, or minimum losses of energy, or minimum energy dissipated in the actuators. We do not expect that one can require minimum error: real systems have usually the level of acceptable accuracy preassigned. Another example: tracking (following the output motion trajectory required): it is just a subset of the problem of positioning. In both examples, model is not known precisely. Thus, feedback controller is required to compensate for real deviations of the output trajectory from the computed one.

By definition, the closed-loop control is doomed to failure since the compensation will always be executed later than the measurement is performed. This is why we will receive a final error if we do not take appropriate measures. These measures can include for example, a development a new set of open- and closed-loop controller which will work with higher sensitivity (at a higher "resolution"). Another measure: combining the closed-loop controller with elements of the more sensitive open loop controller, and creation of the so called, predictive feedback controller.

Definition 1.3. Predictive Control (Active-Reactive Control). A law of transforming the measured error (the deviation of the actual output from the desirable one) into an input variable which is supposed to eliminate the output error and which is based not only upon our measurement but also upon our prediction about this error based upon the assumption that our knowledge of the system is adequate.

One can see that all these cases both pro-active and re-active principles has been employed. The first principle is based upon some (often rudimentary) knowledge of the model. The second principle requires constant measurement of the results of our motion. Our pro-active attitude can be extended to the degree of introducing learning: real motion results can be used to improve our knowledge of the model and thus, to make our subsequent cases of planning more adequate, and our further feedback operation more accurate (learning presumes identification and updating for the sake of adaptation).

The definition implies that the goal is preassigned. Of course, it will be necessary in the future to address the systems which produce their goals internally. However, we will discuss in this paper only a subclass of systems that are working under the goal externally formulated. It would be beneficial first to learn how to deal with this class of intelligent systems, and only as a next step to ascend to a more general class of them.

A terminological subtlety: the specialists in adaptive control often omit the problem of finding the "plan", i.e. input which will create the output identical to the "reference curve", or assignment. This omission became a habit because the problem is usually discussed at a very high level of resolution where "off-line" procedures can be dropped and the whole "control law" can be computed as a feedback control. In a vast multiplicity of problems such an omission is not a good solution, and the "off-line" planning can reduce substantially the computational complexity of control. Thus, planning as a preceding procedure to "execution" control is meant to be done in an open loop fashion while the "execution" closed loop process is dealing with the results of model uncertainties incorporated at the planning stage. If this approach is accepted, the problem of control allows for a recursive solution starting with a very low resolution and ending with a very high one.

Definition 2. Resolution.
Resolution of the control level is the size of the indistinguishability zone for the representation of goal, model, plan and feedback law.

It is important to notice that instead of the word "resolution" the following terms are used intermittently: accuracy, granularity, discrete, and tessellatum. In all cases we are talking about limit of details available at the level of resolution. The idea of resolution was neglected until recent developments in the area of intelligent control because in comparatively simple systems we can succeed in design in one-two steps using only the idea of "accuracy". In complex systems and situations, one level of resolution is not sufficient because the total space of interest is usually large, and the final accuracy is usually high enough. So, if the total space of interest is represented with the highest accuracy, the ε-entropy of the system (the measure of its complexity) is very high.

ε-entropy=log(total volume of space/elementary discrete of space). Thus, the total space of interest is to be considered initially with a much lower resolution. Only a subset of interest is considered at a higher resolution. The subset of this subset is considered with even higher resolution, and so on, until the highest resolution is achieved. This consecutive focusing of attention results in a multilevel task decomposition, and finding the intermediate (nested) plans at several resolution levels of the multiresolutional system.

Definition 3. Multiresolutional system.
Multiresolutional system is defined as a data (knowledge) structure for representing the model of the system at several resolution levels.

In a multiresolutional system, the active part of representation (currently used for control purposes) is repeated many times: as many as we have resolution levels in the system. Instead of the word "multiresolutional system", a word "heterarchy" can be used which is understood as follows: heterarchy - is a hierarchical organization of a heterogeneous information (knowledge). "Hierarchy" is a more general term, it can be related both to "homogeneous" and "heterogeneous" representations. There is a lot of confusion about these terms. For some people, the term "hierarchy" has a meaning of a "tree of subordination", or "tree of decomposition", or "tree of search". The hierarchies we are going to talk about are not homogeneous. So, better to call them "heterarchies". They are not just "trees": they have relational links at a level. This is why eventually we ended up with a term "multiresolutional systems": it will help to avoid many of possible confusions.

In order to construct a multiresolutional system of representation, the process of generalization is consecutively applied to the representation of the higher levels of resolution. Generalization usually presumes clustering subsets and substitution of them by entities of the higher level of abstraction. This is why instead of the term "resolution levels" we use sometimes an expression "abstraction levels" (which is the same as "generalization levels","granularity levels", etc.) When generalization is continuously done in the course of time-varying of variables, it becomes a key tool of learning.

Definition 4. Learning.
The process of generalization upon the statistics of the time-varying functions is called learning of a control system.

In order to satisfy our intuitions about learning, we should add that improvement of numerical data based upon these statistics is also learning: we will call it "quantitative learning". Learning is becoming more significant when not only numbers are changed in our representation, but the representation *per se*. Then we call in, "conceptual learning". Learning also presumes memorizing these new updated results).

Learning results in constant updating of the multiresolutional system of representation, and thus, in improvement of plans and feedback control laws. Quantitative learning reflects the updated information on numerical data: i.e. it leads to adaptation without changing the structure. In this sense, the process of adaptation (in *adaptive systems*) is based upon learning. Conceptual learning, reflects the changes in the structure of knowledge (vocabulary) which happens as a result of cumulative changes in collected information. Sometimes, changes in structure are required for "adaptation". Thus, conceptual learning can be also part of the process of adaptation[1] .

Definition 5. Intelligent Control.
Intelligent Control is directing to a goal of a complex system.

The word "intelligent" in this definition implies that we expect to achieve some resemblance to intelligence demonstrated by living creatures, primarily by the humans. The primary distinctive traits of human intelligence is our ability to generalize (G), our ability to focus attention (FA), and our ability to synthesize new combination in our search for the alternatives of solutions, i.e. to perform a combinatorial search (CS). Later we will link any intelligent activity to the sum G+FA+CS, or the GFACS package. The following attributes of Intelligent Control are known and usually can be demonstrated in all intelligent controllers: multiresolutional system of goals, multiresolutional system of model representation, multiresolutional system of plans, and multiresolutional system of feedback control laws. However, we would like to specify multiresolutional control systems just as a subclass of intelligent controllers since we should allow for a possibility that other tools of intelligence can be discovered later.

Definition 6. Multiresolutional Intelligent Control Systems.
Intelligent Control Systems which are directing a complex system to a goal by using multiresolutional information processing including organization of external information, knowledge representation, and decision making processes.

Some of the important properties of the 'intelligence' are implicit for the known hierarchical control structures , e.g. the property of nesting which holds not only separately for the structures of 'perception' (P), 'knowledge of the world' (K), and 'decision making' (DM), but holds for the control loops as a whole. Control layers (actually, "control loops') of the higher resolution are nested within the layers of the lower resolution recursively. Thus if the lowest level of abstraction (the highest level of resolution) sends its output directly to the actuators: subsystems that transform the commands from DM into actions changing the world, and thus leading to the changes of input information (from the sensors).

---

[1] Clearly, one should not consider "adaptation" and "learning" competing words. Adaptation presumes an assignment to maintain some kind of adjustment to particular environmental changes. To perform this assignment. one must learn (not only). Learning can be performed with no adaptation followed if it was not required.

We can write it down for the highest resolution level as follows:

$$(P; K; DM)_1 \longrightarrow A_1 \tag{1}$$

the arrow means 'sends its output to'; index 1 means the lowest level of abstraction.
Then the second level looks as follows:

$$(P; K; DM)_2 \longrightarrow A_2 \iff (P; K; DM)_2 \longrightarrow [(P; K; DM)_1 \longrightarrow A_1] \tag{2}$$

Comparison (2) with (1) shows that the whole loop (1) is playing a role of an actuator system for the second level . From the third level equation we can see that similar property holds:

$$(P; K; DM)_3 \longrightarrow A_3 \iff (P; K; DM)_3 \longrightarrow [(P; K; DM)_2 \longrightarrow A_2] \tag{3}$$

....

$$(P; K; DM)_n \longrightarrow A_n \iff (P; K; DM)_n \longrightarrow [(P; K; DM)_{n-1} \longrightarrow A_{n-1}] \tag{4}$$

where '$\iff$' means 'is equivalent', or 'can be interpreted as').

Maintenance of the multiresolutional system of representation is done by learning. Levels of resolution are selected to minimize the complexity of computations (minimizing the value of ε-entropy). Planning and determining of the feedback control laws is also done by joint using of generalization, focusing of attention, and combinatorial search: this will be demonstrated in this paper. Many of the existing systems demonstrate GFACS features and capabilities partially, or in full: fuzzy logic controllers are tools of generalization and focusing attention; neural networks are tools of generalization , focusing attention, and combinatorial search; combinatorial search has many particular instantiations: A-star, exhaustive search, complete, or approximate dynamic programming, etc.
In the meantime, each of the layers (P;K;DM) can be considered from the methodological point of view a set of procedures of generalization, focusing attention, and combinatorial search. We have mentioned already that unless these procedures exist - the whole phenomenon of intelligence doesn't exist. The whole phenomenon of 'supervision' does not exist without prior generalization. The whole phenomenon of task-decomposition task distribution, and coordination does not exist without focusing attention The whole phenomenon of finding solution does not exist without combinatorial search (the latter can vary from simple search in the table of precomputed mappings up to the exhaustive search in the space of units combined by a separate routine).
At this point we would like to give a definition for intelligence which would allow for better interpretation of the systems and phenomena which are described in this paper.

Definition 7. Intelligence.
Intelligence is a property of the system which emerges when procedures of focusing attention, combinatorial search, and generalization are applied to the input information in order to receive the output results. This property leads to a drastic improvement in the ability to deal with complex situations. Living creatures and machines with intelligence are dealing with complex situations with a substantially higher probability of success and survival.

One can easily deduce that once a GFACS package is defined, the other levels of the structure of intelligence are growing as a result of the recursion. Having only one level GFACS (because of the insufficient funds) leads to a rudimentary intelligence that is implicit in the air-conditioner (which has some very rudimentary intelligence). Having many levels of GFACS leads to a model of a powerful intelligence.

Definition 8. Incomplete and Inadequate Knowledge Representation.
The system has incomplete and/or inadequate knowledge representation (e.g. *a model*) if this representation does not allow for direct support of an algorithm of design and/or control within a conventional methodology of reasoning which assumes predicate calculus of the first order, differential and integral calculi, etc. It is due to this inadequacy and incompleteness, a variety of nonconventional techniques is required (including techniques of cognitive control and others).

2.2 Issues and Problems

We will more closely discuss the issues which have already been mentioned in Section 1 and in the beginning of this Section. Now we will address them in more detail.

Issue 1. Which one of the two major approaches should we favor: hierarchy or non-hierarchy?

Here we have to immediately ask ourselves: what is the problem to be solved by an intelligent machine we are designing now? Obviously, there is a definite problem (or a set of problems) to be solved, and/or a goal (or a set of goals) to be achieved. We would like to dismiss a non-goal oriented approach to this question. This paper does not consider machines that perform *aimless wandering*. It can be assumed that randomized motion might be of substantial importance when some problems are being solved (especially for robots-guards, and robots for reconnaissance). However, it won't be *aimless wandering*: we cannot substantiate a claim that there exist such an issue at least as far as engineering systems are concerned (however, probably aimless wandering does not exist in other areas either). In will be a kind of motion which the designer intends to obtain with the probabilistic parameters the designer expects to receive from the machine which is being designed.

Then the next question emerges associated with "hierarchies": should we mimic the nature or not? It would not be prudent *not* to copy natural systems. The effects of their operation often depends on the reasons we are presently unable to perceive and/or recognize. In [7] a behavioral architecture for a bird is described (p. 126) which obviously implies multiplicity of complicated learning processes which create a set of layers related to proprioception, then to senses, then to food recognition, then to rivals and mates recognition, then to more sophisticated behaviors. It is easy to recognize the hierarchy within the so-called subsumption architecture (examples of which can be found in recent robotic literature). Clearly, the layers of "entities" emerge as a result of a natural learning process, and they correspond to the object-oriented (entity-relational) hierarchies of world representation which are required for intelligent robots with sophisticated behavior.

So, the conclusion that can be made is: a sophisticated goal-oriented behavior in the incompletely known environment can be based only upon the hierarchical (more precisely: heterarchical, object-oriented, entity-relational) world representation, which can be learned gradually, or designed as a whole (with a partial learning required in the future), and which determines that the rule bases, control mappings, and behaviors generated will have a structure similar to the hierarchical (heterarchical) system of representation which presume the bottom-up clustering and the vertical links of inclusion, as well as relations at the level (horizontal links).

Issue 2. Should we design system with or without World Representation?

As we have stated earlier, and later we will add validity to this subject, no intelligent system can exist without World Representation. From the discussion of *Issue 1*, the learning process can be deduced which leads to the full fledged hierarchy (heterarchy) of World Representation. Another question is: how exhaustive should be this representation? Sure, this representation can be very modest especially in the systems with partial autonomy and teleoperated supervision. But still, representation will exist and it should be treated as a representation. At the level 1 we are dealing usually with a simple state space representation. The best way of treating this problem is to consider a system with learning. It is clear that a robot which learns from experience collects elementary facts, generalizes them into entities of the lower resolution, notices correlations among the implications, discovers the rules and also combines them into rules of the lower resolution. Thus it always ends-up with the hierarchical (heterarchical) knowledge representation.

On the other hand, the system of World Representation does not necessarily need to be concentrated in one particular place as a "lumped" hierarchy, or a homogeneous hierarchy (homorarchy as opposed to heterarchy). The hierarchical representation can be distributed over the multiprocessing system allocating the "islands" of representation in the vicinity of processor that will use them frequently (as connectionist structures usually suggest). However, whether lumped or distributed - this will be a system of representation and it should be treated as a system of representation.

Issue 3. Why multiresolutional representation?

At a particular level of sophistication, we are coming to the multiresolutional (multigranular, object-oriented, entity-relational) system of representation. Numerous examples of multiresolutional representation can be found in [43]. From our discussion earlier we found that these types of representation emerge in a natural way as a part of learning process. These types of representation are not a result of some arbitrary decision: to make the system multiresolutional, or entity-relational, or object-oriented. On the contrary, all these types of information organization emerge because they simplify dealing with information, make searches, and other computational procedures substantially more efficient. *Levels of abstraction (generalization) are introduced in order to make computation more efficient.* The validity of this principle is demonstrated in [18, 56, 95]. This principle is reflected both in the system of World Representation and in the system of Behavior Generation Algorithms (including execution control, compensation, planning, and so on), and also in Sensory Processing, e.g. signal detection, pattern recognition, etc.

Issue 4. What if we face a lack of knowledge: can we simulate in order to make a decision?

We always lack knowledge, and we always simulate to make a decision. Generation of alternatives for the subsequent decision making presumes the need in a simulation. Planning incorporates forecasting as a component, while forecasting always presumes simulation (think about predictive controllers). This simulation can have a form of building inferences based upon incompletely known premises. These inferences are actually predictions into the vaguely known future based upon partially known past. When these inferences are completed the solution can be found for the trajectory of motion to be executed.

These simulations can also have a form of playing ahead this vaguely known future based upon the same vaguely known past. There is no difference between inferring the control recommendations deductively, or (when deduction is difficult, or impossible) inferring them based upon comparison of different versions of processes reproduced by testing the model for the expected circumstances in the future. *Comparison of simulated versions of the future process is equivalent to deduction of the solution implicitly using judgments about nonsimulated future processes.*

If the processes in the system can be described analytically, the operator of the plant G can be often presented explicitly, and inverse G^{-1} can be obtained mathematically. Thus, the required input for the desirable output can be found using one of available techniques of inverting the output function to the input of the operator. This is how we operate at the level of execution controller design. Very often, only a set of general statements can be produced about the plant which allows for inverting them logically and thus, for deducing the required input which can generate the desirable output. This is how we usually operate at the level of planning the overall robot functioning.

However, also knowledge of the plant is very often presented in a form that allows neither for analytical nor for logical knowledge inverse: this form is complicated for doing elegantly analytical , or logical inverse. These are the cases when we employ so called "simulation": testing different alternatives of functioning which is becoming a basis of actual knowledge inverse. (The practical algorithm for doing this is described in [36, 37, 97]). It is possible to show that the validity of control simulation. It depends only on accuracy of the information at hand.

Issue 5. Should Planning/Control be Active, or Reactive?

Planning at all levels (including Control at the lower end) must be both: active and reactive. Open Loop Control, or Feedforward Control (OLC) which is based on our model of the system (including our expectation of the world for the particular planning/control interval) presents the "active" part of it. Closed Loop Control, or Feedback Control (CLC) which compensates for the lack of knowledge demonstrates how we "react" to real deviations from our expectations. On the other hand, one can say that when we plan the OLC operation we are actually computing our *"reaction" to the expectations* we have (or compute): so, active control is in a sense also a reactive one. In general, each of the modeled systems is working reactively with respect to the feedback and with respect to the goal. So, this is actually not an issue: when we use OLC and CLC together (the OLC/CLC concept discussed in Section 7 of this paper) this semantic quarrel is becoming unnecessary.

The practical issues of Architecture Design include determining for the hardware: how the individual actuators should produce the output, and how the individual sensors should be attached. The system of nested algorithms then is developed which are operating in parallel and communicate with each other top-down and bottom-up in the multiresolutional hierarchy.

Issue 6. Should the Hierarchy of Control be with top-down goal assignment and task distribution, or it should be with bottom-up behavior generation and synthesis?

This controversy always emerges when RCS/NASREM theorists of multiresolutional hierarchies discuss the architectural issues with proponents of the "multiagent behavior generation". It is possible to demonstrate that the latter leads to the multilevel aggregation of the multi-agent behavior into a hierarchy of behaviors with a single general vertex at the top which will be interpreted as a complex behavior leading to the same goal as the RCS/NASREM hierarchy.

The following subsection (2.3) suggests that these two different approaches (multiresolutional hierarchical and multiagent behavior-synthesizing) lead to the same control structures.

Issue 7. Should the systems be designed using the concept of autonomous agents?

We have presented an approach to the area of Architectures for Intelligent Control Systems (AICS) based on its multiresolutional representation. One can see that Centralization and Decentralization of control are implied by Aggregation-Decomposition, and are actually determined by the multiresolutional character of representation. However, one can choose to see the whole system of representation, or only a particular box in it. In the latter case one can contemplate an autonomy of the subsystem, reflecting the links with the rest of the hierarchy within formulation of the goal and function of the *autonomous agent*. This allows for talking about *behavior* of the autonomous agent and can be convenient in many cases. (We can talk also about external behavior of the system, its internal behavior, behavior of the subsystem, etc.)

The autonomous agents can be associated with the structural subsystem of the machine (like *propulsion, steering, braking*), or can be associated with the functions to be performed frequently (*following the edge of the road, following the wall, avoiding obstacle*). At the present moment it is not clear whether associating autonomous agents with the functions to be performed frequently is beneficial. Another example of introducing multiple autonomous agents is development of different personalities of the decision makers and having them make the decision together. See the description of a pilot with dual behavior in (see Section 10 and references [27] and [35]).

Descriptions of functioning called "behavior" can often be considered a convenient way of systems representation (more precisely, control systems representation)[2]. As any particular representation (with its particular language) it also tends to be multiresolutional when it grows. You can find in a book written 30 years ago (1960) by prominent behavior scientists that all structures of behavior are multiresolutional, that in the multiresolutional behavior structures we always are dealing with goal orientedness at the upper level and with reactive character of control at the lower level [86]. (Authors describe actions of a human hammering a nail, as an example).

Ten years later (1970) a similar multiresolutional structure of behavior was illustrated for animals [87]. Ten more years later (1981), this has been explored as a basis for robotic behavior representation by J. Albus [7]. Then again, several years later, there was an upsurge of research activities linked with a desire to find alternative ways of building the AICS-intelligence. Most of these alternative ways have been presented in 1990 [88].

The following issues are raised in these papers:

---

[2] One should not necessarily link any mentioning of "behavior" with "behaviorism" that has proven to be a narrow approach in psychology since it was based upon denial of the idea of "representation".

a) Intelligence does not need to have "an explicit" knowledge representation (or explicit "model of the world"). It can have an "implicit" knowledge representation implanted in the controllers of its subsystems. Clearly, we are talking about distribution of "knowledge", spreading the knowledge available all over controller components within the controller.

b) The subsystem controllers can be considered a separate relatively independent units ("autonomous agents") with goals moved down to these units. (Clarification: the units can be totally coupled and/or decoupled, they can have different "degrees of independence"). All these issues are being taken care of at the stage of "task decomposition" in its distributed reincarnation.

c) Since often there is no "explicit" knowledge representation and goal assignment, the system is operating based upon reactive (reflexive) activities of the autonomous agents, and its behavior will be OK. This is supposed to be done by shifting the reasoning activities to the level of "raw sensor data".

All these issues ascend to the problem whether intelligence (AICS-intelligence) is possible with no World Representation. We believe that this problem is an artificial one. Decisions are generated based upon Knowledge of the World. This Knowledge (World Model) can be concentrated within a particular (separate) storage, and/or can be distributed over the multiplicity of subsystems - in both cases it still remains a World Model. It can be stored in the computer memory, or allocated in the mechanical structure of the device - in both cases it still remains a World Model. It can be stored in advance for the subsequent use, or obtained immediately before the decision making is done (by sensing) - in both cases it still remains a World Model.

Issue 8. What kind of behavior should we prefer: reactive or active?

There is a semantic trap though in raising the dilemma "reactive vs goal-oriented" intelligence. We should rather talk about the different and more relevant dilemma "reactive vs active". The existence of this dilemma is overlooked. The stored decision tables are reactive, although the construction of can be based upon different degrees of intelligence. Indeed, all features of the scenario for AICS-intelligence were supposed to be represented in the development of this table. Does it yield "intelligent behavior"? This is more difficult to answer: it does if AICS-intelligence operates with its active GOAL--> HIERARCHY OF TASKS decomposition, with interpretation of sensor hierarchy, etc. It would be better to talk about "ON-LINE INTELLIGENCE vs OFF-LINE INTELLIGENCE". On the other hand, an interesting question can be raised[3] : is it possible to have a system reactive and yet taking in account on-line the goal? (Can you store all reaction rules for all unimaginable multiplicity of possible goals? Our experience shows that it can work in a limited set of circumstances [27, 35] but having the whole hierarchy of on-line decision making would lead definitely to a more consistent behavior).

It is clear that the purely reactive activities are insufficient for intelligent goal-oriented behavior. "Reactive" means a direct coupling between the sensory input and the resulting effector, based on a historyless (or almost historyless) transfer function. (Almost historyless means that a reactive system can have a little bit of state. It's hard to be strict about what exactly "a little bit" means, but basically it means that the system is not collecting its history into a model or using a model to do a forward projection. Of course, many tasks require keeping histories, maintaining internal representations, and occasionally even planning, i.e. a hierarchy of decision making. For those, reactive controllers do not suffice).

This is a very grim picture of a reactive controller. Even a simplistic and shallow PD-controller computes a prediction measuring derivative of the process at the "now"-point, and thus is trying to roughly predict what can happen at the next moment of time. Why should we bound ourselves by promise not even trying to predict? As soon as one linked himself (or herself) with a "reactive" concept he (or she) is not to predict anymore. Even within a framework of reactivity, one can make a look-up table storing responses to both signal and its derivative in a point. I even think that it is more biologically justified.

Purely reactive behavior of different degree of sophistication has been analyzed in depth (see [27, 35]) for the case of two autonomous agents acting cooperatively ("precise planner", and "instantaneous planner"). From the example of a robot with dual behavior (see Section 10), one can see that it results in

---

[3] The question was raised by M. Kokar (Northeastern University).

a diversified robot behavior. The trajectories seem to look "intelligent" because of their relative unpredictability. However, this is exactly why Turing test is improper tool of evaluating intelligence: the superficial sophisticatedness of behavior can easily deceive a viewer. The pure reactive systems pretend to be intelligent: they are not.

Issue 9. What should we do: fit, satisfice, or optimize?

On the contrary, more "mundane" problems turn out to be sadly omitted in the recent AICS research. Researchers and designers are often not concerned with optimization of control; optimization seems to many of them (apparently) a problem of far away future for the "autonomous agents". Fitness to the set of specifications can be performed only in the terms of being within constraints which leads us to a so called "satisficing" systems (the term was introduced by H. Simon). However, the reality of engineering problems is in minimization of a cost-function (or a cost-functional). Even in a single-criterion case the problem of optimization is carefully avoided in the domain of intelligent machines - it is not an easy problem to solve.

Indeed, optimization is a doubtful issue and uneasy task especially for a friendly community of autonomous agents: if they start thinking about optimization they might become engaged eventually into a struggle for existence - then what will happen to their cooperation? Arbitration is a good idea even for complex Pareto cases - however no existing system with arbitration is known. Frequent assumption is that only simple control systems are concerned with optimization. "Complex" controls are assumed to be hard to optimize, their optimization can be vaguely presumed in their *mode of operation*.

One can imagine control solutions which are concerned with the multiplicity and complexity of the modes of operation, and yet - tend to be optimal. By postulating a rule: if the control should function in complex modes one does not need to optimize it, we won't do any good to the emerging technology of AICS. I would better postulate: even if your control should function in complex modes you must try to optimize it. I do not see any obstacles in doing this. Does it mean that these problems should not be addressed by control specialists? They simply do not know yet about new problems that arise in the AICS area.

Issue 10. Does functionality depend on degree of distribution?

A number of researchers concentrate on forming behaviors of the autonomous robots by synthesizing autonomous agents which are single behavior oriented units. A major difference of these architectures is in the key idea (as P. Maes put it in [88]): "...of emergent functionality". The functionality of an agent is viewed as an emergent property of the intensive interaction of the system with its dynamic environment. We already saw that functioning by itself may seem to look pretty intelligent which does not mean that real concern of the goal achievement is behind the behavior (See the example of ID+AP induced behavior of a robot in Section 10). The specification of the behavior of the agent alone does not explain the functionality that is displayed when the agent is operating. Instead the functionality to a large degree is founded on the properties of the environment" (see [88], p.1). This "environment induced" functionality seems to me a pretty passive functionality, sorry about that. The role of the goal is becoming reduced within this type of approach. So, as a result we are floating as a piece of driftwood on the waves of environment...

The problems "hierarchy vs subsumption", or "behavior vs hierarchy" is rather a problem "lumped vs distributed" (or centralized vs decentralized) which is hardly a problem for the control theory. Indeed, the decentralized control is a centralized control with zero, or very small off-diagonal terms. What is actually being proposed: to use controllers with distributed rather than with lumped representation of knowledge, with distributed perception, and distributed decision making processes! Is this not contradictory to the multiresolutional control? On the contrary, the higher the degree of distribution is the bigger are the advantages that hierarchies bring in (complexity-wise, time-wise, and many other things-wise). The canonical NASREM (NIST) diagram with its "PERCEPTION-WORLD MODEL-DECISION MAKING" boxes still exists in the same sequence, just in a distributed form.

It is not difficult to demonstrate that behavior-based approaches merge easily with the multiresolutional model. One of the ways has been described in the previous couple of paragraphs: each level of the multiresolutional controller is built in correspondence with the hierarchy of the language of behavioral representation accepted by a particular designer. Another is presented in the R. Arkin's paper in [88] where uses directly Planner-Navigator-Pilot hierarchy (which is being persistently recommended since 1983 [18, 89]) and attaches to it the distributed reactive execution controller. These ideas are very appealing of merging the hierarchy with the behavioral concepts. As a matter of fact, this is a more natural way rather than making all levels artificially behavioral. Indeed, the higher the level of abstraction is, the more natural is the lumped solution.

We believe that the contradiction "behavior" vs "hierarchical" architecture is an artificial one. The hierarchy of representation will exist in any more or less intelligent system because complexity in AICS is fought by building a multiresolutional representation with the help of GFACS tools. This multiresolutional system of knowledge representation can be lumped in a single "knowledge base" or distributed among other subsystems - this does not change neither the principle of knowledge representation nor the very fact of building the system of representation. And functioning of robots with AICS of different type will be judged upon their behavior and the degree of intelligence in it.

2.3 Conceptual Framework for Intelligent Systems Architecture

A general conceptual frame for the Intelligent Systems Architecture can be described as follows.Functioning of the Intelligent System is presented in a form of the "behavior[4] generating operator":

$$B \longrightarrow S_{minF}(T, R) \tag{5}$$

where   S- is the operation of *search* in its broad interpretation ("find the solution, e.g. find the behavior which satisfies the given and emerging set of conditions")[5] . The subscript minF means that a cost-functional F should be minimized (e.g. time of operation, or losses of energy). However, instead of minF we can request for the "satisficing constraints": e.g. $A_1 \leq F \leq A_2$.

B - is a behavior of the system understood as activities at the output of the system; they can be represented, or executed,

T - is a task which is usually formulated as follows: find and execute a particular set of the "behavior satisfying conditions (BSC)" which include statement of the result of behavior (a goal), and a list of constraints which should be satisfied until the result is achieved,

R-is representation of the World Model in one of the available forms, e.g.
- state space variables (convenient for dynamical models including analytical ones, computational algorithms, and simulation),
- object-oriented programming (convenient for dealing with images, fuzzy logic control, task distribution),
- Neural Networks, and other connectionist methods of representation, etc.

Representation of the World Model incorporates knowledge available about the system, and the environment in which this system should operate. This knowledge is insufficient (does not contain everything we need) and excessive (contains a lot of unnecessary information). Only after completion of design and beginning of functioning, one can thoughtfully decide the real measure of insufficiency and excessiveness in representation. One of the factors which determines this measure is the resolution of representation. Usually systems of representation contain representation at different resolution levels.

---

[4] At this point we can switch from the intuitive understanding of the word behavior to a more rigid definition. *Behavior* is defined as mapping the assignment, structure properties, specifications of the machine, and properties of the environment into the *appropriate functioning of the machine*. The word "appropriate" alludes to the definition of "intelligence" from [98]. Indirectly, this suggests that using the term "behavior" for the "output of the system" implies some degree of "intelligence".

[5] Search can be done by a multiplicity of methods, e.g. by browsing within a library (or within a "look-up-table"), by synthesis of a combination minimizing the "estimator of goodness, for example in a form F=g+h+c (g- is a cost of behavior until "now", h- is an expected cost of behavior from "now" until the end of operation, c- is cost of computation).

This algorithm for behavior generation can be introduced at any level i of resolution. Thus, two additional factors should be taken in consideration: resolution of representation with accuracy ρ_i and time scale at the particular resolution level τ_i, and the behavior generating algorithm can be written as follows:

$$B_i \longrightarrow S^i_{minF}(T_i, R_i, \rho_i, \tau_i) \tag{6}$$

which can be interpreted as follows: at a particular level of resolution with accuracy of representation ρ_i and time scale τ_i behavior of the system is found by using algorithm S^i over the representation R_i under particular task T_i and requirement to minimize the cost-function F. This means that the nature of representation, task formulation, and even the algorithm of behavior generation can be tailored to fit the particular level of resolution. As a result we receive a *set of behaviors* which can be interpreted as follows:
-at the highest resolution level 1 (the lowest level of abstraction) the output behavior (plan) is presented as

$$B_1 \longrightarrow S^1_{minF}(T_1, R_1, \rho_1, \tau_1) \tag{7}$$

where S^1_{minF} is the algorithm of planning;
....
-at the lowest resolution level k (the highest level of abstraction) the output behavior (motion trajectory) is presented as

$$B_n \longrightarrow S^n_{minF}(T_n, R_n, \rho_n, \tau_n) \tag{8}$$

where S^n_{minF} is the algorithm of planning.

This framework allows for describing all Architectures under consideration in a unified comparable form.

3. Overview of the general results.

Intelligent control systems in robotics broadly employ the concept of control hierarchy. Control hierarchies that came from the 60-s [1-3] were based on the idea of system partitioning. All components of GFACS are never mentioned but always implied. G. Saridis' conceptual snapshot of the situation in the area of hierarchical control [4] reveals some of the major features typical for the hierarchical control systems: controller at the top of the system, controls the process as a generalized whole, control devices at the bottom focus their attention and control the subprocesses at a high resolution, the latter should be coordinated. On the other hand, controller at the top is imprecise, it deals with the process at the level of linguistic descriptions; controller in the middle is more precise, but it is still a fuzzy controller; controllers at the bottom have the required precision.

J. Albus noticed that the structure of a hierarchical controller is similar to the structure of brain functioning, and that the hierarchy is generated as a result of "task decomposition" with a gradual focusing attention upon smaller and smaller details [5]. G. Giralt, R. Sobek, and R. Chatila are applying the task decomposition to the problem of mobile robot control [6]. It becomes clear that a hierarchy of functioning evokes not only a need in hierarchical decomposition of tasks, but also a hierarchical decomposition of maps (representations). J. Albus outlines for the area of robotics [7] the structures of brain functioning/hierarchical control as the three interacting hierarchies of task decomposition, world model, and perception (heterarchical control structure). Motivated by these developments A. Meystel [8] proposes a control architecture "Planner-Navigator-Pilot" for robots in which three levels of resolution exist explicitly with a combinatorial search performed at each of them. This architecture is dominating the area in the 80-s (see applications in 12-14, 17, 19, 20, 24). In this architecture not only a planning is performed at each level of resolution but also a feedback compensation as soon as corresponding data arrive. This feedback compensation at high levels of abstraction (low levels of resolution is called "replanning").

G. Saridis arrives with the principle of increase of intelligence with reducing precision bottom-up in the hierarchies of control [9]. It becomes clear that there are some general properties of knowledge processing in the control hierarchies, and that these properties are not determined by the phenomenon of system partitioning: they rather imply partitioning of representation which happen at the highest levels by the laws of linguistics [10], at the middle levels by the laws of fuzzy control [11], and they allow for integration of upper level with the lower ones [12-14]. A hypothesis is proposed [15] that control commands can be obtained for all levels as a time-tagged hierarchies of actions (procedural knowledge) which can be obtained by a corresponding processing of the snapshots of the World (declarative knowledge). Different strategies of mathematically rigid controllers are proposed [16-17], and eventually, a sketch of the theory of nested multiresolutional control appears in 1986 [18].

Known applications are related to the areas of autonomous and teleoperated robots [19-24, 27-32] as well as for the are of material processing [36, 37]. In the meantime the structure of the theory is becoming more clear [25, 26, 33-35] as well as the problems that should be solved. This paper is a further development of earlier papers [18, 26, 29, 30]. It formulates theoretical methods of design and control in systems which allow for multiresolutional world representation and nested decision making. Motion planning, and motion control which are usually treated separately, are becoming a continual process in this approach ("planning/control continuum"). Theory of joint planning-control systems and processes ascends from the theory of decision making applied to control systems. In all cases, the goal is assumed to propagate top-down in a form of task decomposition and distribution.

The ideas of nested hierarchical (multiresolutional, multiscale, multigranular) control are deeply rooted within numerous efficient mechanisms of knowledge representation.Hierarchies of 60-ties [1-3] were focused upon as an organizational tool, and M. Minsky's "frames" (1975) can be considered the first explicitly discussed generator of nested knowledge [38]. Broadly utilized in the practice of programming as a part of LISP, nesting became also an important tenet of the so called "entity-relational approach" in the database area of computer science. B. Mandelbrot announced that the Nature as a whole is built upon "fractally hierarchical patterns" (see p. 93 in [39]). Mathematical treatment of nested representations was explored by H. Samet during 80-ties (collected in [40,41]). Nested Hierarchical (multiresolutional, multiscale, multigranular) representation has generated a rich flow of research results in the area of vision (see a collection of works [42] and a bibliography in [43]). S. Tanimoto, L. Uhr, and others often use a term "pyramidal" for Nested Multiresolutional representation [44, 45].

Nested multiresolutional algorithms were introduced in the area of computational mathematics as "domain decomposition methods", or "multigrid methods" [46-48]. Hierarchical Aggregation of Linear Systems with Multiple Time Scales was discussed in the paper of the same title [49]: this was a thorough mathematical treatment of Nested Hierarchical Markov Controllers. Multiscale statistical signal processing was recommended in [50]. Recently, an effort to formulate a Mustiscale Systems Theory has been done [51, 52].

One can see that most of the research results are related to development of models and to signal processing. Among the early papers related directly to multiresolutional (multiscale) controllers we can mention only [18, 21, 53, 54]. Strong connection of nested hierarchies of representation was early appreciated by the researchers (see survey [55]). An assumption was proposed in [55] called the *time-scale separation hypothesis*, which stated that some of the motion trajectories can be considered independently. An important problem was raised about controlling a class of systems which is represented inadequately: the models are known but we have no means of determining how adequate these models are. This is a frequent situation, and most of the research thrust of this paper is directed toward this type of systems: with incomplete and inadequately represented information. We will try to equally reflect the tendency to visualize a possibility of a general theory of Nested Hierarchical (multiresolutional, multiscale, multigranular, pyramidal) control, as well as to be perceptive to opportunistic domain decomposition, local techniques oriented toward computational efficient schemes. We believe that the balance between these two tendencies can be especially fruitful in the engineering practice of design and development.

4. Evolution of the Multiresolutional Control Architecture (MCA): Its Active and Reactive Components

4.1 General Structure of the Controller

Any machine and/or technological process can be easily identified with Figure 1, a where the following three parts can be distinguished:

A - a source and a storage of the World Model: it contains all Knowledge necessary for modeling the operation as well as the means of communication to acquire this knowledge from the external source (say, an expert, or a collection of models).

B - a computer controller which contains all necessary means for processing information delivered by sensors from the machine, evaluate the situation, and compute plans and immediate commands for controlling the machine.It organizes and stores the newly arrived information, it identifies the entities of the World to be controlled, it compares the assignment with the current situation and outlines the output to be obtained in order to achieve the desired goal of operation.

C - the machine performing the process of interest with actuators that transform control commands into actions, and with sensors that inform the computer-controller about the process. Figure 1, b shows a simplified version of the diagram from Figure 1, a.W - is World, or the process to be controlled, S - a set of sensors, P-a system for dealing with sensor information ("perception"). K - a system of Knowledge representation, interpretation, and analysis, P/C- a subsystem for planning and control which determines the required course of actions, and A - a system of actuators which introduce the desired changes into the World. We will call this simplified version of the information flow structure: *a six-box diagram.*

Let us consider a technological example illustrating that the system is perceived by a designer (or by a control engineer) at multiple levels of resolution. Autonomous operation of a mobile robot (AMR) invokes several different operational paradigms reflected even at the early stage of "customer specifications". Let us imagine a possible scenario of operation exemplified in the specification.

"The following scenario is to be used in the final test after AMR manufacturing:

• A map of the region is given to the robot (the interface is subjected to subsequent negotiations).

• The starting position and the final position (destination of arrival) are marked on the map as zones.

• The map contains representation of the environment in the terms of permanent features of significance (roads, woods, lawns, bridges, buildings, fences, etc.) as well as the objects and sites which existence can be evaluated by a probability measure (marshes subjected to a possibility of totally dry out as well as to increase after a prolonged period of precipitation), construction sites, etc.

• The map does not contain representation of the objects which are significant but not permanent by their nature (parked trucks, road repair sites, etc.), and the objects which affect the motion process insignificantly: bricks, stones, logs, puddles, etc. All these objects should be identified by AMR if necessary and taken in account in motion control.

174

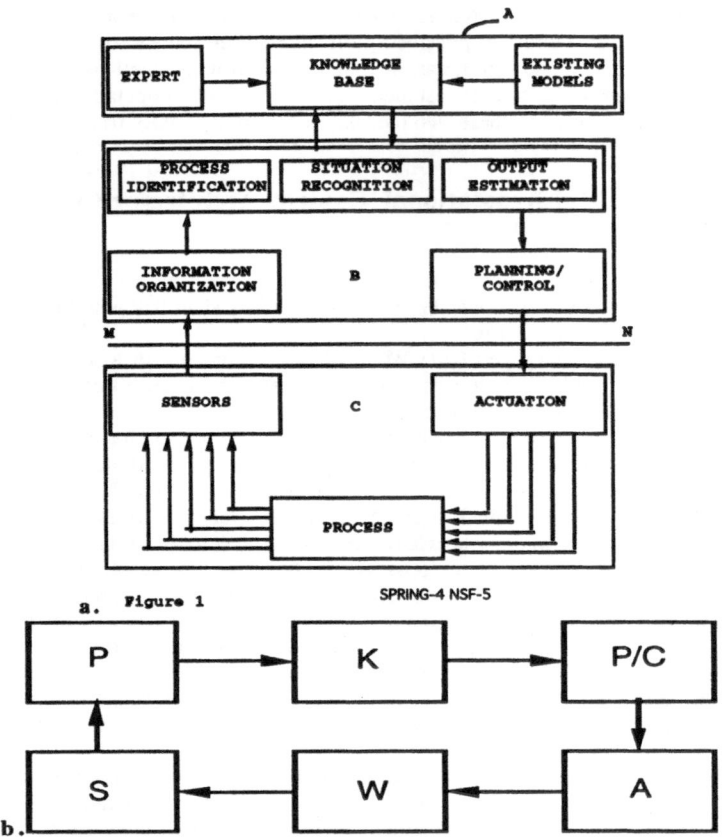

a. **Figure 1**

SPRING-4 NSF-5

b.

Figure 1.General Model of the Machine with its Computer-Controller

• If map should be updated, AMR should be capable of doing this.
• Having the map, and the ability to perceive and identify all above objects, AMR should be able to plan its motion, and to execute the plan so that to achieve the goal in minimum time, and return back.
• During the returning part of the mission, the map might be different (because of updating), so replanning should be performed.
• Moving obstacles should be dealt with according to the expert determinations collected in the set of rules, so that the overall goal of motion would not be seriously affected.

This example is described in many sources (see [18, 20, 24, 26, 56]). One of the common models of dealing with this example is a commonsense model Planner-Navigator-Pilot. It is supposed to deal with the assignment as follows:

1. UPPER LEVEL:PLANNER
   • CONSIDERS ALL AREA AT LOW RESOLUTION AND DELIVERS A ROUGH PLAN.
   • PERFORMS A ROUGH PLANNING OF THE TIME-PROFILES OF INPUT VARIABLES
     WHICH ARE SUPPOSED TO ENSURE THE DESIRABLE OUTPUT TIME-PROFILES.

2. MIDDLE LEVEL 1[6] : NAVIGATOR
- COMPUTING MORE PRECISE TRAJECTORY OF MOTION TO BE EXECUTED,
- DETERMINES MOTION COMPENSATION AND REFINEMENT OF THE INITIAL
  PLAN IF THE PROCESS DEVIATES FROM THE PREPLANNED INPUT/OUTPUT
  TIME-PROFILES AND/OR THERE ARE OTHER INDICATIONS THAT THE PROCESS
  SHOULD BE INTERFERED WITH.
3. LOWER LEVEL: PILOT
- DEVELOPS ON-LINE TRACKING OPEN-LOOP CONTROL TAKING IN ACCOUNT
  DEVIATIONS FROM THE EXPECTED SITUATIONS WHICH CAN BE OBSERVED
  ONLY IN THE IMMEDIATE VICINITY.
4. THE LOWEST LEVEL: EXECUTION CONTROLLER
- EXECUTES THE PLANS AND COMPENSATIONS COMPUTED
       BY THE UPPER AND MIDDLE LEVELS, AND THE PILOT, TRIES TO PROVIDE
       ITS ACCURACY.
One can see that the concept of this controller is equivalent to the "PLANNER-NAVIGATOR-PILOT" concept known from [8, 19, 26, 55]. This example demonstrate a 3-4 levels multiresolutional control architecture (MCA).

4.2 Multiresolutional Control Architecture (MCA)
The MCA diagram is shown in Figure 2. The flows of information between the adjacent levels are demonstrated by inclusion of all P, K, and P/C elements into unifying boxes.The following properties are characteristical for MCA.

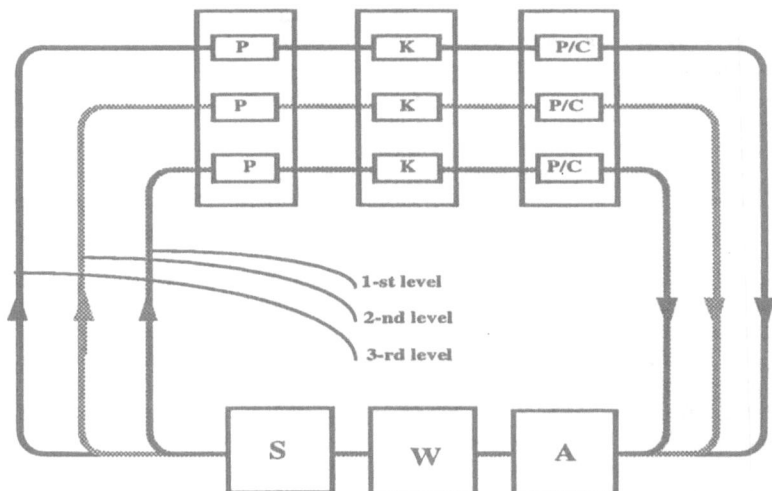

Figure 2. MCA: each level has its own feedback loop:
the lower levels are lumped into the "execution part"

•*Property 1. Computational independence of the resolutional levels.* Each of the loops in Figure 2 can be considered and computed independently from others. Each of them describes the same control process with different accuracy and different time scale which entails the difference in the vocabularies of levels.

---

[6] Later we will discover that the number of resolution levels varies depending on the complexity of the system. Therefore, one can expect having more than 1 "middle level".

176

•*Property 2. Resolutional level represent different domains of the overall system processes.* Since all loops are performing the same operation at different resolutions they are dealing with different subsets of the World (starting with the "small, fine grained, and quick" World at the bottom, and ending with "large, coarse grained, and slow" World at the top).

•*Property 3. Different levels of resolution are dealing with different bands of frequencies within the overall process.* The resolution of the level is associated with the frequency of sampling which not only mean that the frequent sampling is associated with the higher accuracy of the processes representation, but also that the frequencies of the process which are lower than the frequency of the sampling are not likely to be reflected in the control processes of this level.

•*Property 4. Loops at different levels of resolution are 6-box diagrams nested in each other.* Loops are nested one into another (see Figure 2), the lower resolution loops presume a possibility of refining representation of their processes by using the higher resolution loops. Each of the loops contains Perception, Knowledge Representation, and Planning/control subsystems with the external World attached to them via Actuators and Sensors. In the meantime, they operate with different *scope of attention* : each process at higher resolution has a scope of attention narrower than the adjacent level of lower resolution.

•*Property 5. The upper and the lower parts of the loop correspond to each other.* The upper part of the MCA (P, K, P/C) corresponds to the lower part (S, W, A). The hardware realities of S-W-A are represented in computer architecture as P-K-P/C.

•*Property 6. Behavior of the system is a superposition of behaviors generated by the actions at each resolution level.* Action of the system is being generated simultaneously at several levels of resolution (granulation), e.g. if the teleoperated, or an autonomous mobile robot is considered then the list of levels bottom-up will be: 1) output motion level, (the lowest abstraction level, or the most accurate level), 2) maneuver level, 3) plan of navigation level, 4) scenario of the operation level, 5) mission planning level (the highest abstraction, or the lowest resolution level).

•*Property 7. Algorithms of behavior generation are similar at all levels.* All levels execute a particular (pertaining to the level) algorithm of finding the best set of activities (control trajectory); each higher level constitute prediction for the each lower level. At each level the action generating algorithms should perform: ASSIGNMENT GENERATION, PLANNING, PLANT INVERSE, DECOMPOSITION, COMPENSATION, and EXECUTION COMMAND GENERATION.

•*Property 8. The hierarchy of representation evolves from linguistic at the top to analytical at the bottom.* Dealing with these several levels of planning/control process presumes a nested system of representations which can be analytical at the level of high resolution, and linguistic (knowledge based, or rule based) at the level of low resolution.

•*Property 9. The subsystems of the representation are relatively independent.* This means that distributed control solution are not inhibited, and within each particular level a multiplicity of control agents can exist: it does not contradict the overall structure of the system and its knowledge representation. The principle can be formulated also as follows: "First formulate the task of a level, then decompose it for a performance within the level"

The example with autonomous mobile robot has a fundamental significance for the approach to design and control MCA (55). If we consider a state space of the particular process (however complicated this process could be) the goal of control can be formulated as arrival from the initial point (state) to the final point (state) in this space. Thus, the moving point can be identified with an imaginary little robot traveling from one location to another, avoiding obstacles of the disallowed zones, preferring low cost areas to the high cost areas, accelerating and decelerating, etc. Thus in each machine and/or process the levels of multiresolutional control can be similar to what we stated for the mobile robot: 1) output motion level, (the lowest abstraction level, or the most accurate level), 2) maneuver level, 3) plan of navigation level, 4) scenario of operation level, 5) mission planning level (the highest abstraction, or the lowest resolution level). As one can see, this list fully apply to the general problem of traveling within the state space.

Before we are able to treat the system in Figure 2 as a system of control we have to make several transformations. From Figure 2 one can see that all three control loops merge in order to enter the system to be controlled. This requires for a conceptual leap be made: these three control loops can be considered independent loops as demonstrated in Figure 3, a. The Word Model performs the superposition of the control loops working simultaneously: they can be mutually dependent, or independent, the nature of superposition does not change. At the next step the World Model is being decomposed in three different submodels each working within its own loop: the reality of the system to be controlled can be visualized as if three separate subsystems exist to match their control levels (Figure 3,b).

a) Step 1   b) Step 2

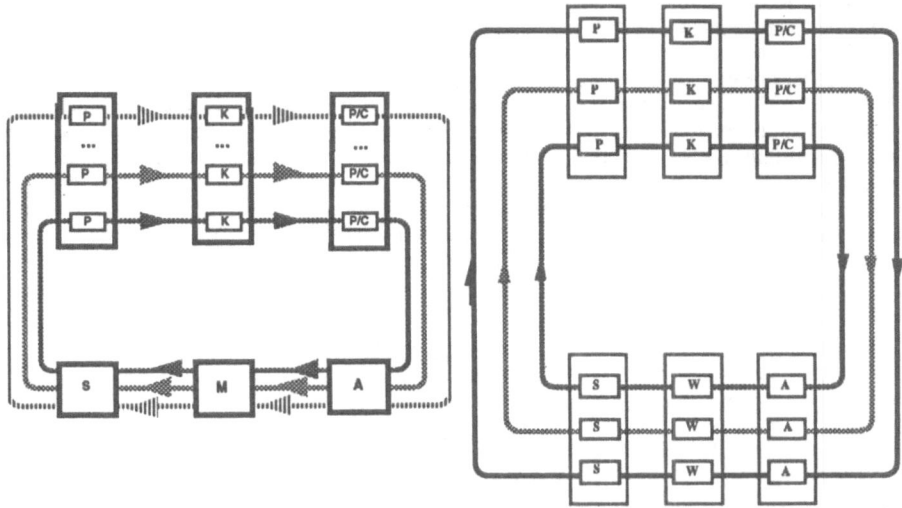

Figure 3. Multiloop Multiresolutional Controller

It is a convenient way to model the World and to arrange the computer controller as if they can be "component-to-component" mapped one into another. After the Step 2 is done we are able to deal with several World Models (as many as we have levels of resolution in MCA). Each WM actually exists for the observer and interpreter associated with a particular level of resolution. Each control loop can be driven by a multiplicity of control agents; its control can be centralized, or distributed.

Usually, the multiresolutional hierarchical systems are convenient as a form of organization for large systems which contain a number of goal-seeking decision units (subsystems). The problem usually include coordinating the actions for optimizing the process of goal achievement (see [57]) .

5 . Nested Control Strategy: Generation of a Nested Hierarchy for MCA

5.1 GFACS Triplet: Generation of Intelligent Behavior

We spoke already about the procedures prerequisites of the intelligent behavior: Generalization, Focusing Attention, and Combinatorial Search. Multiresolutional Consecutive Refinement and Centralized/Decentralized Search in the State Space are the primary algorithms applied in the Intelligent machines and they are based on GFACS. Search in the state space (see [32, 36, 37, 58]) is done by synthesizing the alternatives of motion and scanning the set of available alternatives. One of the

178

strings is selected when the desirable property is met. The vicinity of the solution is considered at the adjacent higher resolution level where the search is executed only within the vicinity. It is not a method of centralized control: it is a general technique which is applicable also for decentralized solutions. Indeed, even if one selects a hierarchy based not upon structural subsystems but based upon functional "behaviors" of "autonomous agents" the latter are not supposed to be doomed to reactivity: each of them is capable of searching is his (or her) own state space.

The concept of the multiresolutional consecutive refinement (Multiresolutional Search in the State Space, or MS^3-search) can be introduced as shown in Figure 4. Clearly, all basic principles characteristical for AICS-intelligence are employed in the diagram. A triplet of consecutively performed operations of "focusing attention", "combinatorial search", and "generalization" is performed consecutively with increasing of resolution at each repetition of the triplet until the resolution of the level is not equal to the accuracy of the decision required. (This triplet is characteristical for all algorithms implementable in the systems of control for intelligent machines).

The process of planning starts with *focusing of attention* which is selection of the initial map with its boundaries. *Combinatorial search* is performed as a procedure of choosing one string (minimum cost) out of the multiplicity of all possible strings formed out of tiles at this particular level of resolution. *Generalization* is construction of an envelope around the vicinity of the minimum cost string. This envelope is being submitted to the next level of resolution where the next cycle of computation starts. Focusing of attention presumes proper distribution of nodes in the state space so that no unnecessary search be performed. Combinatorial search is forming the alternatives. Generalization is generation of the map for the subsequent search at the higher level of resolution.

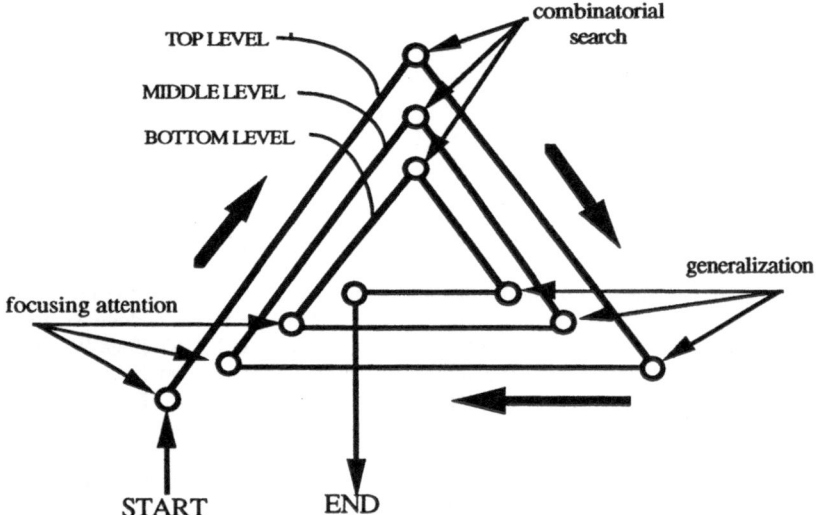

Figure 4. Conceptual Structure of the Multiresolutional Consecutive Refinement

5.2 Off-line Decision making   procedures of planning-control in MCA

In the most general form, the controller can be represented externally as a box with three inputs, and only one output. Internally, this controller can be a set of agents with some distribution of the overall job among them - it does not affect the algorithm of decision making. These inputs can be specified as follows (see Figure 5,a):

-Task: the goal (G) to be achieved, (and conditions to be satisfied including the parametric constraints, the form of the cost function, its value, or its behavior).

-Description of the "exosystem", or map of the world including numerous items of information to be taken in account during the process of control; map of the world (M) is often incomplete, sometimes, it is deceptive.

-Current information (I) is the information set delivered by the sensors in the very beginning of the process of control, and continuing to be delivered during the process of planning/control operation.

The processes within the controller are illustrated in Figure 5,b. Input T (Task) determined the points SP "starting position") and G ("goal") which are situated within the input M (Map). Input I (external information) works for a limited zone (the vicinity of SP), and the information set in this zone is more reliable, or even

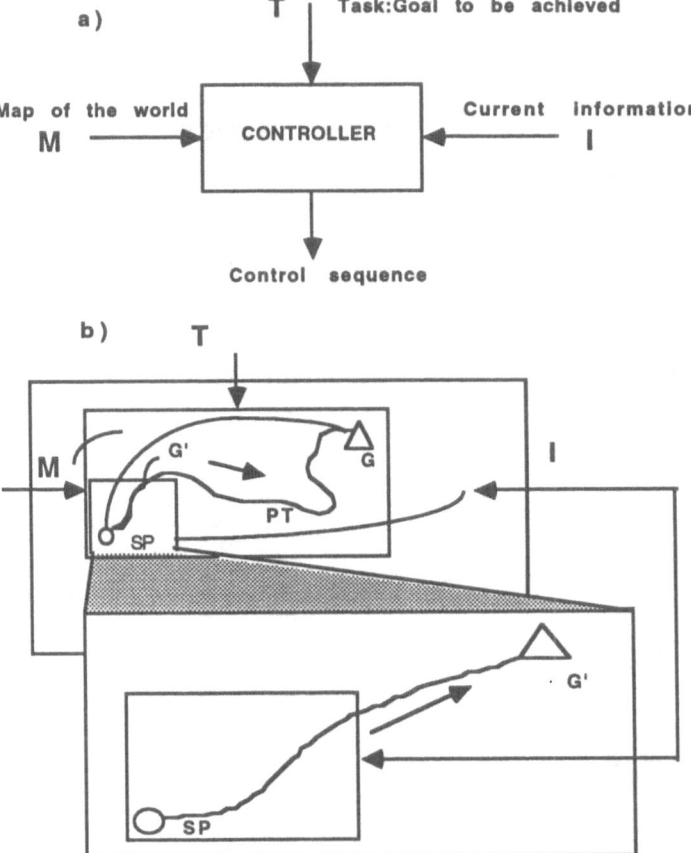

Figure 5. Planning/control subsystem and its off-line processes of decision making

exhaustive.Thus, the part PT* of the overall planned trajectory PT, can be determined with higher accuracy. In other words, our selection of PT can be changed in the future if the input I updates the map M in the way that PT will not be the best trajectory. However, the PT* part of the plan won't be changed: no new information is expected.

The need in "tracking" shows that the execution controller is presumed. Both planning system and execution controller are supposed to work together to solve the tracking problem. The similarity between our "positioning control" and "tracking the target", is becoming even more noticeable after we realize that the tracking trajectory is being constantly recomputed during the process of the above mentioned recursion.

One can see that the algorithm presumes each action to be performed as a unit. Certainly, each action can be a subject of further distribution at a level: even multiple agents associated with this particular level could be a "group performer" of the assignment. This won't change the algorithm of planning/control. It will just generate a set of additional algorithms which will drive the agents toward performance of the unitary action described as an assignment.

5.3 Generalized Controller

The problem of Generalized Controller design is illustrated by diagram Figure 6. Generalized Controller can be proven to consist of two major parts: Open-Loop-Controller (OLC), or *feedforward controller*, and Closed-Loop-Controller (CLC), or *feedback compensation controller*. OLC is designed to

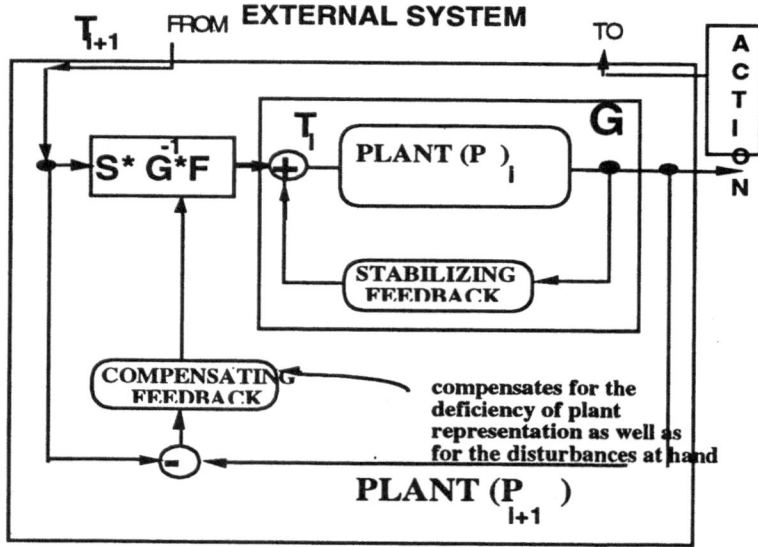

Figure 6. Realization of the Generalized Controller

generate a control input applied to the *plant* (G) to be controlled. Planning of the desired output is done by using S-operator which is finding the output trajectory leading to the desirable final state with minimum cost of this operation. The only way to determine to required input to the plant is to construct the *inverse* (G^{-1}) of the plant and then apply to the input terminals of G^{-1} the *desired output* of the plant. Then, at the output terminals of G^{-1} the required input to the plant will be obtained, so that $G \cdot G^{-1} = I$, and whatever you submit to the input (the desirable trajectory) must appear at the output (actual output trajectories) to the degree of accuracy of our knowledge of G.

Certainly, the computational structure G^{-1} is obtained in assumption that the model G of the plant is known adequately which include also knowledge of the external environment which never is the case. Thus, when the computed required input is actually applied one should carefully compare the actual output with the desired output, and the difference (error) use as an input to another computational structure: compensating feedback which forms immediately the feedback compensation loop. It has been proven that in order to minimize the output error CLC should be also based on the G^{-1} computational structure.(The following subtleties should be taken in account before

one uses these recommendations: the computational structure G^{-1} should be determined only after G is stabilized (if unstable). In rare cases when G is minimum phase, the structure G^{-1} is unstable and should be stabilized in OLC computation is to be done on-line).

At this point, we should declare that we will call *planning* the procedure of search for the input which should be applied to the plant when the desired output of the plant is given. One can see that this procedure is performed for the operation as a whole: it will be distributed among the multiple agents of the level (if they exist) later. Another term for this input is the *feedforward control*. If we compare the real output with the desired output, the error of OLC can be found and the compensation can be computed and applied. The output of the feedback loop we will call *compensation*. Compensation feedback controller we will call *closed loop control*, or CLC. It is clear that planning can be done off-line as well as on-line, in advance as well as in real-time.

The notations of Figure 6 are related to the arbitrary level of any control system: T_{i+1}-task from the level above, G-the transfer function of the plant P_i (stabilized), T_i-task for the plant P_i, this task can be obtained from the task of the level above T_{i+1} by applying the operator of planning/control

$P/C = S*G^{-1}*F$, where S is the generator of the string of input commands, G^{-1} is the inverse of the plant's transfer function, F- is the feedback operator (compensation). Plant of the level of resolution together with the planning control operator P/C can be considered plant of the upper level P_{i+1}.

5.4 Universe of the Trajectory Generator: 2-nd level

The Execution Controller, or the e-th level of control (Figure 7,a) is a machine for producing sequences of preassigned output states. The next adjacent level above, or a Trajectory Generator (2-nd level) has at its input a *trajectory of motion required to perform a particular maneuver* (TMPM). The maneuver is understood as a recognizable sequence of elementary trajectories[7] . Then the whole diagram shown in Figure 7,b (the e+1-th level) is considered "The Machine for Executing TMPM" (e.g. this is a set of PID Execution Controllers for turning the all four wheels of our robot: all operators of CLC for G_1 are assumed to be Multi-Input-Multi-Output, or MIMO). MIMO TMPM Machine has at its output not a motion of a single wheel, and not a set of motions of the four wheels but a trajectory of motion of the robot during some unspecified particular maneuver which is unknown at this level. In other words, these are the *actual time profiles* of speed, position, and orientation of the robot. In order to receive this output the desired TMPM has been obtained using $(S*G^{-1}*F)_1$ operator.

It is clear that as any MIMO, the TMPM machine can be decomposed into a set of subsystems, e.g. using feedback linearization and decoupling, or heuristic decomposition into a set of multiple autonomous agents. One can see that this decomposition won't change anything in the assignment of the TMPM: it will affect only the concrete arrangement of the process of performing this assignment. It is easy to imagine the situation in which performing of a particular maneuver will be distributed among agents of steering and propulsion. In the system of another mechanical construction TMPM can be obtained as a result of a group activities of the actuators associated with each wheel. This will generate other interesting (and sometimes difficult) problems of coordinating the autonomous agents. Theory of control addresses the issue of coordination.

---

[7] Two new terms are used in this phrase: "recognizable", and "elementary trajectories". The latter term implies that there exist a vocabulary of elementary trajectories so that each real trajectory can be described as a concatenation of "elementary" ones. Some of these concatenations can be considered entities, they are used more often, we refer to them in describing functioning of the machine (right turn, K-turn, etc.). These are entities of the next level of (lower) resolution, we can recognize them, we will call them maneuvers.

a) CONTROLLER AT THE EXECUTION LEVEL ("EXECUTION CONTROLLER")

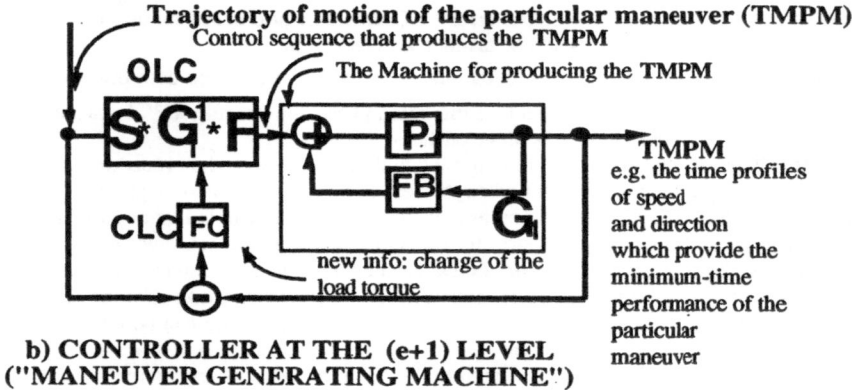

Trajectory of motion of the particular maneuver (TMPM)
Control sequence that produces the TMPM

OLC — The Machine for producing the TMPM

S*G¹*F   **P**   **FB**   **G**ᵢ

CLC FC

new info: change of the load torque

TMPM
e.g. the time profiles
of speed
and direction
which provide the
minimum-time
performance of the
particular
maneuver

b) CONTROLLER AT THE (e+1) LEVEL ("MANEUVER GENERATING MACHINE")

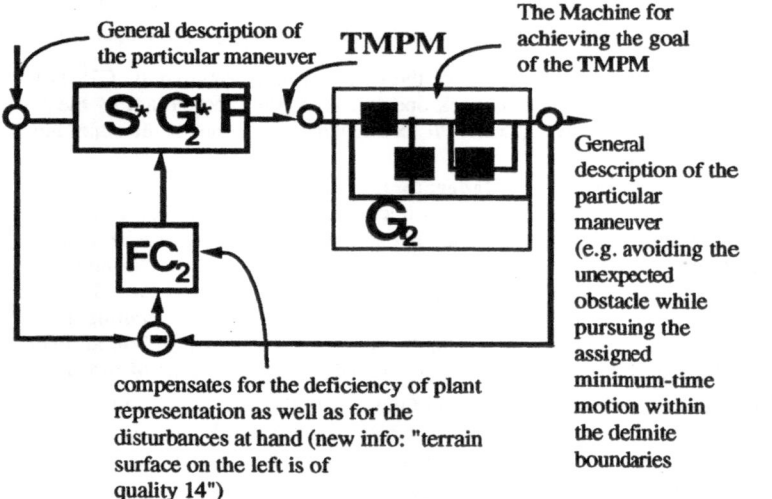

General description of
the particular maneuver   **TMPM**

The Machine for
achieving the goal
of the TMPM

S*G²¹*F   **G₂**

FC₂

General
description of the
particular
maneuver
(e.g. avoiding the
unexpected
obstacle while
pursuing the
assigned
minimum-time
motion within
the definite
boundaries

compensates for the deficiency of plant
representation as well as for the
disturbances at hand (new info: "terrain
surface on the left is of
quality 14")

Figure 7. Controller at the levels of : a) trajectory generation, b) maneuver generation

OLC is not expected to work properly because the actual load torques are not equal to the expected ones, a number of nonlinearities contaminate the picture such as nonlinearity of the friction, slipping factor between the wheels and the ground, and high order components of the Taylor expansion that were neglected in the model of transmission. Thus, the actual time-profiles of position, speed and orientation differ from expected ones.

The maneuver controller has at its input a call for a particular maneuver (Figure 7,b). MGM has at its output not a trajectory of robot motion as a time profile but rather a description of some particular maneuver which is a part of the motion schedule unknown at this level. In other words, the input is presented as a verbal (logical) description in the terms of standardized behavior which is expected from this robot. This OLC also is not expected to work properly because actual environment differ from the typical environment presented in the Look-up Table of the available TMPM goals. Thus, the actual sequence of the maneuver primitives can differ from expected ones. This difference is being measured and used to generate the CLC part of the controller at this level. compensates for the deficiency of plant representation as well as for such disturbances as "terrain surface on the left is of different quality than expected".

5.5 Representation of the planning/control problem in MCA

The planning/control problem in nested multiresolutional systems is considered as a problem of guiding the motion at each level of resolution so that the next level below treats it in a smaller state-space envelope with higher accuracy. "Planning" corresponds to the lowest resolution while the "execution control" is performed at the highest one.

First, the dynamical system should be represented as a formal structure (e.g. mathematical, linguistic, or otherwise symbolic structure) or a "knowledge base" structure representing relationships between the states "x" and their rates of change "x'"

$$\text{KB}[\ X(t),\ X'(t),...],\quad t=t_0,t_1,...,t_f, \tag{9}$$

where $X(t)$ and $X'(t)$ are the sets of time profiles of the evaluations for the variable and its rate of change in any available form[*] . Time profiles are the strings of values of the variables and their rate of change directly stored (or otherwise computable by supporting procedures) for the whole duration of the process of interest starting with initial time t_0 and ending with the final time t_f .

The knowledge base which is denoted symbolically as the structure (9) can be accepted in a form of difference, or differential equations, in a form of set of logical statements, or any another form assuming formal (and computer) manipulations (see in [59] an illustration of this statement). Here, we would like to stress the fact that the form (9) should be understood in a much broader sense than just an analytical form based upon, say, differential equations with all strengths and weaknesses of its apparatus. One should understand structure (9) just as an abbreviation for any array of coded knowledge about the world realities denoted within (9).

The structure (5) should be supplemented by information on the state inequality constraints which in turn, depend on the actual state and the result of measurements and processes of recognition and interpretation

$$X_c = \{x_{ci}(x,x',...,t): |\ |\ x_{ci}(x,x',...,t\ |\ | \leq x_{Mi},\ i=1,2,...,k\},$$
$$X'_c = \{x'_{ci}(x,x',...,t): |\ |\ x'_{ci}(x,x',...,t\ |\ | \leq x'_{Mi},\ i=1,2,...,k\}, \tag{10}$$

A subset of variables $U(x,t)=X_{inp}(t)...X(t)$ is considered to be controlled independently and is called *controls*. The structures (9) and (10) should also be supplemented by the information on the state dependent admissible control set

$$U_A = \{u_i(x,t): |\ |\ u_i(x,t)|\ | \leq u_{Mi},\ i=1,2,...,n\}, \tag{11}$$

and on the cost-functional

$$J= \int_0^{t_f} L[x(t),x'(t),...,t]\ dt \tag{12}$$

which characterizes the final cost of the process, and is supposed to be properly interpreted depending on the situation. For example, $L[x(t),x'(t),...,t]$ should be understood as a sub-knowledge base storing (or computing) the cost of being $x_i(t),x'_i(t),...,t_i$ at a particular moment of time "i " which allows for computing the cumulative cost of the overall process.

The control problem can be formulated as follows. For a given map of the state-space[9] , for a

---

[*]   One can see that instead of writing a symbolic form F[$x(t)$, $x'(t)$,..., $u(t)$] as we are doing for "functions", we write KB[$x(t)$, $x'(t)$,..., $u(t)$] thus reminding that this is not a *function* in a mathematical sense, this is a "knowledge base" i.e. a collection of knowledge more general than a function. Function can be a particular case of a "knowledge base".

[9] Since the cost is different in different parts of the map, we will call it a *"variable-cost-map"*.

given "initial point" , and "final point" (goal) of the motion:

1. declare part of the variables x(t) to be the output variables $X(t)...X_{out}(t)=y(t)$;

2. find the desirable output is proposed $y^*(t)$ which is called the *output plan*; the time profiles of the output vector components (*output plans*) can be found from the knowledge of the "starting position" $y_0(t)$ and final goal $y_f(t)$ using operator denoted S in the previous subsection; the time profiles of the control vector components can be found from the output plans and the inverse transfer function of the plant denoted G^{-1} in the previous subsection.

3. find the *open-loop*, or *feedforward* control vector $u(t)$, or the *input plan* which minimizes, maximizes, or keeps within some inequality bounds the value of J; Since neither S, nor G^{-1} are perfect, the real y(t) will differ from the desired $y^*(t)$; the difference $\Delta=y(t)-y^*(t)$ should be compensated by the operator of feedback compensation F.

We will consider here only additive law of compensation other laws can be considered too, it does not affect the substance of our approach. So, in order to have the control problem solved, the structures (9-12) should be supplemented by the following structures

-for the planned control:

$$u^*(x,t)= G^{-1*} \bullet y^*(t) \tag{13}$$

-for the compensated control:

$$u(x,t)=u^*(x,t)+F \bullet \Delta \tag{14}$$

-for the plan:

$$y^*(t)=S[y_0, y_0'; y_f, y_f'; J]. \tag{15}$$

-for the error of control (deviation from the plan):

$$\Delta=y(t)-y^*(t) \tag{16}$$

The system of structures (9)-(16) is assumed for a particular accuracy of representation. In the previous sections we saw that accuracy (level of generalization) was critical for determining the vocabulary of the level, therefore the number of variables and their contents was determined by the accuracy of representation too. Thus, we can expect that each resolution level will entail its own system of structures (9)-(16). In this paper, we are interested in finding how the systems (9)-(16) from all levels are related to each other.

The system of representation of the available (as well as of the required) information, is looming in the above considerations. Structures (9)-(16) built for real examples, strongly depend on the accuracy of representation, i.e. on the accuracy of the assumed processes of sampling, digitization, etc., and then, decoding, interpreting, storing, and executing the information. It is clear that the quality of representation should allow for planning, and executing the motion, and on the other side, to distinguish the difference Δ between the motion we were able to execute and the plan of this motion which was good enough before the process started.

Here we have a dual situation. On one hand we are still within the realm of *representation for control* (9)-(16). However, when the complexity of Plant and World is growing, the information structures becomes very complicated and its computational complexity can grow impermissibly. Thus, when we try to find a controller, say a feedback controller, the devices for receiving and interpreting information about the states and outputs ("perception") are becoming impermissibly complicated, and the controller is becoming non-relilizable as a technological object: too cumbersome, too complex, too unreliable, probably, too expensive. This entails substantial problems by itself: special "knowledge representation" subsystems are required for perception, for knowledge base, and for the subsystem of planning/control.

Thus, the complexity of knowledge representation implies the complexity of the perceptual information structure and the controller. The processes of task formulation, constraint determining, cost-function computation, and u-vector generation, require special interactions between

a) information structure utilized by controller, and

b) information structure implanted into the subsystem of perception.

Both interactions are performed via the information structure of Knowledge Base. Thus, this process can be considered an internal process of the unified information structure P-K-P/C (see Figure 1b). This information structure ("knowledge system") is to satisfy both destinations: to incorporate both representation for perception and representation for control.

5.6 Search as the General Control Strategy For MCA

Search is a conventional technique of finding solution in the production system. Whatever representation is implemented an algorithm of "search" is to be apply to select the "best" trajectory out of the multiplicity of "candidate" trajectories. Other existing control strategies pursue the same goal. In a particular case when a tree or a string should be found, a number of search algorithms is being used. One of the very efficient search-algorithms, "best-first" algorithm of search [60] is based upon the following sequence of operations:

Search Algorithm (general scheme) for finding minimum cost path on the graph
1. Define initial and final nodes of search.
2. Determine "successors" from the initial node which are to be considered as the "next" standpoint.
3. Determine cost for all of the successors.
4. Select the minimum cost successor as a new initial point.
5. Loop to the step 2.

The cost of each of the successors can be determined as a clear cumulative cost of achieving this particular successor from the initial point.

$$C_f = C_g + C_h \text{ , (or f=g+h as in [61])}$$

where C_g- is the cost from the initial node to the one of the set of generated nodes-candidates,

C_h-is an evaluation of the cost from the node-candidate to the goal.

It was shown [61], that when no additional information is available, one should determine the minimum possible value of distance between the candidate node, and the goal, using the accepted metric of the space of the search. This strategy leads efficiently to the optimum solution.

In this case the algorithm is called The Dijkstra Algorithm, and it propagates in all directions of the state space. The Dijkstra algorithm tends to check all possible paths before it selects the best one. A heuristic can be introduce which prunes the number of trajectories explored by including in the expression for the cost one additional component. In this algorithm (it is called A*-algorithm,the cost is computed as a sum of "clear" cost of achieving (or as a total relevance with) the successor, and of "less clear" cost of moving from the successor to the final node (or relevance between the successor and the final node). The search is propagating not as broadly as the Dijkstra-search, it tends to be "attracted" by the goal.

One of the first approaches proposed for decision-making processes within nested hierarchical structures, was an approach of Y. C. Ho and K.-C. Chu [62-64]. Search for a trajectory satisfying numerous constraints, and minimizing a cost-functional invokes variational methods, and ascends to dynamic programming (DP). The latter was not used as often than it could because of the well known "curse of dimensionality". Many efforts to apply DP were obstructed by subtle computational "glitches". Nevertheless, when not overestimated DP seems to be the most appropriate and perspective method because of the following considerations (see [65-70]):

1) most of the systems we are dealing with in the MCA area, are substantially nonlinear, coupled, and cumbersome ones: off-line precomputation of table-look-up would be expected for control of such a system anyway;

2) we can consider DP as an idea of synthesizing trajectories in the state space, as an idea of a graph-search [70]; this allows for enhancement DP by a number of heuristical methods which are intended to make the algorithms more computationally efficient (e.g. [16, 17]).

Selection of the proper system of cost-functionals is becoming important. We do not have too many cost assignment strategies on hand which can be considered "tractable", and confirmed by a substantial experience of broad application. One of the possible alternatives is the strategy of cost assignment. The total cost of the node selection ("feasibility of the node expansion") C_f is divided in two parts. In a uniformed cost space, C_h is considered the shortest line to the goal ("air-flight path" [61]). In the state space with variable cost, or variable cost space (VCS, simple example of it is considered in [32]) the value of C_h should be accepted as if all units of space from the current point to the final point will have the minimum cost of the unit among all available unit costs.

The method of "Nested Dynamic Programming" (NDP), or method of "nested consecutive refinement" which was proposed recently for systems of nested multiresolutional control [16-19, 21-23, 25-28]. Similar ideas were contemplated earlier for increasing efficiency of dynamic programming [65]. NDP can be considered an extension of continuation methods [71] and as using dynamic programming in the multigrid environment.

Multigrid methods are also employing the ideas of consecutive refinement [46-48]. The solution is initially found by numerical solution of a complicated differential equation at a low resolution (coarse granularity, or with a low resolution grid). Then a vicinity of this solution is being determined and only in this vicinity, a new finer grid is built, and the problem is being solved again.

The idea of NDP which encompass ideas of all these methods is illustrated in Figure 8. At the highest level the system is represented with low resolution (coarse granularity). Nodes of the graph representing the system are shown without edges which connect them. After the solution is found, an envelope is determined which contains the solution and a definite vicinity of it.

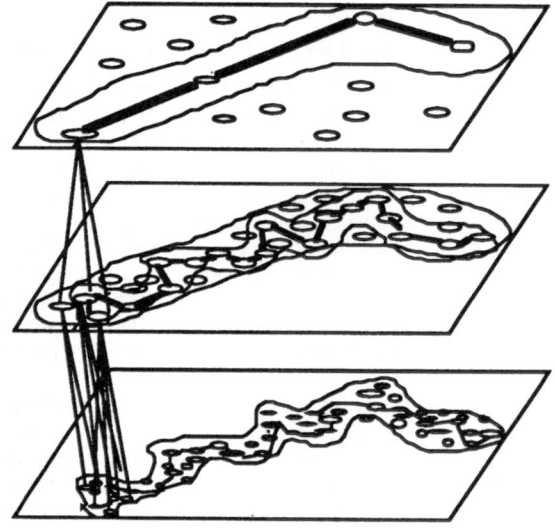

Figure 8. Illustration to the NDP method

One can expect that the width of the envelope should not exceed the width of the unit of the grid at this particular level. In fact, taking in account the uncertainties of knowledge representation, a more cautious approach should be recommended. Any width of the envelope should be evaluated by the probability of missing the optimum solution. This is why in reality one should start with a substantially wider envelope: three times the width of the unit of indistinguishability of the particular level of resolution.

Only the area within the envelope is refined: the higher resolution representation of the system is built for this area. Then the algorithm of solving the problem (a search for minimum cost trajectory is applied), and the trajectory is obtained (see the middle level in Figure 8). The envelope is built around this solution which is based upon the same principles. Only area of this envelope is submitted for the consecutive refinement at the level of representation with higher resolution. Only after the process of consecutive refinement is completed and the problem is solved for a level, the decoupling of the MIMO can be initiated, and the problem can be distributed among the autonomous agents.

6. Elements of the Theory of Nested Multiresolutional Control for MCA

6.1 Commutative diagram for a nested multiresolutional controller

Considering subsystems as categories C, and the interaction among them as functors F, the commutative diagram of the MCA can be shown in Figure 9 (indices mean: s-sensing, p-perception, k-knowledge, pc-planning/control, a-actuation, w-world). Feedbacks are not shown: boxes are connected by "functors" which characterize the structure conservation in a set of mappings of interest. The bold horizontal line separates two major different parts of the system: what is below, is a world of real objects, and what is above the bold line, is the world of information processing. All of the "boxes" in Figure 9 are fuzzy-state automata. They are easily and adequately described in terms of the automata theory, and provide consistency of the descriptions, computer representations, and control operation. Then, the search can be done by NDP technique, discretization of the space is being determined by the level of resolution, and the rules which are formulated within the given context.

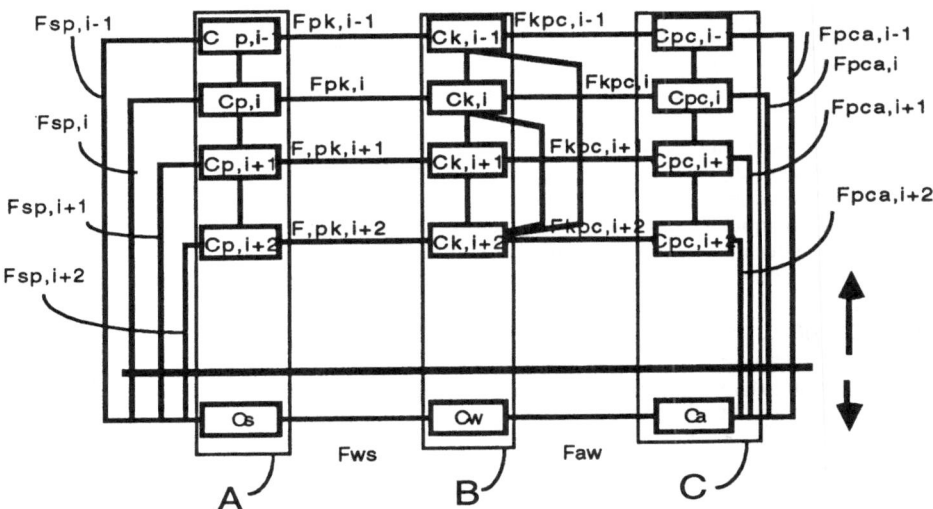

Figure 9. Category-theoretical Representation of MCA

6.2 Tessellated Knowledge Bases

An important issue is the fact of dealing with a discretized representation. This discretization starts at a level where the entities of objects cannot yet be recognized. At this stage we call it "tessellation". The system structures (5) through (12) with its variables and mappings is described in a Hausdorff n-space $(R^n)_h$ (subscript "h" is later omitted everywhere) which means that any two different n-dimensional points in it have disjoint neighborhoods. One may assume that our space is obtained as a result of artificial discretization (tessellation, partitioning) from an initial continuous space of the isomorphic world description. This space can be considered as a state-space for a dynamical system in which the trajectory of motion can be represented as the structures (5) through (12) will require. This space is considered for a given moment of time ("snapshot of the world"). Then, a sequence of snapshots can be obtained. The conventional state-space is a superposition of all snapshots which describe the process.

This tessellation is done in such a manner that the "points of interest" (symbolizing the variables, words, or wwf's) are placed to the center of the "tile" of tessellation (elementary segment, grain, discrete, pixel, or voxel of the space). The terms *segmentation, granularity, discretization*, and *tessellation* will be used intermittently. Property of being centered holds if the tile can be derived as a system of subsets with nonempty intersection. This property should be considered rather in a symbolic way: in fact, we cannot discern any another point within the tile, this is a minimum grain at this level of resolution.

A high resolution levels the discretization is very "fine-grained": in order to describe an entity, a group of "grains" is required. At low resolutions, each "grain" can be an object.

Discretization continues at all resolutions including the lowest ones. Categories as well as their objects and morphisms can be partitioned.The overall description of the world pertains to a definite instant of time. Thus, any particular $C_r(t)$ can be considered as a "snapshot" of the world C_w. This snapshot is supposed to be analyzed as a whole.

Decomposition of the categories of representation is done through decomposition of objects and morphisms represented in this category. Decomposition implies dividing objects and relationships in parts, so the components of objects are also becoming of interest which was not the fact before the process of decomposition. This, in turn, implies higher resolution of world representation than before the process of decomposition. If we continue with decomposition, the further representations are expected to disclose new details about the objects and their parts, as well as about relations and their parts. There is a limit to decomposition of the world: this is decomposition in the tiles of a very small diameter. We will call these tiles ε-tiles. One cannot distinguish any unit of information smaller than the ε-tile. Centers of the tiles at all levels of resolution can be connected into graphs depending on their relationships. This graph will be called ε-graph at the level of ε-resolution.

In the hierarchy of decompositions, all of the levels describe the same world with higher and higher level of resolution. Obviously, this statement is irrelevant to the kind of representation chosen. Later we will illustrate dealing with the multiresolutional (multigranular, multiscale) world representation together with the multiresolutional algorithm of search.

Graphs of the adjacent levels can be transformed into each other. Transformation of the lower resolution graph into the higher resolution graph is called decomposition, or instantiation. Transformation of the higher resolution graph into the lower resolution graph is called aggregation, abstraction, or generalization. These terms are not synonyms: there are subtle distinctions among aggregation as grouping, abstraction as finding the distinct property, or a function for the group, and generalization which makes the group a carrier of a particular law. In this paper, we will consider only simple aspects of these phenomena.

Partitioning of representation in which each "grain" is an object boils down to "grouping", or "clustering" of these objects. This is done for the sake of making the computations more efficient, and this brings us to *generalization*.

6.3 Generalization

So, information can be represented as an ε -net at a definite resolution, and as a system of nested ε-nets with different scale where scale is defined as a ratio between resolution at a level and resolution at the lowest level. Clearly, each of the larger tiles at the upper level (lower resolution level) is a *"generalization"* for the set of smaller tiles at the higher resolution adjacent level of the hierarchy. Selection of one of the tiles at the upper level (focusing attention) entails selection of a definite subset of smaller tiles at the lower level.

The procedure of "generalization" is not defined as a context independent operation. We will assume a loose interpretation for generalization as if we are dealing with an operator of unifying separated units of knowledge in a set. One of the procedural generalization ideas broadly accepted in the control area is "averaging" (see more about it in [75]). We would like to stress the fact that the inclusion $X \supset X \supset x$ shown in the multiresolutional hierarchy of the tile embeddings has more important and broad meaning than just "scaling of the size".

The inclusion predicate ... has a meaning of "a class belonging to a larger class".One can talk about state space, space of weighted properties, and so on, and the notion of "belonging to a class of come spatial neighborhood" is becoming closer to a meaning of "generalization" as if it is understood in the discipline of logic. Then discretization of the state space will contribute simultaneously to a) minimum required interval of consideration, b) hierarchy of classes of belonging to a *meaningful neighborhood*.

The relations among the highly generalized tiles at the upper levels can be considered as generalized representation of the corresponding less general relations among the tiles of the corresponding subsets at the less general lower levels. This implies that not only properties of this tile per se, are generalized but also its relationships with the other tiles at the particular level.

Hierarchies created in this way, satisfy the following principle: *at a given level, the results of generalization (classes) serve as primitives for the above level.* Then each level of the hierarchy has its own classes and primitives; thus, it has its own variables (vocabulary), and the algorithms assigned upon this vocabulary can be adjusted to the properties of the objects represented by the vocabulary. This determines the mandatory rule of combined consideration of two (at least) adjacent levels of the hierarchy. Set of relationships among the variables at the i-th level describes the implications which are being utilized for decision-making at this particular level.

On the other hand, the set of relationships at the (i+1)-th level describes the relationships among classes and thus can be characterized as set of *meta-implications* (meta-rules, meta-clauses) which are governing the process of applying the implications (rules) at the i-th level. In the light of this consideration, each two adjacent levels can be understood as a complete E. Post production system (analog to "general problem solver" or "knowledge based controller") in which the meta-rules applied to the alphabet of the lower level act as a semi-Thue process of the grammar [77].

6.4 Attention and Consecutive Refinement

Clustering of several subsets into a single set entails inclusion which is commonly understood as *inclusion by generalization*. There is another type of inclusion which is somewhat opposite to inclusion by generalization. When a part X_p is detached from the "whole" X (which may happen to be category, object, and/or morphism) and the relation of inclusion holds

$$X \supset X_p \qquad (17)$$

this separation of X_p from X we will name *focusing attention* upon X_p. Sampling is one of the common methods of focusing attention.Usually we focus the attention when the subset of attention can be considered important, or is typical for the whole set. The latter case links focusing attention with mechanism of generalization.

Let us concentrate on the mechanisms which determine the resolution of the level. It was shown by Ho and Chu [62-64] that generalization induces partition of the world representation which is coarser than the initial representation: it represents less details at the upper level. Following Ho and Chu we will consider the field for the overall world representation. We have already demonstrated that the field for the world representation is the state space where the search for the future solutions is actually being performed. Thus, if the field for the lower level is F, and for the upper level is J (see Figure 10) then the following holds

$$F \supset J \qquad (18)$$

or, in other words, J is nested (included) in F which means that both describe (or: represent symbolically) the same part of the world but with different accuracy. Since J carries in a sense less information than F, the direction of nesting reflects this relation of inclusion. This nesting can be named *nesting by generalization*. We will show that another type of nesting: *nesting by focusing attention* must be considered for the systems of representation. These two types of nesting appear in representations simultaneously, and create sometimes confusing situations.

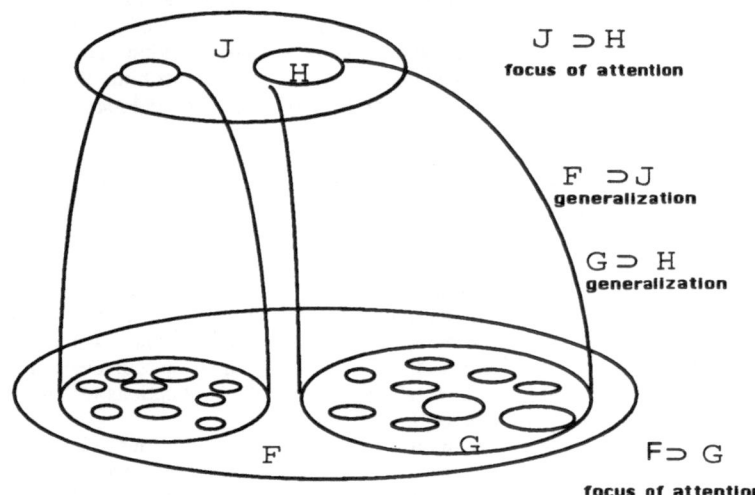

Figure 10. Set-theoretical interpretation of attention and generalization

Indeed, the notion of nesting is understood in a sense of "being a part of, being a subset of". Consider H a subset (part) of J (where J is a field built for the representation of the upper level),G a subset (part) and of F which is a representation of H, or G R H (where F is a field built for the representation of the lower level). The relation R is a relation of generalization. This relation means that all elements of the set H are actually classes for the set G . Obviously (see Figure 15)

$$J \supset H, \text{ and } F \supset G, \tag{19}$$

but on the other hand, taking in account resolution

$$F \supset J, \text{ and therefore } F \supset H, \text{ and } H \ R \ G, \text{ hence } J \supset G. \tag{20}$$

Here we have a situation when the subset of the lower level representation is nested within the upper level representation, and simultaneously the subset of the upper level representation which represents the same subset of the world, is nested in the lower level representation. This is not a contradiction since different kinds of nesting are under consideration. The diagram in Figure 10 illustrates this possibility of *nesting by generalization* (n_g) versus *nesting by focusing attention* (n_{fa}). In this paper, the representation implications are discussed for both of these nestings.

6.5 Accuracy and resolution of representation

To all words (nodes of the imaginary graph) and all relationships among the words (edges of this graph) with the corresponding weight of this edge, the values of *accuracy* are assigned. Accuracy is understood as the value of the relative error of the quantitative characteristics assigned to the node or the edge of the graph. This evaluation can be based on estimation of probabilities, zones of uncertainty, and/or in any another way. The approach to the world representation in this paper does not depend on the particular way of dealing with uncertainty evaluation. The value of uncertainty is assigned to the numerical values associated with the properties of the tile of representation which is considered as a minimum unit at a level.

Thus, the uncertainty of hierarchical system originates at all levels of the representation and propagates through the rest of the hierarchy thus generating the tight bounds of *uncertainty cones* [80]. It is of interest for us that the idea of bounded uncertainty cones is actually a generalization of a variety of convenient and applicable uncertainty evaluators (such as circle stability criterion [81]). The major statement of these works is that assigning a value of conic uncertainty at some level of the hierarchy would eliminate computational complexities. What actually happens inside the cone should be out of interest unless there is a special reason to be involved in the "microstructure" of the stochastic or dynamical processes "wrapped up" in this cone. Similarly, we are not interested in what actually happens within the tile at a given level. The "microstructure" of the tile is considered to be a prerogative of the lower level of the hierarchy.

This is similar to assigning a definite ε-net. However, this also suggests that the minimum value of the radius of the tile, can be time dependent. In order to avoid this predicament in the following treatment, we will try not to get involved with the particular stochastic and dynamical microstructure of the ε-tile at the level of the representation hierarchy.

We will define the radius of the tile by the width of the fictitious uniform distribution which has the same value of the conditional entropy as the real distribution of the variable x. Then, using the value of conditional entropy

$$H = - \int_{-\infty}^{+\infty} p_i(x) \ln p_i(x)\, dx \tag{21}$$

for an arbitrary distribution law, the actual error can be equated to the fictitious error interval $(-\Delta, +\Delta)$ with uniform distribution law as follows

$$H = - \int_{-\infty}^{+\infty} p_u(x) \ln p_u(x)\,dx = - \int_{-\Delta}^{+\Delta} (1/2\Delta)\ln (1/2\Delta)\, dx = \ln(2\Delta) \tag{22}$$

where $p_u(x)=0$, at $x< -\Delta$, $x> +\Delta$, i.e. $|x| > \Delta$,

$p_u(x)=1/2\Delta$, at $-\Delta< x< +\Delta$, i.e. $|x| < \Delta$.

Thus

$$\Delta = -1/2\ e^{-\int pi(x) \ln pi(x)\, dx} \tag{23}$$

which for a Gaussian gives

$$\Delta = s\,(pe/2)^{.5} \tag{24}$$

Referring to the definition of ε-net, we would suggest that Δ from (24) is the minimum size of the tile radius to be assigned. For systems with stationary random processes, evaluation of σ will be sufficient for determining ρ and not being involved in analysis of stochastic processes anymore. Certainly, this depend on the nature of what is considered to be a random component of the variable. This problem will be treated in a separate paper. Initial treatment of this problem given by Ho and Chu [62] demonstrates the strength of recommendations that can be implied.

Labeling the class presumes dealing with this class as a primitive at the given level of consideration. Moreover, this class (now, also a primitive) is being again clustered in classes of the higher level. In order to deal with class as with a primitive we have to neglect the inner content of this class (which might be reflected as new properties of the new primitive but with no mentioning the initial primitives with their properties). The levels of the hierarchy of representation (since they are created by a mechanism of generalization) are dealing with the same world given with different level of richness in submitting specific details, level of coarseness (fineness). We will name this characteristics of world representation at the level, *resolution* which is a measure of distinguishability of the vectors in the state space.

The problem of resolution can be restated as a problem of the covering radius which is addressed in [82]. This brings us to the idea that the second of conditions for assigning ρ is determined by the operation of vector quantization [83, 84]. After assigning to the cluster a new class-label (a word to be included to the vocabulary), this class-label is becoming a new primitive with its own properties, the numerical values are assigned to these new properties, and these numerical values are characterized by the accuracy depending on the accepted evaluation theory for this particular property (including probabilistic approaches, fuzzy set theory, etc.). Clearly, this accuracy evaluation is formally independent from the act of neglecting the inner contents of the new primitive. This means that accuracy and resolution are formally independent. The word "formally" means: in the view of procedure to be performed.

Thus, accuracy presumes the error evaluation in terms of the existence of difference between the world and its description within the accepted concept of the model. The smaller the vocabulary S_i is, the more different phenomena are neglected. This neglect may entail the increase in error and may not. However, the smaller $Card(S_i)$, or size of the vocabulary is, the higher is the level of generalization, and the larger is the radius ρ of the tile in the ε-net. Thus, the following relation should hold

$$\rho = \frac{1}{Card(S_i)} > \Delta \tag{25}$$

where ρ determines the value of allowable error (inaccuracy) and $Card(S_i)$ determines the value of resolution.

Accuracy and resolution can mistakenly be understood as nondeterministic properties of representation. We will separate the characteristics of accuracy and resolution from the nondeterministic information. Remember that whatever experimental results have been received after observation they are subjected to a procedure of identification, and there are two parts in them: recognizable, and the remainder. Let the vector of observations be

$$X_0 = X_{od} + \xi \tag{26}$$

where X_{od}- deterministic model after recognition,

          ξ - stochastic component (a difference between the results of observation and the results of recognition; the remainder).

Definition. A component of observation is named stochastic component if it is

        a) not identified (yet) with any of models stored in the knowledge base,

        b) substantially larger than measure of accuracy and resolution ($|\xi| > \rho$),

        c) presumed to potentially affect the results of decision-making.

Our approach differs from the classical only in a sense of (31) which implies two recommendations.

Recommendation 1. Control problem is to be solved as a deterministic open-loop problems dealing with models X_{od} and a stochastic closed loop (feedback) problem based upon estimating likelihood or plausibility of the decision, or the policy by measure of uncertainty demonstrated in $\{\xi_i\}$.

Recommendation 2.. Learning will be understood as a tool for extracting new recognizable models from $\{\xi_i\}$ rather than for updating knowledge of probabilistic characteristics of the set $\{\xi_i\}$.

The following structure of dealing with unrecognized (unmodeled) information is implied by these two recommendations. Decomposition (26) is considered to be repeated recursively for the nested multiresolutional structure which has been obtained earlier. At each level the component is being decomposed in two parts: which can be recognized and included in the deterministic part of the next level, and which at the next level remains still unrecognizable (with $E[\xi_i]=0$)

$$X_{0i} = X_{odi} + x_{r,i+1} + \xi_i \tag{27}$$

where X_{odi}- deterministic model after recognition at the level i,

$x_{r,i+1}$ -part of the stochastic component at the i-th level which will be recognized

after observation at the (i+1)-th level,("trend").

ξ_i -part of the stochastic component at the i-th level which remains unrecognized, $E[\xi_i]=0$.

This recursive analysis of the stochastic information can be illustrated as follows

$$X_{0i} = X_{odi} + x_{r,i+1} + \xi_i$$
$$X_{0,i+1} = X_{od,i+1} + x_{r,i+2} + \xi_{i+1} \tag{28}$$
$$\dots\dots\dots\dots\dots\dots\dots\dots\dots\dots\dots\dots\dots$$
$$X_{0,n-1} = X_{od,n-1} + x_{r,n} + \xi_{n-1}$$
$$X_{0,n} = X_{od,n} + \xi_n$$

and n is the level where the recursion stops (no consecutive levels are expected to be built).

This decomposition of information (which is possible within the nested multiresolutional structure) allows for multiple reference system. The key motivation for the multiple referencing is simplification of information representation per level. Multiple referencing is indirectly presented in the requirement that $E[\xi_i]=0$. This means that the origin is placed in such a point in the state space as to provide $E[\xi_i]=0$. Then, the rest of the information allocated for decision-making at this level is referenced to these origin.

Another important implication of *multiple referencing* in dealing with nondeterministic information, is related to the topic of learning. As mentioned above, the system is supposed to deal with partially, or completely unknown world. Thus, learning is presumed. Any learned information is being identified with memory models (patterns) which determine the initial referencing. The residual information is supposed to be collected, and later it is expected to generate a new pattern upon the multiplicity of realizations. If generation of a new pattern seems to be impossible (no regularities are discovered), the change in the initial referencing might be undertaken. This philosophy of dealing with new information is to be utilized for procedures of *map updating*.

We can see also within the body of this problem of nested referencing, a direct link among the quantitative characteristics of the system and its linguistic description, and the components of this description. At this time, however, we will restrain from further statements since we do not have enough factual observations.

In other words, the tile of the tessellation determines the *resolution of knowledge* which is defined as a minimum discrete of information, or minimum wwf which can be stated unmistakably.The minimum centered tile will have diameter ε and the net of centers emerging from this tessellation is called ε -net [44]. Let us consider the important process of resolutional hierarchies generation.

Let R be a knowledge (metric) space and ε any positive number. Then a set $R \supset A$ is said to be an ε -net for a set $R \supset M$ if for every $x \in M$, there is at least one point $a \in A$ such that $\rho(x,a) \le \varepsilon$. The idea of nested tessellations is coming together with a definition of a single tile $T(\varepsilon)$ based upon nested sphere theorem which can be rephrased as nested tile theorem. This theorem defines a chain of inclusions

$$T(x_0, \varepsilon_0) \supset T(x_1, \varepsilon_1) \supset T(x_2, \varepsilon_2) \supset ... \supset T(x_n, \varepsilon_n) \tag{29}$$

where $x_1, x_2,..., x_n$ are the coordinates of the centers of the tiles,

$\varepsilon_1, \varepsilon_2, ..., \varepsilon_n$ are the radii of the nested tiles.

The definition of the nested ε -net: a net with elementary tiles satisfying this condition. In the equation of relationships among the tiles

$$\varepsilon_0 = \frac{\varepsilon_1}{\sigma_1} = \frac{\varepsilon_2}{\sigma_2} = \ldots = \frac{\varepsilon_n}{\sigma_n} \qquad (30)$$

coefficients σ_1, σ_2, ..., σ_n are scales of the nested ε -net hierarchy.

6.6 Complexity of tessellation: ε -entropy

In the context of this paper, discretization of the space does not allow for the problem of aliasing because there is no information between the adjacent tiles of tessellation (digitization, quantization), and the information about the tile properties (values for them) is the set of average values over the tile. The term "average", acquires a somewhat unusual meaning of *"class generating property"*.

Let Σ be the alphabet into which each number is to be coded. Let S be the set of finite strings in this alphabet. Coding is to be considered as an injective function $\varphi: \Sigma \rightarrow$ S which is characterized by the length of a string for each coding. Later we will find a condition for minimum coding in the information structure for control.

Complexity of the table is being evaluated by computing the ε -entropy (introduced by A.N. Kolmogorov [93, 94]) for a space where the corresponding e-net has been constructed

$$H_\varepsilon(S) = \log N_\varepsilon(S) \qquad (31)$$

where S-is a space with ε -net assigned,

N_ε -number of elements (nodes) in ε -net, or tiles in the tessellation.

If the category of consideration can be represented as a power set, then trivially

$$H_\varepsilon(S) = N_\varepsilon(S) \qquad (32)$$

Equivalence between the automata representation and look-up table was stated in the literature [2, 50]. The key role is determined by the function which describes the state transitions given inputs as segments of signals determined upon the time interval (t_j, t_{j+1}), $t_j \in$ T. Thus, a set of tables is required where these solutions are listed under the set of different initial conditions and different constraints. Accuracy of this representation is determined by the ε-net for a look-up table (LUT). Since any couple "input-output" is a logical clause, the equivalence between LUT and production system can be verified.

Various theories of planning are based upon the idea of space tessellation. A multiplicity of different strategies of space tessellation is created, and numerous techniques of evaluating the distance, and the complexity are known from the literature. It is essential to understand that all of these techniques ascend to the theory and formalisms of ε -nets.

A question arises, how many zoomings should be done in a particular situation. Let the total size of the space of operation available is A, and the minimum value of the tile at the lowest level is Δ which is determined by the accuracy of motion at the lowest level which is *real* motion. Assuming that the number of levels is n, and the scale among the levels is m, we can determine an estimate for the required computer power as P=mn. It is easy to show that minimum computer power is required when the scale is equal to **e**. Indeed, the total number of the minimum size tiles is determined by the equation

$$A/\Delta = mn, \qquad (33)$$

After transformation we have

$$n = \log m(A/\Delta). \qquad (34)$$

Minimizing mn we should minimize

$$m \log m(A/\Delta) = \min. \qquad (35)$$

After differentiation and simple transformations we have

$$m = e. \qquad (36)$$

In reality of discrete system a condition should be imposed of m and n of being integers which changes the results of minimization computed for definite values of A.

7. MCA in Autonomous Control System

7.1 The Multiresolutional Generalization of System Models

In this section we outline the premises of the algorithms recommended. Levels of generalization and multiresolutional representation, as discussed here, are considered to be depictions of the same object with different degrees of accuracy. We will formalize the preceding statement in mathematical form by applying concepts of the usual state space representation for the (not necessarily linear time invariant) system:

$$\dot{x}(t)=A(x,u,t)x(t)+B(x,u,t)u(t)$$

$$y(t)=C(x,u,t)x(t) \tag{37}$$

where

$$x\in R^n,\ u\in R^m,\ y\in R^p,\ t\in R^+$$

Thus it is possible to form a solution of these equations as mappings describing the state transition and output functions:

$$\Phi:R^n\times R^m\times R^+\to R^n\times R^+$$

$$\Psi:R^n\times R^m\times R^+\to R^p\times R^+$$

$$\tag{38}$$

so that for any input function "u" on the interval $[t_0,\ t_f]$ it is possible to determine the corresponding output function "y" on the same interval. If it can be shown that there exists a pair of functions

$$\Phi':R^{n'}\times R^{m'}\times R^+\to R^{n'}\times R^+$$

$$\Psi':R^{n'}\times R^{m'}\times R^+\to R^{p'}\times R^+$$

$$\tag{39}$$

for which n' is strictly less than n, and for which the same input function "u" generates the output function "y'" such that inequality

$$\left\|\int_{t_0}^{t_f}(y'(t)-y(t))dt\right\|<\varepsilon$$

$$\tag{40}$$

holds for all admissible inputs in the input function space where ε is a value which depends on the level of resolution under consideration. Then, it is claimed that

$$[\Phi',\Psi']\ \text{is an e–generalization of}\ [\Phi,\Psi] \tag{41}$$

The strictness of this formulation may be relaxed by considering a stochastic measure for associating a confidence level with the generalization to construct the concept of ε-generalization nearly everywhere. Thus,

196

$$P\left[\left\|\int_{t_0}^{t_f}(y'(t)-y(t))dt\right\|<\epsilon\right]<\tau$$

(42)

is a statement of the belief that the constraint holds with a probability defined by the preassigned threshold τ.

This formulation can be extended to an ordered collection of epsilons, $\{\epsilon_1, \quad \epsilon_2,...,\epsilon_k\}$ thereby defining a hierarchy of models which describe the same input-output behavior with increasing degrees of accuracy. The necessity of considering all elements of the input and output vectors as time varying functions may also be relaxed so that at some level 'i', u_{ki} $[t_0, t_f]$ could be considered constant in the interval, whereas at some lower level (at higher resolution) the same input may be represented as a time varying function.

The ability to formulate the world models with this multiresolutional generalization is an essential device for coping with the complexity of planning the system operation in MCA.

7.2 Perception Stratified by Resolution

In the subsystem P (perception) the information mapped from the world is being stored and organized. The process of organization presumes the process of recursive generalization.: the "whole" at the input in P is many times stratified by resolution.We will call "phaneron" the array of information which is coming into P from the whole multiplicity of sensors. (The term "phaneron" was introduced by C.S. Pierce for the totality of information which can be called *phenomenal world*). Phaneron is not structured at the moment of arrival, it should be recognized, identified within ER-structure. These processes are broadly discussed in literature, and the importance of such phenomena as "attention", and "resolution" was emphasized many times in literature.

Separation in levels appeared to be a natural phenomenon linked with the properties of attention, and its intrinsic links with the process of generalization. In fact, generalization is required to provide the efficiency of computing resources use and allocation, and attention is one of its tools. Thus, the new class labels which are created by the process of generalization, are being considered as new primitives of the upper level of world representation. This rule: the class labels of the lower level are considered as primitives for the higher level, is one of the laws of the mechanism of nested hierarchy.

The results of this identification (a *snapshot of the world*), contain information part of which can be different in the previous snapshot, and part won't change (e.g. about relations among objects and/or their properties). Thus, the identification can be done only in the context, i.e. in constant interaction with another body of information which is not specified in detail in this paper and could be called a "Thesaural Knowledge Base. This affects the set of preprocessing procedures which are being separated from the rest of the intelligent module primarily because of the first experience of manufacturing of the computer vision systems . Simultaneously with the process of finding phaneron structure (or image interpretation) the problem of proper allocation of the information contained within phaneron should be done.

As one can see from Figure 11 the systems of phaneron at different levels are nested by generalization

$$P_1 \overset{g}{\supset} P_2 \overset{g}{\supset} ... \overset{g}{\supset} P_{i-1} \overset{g}{\supset} P_i$$

(43)

The results of structuring in a form of nested multiresolutional system are delivered to the Nested Hierarchy of the Knowledge. Knowledge for a mobile robot is represented by a system of maps with tables of interpretation.

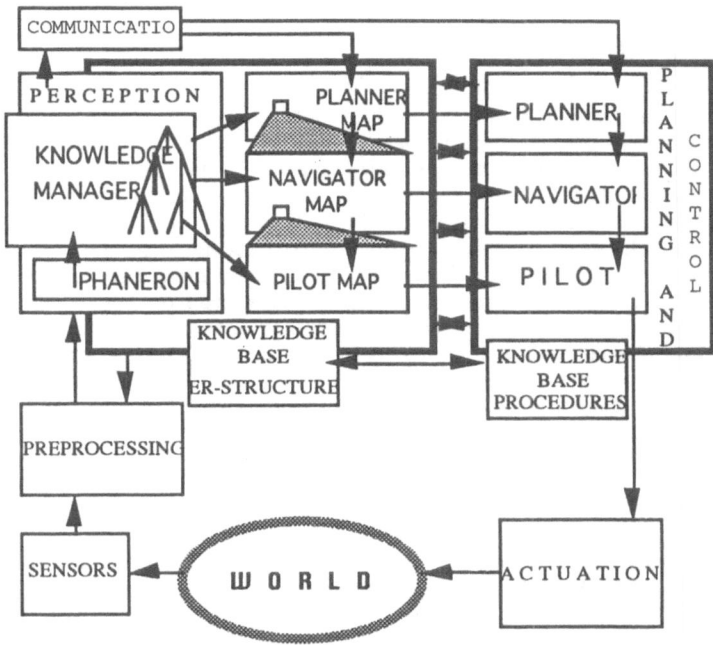

Figure 11. An example of MCA structure: three-level operation is illustrated

7.3 Maps of the world stratified by resolution

Decisions are based not only upon the phaneron (unless we are talking about decisions of the lowest level). Decision-making process utilizes the map of the world which consists of the organized multiplicity of phanerons (or their generalizations) known from the previous experience

Thus, the information of the world which is required for decision-making, should be explicated and prepared for the subsequent decision-making processes. Map is defined as a subset of the state-space which is to be taken in consideration during the process of decision-making. Thus, map is a part of policy delivered from the leader to the follower. Map of the upper level contains the maximum subset given at the lowest resolution.

Maps for the next levels of the nested map structure (top down) are obtained using the second apparatus of nesting: focusing of attention. Thus, only one part of the map is being activated: which is required for the current decision-making and is illustrated in Figure 11. The system of maps has dual nesting; firstly, we have a subsystem with nesting by generalization

$$M_1 \overset{g}{\supset} M_2 \overset{g}{\supset} \ ... \ \overset{g}{\supset} M_{i-1} \overset{g}{\supset} M_i \tag{44}$$

which we name "CARTOGRAPHER", and assign to this subsystem maintenance of the information rather than active participation in the process of decision-making, and secondly, we have nesting by focusing attention superimposed upon the nested hierarchy by generalization, and this information extracted from the cartographer $\{m_i\}, i=1,2,...$ is delivered to the "C" system (Figure 11) in a form of "maps for decision making".

$$\begin{array}{cccccccc} fa & g & fa & g & g & fa & g & fa \\ & \supset & & \supset & \supset & & \supset & \\ (m_1 \in M_1) \subset (m_2 \in M_2) & \subset & ... & \subset (m_{i-1} \in M_{i-1}) & \subset & (m_i \in M_i) \\ fa & & fa & fa & & fa \end{array} \qquad (45)$$

All fa-predicates are performed on the basis of MCA-algorithm results and thus belong to the subsystem of control, all g-predicates are prerogative of the system of map maintenance. Later we will show how these predicates are built in the algorithms of planning/control and map maintenance.

The problem of map maintenance is of scientific and practical importance. The upper level map ("planner's map") should be maintained for a long time due to the largest scope, and the "slow rhythm" of this level. Changes in the upper level map are not frequent. Maps of the subsequent levels are to be regularly updated with increasing frequency but decreasing volume. The lowest level map ("pilot's map") may or may not need a map maintained as a part of the nested hierarchy. (Actually, from our first experience of dealing with MCA we found that intelligent module cannot afford maintenance of the pilot map , i.e. of the lowest level of world representation), and therefore all processes related to the real time operation have *ephemeral structure* with a number of logical filters determining whether this ephemeral information contains anything to be included in the maps of the upper level

8. Development of Algorithms for MCA

8.1 Extension of the Bellman's optimality principle

The principle of optimality of Bellman [67] can be stated as follows for stochastic problems: at any time whatever the present information and past decisions, the remaining decisions must constitute an optimal policy with regard to the current information set [85].

Three types of inclusion participate in (37) through (45): by generalization (g), by focus of attention (fa), and by time (t). The last points to the fact that the time sequence of information is nested, i.e. each next contains its predecessor. Bar-Shalom has shown [85] that in the case of incompletely observed Markov process, stochastic dynamic programming can be applied as well as similar search type control algorithms.

$$\begin{array}{ccccc} fa & g & fa & g & fa \\ [X^\circ_{r3} \in C_{r3}] & \supset & [X^\circ_{r2} \in C_{r2}] & \supset & [X^\circ_{r1} \in C_{r1}] \end{array} \qquad (46)$$

the inclusion C_{r2} ... X°_{r1} should be considered to be the zone of convergence for X°_{r2} and the inclusion $C_{r3} \supset X^\circ_{r2}$ should be considered to be the zone of convergence for X°_{r3}.

8.2 Nested Multiresolutional Search in the State Space

Optimum decisions found at different levels are nested if they are found under cost-functionals which constitute a nested system. (In particular, minimum-time controls found at different levels of resolution, are nested). Multiresolutional decision making process allows for using efficiently the full computer capacity which is limited at each level of such hierarchy (with no branching). In this case, the tree hierarchy of intelligent control converts into MCA. If MCA is acting under the above mentioned constraints the process of control allows for decoupling in parts dealing separately with information of different degree of resolution (easily interpreted as certainty, belief, etc.)

The problem of motion planning was given substantial attention in the literature on AI and robotics. However, problems of optimum planning (or optimum navigation, or optimum guidance) as well as optimum control until now do not have consistent and tractable solutions. Either solution is consistent or tractable but never both. Motion planning is frequently understood in the context of "solvability" of the problems of positioning or moving the object rather than in the context of finding the desirable trajectory of motion. Nevertheless one cannot argue that the real problem of concern is finding the location and/or trajectory of motion which provides a desired value of some "goodness" measure (e.g. the value of some cost-function).

We will demonstrate that the decision-making process in MCA is an incompletely observed Markov process. The space is being discretized, the centers of tiles in the tessellation are connected and form a graph, cost of each edge of the graph is assigned. This allows for using an approach to discretization of the control system which is different from the existing approaches of systems "digitalization". We do not discretize the time by introducing a "sampling period". The system is being considered to move in a discretized state-space from one node on the graph to another one paying the cost of the edge upon its traversal.

Method of Nested Dynamic Programming (NDP) follows from the commutative diagrams actually shown in Figure 10. It follows also from (42) through (45) and states that the optimum control should be found by consecutive top-down and bottom-up procedures, based on the following rules.

Rule 1. NDP should be performed first at the most generalized level of information system with complete (available) world representation.

This will obviously lead to a very fuzzy solution from the view of the lowest level of system ("actuator"). However, this enables substantial advantages later: the substantial part of the world will be excluded later from consideration at the lower levels.

Rule 2. NDP is being performed consecutively level after level top down. The subspace of the search at each of the consecutive lower levels is constrained by the solution at the preceding upper level recomputed to the resolution of the next lower level.

The optimum solution for the upper level, is considered at the lower level as the stripe (zone, envelope of search) for further independent decision-making. However, due to the additional information which appears at a given level during the performance, the optimum solution of the lower level may require to seek beyound the stripe of independent decision making. Rule 3 is to be applied in this case.

Rule 3. When during the actual motion, due to the new information, the optimum trajectory determined at a given level must violate the assigned boundaries, this new information should be submitted to the upper level (proper generalization must be performed, and the information structure must be updated). This generates a new top-down NDP process.

The nested resolutional (by generalization) world representation (e.g. category of "knowledge" C_{gk} corresponds to the nested resolutional (by attention) world representation as follows

$$
\begin{array}{ccccccc}
\cdots \supset & C_{gk,i-1} & \supset & C_{gk,i} & \supset & C_{gk,i+1} & \supset \cdots \\
& \downarrow & & \downarrow & & \downarrow & \\
\cdots \subset & C_{ak,i-1} & \subset C_{ak,i} & & \subset & C_{ak,i+1} & \subset \cdots
\end{array}
\tag{47}
$$

which is the major basis of nested decision-making processes upon these hierarchies. From (47), a rule of ordering the decisions follows on the basis of nesting and the policy of decision making. We will formulate this rule as follows: given a nested world representation

$$
S_1 \supset S_2 \supset S_3 \supset \cdots \supset S_i
\tag{48}
$$

and a set of cost-functionals for these representations, based upon common policy of decision-making, the set of decisions will constitute a nested hierarchy

$$
D_p(S_1) \supset D_p(S_2) \supset D_p(S_3) \supset \cdots \supset D_p(S_i)
\tag{49}
$$

Figure 12 illustrates three levels of world representation which is required for a decision-makers in the nested multiresolutional team for an autonomous vehicle.

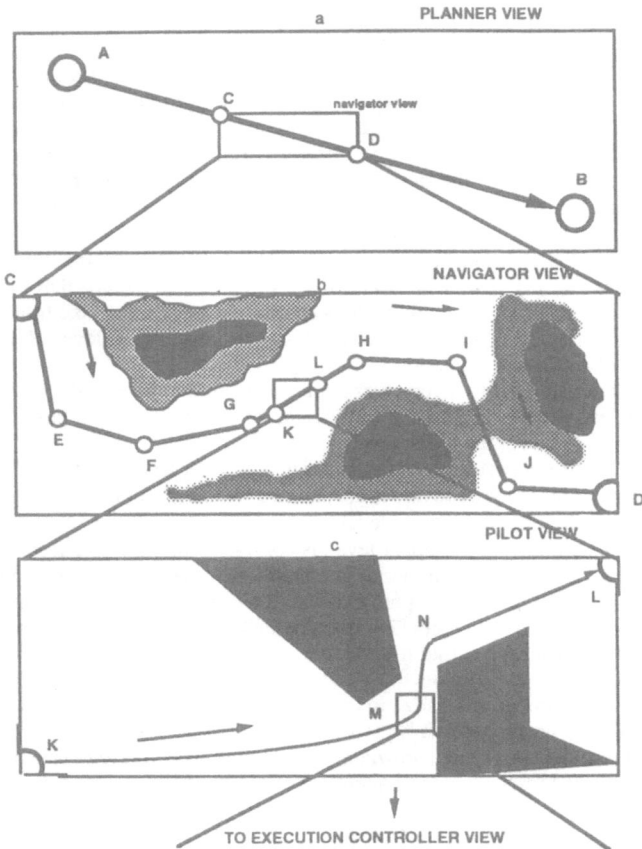

Figure 12 Example of three-level world representation

The world represented at the upper level ("a") can be empty, and only initial point A and the goal B are shown. At this level of consideration, no particular object of the world is shown, all of them are below the level of generalization. The only consideration which is presumed to be reasonable at this level, is a judgment of the general properties of the space: "how good the state space is for the motion" or which is the same, what is the cost of moving in the particular tile of the tessellation at this level.

It is also presumed, that the world representation occupies the maximum of computer facilities available for dealing with representation. (For systems with sequential, or pseudoparallel operation of decision-makers this means using full computer capacity at the level. For multiprocessor systems this presumes that corresponding value of computing power will be assigned in advance, still the total power is limited).

In order to reduce ourselves to the next level of the nested hierarchy, we will focus the attention of the next decision-maker upon a definite subset (shown in Figure 12,a as a rectangle), and will "zoom" this rectangle into a higher level of resolution shown in Figure 12,b. (Now again the whole computer power is dedicated to dealing with this magnified representation). Here we have smaller subset of the overall state space, but given with more details. These details can be represented in a twofold manner depending on the vocabularies accepted.

Firstly, we can continue with our upper level tendency to consider only cost of achieving one state-tile from another. At this levels we consider tiles which are "finer": their size is smaller, the values of states are presented with higher accuracy, the value of cost is given with high accuracy too. On the other hand, the decision-maker at this level might be able to understand another way of world representation. Indeed, the obstacles can be shown as geometrical entities, and the decision-maker will contemplate "how to avoid" the. Let us state clearly, that the difference between these two ways of representation is not essential at all. In fact, representing the obstacle we are representing the areas of the space which have infinitely large cost of "being achieved", and therefore will never be selected by a reasonable decision-maker (will be "avoided").

After second zooming (Figure 12,c) more details are being shown for the smaller subset of the state-space. More details emerge for a decision maker. However, the ultimate nature of the representation does not change at all: they remain to be refinements for the cost of achieving the smaller tile from its neighbors whereas definite values of the states are assigned for this particular tile. If the arrangement of neighbor detection is assigned as shown in Figure 12, i.e. by the principle of "location coordinate", then the other states, like rates of this coordinate change (speed, acceleration, jerk) can be assigned combinatorially to this particular neighbor-tile.

9. Complexity of Knowledge Representation and Manipulation

9.1 Multiresolutional Consecutive Refinement: Search in the State Space

Search by scanning the string of available alternatives ("browsing"), and selection when the desirable property is met, is one of the straightforward algorithms of *combination generation*. If the results of search are constantly enhancing the input vocabulary, then during the exhaustive search with recursive enhancement of the vocabularies, all possible unions, intersections, and complementations of the sets are being obtained.

On the other hand, any combinatorial algorithm is an operator of generating *solution alternatives* for a decision-making process; see [32, 36, 37, 90, 91]. Then a number (value) is assigned to each of the combinations generated (preferability, closeness, propensity, cost-effectiveness, etc.) which will enable the decision-maker to make his choice under the accepted strategy of decision making. According to the existing terminology, the chain of consecutive decisions is called *policy of decision-making*, or policy of control).

We consider n-dimensional continuous Euclidian state space E^n and a closed domain (of interest) in it Ω ($E^n \supset \Omega$) with volume V and diameter d. This domain can be divided (decomposed) in a finite number of nonintersecting subdomains so that their union be equal to the domain of interest (Ω_j- is a closed j-th subdomain with a diameter d_j, $j \in H$, where H - is a set of all subdomains). Obviously, in this case

$$\bigcup_{j \in H} \Omega_j = \Omega. \tag{50}$$

If each pair of subdomains can have only boundary mutual points ($\Omega_{j2} \cap \Omega_{j1} \neq \varnothing$) then they are called *adjacent* subdomains. Relation of adjacency will play a key role in representing the context of the problem to be solved.

Each of the subdomains can in turn be decomposed in a finite number of sub-subdomains and so on. This sequential process we will call *decomposition* while the opposite process of forming domains out of their subdomains we will call *aggregation*. Every time we will decompose (or aggregate) all domains simultaneously, and the results of decomposition (or aggregation) we will call *a level* (of decomposition, or aggregation). We will consider the *relation of inclusion* which emerges as a part of processes of decomposition-aggregation to be the key property which should play an essential role in representing the context of the problem to be solved. We will illustrate this in Figure 13.

202

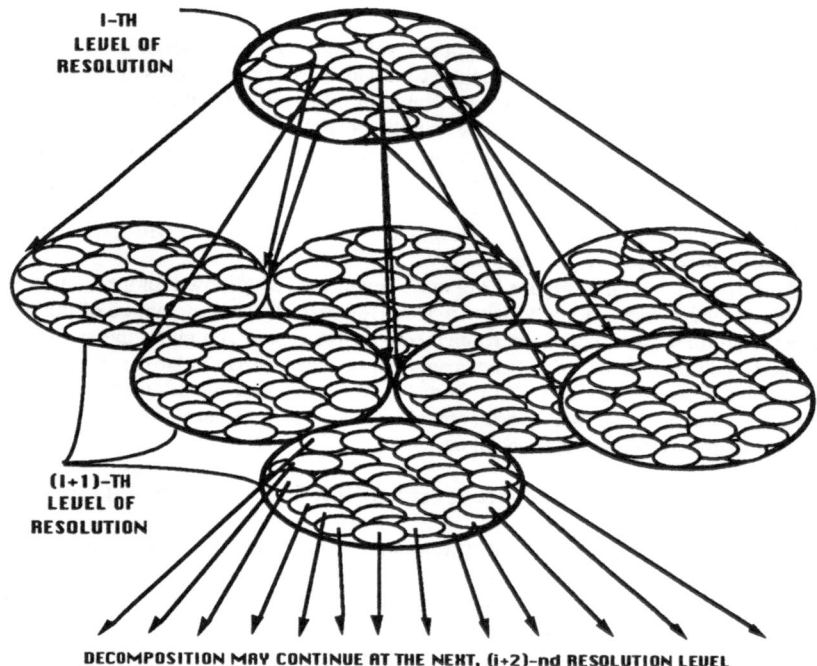

I-TH
LEVEL OF
RESOLUTION

(I+1)-TH
LEVEL OF
RESOLUTION

DECOMPOSITION MAY CONTINUE AT THE NEXT, (i+2)-nd RESOLUTION LEVEL

Figure 13. Multiresolutional decomposition of a domain

We won't allow for the infinite process of decomposition: we introduce a notion of *accuracy (resolution)* by determining the smallest distinguishable (elementary) vicinity of a point of the space with a diameter $d_j = \Delta$. This diameter can be considered *a measure*, and this vicinity can be considered a single *tessela*, or a single *grain* which determines tessellation (granulation) of the space, or which is the same, the *scale* of the space. Each tessela contains an infinite number of points however for the observer they are indistinguishable. However, each subdomain of decomposition contains a finite number of elementary subdomains (grains) which have a finite size and determine the accuracy of the level of decomposition (see Figure 13). The observer will judge this elementary domain by the value of its diameter and by the coordinates of its center. Let us demonstrate how it can be done. Obviously, our construction alludes to the Hausdorff space. We will not introduce the Hausdorff space as a tool in this paper. We will discuss the Euclidian space with limited accuracy of measurements available (which is a practical example of a Hausdorff space, see [13]).

We did not make any assumptions about the geometry of subdomains which begs for a definition of the *diameter* of the subdomain. A standard definition for the diameter $d(\Omega_j)$ of the subset Ω_j of a metric space [92] is presented via notion of distance D in a metric space as

$$d(\Omega_j) = \sup_{x,y \in \Omega_j} D(x,y) \tag{51}$$

and holds for all levels of decomposition except of the lowest where is understood as the elementary subdomain in which points x, y cannot be specified, and the distance cannot be defined. Thus, formula (51) cannot be used for computing the diameter of the lowest level of decomposition.

Let us take one arbitrary point from the elementary subdomain Ω_j, $j \in H$ and call it the center of the elementary subdomain (it does not matter which one of the points is chosen: at the highest level of resolution different points of the elementary subdomain are indistinguishable anyway). In order to distinguish the centers of the elementary subdomains from other points of the space we will call them nodes and denote q_j, $j \in H$. The relationships $\Re_{ij}{}^k$ among each two particular nodes i and j of a particular level k of decomposition are very important for solving the problem of control. We will characterize this relationship by the *cost* of moving the state from one node (i) to another (j). Cost is a scalar number C_{ij} determined by the nature of a problem to be solved. Graphically, cost will be shown as a segment of a straight line connecting the nodes, this segment we will call an "edge". A set of all nodes and edges for a particular subdomain we will call "a graph representation of knowledge for this subdomain".

For the lowest level of decomposition where each subdomain is the elementary subdomain we will introduce the diameter of the elementary subdomain (or the diameter of a tessela, or the diameter of a grain) as the average diameter. At this time we will not focus on the statistical data which are the basis for the averaging. It can be a multiplicity of measurements of a single elementary subdomain Ω_j.

It can be also a multiplicity of measurements performed on a population of the elementary subdomains belonging to the same subdomain of the upper adjacent level. It may happen that elementary subdomains belonging to the different subdomains of the adjacent upper level will have different diameters. One can expect that in order to compute the statistical data for measuring at the level of required accuracy, the information should be submitted concerning with the next adjacent (higher accuracy) level of decomposition. We intentionally omit these continuations which are to be addressed in a special paper.

Assume Δ is the average diameter of the elementary subdomain at the highest level of resolution. Then, the order of the average volume for the elementary subdomain is evaluated

$$V_{av} = O[\Delta_n] \tag{52}$$

as well as the quantity of the elementary subdomains at the level (of highest resolution), or the cardinal number of the set H:

$$N = {}^V/_{V_{av}} \tag{53}$$

Cost also can be characterized by the value of accuracy which is determined by the accuracy of the particular level of resolution. Statistically, cost at a level k of moving from a node i to the node j, is understood as the average of all costs of moving from each subdomain of the node i to the each subdomain of the node j. Even more important are relationships of inclusion between the subdomains of the adjacent levels of decomposition.

9.2 Multiresolutional Consecutive Refinement: Multiresolutional Search of a Trajectory in the State Space

We are looking for a trajectory in the state space which satisfies the specifications. Trajectory is a string of adjacent elementary subdomains denoted as $\cdot \Omega j \dot{O}$. It can be also represented as a sequence of the subdomain indices $T = \langle \varphi(\mu) \rangle$ where μ- is a number of elementary subdomains in a string ($\mu = 1, 2, ..., z$). A well posed problem should have the initial point (SP) and the final point (FP) of the trajectory assigned. If T has a start point $SP \in \Omega_{j(1)}$ and a finish point $FP \in \Omega_{j(z)}$, then T is called a *feasible trajectory*. A feasible trajectory in the graph for the particular subdomain is represented as a *path on the graph* of this subdomain.

The off-line method is introduced for finding the best trajectory of motion which should be followed by the control system. Search in the state space (S^3-search, see [36-37, 58]) is done by synthesizing the alternatives of motion and scanning the set of available alternatives[10] . One of the

---

[10] Each alternative is considered a string of consecutive states which can be translated into corresponding string of commands (control input).

strings is selected when the desirable property is met. The whole process was illustrated in Figure 8. The state space is represented as a set of k maps with different resolution (increasing top-down). The maps are discretized by building a random graph (grid). The density of this grid can be introduced as the ratio

$$\rho_k = N_k / V_k \qquad (54)$$

where N_k is the number of points at a level k, and V_k is the total volume of the state space (under consideration at this level k). Density ρ_k can be considered a measure of accuracy for the k-th level.

The points are put in a uniform manner (random graph with uniform distribution) and the idea of the volume of the vicinity can be introduced which is the value inversely proportional to the density

$$\sigma_k = 1/\rho_k = V_k / N_k. \qquad (55)$$

This discretization can be characterized by the value of average distance between the two "adjacent nodes" of the graph. The space is characterized by the rate of losses which are supposed to be associated with motion from one point to another (rated cost, or unit cost). These losses (rated cost) are in general case different in the different domains, or under different circumstances. They can be determined by the time required to traverse the distance between two consecutive points, energy dissipation, energy consumption, dynamics, etc. Thus we are dealing with the general problem problem of optimum (minimum cost) control of motion in a variable cost space.

Domains of the state space Ω_k and their densities of points ρ_k at different levels are different so that the following inequalities hold

$$\Omega_1 \supset \Omega_2 \supset ... \supset \Omega_k \supset ... \supset \Omega_m; \ k=1,2,...m, \qquad (56)$$

$$\rho_1 < \rho_2 < ... < \rho_k < ... \varphi_m; \ k=1,2,...m, \qquad (57)$$

while

$$V_1 > V_2 > ... > V_k > ... > V_m. \ ; \ k=1,2,...m, \qquad (58)$$

where V_1 is the total volume of space under consideration. The heuristic of *contraction* is introduced in (56).Since solution is searched for within the volume of the state space V_k designated for search, we have to have some justification for the contraction: we should reduce the probability that contraction eliminate some or all of the opportunities to find *the* optimum path trajectory. The following heuristic strategy of contraction is chosen. After the search at the lowest resolution level is performed, the optimum trajectory is surrounded by an envelope, which is a convex hull which has a width w determined by the context of the problem. Then, the random points generation at the next level of resolution is performed only within this envelope of search. This strategy is demonstrated to be acceptable in many practical cases. However, the problem of consistency representation under the contraction heuristic has to be addressed in the future.

Let Ω-state space in which the start and final points SP and FP are given. The path from SP to FP is to be be found with the final accuracy ρ. Let us consider $\Omega = \Omega_1$ and $\rho = \rho_m$. We will introduce three operators.

I. Operator of Representation ($\mathfrak{R}$). This operator is based upon generalization.

$$\mathfrak{R}:(\Omega, \rho) \to M, \ or \ M = \mathfrak{R}(\Omega, \rho)$$

where M- is the map representing the state-space Ω, ρ is the level of resolution of this map determined by the density of the search-graph.

II. Operator of state space search (S^3), or operator of combinatorial search at a level.

$$S^3: (M, SP, FP, J) \to P \ , \ or \ P = S^3(M),$$

where P- is the optimum path connecting the start point SP and the finish point FP, J- is the cost of operation which should be minimized as a result of search S^3.

III. Operator of contraction (C), or operator of focusing attention at a level.

$C:(P, w) \rightarrow \Omega$, or $\Omega = C(P)$, where w- is the parameter of the envelope, (e.g. the "width" of the envelope).

The multiresolutional control algorithm (one particular instantiation of GFACS) can be described as follows.

For $k=1, ..., m$ do the following string of procedures:

a) $\Omega_k = C(P_{k-1})$, or at $k=1$ assume $\Omega_k = \Omega$,

b) $M_k = \Re(\Omega_k, \rho_k)$,

c) $P_k = S^3(M_k)$.

The algorithm of control can be represented as a diagram

$$
\begin{array}{ccc}
w & \rho_k & SP,\ FP,\ J \\
\downarrow & \downarrow & \downarrow \\
P_{k-1} \rightarrow C \longrightarrow \Re \longrightarrow S^3 \longrightarrow P_k \\
\Omega_k \quad M_k
\end{array}
\tag{59}
$$

or a recursive expression

$$P_k = S^3(\Re(C(P_{k-1}, w), \rho_k)\ SP,\ FP,\ J) \tag{60}$$

Algorithm (59) shows how the united plan of a level should be obtained. It does not demonstrate how the decomposition of this assignment (plan) is done among the multiple autonomous agents operating at a level of resolution. We would like to focus on this property of the plans obtained by (59) since this is a fundamental property of intelligent information processing in multiresolutional systems. This property can be formulated as follows: plan propagation top-down as well as performance propagation bottom-up does not depend on the particular solution of decomposition accepted at a level. The same top-down and bottom-up processes of propagation will be valid at different decompositions of a level into a set of autonomous agents. In other words: the plan/performance propagation is unique, while its decomposition and/or decoupling is not unique.

9.3 Evaluation and Minimization of Complexity of the MCA

Thus, we can address the problem of complexity only for the process of top-down propagation of the planning/control process and the bottom-up propagation of sensing and performance evaluation. However, it will be clear from the subsequent material that whatever solutions concerning with the agents selection are made, they will be affected favorably by recommendations of this subsection.

Let C_k- complexity of the GFACS algorithms at the level k, or just complexity of the combinatorial search; φ-a function of complexity depending on the number of nodes N_k, then the total complexity of the control system will be

$$C = \sum_{k=1}^{m} C_k = \sum_{k=1}^{m} \varphi(N_k) = \sum_{k=1}^{m} \varphi(V_k \cdot \rho_k) \tag{61}$$

We have to transform (61) in such a way as to determine how is the complexity of GFACS (or just of a combinatorial search) affected by the number of resolution levels and the law of changing the value of resolution from level to level (refinement ratio).

Let us introduce a set of variables (refinement ratios, or ratios between the densities of the adjacent levels):

$$x_{k+1} = \frac{\rho_{k+1}}{\rho_k} > 1, \; k = 1, 2, \ldots, m-1 \tag{62}$$

Then

$$\rho_2 = x_2 \rho_1, \tag{63}$$

$$\rho_3 = x_3 \rho_2 = x_2 x_3 \rho_1, \tag{64}$$

Let us introduce a new set of variables (contraction ratios):

$$y_{k+1} = \frac{V_k}{V_{k+1}}, \; k = 1, 2, \ldots, m-1, \tag{65}$$

which means that the value for areas can be rewritten as follows

Taking in account the new expression for V_k and assuming $y_1 = 1$ (no contraction at the first level) the equation for complexity can be obtained with the set of refinement ratios $\{x_i\}$ and the set of contraction ratios $\{y_i\}$ both reflected

$$C = \varphi(x_1) + \sum_{k=2}^{m} \varphi \left(\prod_{i=1}^{k} \frac{x_i}{y_i} \right), \tag{66}$$

and the general expression for complexity can be introduced as

$$C = \varphi(x_1) + \sum_{k=2}^{m} \varphi \left[Ax_k \left(\prod_{i=1}^{k-1} x_i \right)^{\frac{1}{n}} \right] \tag{67}$$

Optimization of the controller structure for the class of systems under consideration can be done by minimization of the form (67).

Computational Complexity of the multiresolutional controller as a function of the number of levels of resolution is a curve which can be characterized by a very high value at $m=1$, rapid drop when the number of levels is growing, and then slow asymptotic approach of the constant value (Figure 18). It has a weakly demonstrated minimum below the level of asymptotic complexity. This minimum does not create clearly distinguishable preference for selection of the number of levels, and we believe that the choice should rather be done based upon the value of m which corresponds to a definite stage of reduction of complexity $C(m)$ before the actual minimum is achieved. As a tool for selecting the preferable value of m we would recommend to target the "practical" value of complexity $C_{pr} = (1.05-1.15)C_{\infty}$. Figure 14 shows that when α is reduced, the values of m_{opt} are becoming smaller, and the minimum

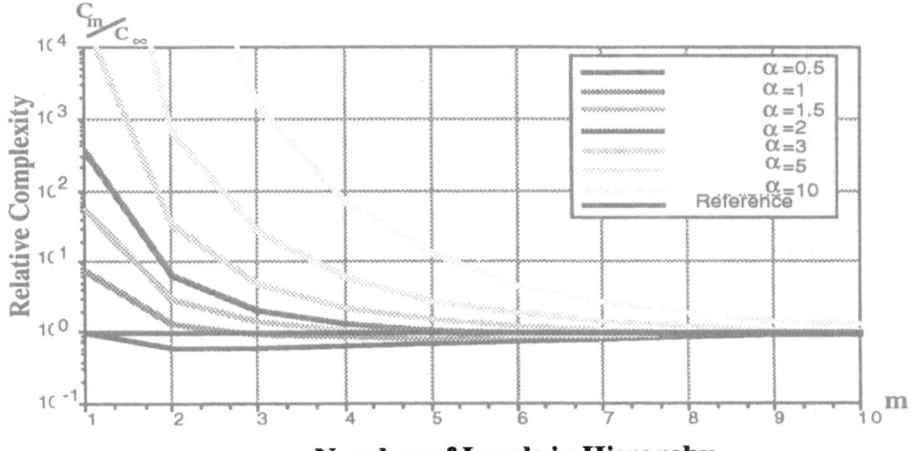

Number of Levels in Hierarchy

Figure 14. Dependence between relative complexity and and the number of levels
at different order of search complexity (α=.5 through 10) and at the number of variables n=3

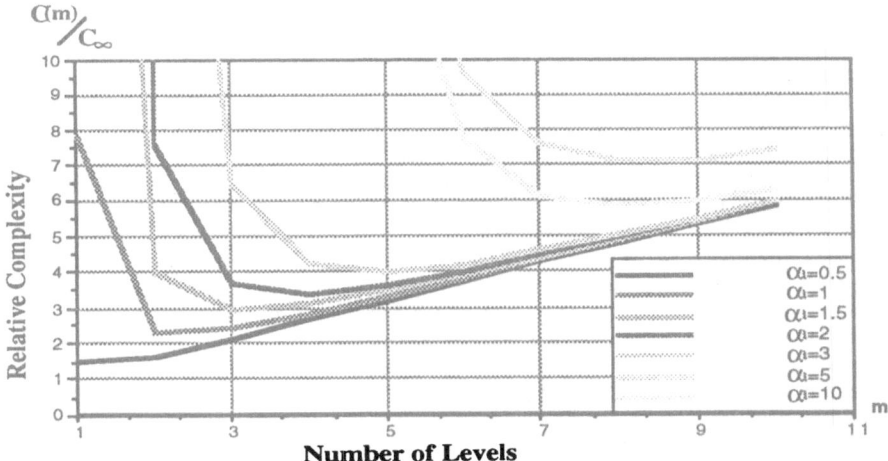

Number of Levels

Figure 15. Relative Complexity of the Multiresolutional System as a function of Number of Levels if
the cost of adding new levels linearly increases with their numbers

is sharper. For the algorithm of search with high complexity one needs larger number of resolution levels
even to achieve the "practical" choice value C_{pr} . Finally, we decided to evaluate a more realistic case
when increase in the number of levels leads to the additional costs. For the case with linearly growing
costs, the case is demonstrated in Figure 15 which shows that the minimum is becoming sharper and
shifts to the zone of smaller number of resolution levels (e.g. m_{opt}=1 at α=.5; m_{opt}=2 at α=1, m_{opt}=3
at α=1,5; m_{opt}=4 at α=2, and so forth).

Figures 14 and 15 demonstrate that in the systems with multiresolutional representation and nested hierarchical control the following statements can be made concerning the issue of complexity:

1. Complexity of MCA is drastically reduced in comparison with a single resolution controller. It implies that MCA are capable of practically address the set of problems which are considered theoretically as prohibitively complex ones.

2. As the number of level of resolution grows, the complexity falls, however with the

further creation of additional resolution levels this growth is becoming slow. One can expect that since there always is a price to pay for introduction of additional levels, we should not continue adding new levels as soon as some reasonable reduction of complexity is achieved.

10. Case Studies

10.1 A Pilot for an Autonomous Robot (two levels of resolution)

In order to begin planning the path of a robot in a space with obstacles, it is necessary to have a geometrical representation of the extent and the elements of the search space, including the robot. This job is accomplished by creating geometrical models of the workspace, robot, and obstacles. Templates for these objects are components of the simulation package SIMNEST developed by Drexel University with NIST participation for design and control of multiresolutional systems [96, 97].

The state of the autonomous robot includes the following variables: 1] x - position, 2] y - position, 3] orientation, 4] speed, 5] steering angle. We will consider this to be the level of 'complete informedness' or maximum resolution (even though we can include more complex issues such as the effect of roll on cornering dynamics or skidding) because it is sufficient to illustrate the principle of successive generalization required for obtaining a representation which fits into our concept of planning/control algorithm of consecutive refinement.

Depending upon the acceleration imparted to the robot by the propulsion and steering actuators, it is possible to formulate a model of the system in which the next position of the car, given its current state, is a function of the effective steering angle and velocity, 'a' and 'v', during the interval of modeling. The approach to system modeling at this stage, begins to come under the influenced of the desire to solve the motion planning problem using the principles of consecutive refinement. This means that the robot should be represented at more than one level of abstraction, and a rough plan should be created with an imprecise model before an attempt is made to synthesize the final plan. Since this example illustrates the use of two nested levels of planning, it will only be necessary to construct these two versions of the robot model.

The technique of planning which is being exemplified here involves top-down search with successive refinement. In generating the graph of the first level, the model that is used is one that allows the inclusion of the least amount of detail that is considered sufficient to provide rough predictions of qualitative dynamical behavior. This is accomplished by representing the robot as an object whose velocity is conditioned by its change of direction. In our model only the resistance to change of direction is shown; other kinematic or dynamic issues are not represented. In order to utilize this model for planning, the first level search procedure begins by discretizing the horizontal workspace of the robot. The rule of discretization is novel, in that a regular grid is not utilized. Instead, the xy plane is partitioned into regular subsets, and a random coordinate pair is used to represent each subset. This approach eliminates the idiosyncrasies associated with regular grids without utilizing a biasing heuristic. The number of points used to represent the space is a function of the average inter-point spacing and the narrowest passageways in the map. The influence of these parameters can be determined by experiment.

The connectivity of the randomly discretized version of the workspace is decided by the definition of vicinity at a given level of the representation. If the concept of vicinity is relaxed to include

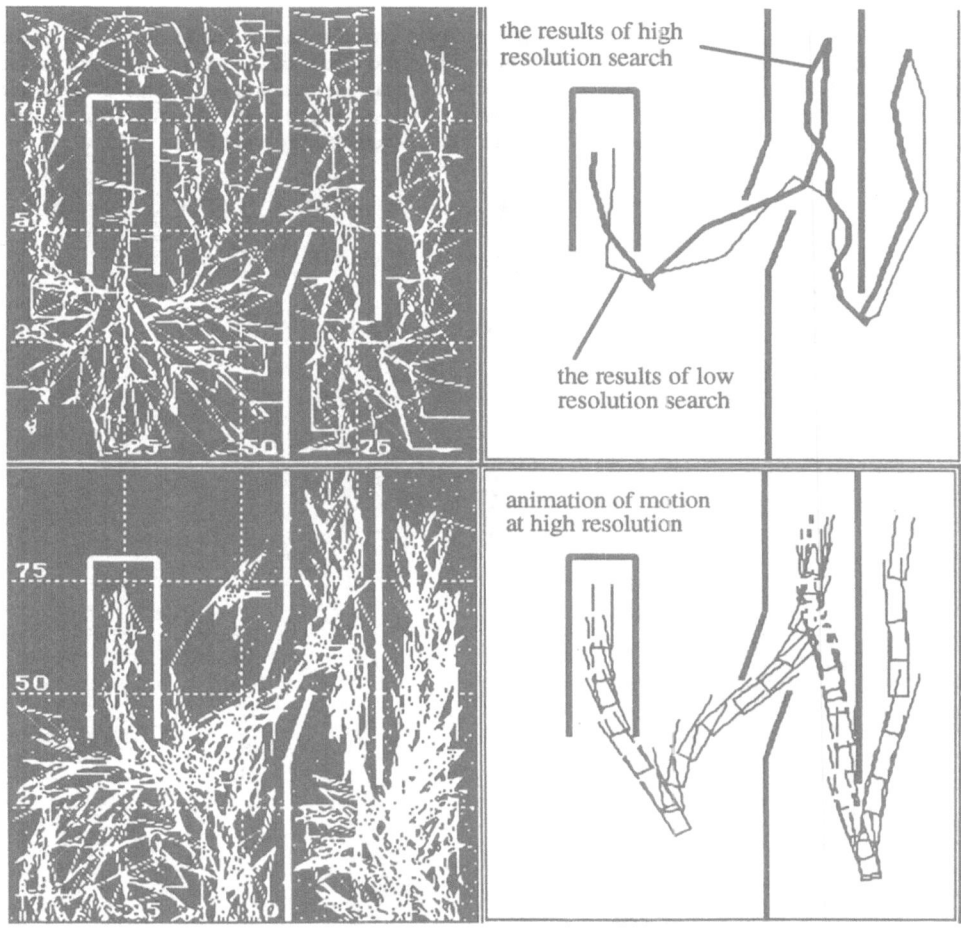

the results of high
resolution search

the results of low
resolution search

animation of motion
at high resolution

Figure 16. Experiment in a configuration 1 (a-search in the whole space at low resolution, b-search in
the reduced space at high resolution)

the whole space, the graph could hypothetically be fully connected. On the other hand, curtailing the
measure of vicinity may lead to a graph which is not connected. This problem can be addressed by using
heuristics, or by trial and error. There remains the issue of whether edges in this graph traverse forbidden
regions in the workspace or whether the robot will intersect obstacles during such traversals. Such
potential intersections are tested individually during the construction of the graph. The concept of the
robot being represented as an expanded point is exercised and the degree of expansion is fixed by iterations
beginning with the minimum dimensions of the robot.

The trajectory determined at the first level of the search, embodied by a string of xy
coordinates, is then provided to the next level for refinement.

The second level of refines the results of the first search by utilizing a similar search
procedure but it operates on a reduced space in the vicinity of the first (approximate) plan and the graph
generation process takes into account the body configuration, kinematics and dynamics of the robot in

210

greater detail.

The robot model that is used at this second level includes the kinematic constraints of the robot as well as dynamical constraints on acceleration. Thus, the model of the second level utilizes the assumption that the steering angle and velocity of the model remain constant in the interval of modeling to predict the new coordinates of the robot. Also, a maximal rate of change of steering angle and velocity is used to describe the dynamics of steering and acceleration. A neighboring point is

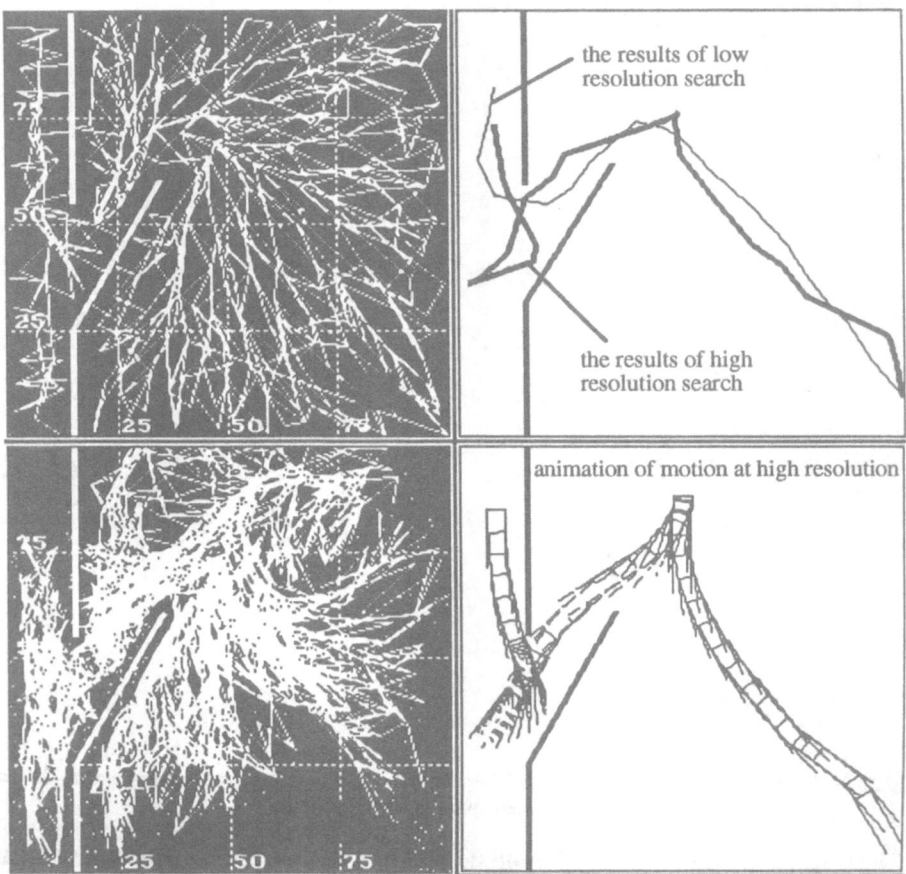

Figure 17. Experiment in a configuration 2 (a-search in the whole space at low resolution, b-search in the reduced space at high resolution)

deemed to be unreachable from another if the transition requires a change of steering larger than allowed for a single transition.

The trajectory, which is described by a string of points, is enclosed by sequence of cells whose width is a function of the characteristics of the trajectory itself. As can be seen in the figure, straight segments are enclosed in narrow cells such as cell (i) whereas a change in direction establishes the need for greater width, as in cell (i+1). The area of the cells must be increased if complex maneuvers are expected (such as turns or direction reversal) because new constraints will be included in the new level of description and the abrupt changes which can be made with the slack constraints of the rough description may no longer be possible. The selective refinement of the scope of search is thus a heuristic for guaranteeing consistency in this paradigm. One further heuristic is applied at this stage; the cell size is increased if the baseline cell intersects an obstacle, because avoiding the obstacle may involve departing significantly from the nominal trajectory.

The number of dimensions in the search space is increased at the second level, where velocity becomes an additional axis. Thus, each cell from Figures 16 and 17 is decomposed into enough subcells to achieve the assigned average spacing of this second level and each randomly generated point has associated with it an x-position, a y-position, and a velocity.

A method of successor validation is utilized to determine children for each parent in order to preserve the fidelity of the representation of this level. This process begins with an evaluation of the steering angle 'a' which is required to attain each candidate successor point from each putative parent node. Inverse kinematic routines are used to determine this information as well as the corresponding center and radius of rotation. Constraints are applied in the following manner:

    i] If the required steering angle exceeds the maximum steering angle, the successor is invalid

    ii] If the required change in steering angle exceeds the maximum allowable value, the successor is invalid.

    iii] The time of transition is constrained by the maximum velocity allowed for the required radius of rotation. This value is determined from the maximal centripetal force which can be tolerated without slippage.

The edge cost is determined by the distance to be moved along the circumference of the circle and by the random velocity which is the third coordinate of the parent node. It has been determined experimentally that in order for the inverse kinematics routine to be numerically stable, the density of points in the XY-plane should be such that, on average, at least three points are located in a unit equivalent to the cross-sectional area of the robot.

Figures 16 and 17 depict the process of planning via snapshots of the screens presented to the user during planning. The process of search is shown in the left part of each Figure. Upper part shows the search in full space at low resolution. The lower part shows search in a reduced search space but at higher resolution. The final trajectory of the of the robot is shown in the right part of each figure for a workspace including a garage, wall, and gate.

The order of synthesis of this result can be seen beginning with Figure 16 which is a depiction of the search tree at the low level of resolution, overlaid on the description of the workspace. The kinematics of the robot are clearly absent from this consideration as can be seen by the result of search at the first level in the upper right part of the Figure 16 (the thin line trajectory) but are evident in the bold line trajectory in the same Figure which is the result of search at the next level: one can see the maneuvering of the robot. The search at this high-resolution level is depicted in the lower parts of Figures 16 and 17 where the reduced search tree of that level is shown.

These results demonstrate that it is possible to synthesize complex maneuvers such as reversing and K-turns without using expert rule-base generated by a human being. Comparatively complex maneuvering is performed just by constructing a multiresolutional representation of the system and searching for successive approximations to construct an e-optimal solution of the problem.

10.2 PILOT with two agents for control (A case of behavioral duality)

The PILOT level will be discussed which provides the real-time guidance of the system and is responsible for the generation and tracking of dynamically feasible trajectories which avoid local obstacles while following the planned path given by the upper level (NAVIGATOR). In this subchapter we present algorithms of two agents which incorporate different strategies of decision making (DM). Depending on the set of circumstances including the level of "informedness", initial data, particular environment, etc., the "personality" of the PILOT is selected from among two DM alternatives:

    1) DM-agent based on a set of rules derived from the results of exhaustive search.

2) DM-agent inclined to make a choice of solution in a rather reckless manner base upon short term alternatives not regarding long term consequences.

Simulation shows that these personalities support each other in a beneficial way (see more in [27 ,35 , 99 ,100).

The idea of multi-agent control is based upon designing a multiplicity of special purpose algorithms each with clearly defined "operating limits", and to use a meta-control, or dispatcher to choose among these available algorithms, or both of them at once depending on the circumstances. Each of these algorithms results in a definite *behavioral portrait* of the Autonomous Mobile Robot (AMR).

We start with defining define the Problem Space. It consists of the state space of the system being controlled as well as dimensions describing: (a) completeness of knowledge of the world; (b) self model; (c) goal. These "extra" dimensions are used to support the Meta-level (the "arbiter") in its decision making about which algorithm to apply and with which cost functions. This continuous selection of the appropriate algorithm from among a set of possible algorithms results in a *behavioral plurality*. AMR control algorithm is predicated on some assumptions about the level of informedness of the environment which is actually a continuum of possibilities. We consider two extreme cases:

(a) Known world. In the case of *a priori* known environments, the problem of motion control reduces to that of planning a complete trajectory, and then tracking it, with possible compensation for error during the execution. Since the world is given, the complete trajectory is to be found off-line, and thus more time consuming methods may be employed. A partially known world presumes the need of dealing with unforeseen (moving, or temporary) obstacles. For this case too, tentative preplanning reduces the computation required for each trajectory, while ensuring that the path found will not become invalid too soon. Our "Accurate Preplanning" (*A P*) algorithm addresses this scenario.

(b) Unknown World. The other extreme is a completely unknown world. Real world is never *completely* unknown, however no information might be given to the robot, other than as a part of "built-in" assumptions. In this case, planning the complete trajectories, at any stage of the motion, is not reasonable since they will quickly become invalid. There are far fewer existing "solutions" to the problem of motion planning in unknown worlds. Our "Instantaneous Decision Preplanning" (*ID*) algorithm addresses this scenario.

Meta-level requires also knowledge of self. This includes estimates of computing power available, characteristics of each decision making method (how long it takes, how accurate its results are, etc.), and models of the physical robot. Using information about current velocity and computing power, for example, the Meta-level can select the algorithm that provides an adequate frequency of decision making to stay within the desired error bounds.

Finally, Meta-level must have access to the goal. This includes the actual goal instance (e.g. position to attain), criteria to use in comparing alternatives (cost functions), and error allowable in achieving the goal. It is a general and explicit representation of the goal, from which any "goals" required by individual algorithms are derived.

The two personalities are represented by the two algorithms: 1)Accurate Preplanning algorithm (AP), and 2) Instantaneous Decision Maker (ID). Let us first describe the AP-algorithm. In a completely known world, the problem of planning a dynamically feasible (e.g. minimum time) trajectory in the presence of obstacles can be formulated as an optimal control problem, with the obstacles acting as constraints on position. One way to approximate a solution to this problem is by using search in a state space consisting of the state variables of the vehicle. Positions of obstacles can then be superimposed on this space making covered positions "illegal" states. Using a model of the vehicle, a successor generating function was defined, which generated the next possible states, given any initial state. By applying the successor generating function recursively, we can create an entire graph representing the state space and all available chains of solutions

Often various heuristics are used to reduce the computational complexity involved in such an exhaustive state space search. Our approach has been to use *generalized* results of search done off-line to create a rule base to use on-line. This generalization of the search results can be seen as pruning after the fact, where each of the pruned regions are transformed into a description of the (class of) situation common to the pruned set, and a trajectory representative of that set. To properly generalize the results of search we must define the proper "vocabulary" with which to describe situations.

The rule base has been defined as a set of rules in a form:

(Situation description) ==> (Trajectory)

This language supporting such situation descriptions is what is called the Rule Language. Along with knowledge of the behavior of our system gained from experience, we have developed a language with which to implement the control rules. The "word classes" were derived directly from the variables of the analytical, differential equations formulation of the problem. Following are the descriptions of the word classes used to form the *A P* rules.

- Goal Angle   (Goal_ang)

Goal Angle describes the direction to the goal relative to the current position of the robot. All measurements such as *Goal Angle*, are done in a polar or rectangular coordinate frame, with the origin being the current robot position. Its values are not distributed uniformly over the 360° range because we found that goal angles within certain adjacent regions resulted in the same control recommendations, regardless of the values of the other precondition variables.

- Goal Range (goal_range)

Goal Range   describes the distance to the *subgoal* from the current robot position. The subgoal is determined by selecting a position in an immediately surrounding passageway that allows achievement of the NAVIGATOR's currently planned subgoal, with minimum cost.

- Phantom Points (phantoms)

This can be thought of as a discrete "artificial potential field" *surrounding the robot* . They may also be seen as "virtual proximity sensors". They are used to help classify obstacle configurations in constrained situations. Since the underlying representation of obstacles is a local grid surrounding the robot, this structure is fast and easy to use. For each Phantom Point, we only need to reference the obstacle array to determine if it is "stimulated". Alternatively, the *Phantom Points* could be implemented as *actual* proximity sensors, thus bypassing the need for an underlying representation of obstacle positions.

- Feasibility of Trajectory (traj_feasible?)

An obvious part of the precondition for a given trajectory-rule is that the trajectory be collision-free. This "word" is a binary variable which is true if the "region swept" by the trajectory of the right side of the rule is obstacle-free. The "region swept" is determined by the width of the vehicle plus some safety margin.

- Current State (curr_state)

Represents the state of the vehicle (system) itself via pertinent state variables, such as steering position, velocity, acceleration. In the *A P* algorithm, this is likely to determine the first few commands of a given trajectory. In some cases, this can be a "dominating variable" in that for a given situation, the trajectories found will be quite different based on the value of curr-state.

With these "words" defined, we can now give the actual form of an *A P*   rule.

If [(goal_ang = a) & (goal_range = r) &
(phantoms = p) & (traj_feasible? (Ti))]   **Then (Trajectory <-- Ti)**

The *A P*   algorithm operates in accordance with the principles of multiresolutional planning/control. The first step is to transform the NAVIGATOR's currently planned subgoal to a locally reachable "passageway". This is done using simple criteria of minimum deviation from path, and safe width of passageway. Next the precondition variables are instantiated. Rule matching is facilitated by having the rules organized in a loose tree form. Common parts of preconditions are extracted and are incorporated into a "dummy" parent rule, which determines which subordinate set of rules should be tested. This organization makes matching fast, but at the expense of more difficult addition of rules. Clearly, an automatic "classification tree" could be formed from the full set of rules.

Most of the above rule vocabulary components depend on an underlying representation. We found a simple binary grid to work well. Obstacle boundaries are projected onto the grid, and thus each cell is marked as either "free-space" or "obstacle". This grid is the basis of the PILOT's "context map". Although the sensors provide less than a 360° view, the PILOT Context Map is 360° so that certain trap situations can be avoided. The mapping of obstacle boundaries onto the grid can be facilitated by using commonly available built-in graphics functions for drawing lines.

Goal Angle and *Goal range* are computed using integer division and a simple table look-up. A minimum division is defined, and then a table is defined which maps each division to a linguistic variable. An alternative would be to use the integer division only. However, it would then be necessary to have substantial redundancy in the rule base. *Phantom Points* are easily implemented with the grid representation of obstacles. A "ray" of the phantom point is activated (or set to true) if any point along the ray corresponds to a cell tagged as "obstacle" in the underlying grid. *Feasibility of Trajectory* is tested using the phantom points structure. For a given trajectory Ti, the robot position is "conceptually" moved to each position of the trajectory. For each position, the phantom points are checked against the obstacle-grid. A trajectory is collision-free if there are no phantom points activated during such an iteration. *Current State* is normalized into a linguistic value by simple integer division of the full range of values that are used to measure the state.

Different premises are employed within the Instantaneous Decision Algorithm (ID). In situations where the robot's information about the world is changing rapidly, the trajectory generated by the *A P* algorithm will quickly become invalid. This can be compensated for, to some extent, by adding a mechanism to "predict" how the world (robot's map of the world) may change. In general, in such environments it is computationally wasteful to look for an entire trajectory when only one or two commands will be used before replanning.

Our objective with the *ID* algorithm is to develop a control module which quickly, (and rather superficially) classifies the situation and gives a *single command*, rather than a complete trajectory of commands. Operating at a relatively high frequency, this decision procedure is expected to produce reasonable paths for environments that are rapidly changing. (Actually it may not be the *environment* which is changing, but the robot's *knowledge* of the environment.)

A fundamental premise that the *ID* algorithm rests on, is that for DM in *immediate time-space vicinity* of the present state, we are able to decouple, and consider separately, different components of a situation description, and how they will affect the decision. Consider the following general form of the problem at hand.

Let Wi = Description of world known to robot, C = a control to the system, then we are interested in solving F (C, Wi) = Wk and inverting it to obtain G (Wi, Wk) = C which in a linguistic form can be stated as:

IF (current state is Wi, AND desired state is Wk)
THEN (apply control = C).

Now obviously the function G is a complex function of the many variables necessary to "describe" the state of the world. (By state of the world we mean description of the system being controlled, as well as all aspects of the "external" environment which affect operation of the system e.g. obstacles). The *ID* algorithm is approximating G over very short time intervals and assuming that this will lead to acceptable "global" behavior.

The accuracy of this approximation will be dependent on (a) the increment of time at which we are determining a value of G, (b) the particular topology of G itself (which is unknown), and (c) the discretization of the variables of G (since we are working in a discrete world). Within definite limits, we as designers have some control over the increment of time (DM frequency) at which we approximate G, as well as over the discretization of the variables (to provide for a desirable level of computational complexity).

Some of the linguistic variables used in *ID* are common to the *A P* algorithm. Goal angle, Goal range, Phantom points, and Current state are all understood in *ID* as they were for *A P*. In addition to these, *ID* uses a construct called "Dynamic Avoidance Regions" (DARs). This is simply another variable to characterize obstacle positions. Figure 18 shows the geometric representation of the front and rear DARs associated with the underlying grid representation of local obstacles.

There are two considerations in determining the discretization of a linguistic variable.

(1) <u>Errors from observation:</u> The information given by a linguistic variable can be at most as accurate as its input. Therefore, the range of errors involved in assigning a linguistic value to an observation should be smaller than the support set of the linguistic value.

(2) <u>Compromise between accuracy of motion and computational complexity.</u> Regardless of the algorithm used, it is reasonable to assume that computational complexity will be positively correlated with size of vocabulary. The more words (values of variables) the more rules, and thus the higher the time for rule matching. On the other hand, a finer discretization of the variables is needed to provide accurate motion control. Therefore, we seek a compromise that provides a vocabulary size just large enough to produce the required accuracy of control. Once we have been given the required accuracy of control (say in terms of position error) we can derive the discretizations required for each of the linguistic variables.

General ID algorithm consists of the following steps:

        1.   Normalize Precondition Variables.

        2.   Rank Desired Direction vectors.

        3.   Translate ranking of directions into motion command.

Step 1 could also be called "assigning observed values to linguistic variables". We use the word "normalize" since the values assigned to the linguistic variables are actually derived from an intermediate representation, and not directly from "observed" values. For the simple variables such as *Goal angle* and *Goal range*, normalization is simply an integer division of the currently represented goal position relative to the robot. The table look-up is necessary since the linguistic variables of *Goal angle*, for example, do not represent uniform divisions of the range of the variable.

Values of the *Phantom points* variable are instantiated using the underlying grid representation of local obstacles. Since this grid is continually redefined at each decision cycle, with the robot at the origin, each *Phantom-ray* can have an associated obstacle-grid coordinate set. These coordinates can then be easily referenced to see if a particular *Phantom-ray* is activated. It was shown earlier, that *Phantom points* refers to a class of variables, each taking on a Boolean value. Thus normalizing the *Phantom points* requires setting each of these variables to true or false.

The DAR's are handled similarly. Each variable of the DAR class is defined by the set of obstacle-grid coordinates to which it corresponds. Thus to determine whether a particular DAR word is "activated" (contains and obstacle) we need only to check each obstacle-grid coordinate belonging to the set of coordinates defining that word. "Front" and "rear" DARs are treated as two separate classes of words in the current algorithm. Both "front" and "rear" DARs could be handled as a single class however this should not give any computational advantage.

Step 2 of the algorithm alludes to the idea of "Desired Direction" vectors. These are simply a result of a discretization of the space of possible resultant directions (and magnitudes) of motion. The ranking of desired directions, based on the current situation is achieved using a special type of table look-up. The numbers in the row effectively impose an ordering of the possible directions of motion as a measure of "compatibility" with each direction. Any situation is described by one word from each class. The rows corresponding to the words describing the current situation are extracted from the table, and are "combined" using some combination operator, in this case addition. For the "special" classes, DARs and Phantom points, the multiple words for a single description are precombined.

216

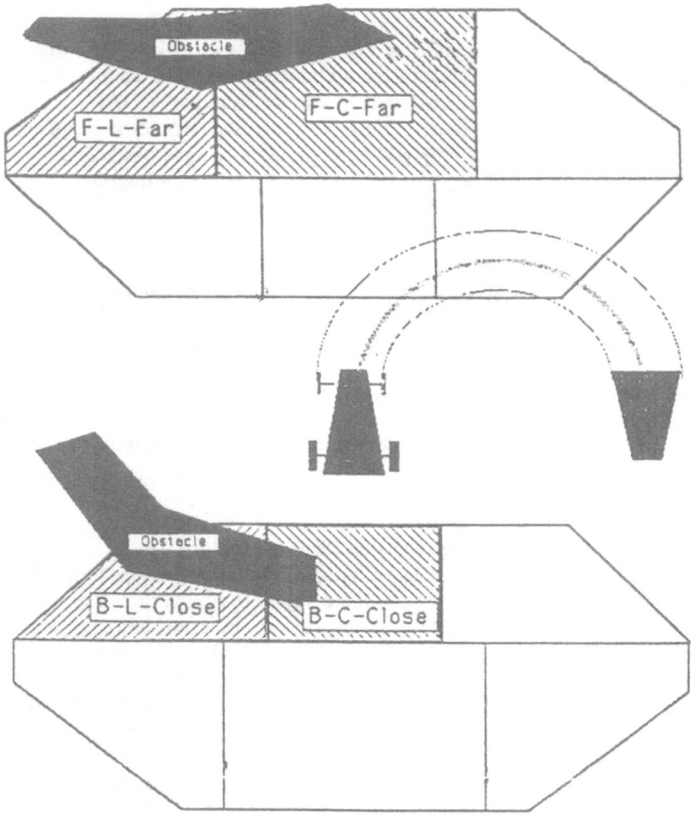

Figure 18. Fuzzifying the World in the set of "zones of response" (DAR)

The set of DAR or Phantom point words instantiated for a particular situation are superimposed to result in a single magnitude for each of the desired directions, for that word class (i.e DAR or Phantoms), using a **min** function as follows:

Let d_i for i = 1 to n be the set of desired directions represented by

the table.

W_a be the set of words of class W that are true for the

given situation.

v_k (w) = value of word w for direction k.

then the ranking of the d_i due to the word class W is

defined as:

For i = 1 to n

d_i = MIN (for all variables v_i (w) such that w is in W_a)

Once a single row is obtained for each word class, a final ranking of the directions is found by column-wise addition of the rows selected for the current situation. This results in a *ranking* of the possible directions of motion, which is the input for the next step of the algorithm.

Step 3 incorporates knowledge about the kinematic and dynamic characteristics of the particular vehicle being controlled. The entire ranking of directions is given to this step instead of a single "best" direction, since that best direction may not be directly achievable. Having the entire ranking of directions enables this step to compromise between the direction the robot *should* go, and the direction it *can* go, in a more informed way.

The transformation of direction into command is done using a simpler form of table look-up. In the simple case, the "best" direction, and the normalized current state are indices into the table that contains the command to achieve that direction from a given state. Since our robot is a steered vehicle, it is biased toward moving forward which creates a number of "ambiguous" cases. Directions that are near 90° from the heading of the robot, for example, can be achieved either by a hard turn and forward motion, or by backing up and then moving forward, in a "three point turn" fashion. For these cases, both commands can be stored in the cell of the matrix, along with the "temporary" direction associated with each of them. Then the command with the highest value of direction ranking will be used. We currently just assign one of the choices for ambiguous cases. This results in the vehicle having a bias to either turn around forward, or "3-point" turn, when the goal is to the side.

Simulated behavior of AMR is shown in Figure 19,a for operation with AP-Pilot, in Figure 19,b for operation with ID-Pilot, and in Figure 19,c for a multiagent Pilot (using both AP and ID). One can see that AP-Pilot provides an acceptable trajectory with one serious deficiency: it works longer than ID. Simulated behavior of the AMR with single-agent and double-agent behavior contains several zones in which no precomputed rule could be found. In this case, AMR must stop an use search to determine the correct action, thus halting real-time operation.

On the contrary, the deficiency of ID-Pilot can be found in its definite lack of long-term optimality with no "wasteful thinking". The compromise is obviously achieved when both of the "personalities" are used within one system. Here the best solutions (trajectory-rules) are employed for the cases when these solutions exist. When no rule applies, the ID-Pilot takes over and issues a single command enabling continued real-time operation.

218

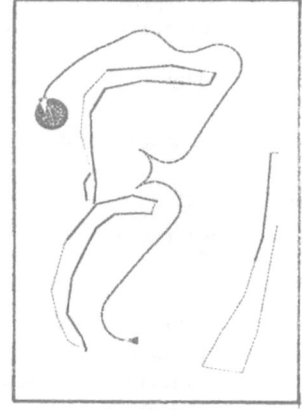

Simulated behavior of ID-pilot.

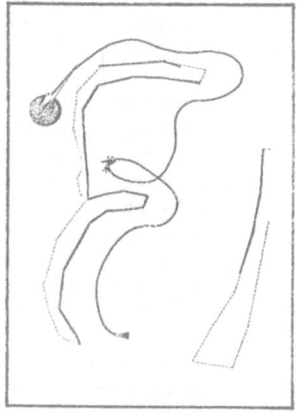

Simulated behavior of AP-pilot.

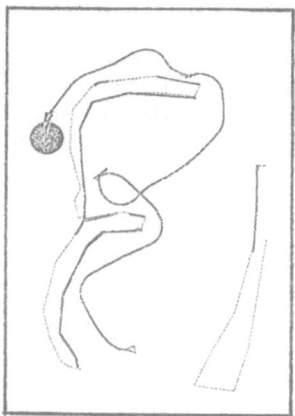

Simulated behavior of (AP + ID)-pilot.

Figure 19. Simulated AMR behavior

11. Conclusions

1. The need in efficient knowledge organization promulgates multiresolutional structures in architectures for intelligent control. Multiresolutional Architectures are dominating in robotic control systems. So called, "behavioral" (usually understood as "multiagent") solutions exist practically only at the lowest level, and when the robot is becoming more complicated, or its mission more sophisticated - multiresolutional control architectures emerge since they reflect the organization of knowledge to be manipulated. Multiagent solutions do not contradict the multiresolutional hierarchies (or heterarchies) — this is just one of forms of distributed control organization. In principle, decomposition of the robot behavior in the multiagent operation can be done at each level of the multiresolutional planning/control system.

2. Theory of Multiresolutional Architectures for Intelligent Control Systems (AICS) is a synthesis and a further development of theories of multiresolutional (multiscale, nested, multigranular) image representation, and multiresolutional (multiscale, multiple time-scale, wavelet) signal representation into the domain of Control Theory. Multiresolutional AICS should allows for solving numerous problems of control in systems with incomplete, and/or inadequate information, large systems, autonomous control system, particularly in the area of intelligent machines. There is no contradiction between the Multiresolutional Architectures and the "behavioral" approach: operation of the multiresolutional system is called "behavior" at each level of resolution.

3. Multiresolutional controllers are solving planning/control problems via consecutive refinement of the solution top-down starting with low resolution larger picture of the situation where the processes are fuzzy and slow, and can be resolved off-line, and ending with the high resolution fast processes which should be computed on-line. Consecutive refinement is done with simultaneous contraction of the zone of the state space in which the search for the solution is being performed. After planning/control behavior of the level is defined, it can be decomposed into activities of several control agents (distributed control) and these agents can be autonomous as well.

4. Algorithms of Nested Multiresolutional Control have been developed which allow for time-efficient computation of control sequences at all levels; some of them has serious computational advantages in comparison with algorithms of dynamic programming. These advantages include working with a non-uniform multiresolutional grids and consecutive contraction of the zone of computations in the state space. All these algorithms can be successfully applied to control of the autonomous agents after decomposition is resolved at the level of consideration.

5. Analysis of the complexity of Multiresolutional Control systems shows that when the number of hierarchical levels grows, complexity initially falls, and then starts growing again. Thus, in many cases, an optimum number of levels can be selected. A law of change in refinement ratio from level to level can be found for each particular case. This law entails definite values of the contraction factors which can be computed for a given level of reliability of the design results. Multiresolutional Systems of Knowledge Representation and Processing reduce the complexity of computations no matter whether the control system is centralized or decentralized. Thus, if one finds beneficial to use a distributed controller with multiple autonomous agents at the level, the complexity of the system should be reduced by a prior organization of this system in a multiresolutional fashion.

6. The results shown in this paper demonstrate that the multiresolutional search in a tessellated space can be successfully utilized for planning and control in robotics for both centralized and multiagent concepts of control. Tessellation of the space is done by discretization of the range of a variable at the levels of higher resolution, and it is done by a proper organization of the vocabulary at the lower resolution. The potential of this approach in the reduction of the complexity inherent in search

220

techniques has been illustrated by a practical examples. Maneuvering of the robot with no human involvement generated rules have never been obtained before. It demonstrates that multiresolutional search can be a powerful tool for a truly autonomous robot planning and control.

References.

1. M. D. Mesarovic, "On Self Organizational Systems", Eds.: M. Yovits, G. T. Jacobi, G. D. Goldstein, Self-Organizing Systems-1962, Spartan Books, Washington, DC 1962, p.p. 9-36
2. G. Pask, "Interaction Between a Group of Subjects and an Adaptive Automation to Produce a Self-Organizing System for Decision Making", ibid., p.p. 283-312
3. M. Mesarovic, D. Macko, Y. Takahara, Hierarchical Multilevel Systems Theory, Academic Press, New York, 1970
4. G.N. Saridis, Self-Organizing Control of Stochastic Systems, Marcel-Dekker, New York, 1977
5. J.S. Albus, "Mechanisms of Planning and Problem Solving in the Brain"," Mathematical Biosciences", v. 45, p.p. 247-293, 1979
6. G. Giralt, R. Sobek, R. Chatila, "A Multilevel Planning and Navigation System For a Mobile Robot: A First Approach to Hilare", Proc. of IJCAI-79, Vol. 1, Tokyo, 1979
7. J. S. Albus, Brains, Behavior, and Robotics, BYTE Books/McGraw-Hill, Peterborough, NH 1981
8. A. Meystel, "Intelligent Control of a Multiactuator System", in IFAC Information Control Problems in Manufacturing Technology 1982, ed. by D.E. Hardt, Pergamon Press, Oxford, 1983
9. G.N. Saridis, "Intelligent Robotic Control", IEEE Transactions on Automatic Control, Vol. AC-28, No. 5, 1983
10. G.N. Saridis, J.H. Graham, "Linguistic Decision Schemata for Intelligent Robots", Automatica, Vol. 20, NO 1, 1984
11. C. Isik, A. Meystel, "Knowledge-based Pilot for an Intelligent Mobile Autonomous System", Proc. of the First Conference on Artificial Intelligence Applications", Denver, CO, 1984
12. G. Giralt, R. Chatila, M. Vaisset, "An Integrated Navigation and Motion Control System for Autonomous Multisensory Mobile Robots", In Robotics Research, Ed. by M. Brady and R. Paul, MIT Press, Cambridge, MA 1984
13. R. Chavez, A. Meystel "Structure of Intelligence For An Autonomous Vehicle", Proc. of the IEEE Int'l Conf. on Robotics and Automation, Atlanta, GA, 1984
14. J.S. Albus, C.R. McLean, A.J. Barbera, M.L. Fitzgerald, "Hierarchical Control for Robots, and Teleoperators", Proc. of the Workshop on Intelligent Control, Troy, NY, 1985
15. A. Meystel, "Autonomous Mobile Device: A Step In the Evolution", In Applications in Artificial Intelligence, Ed. by S. Andriole, Petrocelly Books, Princeton, NJ, 1985, p.p.369-418
16. A. Guez, A. Meystel, "Time-Optimal Path Planning and Hierarchical Control via Heuristically Enhanced Dynamics Programming: a Preliminary Analysis", Proc. of the Workshop on Intelligent Control, Troy, NY, 1985
17. A.Meystel, A. Guez, G. Hillel, "Minimum Time Path Planning for a Robot", Proc. of the IEEE Conf. on Robotics and Automation, San Francisco, CA, 1986
18. A. Meystel, "Planning in a Hierarchical Nested Controller for Autonomous Robots", Proc. of the 25-th IEEE Conference on Decision and Control, Athens, Greece, 1986
19. A. Meystel, Primer on Autonomous Mobility, Drexel University, Philadelphia, PA, 1986
20. A.M. Parodi, J.J. Nitao, L.S. McTamaney, "An Intelligent System for Autonomous Vehicle", Proc. of IEEE Intl. Conf. on Robotics and Automation, San-Francisco, CA 1986, p.p.1657-1663
21. A. Meystel, "Nested hierarchical controller for intelligent mobile autonomous system," Proc. Int'l Congress on Intelligent Autonomous Systems, Amsterdam, The Netherlands, 1986
22. A. Meystel, "Knowledge-based Controller for Intelligent Mobile Robots", In Artificial Intelligence and Man-Machine Systems, Ed. by H. Winter, "Lecture Notes in Control and Information Systems",v. 80, Springer Verlag, Berlin, 1986
23. A. Meystel, "Nested Hierarchical Intelligent Module for Automatic Generation of Control Strategies", Languages for Sensor-based Control in Robotics, Ed. by U. Rembold, K. Hormann, Springer-Verlag, Berlin, 1987
24. A. Waxman, et. al., A Visual Navigation System for Autonomous Land Vehicles, IEEE J. of Robotics & Automation, Vol. 3, No. 2, 1987, pp. 124-141

25. A. Meystel, "Planning in a hierarchical nested autonomous control system," Proc. SPIE Symp., Vol. 727, Cambridge, MA, Oct 30-31, 1987
26. A. Meystel, "Theoretical Foundations of Planning and Navigation for Autonomous Mobile Systems", *International Journal of Intelligent Systems*, Vol.2, No.2, 1987
27. A. Meystel, et al.,"Multiresolutional pyramidal knowledge representation and algorithmic basis of IMAS-2," Proc. of Mobile Robots, SPIE, Vol. 851, pp. 80-116, 1987
28. A. Meystel, "Nested hierarchical controller with partial autonomy," Proc. Workshop on Space Telerobotics, V1, NASA, JPL, pp. 251-278, Pasadena, CA, July 1, 1987
29. C. Isik, A. Meystel, "Pilot level of a hierarchical controller for an unmanned mobile robot," *IEEE J. of Robotics & Automation*, Vol. 4, No. 3, 1988, pp. 244-255
30. A. Meystel, "Intelligent control of robots," *J. of Robotic Systems*, Vol. 5, No. 4, 1988, pp. 269-308
31. A. Meystel, "Mobile robots, autonomous," in: Int'l Encyclopedia on Robotics; editor, R. Dorf, Wiley, 1988
32. G. Grevera, A. Meystel, "Searching for a path through Pasadena", Proc. of the IEEE Symposium on Intelligent Control, Arlington, VA,1989
33. A. Meystel, "Multiresolutional Control System", Proc. of the IEEE Symp. on Intelligent Control, Arlington, VA 1989
34. R. Bhatt, D. Gaw, A. Meystel, "Learning in a Multiresolutional Conceptual Framework", Proc. of the IEEE Int'l Symposium on Intelligent Control 1988, IEEE Comp. Soc. Press, 1989
35. A. Meystel, Knowledge-Based Nested Hierarchical Control", in Knowledge-based Systems for Intelligent Automation, Vol. 2, Ed. by G. Saridis, JAI Press, Greenwich, CT 1990, p.p. 63-151
36. A. Meystel, S. Uzzaman, G. Landa, S. Wahi, R. Navathe, B. Cleveland, "State Space Search For An Optimal Trajectory", Proc. of the IEEE Symposium on Intelligent Control, Vol. II, Philadelphia, PA, 1990
37. B. Cleveland, A. Meystel, "Predictive Planning+Fuzzy Feedback Compensation=Intelligent Control", Proc. of the IEEE Symposium on Intelligent Control, Vol. II, Philadelphia, PA, 1990
38. M. Minsky, "A Framework for Representing Knowledge", in P. Winston (ed.), The Psychology of Computer Vision, McGraw-Hill, 1975, p.p. 211-277
39. B. Mandelbrot, The Fractal Geometry of Nature, W. H. Freeman and Co., 1977, 1983
40. H. Samet, The design and analysis of Spatial Data Structures, Addison-Wesley, 1990
41. H. Samet, Applications of Spatial Data Structures, Addison-Wesley, 1990
42. A. Rosenfeld (ed.), Multiresolutional Image Processing and Analysis, Springer-Verlag, New York, 1984
43. A. Meystel, "On the Phenomenon of High Redundancy in Robotic Perception", in Highly Redundant Sensing in Robotic Systems, ed. by J. Tou, J. Balchen, Springer-Verlag, Berlin, 1990, p.p. 177-250
44. S. Tanimoto, "Paradigms for Pyramid Machine Algorithms", in Pyramidal Systems for Computer Vision, eds. V. Cantoni, S. Levialdi, Springer-Verlag, Berlin, 1986
45. L. Uhr, Algorithm-structures Computer Arrays and Networks: Architectures and Processes for Images, Precepts, Models, Information, Academic Press, New York, 1984
46. W. Hackbush, Multi-grid Methods and Applications, Springer-Verlag, Berlin, 1980
47. A. Brandt, Guide to Multigrid Development, in Multigrid Methods, Lecture Notes in Mathematics 960, eds. W. Hackbush and U. Trottenberg, Springer-Verlag, Berlin, 1982
48. A. Brandt, D. Ron, D. Amit, "Multi-Level Approaches to Discrete States and Stochastic Problems", in Multigrid Methods II, Lecture Notes in Mathematics 1228, eds. W. Hackbush and U. Trottenberg, Springer-Verlag, Berlin, 1986
49. M. Coderch, A. Willsky, S. Sastry, D. Castanon, "Hierarchical Aggregation of Linear Systems with Multiple Time Scales", *IEEE Transactions on Automatic Control*, Vol. AC-28, No. 11, November 1983, p.p.1017-1029
50. S. G. Mallat, "A Theory for Multiresolutional Signal Decomposition: The Wavelet Representation", *IEEE transactions on Pattern Analysis and Machine Intelligence*, Vol. 11, No. 7, July 1989, p.p. 674-693
51. M. Basseville, A. Benveniste, K. C. Chou, A. S. Willsky, "Multiscale Statistical Signal Processing: Stochastic Processes Indexed by Trees", MTNS 89, June 19-23, Amsterdam, 1989
52. A. Benveniste, R. Nikoukhah, A. S. Willsky, "Multiscale System Theory", internal publication No. 518, IRISA, INRIA, February 1990
53. U. Ozguner, "Near-optimal Control of Composite Systems: the Multi Time-Scale Approach", *IEEE Transactions on Automatic Control*, Vol. AC-24, No.4, April 1979, p.p. 652-655
54. B. Litkouhi, H. Khalil, "Multirate and Composite Control of Two-Time Scale Discrete-Time Systems", *IEEE Transactions on Automatic Control*, Vol. AC-30, No. 7, July 1985, p.p. 645-651
55. W. Findeisen, et al., Control and Coordination in Hierarchical Systems, Wiley, New York, 1980

222

56. A. Meystel, Autonomous Mobile Robots : Vehicles with Cognitive Control, World Scientific, 600 pages, 1991
57. A. Meystel, "Coordination in a Hierarchical Multiactuator Controller", Proc. of the NASA Conference on Space Telerobotics, Pasadena, CA 1989
58. G. Grevera, A. Meystel, "Searching in a Multidimensional Space", Proceedings of the IEEE Symposium on Intelligent Control, Vol. I, Philadelphia, PA,1990
59. P.K.S. Wang, "A Method for Approximation of Dynamical Processes by Finite State Systems", *Intern J. Control*, Vol.8, No.3, p.p. 285-296, 1968
60. J. Pearl, "Heuristics," Addison-Wesley, Reading MA, 1984.
61. P.E. Hart, N.J. Nilsson, B. Raphael, "A Formal Basis for the Heuristic Determination of Minimum-Cost Paths", *IEEE Transactions on Systems, Science, and Cybernetics*, Vol. SSC-4, No. 2, July 1968
62. Y.C. Ho, K.-C. Chu, "Team Decision Theory and Information Structures in Optimal Control Problems", Parts I and 2, *IEEE Transactions on Automatic Control*, Vol. AC-17, No.1, Feb. 1972
63. Y.C. Ho, K.-C. Chu, "On the Equivalence of Information Structures in Static and Dynamic Teams", *IEEE Transactions on Automatic Control*, Vol. AC-18, p.p.187-188, 1973
64. Y.C. Ho, K.-C. Chu, "Information Structure in Dynamic Multi-person Control Problems", *Automatica*, Vol. 10,p.p. 341-345, 1974
65. R. E. Larson, "A Survey on Dynamic Programming Computational Procedures", *IEEE Transactions on Automatic Control*, Dec., 1967, p.p. 767-774
66. J.-P. Forestier, P. Varaiya, "Multilevel Control of Large Markov Chains", *IEEE Transactions on Automatic Control*, Vol. AC-23, No.2, April 1978
67. R. Bellman, Dynamic Programming, Princeton University Press, Princeton, NJ 1957
68. A. Haurie, P. L'Ecuyer, "Approximation and Bounds in Discrete Event Dynamic Programming", *IEEE Transactions on Automatic Control*, Vol. AC-31, No.3, March 1986
69. Cheney, W., and Kincaid, D., Numerical Mathematics and Computing, Brooks/Cole Publishing Co., Belmont, CA, 1980.
70. S. Gnesi, U. Montanari, A. Martelli, "Dynamic Programming as Graph Searching: An Algebraic Approach", *J. of the ACM*, Vol. 28, No.4, Oct. 1981
71. N.N. Moiseev, F.L. Chernous'ko, "Asymptotic Methods in the Theory of Optimal Control", *IEEE Transactions on Automatic Control*, Vol. AC-26, No. 5, Oct. 1981
72. H. Herrlich, G.E. Strecker, Category Theory, Allyn and Bacon, Inc., Boston, 1973
73. M.A. Arbib, E.G. Manes, "Foundations of Systems Theory: Decomposable Systems", *Automatica*, Vol. 10,p.p.285-302, 1974
74. M.A. Arbib, Theories of Abstract Automata, Englewood Cliffs, NJ, Prentice Hall, 1969
75. J. A. Sanders, F. Verhulst, Averaging Methods in Nonlinear Dynamical Systems, Springer-Verlag, New York, 1985
76. G. Takeuti, W.M. Zaring, Introduction to Axiomatic Set Theory, Springer-Verlag, New York, 1982
77. M.D. Davis, E.J. Weyuker, Computability, Complexity, and Languages, Academic Press, NY, 1983
78. R.E. Kalman, P.L. Falb, M.A. Arbib, Topics on Mathematical System Theory, McGraw-Hill, New York,1969
79. M. Genesereth, N. Nilsson, Logical Foundations of Artificial Intelligence, Morgan Kaufmann, Los Altos, CA, 1987
80. M.G. Safonov, "Propagation of Conic Model Uncertainty in Hierarchical Systems", *IEEE Transactions on Automatic Control*, Vol. AC-28, No.6, June 1983
81. P.E. Crouch, F. Lamnabhi-LaGarrique, "Local Controllability about a Reference Trajectory", Proc. of the 24-th Conference on Decision and Control, Ft. Lauderdale, FL, Dec. 1985
82. G.D. Cohen, M.G. Karpovsky, H.F. Mattson,Jr.,J.R. Schatz, "Covering Radius - Survey and Recent Results", *IEEE Transactions on Information Theory*, Vol. IT-31, No. 3, May 1985
83. J. H. Conway, N. J. A. Sloane, "A Lower Bound on The Average Error of Vector Quantizers", *IEEE Transactions on Information Theory*, Vol. IT-31, No.1, Jan. 1985
84. A. Gersho, "On the Structure of Vector Quantizers", *IEEE Transactions on Information Theory*, Vol. IT-28, No.2, 1982
85. Y. Bar-Shalom, "Stochastic Dynamic Programming: Caution and Probing", *IEEE Transactions on Automatic Control*, Vol. AC-26, NO. 5, Oct. 1981
86. G. Miller, E. Galanter, K. Pribram, Plans and the Structure of Behavior, Holt, Rinehart and Winston, NY, 1960

87. R. Hinde, Animal Behavior: A Synthesis of Ethology and Comparative Psychology, McGraw-Hill, NY, 1970

88. Ed. by P. Maes, Designing Autonomous Agents, MIT/Elsevier, Cambridge, MA, 1990, (194p).

89. A. Meystel, "IMAS: Evolution of Unmanned Vehicle Systems," *Unmanned Systems*, Vol. 2, No.2, 1983

90. A. Meystel, "Multiresolutional Recursive Design Operator for Intelligent Machines", Proc. of the IEEE Int'l Symp. on Intelligent Control, Arlington, VA 1991, p.p. 79-84

91. A. Meystel, "Multiresolutional Feedforward-Feedback Loops", Proc. of the IEEE Int'l Symp. on Intelligent Control, Arlington, VA 1991, p.p. 85-90

92. A. N. Kolmogorov, S. V. Fomin, Introductory Real Analysis, Dover Publ., New York, 1970

93. A.G. Vitushkin, Evaluation of Complexity of Look-up Tables, Publ. Phys-Math, Moscow 1959 (in Russian)

94. N.F.G. Martin, J.W. England, Mathematical Theory of Entropy, Addison-Wesley, 1981

95. Y. Maximov and A. Meystel, "Optimum Design of Multiresolutional Hierarchical Control Systems," Proc. of the IEEE Int'l. Symposium on Intelligent Cont., Glasgow, U.K., 1992.

96. J. Albus, A. Meystel, S. Uzzaman, "Nested Motion Planning for an Autonomous Robot", Proc. IEEE Conference on Aerospace Control, Westlake Village, CA 1993

97. A. Meystel, S. Uzzaman, "Planning via Search in the Input/Output Space", Proc. IEEE Conference on Aerospace Control, Westlake Village, CA 1993

98. J. S. Albus, " Outline of the Theory of Intelligence", *IEEE Transactions on Systems, Man, and Cybernetics*, Vol.SMC-21, No.3 , 1991

99. R. Bhatt, D. Gaw, A. Meystel, "A Real Time Guidance System for an Autonomous Vehicle", Proceedings of IEEE Int'l Conference on Robotics and Automation, Vol. 3, Raleigh, NC 1987

100. R. Bhatt, et al, "A Real Time Pilot for an Autonomous Robot", Proceedings of IEEE Int'l. Conference on Intelligent Control, Philadelphia, PA, 1987.

8
DISTRIBUTED INTELLIGENT SYSTEMS IN CELLULAR ROBOTICS

TOSHIO FUKUDA(*), TSUYOSHI UEYAMA(**),
KOUSUKE SEKIYAMA(*)
(*) Department of Mechano-Informatics and Systems
Nagoya University, Furo-cho, Chikusa-ku, Nagoya, Japan
(**) Nippondenso Co., Ltd
500-1, Nisshin-cho, Komenoki, Minaniyama, Aichi, 470-01, Japan

1 INTRODUCTION

Recently, as the robotic systems are being improved, these systems are expected to be applied in more global fields. In addition to the conventional research for the improvement of the function of individual robot, distributed robotic systems are being researched. The distributed robotic systems consist of a number of autonomous robots. In the systems, some tasks are carried out through the cooperation, coordination and competition among the robots. The distributed robotic systems will work more efficiently than the sum of the ability of each. Therefore, the distributed robotic systems are expected as an advanced robotic technology. Following advantages can be considered such as flexibility, robustness, fault tolerance, and extendibility of the system. On the other hand, to realize the distributed robotic systems, there are many subjects to research. In recent years, the research works on the distributed robotic systems have been carried out actively and attractively, which include the research on cooperative/coordinate mobile robots [1][2], multi-arm manipulator robots [3][4], several kinds of robots [5][6], and group/swarm robotic systems [7][8]. Related research works also include distributed artificial intelligence [9][10], distributed computer systems[11], artificial life[12] and so on. The research on distributed artificial intelligence will indicate any ideas to the research on the intelligence of the distributed robotic systems. The distributed computer systems will show the control and communication architecture for the distributed robotic systems. And the research on the artificial life will present how to organize the distributed robotic systems, since the systems as seen in the society of ants or human beings will be considered as one of the distributed autonomous systems. Therefore, the distributed robotic systems will be developed by the coordinate research from viewpoint of the biological fields.

In addition to these advantages, we have proposed a cellular robotic system (CEBOT)

S. G. Tzafestas and H. B. Verbruggen (eds.),
Artificial Intelligence in Industrial Decision Making, Control and Automation, 225–246.
© 1995 Kluwer Academic Publishers.

as an advanced distributed intelligent system in 1987 [5]. The concept of the CEBOT is based on the biological organization composed of a enormous number of cells. That is, in cellular robotics, the robot itself is decomposed into functional units which have intelligence, and the CEBOT consists of a number of autonomous robotic units called cells. The concept is described in section 2. Section 3 presents prototypes of the CEBOT. Section 4 introduces a distributed genetic algorithm as an example of distributed intelligence.

2 CONCEPT OF CELLULAR ROBOTIC SYSTEM

The CEBOT, an distributed autonomous robotic system, which was proposed by Fukuda in 1987 and has been researched and developed, is a self-organizing robotic system consisting of a number of autonomous robots called cells. The cell is a robotic unit having basic functions, such as a move, rotation, extension, bending, and so on[5], that is, the CEBOT is regarded as a heterogeneous system. The CEBOT, therefore, realizes self-organization in terms of both hardware and software. By organic integration of a number of cells, the structure appropriate to the environment and tasks is dynamically reconfigurable. Consequently, this system composes functional cellular modules, and then composes individuals, which will become robots, by combining such modules, in the same flow as living organisms which form individuals from cells, then clusters, and finally social systems. Furthermore, as in a social organization of creatures, multiple robots form a group, and a social system is formed through an assembly of such groups of the robots. Figure 1 gives analogy between creatures and the cellular robotic system.

To realize the system, there are many attractive research works. The research topics mainly contain three-fold, that is, a mechanically flexible system, intelligent communication system and intelligent control system. For each topics, we have reported an experiment of automatic docking between cells, research of cellular structure determination, a communication experiment among cells, research of optimal knowledge allocation, and so on [5][13 - 18]. In this chapter, two different sorts of the prototypes, a prototype Mark IV and a cellular manipulator, are introduced and a cooperative search algorithm based on a modified genetic algorithm is presented.

According to the concept, the CEBOT can be applied to many fields, as follows:
(1) Medical field, where the idea of CEBOT is extended to the micro robotic cell. According to the development of IC processing, we will realize a miniature cell. The system can inspect in a living body or operate on organs inside the bodies with a hardware or software linkage distributively. Figure 2 shows a conceptual figure, in which miniature cells operate an organ in a heart and micro intelligent pills are administered.
(2) Agriculture field, where the ability of adaptive structure or self-organization allows most useful function that is a reconfigurable ability depending on the growth of plants.
(3) Construction field, where the robotic system can change the structure in order of building constructions.
(4) Inspection/maintenance field, where the ability of this application is related to distributed sensing and control. The purpose of this application can be adopted to huge plants, where the centralized sensing and control are impossible, and unknown environment, and so on.
(5) Flexible Manufacturing System, in which the flexibility and the multiple combination of the system can respond to demands of productions rapidly. Figure 3 represents the concept of the CEBOT applied to the manufacturing system.
(6) Space field, where the advantage of CEBOT is the cells' size, decentralized control, fault-tolerance, and so on.
Since the system is an open system, it has an interaction to environments and human beings

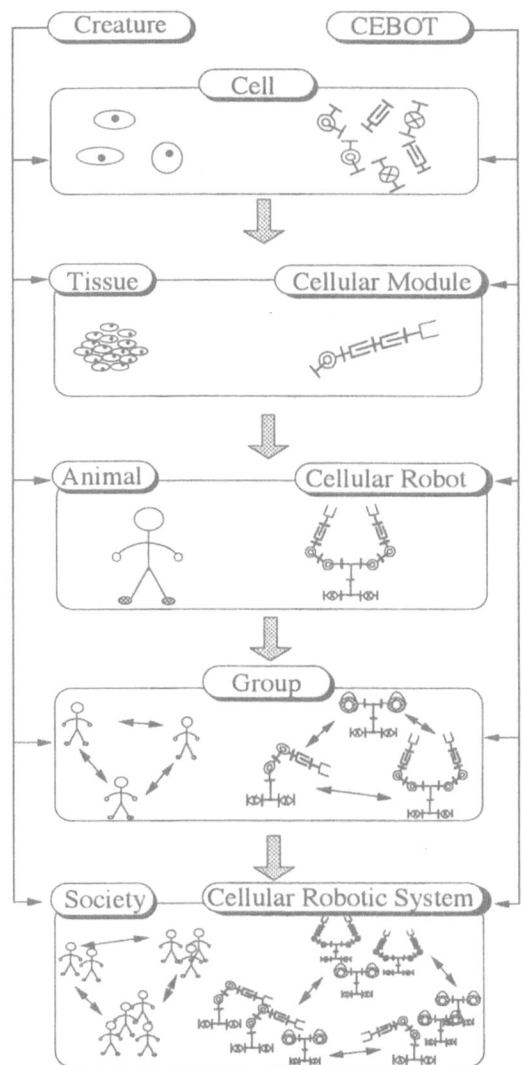

Fig. 1 Analogy between creatures and the cellular robotic system

dynamically. So it can organize
(7) an adaptive human interface by using the adaptive or organizing abilities of itself.

3 PROTOTYPES OF CEBOT

In this section, two different sorts of prototypes are introduced. One is a robotic system which consists of several types of cells, such as moving cells, bending cells, sliding cell and so on. In this system, the cellular robot organizes automatically by using the

228

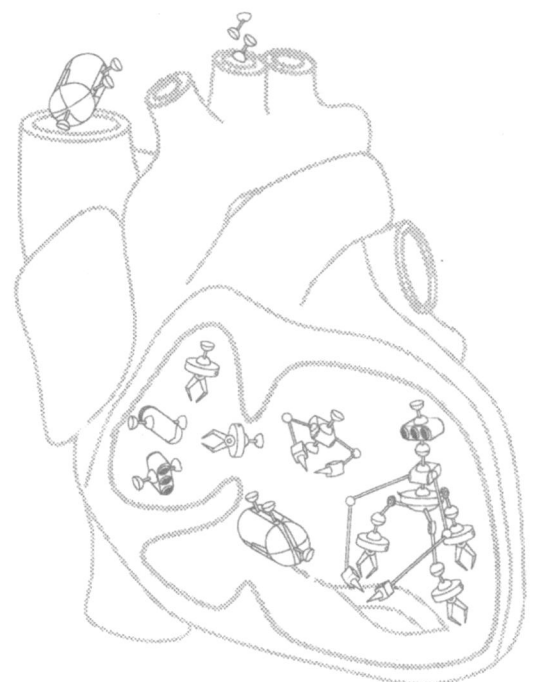

Fig. 2 Concept of CEBOT as micro CEBOT applied in a heat

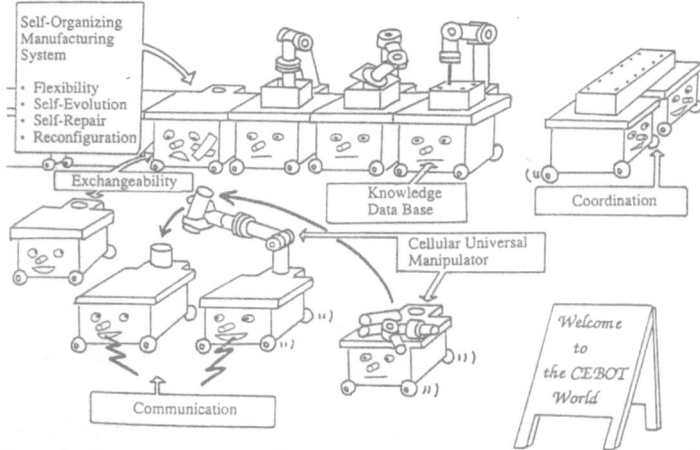

Fig. 3 Concept of CEBOT applied in manufacturing system.

movability of the moving cells. On the other hand, the other is a cellular manipulator which is assembled by other manipulators in the system. By simplifying the function of the cells in the cellular manipulator, it can be miniaturized and the system will be applied effectively in the hazardous environment.

3. 1 Prototype CEBOT Mark IV

The prototype CEBOT Mark IV is shown in figure 4, in which the separation state and

(a) Separation state

(b) docking state
Fig.4 Prototype CEBOT Mark IV

the docking state are shown. In figure 4 (a), a bending cell, a moving cell and a sliding cell are shown form the left side. The system configuration and the docking mechanism are presented in the following sub-sections.

3. 1. 1 System Configuration

To carry out the docking among cells, every cell must have sensors to recognize another cell, and measure the relative distance and relative angle between connection cell. Each cell has connectors for docking with another cell. The moving cell of the CEBOT MarkIV can move by wheels, and has two infrared LEDs and two infrared photo-diode sensors at the front and back of cell individually. The connectors for the docking are installed at the front and back of the cell. The cells have wireless modems of RS232C to communicate among the cells. By using the communication system, the system can coordinate, cooperate and negotiate among the cells. The basic system configuration is shown in Fig. 5. Here each cell has a CPU to behave autonomously. These cells are almost equal in its size and weight. For example, the frame size of a moving cell is 190mm(L)x110mm(W)x70mm(H) and the weight is 4.1 Kg. On the top of each cell, any equipment, such as a CPU luck, motor drivers, wireless modems and so on, are equipped.

3. 1. 2 Docking Mechanism

The docking mechanism between the cells is shown in Fig. 6. The docking between the cells is carried out by using a pair of hooks mechanically. Comparing the voltage of the two photo-diode sensors equipped with the front face of a moving cell, a moving cell can measure the relative distance and the relative angle between the cells. Here the infrared LEDs of the another cell are turned on during the approach and connection of the moving cell. According to the data of the photo-diode sensors, the moving cell approaches and docks with another cell. Figure 7 shows the initial state of the cells for the docking experiment between the cells. The experiment result is shown in Fig. 8. In the experimental result, the vertical axis represents the voltage of the right and left photo-diodes. The horizontal axis represents the iteration times of the approach to connect with the another cell.

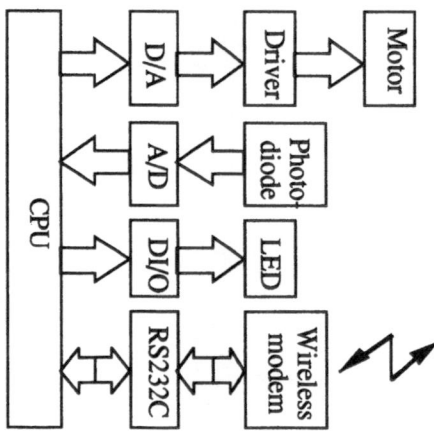

Fig. 5 System configuration

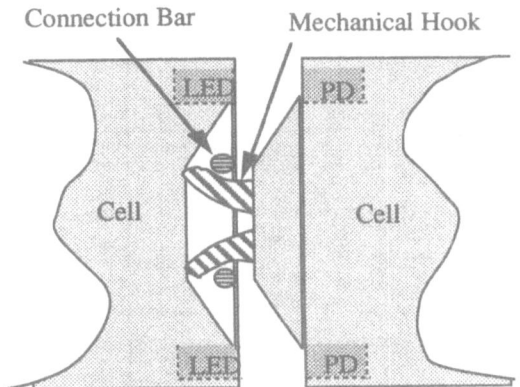

Fig. 6 Docking mechanism of CEBOT MarkIV

In this experiment, the relative distance between the cells is 20cm at the initial condition. According to the approach of the cell, the voltages are increased and the cells connected at 150 iteration times for approach. At the point of **2**, the moving cell went back and retried to dock with the another cell, because the difference of the voltages between the right and the left photo-diode was increasingly larger, therefore, the moving cell detected that it was going to fail docking. At the point **1**, the docking between the cells succeeded.

3. 2 Cellular Manipulator

The Cellular Manipulator System is a system that can automatically produce the manipulator consisting of cells. Each cell has a simple function such as rotation, bending, and extension. The configuration of the cellular manipulator depends on a given task and its environment. That is, the combination of the cells can be adapted to the task in various way. Figure 9 shows the concept of the cellular manipulator system. The experimentally developed cells are shown in Fig.10. The connection between the cells is carried out by the other manipulators, therefore, the manipulation is an important problem to assemble the cellular manipulator. We have presented the off-line planning work which generate the construction of a cellular manipulator according to a given task, and assemble this cellular manipulator by cooperative work of two manipulators [17]. We have also presented the assemble method with on-line planning, since the real environment is very difficult to be

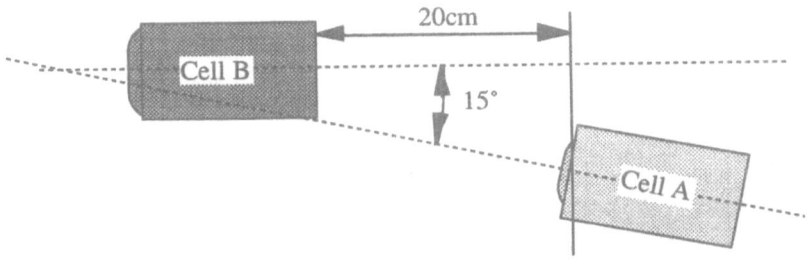

Fig. 7 Initial State of Cells

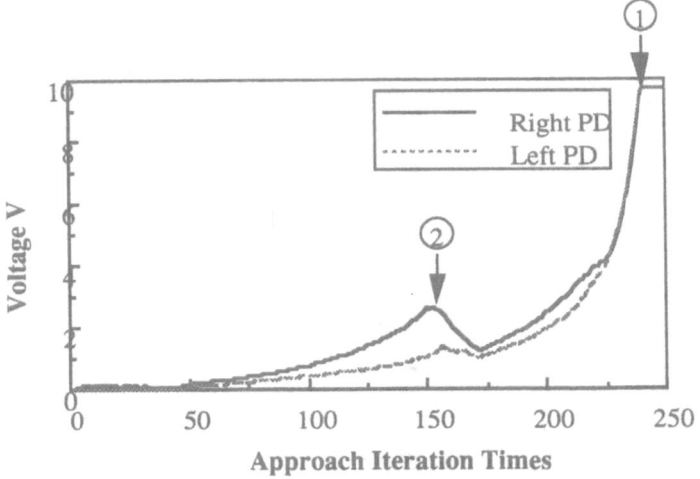

Approach Iteration Times

Approach Iteration Times

Fig. 8 Experimental results of docking between cells

represented by a simple model, and also there may be unpredictable interference due to noise or unknown reason. The total system is presented in figure 11, in which three manipulators, two CCD cameras, and one force/torque sensor are included. The two manipulators are used to assemble the cellular manipulator and one has a CCD camera fixed to the hand, which is used to sense the state of the cells precisely. The other camera is set on the ceiling used to recognize the global environment of the system. The force/torque sensor is used to recognize the contact status between the cells, when the assemble task of the cellular manipulator is carried out. We have researched the method to assemble the cellular manipulator precisely by using active sensing [18]. Since the visual information and force information are both necessary for human being to implement any manipulation tasks, the system includes both sensors. Referring to the cooperation of hand and eye of human being, the previous work presented a evaluation criterion to assemble the cellular manipulator

Fig. 9 The Concept of Cellular Manipulator

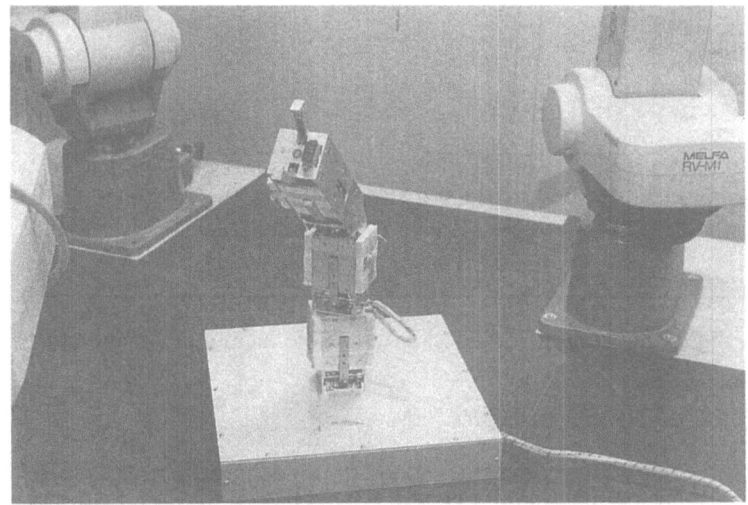

Fig. 10 Photo of Cellular Manipulator

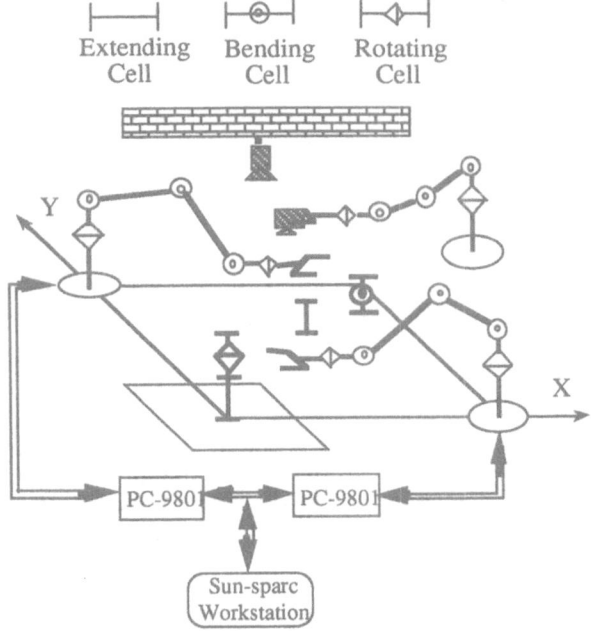

Fig. 11 Cellular Manipulator System

cooperatively with the force/torque sensing and the visual information. Figure 12 depicts the flow of the sensing information to assemble a cellular manipulator.

234

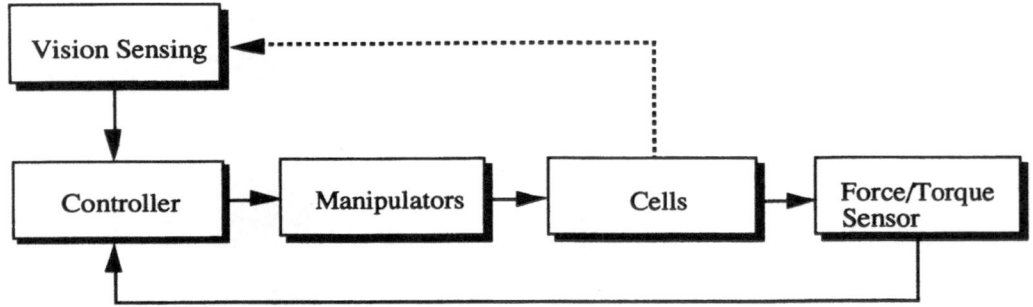

Fig.12 Sensing information to assemble a cellular manipulator

4 DISTRIBUTED GENETIC ALGORITHM

This section deals with a path planning to organize the structure of CEBOT as an issue of the distributed decision making. Since the CEBOT is configured by the connection of a large number of cells dynamically, the path planning of cells for the structure configuration must be determined among cells coordinately and optimally. We describe a method of a distributed decision making for the path planning, where the method employs a genetic algorithm [19][20]. Genetic algorithms are search algorithms based on the model of ecosystem, which adopt the principle of natural selection and natural genetics. The decision making is calculated according to survival rule of planning candidates. We present a *distributed genetic algorithm*, in which the selection of the survival to the next generation is evaluated by the communication between two cells randomly [21]. Repeating the communication between two cells randomly in the system, the cells to construct the CEBOT are selected and the path for each cell is planned by the *distributed genetic algorithm*. First, we discuss distributed decision making briefly. Then the application method of the genetic algorithm is described and the distributed genetic algorithm is explained. Finally, the simulation results are presented for the planning in two cases as follows: (1) Case for using a general genetic algorithm, (2) Case for using the proposed genetic algorithm.

4. 1 Distributed Decision Making

In this section, we adopt the distributed planning as one of the issues of the distributed decision making. The distributed planning is one of the distributed decision making in the distributed artificial intelligence. The distributed planning is the most important and interesting issues for the realization of the distributed system and robotic society. As methods of distributed decision making, the black board model, the contract net protocol and the scientific community metaphor were proposed and researched in the distributed artificial intelligence field [9][10]. The CEBOT can realize this idea for distributed decision making to make a decision in the system by using flexible communication architecture. In this section, we present a distributed genetic algorithm [21], in which a genetic algorithm [9][10] is adopted to distributed robotic system, as one of distributed decision making methods. Then this idea is applied to the path planning for the structure configuration of cellular robots.

4. 2 Structure Configuration Problem

Since the CEBOT consists of a large number of cells, the structure depends on given tasks, its environment and its condition. For the executing given tasks, these cells work cooperatively, or configure cellular modules or cellular robots. And the structure must be determined by themselves. In the next step, these cells have to configure the structure for realization of the system cooperatively. In this case, they should select the optimal path to connect each other. The sequence is represented as follows:
(1) Recognition of task or objects,
(2) Structure determination,
(3) Request of cells and broadcasting,
(4) Configuration of cellular modules or cellular robots.
In this paper, we describe the path planning as follows:
(1) how to treat the structural representation,
(2) how to determine the path cooperatively.
The concept of Cellular Robotic System is based on the biological configuration. In biology, especially the research of the *developmental biology* has been carried out for an analysis of a structural formation. Since the CEBOT is a dynamically reconfigurable robotic system, this research presents the idea for the structural configuration of cellular robots. The goal of structure configuration is given for each cell, that is, the knowledge of each cell is concerned with possessing the positional information in the structure. By using this knowledge, each cell can generate the candidate for the path planning to configure the structure separately. The idea of positional information is important for the reconfigurable robotic system, because the position value gives the priority of cells to the position in the structure. The CEBOT has several kinds of cells and can realize the various structure or combination of these cells by changing the kind of cells. For example, in case of the position corresponding to a moving cell, we should select a moving cell. In this paper, we only consider the kind of cells in each position of the goal structure and assume that each cell has the same priority for each position in a string.

4. 2. 1 How to treat the structural presentation

In this chapter, we assume that the desirable structure is given. It is necessary to determine the path of the robotic system to make this desirable structural configuration. We adopt genetic algorithms as search algorithms for searching the path planning to the optimal structure configuration. Using genetic algorithms, it is necessary to encode a problem. In this character, we can apply genetic algorithms to the structure configuration. Since the Cellular Robotic System constructed by a large number of different kinds of cells, we can treat the string as a structure presentation. The string is represented by the character of cell kinds in a line. This paper deals with one dimensional structure of a Cellular Robot. We must encode a two dimensional structure into a string. Then we can treat a string as a structure configuration with connection data, where the connection data are presented by a connection matrix. The matrix is described in section 4.3.2.

4. 2. 2 How to determine the path planning cooperatively

In the distributed intelligent robotic system, the decision making can be treated in each robotic unit separately. In the Cellular Robotic System, we must consider the character for the decision making, that is, these systems can realize the distributed decision making. On the other hand, these systems must realize any algorithms for the arrangement of decision making among robotic units or cells. In this paper, we assume that each cell only communicates the strings with another cell, which include the position information for structural configuration. In this communication like swarm intelligence, the strings are the improved according to rules. The rule in genetic algorithm depends on natural selection and natural genetics. With the communication among cells, the system consisting of a large number of cells can improve the solutions for this problem cooperatively. Figure 13 shows an example of the structure configuration problem with path planning.

4. 3 Application of Genetic Algorithm

4. 3. 1 Conceptual idea of genetic algorithm

Genetic algorithm is a search algorithm based on the model in ecosystem, in which the mechanics of natural selection and natural genetics are important factors for improving the candidate of solutions. Here, these candidates are coded into string structures likewise genetic code. The survival, which is selected based on the principal of the survival of the fittest, is combined among string structures with a structured yet randomized information exchange to reform fitness values of the survival with historical information. In renewal generation, a new set of the candidate strings, which is a kind of artificial creatures or strings, is created by using bits and pieces of the fittest in a previous set.
Genetic algorithms have three operations to abstract and rigorously explain the adaptive process of natural systems, as follows:
(1) Reproduction operation,
(2) Crossover operation,
(3) Mutation operation.
The reproduction process is the operation to select the survival in a set of candidate strings. In this process, the fitness values are calculated for each candidate string using a fitness function, which depends on a goal for searching problems. According to the values, the selection rate is determined for the present candidate strings, and the survival is selected in any rate depending on the selection rate. The crossover process is the reform operation for

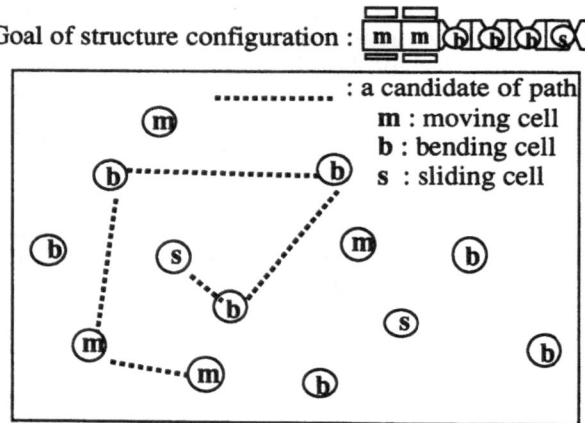

Fig.13 Example of structure configuration problem with path planning

the survival candidates. In natural system, a set of creatures reproduces a new set of the next generation by crossing among the creatures. In the same way, the crossover process is performed by exchanging bits or pieces of strings using the information of old strings. The bits or pieces are crossed in couples of strings selected randomly. In the artificial genetic approach, the mutation process is held to escape the local minimum in search space. In the following, we show an adaptive method of the path planning for a genetic algorithm.

4. 3. 2 Fitness function

As a simplification of the path planning problem for the structure configuration, we assume that the position of cells is given in the absolute coordinates system and all cells can know the position of other cells. And we assume that there are three kind of cells, moving cell, bending cell and branch cell, and only moving cells have the movability. In this fitness function, we consider the situation that a given structure is two dimensional formation on two dimensional plane and the structure is configured according to the order of the kind of cells. An example is shown in fig. 14. In this chapter, we consider only the total distance between cells to configure the structure. We define the fitness function as follows:

$$L_k = \{\sum_{i=1}^{n-1} \sum_{j=1}^{n-1} \sqrt{(x_i - x_j)^2 c_{ij} + (y_i - y_j)^2 c_{ij}}\}/2 \tag{1}$$

where, L_k represents the fitness value for the string k, i and j refer to the cell number and the connects with cell j virtually in the planning of string k respectively. x_i and y_i refer to the position of cell i on the absolute coordinates. n represents the number of cells in the system. (see Fig. 14) c_{ij} refers to an element of a connection matrix C, which is defined as follows:

$$C = [c_{ij}] = \begin{cases} 1 : connection \quad between \quad cell\ i \quad and \quad j \\ 0 : otherwise \end{cases} \tag{2}$$

The values of the fitness function represent the length of path to configure the structure. In this chapter, we select the string, which has the minimum value, as a best candidate.

4. 3. 3 Reproduction Rule

We adopt the roulette rule as a reproduction rule to select the survival in proportion to

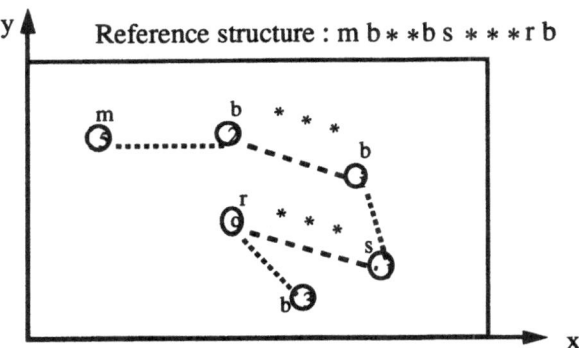

Fig.14 Example of the position of cells

the fitness rate. (see fig. 15) The roulette rule has the characteristic that the survival rate is in proportion to the fitness rate and we can select the number of the fittest candidates effectively. In this paper, the fittest is the smallest value in the fitness values for the strings. We calculate the survival rate in proportion to the difference between the maximum value and each value as follows:

$$L_k' = L_{max} - L_k \tag{3}$$

where $L_{max} = \{L_k \mid max(L_k)\}$

$$C_k' = L_k' - \overline{L}_k \tag{4}$$

where, C_k' is a count value for the reproduction. $\overline{L}_k = (\sum_{k=1}^{m} L_k)/m$, m is the total number of strings. Then according to the count value C_k', we calculate the actual count (integer)value C_k ($\sum_{k=1}^{m} C_k = m$) in proportion to a survival rate. Here the survival rate is a weight value of a roulette rule in the reproduction.

4. 3. 4 Crossover operation

We adopt a position based crossover [20] as a crossover operation for the path planning to configure the structure, because the strings are encoded by the order of characters, which represent the kind of cells. The string code is a set of number of cells. The position based crossover is based on the random selection of a set of positions. And the position of cells in the selected position in one parent is fixed on the corresponding cells in the other parent. The example of this process is shown in figure 16 arranged for our search problem, where the reference structure presents a constraint for the position of each cell. In this paper, we assume that two cells exist in this system. These cells are encoded by the order in the string, as follows :

reference structure (bold characters) : m b b b m l m b b b m
                 coding number : 1 2 3 4 5   6 7 8 9 10

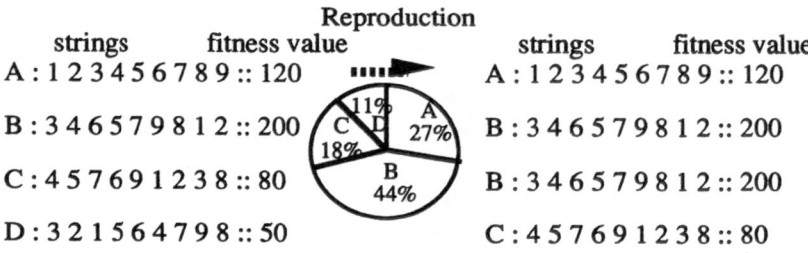

In this example, we assume that the maximum fitness value is best
and the survival rate of the strings is in proportion to the fitness values.

Fig. 15 Example of roulette rule

```
Reference structure :  m   m   b   b   b   b   b   r   b   b   b   b   b   m

     Parent 1 :  1   2   3   4   5   6   7   22  9   10  11  12  13  14
     Parent 2 :  2   15  4   11  7   6   12  8   17  13  10  5   9   28

 Select Position :         *       *               *           *   *

     Child 1 :  1   15  3   11  4   6   7   8   9   12  10  5   13  2
     Child 2 :  15  2   7   4   6   17  13  22  10  5   11  12  9   28
```

where a set of moving cells "m" : {1, 2, 14, 15, 16, 28}
      a set of bending cells " b" : {3, 4, 5, 6, 7, 9, 10, 11, 12, 13, 17,
                            18, 19, 20, 21, 23, 24, 25, 26, 27}
      a set of branch cells "r"   : {8, 22}

Fig. 16 Example of position based crossover

Each cell should be settled on the position of the same kind of cells in the string. Parent 1 and 2 are a couple of the candidate strings which is selected randomly. The mark of * at the position refers to the points chosen randomly. In the points of *, the cells are fixed in that position. For example , that character of in parent 1 is presented in child 2. Here we also consider the position constraint caused by position information in the string and select the rest of cells from the left side. Child 2 and 2 present the new produced strings by crossover.

4. 3. 5 Mutation operation

For the mutation process, first we select a cell in a string at random and the other cell is selected out of the string, which is the same kind of cell and sets before the first. Then these cells are interchanged. In this paper, we treat the mutation operator inside the same kind of cells. In the other hand, we can consider another type of mutation operation that is real type of mutation in nature, which will cause a mutant of structure configuration.

4. 4 Distributed Genetic Algorithm

4. 4. 1 Concept of distributed genetic algorithm

By using the advantage of the distributed intelligent robotic systems, we can suppose that the distributed decision making can be realized in the system. With the massively parallel and massively distributed structure of the distributed robotic systems, the method of the cooperative control and communication is one of the important issues in the system. For these issues, we have to consider the method of the cooperative decision making to control effectively. According to the point, we propose a solution for the distributed decision making in the distributed robotic system with swarm intelligence calculating the evaluation function and only the communication ability of the simplest signals. According to the above described assumptions, since we can consider the distributed decision making by treating the character in a genetic algorithm for search problems, we adopt a genetic algorithm to determine the path planning among cells or autonomous robotic units by assuming that the cells can communicate.

In conventional genetic algorithms, we must take a global statistics of fitness values for reproduction. Therefore, in the distributed intelligent robotic system, any robotic units or cells must communicate with other units or cells. Here, the communication problem is very important issue, because the communication among many robotic units like the Cellular Robotic System has to be controlled in any communication architecture to cooperate effectively. On the other hand, these distributed intelligent systems have the ability of decision making separately. By using this characteristic, we can realize a simple distributed decision making like a swarm intelligence. In this chapter, we present a distributed genetic algorithm for distributed robotic systems by extending the concept of a genetic algorithm. The proposed genetic algorithm makes a reproduction based on the local statistics in the whole system. The conceptual figure is shown in figure 17. By separating the area into several sub-areas, the communication frequency among cells in the distributed condition is reduced compared to the centralized condition. Especially in this paper, we assume that only two cells communicate to reproduce the candidate strings. Because the communication between cells is the simplest situation in any communication architecture. Then in this paper, we assume that each cell has two candidate strings, where the idea is based on the principle of natural generation by supposing the strings as chromosomes in generative cells. (see figure 18) The sequence of the distributed genetic algorithm is shown in figure 19. Here we can see that the differences between a general genetic algorithm and a distributed algorithm are in the random selection of pairs of cells and the research size of genetic operation.

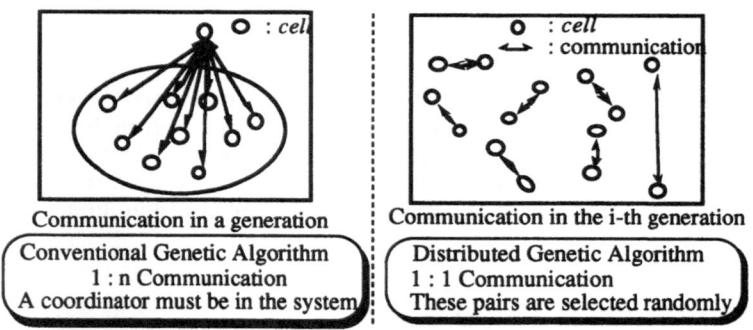

Fig. 17 Conceptual figure of distributed genetic algorithm

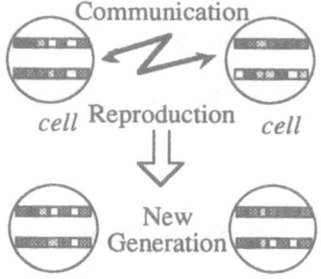

Fig. 18 Conceptual figure for proposed genetic algorithm

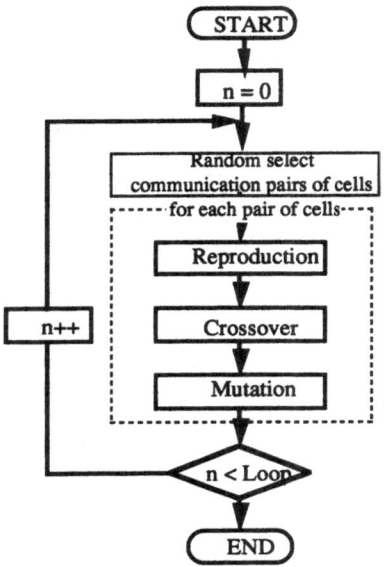

Fig.19 Sequence of distributed genetic algorithm

4. 4. 2 Communication frequency

By assuming the followings, we compare the communication frequency in the distributed condition with that in the centralized condition.
(1) In the case of the centralized condition, only one master cell [22] exists in the system and the cell communicate with other cells to reproduce the strings, where the strings refers to the candidates of the path planning.
(2) In the case of the distributed condition, couples of cells are selected randomly and only 2 cells communicate each other.
(3) In the case of the distributed condition, the fitness function is calculated simultaneously.
According to these assumptions, we calculate the communication frequency as follows:
In the centralized condition for a general genetic algorithm,

$$C_c = (n\text{-}1)\, m \qquad\qquad (5)$$

where, C_c refers to the communication times and n refers to the number of cells. m refers to the reproduction times.
In the distributed condition for a distributed genetic algorithm,

$$C_d = [n/2]\, m \qquad\qquad (6)$$

where, the C_d refers to the communication times. [ ] presents the gauss' notation.

4. 5 Simulation Results

In this simulation, we only consider two dimensional structure of the Cellular Robot on two dimensional plane. We have some assumptions as follows : (1) Each cell has a positional information in the structure itself (2) Each cell knows the information of the location of all cells. (3) These cells can communicate each other. And we assume that the number of cells, which exist in the environment, is the double number of cells for the reference structure. In this simulation, we use a reference structure represented in fig.20, where we have two reference structures in the environment. The fitness value is calculated as a total path length to configure the two reference structures. The length is calculated easily by using the connection matrix. The structure is encoded in a string as follows :

module 1   module 2

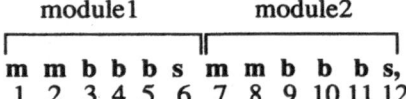

m m b b b s m m b   b   b s,
1   2   3 4 5 6   7   8   9 10 11 12

where **m** refers to a moving cell, **b** refers to a bending cell and **r** refers to a branch cell. The number drawn under the string of the structure refers to the order of the character in the string. According to the number, we derive a connection matrix in eq. (7). Here the number of each kind of cells is eight in the case of moving cells, ten in the case of bending cells and four in the case of sliding cells. In this case, the size of the search space is $_{10}P_5 \cdot _8P_4 \cdot _4P_2$ ($\approx$ 4.35x10^7).

We show the two different types of simulation results as follows:
(1) Case for using a general genetic algorithm,
(2) Case for using a distributed genetic algorithm, in which two cells exist in each area. In the simulation, we adopt the above described method for the reproduction, the crossover and the mutation of the genetic algorithms.

$$
C = \begin{pmatrix}
0 & 1 & 0 & 0 & 0 & 0 & 0 & 0 & 0 & 0 \\
1 & 0 & 1 & 0 & 0 & 0 & 0 & 0 & 0 & 0 \\
0 & 1 & 0 & 1 & 0 & 0 & 0 & 0 & 0 & 0 \\
0 & 0 & 1 & 0 & 1 & 0 & 0 & 0 & 0 & 0 \\
0 & 0 & 0 & 1 & 0 & 1 & 0 & 0 & 0 & 0 \\
0 & 0 & 0 & 0 & 1 & 0 & 0 & 0 & 0 & 0 \\
0 & 0 & 0 & 0 & 0 & 0 & 0 & 1 & 0 & 0 \\
0 & 0 & 0 & 0 & 0 & 0 & 1 & 0 & 1 & 0 \\
0 & 0 & 0 & 0 & 0 & 0 & 0 & 1 & 0 & 1 \\
0 & 0 & 0 & 0 & 0 & 0 & 0 & 0 & 1 & 0
\end{pmatrix}
\qquad (7)
$$

4. 5. 1 Searching results using general genetic algorithm

module 1 :

module 2 :

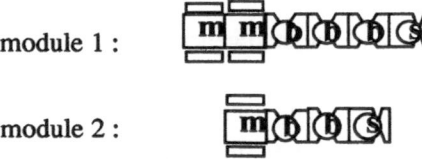

"m" refers to a moving cell, "b" refers to a bending cell and "s" refers to a sliding cell.

Fig. 20 Reference structure for simulation

The simulation results in figure 21 shows the transition state of the best condition in the candidates of solutions, where the final generation is 99th and the mutation probability is 0.05. Figure 22 shows the simulation results of the best fitness value and the average value for each generation. In figure 22, the vertical axis refers to the fitness value and the horizontal axis refers to the generation number. Here the final fitness value is 208.1. In this simulation, the optimal solution is obtained at the 54th generation.

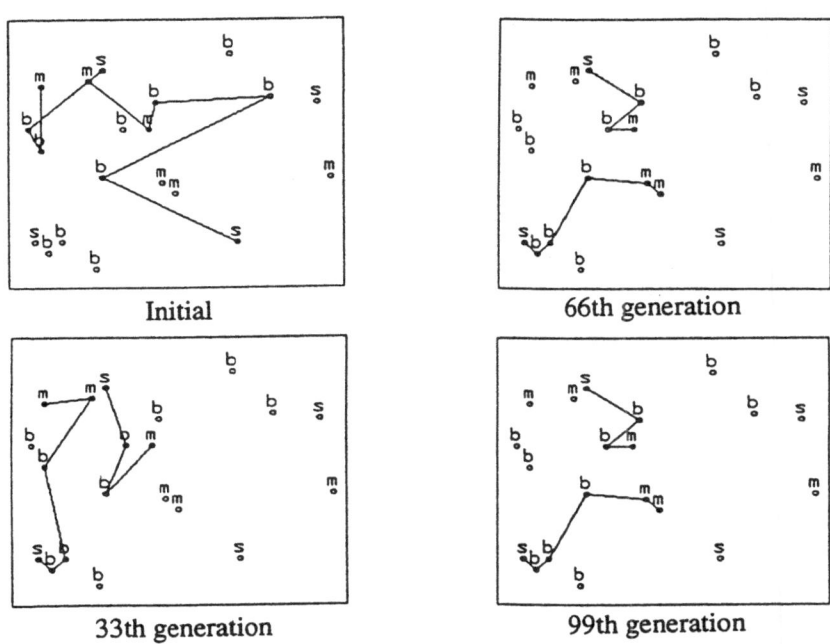

Fig.21 Simulation results of path planning by general genetic algorithm

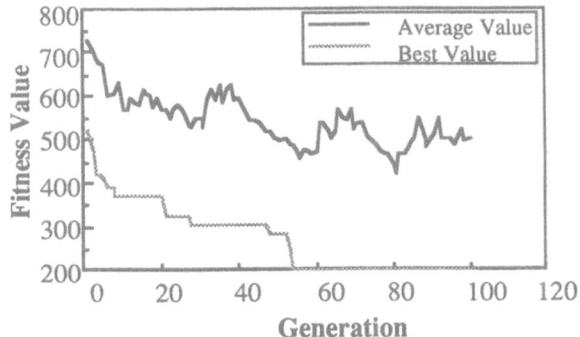

Fig.22 Simulation results of the fitness values by general genetic algorithm

4. 5. 2 Searching results using distributed genetic algorithm

In this simulation, only two cells communicate for the reproduction in the genetic algorithm. The couples of two cells are selected randomly in each generation. Fig. 23 shows the transition state of the best condition where the mutation probability is 0.05 and the final state is in the 99th generation. Fig 24 shows the simulation results of the best fitness value and the average value for the all candidate strings in each generation, where the final best value is 208.1. In this simulation, the optimal solution is obtained at the 16th

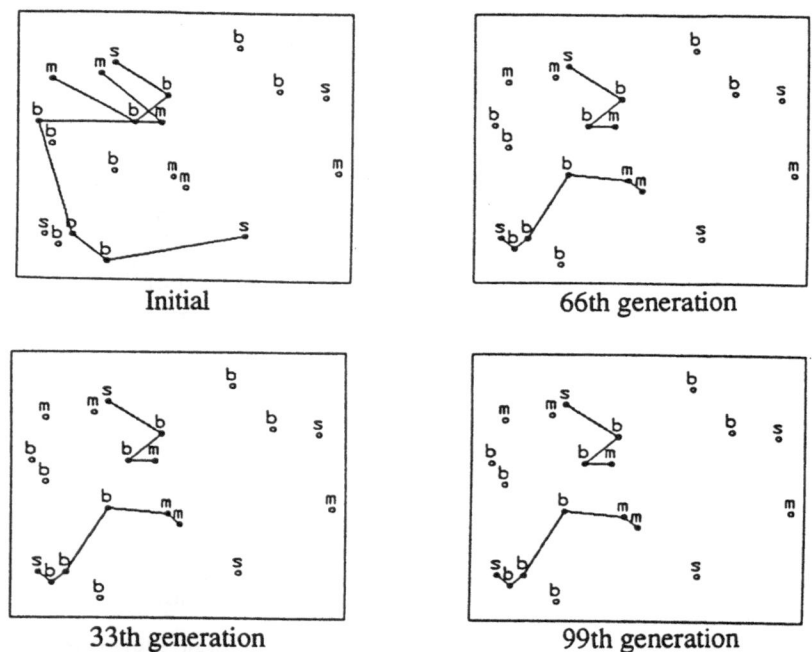

Fig.23 Simulation results of path planning by distributed genetic algorithm

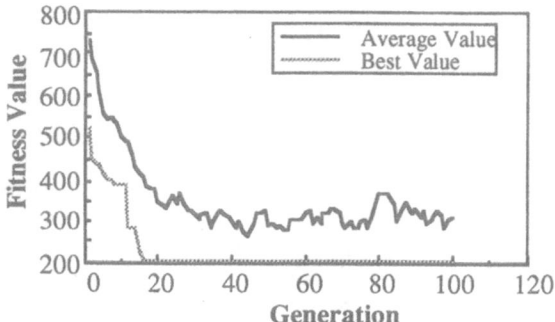

Fig.24 Simulation results of the fitness values by distributed genetic algorithm

generation.

Comparing the results of figure 22 and 24, the distributed genetic algorithm proposed in this paper is more effective than the general genetic algorithm to search the path for the structual configuration. Since the distributed genetic algorithm selects and makes a crossover in a local area, it is regarded that the distributed genetic algorithm can select the candidates more effectively than the conventional genetic algorithm. Therefore, the distributed genetic algorithm can search the feasible solution in a huge search space more than the conventional genetic algorithm. That is, we can consider that the effectiveness is equal to the effectiveness of the group discussion.

5 CONCLUSIONS

This chapter presented the recent research works on distributed intelligent system in cellular robotics. First of all, the concept of the CEBOT was described and two prototypes of the CEBOT were introduced. As a cooperative search method in distributed robotic system, the distributed genetic algorithm was presented. As a search problem, the path planning problem to organize the cellular robot was adopted. And this chapter showed that the system can search the solution in the searching space by communicating each other based on the distributed genetic algorithm. This idea represents the basic idea for the distributed decision making in the distributed intelligent robotic system.

As mentioned in this chapter, the concept of the CEBOT presents the strategy to develop advanced intelligent robotic systems, such as the distributed robotic systems. In the decentralized robotic system, the collective or group intelligence and collective or group function of the system provide the system high performance ability. Therefor, the research of the cooperation and coordination among robots is one of the most important issues to realize the decentralized robotic system such as human beings in society.

REFERENCES

[1] R.C. Arkin, "Cooperation without Communication Multiagent Schema-Based Robot Navigation," Journal of Robotic Systems, pp.351-364, (1992)
[2] F.R. Noreil, "An Architecture for Cooperative and Autonomous Mobile Robots," Proc. of IEEE International Conference on Robotics & Automation 1992, pp.2703-2710, (1992)
[3] N. Xi, T.J. Tarn, and A.K.Bejczy, "Event-Based Planning and Control for Multi-Robot Coordination," Proc. of IEEE International Conference on Robotics & Automation 1993, Vol.1, pp.251-258, (1993)
[4] X. Cui, and K.G. Shin, "Intelligent Coordination of Multiple Systems with Neural networks," IEEE Transactions on Systems, Man, and Cybernetics, Vol. 21, No. 6, pp.1488-1497, (1991)
[5] Fukuda, T. and Nakagawa, S. "A Dynamically Reconfigurable Robotic System (Concept of a system and optimal configurations)," IECON'87, pp. 588-595, (1987)
[6] A. Matsumoto, H. Asama, Y. Isida, K. Ozaki, and I. Endo, "Communication in the Autonomous and Decentralized Robot System ACTRESS," Proc. of IEEE International Workshop on Intelligent Robots and Systems, pp.835-840, (1990)
[7] G. Beni, "The Concept of cellular Robotic System," Proc. of IEEE International Symposium on Intelligent Control, pp.57-62, (1988)
[8] P. Dario, V. Genovese, F. Ribechini, and G. Sandini, ""Instinctive" Cellular Robots," Proc. of International Conference on Advanced Robotics, pp.551-555, (199

246

[9] A.H. Bond, and L. Gasser (eds.), "Readings in Distributed Artificial Intelligence," Morgan Kaufmann, (1988).

[10] L. Gasser, and M. N. Huhns, "Distributed Artificial Intelligence Vol, II," Morgan Kaufmann, (1989)

[11] K. Hwang, and F.A. Briggs, "Computer Architecture and Parallel Processing," McGraw-Hill Book Co., (1988)

[12] C.G. Langton (ed.), "Artificial Life," Proc. of Interdisciplinary Workshop on the Synthesis and Simulation of Living systems, (1988)

[13] Fukuda, T. and Nakagawa, S. "Approach to the Dynamically Reconfigurable Robotic System," Journal of Intelligent and Robotic System Vol. 1, Kluwer Academic Publishers, pp. 55-72, (1988)

[14] Fukuda, T., Kawauchi, Y and Buss, M. "Communication Method of Cellular Robotics CEBOT as a Selforganizing Robotic System," Proc. of IEEE International Workshop on Intelligent Robots and Systems (IROS'89), pp. 291-296, (1989)

[15] Fukuda, T., Kawauchi, Y., and Asama, H. "Analysis and Evaluation of Cellular Robotics(CEBOT) as a Distributed Intelligent System by Communication Information Amount," Proc. 1990 IEEE International Workshop on Intelligent Robots and System (IROS'90), pp. 827-834, (1990)

[16] Fukuda, T., Ueyama, T., Kawauchi, Y. and Arai, F. "Concept of Cellular Robotic System (CEBOT) and Basic Strategies for its Realization," Computers Elect. Engng Vol. 18, No. 1, pp.11-39, Pergamon Press, (1992)

[17] Fukuda, T., Xue, G., Arai, F., Kosuge, K., Asama, H., Omori, H., Endo, I., and Kaetsu, H., " Self Organizing Manipulator System Based on the Concept of Cellular Robotic System", Proc. of IEEE/RSJ International Workshop on Intelligent Robots and Systems (IROS'91), pp. 1184-1189, (1991)

[18] Fukuda, T., Xue, G., Arai, F., Kosuge, K., Asama, H., Omori, H., Endo, I. and Kaetsu, H. "Assembly of Self-Organizing Manipulator Using Active Sensing," Proc. of International Symposium on Distributed Autonomous Robotic Systems (DARS'92), pp. 69-76, (1992)

[19] David E. Goldberg "Genetic Algorithms in Search, optimization & Machine Learning," pp. 1-23, Addison Wesley.

[20] Lawrence Davis (ed) "Handbook of Genetic Algorithms," pp. 332-349, Van Nostrand Reinhold.

[21] Ueyama, T., Fukuda, T. and Arai, F. "Coordinate Planning Using Genetic Algorithm - Structure configuration of Cellular Robotic System -," Proc. of the 1992 International Symposium on Intelligent Control, pp.249-254, (1992)

[22] Ueyama, T., Fukuda, T., and Arai, F. "Configuration of Communication Structure for Distributed Intelligent Robotic System - Evaluation of the organization of Cellular Robotic System as Group Robotic System -," Proc. of 1992 IEEE International Conference on Robotics and Automation, pp.807-812, (1992)

9

DISTRIBUTED ARTIFICIAL INTELLIGENCE IN MANUFACTURING CONTROL

S. ALBAYRAK, H. KRALLMANN
Institute for Quantitative Methods
Department of Informatics
Technical University of Berlin
10587 Berlin, Germany

1 Introduction

Manufacturing companies have not been able to sufficiently solve the tasks of manufacturing control with computer support since these are, at least partially, very complex and poorly structured. In addition, many tasks are so closely interconnected with one another that it has hindered adequate computer support. On the other hand, achieving the company goals was often determined by the solution of the manufacturing control tasks.

In addition to the problem in the solution of manufacturing control tasks mentioned above, there is also the problem that the manufacturing company must reorganize their manufacturing structure, i.e. reduce the manufacturing depth and introduce lean organization forms. This reorganization led to a situation where many companies have virtual production structures. These make the manufacturing control task even more complicated, since the individual virtual structures are different: act autonomously and closely cooperate with one another. The DAI methods form a possible basis for the computer-supported solution of these tasks.

The manufacturing control tasks are described in Section 2. Section 3 provides an overview of the state of the art in DAI technology in manufacturing control. Section 4 introduces DAI provides an extensive description of the blackboard architecture. In section 5 we conclude with a presentation of the VerFLEX-BB system which is based on blackboard architecture.

247

S. G. Tzafestas and H. B. Verbruggen (eds.),
Artificial Intelligence in Industrial Decision Making, Control and Automation, 247–294.
© 1995 *Kluwer Academic Publishers.*

2 Tasks of Manufacturing Control

Manufacturing control is the heart of a manufacturing company. It includes all decisions and measures which are necessary immediately before, during, and after executing the manufacturing program preset by the planning level, in order to guarantee a smooth, punctual and economical manufacturing process. Therefore, material flow control, order execution and quality management are subtasks of the process control (see Fig. 1).

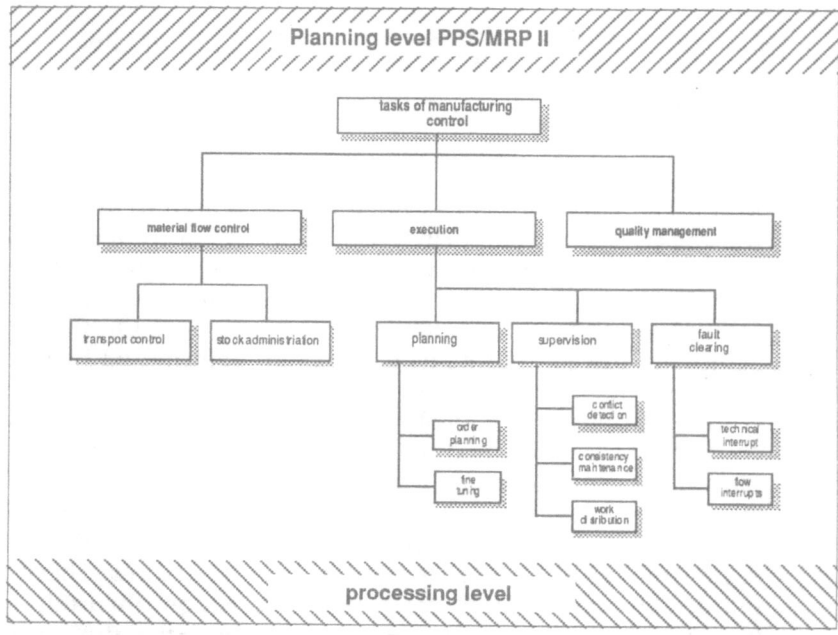

Fig. 1: Process control tasks

The tasks of the process control are increasingly characterized by complex and alternating market situations, a restructurization of companies and a reduction of the processing depth.

The problems of present and future process control can be described as follows:

- On the one hand, the tasks of process control are very **complex** and **poorly structured,** on the other hand, the tasks are **interconnected in a number of ways** so that it is impossible to deal with them separately. In reality one expert (specialist) each responds for the solution of one task. During the task solution process, the experts cooperate in an intensive and goal-oriented way. Furthermore, the real-time conditions prevailing in manufacturing require high efficiency and acceptability of the computer supported solutions. Due to the above-mentioned reasons, the application of conventional methods to the computer-supported solution is nearly impossible.

- Today, modern production systems of process manufacturing consist of different complex process plants and equipment. Buying such equipment represents a considerable investment. Concerning process manufacturing, these plants must be always available and must be run in layers. However, with the growing complexity of the plants, the error proneness is also increasing, and, at the same time, the requirements of the staff-members which run and maintain the plants are growing. So the manufacturing companies are forced to train specialists to run and **maintain** the plants. These **specialists** will be responsible for the maintenance of one type of plant located at different sites.

- In order to attain a complete capacity utilization of these highly developed and expensive plants, many firms start to modularize and to segment the manufacturing (e.g. plant within the plant, team work concepts, lean management etc.), which means the companies are forced to reduce their manufacturing depth. The **reduction of the manufacturing depth** is necessary and sensible due to capital-intensive production plants, since several production plants at different production sites, probably often standing idle, are far too expensive. Instead of this, the production is carried out on different sites, the final and semifinished products are transported, if necessary, to other sites and will be assembled there. A positive effect of such flat manufacturing structures is, e.g., that the capacity of the production plants is better exploited. However, the requirements to organization and realisation of material supply and transportation of single process modules are higher. Small intermediate buffers and several process modules as well as varying transport times contribute to the complication of controlling even just a small amount of process modules. Due to the distance in time and space between the process modules the quality of the subproducts manufactured becomes more and more important.

A computer supported solution of process control tasks requires **distributed software structures** which should help to model adequately the single tasks as well as their interaction. Due to the real-time environment prevailing in case of the manufacturing control, cooperative problem solving methods specific to the problem class also have to be developed. This can be realized using techniques from the field of **Distributed Artificial Intelligence (DAI)**.

Since the cooperation between the manufacturing control experts is very communication intensive and with reduced manufacturing depth, the distributed manufacturing is increasing in importance, the demand for very fast and efficient networks like B-ISDN grows in order to deal with the enormous communication requirements of distributed problem solving. One could mention here, by way of example, the transfer of graphic presentations (multimedia elements) as help in the managing of technical faults (distributed diagnosis). The acceptance of the solutions can be additionally increased by using multi-media techniques. The use of multi-media communications (multi-media conferencing) improves group work among experts. In addition to the technical possibilities, the organization forms must be adapted to the process flow structures, i.e. leaner organization forms must be realized.

Here, the following simple scenario demonstrates the complexity and interconnectedness of the manufacturing control tasks [Albayrak 90b].

The following example, "machine fault" makes clear that the knowledge necessary to solve the individual subproblems is embodied by the different experts and that the solutions can first be found using an intensive and goal-oriented communication among experts (cf. Fig. 2).

If, on the machine level, a technical fault arises, the expert responsible for the setting-up and well-functioning of the machine is called. He tries to locate the reason for the fault and determine its scope. The expert decides, whether he can repair the machine and how long that would take, or if the manufacturer of the machine has to be called. The expert gives all this information to his superior, commonly the shop floor manager. The shop floor manager, in turn, evaluates the consequences on other orders in his work shop unit, depending on the duration of the breakdown and takes adaptive measures, possibly after consulting manufacturing planning and control. On the planning level the breakdown is evaluated and, once delivery dates are checked and other departments are consulted, orders connected with the breakdown are assigned new priorities and - if necessary - put in a different order, which implies re-planning on the work shop level.

The knowledge necessary for problem solving in realization of the manufacturing process is distributed amongst various experts, as shown in the example above. In order to realize a computer-based solution for this group of tasks, it is necessary to model the individual agents who are working on the problem and their cooperation with one another.

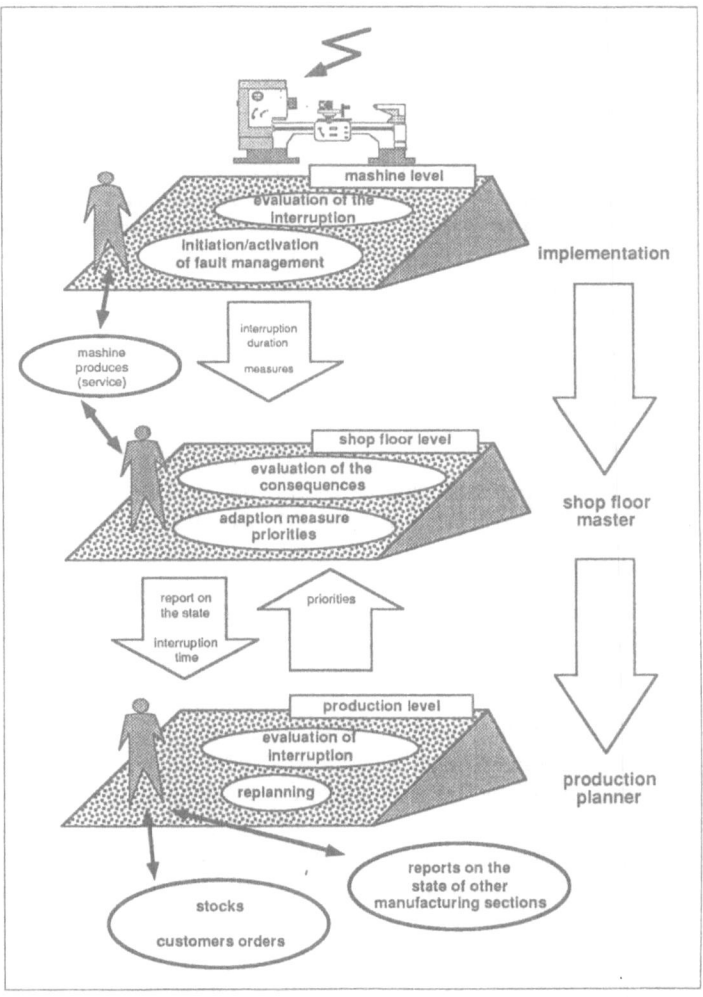

Fig. 2: Fault management procedure [Albayrak 90b]

3. The State-of-the-Art of the DAI technique in Manufacturing Control

3.1. ISIS/OPIS

Both the ISIS System (Intelligent Scheduling and Information System) and the following advanced system OPIS (Opportunistic Intelligent Scheduling System) are systems for order sequence planning for shop floor manufacturing and were developed at Carnegie Mellon University [Fox 84], [Fox 86]. ISIS formalizes the different marginal conditions which restrict the number of possible solutions and which, in this way, lead to a solution ("constraint-directed reasoning"). These restrictions are propagated through a planning process which runs on three levels. The restrictions represented by ISIS (see Fig. 3) are:

- corporate goals like product quality, manufacturing costs, prompt delivery and low stocks,

- physical restrictions like machine capacities and machine types, characteristics of the machines, processing time and quality,

- organizational restrictions like alternative work sequences, machines, requirements on tools and material, qualification,

- restrictions on the availability, e.g. of machines and material,

- causal restrictions like the stipulated process sequences,

- preferences like low costs, short run-through times, high machine utilization.

ISIS names, besides the availability of material and machines, preferences for manufacturing steps and organizational goals as well as restrictions. The manufacturing steps and organizational goals are not really restrictions, as they do not restrict the amount of possbile solutions but just provide a selection opportunity of the solutions within the solution space, e.g. select from the amount of possible solutions the one which leads to low costs, short set-up times, low material stocks and high machine capacity utilization etc.

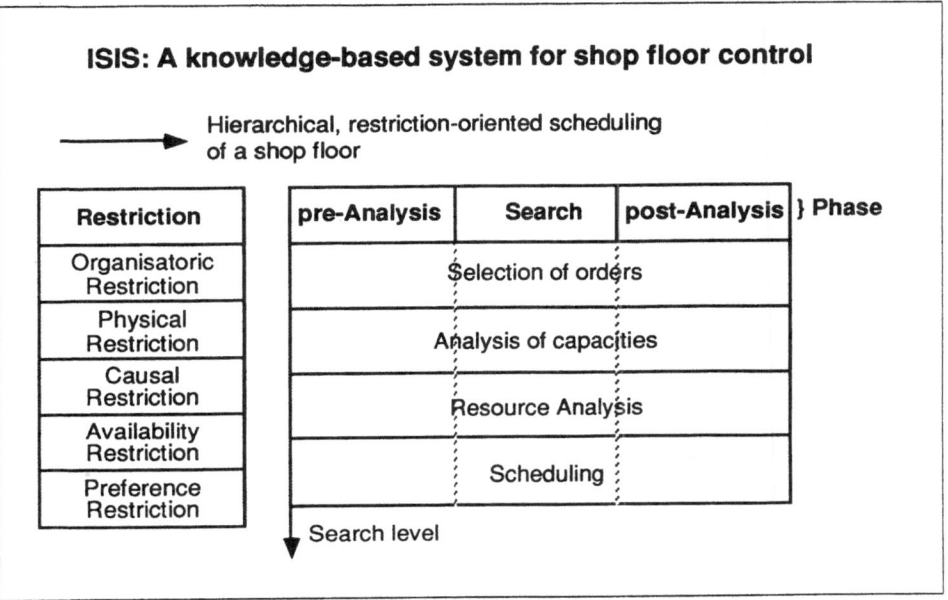

Fig. 3: Basic concept of ISIS-System [Fox 84]

ISIS was implemented in the object-oriented language SRL based on Franz LISP. ISIS was applied on trial to the planning of turbine manufacturing (Westinghouse Company, USA). It could not be employed in practice as the size of the knowledge base caused problems with the run-through time and thus ISIS could not be integrated in the existing DP environment due to missing interfaces [Fox 86].

OPIS, ISIS´ successor has a much better run-through time and knowledge representation due to the facts that

- the reasoning mechanism became more flexible by using Blackboard structure [Nii 86a] and

- the plans are regarded from two different resourced-based points of view.

For the above-mentioned reasons neither ISIS nor OPIS was applied in practice. Kempf writes:

"[...] Even the most publicized AI-based manufacturing schedulers, the ISIS/OPIS system produced at Carnegie Mellon

*University starting in 1980 have yet to enter productive service
and are not considered to be on the verge of doing so" [Kempf 0]*

We are not in a position to report on OPIS´ state of development
based on the publications on hand. The Carnegie Mellon University
(CMU) is working on different approaches to the scheduling
problem, e.g. Distributed Scheduling [Fox 90] and Case-based
Scheduling [Sycara 91].

3.2.   SOJA/SONIA

The SOJA (Systeme d´Ordonnacement Journalier d´Atelier)
[Stauve 87] and its successor SONIA [Collinot 88], [Suave 89] are
systems for knowledge-based manufacturing planning and control,
which especially support the short-term allocation planning on the
shop-floor level.

SOJA creates an allocation schedule based on predictions derived
from capacity calculations and the reservation of resources. Suave
defines the goal of SOJA:

> *"[...] The aim of SOJA is to generate an admissible solution of
> the estimated scheduling problem that implies: selecting the
> operations to be performed over a short period (a day) assigning
> resources to these operations with respect to preference criteria
> and the dispatching of freedom degrees. Scheduling them
> satisfying pure constraints." [Sauve 87]*

Accordingly, SOJA takes the static prediction base into
consideration, however it does not consider the dynamic processes
of manufacturing.

SONIA includes SOJA´s functions as well as a manufacturing
control component, that is a reactive component. SONIA integrates
predictive modules which create a momentary allocation plan as
well as reactive modules which, in case of a deviation of the current
state in the shop floor from the planned goal state (e.g. working
sequence cannot be finished due to machine failure), restructures or
newly plans the working sequences and orders.

SONIA is realized in form of a Blackboard system [Sauve 89],
every system component (with the exception of the simulator)
standing for one knowledge source. This increases the modularity as
well as the expendability of the system architecture, as the unique
data structure in form of the Blackboard is used for communication
so that a conform protocol can be defined for the access to the data of
all knowledge bases. More knowledge sources can be integrated or

removed without complicated changes since knowledge sources do not make predictions/assumptions concerning the existence of other knowledge sources and only the Blackboard is used as a communication medium.

SONIA is being tested by means of simulators. At the same time industrial applications based on SONIA are being developed by Laboratories de Marcouissis in cooperation with Alcatel AMTC, Alcatel TITN and partners of the Esprit 2 project IMPACS (Integrated Manufacturing Planning and Control Systems).

3.3. YAMS

YAMS (Yet Another Manufacturing System) was mainly developed by H. Parunak [Parunak 87] and is a system for the control of business manufacturing plants. The contract network protocol is used for the adaptive distribution of the processing of subproblems.

YAMS is to be applied to the control of several production plants which are largely remote from each other. As many operations during the manufacturing process are subject to real-time conditions, production plants, manufacturing cells and often the machines need a processor of their own. Fig. 4 shows the hierarchic structure of YAMS including the important levels on which the main components are organized. These features of the manufacturing control are the basis of YAMS' distributed character, where communicating knowledge-based systems are assigned to the processors.

In case of the problem of manufacturing control, the solution space is defined by many factors, e.g. the available equipment of a machine, the products to be manufactured and the available resources (time, inventory, stocks etc.). In conventional systems of manufacturing control the knowledge of such parameters and factors which make efficient problem-solving possible are integrated in the code (manufacturing control structures), so that the system is designed specifically for one factory taking into consideration its current situation; changes have to be made in the code. Normally, these changes are not only difficult to realize and complicated, but it is also difficult to predict the consequences and repercussions.

The control system YAMS uses on every processor the same control structures, characteristics of the local environment of a processor (e.g. equipment and control procedures of a machine) are stored in local data bases. Thus it is easy to integrate new machines and manufacturing cells in the total system and to perform easily and quickly the changes in the local data base which respond to changes in the local environment which can be handled by the control.

Production plants are embedded in an open environment. Due to events and changes in the environment which occur asynchronically, the duration of the single operations cannot be defined from the beginning. The resources needed for one operation are in complex interconnection with the current state of a machine. Especially, whole machines might break down during operation so that other (and eventually less appropriate) machines have to assume their tasks. That means that static models of manufacturing control as they are realized by conventional manufacturing systems are not sufficient because the system has to react in a quick and flexible way to asynchronic changes in the environment.

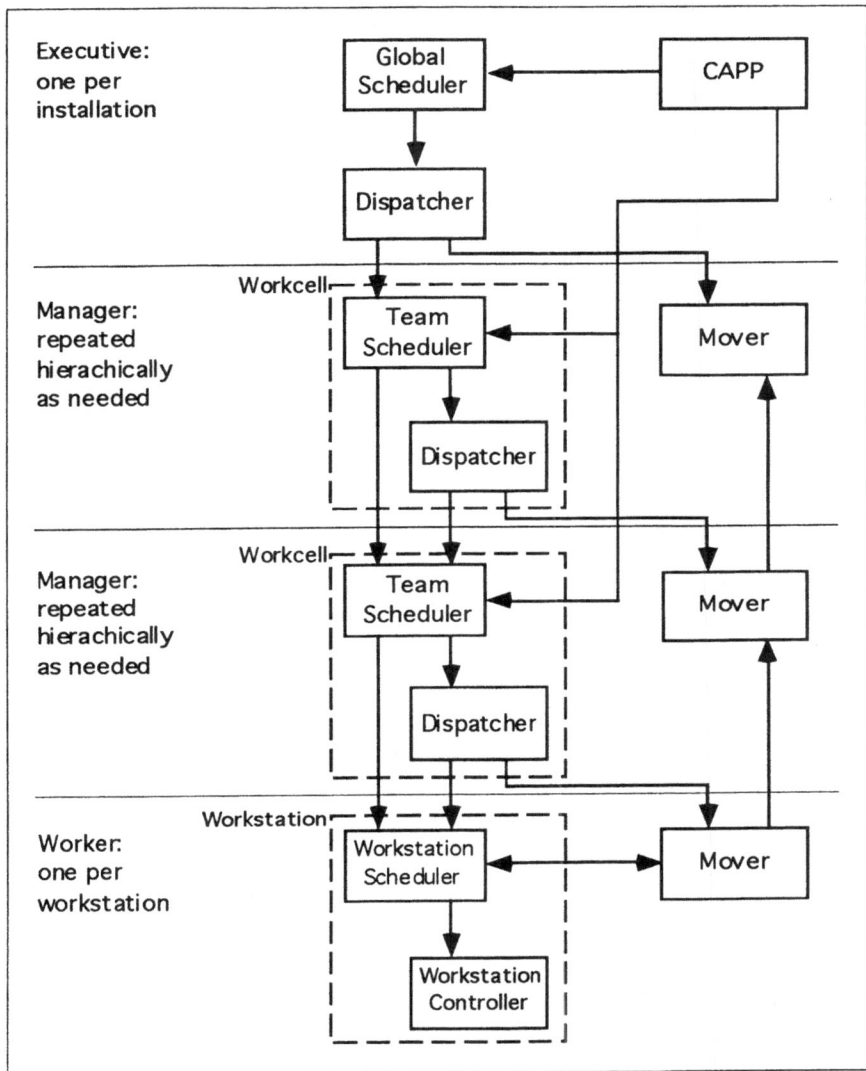

Fig. 4: Hierarchy-structure YAMS [Parunak 88]

The features of the manufacturing control described above suggest to base YAMS on the negotiation concept in form of contract network protocols, the involved processors or the assigned knowledge-based systems being modelled as nodes in the contract network. YAMS represents the plants to be controlled by a hierarchy of production plants, manufacturing cells and machines (see Fig. 4). Each node of the hierarchy corresponds with a node of the contract network. The hierarchy does not define the interaction structure, but expresses only the content of the components of the production plants. Basically, each node can communicate with other nodes and represents the orderer as well as the performer.

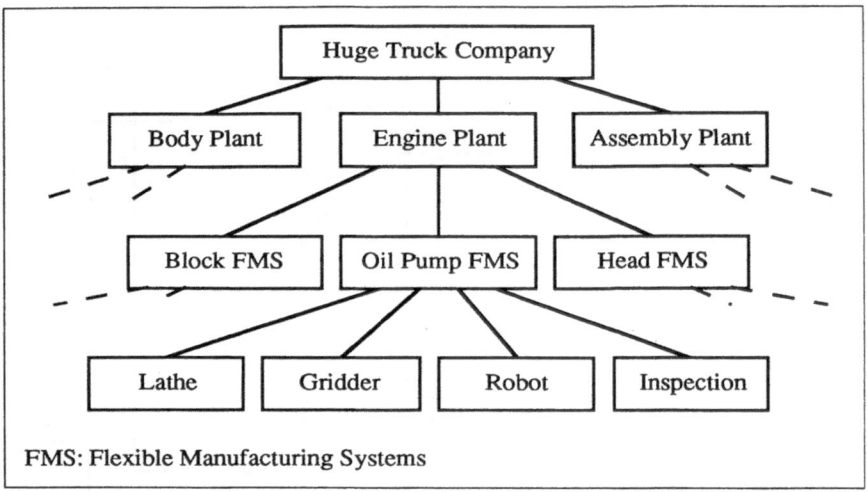

Fig. 5: Example of a component hierarchy in production sites [Parunak 87]

Every node has process plans of the operations which it can execute. Parts of the process plans of a node can possibly not be executed by the node itself but have to be transferred to other nodes. The nodes use the contract network protocol in order to find the most appropriate performer. The existence of process plans presupposes that the *decomposition* of the total problem in single subproblems had been done a priori so that each node has already at the beginning of the execution the corresponding process plans at disposal.

For example, the engine factory (see Fig. 5) accepts the order for manufacturing an engine. By the according process plans the node decides that, among other things, an engine block has to be manufactured. As the node itself cannot process this subproblem, he makes a tender for it per broadcast for processing. Then those nodes apply for it which are assigned to the manufacturing cells for engine blocks and cylinder heads as both have the required capacities for manufacturing an engine block. The order is placed with the manufacturing cell for engine blocks since it has better equipment. If at the moment the capacities of the manufacturing cells of engine blocks are fully utlized or, in the extreme case, if this manufacturing cell had failed, the order for manufacturing the engine block is placed automatically with the manufacturing cell for cylinder heads without entering this explicitly in the control structures of this case.

4.   Distributed Artificial Intelligence

Distributed Artificial Intelligence (DAI) is a rather young subfield of Artificial Intelligence (AI). While AI deals, among other things, with the development of monolithic systems for modeling human reasoning capabilities, DAI focuses on cooperative problem solving by several problem solvers. The single problem solvers are modeled in form of so-called "agents" which interact.

An agent is an intelligent problem solver. The degree of complexity of the problems to be solved ranges from relatively low to very high, up to now, these could only be solved by specialists. The behavior of the agents is characterized by the fact that they follow goals during the problem solving process, that they apply solution methods which are problem-adequate and that they have the ability to cooperate.

A minimal DAI scenario contains at least two agents which cooperate during the problem solving process. At least one of them should possess "intelligent" behavior according to AI systems (see e.g. [Rosenschein 86]).

Central terms of DAI are *cooperation, communication* and *coordination. Cooperation* means the coordinated action of two or more agents in order to attain the individual goals optimally or to understand better the common total target, this presupposes *communication* between the agents. The term *coordination* means the aspect of the control of cooperative problem solving, i.e. the temporal control and supervision of the cooperation regarding efficient problem solving behavior.

Two approaches comprise the field, they form the poles of DAI: Distributed Problem Solving and Multiagent Systems (see e.g. [Martial 92]).

The method of Distributed Problem Solving (DPS) is such that after the decomposition of a problem into subproblems, specialized agents solve one of these subproblems each. The cooperation is related, on the one hand, to the information exchange and the composition of the subsolutions to a total solution and, on the other hand, to the activation of other agents for the problem solution. The coordination of the agents is controlled via a central instance.

The agents of *Multi-Agents-Systems* (MAS) are almost autonomous, they can coordinate their knowledge, their targets and their plans for common actions and for problem solving independently. They must not necessarily follow a common aim, they can have aims which are local and which interact with each other.

> *"[...] research in Distributed Problem Solving [...] considers how the work of solving a particular problem can be divided among a number of modules, or "nodes", that cooperate at the level of dividing and sharing knowledge about the problem and about the developing solution.*

> *"[...] in Multiagent [...] systems,.research is concerned with coordinating intelligent behavior among a collection of (possibly pre-existing) autonomous intelligent "agents" how they can coordinate their knowledge, goals, skills and plans jointly to take action or to solve problems."*

[Gasser 88]

The two system approaches, MAS and DPS, (see fig. 6) form the poles of DAI systems and between them there is a wide variety of further systems. Independent of this classification, all DAI systems have the same problem solving method, called cooperative problem solving. Cooperative problem solving has several phases, which follow either, like MAS systems the bottom-up or, like DPS systems, the top-down way of proceding. The single phases of cooperative problem solving are described in section 5.2.

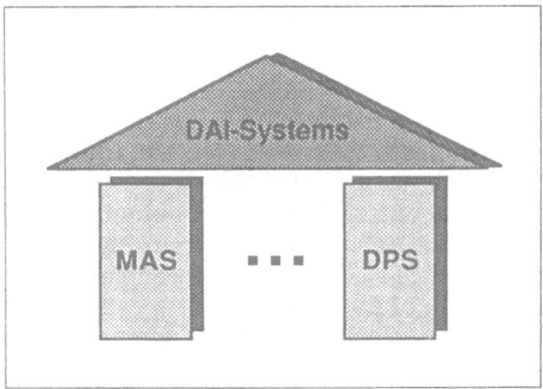

Fig. 6: DAI Subareas (according to [Bond & Gasser88])

4.1. Cooperative Problem Solving

Cooperating systems are systems which have the capacity to solve a problem in a cooperative way. The problem solving process is called Cooperate Problem Solving. The image of Cooperating Problem Solving is a group of human experts which solve a complex problem in cooperation. Every expert solves a defined subtask and can at any moment of the problem solving process of his task involve the other experts as long as they are willing and capable. On the other hand, passing on solutions of subproblems is to make the solution of similar, eventually interconnected subproblems possible. The synthesis of different subsolutions leads to the solution of a complex problem. This process is called cooperating problem solving. The techniques of cooperating systems as well as their history can be found in [Albayrak 92], [Krallmann 92] and [Müller 93].

4.2. Phases of Cooperating Problem Solving

According to Davis a cooperating problem solving process is divided in three phases: problem decomposition, solution of the subproblems and synthesis of the total solution. The relevance of the different phases is related to the concrete problem which can be solved by the group. So, single phases have not to be run through during the solution process. On the other hand it is possible that some phases have to be run through repeatedly.

As an example we use here the control of the traffic light for an entire town which depends on the current traffic situation. The problem decomposition depends on the geographical distribution of the single traffic lights (controlled by one agent each). The problem decomposition, that is the distribution of the processing of subtasks (that is the control of the single traffic lights) is preset and does not require an independent phase. The phase synthesis of the total solution can be omitted as well if the total solution corresponds with the current states of the single traffic lights.

Basically, cooperating problem solving can be divided in three phases:

- problem decomposition & problem distribution
- solution of the subproblems
- synthesis of the total solution

4.2.1. Problem Decomposition

Problem decomposition means the decomposition of the total problem in subproblems and the distribution to different existing agents. If the task formulation requires a decomposition of the problem, there are two basic problems:

- Who decomposes the problem?

- Which way is the problem decomposed?

The two problems are in close connection. Firstly, one can effect the decomposition by a selected agent. This can be oriented, e.g., on the character and the formalization of the problem or on the capacities of the available agents. Secondly, one can have the decomposition done by several or even all agents. Eventually, each agent chooses a subproblem for which he seems to be most capacitated.

There are two main approaches to the decomposition and distribution of subproblems to the most competent agents, called "Connection Problem" in the corresponding literature:

"Contract Net": The agent who is going to solve the subproblem is elicited by tenders. He performs as a supplier. Agents who think they are capable of solving the problem in question apply for processing. The supplier selects from the amount of applicants the one who seems to be most capable and makes a contract with the selected agent.

The problem is delegated directly. Therefore the agent who delegates needs to have exact knowledge about the capacities of the agents to be ordered.

Which of the two procedures is the most favourable, depends mainly on the problem type and the capacities of the agents. In any case, one can say, the problem decomposition and the subsequent distribution of the subproblems is easier, the more the distribution of the total problem is presented inherently distributed and corresponds to the agents´ capacities correspond to this distribution.

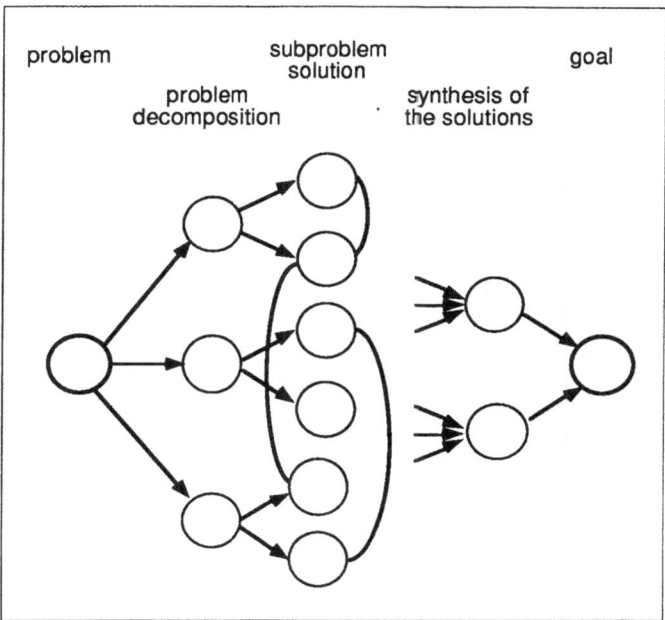

Fig. 7: Phases of distributed problem solving [Smith81]

4.2.2.   Solution of the subproblems

When the agents process the corresponding subproblems, they can interact. These interactions have fundamentally two different causes:

The subproblem solvers realize that their respective subsolutions conflict with one another. One must differentiate between two types of conflicts. On the one hand, there are environments of DAI systems where the agents solve the problems together and where lacks of resources might cause problems. On the other hand, the agents could recognize while processing subsolutions that subproblems are contradictory.

As long as the agent is in a position to solve the assigned subproblems alone, there are no interactions. Normally, however, subproblems cannot be solved autonomously. If one of the agents needs the help of another one, it has to communicate and to cooperate with the other agents in order to reach a total solution.

According to Davis [Davis 81] there are basically two types of cooperation between the problem solvers:

• task sharing

• result sharing

Task sharing means that the problem solvers support each other by assuming subtasks of the partner agent.

Result sharing means that the problem solvers cooperate by exchanging subsolutions.

4.2.3.   Synthesis of the Total Solution

If the amount of single solutions has to be put together to form a total solution, there are two problems which regard the decomposition:

- Who puts the subsolutions together?

- How are they put together?

There are two possibilities of putting total solutions together, either in a centralized or decentralized manner. Which one of these possibilities is more appropriate, depends on the type of problem and on the agents´ capabilities.

4.3.   Blackboard Metaphor, Model and Frameworks

Historically, the idea of the blackboard technique arose out of concerns with organizational problems of AI-programs existent in the early sixties operating on sequential flow of control and closed subroutines. The deficiencies of these AI programs can be characterized as follows [Nii 86a]:
- Operations must happen sequentially

- The closed subroutines do not operate on all available information provided by the system referring to the current state of the problem solution. On the one hand, there is no direct access to the subsolutions generated by other subroutines, on the other hand, the necessary communication between the subroutines requires defined communication protocols.

In order to remedy these deficiencies, the problem solving state must be available in form of a global data structure, while maintaining the isolation of the subroutines. This idea forms the base of the blackboard model. In the following, definitions of the blackboard technology terms are introduced according to [Nii 89]. They are necessary since literature as well as spoken language do not sufficiently distinguish among these terms.

- *Blackboard model* is a specific problem solving model.
- *Blackboard framework* is a specification of the components of the blackboard model or of the implementation.
- *Blackboard shell* is used as a synonym for the term *blackboard framework.*
- *Blackboard application* is a system based on blackboards which solves the problems of one field (domain).
- *Blackboard system* includes the framework as well as the application.
- *Blackboard Architecture* refers to the design of a blackboard system.

4.3.1. The Blackboard Metaphor

Quoting Allen Newell, 1962, the properties of the blackboard metaphor can be described as follows:

> *"[...] Metaphorically we can think of a set of workers, all looking at the same blackboard: each is able to read everything that is on it, and to judge when he has something worthwhile to add to it. This conception is just that of Selfridges Pandemonium: a set of demons, each independently looking at the total situation and shrieking in proportion to what they see that fits their nature."*

[Newell62].

This metaphor is to be interpreted as a group of experts (or problem solvers) analyzing a blackboard which records the individual states of the ongoing problem solving process (where the different states are the initial data in the beginning, passing over to contain the solution by the end of problem solving). Each expert is capable of recognizing when it can contribute to the problem stated on the blackboard and can take then the appropriated action (see Fig. 8).

Fig. 8: Illustration of the blackboard metaphor [Winston 87]

4.3.1.1. Properties of the Blackboard Metaphor

According to [Corkill 91] the properties of the blackboard metaphor are described as follows:

- *Independent expertise*: The experts do not need the direct support of other experts. They do not even have to know about their existence. Therefore, the communication happens only through the blackboard.

- *Different Problem Solving Techniques:* The independence of each expert implies that experts are free to chose their problem solving techniques. This implies further that for each subproblem the best possible problem solving method can be utilized.

- *Flexibility of Representation on the Blackboard:* The form of representation on the blackboard can be application-dependent.

- *Uniform Representation of each Application:* Each expert has to be capable of understanding the contributions to the problem solving process placed by other experts on the blackboard. In order to achieve this aim optimally, the representation of information on the blackboard can be free, but uniform and fixed for the entire application.

- *Positioning Metrics:* The more complexity inherent in an application, the more complex the data placed on the blackboard will be. In order to allow the experts quick and efficient access to the information on the blackboard so that they can have an overview of which changes produced their activations, it is partitioned (either in regions characterized by different abstraction levels or into new blackboards). Each expert supervises only the changes made in its particular region of interest.

- *Event-driven Activation:* Each expert monitors the ongoing problem solving process and makes attempts to contribute with its knowledge whenever possible. Situations in which its contributions are required usually arise, when other experts modify the state of the blackboard, more specific, when events occur in the shared memory. These events can be the placement of a new hypothesis or the change of an existing one on the blackboard. Triggered by those events, the expert will react either by confirming existing hypothesis, or by placing new ones on the blackboard (which, in turn, will trigger other experts for activation). Other forms of expert-activation could be an expert-driven activation. In this approach experts define at the beginning of the problem solving process, in wich type of events they will be interested. The system will then activate the experts based on the type of event that just occurred.

- *Necessity of Control:* In principal, there is the possibility that several experts may react to one single event, especially in case of a computer-supported solution. Thus, a control entity has to be created which selects the one best-suited to react to that event. A very simple control mechanism could activate the first expert that reacts to an event (first come, first serve). A considerably more efficient approach would judge the quality of the contributions (by efficiency, trust, importance for the problem solving process etc.) coming from each of the responding experts. This approach implies, though, that the control mechanism has to be provided with strategy knowledge, that is, knowledge about *which* expert to select in *what* type of situation.

- *Incremental Generation of the Problem Solution:* The problem solution is produced by the following loop:

  - Experts react to events
  - One expert is rated "most appropriate" and selected to execute the task.
  - The contribution results in a modification on the blackboard.

4.3.2. The Blackboard Model

The blackboard model is a special problem solving model. A problem solving model is defined as a conceptual frame which specifies which data (or what knowledge) is to be applied when and how in order to solve a given problem. Thus, the main components of a problem solving model are data (knowledge) and rules which define an application. Very simple models can be found in the classic expert system approach where the knowledge is embedded in the procedures.

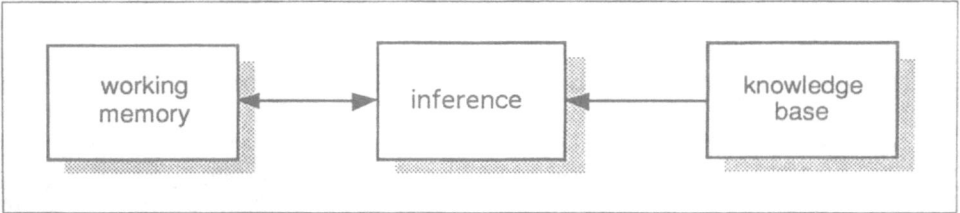

Fig. 9: The classic expert system [Nii 89]

The classic expert systems are problem solving models, too, where the knowledge is stored explicitly separated from the application field (the problem solving mechanism or "expert"). The problem solving mechanism uses the knowledge contained in a knowledge base to construct the solution in the working memory. That means, hypotheses are built successively and verified in the working memory, until the problem is solved (see Fig. 9).

The classic expert systems have several disadvantages:

- The control is implicit, that means, it depends on the structure of the knowledge base. A suitable example is the production system. Its representation formalism makes the control dependent upon the production rules.

- The choice of the knowledge representation is dependent upon the problem solving mechanism and vice versa.

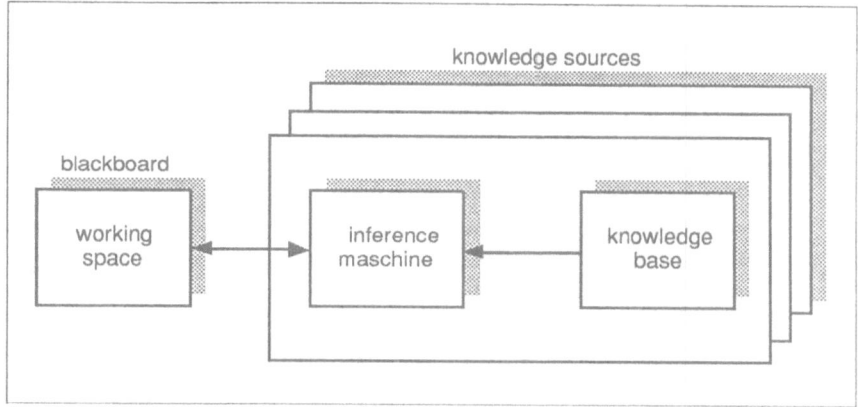

Fig. 10: The rudementary blackboard structure [Nii 89]

In order to eliminate this disadvantage, the problem solving component and the knowledge base are distributed onto several subexpert systems which reach the problem solution in a cooperative way (see Fig. 10). The components of this structure are specified as follows resulting in the standard blackboard structure (see Fig. 11):

- The knowledge sources necessary to solve the subproblems are partioned in the knowledge sources together with an appropriate problem solving component.

- The current state of the problem solving process is illustrated by the global data structure, the blackboard. So the blackboard is the means of communication for the knowledge sources.

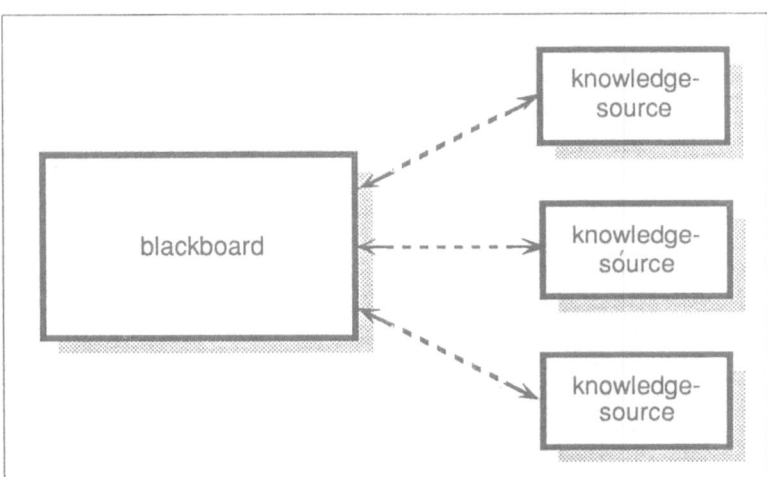

Fig. 11: The standard blackboard structure [Nii 89]

4.3.2.1. Properties of the Blackboard Model

The blackboard model does not contain any control component. As shown earlier in the description of the blackboard metaphor, a control component turns out to be necessary when problem solving strategies have to be integrated. The blackboard represents the solution space of the ongoing problem solving process, that is, initial data, sub and final solutions. Due to this fact, the blackboard is commonly divided into application-specific, hierarchical levels. The transformation on these levels (synthesis of subsolutions to global solutions or the confirmation of subsolutions by already found global solutions), is made by the knowledge sources.

Problem solving using the blackboard model is executed opportunistically, that means knowledge can be applied either using forward-chaining or backward-chaining algorithms, depending on the fact which technique appears to be the most appropriate in the context of the application and the problems to be solved.

4.3.2.2. The Koala Example

The problem to be solved: find a certain koala in the eucalyptus forest. This example is used to illustrate the properties of the blackboard model. Since it does not require specific knowledge, it is used as a standard example in the relevant literature.

Firstly, different types of knowledge are necessary in order to find the koalas, e.g. color and shape of them, the different colors and shapes of their environment, their behavior (locations, movements etc.), seasonal and daytime influences. In order to make a decision how these different types of knowledge can best be utilized, one first considers what the solution could possibly be. While considering several solutions, it becomes apparent that solutions or subsolutions are also composed of subsolutions, e.g. one defines that an object is a koala, if it looks and acts like a koala. An object looks like a koala if it has the head and the body of a koala, etc. This abstraction hierarchy is continued on down to the lowest level, the single pixels of the snapshots taken in the eucalyptus forest. A solution of the total problem is composed of sub and hypothetical identifications which are attained by gradual abstraction of the solution space. This corresponds with the blackboard structure of the model.

The types of knowledge defined at the beginning are applied to the distribution of the solution space, in order to verify the hypothesis of higher levels or to synthesize the results of lower levels. In order to attain this, the different types of knowledge are organized in form of knowledge sources. For example, a knowledge source (the expert in question is called the color specialist because it applies its knowledge to the color design of the screen lines) could formulate hypotheses whether certain lines figure parts of the koalas´ bodies. The color specialist transforms the steps between the different graduation levels, here between the level of the screen lines and the one of the figures of the koalas´ bodies. Another knowledge source (body specialist) parts from the body hypothesis in order to find in the figure more parts of the koalas´ bodies in the correct order. That means it formulates its own hypothesis which have to be verified on a lower level. Fig. 12 illustrates the organization of the system.

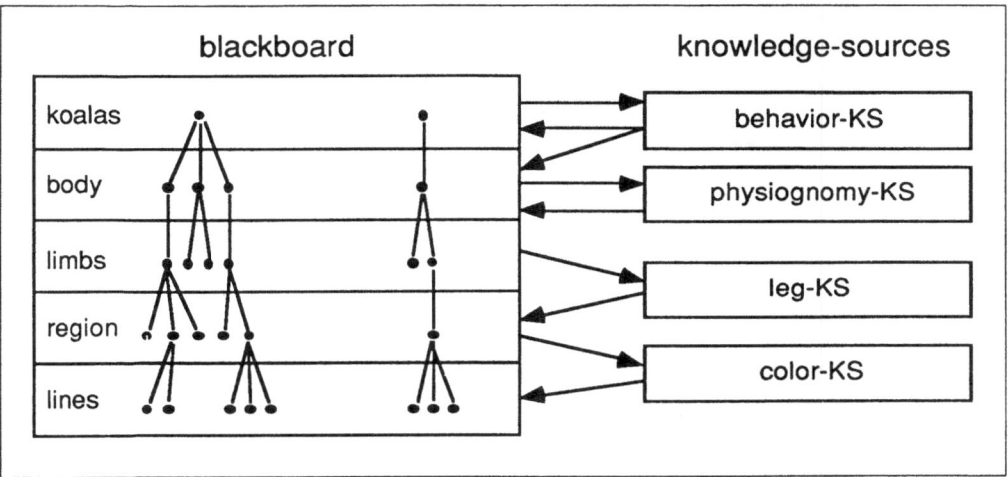

Fig. 12: Blackboard structure und knowledge-sources for the koala problem [Nii 89]

The koala example is used to illustrate two facts: Firstly, here the typical organization of the system, partioning application knowledge and specific knowledge sources is realized; the solution space is partioned in different abstraction levels. Secondly, the characteristic problem solving behavior of the blackboard models is made apparent, the opportunistic reasoning as described above as a combination of hypothesis and verifications while solving gradually the problem.

4.3.3. The Blackboard Framework

The blackboard framework is an extension to the blackboard model, serving as an intermediate step toward any application that is to operate on the blackboard model. Many blackboard applications share common properties. These application-independent properties will be shown in the following, they can be regarded as guidelines for blackboard applications.

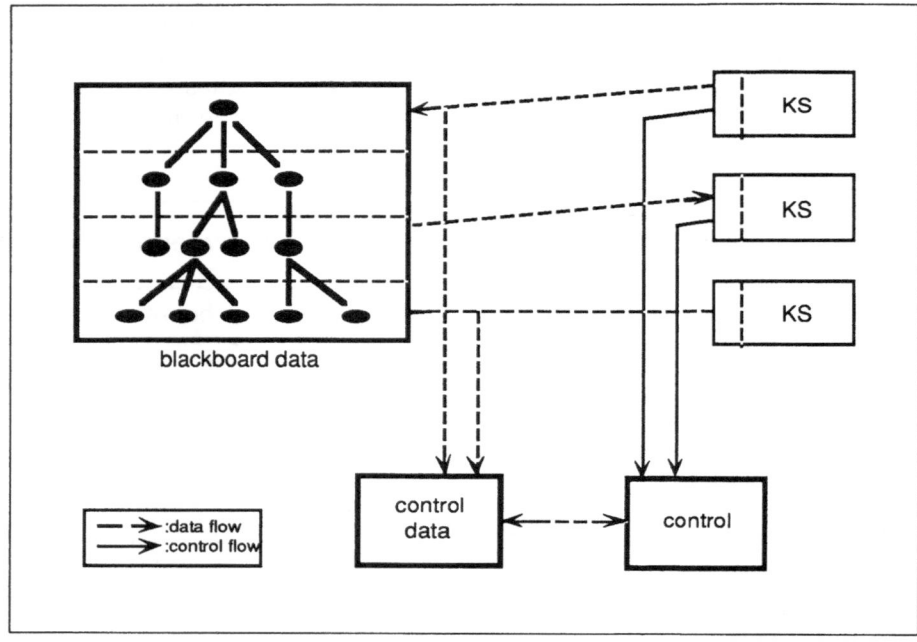

Fig. 13: The blackboard framework [Nii89]

The most significant extensions to the blackboard model are a more precise specification of the knowledge sources and the inclusion of control modules and control data.

4.3.3.1. Knowledge Sources

The main characteristic of knowledge sources is their independence and the fact that they have to be separate from each other. The function of knowledge sources is the contribution of knowledge to the problem solving component. They process internal freedom of representation (mostly in form of procedures and/or rules). They are the only components in a blackboard which modify it. They are mainly divided in: *preconditions* which indicate, upon input of certain data, if the knowledge source is to be activated and *body* which indicates the contribution of the knowledge source in the given context.

Triggering a knowledge source for execution is called Knowledge Source Activation Record (KSAR), it cannot be distinguished from the knowledge source itself. The difference results from the context of KSAR like a program differs from a process. KSAR is the dynamic instance of the knowledge source containing specific data.

4.3.3.2. The Blackboard Structure

The blackboard serves as a communication medium between the knowledge sources. It stores objects that belong to the solution space: initial data, sub- and global solutions, and, considering the existence of the blackboard framework, it also contains control data. The blackboard objects are classified and placed into several, hierarchically organized levels. Properties of these objects are input into the knowledge sources, which in turn, produce more information on the same or on different levels of the blackboard.

4.3.3.3. Control

The framework described here is completed by control modules which, on the one hand, monitor the modifications on the blackboard and, on the other hand, select the knowledge sources reacting in an opportunistic way to modifications. That means control modules define the focus of attention regarding the problem solution. There are several possibilities which select the appropriate focus of attention, e.g.:

- *knowledge-source oriented* "Which knowledge is to be activated?"

- *blackboard-oriented* "Which area on the blackboard needs further completion?"

- *combination method* "Which knowledge source is to be applied to what object?"

  This results in an iterative sequence of problem solving activities:

- A KSAR leads to modifications on the blackboard. This situation is called *blackboard event*.

- Events result in the precondition check of all knowledge sources.

- Depending upon the strategy in use (blackboard-oriented, knowledge source-oriented or both) the control module selects the current focus of attention.

- The selected focus of attention is prepared for execution. If the focus is a knowledge source, the appropriate blackboard object will be selected (so-called knowledge-scheduling approach), if the focus is a blackboard object, the appropriate knowledge source is selected (so-called event-scheduling approach).

- The KSAR generated this way is executed.

The above described cycle makes apparent that, at each moment, any direction of the conclusion can be taken, as the knowledge sources can either analyze hypothesis or construct subsolutions. Due to this fact, the problem solving process occurs opportunistically and dynamically.

4.3.3.4. Strategies of Problem Solving

The behavior of a blackboard system is determined by the criteria implemented inside the control modules. As described above while discussing the blackboard metaphor, these criteria will determine the problem solving strategy. The framework makes no specifications regarding these criteria, as the appropriate strategy will always be application-dependent. The approach to select amongst several available strategies the most appropriate one by blackboard technologies is an important approach.

4.4   History of the Blackboard Model

Hearsay-II [Erman 80] is a system developed in the early seventies at Carnegie Mellon University (CMU) for language recognition. It is regarded as the initial, experimental work for the development of the blackboard model.

After Hearsay-II the blackboard model was applied to other applications as well. Applications have been developed e.g. for the military field (HASP which analyzes sensor data). Victor Lesser is one of the Hearsay-II architects. He moved CMU to the University of Massachusetts at Amherst (UMAS). There he continued his research work at the blackboard model and developed the blackboard model as a base for cooperative problem solving.

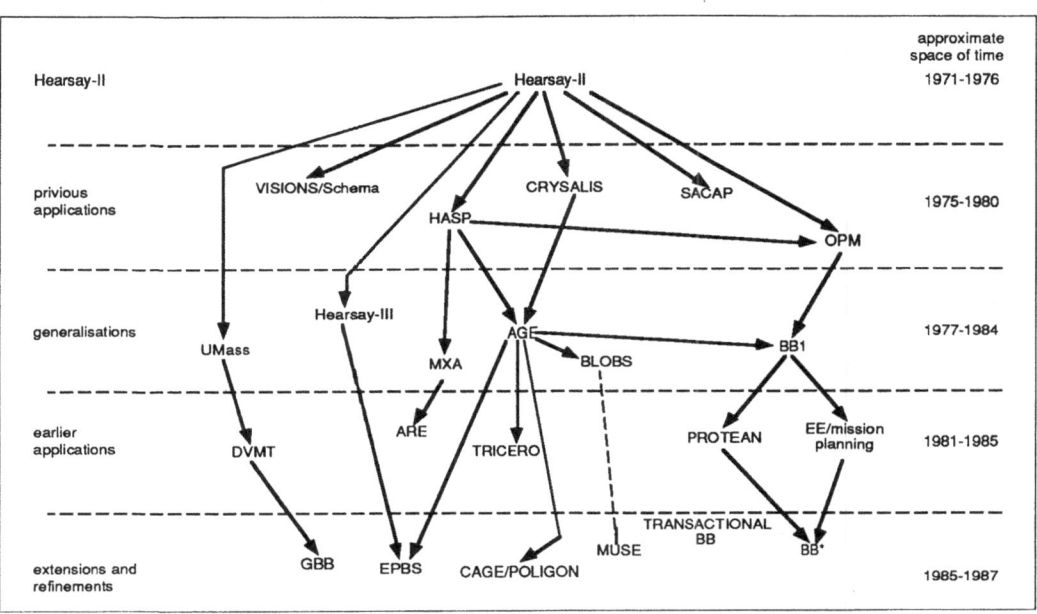

Fig. 14: Evolution of blackboard systems [Nii 89]

At the same time, the AGE project tried to generalize the strategies which were applied to the previous blackboard generations (like Hearsay-II and HASP). Victor Lesser continued his research work at UMAS based on cooperative distributed problem solving. Together with his colleague Daniel Corkill he developed a distributed blackboard application for the monitoring of vehicles: the system "Distributed Vehicle Monitoring Testbed, DVMT". Over the last years the tool or development environment Generic Blackboard System (GBB) has been developed as an extension of the current blackboard model which is to make it possible to compile rapidly blackboard applications with a short run-through-time. The Blackboard Technology Group took over GBB and developed it further. Meanwhile the GBB 2.0 version is distributed commercially. [Nii 89] gives a survey of the history of the blackboard model.

After Victor Lesser left UMAS, three developers of Hearsay-II remained. Barbara and Frederick Hayes-Roth moved to the Rand Cooperation where they worked together at the development of the planning system OPM. Later on Barbara Hayes-Roth developed BB-1 Control Shell. The BB-1 Control Shell was introduced as the first OPM planning system. BB-1 was applied later on in many blackboard frameworks to the realization of different applications. BB-1* is the advancement of BB-1. The third developer of Hearsay-II, Lee Erman, moved to the Information Science Institute (ISI) of the

University of Southern California and there started a project transforming Hearsay-II into a new framework, called Hearsay-III. The research work at UMAS reached an advanced state with regard to cooperative problem solving based on blackboard models.

4.5. Advantages of DAI

There are a number of advantages associated with the use of DAI techniques [Durfee 89]:

- Fast and efficient problem solving by taking advantage of inherent parallelisms.
- Reduced communication due to conventional distributed systems due to the transfer of subsolutions to the corresponding agents.
- Flexible processing through dynamic cooperation of differently qualified agents.
- Increased dependability if problem solvers are able to assume responsibility from the down units. This is realized by several problem solvers working on a common or overlapping sub-problem.

Moreover, DAI models can contribute to the validation of theories in sociology as well as management and organization theory.

DAI is also available to provide valuable assistance with regard to the acceptance of systems. The ability to cooperate flexibly and "intelligently" with other systems and people and to coordinate its activities with them brings closer to the goal of a human-adequate machine interaction capability.

5 VerFlex-BB System: Approach and Implementation

5.1. Distributed Approach to the Solution of the Task Order Execution

The description of the task order execution makes apparent that the structure is divided in several subtasks, all of them being very complex and poorly structured. So it is nearly impossible to apply conventional methods to the solution.

The subproblem solvers interact during the problem solving process, that means a solution is attained by intensive and goal-oriented cooperation of the subproblem solvers. The subproblem solvers provide subsolutions during the problem solving process.

Furthermore, the task *order execution* has to meet more restrictions which further the acceptance of the system and which support effective and efficient problem solving. These criteria are applied to the evaluation of the knowledge-based systems.

- Acceptance

- Run-through-time are the main criteria besides adequate modeling.

The properties which imply the acceptance of the solution of the task order execution are: interface, support, qualitative data, influence in the restrictions and goals.

The requirements to the interface enhance the necessity of adequate and efficient modeling which make it possible to influence in the behavior of the whole system as well as of the single subproblem solvers. Furthermore, the functions and the user interface of the system have to be customized (oriented to the skilled user's way of proceding).

The requirements to support the user underline the necessity that the solution approach is to support the user while controlling the problem solving process, it should give hints how to solve the problem, detect inconsistent and contradictory solutions and offer the corresponding suggestions of how to solve the conflicts. It is of the same importance that the solution approach makes reliable decision values available at any time.

The next requirement is that the strategies underlying to the problem solution (or the subproblem solvers) have to be observed during the solution process and that the restrictions are kept which indicate to the user how he can preset or modify them.

Furthermore, it must be guaranteed that all subproblem solvers have a consistent and constant sight of the shop floor at any time.

There are high requirements to the run-through-time of the problem solver since the (sub)solutions in manufacturing have to be found within a sensible period of time. The solutions have to fulfill the requirements of different goals.

These requirements imply that the single subproblem solvers are based on efficient and flexible problem solving methods so that the quasi-real-time-requirement is met.

The reason for the preferred application of AI methods compared with conventional methods is the poor structure of the tasks, they are really complex and even NP complete. AI methods are necessary, as the application of conventional methods does not lead to a solution in a responsible period of time, in order to attain a prompt solution, even if it is not the optimal solution.

If the task order execution has been solved in an efficient and flexible manner, this means support to the achievement of the superior corporate goals.

This approach, realized by VerFLEX-BB based on the principle of distributed problem solving is characterized by adequate modeling of the subproblem solvers. One agent was developed for each subproblem solution, the agents being based on problem solving methods which are specific to the problem class, independent on the domain and efficient. The problem solving methods fulfill the requirements to efficient and effective problem solving.

The interaction of the agents during the problem solving process is realized via cooperation on the blackboard. The agents provide subsolutions during the problem solving process, this cooperation form is called "Result-sharing" and it is supported efficiently by the blackboard model.

This solution approach includes a user interface which was adequately modeled and which meets all above-mentioned requirements. The user gets support from the system by the interface and has possibilities to set or to modify restrictions, strategies and other heuristics.

There is a fault clearing component included in the solution approach composed of two diagnosis agents: *technical fault clearing* and *interruption clearing*. In case of an interrupted capacity site detected, the origin of the interrupt is found out and measures are suggested how to clear the fault and how many time is needed.

The interruption time predicted by the technical fault clearer is interpreted by the component *conflict analysis*, the consequences are propagated in order to determine which orders suffer from the interruption. And then replanning is possible guided by the duration of the interruption. The resulting decision value *interruption duration* is rather reliable.

The agent *interruption clearing* deals with the interruption flows resulting from the technical breakdown cooperating intensively with the scheduler.

Furthermore, this solution approach contains an agent which has the task to put at any time the current sight of the shop floor at disposal. This agent puts the data at disposal in a way that all other agents can understand and interpret them.

This solution approach is based on ISIS/OPIS [Fox 84], [Fox 86] and SOJA/SONIA [Sauve 87], [Sauve 89]. The main difference is that our solution approach contains two explicit diagnosis agents, which, on the one hand, guarantee for the provision of reliable decision values and, on the other hand, for a sensible clearing of the interruption flow (also if replanning had been done before).

This solution approach meets all requirements which are needed for high acceptance. Especially the problem solving methods which are specific to the type of problem and efficient improve the run-through-time of the single agents.

Fig. 15 illustrates the architecture of the VerFLEX-BB system where this approach has been implemented. Fig. 17 shows the subtasks of order execution and the agents embedded in the architecture of the VerFLEX-BB system where the solutions of the single subtasks are modeled in form of agents. The modelling of the solution of the task *order execution* is really adequate.

The next section introduces the architecture of the agents, each agent of the system has the same architecture (composed of conform components) but a different content.

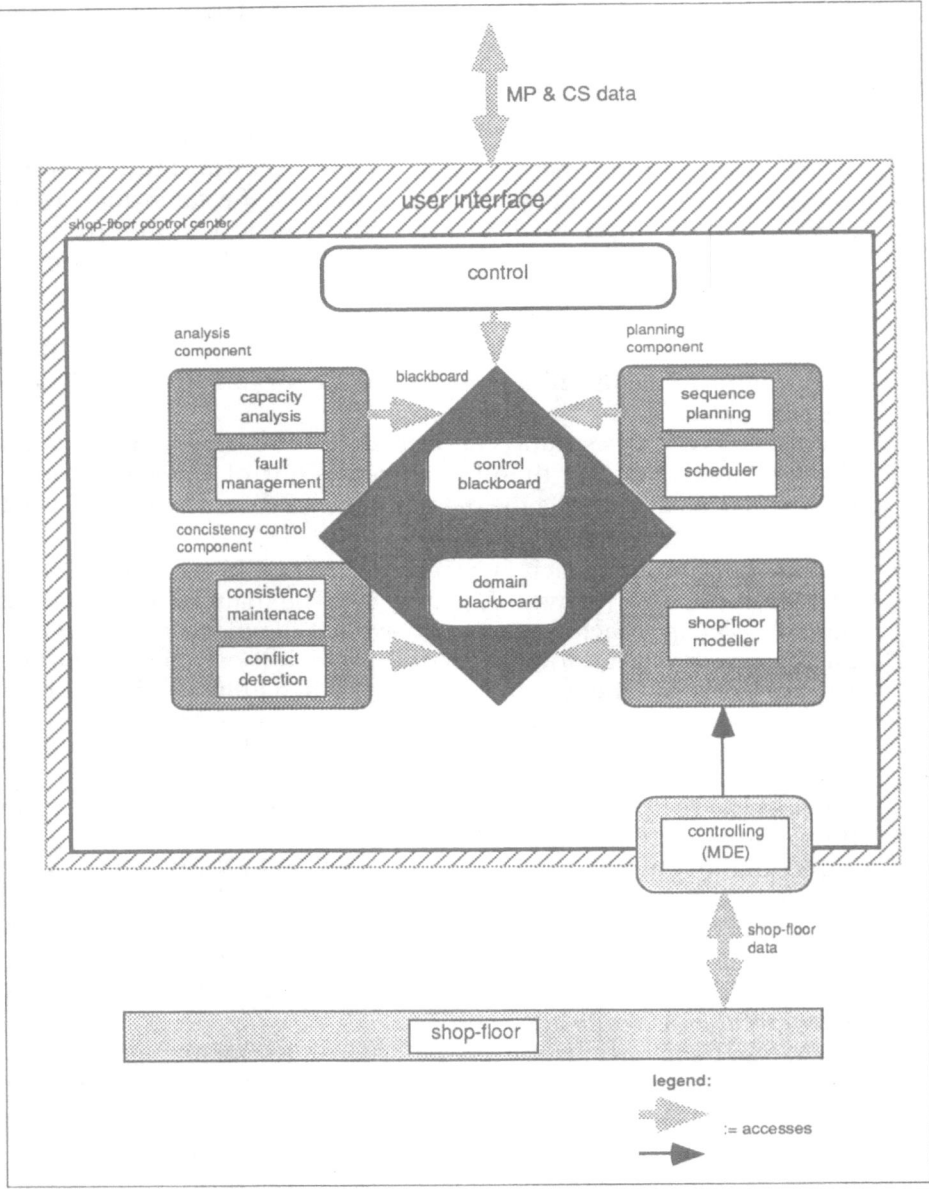

Fig. 15: Architecture of VerFLEX-BB system [Albayrak 92]

5.2.   Why was the Blackboard Model used?

The blackboard model is especially suited to the cooperative solution of the task order execution due to the following reasons:

- All agents have to operate on the same object model. That means these agents must have at any time the same sight of the object. This requirement is fulfilled by the blackboard architecture easily and efficiently. Other applications (Contract-Net) have to propagate the modifications via Broadcast. These messages imply high communication expenditures. The asynchronical message transference and the high communication expenditures which involve delays are in opposition to the rigid real-time requirements.

- The cooperative solution of the order execution is attained by the procedure that the subsolutions of elementary tasks of other agents are put at the disposal of the subproblem-solving agent. This procedure is called result-sharing and can be realized easily and efficiently by our model.

- There is an inherent distribution of the problem, order execution is composed of several subproblems which are on hand inherently. The problem decomposition and distribution of the different subproblems can easily be realized by control knowledge sources.

- Success up to now has been attained by systems or approaches based on blackboard architecture models.

5.3.   The VerFLEX-BB System

The VerFLEX-BB system is to support staff-members of a company (shop floor manager, installation engineer, machine engineer etc.) to solve the task *order execution*. VerFLEX-BB supports on the operative level the attaining process of the corporate goals.

VerFLEX-BB puts one agent at disposal for the solution of each subtask. The cooperation during the problem solving process is done by providing the corresponding subsolutions on the blackboard.

Within the corporate CIM chain, VerFLEX-BB is set before the planning level and behind the planning system (see Fig. 16). Fig. 17 illustrates the modeling of the problem solution. The Figure shows the task *order execution* and the corresponding subtasks and, at the same time, the implementation of the system architecture of

VerFLEX-BB. The VerFLEX-BB system is composed of two parts which use a common knowledge base:

- The *user interface*, it makes "manual" planning possible and influences the behavior of the single agents.

- Agents for the solution of the respective subtasks like technical fault clearing, clearing of interruption flows, scheduling, sequence planning, consistency maintenance, conflict detection and provision of the current view of the shop floor.

The above-mentioned agents enable the VerFLEX-BB system to act based on predictions (planning etc.) as well as to react on events in the shop floor (supervision, replanning and fault clearing, especially technical faults and interrupts). These properties enable the VerFLEX-BB system to solve the problems on hand efficiently.

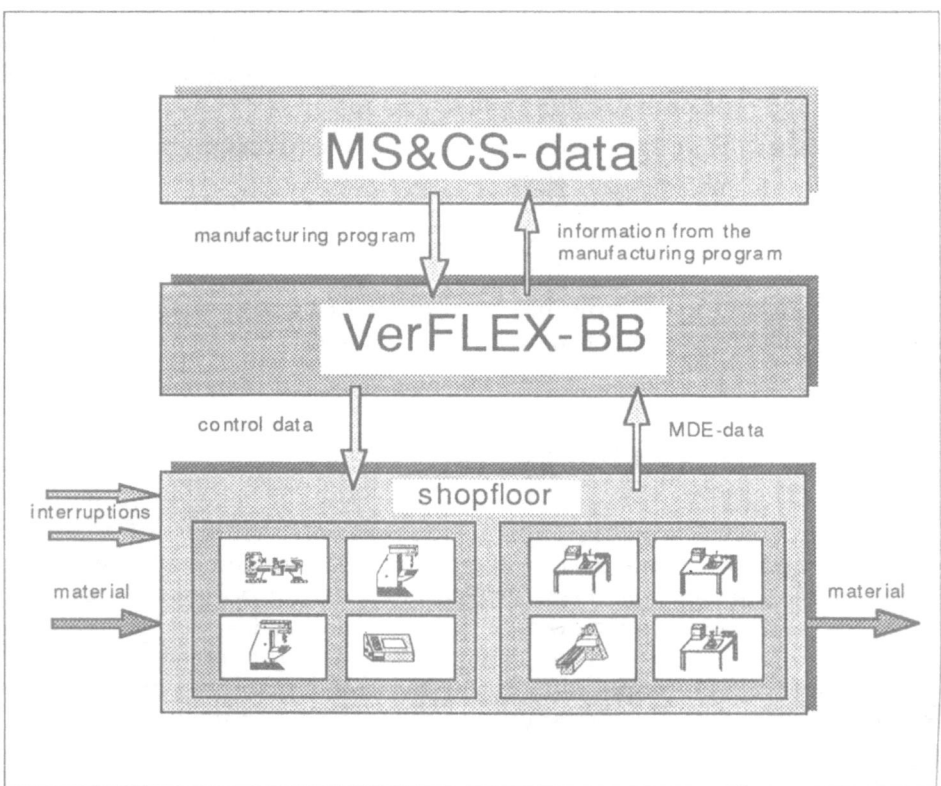

Fig. 16: Position of the VerFLEX-BB system in CIM

The manufacturing sector modeled is composed of 14 allocation units which include 3 to 4 capacity sites each. The planning horizon can vary from one day to four weeks. This value can be adjusted via the user interface.

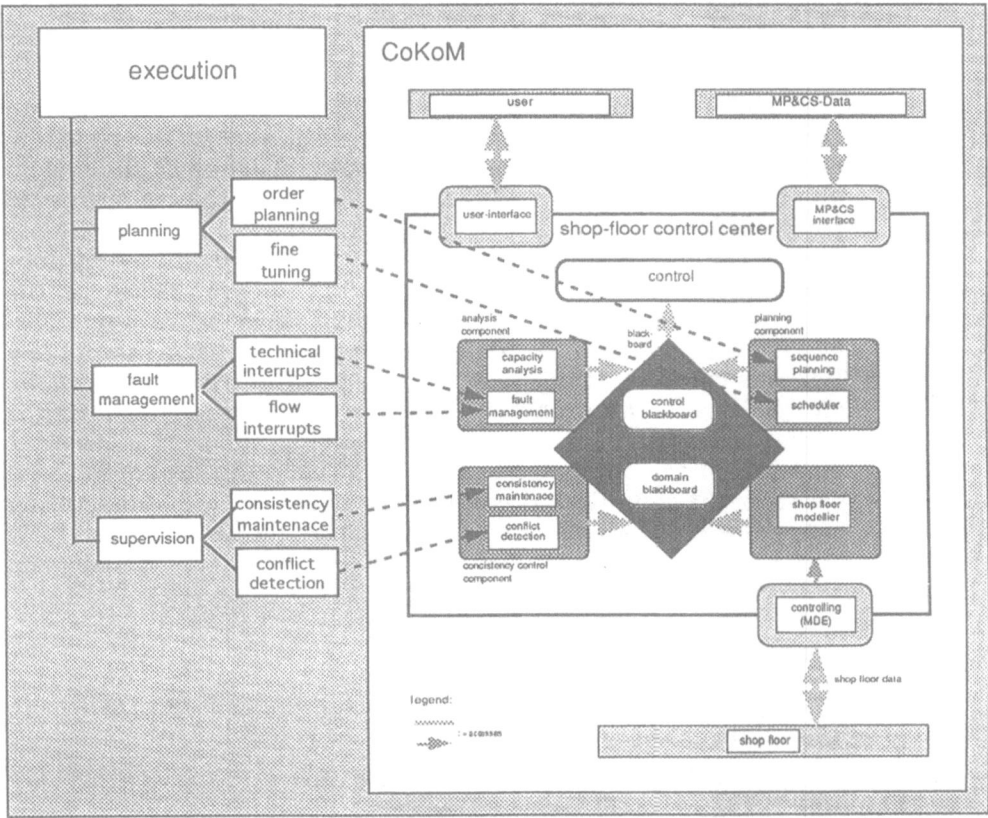

Fig. 17: Task implementation and system VerFLEX-BB

The conductor plates manufactured in the manufacturing sector under consideration are passed on to another manufacturing sector to be processed further and so they influence in its planning. The manufacturing system is composed of three manufacturing sections. In the first, the conductor plates are produced, in the second the casings of the phones are manufactured. In the third manufacturing section conductor plates and casings are put together to telephones.

The VerFLEX-BB system has been created for *shop-floor oriented manufacturing* and for *small and medium series production* because they involve really complicated order execution steps. Companies which produce this way must be very flexible concerning the customers and must support efficiently the goals set on the manufacturing level.

The VerFLEX-BB system was implemented using the following tools: GBB [Johnson 88], Common-LISP, FOXGLOVE (a rule interpreter) and OSF-Motif.

Modelling Components

The modelling components give the VerFLEX-BB system a high degree of flexibility. On the one hand, one can transfer the system to other manufacturing areas, i.e. the operationalization of the system with other domain-specific knowledge as well as the connection to the new BDE system.

If it is connected to the new BDE system, then the shop-floor modeler must be informed about what the structure of the information from the BDE system looks like, since the form in which VerFLEX expects the information is contained in this component. When this component receives information from the BDE system, then it adapts the information, if needed, and then writes it on the domain blackboard.

The modelling components assure that current information about the state of the shop-floor is continually available. In manufacturing it is important that all the agents always have the same current view of the shop-floor. This prevents the development of inconsistent conditions and contradictory solutions.

Analysis Components

The analysis components consist of three agents for the tasks:
- Capacity analysis
- Managing technical faults (diagnosis agent)
- Managing work flow faults (fault manager)

These agents were brought together in one component because they all belong to the problem class, classification.

Capacity Analysis

The capacity analysis agent compares the available capacity and the specific degree of capacity utilization using plan orders, taking into account the current utilization. Here, under utilized and over utilized capacities are localized which has a corresponding effect on the planning component.

The capacity analysis produces GBB heuristics which are deposited on the control blackboard and thus influence the scheduling. A possible heuristic is when the scheduler plans plan orders, he first takes into account the Afos which require bottle-neck resources and thus are to be classified as bottle-neck Afos.

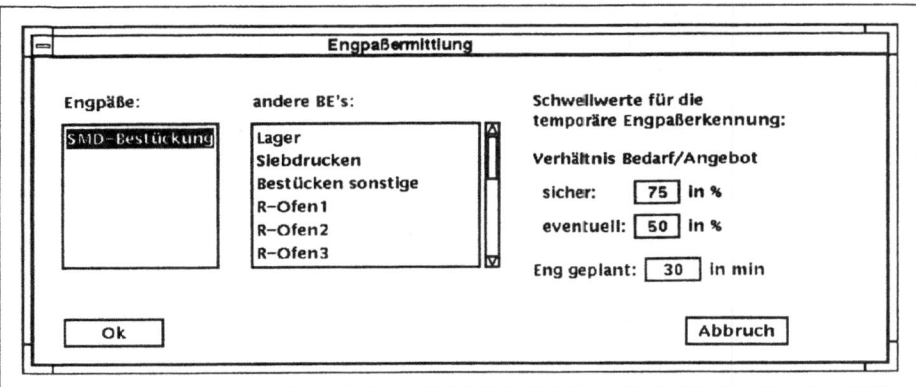

Fig. 18: Control knowledge of capacity analyzer

The domain-specific control knowledge required by the capacity analysis is available in the form of threshold values. These threshold values can be changed for each BE at any time with VerFLEX-BB via the user interface (see Fig. 18) Figure 19 shows for each BE, the time period for each capacity analysis as well as the demand and supply for this time period and the result of the analysis.

```
                                  Temporäre Engpäße:
Kommandos
                          Quittieren

BE      Start               Ende            Bedarf Angebot Engpaß?
100001  18.08.92  10:24  18.08.92  10:24        0       0    nil
129100  18.08.92  10:24  30.08.92  14:57      771   35106    nil
129179  18.08.92  10:37  30.08.92  22:16     3533   53937    nil
112111  18.08.92  13:18  30.08.92  22:44     25C8   53538    nil
120733  18.08.92  15:12  31.08.92   2:46      550   17974    nil
120734  18.08.92  12:13  30.08.92   7:50     1592   17017    nil
120735  18.08.92  15:42  30.08.92  11:19      367   17017    nil
129144  18.08.92  10:24  30.08.92   4:48     1669   11256    nil
129145  18.08.92  12:21  30.08.92   6:50      624   11189    nil
129156  18.08.92  10:24  30.08.92   9:45     1416   11520    nil
112101  18.08.92  16:48  31.08.92   1:28     3604   53400    nil
129173  18.08.92  16:01  30.08.92  23:11      703   17710    nil
129505  18.08.92  17:39  31.08.92   0:30     1607   17691    nil
112201  18.08.92  16:50  29.08.92   4:43      112    5203    nil
112202  18.08.92  18:28  31.08.92   0:45      461   17775    nil
112307  18.08.92  18:32  30.08.92  23:49      320   17597    nil
112305  18.08.92  18:44  31.08.92   0:22      632   52854    nil
```

Fig. 19: Result of bottle-neck analysis

Fault Management

The task of manufacturing control is to process orders punctually and efficiently. Real production processes however are characterized by the fact that there are always faults in the planned flow, e.g. through machine faults, personnel illness and material shortages.

If there is a fault, then this requires a quick reaction, i.e. a decision must be made how to continue. Normally, a decision is made by a responsible manager with the appropriate competence. If, in the course of the decision making process, it becomes apparent that ones own area of competence is exceeded, then the decision is passed on to the next highest level and the necessary decision is made cooperatively taking advantage of the information of other decision makers.

In practice, it is often the case that the decision making process, which requires several decision making levels and information agents, requires a great deal of time, since often the competent person is not immediately available thus leading to longer communication times. However, the longer the reaction time, the more severe the consequences resulting from the fault. Sometimes provisional decisions are made to solve pressing problems. But in another context, the temporary decision can lead to additional complications. In the VerFLEX-BB System this problem is solved in such a manner that through an agent (the diagnosis agent) only technical faults of the type "machine down" are managed and another agent (the fault agent) is responsible for the work flow faults. Work flow faults include "material shortage", "personnel shortage" or as a result of technical faults "subsequent faults".

Planning Components

The planning components provide two agents for the solution of the planning tasks mentioned in Section 3:

- Sequence Planner
- Scheduler

The planning components have the task of performing the detail planning on the orders passed on by the superior planning system. Here, both automatic as well as manual detail planning must be possible. The sequence planner in the automatic detail planning organizes the orders in the sequence in which they are then planned by the scheduler.

The machine utilization problem has [according to Berry 90] a complexity of $(N!)^M$, where N is the number of orders to be planned and M corresponds to the number of Afos. Actually, this problem is significantly more complicated: the formula is based on the assumption that an Afo can be processed on exactly one capacity site (1:1 relation), however in shop-floor practice similar KAPLs are brought together as BEs. This means that when selecting the KAPLs several alternatives are possible (1 Afo: N KAPL). Thus the solution space increases explosively. In order to reduce the complexity of the machine capacity planning, it is divided between sequential planning and detail planning. The reduction of the problem solving complexity by dividing the tasks is an advantage of DAI which was utilized in the planning components of VerFLEX-BB.

5.3.1. Monitoring Component

This component is composed of two agents which can be used for:
- maintaining the consistency of the allocation sites which have been created by the planning component or by the user;
- detecting conflicts between goal state and current state in the shop floor. The necessary data are provided by the shop floor modeler.

The component *conflict detection* has the following main tasks:
- up-dating the planning data base (state of the resources, operations and the corresponding restrictions etc.).
- supervising the restrictions (referring to the sequence of operations including the corresponding start and finishing times within an order).

The supervision component is to detect conflicts which had been caused by the failure of resources (machines, personnel, material) or by overloading the capacities and which could delay single operations as well as the total manufacturing order.

5.3.1.1. Conflict detection

The task of the agent is to detect conflicts between goal state and current state of the shop floor. The agent has the following subtasks:
- up-dating the planning data base
- checking the restrictions
- propagating the interrupts, that means eliciting the consequences of an interrupt.

The conflict detection component up-dates the planning data base according to the BDE news which the shop floor modeler wrote on the domain blackboard. The next step is that he checks whether there is a conflict between the goal state and the current state. If this checking leads to a conflict, e.g. the failure of one machine, an interrupt unit is generated. Thus, the agent corresponding to the type of interrupt is activated (either diagnosis agent or fault clearing agent).

Furthermore, the conflict detection component is in a position to detect faults early while checking the goal state and the current state. For example, using the reports on the advancement of the operations it can give statements how the stock of work at a machine has developed.

If an interrupt has occurred, and if its duration has been elicited, the conflict detection component propagates the interrupt in order to find out which orders will suffer from the interrupt. The results of the propagation are used by the fault clearing component to solve its subtasks, to clear "internal-external" faults and the "consequence faults".

5.3.1.2. Consistency Maintenance

This agent has the subtask to guarantee for the consistency of the allocation plans, which either are generated automatically by the scheduler or manually by the user. The allocation plan generated by the scheduler checks whether there might arise conflicts from the chronological and technological dependabilities of the Afos. In that case the conflicts are solved by removing the orders in question from the allocation plan.

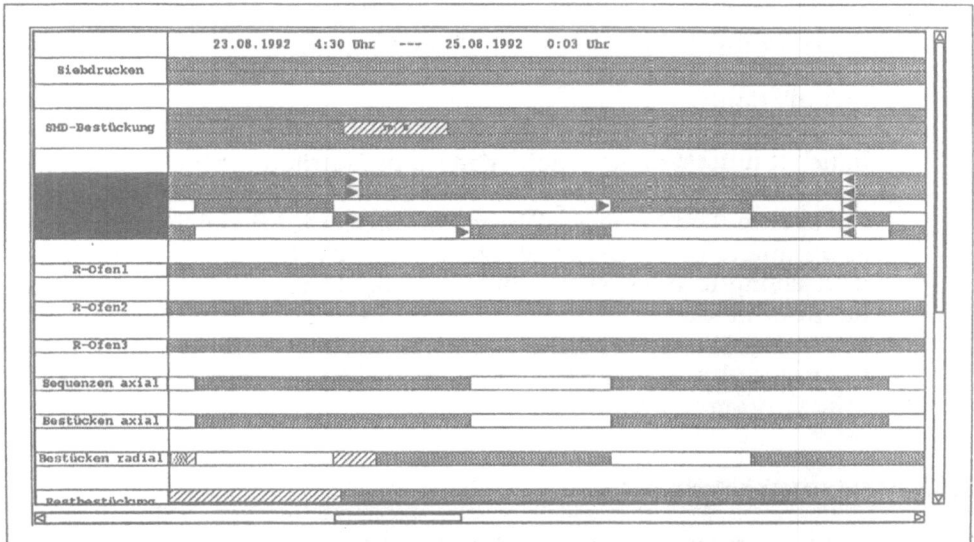

Fig. 20: Admissible solution space in manual planning as determined by the consistency maintainer

The consistency maintenance component guarantees a constant consistency of those allocation plans which were generated manually. Therefore, the dependencies described above are monitored. With manual planning it defines the allowable solution space for each Afo to be planned within the allocation unit, taking into consideration the chronological and technological dependabilities. Fig. 20 shows the solution space of one allocation unit determined by the consistency maintenance component during manual planning. It is not allowed to plan beyond this solution space.

5.3.2. Control Component

In order to coordinate all agents and to guarantee for an efficient application of all system components, control knowledge is required. Control knowledge serves for controlling and activating the agents involved in the planning process. The coordination must already be concipated and developed during the development process of a distributed problem solving system like VerFLEX-BB since this component is responsible for the problem decomposition and for the assignment of the subtasks to the agents.

The coordination is realized in a superior control agent which analyses the current situation of the shop floor and which activates the agent which seems to be best suited for the current problem formulation.

The control agent has two main tasks in this system, on the one hand, it must decide which agent is best suited for the solution of a subtask, on the other hand, he must define the activation sequence of the agents. The VerFLEX-BB system has simplified considerably the assignment of subtasks to the agents by the agent architecture underlying to the system since each agent describes which services he puts at disposal, and so, which subtasks he can solve. The activating sequence of the agents is supported by the advantages provided by the GBB1 KS-Shell (Rating, Trigger-Conditions, Preconditions etc.).

5.3.3.   Blackboards

There are two different *main blackboards*, the domain blackboard and the control blackboard. The domain blackboard itself is composed of the two blackboards *shop floor* and *interface* . The blackboard *shop floor* is composed of four spaces: base, order, schedule, and interrupt. The blackboard interface is composed of three spaces: user, mps,bde.

The space *base* of the blackboard *shop floor* contains the order objects which are needed for the two planning agents. The space *order* contains all objects of the order which the planning agent needs. The space *schedule* contains the objects of the allocation plan of the type KAPL-BELEGEUNG: The objects which are specific to the interface are located on the blackboard *interface*. On the space *BDE* the objects of the BDE component are located. The space user contains all objects which are specific to the interface like *planning horizon*.

The objects of the control blackboard are shown in Fig. 21. The objects stored on the domain blackboard reflect the current sight of the shop floor guaranteeing that all agents have at any time the same sight of the shop floor. This is necessary for that each agent can work with consistent data and can be fulfilled quite easily compared with actor-based approaches by means of the blackboard model.

control blackboard

- problem descriptions, goals (on time, optimization capacity utilization ...)
- strategies, heuristics
- agenda of possible planning steps
- planning steps realized (protocol of problem solving process)

Fig. 21: Control blackboard

5.3.4. User interface

The acceptance of a software system depends decisively on the user-friendliness and, thus, on the user interface. The design of the interface of VerFLEX-BB was guided by the following criteria:

Uniform structure of windows and menus so that the user can handle the system more easily and gets quickly familiar with it.

The user is provided only with the information which is relevant for each operation (like windowing headlines) but additional information is quickly (by mouse-click) available. Information overflow is to be avoided in order to support goal-oriented working (and in order to avoid a distraction of the user by providing him with less important information).

Many activities are to be modeled on the computer exactly like that had been done before manually so that staff-members, who up to that moment had executed them without EDP support, can do them now supported by the computer.

Support to the experienced user (explanation of the keys for the most important commands and for the control of the dialog) as well as to the inexperienced user (helps specific to the situation). Adequate graphical representation of the decisive information.

Fig. 22: The logo window

After the start of the VerFLEX-BB system at first the logo window pops up (see Fig. 22). Next to the logo there is a window which offers seven pull-down menus and a central message window which does not pop up before a new message comes in the middle of the logo window. The seven pull-down menus system, inspection, planning, PPS, VerFLEX-WA, interrupt and help form the user/system interface. The user has the opportunity to plan either manually and to inspect the shop floor, or to influence in the behavior of the single agents.

A. Literature

[Albayrak 90a] Albayrak, S., Drewes, B., Krallmann, H.: Wissensbasierter Fertigungsleitstand auf der Basis einer Blackboard-Architektur, in: Hübers, W. (Hrsg.): Congressband VIII zur 13. Europäischen Congressmesse für Technische Kommunikation ONLINE `90, S. VIII.18.01-18.41, Hamburg 1990

[Albayrak 90b] Albayrak, S.: Verteilte wissensbasierte Systeme in der Fertigung, in: Krallmann, H.; Rieger, B. (Hrsg.): Wissensbasierte Systeme in der Betriebswirtschaft, Erich Schmidt Verlag, Berlin 1990

[Albayrak 92] Albayrak, S.: TUBKOM-Projekt: Blackboard-DEC in: KI 1/92, FBO-Verlag, 1992

[Collinot 88] Collinot, A., Le Pape, C., Pinoteau, G.: SONIA: A Knowledge-based Scheduling System, in: Artificial Intelligence in Engineering, 1988, Vol. 3, No. 2, S. 87-95.

[Corkill 91] Corkill, D.: Blackboard Systems, AI Expert 9/91, S. 41-47, 1991

[Durfee 89] Durfee, E., Lesser, V., Corkill, D.: Trends in Cooperative Distributed Problem Solving, in: IEEE Transactions on Knowledge and Data Engineering, Vol. 1, No.1, S. 63, 83, March 1989

[Ermann 80] Ermann, L, Hayes-Roth, F., Lesser, V., Reddy, D.: The HEARSAY-II Speech Understanding System: Intergrating Knowledge to resolve Uncertainty. in: Computing Surveys, Vol. 12, No. 2, S. 213-253, 1980

[Fox 84]   Fox, M. S., Smith, S. F.: "ISIS, a Knowledge-Based System for Factory Scheduling", in: Expert Systems, the international Journal of Knowledge Engineering, Vol. 1, Learned Information Inc., Medford, NJ, July 1984

[Fox 86]   Fox, M, Smith, S. F.: Constructing and Maintaining detailed Production Plans: Investigations into the Development of Knowledge-Based Factory Scheduling Systems, in: AI Magazine, Fall 1986.

[Fox 90]   Fox, M.S.: AI and Expert System Myths, Legends, and Facts. in: IEEE EXPERT February 1990

[Gasser 88]   Gasser, L.: Collected Draft Papers of the 1988 Workshop on Distributed Artifial Intelligence. Technical Report CRI-88-41, USC Computer Research Institute Los Angeles, CA, 1988

[Huber 90]   Huber, A.: Wissensbasierte Überwachung und Planung in der Fertigung, Erich Schmidt Verlag Berlin 1990

[Johnson 88] Johnson, P. M., Gallagher, K.Q., Corkill, D.D.: GBB Reference Manual, GBB Version 1.2. COINS Technical Report, July 1988

[Kempf 91] Kempf, K., Russel, B., Sidhu, S., Barrett, S.: AI-Based Schedulers in Manufacturig Practice: Report of a Panel Discussion, in: AI Magazine Special Issue, Vol. 11, No. 5, 1991, S. 46-55

[Martial 92] von Martial, F.: Einführung in die Verteilte Künstliche Intelligenz, KI 1/92, FBO-Verlag, 1992

[Newell 62] Newell, A.: Some problems in basic organization in problem-solving problems. In M. C. Yovits, G. T. Jacobi & G. D. Goldstein (Hg.): Conference on Self-Organizing Systems. Washington, D. C., Spartan Books, S. 393-423, 1962

[Nii 86a]   Nii, H. P.: Blackboard Systems: The Blackboard Model of Problem Solving and the Evolution of Blackboard Architectures. AI Magazine, (38-53), Summer 1986

[Nii 86b]   Nii, H. P.: Blackboard Systems: Blackboard Application Systems, Blackboard Systems from a Knowledge Engineering Perspective, AI Magazine, S. 82-106, August 1986,

294

[Nii 89]   Nii, H. P., Feigenbaum, E. A., Anton, J. J., Rockmore, A. J.,: Signal-to-Symbol Transformation: HASP/SIAP Case Study, in (eds.) [Engelmore & Morgan88]

[Parunak 87]   Parunak H.V.D.: Manufacturing Experience with the Contract Net, in: Michael N. Huhns (Ed.), Distributed Artificial Intelligence, pp. 285-310, Pitman Publishing/Morgan Kaufman Publishers, San Mateo, CA., 1987

[Parunak 88]   Parunak, H.V.D.: "Distributed Artificial Intelligence Systems". in: Artificial Intelligence and CIM, 1988

[Puppe 90]   Puppe, F.:   Problemlösungsmethoden   in Expertensystemen, Springer-Verlag 1990

[Rosenschein 86]   Rosenschein, S.: Rational Interaction: Cooperation among Intelligent Agents. PhD Tehsis, Stanford University, 1986

[Sauve 87]   Sauve, B., Collinot, A.: Expert System for scheduling in a flexible manufacturing system, in: Tagungsband International Conference on intelligent Manufacturing Systems, Budapest, S. 229-233, 1987

[Sauve 89] Sauve, B.: "Job Shop Dynamic Scheduling, the Knowledge-Based Approach of SONIA", in: Knowledge-Based Production Management Systems, J. Browne (Ed.), Elsevier Science Publishers B.V. (North Holland), IFIP 1989

[Sycara 91]   Sycara, K., Miyashita, K.: Case-Based Schedule Repair: An Initial Report, in: Garlick, S.: The IJCAI-Workshop on Artificial Intelligence Approaches to prodution Planning: Tools for Master Scheduling and Sequencing 1991

[Winston 87]   Winston, P.: Künstliche Intelligenz, Addison-Wesley 1987

PART 3
NEURAL NETWORKS IN MODELLING
CONTROL AND SCHEDULING

10
ARTIFICIAL NEURAL NETWORKS FOR MODELLING

A.J. KRIJGSMAN, H.B. VERBRUGGEN, P.M. BRUIJN
Delft University of Technology
Department of Electrical Engineering, Control Laboratory
P.O. Box 5031, 2600 GA Delft, The Netherlands

1 Introduction

The technique of subsymbolic information processing is based on the functioning of the human brain. It appeals to the idea of black-box modelling, using biologically inspired models of the human brain. When using these models, tools are provided to implement arbitrary complex functions. No explicit knowledge is needed to apply these techniques, in contrast with that which is necessary in the application of symbolic AI techniques based on logic. In symbolic AI systems, the knowledge is represented explicitly, for example by using production rules.

In a subsymbolic approach, a relation is not explicitly given, but coded into a network structure. Given an input (vector), an output (vector) is produced by using the control algorithm of the neural network involved (which can be a relatively simple feedforward calculation). The main characteristic is the learning ability of such networks (via weight updating).

Many neural networks exist, with a wide variety of topologies, control and learning strategies. In section 2, the mathematical concept of a neuron used to construct Artificial Neural Networks (ANN), both static and dynamic, as discussed in section 3 is given.

The use of neural networks as universal function approximators is highlighted in section 4. The choice of network type depends on the degree of generalization required. For control purposes like modelling and control, we want networks which produce continuously valued outputs. For this reason, a number of neural network architectures are out of the scope. The second limitation to the neural network architecture is the learning mode required. In adaptive and learning control systems, the ability to learn from experience is crucial. For example, in a modelling application, we want to update the weights of the network using the difference between the predicted process output and the actual process value. Therefore, besides a continuously valued neural network, we also need a supervised type of learning. The last demand on the network is the environment, which is described as a discrete time system.

Given these restrictions (continuous, supervised and discrete time), a number of local and global generalizing networks are described in section 5.

In section 6, identification using ANN is described, focusing on the type of models used. A new concept is given which uses hybrid structures for identification. The results of the application of this concept are promising and can be used as the basis for further research.

S. G. Tzafestas and H. B. Verbruggen (eds.),
Artificial Intelligence in Industrial Decision Making, Control and Automation, 297–325.
© 1995 *Kluwer Academic Publishers.*

Experiments which compare the results of various approaches are given. Finally, an evaluation of the comparison and some conclusions are given.

2 Description of artificial neurons

Most artificial neurons are based on the McCulloch-Pitts model of a neuron which is defined by a weighted summation and a nonlinear function. This description is a subset of a more general model for an artificial network, which covers a large set of other models (like the Perceptron [18] or the adaptive linear element: Adaline [21]). An artificial neuron is defined as an arbitrary composition of three elements [12]:

- a weighted summation of the inputs

- a linear dynamic system of the SISO type

- a (nondynamic) nonlinear function to transform the output of the SISO dynamic system into an output signal

An artificial neuron model includes both static and dynamic neurons. Static artificial neurons are defined as artificial neurons in which the linear dynamic system is reduced to a transfer function $H_n(s) = 1$. All other choices for the dynamics of a neuron leads to more complex behaviour of the neuron itself.

Definition 2.1 *Artificial neural networks are defined as (an) arbitrary combination(s) of one or more artificial neurons.*

Definition 2.2 *A static neural network is built up using artificial neurons which include **no** dynamics at the neuron level, and no feedback connections.*

Definition 2.3 *Dynamic neural networks are networks which are **not** static, e.g. memory is included in the network (by using either dynamic neurons or feedback connections).*

In figure 1, a description is given of a possible description of a dynamic artificial neuron. Many artificial neural networks known from literature can be described using this description [12]. One should be aware of the fact that other configurations are possible, for example, they can be obtained by changing the order of the dynamic part and the nonlinear element. The transient behaviour of the neuron in figure 1 can be described in the time domain. The input vector i of this neuron is a $(N + M + 1) \times 1$ vector. This input vector can be decomposed into :

$$i_i = [y_{n,1} \cdots y_{n,N} \ u_{n,1} \cdots u_{n,M} \ 1]^T$$

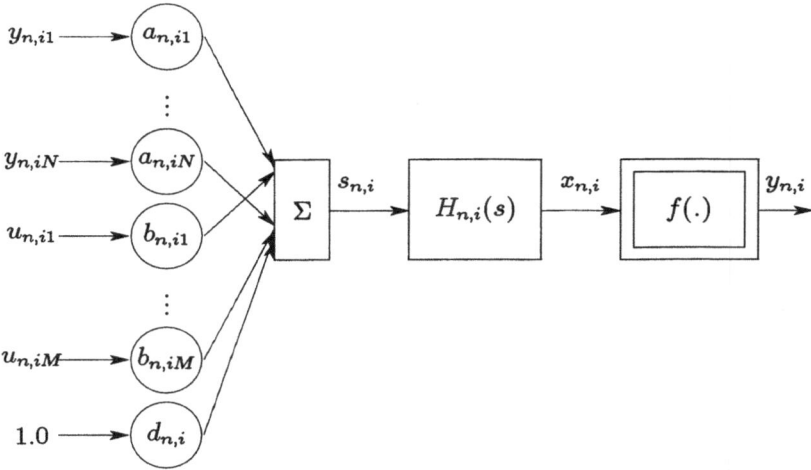

Figure 1: *Basic mathematical model of the i-th artificial neuron in a network. The index $._n$ is used to indicate that a signal or variable is related to an artificial neuron.*

This notation can be seen in analogy with 'normal' (linear) system theory. The result of the summation can now be described as:

$$s_{n,i}(t) = \sum_{j=1}^{N} a_{n,ij} y_{n,j}(t) + \sum_{k=1}^{M} b_{n,ik} u_{n,k}(t) + d_{n,i}$$

The weighted sum $s_{n,i}$ is given as a function of the outputs of all elements $y_{n,j}$ in the network, the external inputs of the network $u_{n,k}$ and the weights $a_{n,ij}$, $b_{n,ik}$ and $d_{n,i}$. The index $._n$ is used to indicate that a signal or variable is related to an artificial neuron. For N elements these elements $s_{n,i}$ form a vector s_n which can be written in matrix form as:

$$s_n(t) = A_n y_n(t) + B_n u_n(t) + d_n$$

in which A_n is a $N \times N$ matrix, and B_n a $N \times M$ matrix.

3 Artificial Neural Networks

Artificial Neural Networks (ANN) are constructed by interconnecting artificial neurons. The components of the artificial neurons (summation, dynamics and nonlinear transformation) can be combined in numerous ways. When the network is constructed using static neurons, the total behaviour of the artificial neurons can be described by a set of *algebraic equations*. Using a first order behaviour ($H_n(s) = \frac{1}{s+1}$), a network can be described by a set of *differential equations*. The most important parameters to describe a network are the interconnection matrix A_n and the dynamic part of the neuron described by H_n. The nonlinear part of the artificial neuron can be chosen free. A common choice for this function is a tangent hyperbolic function, which is a diff*ertiable* function with zero-mean value.

For binary decision problems, non-differentiable, step-like functions like thresholds are more appropriate. Using these three 'parameters' the main networks types in the literature can be described.

An infinite number of networks can be constructed by choosing:

(a) the dynamic behaviour of the artificial neuron. In principle, every neuron can have its own specific dynamic behaviour.

(b) the nonlinear transformation inside an artificial network. Again every neuron can have its own nonlinearity.

(c) the interconnections between neurons. The interconnection type can vary from fully connected to more loosely connected feedforward type of networks.

(d) the combination of both static and dynamic components within one architecture.

In almost all networks presented in the literature, simplifications are made by choosing all neurons of the same type (same dynamics, same nonlinearity), while an organisation in layers is used to be able to handle the complexity of the configuration.

4   Nonlinear models and ANN

The aim of applying neural networks is to build models which are essentially nonlinear. These nonlinear models can then be used to implement nonlinear relations between a multidimensional input space and a multidimensional output space. No restrictions are imposed on the dimension of either spaces. The introduction of neural networks introduces the possibility of nonlinear black-box models.

In the literature, many possible descriptions of nonlinear systems are given. The two most important classes of nonlinear systems are:

- **Wiener class:** Representing a large number of systems with noninfinite memory. This class is derived from the Wiener theory, which is in fact an extension of the theory of linear time-invariant systems [19].

- **Parametric approximating function:** This class of system is derived from the theory of multivariable approximation. The class is extensively used in the literature on neural networks. Parametric approximation functions generally yield less complex models than Wiener models.

It is stated that by defining an identification problem as a static problem, where the dynamics are introduced externally (by means of tapped delay lines), every identification problem can be defined as a multidimensional function approximation. We can use neural networks for this function approximation (see section 6).

Neural networks are special models within the class of parametric approximation functions that receive considerable attention. There are two ways of achieving function approximation using neural networks:

- Neural networks realizing parametric functions $f(u, w)$ that are nonlinear with respect to its parameters w

- Functional networks realizing parametric functions $w.f(u)$ that are linear with respect to its parameters w

Both network types are suitable to implement nonlinearities and they both contain several nonlinearities. For neural networks, it is not an easy and straightforward task to find the optimal set of parameters as the problem is nonlinear in its parameters. For functional networks, it is easy to find the optimal set of parameters, because the problem is linear in the parameters,

Wiener models

In [19] an overview of system modelling based on Wiener theory, is given. Discrete-time systems with finite memory can be represented by a finite multi-dimensional, discrete impulse response or power series with memory.

$$\hat{y}(k+1) = h_0 + \sum_{\tau_1=0}^{n_1} h_1(\tau_1)u(k - \tau_1) + \ldots$$

$$+ \sum_{\tau_1=0}^{n_1} \cdots \sum_{\tau_l=0}^{n_l} h_k(\tau_1 \ldots \tau_l)u(k - \tau_1) \ldots u(k - \tau_l)$$

$$= h_0 + H_1[u(k)] + h_2[u(k)]$$

$$\tau_1, \tau_2, \ldots, \tau_l \geq 0 \tag{1}$$

Equation (1) is called the truncated discrete Volterra series of order l. The model output $\hat{y}(k + 1)$ is a linear combination with respect to the multi-dimensional Wiener kernels $h_1 \ldots h_l$. These kernels are symmetric, i.e. $h_2(\tau_1, \tau_2) = h_2(\tau_2, \tau_1)$, etc.

The regression part of the Wiener model has a fixed form: a set of power expansions of the system input, so that only the linear Wiener kernels can be optimized, given a set of power expansions. Like the continuous-time Volterra series, practical application of this model is limited to only the first- and second-order series expansions.

In order to illustrate the quality of a single-layer network realization of a simplified Wiener model representation, the following simulated nonlinear system:

$$y(k+1) = 0.3y(k) + 0.1y^2(k) + u^2(k)u(k-1) \tag{2}$$

is modelled by a truncated second-order Volterra series, with $n_1 = n_2 = 60$ samples. It can be shown that this series contains $\frac{1}{2}(n_2 + 2)(n_2 + 3) = 1953$ terms:

- 1 kernel h_0

- plus $n_2 + 1$ kernels $h_1(\cdot)$

- plus $\frac{1}{2}\left((n_2 + 1)^2 - (n_2 + 1)\right) + (n_2 + 1)$ kernels $h_2(\cdot)$

5   Networks

The basics of artificial neurons (AN) and artificial neural networks (ANN) have been given. By choosing the network's parameters (size, dynamics of the neuron, etc.) a large number of networks can be created. In this section, four general network types are given which are used as the basis for identification, based on a classification according to the generalization property of the network proposed.

In section 5.1, the use of a network with global generalization properties is given. For these multilayered networks, the parameters to be adjusted are the nonlinear parameters of the networks. When a very local generalization property is required, a CMAC (section 5.3) is very well suited. An intermediate generalization property is given by Radial Basis Function Networks (RBFN, section 5.2), where the adjustable parameters appear linearly in the network description.

5.1   Multilayered Static Neural Networks

The neurons which are used in multilayered neural networks (MNN) are usually of the McCulloch-Pitts type and use a sigmoidal nonlinear function, a continuous monotonically increasing function and a continuously differentiable function. The function approaches fixed values asymptotically as the inputs approaches plus or minus infinity. The function description is:

$$f(x) = \frac{1 - e^{-x}}{1 + e^{-x}} \tag{3}$$

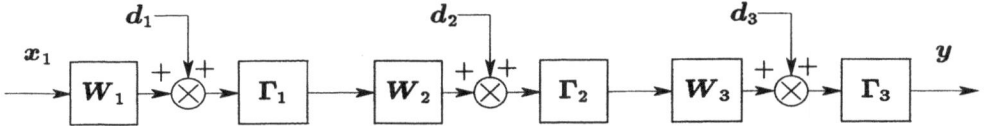

Figure 2: *A three-layered MNN structure. The matrix W_i is a the weight matrix, Γ_i the nonlinear operator and d_i the bias of layer i.*

Every layer in a neural network can be denoted as a weight matrix W_i, a nonlinear operator Γ_i and an offset (bias) vector d_i. The output of a layer produced by an input vector x_i is:

$$N_i[x_i] = \Gamma_i[W_i x_i + d_i] \tag{4}$$

The output of an n layer network can be described by a multiplication of n layers:

$$y = \Gamma[W_n \Gamma[W_{n-1} \cdots \Gamma[W_1 x_1 + d_1] + \cdots + d_{n-1}] + d_n] \tag{5}$$

in which y is the output vector produced by the network. In figure 2 the configuration of a multilayered network with three layers has been depicted.

5.2 Radial Basis Function Networks

Radial Basis Function Networks are set up using the idea of function approximation. The network consist of a nonlinear transformation of the input signals. The outcome of these nonlinear transformations are used for a linearly weighted summation (see figure 3). The network realizes an approximation function $f : \Re^m \to \Re$ which is described mathematically as:

$$y = f(u) = \sum_{i=1}^{n} w_i \phi_i(u, c_i) \tag{6}$$

The basis functions $\phi_i = \phi_i(\|u - c_i\|) = \phi_i(r)$ are radial functions. Some common choices for these functions are:

- $\phi(r) = r$, a linear radial function
- $\phi(r) = r^2$, a quadratic function

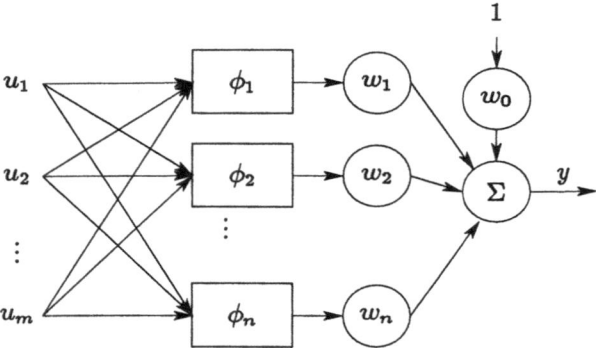

Figure 3: *Scheme of a Radial Basis Function Network. The basis functions are denoted as* ϕ_i.

- $\phi(r) = \exp(-r^2/\rho^2)$, a Gaussian function
- $\phi(r) = r^2 \log(r)$, a thin-plate-spline function
- $\phi(r) = (r^2 + \rho^2)^{\frac{1}{2}}$, a multiquadratic function

The radial functions are (typically) centred on a subset of the N data points (i.e. $c_i = u_i \ i = 1 \cdots n \le N$). The choice of the centres c_i is part of the optimization procedure which must be carried out.

An important property of RBFN is its stability, corresponding to the choice of the radial functions. Because Gaussian basis functions tend to zero outside the region where they are centred, the network output is always bounded, provided that the network weights are bounded. This BIBO stability property is important, and, therefore, in many cases, Gaussian functions are used.

5.3 Cerebellum Model Articulation Controller (CMAC)

In 1975, J.S. Albus published his ideas upon a simple method, called CMAC (Cerebellar Model Articulation Controller) [1, 2]. This algorithm is based on the functioning of a part of the small brains: the cerebellum. The algorithm is based on its ability to learn vector functions by storing data from the past in a clever way.

The concept of storage, makes the CMAC fit into the category of Associative Memories (AMS).
The algorithm can be characterized by three important features, which are extensions to the normal table lookup methods. These important extensions are: distributed storage, the generalization algorithm and a random mapping.

In the mapping from the stimulus space (the input space) to the association cell (the table), two properties are fulfilled: only a small number ρ of the cells will have a nonzero value for a specific input and, secondly, similar vectors in the stimulus space will map to similar association cell vectors. This generalization feature is determined by the generalization number ρ. In normal applications $\rho > 1$, for $\rho = 1$ the algorithm reduces to a simple lookup table.

An important part of the algorithm is the part which determines which association cells are addressed. When a generalization parameter ρ has been determined, quantization must be chosen. This results in R_i intervals on each component s_i of the input vector. Thus the original stimulus space is mapped to a quantized space S_q.

Basis functions in CMAC

The mapping from the association cell space to the response space is determined by basis functions, used to cover the association cells and a single layer network. These basis functions can also be found in RBFN. The input to this single-layered network consists of the ρ basis functions whose output is nonzero. The response y_i is constructed by the normalized sum:

$$y_i = \frac{\sum_{j=1}^{\rho} \Phi(S_q, C_{q_j}) * w_{ij}}{\sum_{j=1}^{\rho} \Phi(S_q, C_{q_j})} \tag{7}$$

in which y_i is the CMAC response for the ith coordinate. The basis function $\Phi(S_q, C_{q_j})$, in which C_{q_j} is its centre. The output of the basis function lies in the interval $[0, 1]$. The weight associated with association cell j for the ith output is denoted as w_{ij}.

The interpolation is determined by the basis function $\Phi()$. Where Albus originally used a binary function:

$$\Phi(S_q, C_{q_j}) = \begin{cases} 1 & \text{if} \|S_q - C_{q_j}\| < \frac{\rho}{2} \\ 0 & \text{otherwise} \end{cases} \tag{8}$$

a large number of other choices exist, like B-splines [9, 10].

Learning in CMAC

The error between the actual CMAC output and the desired output is given as $(\hat{y}_i - y_i)$. The weight update is performed by a Widrow-Hoff updating rule:

$$\Delta w_{ij} = \beta_j (\hat{y}_i - y_i) \Phi(S_q, C_{q_j}) \tag{9}$$

in which $\beta \in [0, 1]$ is the learning rate for the jth association cell.

6   Identification of dynamic systems using ANN

In the previous sections it has been explained why arbitrary functions can be approximated by neural network techniques. Modelling of dynamic systems can be achieved by using static neural networks in a dynamic environment and using tapped-delay lines to construct the input of the network.

The research on the use of neural network techniques in process control attracted much attention. The learning capabilities of these networks makes them very attractive for those situations where process knowledge is not complete or even not available. This lack of knowledge, indicated by the absence of a model or by an approximate model, makes it very hard to design a proper control scheme. Neural networks can be used to cope with this lack of knowledge, by modelling the unknown characteristics of the process.

Identification problem definition

In general control applications, a (mathematical) model of the system is derived in order to design a proper control scheme, which should meet all the requirements given by the user. A model of a system is expressed as a mapping P from an input space $\mathcal{U}$ into an output space $\mathcal{Y}$. In [15] the objective of identification is given as: *"characterize the class $\mathcal{P}$ to which P belongs. Given a class $\mathcal{P}$ and the fact that $P \in \mathcal{P}$, the problem of identification is to determine the class $\hat{\mathcal{P}} \subset \mathcal{P}$ and an element $\hat{P} \in \hat{\mathcal{P}}$ so that $\hat{P}$ approximates P (in the way it was desired). In static systems, the spaces $\mathcal{U}$ and $\mathcal{Y}$ are subsets of $\Re^n$ and $\Re^m$ while in dynamic systems they are generally assumed to be bounded Lebesque integrable functions on the interval $[0, \infty]$. "*

It is important that in both cases the operator P is defined by the specified input-output pairs. The choice of the class of identification models $\hat{\mathcal{P}}$, as well as the method to determine $\hat{P}$ depends on a variety of factors which are related to the accuracy desired as well as to the a priori information that is available concerning the system to be identified.

In many cases, the learning is performed by learning input-output mappings: given an input vector $u(k)$ a corresponding output vector o should be produced by the network. The examples provided to train the network are obtained by observation of the actual process. In figure 4 the SISO case is depicted. This chapter examines SISO processes, but by no means is this identification approach restricted to SISO systems. In dynamic systems, the operator P is defined by the input-output pairs of discrete time functions $u(t)$, $y(t)$, $t \in [0, T]$. The objective of the model is:

$$\|\hat{y}_p - y_p\| = \|\hat{P}(u) - P(u)\| \leq \epsilon, \quad u \in \mathcal{U} \tag{10}$$

for $\epsilon > 0$. In this equation the output of the identification model is given by $\hat{P}(u) = \hat{y}_p$ and $\hat{y}_p - y_p \triangleq e$ is the identification error. The output $\hat{y}_p$ is taken equal to the output of an artificial neuron in the network.

Based on theorem of Stone-Weierstrass [7] and the theorem defined by Kolmogorov [13], it can be stated that neural networks are able to implement *any input-output mapping*. Using this property, neural networks can be applied for the identification of static nonlinear

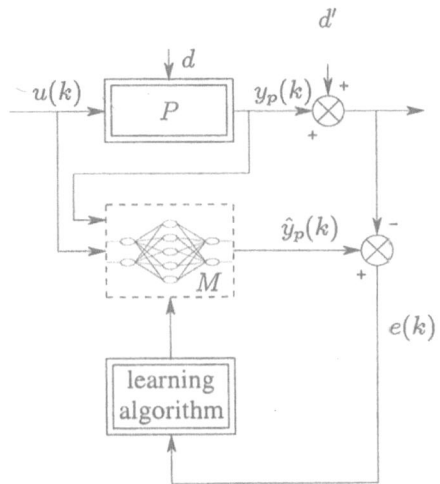

Figure 4: *General scheme for plant identification. In this scheme a SISO system is depicted. The identification error e is used to adapt the model.*

mappings. When the dynamics of the system are defined externally (by using tapped delay lines) the network can be chosen to be a static artificial neural network, which simplifies the identification structure considerably!

The power of neural modelling becomes even more apparent in those situations where the input output relation, which is learned by the neural network, is not restricted to a special form. The input vector as well as the output vector of the network can contain any kind of information: control signal values, process signal values, historical data, sensor data, etc. Neural networks have the ability to approximate large classes of (nonlinear) functions.

The optimization problem can be described by a number of parameters:

- $J(\Theta)$: the cost function expressing the aim of the optimization;
- Θ: the parameters which can be used to optimize $J(\Theta)$;

The optimum solution Θ^* is given by:

$$\Theta^* = \min_{\Theta} J(\Theta) \tag{11}$$

If $J(\Theta)$ is a nonlinear function of the parameter vector Θ then no general analytical solution is known. For this type of optimization problem, nonlinear optimization techniques are available. These techniques solve the optimization problem by *searching* for the optimum in an iterative way.

308

Model descriptions for identification

There are many possible models to describe a system. Here, we want to focus on models using neural networks (or other 'intelligent components'). One of the possibilities is to look on a system as being a black box, observing only the inputs and the outputs, and deriving a nonlinear mapping between inputs and outputs.

It is also possible to use a neural model as a submodel of the complete model of the process. The neural model in this case compensates for the unmodelled behaviour. The unmodelled part can, for example, be used to compensate for an (extreme) nonlinear behaviour, unmeasurable influences, etc. Especially for those circumstances in which a coarse model of the system is available, and only some minor parts of the system are unknown, this is a very powerful approach.

In discrete form 4 of these (SISO) models Σ_1 through Σ_4 are given by input-output equations [16]:

$$\Sigma_1 : \hat{y}_p(k+1) = \sum_{i=0}^{n-1} \alpha_i y_p(k-i) \tag{12}$$
$$+ g[u(k), u(k-1), \cdots, u(k-m+1)]$$
$$\Sigma_2 : \hat{y}_p(k+1) = f[y_p(k), y_p(k-1), \cdots, y_p(k-n+1)] \tag{13}$$
$$+ \sum_{j=0}^{m-1} \beta_j u(k-j)$$
$$\Sigma_3 : \hat{y}_p(k+1) = f[y_p(k), y_p(k-1), \cdots, y_p(k-n+1)] \tag{14}$$
$$+ g[u(k), u(k-1), \cdots, u(k-m+1)]$$
$$\Sigma_4 : \hat{y}_p(k+1) = f[y_p(k), y_p(k-1), \cdots, y_p(k-n+1); \tag{15}$$
$$u(k), u(k-1), \cdots, u(k-m+1)]$$

These models have a fairly simple structure: the first three models have a strict separation between a linear and a nonlinear part. Model Σ_4 is general because it has no assumptions with respect to a separation into a nonlinear and a linear part. We have to be aware of the fact that these four models are only a subset of the infinite number of possible models. In those cases where we have a priori information about the system to be identified, it should be used to construct a more detailed identification model.

7 Hybrid modelling

An extension of the modelling techniques using neural networks, is combination of RBF-type network models with linear functions (hybrid model). The optimisation of the parameters of these models is an off-line procedure.

The hybrid network models have the following form:

$$\hat{y}(k+1 \mid k) = w_0 + \sum_{i=1}^{n} w_{1i}\phi_i[r_i(k)] + w_2^T x(k) \tag{16}$$

$$r_i(k) = \parallel x(k) - c_i \parallel$$

$$x(k) = \begin{bmatrix} y(k) & \cdots & y(k - n_y) \\ u(k) & \cdots & u(k - n_u) \end{bmatrix}^T$$

(16) is a linear regression model, consisting of RBF-type and linear regressors. In this section the Orthogonal Least Squares algorithm (OLS) is used to select the elements of the model. In literature (Chen et al (1,2)), different choices for $\phi(r)$ have been suggested. Gaussian functions: $\phi(r) = \exp(-r^2/\rho^2)$ and inverse multiquadratic functions: $\phi(r) = (r^2 + \rho^2)^{-\frac{1}{2}}$ tend to zero outside the region where they are centred. As a consequence, the output of the NOE model configuration is always bounded, provided that the network weights are bounded. Note however that the linear submodel may give rise to stability problems, so it is necessary to examine the stability of the linear part of the hybrid model. In the case of an unstable linear submodel, the model order should be reduced until a stable linear submodel is obtained. Thus, ultimately a stable nonlinear submodel is left.

The RBF model contains additional free parameters c_i and ρ, whose values are generally unknown. These values might be found during an optimisation stage, but in order to maintain a linear optimisation problem, a better way is to fix c_i and ρ to some predetermined values. The important point with respect to the RBF centres and widths is that these parameters have a geometrical meaning, allowing them to be fixed to reasonable values. In the next section, the OLS algorithm is described as a way to determine a set of optimal RBF-centres, which are initially located on all measured points in the input space X of the prediction model. Other techniques to determine RBF centres and widths can be found in e.g. Chen and Billings [3] and Sbarbaro and Gawthrop [5].

An important property of the hybrid network model is that linear systems can be identified with OLS as well. Then, OLS selects only the linear regressors of the regression matrix, which is a desirable property. The ability to identify linear systems in this way, has been confirmed by experiment. This makes the OLS algorithm, in combination with hybrid network models, a powerful identification tool for both linear and nonlinear systems.

ORTHOGONAL LEAST-SQUARES ALGORITHM

In this section the OLS optimisation technique (Chen et al (1,2,3)) is described, which is a generally applicable technique to iteratively orthogonalise a linear regression problem. Orthogonalisation improves the numerical stability of the LS solution to the regression problem, at the cost of a higher computational complexity.

OLS is an off-line procedure that computes an LS solution to a fixed dataset from the system. This implies that OLS is not directly useful for adaptation, and that the model performance heavily relies on information of the system contained in the dataset. Due to the high calculation load, imposed by the performance of many orthogonalisation steps, application of OLS severely limits the size of the dataset, which is its principal drawback.

Besides the numerical robustness, OLS allows a model reduction procedure, which cannot be performed with normal least-squares computation. With OLS it is possible to find a considerably reduced nonlinear model that effectively describes the relevant nonlinear system dynamics. This reduced model is the result of a model selection procedure that selects significant regressors from a large prespecified linear regression model, containing suitably chosen regressors. However not necessarily, we assume here that the prespecified model contains as many regressors as datapoints, which is convenient in the case of radial basis function type regressors.

The identification problem is stated as a linear regression problem. The model prespecified contains N regressors with respect to N datapoints:

$$y = \Phi w + \varepsilon \tag{17}$$

Hereby, y is the $N \times 1$ desired model output vector, Φ is the $N \times N$ regression matrix, w is the $N \times 1$ weight vector, and ε is the $N \times 1$ vector of modelling errors. y and Φ, determined by the data from the system, must be specified a priori. The N columns ϕ_k of the specified regression matrix are referred to as the regressors and are subjected to selection by the OLS procedure. w is computed as the LS solution, according to y and the corresponding model, which consists of selected regressors ϕ_k. ε is determined by the matrix equation (17) that results from the selected model, and y.

The essential procedure is to select significant regressors $\{\phi_k \mid 1 \le k \le n \le N\}$ from the model (17). The OLS algorithm is constructed such that a set of n regressors that have maximum contribution to the variance of the desired output y in (17) are chosen. The individual contribution of an individual regressor can be determined by orthogonalisation of all regressors: $\Phi = QP$. Q is an $N \times N$ matrix with columns q_k which are orthogonal to each other, and P is an $N \times N$ invertible upper triangular matrix. From (17) we have:

$$y = Q\gamma + \varepsilon$$
$$\gamma = Pw \tag{18}$$

Matrix Φ can be orthogonalised using QP-orthogonalisation, based upon the modified Gram-Schmidt procedure. The basis vector q_k is iteratively computed on basis of the previously orthogonalised and selected regressors $\{\phi_j \mid j \in [1, k-1]\}$:

$$q_k = \phi_k - \sum_{j=1}^{k-1} \frac{q_j^T \phi_k}{q_j^T q_j} q_j \tag{19}$$

From this result γ_k can be calculated (from q_k) as the LS solution to (18):

$$\gamma_k = \frac{q_k^T y}{q_k^T q_k} \tag{20}$$

This calculation is reduced to a scalar computation, because the regressors q_k are orthogonal. From (19), P can be determined using $\Phi = QP$. Once a set of regressors has been made orthogonal, the contribution of an individual regressor to the output variance can be determined by the 'error reduction ratio' ζ. The crossproducts in the second term of (21)) can be neglected for large N and zero mean ε.

$$\frac{1}{N} y^T y = \frac{1}{N} \gamma^T Q^T Q \gamma +$$

$$\frac{1}{N} \left(\gamma^T Q^T \varepsilon + \varepsilon^T Q \gamma \right) +$$

$$\frac{1}{N} \varepsilon^T \varepsilon$$

$$\approx \frac{1}{N} \sum_{k=1}^{n} \gamma_k^2 q_k^T q_k + \frac{1}{N} \varepsilon^T \varepsilon \tag{21}$$

$$\zeta_k = \frac{\gamma_k^2 q_k^T q_k}{y^T y}$$

Algorithm:
With the above described methods, the OLS procedure globally works as follows: Initially, q_1 is equal to the regressor ϕ_i of the full model that maximises ζ_i:

$$q_1 = \phi_i \mid i = \max_{i=1 \ldots N}(\zeta_i)$$

$$\zeta_i = \frac{\gamma_i^2 \phi_i^T \phi_i}{y^T y}$$

$$\gamma_i = \frac{\phi_i^T y}{\phi_i^T \phi_i}$$

The selected regressor now forms the initial model. During the remaining OLS iterations, orthogonalisation is performed. At the beginning of the k-th iteration, we have $Q = [q_1 \ldots q_{k-1}]$, which means that the so far selected model consists of $k-1$ regressors. The $N - k + 1$ remaining regressors of the full model, that have not been selected yet, are all orthogonalised with respect to $q_1 \ldots q_{k-1}$, resulting in $\{q_k \ldots q_N\}$. The element of $\{q_k \ldots q_N\}$ that maximises ζ in (21) is finally added to the model that now consists of k regressors.

The OLS procedure is thus minimising the criterion:

$$J_k = 1 - \sum_{j=1}^{k} \zeta_j \tag{22}$$

$$= \frac{\varepsilon^T \varepsilon}{y^T y} \in [0, 1]$$

If after n stages, J_n is small enough, the OLS procedure is stopped. The threshold for J_k should be related to the level of measurement noise (σ_ε^2) and model complexity (n).

This is especially important when the dataset is relatively small, which is the case for practical application of OLS. The value of the stopcriterion must be chosen carefully in order to avoid noise fitting on one hand, and to achieve sufficient accuracy on the other hand (parsimony principle).

In the case of measurement noise e, J_k should not be smaller than $\frac{e^T e}{y^T y}$. The noise variance can be estimated, according to (see Söderström (6)):

$$\hat{\sigma}_e^2 = \frac{J_N(\hat{w}_N)}{(1 - \frac{n}{N})} \tag{23}$$

with $J_N(\hat{w}_N) = \frac{1}{N} \varepsilon^T \varepsilon$.

Besides the value of J_k, another stopping criterion is the development of $J_k - J_{k-1}$, or the norm $\| q_k \|^2$ of the newly added orthogonalized regressor. If these values are too small, the selection procedure can be stopped.

The use of a complexity criterion instead of (22), depending on model size n, is a possible way to deal with the complexity problem (Chen et al (1,2)), whereby an increased model complexity tends to lead to an increased error variance. However, this complexity problem only applies for $N \to \infty$, which is not the case in our experiments. Moreover, in most cases the OLS procedure appears to select relatively small models, so that we have not been directly concerned with the model complexity problem.

Resuming, for OLS the following algorithm parameters must be specified a priori by the user: number of datapoints N, model order m, number of regressors (for radial basis functions this is the number of datapoints N), threshold for J_k, threshold for $J_k - J_{k-1}$, threshold for $\| q_k \|^2$.

The large prespecified model that determines the linear regression problem of our hybrid network model, and from which OLS selects a number of significant regressors ($\ll N$), consists of N RBF-type regressors ϕ_i and $m + 1 = \dim(x) + 1$ linear regressors x_i, so that Φ is an $N \times (N + m + 1)$ matrix.

Each regressor ϕ_i is determined by a RBF that is centred on one of the N datapoints, i.e. $c_i = x(i)$ for $i = 1 \ldots N$, as a convenient way to establish N different RBF regressors. For simplicity, the values of all RBF-widths ρ are set to ρ_0, which must be specified a priori.

The number of linear terms (m) contained in x, which defines the model order, must be chosen a priori as well. Notice that the complexity of the OLS algorithm, which is determined by the size of the regression matrix Φ, increases linearly with the model order m.

When the finally selected model contains both RBF's and linear terms, we have the total amount of $n + l$ regressors and weights, with $n \in [1 \ldots N]$ and $l \in [1 \ldots m + 1]$.

8 Model validation

The quality of a model has to be validated carefully . The validation is carried out using a part of the data set: the test set. The weight updating has been performed on the other part of the data set: the training set. In the set-up of an identification experiment such as described in section 6 and depicted in figure 4, a quadratic criterion is used for the weight optimization. This error measure can be found in the sum squared error (SSE), which is defined for a data set consisting of N data samples. For a SISO process this performance measure is given as:

$$\text{SSE} = \sum_{k=1}^{N} (y_p(k) - \hat{y}_p(k))^2 \tag{24}$$

The mean squared error (MSE) is corrected for the number of data points:

$$\text{MSE} = \frac{\text{SSE}}{N} = \frac{1}{N} \sum_{k=1}^{N} (y_p(k) - \hat{y}_p(k))^2 \tag{25}$$

To obtain a normalized criterion, a variance performance measure can be calculated:

$$J_\sigma = \sqrt{\frac{\sum_{k=1}^{N} (y_p(k) - \hat{y}_p(k))^2}{\sum_{k=1}^{N} (y_p(k) - \bar{y}_p)^2}} \tag{26}$$

where $\bar{y}_p$ is the mean value of the output. The value of J_σ should tend to zero, as the quality of the model improves.

The standard test used in (linear) identification problems is the cross correlation between the error signal and the input(s) of the network. The resulting error signal should have no correlation with the input signal(s) used to excite the system. In the case of a noise-free system, the error signal should be zero. If a model of a system is adequate, the one-step-ahead prediction error $e_i(k) = \hat{y}_p(k) - y_p(k)$ should be *unpredictable* from all linear and nonlinear combinations of past inputs and outputs [3]. These tests hold:

$$r_{e_i e_i}(\tau) = \text{E}[e_i(k - \tau)e_i(k)] = \delta(\tau) \tag{27}$$

$$r_{u e_i}(\tau) = \text{E}[u(k - \tau)e_i(k)] = 0 \quad \forall \tau \tag{28}$$

$$r_{y_p e_i}(\tau) = \text{E}[y_p(k - \tau)e_i(k)] = 0 \quad \forall \tau \tag{29}$$

The normalized sampled correlation functions $c(\tau)$ are calculated, corresponding to the above expressions. The function $c(\tau)$ can be plotted against the variable τ, looking at the 95% confidence limits. If we consider N data points these confidence limits are found at $\pm 1.96/\sqrt{N}$. The expectations $\text{E}[c(\tau)] \approx -1/N$ and $\text{var}[c(\tau)] \approx 1/N$. The model is considered as appropriate if less that 1 out of 20 values is outside this range.

314

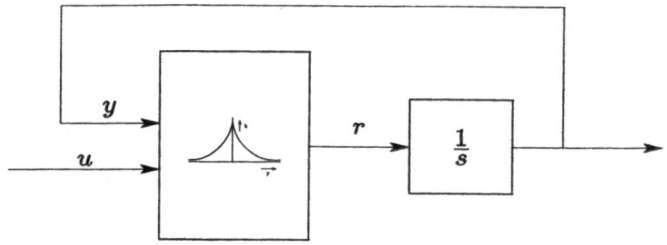

Figure 5: *Implementation of a process with a high nonlinear gain.*

Input signals
Thus far nothing has been said about the type of input signals for identification of nonlinear systems. For linear systems the superposition principle holds, so the response to the sum of various input signals is equal to the sum of the individual responses. This makes (pseudo) white noise signal suitable for identification, because these signals contain information about every frequency. Therefore, information is obtained for each frequency within the input signal band. With nonlinear systems this is no longer true [14]. The consequence is that it is imperative to train neural networks with excitation signals that are in the same frequency range as the future inputs, and do not contain frequencies far beyond this range.

The system can also be tested in an NOE configuration. If the model does not deviate from the actual process data, it might be accurate. The only problem is that sufficient model validation is almost impossible, because all possible states of the system should be checked for. Therefore we use the dimensionless performance measure J_σ as an indication of convergence. If $J_\sigma = 0.2$ this is interpreted as the explanation of 80% of the variance in the data, e.g. the correlation between input(s) and output(s) of the system.

9 Experiments and results using neural identification

To illustrate the modelling capabilities of the various networks, an experiment is carried out which is set up to model a highly nonlinear system.

A very well-known nonlinear system is a titration curve. Such a system is a very complicated and nonlinear system from which a simplified model has been derived. The purpose of the model is only to obtain a nonlinear model and not to model the ph process accurately. This model is depicted in figure 5. The ph level y in this model is described by

$$\frac{Y(s)}{U(s)} = k(y).\frac{1}{s} \qquad (30)$$

$$k(y) = e^{-\alpha||y||} \qquad (31)$$

First of all, the model is discretized and learned using input output information. The discretized version of this system is given by the difference equation using a sampling interval of 1 second:

$$y_p(k) = y_p(k-1) + \exp^{-3.0\|y_p(k-1)\|} u(k-1) \tag{32}$$

The resulting behaviour of the system is dominated by the integration and the nonlinear feedback gain. This gain depends exponentially on the state of the system and the parameter α. In this example $\alpha = 3.0$, without any physical reason. From this model a data set was derived which is depicted in figure 6. The objective for modelling is to minimize the total error SSE on the one-step-ahead prediction of this process output. The learning procedure is started using a heuristic training procedure based on backpropagation. The network dimensions are set up according to the guidelines given by Cybenko and Kolmogorov. Using these guidelines a $\aleph^{2,5,1}$ network is used. The initial result is depicted in figure 7a. The result after a training period of 25 weight updates is depicted in figure 7b. The accuracy which can be obtained using this neural identification depends on the size of the network and the number of learning steps applied to the network. In order to illustrate the superior modelling capability over that of linear identification, an experiment was carried out in which a linear model was estimated. The criterion value for the mean square error value of this data set is J_{σ}linear $= 10^{-6}$. But when the validity of the model was tested in an NOE structure in which the model output is fed back to the input of the model, a large deviation from the actual behaviour was observed (see figure 8a). The same experiment, but now applying the neural network gave a criterion value J_{σ},MNN $= 10^{-4}$. Despite the lower value for the mean square error criterion, the model incorporates the behaviour of the system, especially the nonlinear characteristic. This is really a remarkable result. The explanation for the relatively low value of the criterion is the fact that the data set contains only a few points which really excite the nonlinearity in the system. The influence of these few points in the data set on the criterion value is very low. Again the necessity of a thorough model validation set was made clear. Output error configurations are the most appropriate candidates to obtain this validation.

The results of the CMAC approach are encouraging. Using the same data as described for the MNN, a learning experiment was carried out in which CMAC identifies the observed behaviour by making one-step-ahead predictions, as explained in figure 4. The learning module is trained in an output error configuration, using the predicted output as an input for the CMAC module.

The convergence of the three methods (linear, MNN and CMAC) shows some remarkable differences. The linear identification procedure is very fast. By applying batch-oriented learning over the complete data set, the least-squares error solution is found within a few calculation steps. For the MNN approach every weight update corresponds to a correction with respect to the complete data set using a cumulative error for the weight updating procedure. CMAC is trained in a sequential way, because of the NOE structure. The time in which convergence is obtained is much slower that in the linear case, but faster than for the MNN case. The other structural difference between MNN and CMAC concerns

316

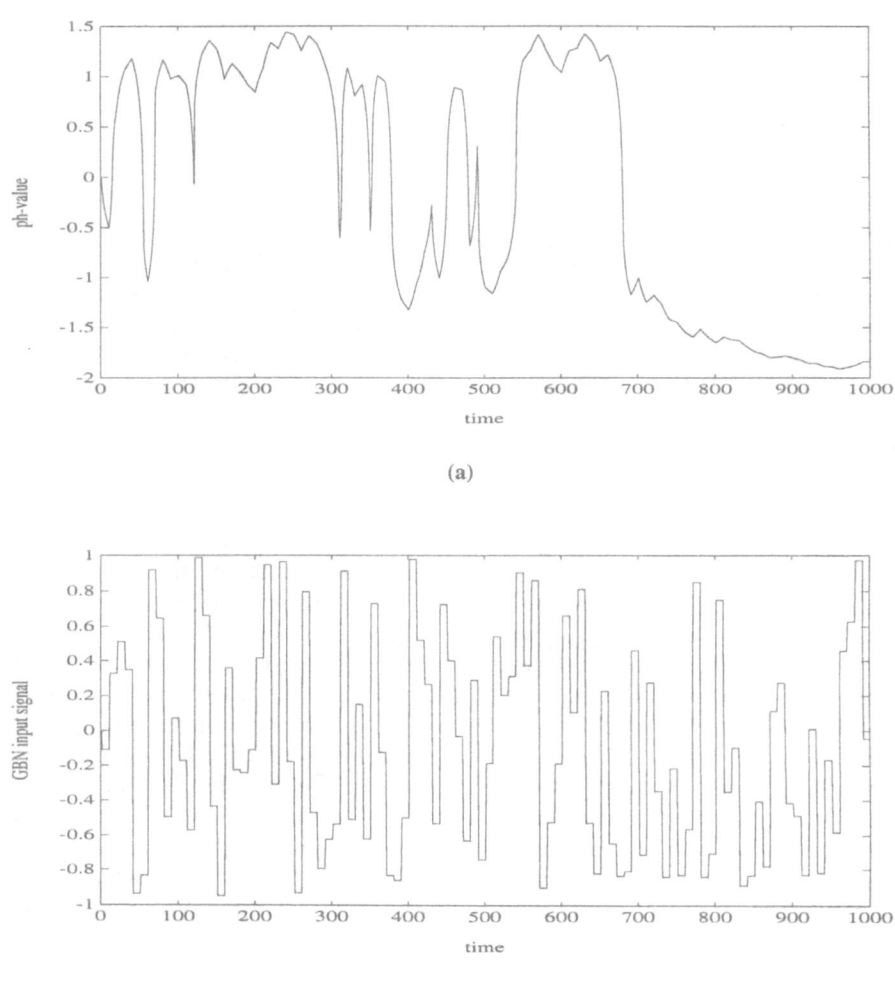

(a)

(b)

Figure 6: *Response (a) of simplified ph model to a Generalized Binary Noise input sequence [20] (b).*

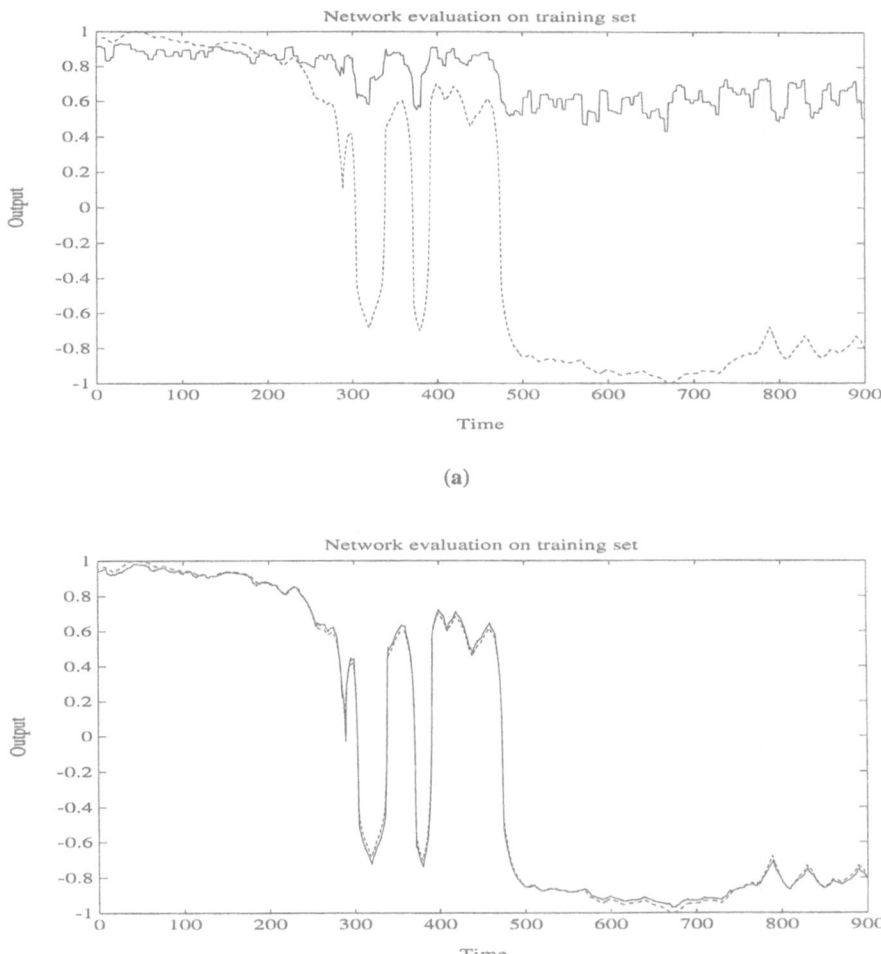

Figure 7: *Identification of a system with highly nonlinear gain. The initial result of the identification procedure is given in figure (a). Figure (b) gives the result of the one-step-ahead prediction after 25 training steps.*

318

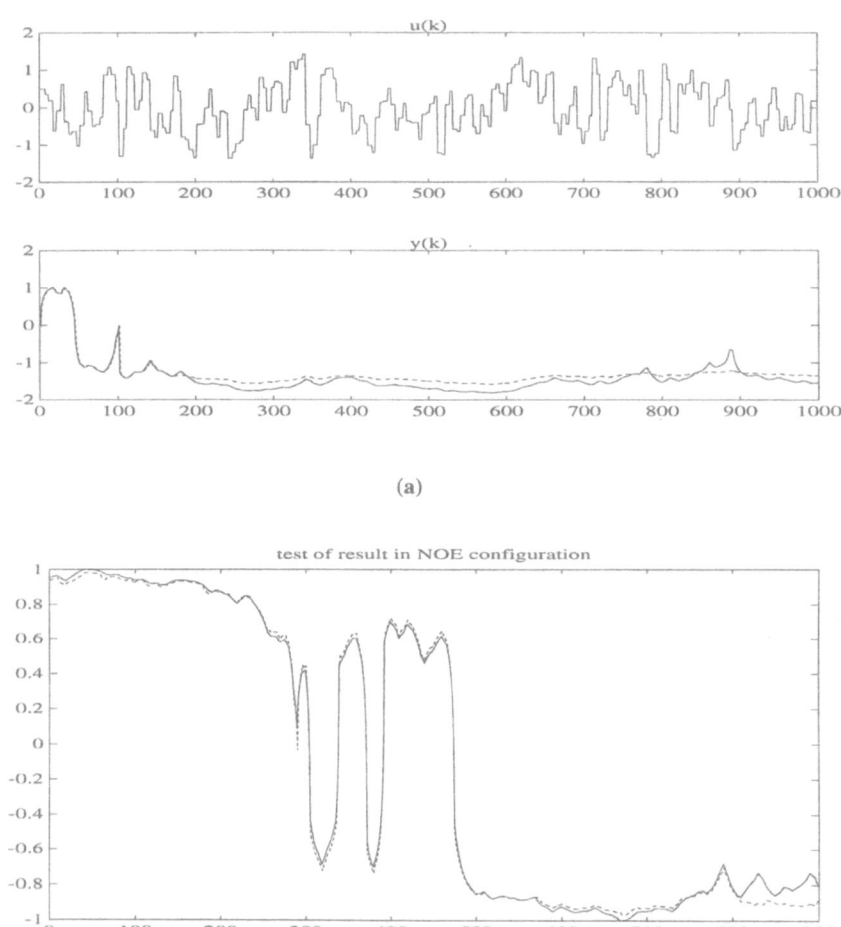

(a)

(b)

Figure 8: *Figure (a) gives the result of a model validation of a **linear** model, while figure (b) gives the result of a **neural** model in an output error configuration.*

the local properties of CMAC. Where MNN are used to get a global model fit (using continuous nonlinear functions), CMAC uses locally generalizing functions. When a recall of the system behaviour is required outside the learning range (other values for the input signal for example) the CMAC model will produce a very poor model fit.

So far the "standard" methods using a full neural model have been discussed. To show the superior behaviour of a hybrid approach, the hybrid RBFN is here examined.

Figures 9- 11 show the identification results obtained from the OLS-procedure as applied to a noise-free data set from the system. Table 1 summarizes the identification specifications. The model is obtained by the evaluation 400 data points. A compact model of 31 network nodes is selected. With this model, an error criterion value of 2.0E-04 is attainable. The correlation functions show that this model is quite good, as can also be concluded from the time-responses. To demonstrate the various uses of a dynamic model, we show the following way in which the time responses of the networks model are produced.

- Prediction error method (all purposes) in which the one-step ahead prediction is compared with the actual output:

$$\hat{y}(k+1) = \sum_{i=1}^{31} w_{1i}\phi_i(r_i(k)) + w_2^T x(k)$$

$$r_i(k) = \| x(k) - c_i \|$$

$$x(k) = \begin{bmatrix} y(k) & u(k) \end{bmatrix}^T$$

$$\phi(r) = e^{-\frac{r^2}{0.04}}$$

- Output error method (e.g. off-line control synthesis) in which the output of the model is fed back to the input of the model:

$$\hat{y}(k+1) = \sum_{i=1}^{31} w_{1i}\phi_i(r_i(k)) + w_2^T x(k)$$

$$r_i(k) = \| x(k) - c_i \|$$

$$x(k) = \begin{bmatrix} \hat{y}(k) & \hat{u}(k) \end{bmatrix}^T$$

$$\phi(r) = e^{-\frac{r^2}{0.04}}$$

- 10 steps ahead (e.g. model-based predictive control): Application of the output error method during 10 successive samples, and application of the prediction error method during one sample at the beginning of a 10 samples horizon.

The settings for the OLS procedure are given in table 1.

It is clear that in this case, where a linear part is apparent, a hybrid model approach is superior to full neural models. The accuracy reached can be expressed clearly by that

training set	400 samples
final error criterion	$\varepsilon^T \varepsilon / y^T y = 2.0\text{E-}04$
width of RBF	$\rho = 0.2$
linear model terms	$y(k) + 0.025u(k)$

Table 1: *Settings for the OLS algorithm.*

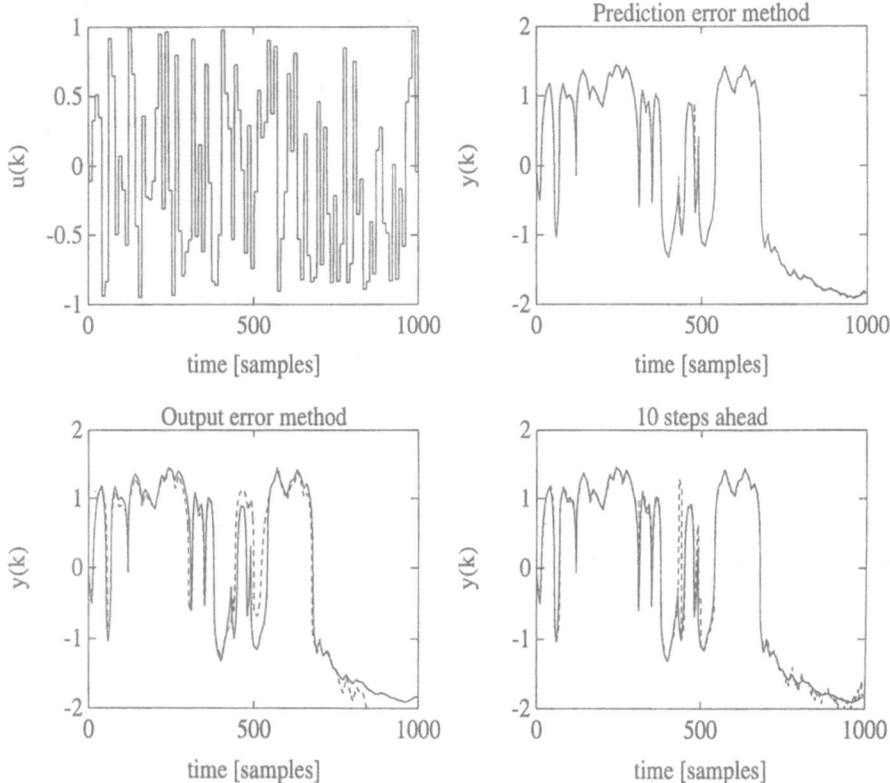

Figure 9: *Time responses: input sequence and output sequence of the system (solid line), compared to the network output (dotted line). The network outputs represent different interpretations: 1. Prediction error method (prediction horizon of one sample) 2. Output error method (prediction horizon is infinite) 3. combination of 1 and 2: prediction horizon of 10 samples.*

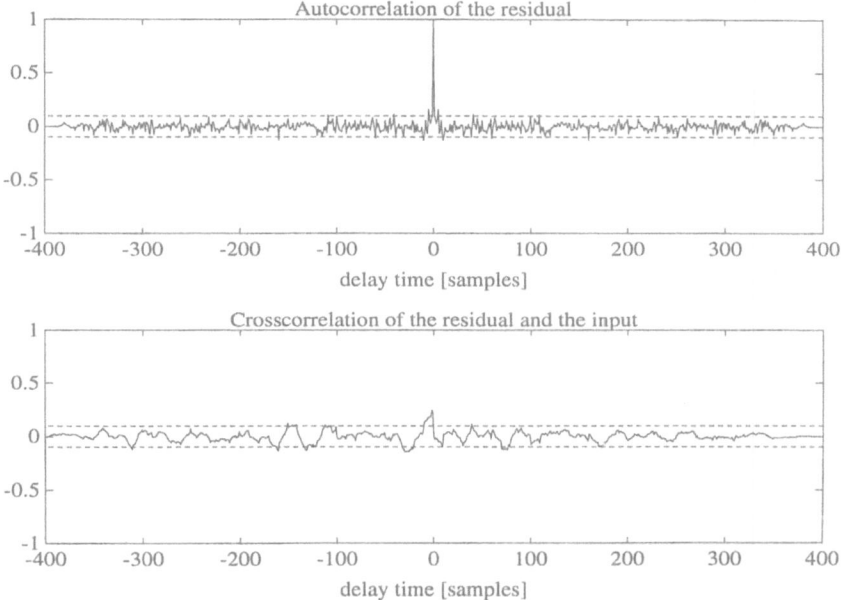

Figure 10: *Correlation functions of the residuals. The dashed lines denote the boundaries of the 95% confidence interval*

Model	J_σ
MNN	0.12
CMAC	0.15
RBFN	0.12
Hybrid model	0.0442

Table 2: *Comparison of J_σ for 4 modelling approaches.*

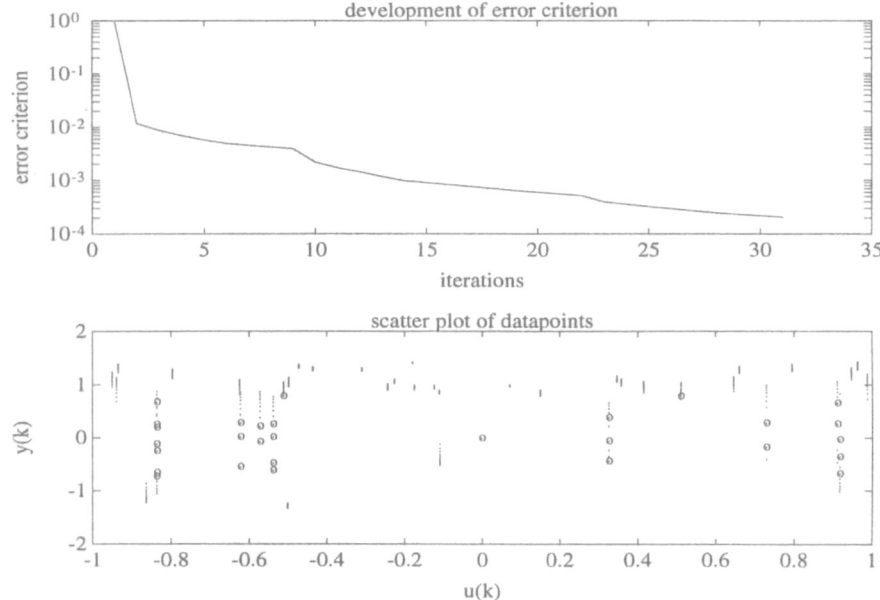

Figure 11: *Upper plot: Development of the error criterion of the OLS procedure. Lower plot: Scatter plot of data points (·) of the system and centres (o) of the Gaussian radial basis functions, found using the OLS selection procedure and the first 400 data points.*

of comparison of the J_σ value. This comparison is summarized in table 2. The use of a hybrid model (linear + RBFN) produces results superior to those obtained from full neural network models. The greatest advantage of this approach (OLS procedure) is the fact that when no linear behaviour is included no linear part is selected. The same holds for the other situation: if a linear model is identified, no nonlinear parts (centres for radial basis functions) are selected by the OLS procedure.

10 Conclusions

In this chapter artificial neural networks have been introduced as one of the 'products' of the AI research, as the counterpart of symbolic methods. The attractive idea of systems which can learn is used as the basis for the identification of unknown nonlinear systems.

ANN are universal approximators. By transformation of the identification problem of dynamic systems, into the learning of a static multidimensional mapping, neural networks can be used for identification.
Neural networks can be classified according to their ability to generalize. CMAC is a local generalizing network, where MNN are networks for global generalization.
RBFN networks are positioned in between. The results of RBFN, MNN and CMAC are in many cases comparable. The choice for either one of the networks depends on the demand for local generalization. When a strong local generalization is required CMAC is the best candidate. When local generalization is desired, but with a somewhat larger generalization width, RBF networks are the proper candidates. For global generalization MNN is a good alternative, combined with an improved weight optimization scheme.

Batch-oriented approaches are more reliable and offer faster convergence than in-line and recursive identification using neural networks. In combination with a heuristic weight optimization algorithm based on backpropagation good results are obtained for training MNN.
This heuristic scheme is a combination of a number of heuristics reported in the literature as improvements on basic gradient learning schemes. The heuristics are based on well-known rules of thumb and the theoretical results of linear system identification techniques. Common optimization problems, like local minima, will be met in neural network weight optimization as well.

The introduction of hybrid models, using a combination of a linear network and an RBFN, has made compact models possible. These models have proved to be reliable. The disadvantage is the off-line data preprocessing which has to be done to obtain the model. Once the model has been obtained, a linear (recursive) parameter update scheme can be used in real time, to cope with small parameter variations.
For all models obtained using neural network identification, model validation is of crucial importance. The linear part of the hybrid model can be checked for its stability in an easy way by examining pole/zero locations.

References

[1] Albus J.S. (1975). *A New Approach to Manipulator Control: The Cerebellar Model Articulation Controller (CMAC)*. Transactions of the ASME, pp. 220-227, September.

[2] Albus J.S. (1975). *Data Storage in the Cerebellar Model Articulation Controller (CMAC)*. Transactions of the ASME, pp. 228-233, September.

[3] Billings S.A., H.B. Jamaluddin and S. Chen (1992). *Properties of neural networks with applications to modelling nonlinear dynamical systems.* International Journal of Control, Vol. 55, No. 1, pp. 193-224.

[4] Chen S., S.A. Bilings and P.M. Grant (1990). *Nonlinear system identification using neural networks.* International Journal of Control, Vol. 51, No. 6, pp. 1191-1214.

[5] Chen, S. et al. (1991). *Orthogonal least squares learning algorithm for radial basis function networks.* IEEE Transactions on Neural Networks, Vol. 2, No. 2, p.p. 302-309.

[6] Chen, S. and S.A. Billings (1992). *Neural networks for nonlinear dynamic system modelling and identification.* International Journal of Control, Vol. 56, No. 2, p.p. 319-346.

[7] Cotter N.E. (1990). *The Stone-Weierstrass Theorem and Its Application to Neural Networks.* IEEE Transactions on Neural Networks, Vol. 1, No. 4, pp.290-295, December.

[8] Cowan, C.F.N. and Adams, P.F. Non-linear system modelling: concept and application. *Proceedings IEEE International Conference on Acoustics, Speech and Signal Processing.* May 1984.

[9] Cox M.G. (1971). *Curve fitting with piecewise polynomials.* Journal of Inst. Mathematical Applications, Vol. 8, pp.36-52.

[10] Cox M.G. (1984). *Practical spline approximation.* In: P.R. Turner (Ed): Topics in Numerical Analysis, Lecture notes in Mathematics 965, New York. Springer Verlag, pp.79-112.

[11] Haber, R. Structural identification of quadratic block-oriented models based on estimated Volterra kernels. *International Journal of Systems Science.* Vol. 20, No. 8, p.p. 1355-1380, 1989.

[12] Hunt, K.J., D. Sbarbaro, R. Zbikowski and P.J. Gawthrop (1992). *Neural Networks for Control Systems - A Survey.* Automatica, Vol. 28, No. 6, pp. 1083-1112, Pergamon Press Ltd, U.K.

[13] Kolmogorov A.N. (1957). *On the representation of continuous functions of many variables by superposition of continous functions of one variable and addition.* Doklady Akademy Nauk SSSR (N.S.), 114:953-956. Translation in American Mathematical Society Translations, Series 2, 28, pp. 55-59, 1963.

[14] LEONTARITIS I.J. AND S.A. BILLINGS (1987). Model selection and validation methods for non-linear systems. *Int. Journal of Control,* vol. 45, no. 1, pp. 311-341.

[15] Narendra K.S. and K. Parthasarathy (1990). *Identification and Control of Dynamical Systems Using Neural Networks,* IEEE Transactions on Neural Networks, Vol.1, No. 1, March, pp.4-27.

[16] Narendra K.S. and K. Parthasaraty (1991). *Gradient Methods for the Optimization of Dynamical Systems Containing Neural Networks.* IEEE Transactions on Neural Networks, Vol. 2, No. 2, pp. 252-262, March.

[17] Rodd M.G., H.B. Verbruggen and A.J. Krijgsman (1992). *Artificial Intelligence in Real-Time Control* Engineering Applications of Artificial Intelligence, Vol: 5, No: 5, pp. 385-399, Pergamon Press.

[18] Rosenblatt F. (1961). *Principles of Neurodynamics: Perceptrons and the Theory of Brain Mechanisms.* Spartan Books, Washington DC.

[19] Schetzen M. (1981). *Nonlinear system modeling based on the Wiener theory.* Proceeding of the IEEE, Vol. 69, No. 12, pp. 1557-1573.

[20] Tulleken H.J.A.F. (1992). *Grey-box Modelling and Identification Topics.* PhD Thesis, Delft University of Technology.

[21] Widrow B. and M.E. Hoff (1960), *Adaptive Switching Circuits,* In *Convention Record, Part 4.,* IRE Wescon Connection Record, New York, pp. 96-104.

11

NEURAL NETWORKS IN ROBOT CONTROL

SPYROS G. TZAFESTAS

Intelligent Robotics and Control Unit,
Department of Electrical and Computer Engineering
National Technical University of Athens
Zographou 15773, Athens, Greece

1. INTRODUCTION

Neural nets (NNs) are large scale systems involving a large number of special type nonlinear processors called "neurons"[1-4]. Biological neurons are nerve cells that have a number of internal parameters called synaptic weights. The human brain consists of over ten million neurons. The weights are adjusted adaptively according to the task under execution such that to improve the overall system performance. Here we are dealing with artificial NNs the neurons of which are characterized by a *state*, a list of *weighted inputs* from other neurons, and an *equation* governing their dynamic operation. The NN weights can take new values through a learning process which is accomplished by the minimization of a certain objective function through the step-by-step adjustment of the weights. The optimal values of the weights are stored as the strengths of the neurons' inteconnections. The NN approach to computation is suitable for problems for which more conventional computation approaches are not effective. Such problems involve systems or processes that cannot be modelled with concise and accurate mathematical expressions, typical examples being machine vision, speech and patern recognition, control systems and robotic systems. The implementation of NNs was made possible by the recent developments in fast parallel architectures (VLSI, electrooptical, and other). The principal features of NNs are:
- associative storage and retrieval
- signal regularity extraction
- convergence rate independent of number of nodes.

The aim of this chapter is to investigate the applicability of NNs in the design of robot control systems which involve significant model uncertainties and external disturbances. A NN controller performs some specific type of adaptive control, with the controller taking the form of the feedforward and feedback network. Learning is here identified with adaptation (the adaptable parameters being the strengths of the interconnections between the weights).

The three primary features that any NN controller should possess are:
- utilization of large amounts of sensory information
- collective processing capability
- learning / adaptation / robustness

Usually learning and control in neurocontrollers are acheived simultaneously, and learning continues as long as perturbations are present in the plant under control and/or its environment.

327

S. G. Tzafestas and H. B. Verbruggen (eds.),
Artificial Intelligence in Industrial Decision Making, Control and Automation, 327–387.
© 1995 *Kluwer Academic Publishers*.

328

The structure of the chapter is as follows. Section 2 is dealing with the principal neurocontrol architectures, starting with a general presentation of them, and then presenting in more detail a neurocontroller for linear deterministic systems and a neurocontroller for CARMA stochastic systems. Section 3 starts with a general look at robotic problems, proceeds to a review of the use of NNs in robotics, and then presents three particular robot neurocontrol techniques. Section 4 provides a set of numerical examples that illustrate the methods of Sections 2 and 3, and finally the Appendix briefly outlines the back-propagation and Hopfield networks that are most commonly used in robotic control.

2. NEUROCONTROL ARCHITECTURES

2.1. General Issues

The need to control complex systems under strong uncertainties has led to a re-evaluation of the adaptive control methods currently available. Two major issues in this are the increasing complexity of the systems to be controlled and the increasingly demanding design requirements with less precise a-priori knowledge of the systems under control and their environment. With the emergence of NNs, adaptive control schemes can be produced that possess stronger robustness and higher intelligency. Two excellent surveys of adaptive control can be found in [5,6]. Several approaches to the application of NNs to control system design are presented in [7-21]. These approaches can be divided according to the type of learning in two major classes, viz neurocontrol via supervised learning and neurocontrol via unsupervised learning. In the first class of NN control a teacher is assumed to be available, capable of teaching the required control performance. That is, in order to train the neurocontroller a controller must be already available capable of successfully controlling the system. This is a good approach in the case of a human trained controller, since it can be used to automate a previously human controlled system. However, in the case of automated linear and nonlinear teachers, the teacher's design requires the a priori knowledge of the dynamics of the system under control.

The structure of supervised NN control system has the general form of Fig.1 and involves three main components, namely a *teacher*, the *trainable controller* and the *system under control* [21].

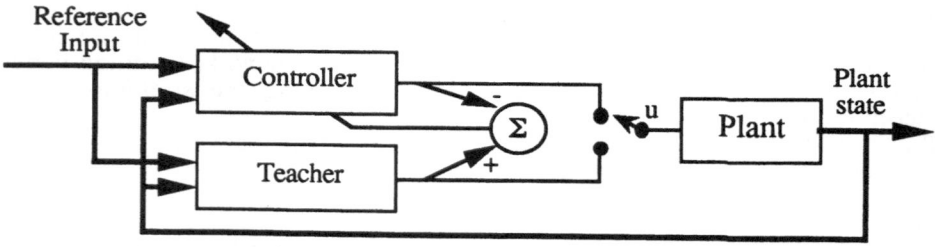

Fig.1 Structure of supervised NN control system

The *teacher* teaches the trainable controller by presenting examples of the control signals (u) needed for successfully controlling the plant. The teacher can be either a human controller or another automated controller (algorithm, knowledge based process, etc). The *trainable controller* is a NN appropriate for supervised learning, prior to training. The plant under control can be linear or nonlinear with unknown or partially known dynamics. Its states are measured by specialized sensors and are sent to both the teacher and the trainable controller. During the control of the plant by the teacher, the control signals and the state variables of the plant are sampled and stored for the NN controller training. After the end of successfull training the NN has learned the right control action, and replaces the teacher in controlling the plant.

In unsupervised NN learning control no external teacher is available and the dynamics of the plant under control are unknown and/or involve severe uncertainties (as is the case of fast robotic manipulators). In general, unsupervised NN learning control can be performed by several ways, the principal of which are described in the next subsection.

2.2. Unsupervised NN Control Architectures

In the following the four principal architectures for unsupervised NN control are presented. These architectures are: i) Direct inverse modelling architecture (DIMA): Version I, ii) Direct inverse modelling Architecture (DIMA): Version II, iii) Feedback error learning architecture (FELA), and iv) Adaptive learning architecture (ALA).

i) DIMA: Version I

This is the simplest architecture and is used in most applications (Fig.2). Actually, this direct inverse modelling is not used in the human's central nervous system.

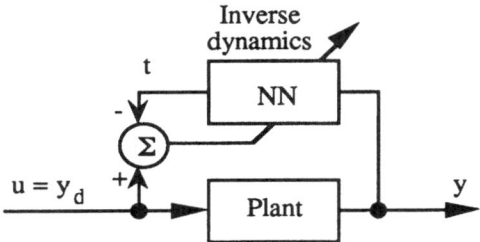

Fig.2. Direct inverse modelling architecture: Version I

One disadvantage of this architecture is that the system cannot selectively be trained to respond correctly in output regions of interest, since normally it is not known in advance which plant inputs u correspond to desired outputs. Obviously, here the neural net should act as the inverse of the plant producing a u(t) that makes y equal to the desired one.

ii) DIMA: Version II

This architecture is similar to Version I and is shown in Fig.3. Here, the NN controller can be trained to work properly in desired output regions only. This is so since the NN is now fed directly by the desired output y_d.

330

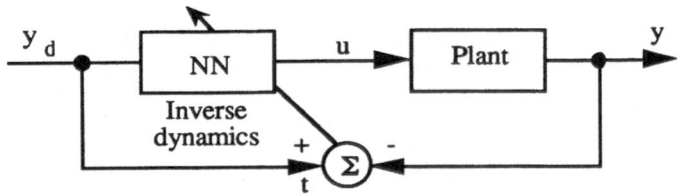

Fig.3. Direct inverse modelling architecture: Version II

iii) FELA: Feedback Error Learning Architecture

The structure of FELA is shown in Fig.4 where the direction of information flow within the inverse dynamics (NN) block is opposite to that of the DIMA I architecture.

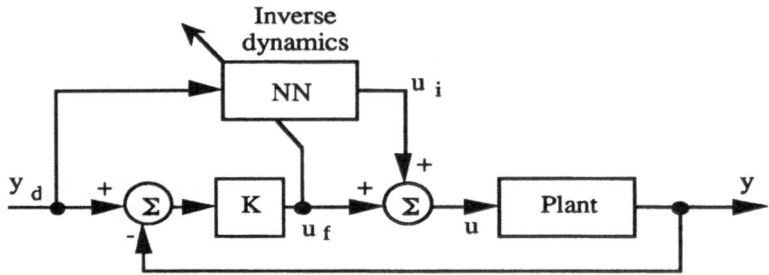

Fig.4. Feedback error learning architecture

This architecture is used in the central nervous system, and includes the *motor cortex* (feedback gain K and summation of feedback (u_f) and feedforward (u_i) commands), the *transcortial loop* (negative feedback loop) and the *cerebrocerebellum parvocellular red nucleus system* (inverse dynamics block). The total control input u to the plant is equal to

$$u(t) = u_f(t) + u_i(t) \tag{1}$$

The inverse dynamics NN block receives the desired output trajectory $y_d(t)$ and monitors the feedback signal $u_f(t)$ as the error signal. The feedback signal tends to zero as learning proceeds. The name "feedback error learning" is due to the utilization of the feedback signal u_f (input command) as the error signal of heterosynaptic learning. According to Kawato [13,14] the FELA scheme possesses the following advantages over the other architectures:
a) Learning and control are performed simultaneously,
b) Back propagation of the error signal through a forward model of the plant is not required,
c) The learning is goal-directed and can resolve any ill-posedness of the problem at hand.

iv) ALA: Adaptive Learning Architecture

This architecture has the form shown in Fig.5 and involves a NN estimator which on the basis of the status of the plant and important parameters available to the controller (and to itself) changes its state, the components of which correspond to the parameters of the controller.

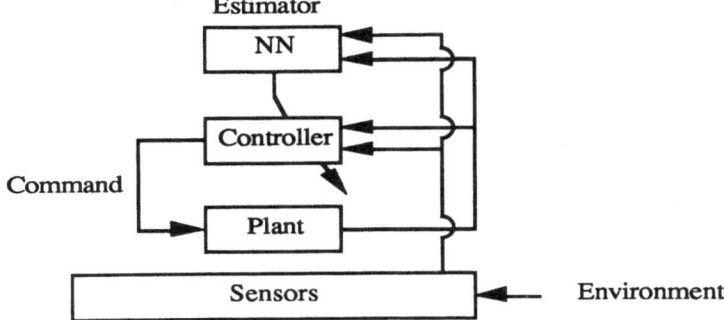

Fig.5. Adaptive learning NN architecture

The plant controller modifies its parameters according to the received estimator status, and then generates and sends a command to the plant under control. The state space of the NN estimator may be continuous or discrete, and so both continuous-time and discrete-time situations can be dealt with. The convergence speed of the estimator toward the optimal control parameters was shown experimentally to be high and independent of the number of parameters under adaptation (i.e. of the NN dimension).

2.3 DIMA II. Neurocontroller for Linear Systems

Here, a neurocontroller of the DIMA II type [20] will be fully analyzed for the case of a discrete-time linear process of the form:

$$G_p(z) = \frac{\beta_0 + \beta_1 z^{-1} + ... + \beta_m z^{-m}}{\alpha_0 + \alpha_1 z^{-1} + ... + \alpha_m z^{-m}}$$

where z is the discrete complex frequency ($z = \exp(sT)$).
    The general structure of the closed-loop system has the form of Fig. 6.

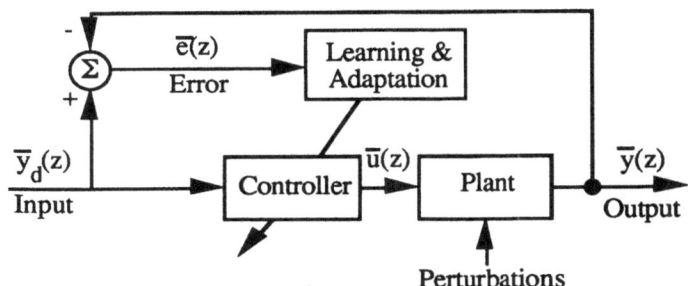

Fig 6. Structure of DIMA II linear control system

This structure is motivated by the following observation. Consider the open-loop control scheme of Fig. 7.

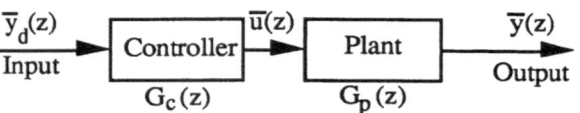

Fig. 7. Linear open-loop control scheme

We have

$$\bar{y}(z)=G_c(z)G_p(z)\bar{y}_d(z) \tag{2}$$

Therefore, if the controller is selected as

$$G_c(z)=[G_p(z)]^{-1} \tag{3}$$

then

$$\bar{y}(z)=G_p^{-1}(z)G_p(z)y_d(z)=\bar{y}_d(z) \tag{4}$$

Equation (4) says that "if the controller transfer function $G_c(z)$ is selected to be the inverse $G_p^{-1}(z)$, of the plant's transfer function $G_p(z)$ then the output y(t) follows (tracks) precisely the input $y_d(t)$". This can be precisely done if the dynamics $G_p(z)$ is exactly known. If not, then a learning mechanism is required as is shown in Fig. 6. The adaptive controller learrns the system dynamics $G_p(z)$ trhrough the error $\bar{e}(z)$ and updates adaptively the dynamics $G_p(z)$ of the controller.

Suppose, for simplicity, that $G_p(z)$ is of first order, i.e.

$$G_p(z)=\frac{\beta_0+\beta_1 z^{-1}}{\alpha_0+\alpha_1 z^{-1}} \tag{5}$$

Then, in order to obtain $\bar{y}(z)=\bar{y}_d(z)$ one must select

$$G_c(z)=\frac{a_0+a_1 z^{-1}}{b_0+b_1 z^{-1}} \tag{6}$$

with

$$a_0=\alpha_0 , \ a_1=\alpha_1 , \ b_0=\beta_0 , \ b_1=\beta_1 \tag{7}$$

The error $\bar{e}(z)$ is given by

$$\bar{e}(z)=\bar{y}_d(z)-\bar{y}(z)=\bar{y}_d(z)-G_c(z)G_p(z)\bar{y}_d(z)=[1-G_c(z)G_p(z)]\bar{y}_d(z) \tag{8a}$$

Therefore selecting $G_c(z)$ as in eq. (3) the error becomes zero.

In the NN controller the block $G_c(z)$ is replaced by a neural net, the weights of which are selected so as to minimize the quadratic error function

$$J(e)=\frac{1}{N}\sum_{k=1}^{N}e^2(k) \tag{8b}$$

where k is discrete time (t=kT, T=sampling period). This ensures that the controller's dynamics tends to the inverse dynamics of the plant under control, and is performed through the steepest descent algorithm

$$\theta(k+1)=\theta(k)+\Delta\theta(k) \tag{9a}$$

where

$$\Delta\theta(k)=-\mu e(k)S_\theta(k) \tag{9b}$$

Here $\theta(k)$ is the current value of the parameter θ, $\theta(k+1)$ is the updated value, $e(k)$ is the error signal $e(k)=y_d(k)-y(k)$, μ is an adjustable learning gain (that specifies the learning rate), and $S_\theta(k)$ is a function of the parameter state. Now, let us show the implementation details for a second order system, where:

$$G_p(z)=\frac{\beta_0+\beta_1 z^{-1}+\beta_2 z^{-2}}{\alpha_0+\alpha_1 z^{-1}+\alpha_2 z^{-2}} \quad , \quad G_c(z)=\frac{a_0+a_1 z^{-1}+a_2 z^{-2}}{b_0+b_1 z^{-1}+b_2 z^{-2}} \tag{10}$$

In this controller the a_is must follow the α_is and the b_is must follow the β_is. The error criterion is given by (8b) where, for faster computations we select N=1 (one step optimization). Then, using (8a) the criterion $J(e)$ becomes:

$$J(e)=[1-G_c(z)G_p(z)]^2 y_d^2(z) \tag{11}$$

Here, $\theta(k)=[a_0, a_1, a_2 ; b_0, b_1, b_2]^T$ and

$$\Delta\theta(k)=-\mu\partial J(e)/\partial\theta \tag{12a}$$

where

$$\frac{\partial J}{\partial\theta}=\frac{\partial}{\partial\theta}\{[1-G_c(z)G_p(z)]^2 y_d^2(z)\}$$

$$=2[1-G_c(z)G_p(z)]^2 y_d^2(z)\frac{\partial}{\partial\theta}[-G_c(z)G_p(z)]$$

$$=-2\bar{e}(z)\bar{y}_d(z)\frac{\partial G_c(z)}{\partial\theta}G_p(z) \tag{12b}$$

Thus, choosing the parameter state signal $\bar{s}_\theta(z)$ as

$$\bar{S}_\theta(z)=\frac{\partial G_c(z)}{\partial\theta}\bar{y}_d(z), \tag{13}$$

the parameter correction term (12a) becomes

$$\Delta\bar{\theta}(z)=-2\mu G_p(z)\bar{e}(z)\bar{s}_\theta(z) \tag{14}$$

where

$$\bar{S}_\theta = [S_{a0}, S_{a1}, S_{a2} ; S_{b0}, S_{b1}, S_{b2}]^T$$

The convergence of $G_C(z)$ to $G_p^{-1}(z)$ takes place for $\theta(k)=\theta^*$ where $\nabla_\theta J\,|_{\theta=\theta^*}=0$. The components of $\bar{S}_\theta$ are determined from (13) and the second equation (10) as follows.

$$\bar{S}_{a_0}(z) = \frac{\partial}{\partial a_0}[G_c(z)\bar{y}_d(z)] = \frac{\bar{y}_d(z)}{b_0+b_1z^{-1}+b_2z^{-2}} \tag{15a}$$

$$\bar{S}_{a_1}(z) = \frac{\partial}{\partial a_1}[G_c(z)\bar{y}_d(z)] = \frac{z^{-1}\bar{y}_d(z)}{b_0+b_1z^{-1}+b_2z^{-2}} \tag{15b}$$

$$\bar{S}_{a_2}(z) = \frac{\partial}{\partial a_2}[G_c(z)\bar{y}_d(z)] = \frac{z^{-2}\bar{y}_d(z)}{b_0+b_1z^{-1}+b_2z^{-2}} \tag{15c}$$

$$\bar{S}_{b_0}(z) = \frac{\partial}{\partial b_0}[G_c(z)\bar{y}_d(z)] = -\frac{\bar{u}(z)}{b_0+b_1z^{-1}+b_2z^{-2}} \tag{15d}$$

$$\bar{S}_{b_1}(z) = \frac{\partial}{\partial b_1}[G_c(z)\bar{y}_d(z)] = -\frac{z^{-1}\bar{u}(z)}{b_0+b_1z^{-1}+b_2z^{-2}} \tag{15e}$$

$$\bar{S}_{b_2}(z) = \frac{\partial}{\partial b_2}[G_c(z)\bar{y}_d(z)] = -\frac{z^{-2}\bar{u}(z)}{b_0+b_1z^{-1}+b_2z^{-2}} \tag{15f}$$

where $\bar{u}(z) = G_c(z)\bar{y}_d(z)$.

Thus, the controller parameter corrections are given by

$$\Delta a_i(k) = -\mu_{a_i}e(k)S_{a_i}(k) \quad , \quad \Delta_{b_i}(k) = -\mu_{b_i}e(k)S_{bi}(k) \tag{16}$$

for $i = 0, 1, 2$ where the gains μ_{ai} and μ_{bi} include the effect of $G_p(z)$. The weights $[a_0, a_1, a_2 ; b_1, b_2, b_3]$ of the NN controller are modified using the above algorithm [(9a), (12a)]. This updating process continues until $G_c(z)$ becomes approximately identical to $G_p^{-1}(z)$. The neurocontroller structure and the generation of the parameter state signals S_{ai} and S_{bi} are shown in Fig.8. The architecture of the complete neuro-controlled system has the form shown in Fig.9.

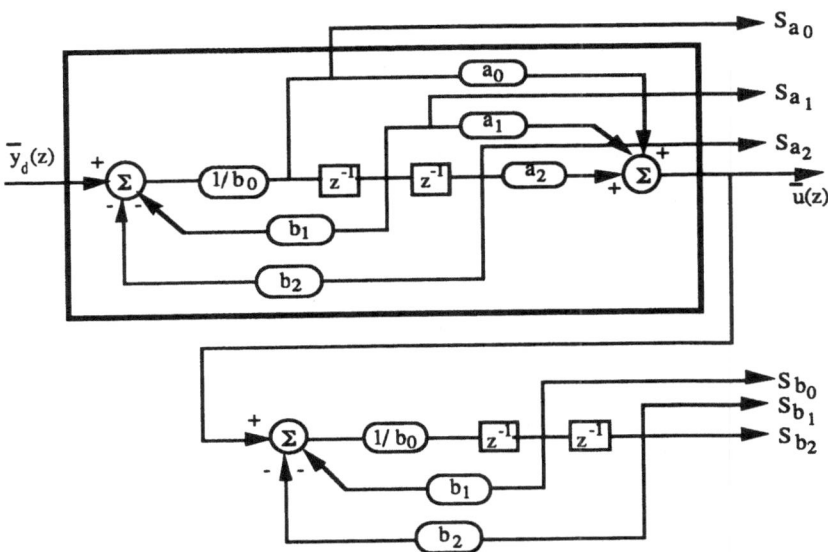

Fig.8. NN controller structure and generation of the S_{a_i}'s and S_{b_i}'s.
The weights of the NN are a_i and b_i (i=0, 1, 2).

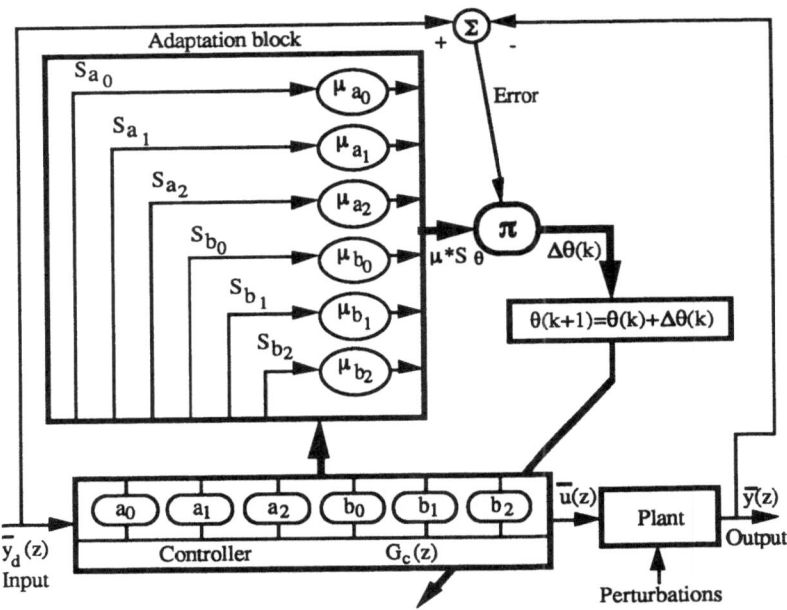

Fig.9. Structure of the overall DIMA II neurocontroller
for the second-order linear plant $G_p(z)$.

2.4. Adaptive Learning Neurocontrol for CARMA Systems

The Controlled Autoregressive Moving Average (CARMA) system is a noisy stochastic system described by the model

$$A(z^{-1})y(k) = B(z^{-1})u(k-1)+C(z^{-1})\xi_k \qquad (17)$$

where here z^{-1} denotes the unit time delay operator: $z^{-1}y(k+1) = y(k)$, $\{\xi(k)\}$ is a Gaussian stochasitc noise process with zero mean and covariance $E[\xi(k)\xi(j)]=r\delta_{kj}$ (δ_{kj} = Kronecker's delta) and

$$A(z^{-1}) = 1+a_1z^{-1}+...+a_{na}z^{-na} \qquad (18a)$$
$$B(z^{-1}) = b_0+b_1z^{-1}+...+b_{nb}z^{-nb} \qquad (18b)$$
$$C(z^{-1}) = 1+c_1z^{-1}+...+c_{nc}z^{-nc} \qquad (18c)$$

are polynomials with real coefficients. The polynomials $B(z^{-1})$ and $C(z^{-1})$ are assumed to be stable (minimum phase system). The polynomial degrees $n_a \geq n_b \geq n_c$ are known, but the parameters a_i, b_i and c_i are unknown.

The neurocontroller here consists of a two-layer NN for learning the inverse CARMA dynamics and a NN for learning the (forward) dynamics of the system [22-23].

Inverse dynamics and learning rule: The model (17) is solved for u(k-1), i.e

$$u(k-1)=\bar{A}(z^{-1})y(k)-\bar{B}_1(z^{-1})u(k-1)-\bar{C}(z^{-1})\xi(k) \qquad (19)$$

where

$$\bar{A}(z^{-1}) = \frac{1}{b_0}+\frac{a_1}{b_0}z^{-1}+...+\frac{a_{na}}{b_0}z^{-na} \qquad (20a)$$

$$\bar{B}(z^{-1}) = \frac{b_1}{b_0}z^{-1}+\frac{b_2}{b_0}z^{-2}+...+\frac{b_{nb}}{b_0}z^{-nb} \qquad (20b)$$

$$\bar{C}(z^{-1}) = \frac{1}{b_0}+\frac{c_1}{b_0}z^{-1}+...+\frac{c_{nc}}{b_0}z^{-nc} \qquad (20c)$$

Assuming for the moment that the noise sequence $\{\xi(k), ... ,\xi(k-n_c)\}$ is known, and that we are given the data $\{y(k), ... ,y(k-n_a)\}$ and $\{\{u(k-2), ... ,u(k-n_b-1)\}$ the model (19) can be written in the following NN form

$$\hat{u}(k-1)=\sum_{i=1}^{na+1}w_ix_i+\sum_{j=1}^{nb}w_{na+1+j}x_{na+1+j}$$

$$+\sum_{m=1}^{nc+1}w_{na+nb+1+m}x_{na+nb+1+m} \qquad (21)$$

where

$$x_{i+1} = y(k-i) , i=0, 1, \dots , na-1$$
$$x_{na+2} = u(k-2) , \dots , x_{na+nb+1}=u(k-nb-1) \tag{22}$$
$$x_{na+nb+2} = \xi(k) , \dots , x_{na+nb+nc+2} = \xi(k-nc)$$

and the time values of the weights (parameters) are

$$w_1=1/b_0 , w_2=a_1/b_0 , \dots , w_{na+1} = a_{na}/b_0$$

$$w_{na+2}=-b_1/b_0 , \dots , w_{na+nb+1} =b_{nb}/b_0 \tag{23}$$

$$w_{na+nb+2}=-1/b_0 , \dots , w_{na+nb+nc+2} = -c_{nc}/b_0$$

Clearly, the NN output $\hat{u}(k-1)$ is an estimate of $u(k-1)$. One can easily observe that the model (21) is a two-layer NN with inputs x_i (i=1, 2, ... , $n_a+n_b+n_c+2$) , output $\hat{u}(k-1)$ and synaptic weights w_i (i=1, 2, ... , $n_a+n_b+n_c+2$).

The criterion J_I to be minimized is

$$J_I=\frac{1}{2}[u(k-1)-\hat{u}(k-1)]^2 \tag{24}$$

where the true signal $u(k-1)$ is the teaching signal, and the learning rule is

$$w_i (p+1)=w_i (p)+n_I.\delta.x_i \tag{25a}$$

where p is the learning index, n_I is the learning rate constant, and

$$\delta=u(k-1)-\hat{u}(k-1) \tag{25b}$$

is determined from

$$\frac{\partial J_I}{\partial w_i} = \frac{\partial J_I}{\partial \hat{u}} \frac{\partial \hat{u}}{\partial w_i} = -\delta.x_i \tag{25c}$$

Neurocontroller: Given the measured input-output sequences $\{u(k-2), \dots , u(k-nb-1)\}$, $\{ y(k-1) , \dots , y(k-na) \}$, the noise sequence $\{ \xi(k) , \dots , \xi(k-nc) \}$ and a setpoint $y_d(k)$ at time k, the control signal $u(k-1)$ is generated using the same NN (21), i.e

$$u(k-1) = \sum_{i=1}^{na+nb+nc+2} w_i x_i^0 \tag{26a}$$

with
$$x_1^0=y_d(k) , x_i^0=x_i , i = 2, 3, \dots ,na+nb+nc+2 \tag{26b}$$
The NN (26) gives the present neurocontroller.

NN noise estimator: The noise sequence is not known and thus it must be estimated. Let

$$Y_k \equiv \{y(0), \dots ,y(k)\} , \quad U_k \equiv \{u(0), \dots ,u(k)\}$$

The noise is approximately reconsrtucted from $\hat{\xi}(k)$, i.e.

338

$$\xi(k) \approx \hat{\xi}(k) \equiv y(k) - \hat{y}(k/k-1) \tag{27a}$$

where

$$\hat{y}(k/k-1) \equiv E[y(k) / Y_{k-1}, U_{k-1}) \tag{27b}$$

$$\hat{y}(k/k-1) = \hat{\phi}^T(k)\hat{\theta}(k-1) \tag{27c}$$

$$\hat{\phi}^T(k) = [-y(k-1),...., -y(k-na) ; u(k-1),...., u(k-nb-1) ;$$
$$; \hat{\xi}(k-1),...., \hat{\xi}(k-nc)] \tag{27d}$$

$$\hat{\theta}^T(k-1) = [\hat{a}_1(k-1),...., \hat{a}_{na}(k-1) ; \hat{b}_0(k-1),...., \hat{b}_{nb}(k-1) ;$$
$$; \hat{c}_1(k-1),...., \hat{c}_{nc}(k-1)] \tag{27e}$$

Here, $\hat{\theta}(k-1)$ is the value of the parameters (weights) provided by the NN (21). The model (27c) can be represented by the following NN:

$$\hat{y}(k/k-1) = \sum_{i=1}^{na} s_i y_i + \sum_{j=1}^{nb+1} s_{na+j} y_{na+j} + \sum_{m=1}^{nc} s_{na+nb+1+m} y_{na+nb+1+m} \tag{28}$$

where y_i $(i = 1, 2, ... , na+nb+nc+1)$ are the NN inputs and s_i $(i = 1, 2, ... ,na+nb+nc+1)$ are the weights of the NN for the system dynamics. We, therefore have system

$$y_1 = y(k-1), \quad y_2 = y(k-2),...., \quad y_{na} = y(k-na)$$

$$y_{na+1} = u(k-1), \quad y_{na+2} = u(k-2),..., \quad y_{na+nb+1} = u(k-nb-1) \tag{29}$$

$$y_{na+nb+2} = \hat{\xi}(k-1),....., \quad y_{na+nb+nc+1} = \hat{\xi}(k-nc)$$

and true parameter values are:

$$s_1 = -a_1, \quad s_2 = -a_2,....., \quad s_{na} = -a_{na}$$

$$s_{na+1} = b_0, \quad s_{na+2} = b_1,....., \quad s_{na+nb+1} = b_{nb} \tag{30}$$

$$s_{na+nb+2} = c_1, \quad s_{na+nb+3} = c_2,....., \quad s_{na+nb+nc+1} = c_{nc}$$

The predictor NN (28) is based on the information $\{ Y_{k-1}, U_{k-1} \}$. Using the criterion

$$J_R = \frac{1}{2}[y(k) - \hat{y}(k/k-1)]^2$$

where $y(k)$ is the measured (true) value of the output at time k, one finds the weight updating (learning rule):

$$s_i(p+1) = s_i(p) + \varepsilon_1 \delta_y y_i, \quad \delta_y \equiv y(k) - \hat{y}(k/k-1) \tag{31}$$

where ε_1 is a learning rate constant.

Setting $\delta_y = \hat{\xi}(k)$, the value provided by the NN (28), the signals $x_{na+nb+1+m}$ $(m=1, 2, ...,nc)$ of the NN (21) can be replaced by

$$x_{na+nb+2}=\hat{\xi}(k) \quad , \quad x_{na+nb+3}=\hat{\xi}(k-1) \quad ,....., \quad x_{na+nb+nc+2}=\hat{\xi}(k-nc) \tag{32}$$

Similarly, the noise terms in the neurocontroller (26a) become:

$$x^0_{na+nb+2}=0 \quad , \quad x^0_{na+nb+3}=\hat{\xi}(k-1) \quad ,....., \quad x^0_{na+nb+nc+2}=\hat{\xi}(k-nc) \tag{33}$$

The term $x^0_{na+nb+2}$ is zero since the NN controller is based on the information $\{Y_{k-1}, \mathcal{U}_{k-2}\}$. The block diagram of the overall NN adaptive control system is shown in Fig. 10.

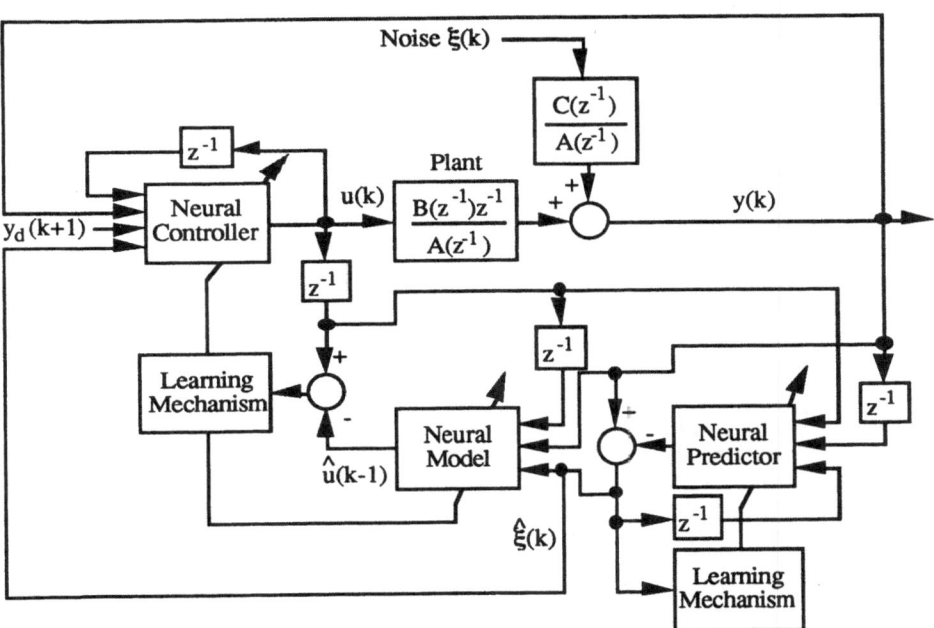

Fig. 10. Overall structure of the NN adaptive CARMA system

3. ROBOT NEUROCONTROL

3.1. A Look at Robotics

The two forms of NNs that are usually used in robotics are the multilayer perceptron and the Hopfield network. These networks are briefly described in the Appendix. Other networks proposed for use in robotic problems include the *competitive* and *cooperative networks* [24] and the *reward / punishment* network [12].

Robotics is concerned with the study of industrial and other robots. Most industrial robots can be regarded as open-loop link mechanisms consisting of several links connected together by joints. Typical joints are revolute joints and prismatic joints. A robotic system usually involves three subsystems, namely a motion subsystem (the arm), a recognition subsystem, and a control subsystem. The motion subsystem is the physical structure that carries out desired motions corresponding to human arms or legs. The recognition subsystem uses several sensors to collect information about any object being acted upon, about the robot itself and about the environment of the robot. The control subsystem influences the motion subsystem to achieve a task using the information from the recognition subsystem. Major problems of robotic control include kinematics, dynamics, path planning, control, sensing, programming and intelligence [25].

Robot kinematics is devoted to the study of the geometry of motion of a robotic manipulator without regard to the forces / moments that cause the motion. Therefore, kinematics deals with the analytical description of the robot as a function of time. Kinematics is distinguished in direct or *forward* kinematics and *inverse* kinematics (or arm solution kinematics). Direct kinematics deals with the determination of the cartesian position of the end effector given the values of the joint variables. Inverse kinematics deals with the inverse problem of finding the values of the joint variables given the position / orientation of the end effector.

Robot dynamics deals with the mathematical formulation of the dynamic equations of the robot arm motion, i.e. with the problem of relating the applied forces and torques (by the joint actuators) to the joint motion (positions, velocities and accelerations with respect to time). The three principal dynamic robot modelling techniques are the *Langrange-Euler* (L-E) approach, the *Newton-Euler* (N-E) approach and the *Generalized d'Alembert* (G-D) approach. The L-E formulation is less efficient, although the recursive algorithm of Hollerbach has improved efficiency. A compact presentation of all these models can be found in [26].

Robot path planning is the process of determining a smooth and safe motion trajectory along which the robot end-effector must be moved for going from an initial location to a desired final location. To this end, one needs to identify a selected set of intermediate points along the trajectory and then use straight-line or circular or higher order interpolation techniques (e.g. spline functions) to link these points. A representative path planner example can be found in [27].

Robot motion control is achieved by using a suitable controlled actuator (e.g. electric motor) at each joint of the robot. The aim of the control is to ensure that the robot will maintain a desired dynamic response (e.g. follow a prespecified path). In actual industrial applications (such as grinding or inserting a peg into a hole) the robot must exert a desired force or torque on an object while moving in the prescribed way. Obviously this needs a combined position and force-torque control [28]. The robot control design problem is, in general, very difficult since the dynamics of the links are nonlinear and coupled. The main robot control techniques used so far are: PID local control, computed-torque control, resolved motion rate control, robust (sliding mode) control, self-tuning control, and model reference adaptive control [29-34].

Robot sensing is the process of perception and translation of appropriate object properties into the information needed by the robot to carry out a given operational task. Object properties include geometric, optical, electric, mechanical, acoustic, thermal and chemical properties. The two basic stages of robot sensing are *transducing* (i.e. converting by hardware the properties at hand into a signal) and *processing* (i.e. transformation of the signal into the appropriate information). The transducing stage is performed by external sensing mechanisms called *sensors*. Types of sensors include position, velocity, acceleration, range, proximity, force, tactile and visual. Most industrial robots involve minimal sensory feedback.

Robot task planning is the robotics area in which the application of artificial intelligence (AI) techniques is most promising. Robot task planning is the planning of robot motions to perform some prespecified task. A plan is a representation of action for achieving a goal (here performing a task). Robot programming is the area of developing suitable high level languages (procedural and object oriented) for the easy and effective programming and reprogramming of a robot. Today a large variety of robot languages exist with relative advantages and disadvantages. In future more flexible robotic systems will have to be created to be able to operate in changing and harsh environments. Robots will have to be equiped with some kind of *intelligence* to cope for uncertainties in the environment. One direction to intelligence is offered by the neural nets.

3.2. Neural nets in robotics: General Review

A mathematical problem is said to be *well-posed* when its solution exists, is unique and depends continuously on the initial data. In most robotic control problems the solution is not unique, i.e. they are *ill-posed* problems. The key feature of NNs lies exactly here, i.e. NNs are able to reduce the computational complexity and deal with ill-posed robotic problems. Three representative ill-posed robotic problems are [35,36]: i) Trajectory formation (determination), ii) Inverse kinematics or redundant robots, iii) Inverse robot dynamics in the presense of agonist and antagonist actuators.

The analytic solutions of inverse kinematics yield accurate numerical results whereas the NN solution is in general nonaccurate. Works devoted to the inverse robot kinematics problem through NNs include [37-39]. Due to their approximate nature, NN inverse kinematics solutions can be best used as initial estimates of iterative-type solutions.

In robot dynamics, the NN learns the inverse dynamic relationship of robot directly. The architectures described in Section 2.2 can be used here. Most notable is the approach of Kawato and coworkers [13,14,35,36] called feedback error learning NN which is based on neurophysiology. If the form of the inverse dynamics is assumed known a-priori (e.g. Lagrange-Euler) then the NN can be used to estimate the unknown parameters involved.

Neural nets were used for robot path planning in [40-45]. The path planning problem with obstacle avoidance was considered in [41,42] using Hopfield nets and truss structures. Whenever a new obstacle is found, the weights are updated by adding a term to the NNs energy function. Liu in [43] used a multilayer perceptron to classify various robotic hand grippers on the basis of key features of the object to be grasped. The case of mobile robot path planning was studied in [44] with the NN trying to minimize the length of a path to a goal position. Finally Eckmiller in [45] has developed a neural net called "neural triangular lattice" for storing and retrieving robot trajectories. The main body of works dealing with the application of NNs in robotics concern the control problem. Further works to the ones cited in Section 2 [7-23] are presented in [46-54]. Albus [46-48] has developed a general NN robotic control technique called Cerebellar Model Articulation Control (CMAC) which can be considered as a distributed look-up table. Pao and Sobajic [49] implement a 2-d.o.f. robot positional control system using a multi-layer perceptron. Guez and coworkers [50-52] use the Hopfield NN model to implement model reference adaptive control (MRAC). Miller [53] employs the CMAC technique in conjunction with the computed torque control method. Finally Elsley [54] implements an inverse Jacobian control using multilayer perceptrons. The resulting control is better than that of the traditional inverse Jacobian control, especially on long movements.

We close this review section by mentioning the works [55,56] where NNs are used in sensing and simultaneous multiple robot control. The work in [55] studies the utilization

342

of NNs in tactile perception, through the use of the deconvolution properties of the Hopfield NN. In the direction of intelligent control, the works of Albus [46-48] are worth mentioning. At each successive level of the CMAC hierarchy, a CMAC module decomposes a high level command to a set of lower level commands. The same is true in the architecture proposed by Kawato and coworkers in [14,35] which is appropriate for the control and learning of voluntary movement. Kawato's hierarchical architecture is shown in Fig.11. This architecture is composed of four parts: (i) the main descenting pathway and the transcortial loop designated by heavy lines, (ii) the spinocerebellum-magnocellular red nucleus system (SPINO-CBM & MGN RN) as an internal neural model of the dynamics of the musculoskeletal system, (iii) the cerebrocerebellum - parvocellular red nucleus system (CEREBRO - CBM & Parvo RN) as an internal neural model of the inverse dynamics of the musculoskeletal system, and (iv) the sensory-association cortex (SENSORY-ASSN CX, AREAS 2,5,7) as an associative memory for iterative learning control. These subsystems of the CNS are phylogenically and ontogenetically older in this order.

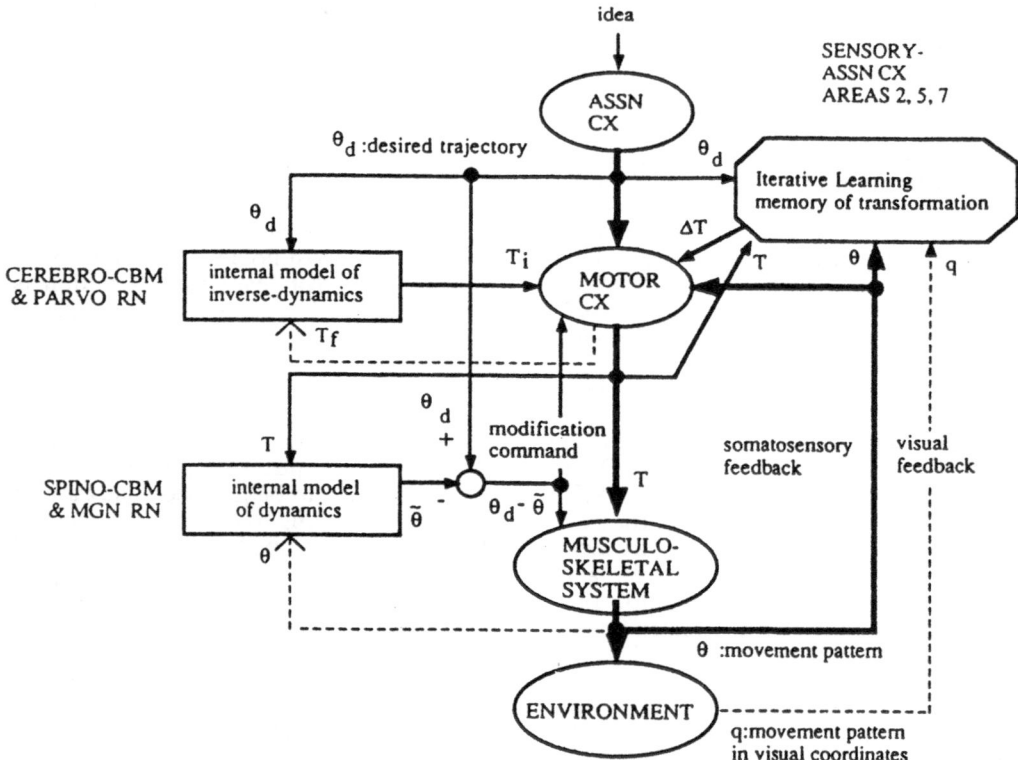

Fig.11. Kawato's hierarchical NN model for control and learning of voluntary movement (Inputs used nonlinear input-output relations, T and θ_d, are designated by solid lines; Inputs for synaptic modification, θ and T_f, are designated by dotted lines; ASSN CX=association cortex, MOTOR CX = motor cortex).

3.3. Robot Control Using Hierarchical NNs

Here an unsupervised neurocontrol scheme will be described using a hierarchical (4-layer) NN structure assuming only the a-priori knowledge that the robot is an open serial linkage of rigid links with a given number n of degrees of freedom [21].

The dynamic model of a robot consisting of an open kinematic chain of rigid links is described by the equation [26]:

$$D(q, t)\ddot{q}(t) + h(q, \dot{q}, t) + B\dot{q}(t) + g(q, t) = u(t) \qquad (34)$$

where $q(t), \dot{q}(t), \ddot{q}(t)$ is the nx1 vector of angular positions, velocities and accelerations respectively, $D(q,t)$ is the nxn robot's inertial matrix, $h(q, \dot{q}, t)$ is the nx1 vector of centrifugal and Coriolis terms, B is the nxn diagonal matrix of viscous friction terms, $G(q,t)$ is the nx1 vector of gravitational terms, and u(t) is the nx1 vector of driving torques or forces. This model is strongly nonlinear, coupled, time varying and robot configuration dependent.

The usual local PID control (i.e. control of each joint individually) is relatively successful at low speeds. For high speed robots one must use controllers that compensate for the robot's nonlinearities and model/environment uncertainties. Under the assumption that the robot parameters are exactly known, the nonlinearities are cancelled by using the well known computed torque (or feedback linearization) technique. In this technique u(t) is given by

$$u(t) = D(q, t)\upsilon(t) + h(q, \dot{q}, t) + B\dot{q}(t) + g(q, t) \qquad (35)$$

where $\upsilon(t)$ is a certain control law such as the proportional plus derivative (PD) law

$$\upsilon(t) = K_p e + K_\upsilon \dot{e} + \ddot{q}_d(t) \qquad (36)$$

where K_p and K_d are diagonal matrices $q_d(t)$ is the desired robot trajectory and e(t) = $q_d(t)$ - q(t) is the trajectory tracking error. Inserting u(t) of (35) in Eq. (34) gives

$$\ddot{q}(t) = \upsilon(t) \qquad (37)$$

which, if the control law (36) is employed, gives the error dynamics

$$\ddot{e}(t) + K_\upsilon \dot{e}(t) + K_p e(t) = 0 \qquad (38)$$

Now choosing $K_\upsilon = \text{diag}[2\zeta_i\omega_i]$ and $K_p = \text{diag}[\omega_i^2]$, eq. (38) represents an asymptotically stable system (damping ratio ζ). Now, Guez and Selinsky's method [21] starts from the observation that the robot dynamics can be expressed as a weighted linear combination of suitable functions which in general are a-priori known. These nonlinear functions can be trained into a series of multilayer (say 3-layer) feedforward network modules prior to use and then combined online in a final layer. In this way information available prior to use is employed at one level of the hierarchy, while information that can be obtained online is learned at another level.

The 4-layer neurocontroller architecture is shown in Fig.11A and has the form

$$y(t) = \sum_{i=1}^{N} w_i x_i \qquad (39)$$

where $w_i(i=1,2,...,N)$ are the synaptic NN weights and $x_i(i=1,2,...,N)$ are the nonlinear signal functions. The structure of the overall unsupervised learning control scheme is shown in Fig.12.

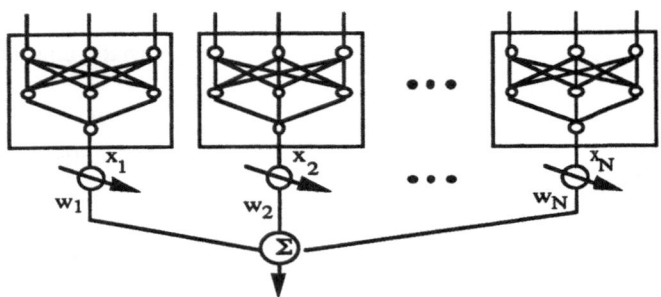

Fig.11A. NN controller architecture

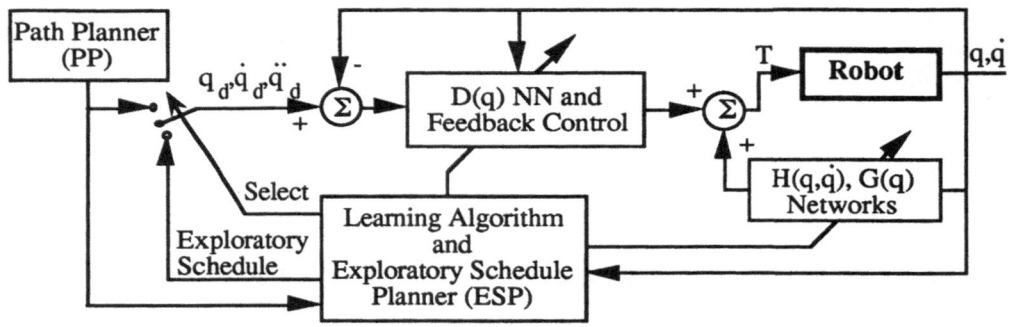

Fig.12. Architecture of overall NN learning control scheme

During the initial purely learning phase and whenever the robot can be out of production, appropriate explanatory schedules are generated to isolate and identify the nonlinear compensation terms. During the purely learning phase, the exploratory schedule planner (ESP) provides the desired trajectory $q_d(t)$. In the learning production phase, the path planner (PP) produces the trajectory for the production task. The nonlinear compensation terms are learned during appropriate sections of the trajectory. After receiving the desired trajectory, the control that is to applied to the robot is calculated using the feedback control from the linear controller and the control from the learning nonlinear compensation terms. The gravity terms $g(q)$ are first learned and then the term

$$\mathbf{H}(\mathbf{q},\dot{\mathbf{q}}) = \mathbf{h}(\mathbf{q},\dot{\mathbf{q}},t) + \mathbf{B}\,\dot{\mathbf{q}}(t) \tag{40}$$

is learned. The $g(q)$ terms are learned during the positioning control phases of the trajectory. At steady state positioning $(\dot{\mathbf{q}} = \ddot{\mathbf{q}} = 0)$, Eq.(34) becomes $u=g(q)$.

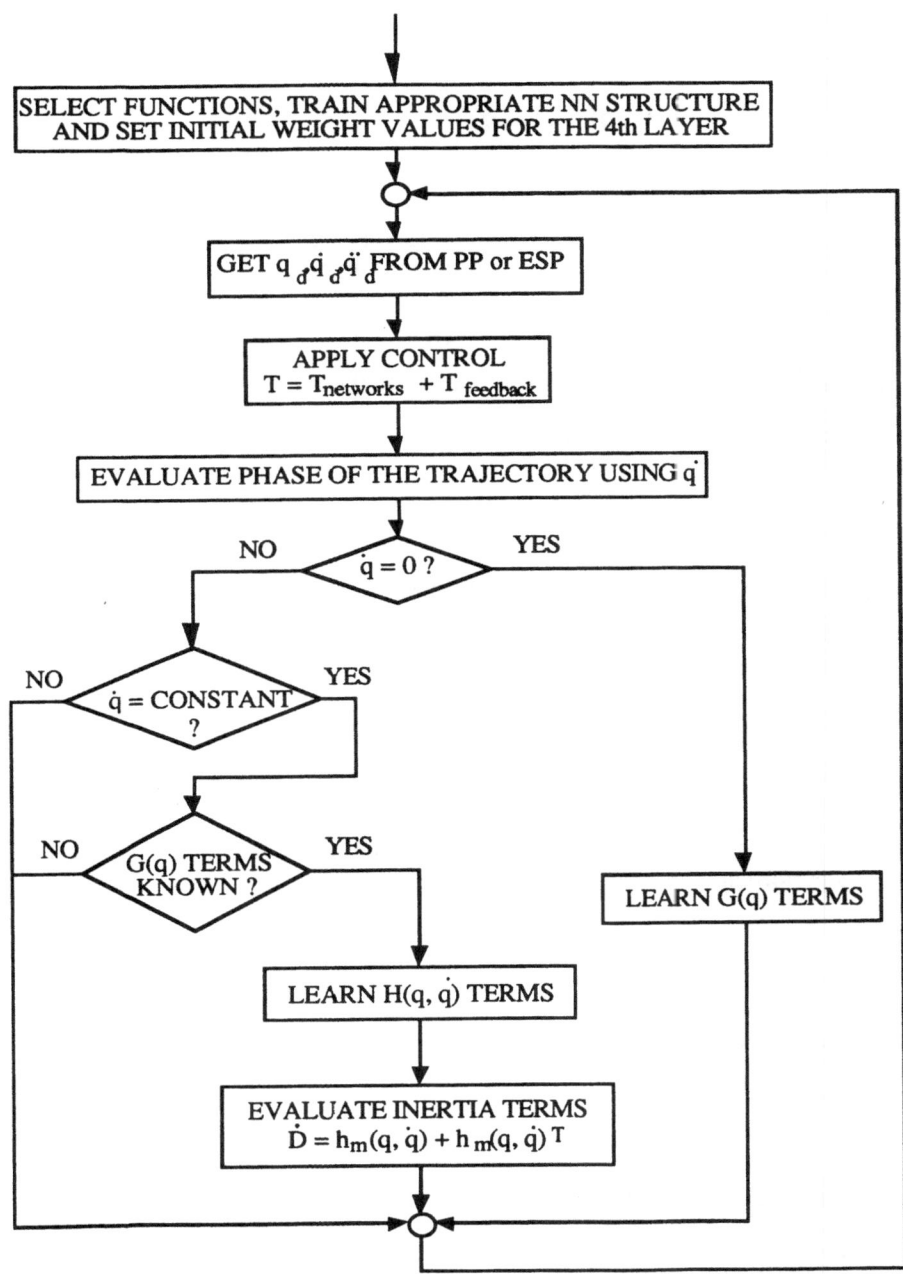

Fig.13. Flow chart of the Guez and Selinsky NN learning scheme

Therefore, $g(q)$ is learned from the applied torque at steady state. The terms $H(q,\dot{q})$ are learned during the velocity control phases ($\dot{q}=$ constant, $\ddot{q}=0$) of the trajectory for which Eq.(34) gives

$$u = g(q) + H(q,\dot{q}) \tag{41}$$

Thus

$$\hat{H}(q,\dot{q}) = u - \hat{g}(q) \tag{42}$$

where $\hat{g}(q)$ is the learned (estimated) $g(q)$ and $\hat{H}(q,\dot{q})$ is the learned value of $H(q,\dot{q})$.

The inertia terms $D(q,t)$ can be computed on-line using the well known a-priori relationship [57]:

$$\dot{D}(q,t) = h_m(q,\dot{q}) + h_m^T(q,\dot{q}) \tag{43}$$

where $h_m(q,\dot{q})$ is an nxn matrix satisfying the relation $h(q,\dot{q},t) = h_m(q,\dot{q})\dot{q}$. After the elapse of sufficient learning time, the NN provides the values $\hat{D}(q,t), \hat{H}(q,\dot{q})$ and $\hat{g}(q)$. Thus, the computed torque control (35) gives

$$u = \hat{D}(K_p e + K_\nu \dot{e} + \ddot{q}_d) + \hat{H}(q,\dot{q}) + \hat{g}(q) \tag{44}$$

This scheme starts with some initial estimates of D, $\hat{H}$, $\hat{g}$ and uses learned knowledge of the robot dynamics to build a controller capable of controlling the robot at high speeds over the whole work space. The flow chart of the resulting overall unsupervised learning scheme is as shown in Fig.13.

3.4. Minimum Torque-Change Robot Neurocontrol

Here, the technique of Kawato and coworkers [14,35,36] will be reviewed which involves a NN that is based on the minimum torque-change criterion and coherently resolves all three ill-posed robotic problems, i.e trajectory formation, inverse kinematics and inverse dynamics problems. Actually, one way to face ill-posed motion problems is to use a smoothness performance criterion. Two such criteria are:

(i) *Minimum jerk criterion (rate acceleration change)*

$$V_J = \frac{1}{2}\int_0^{t_f} \{(\frac{d^3 x}{dt^3})^2 + (\frac{d^3 y}{dt^3})^2 + (\frac{d^3 z}{dt^3})^2\}dt \tag{45}$$

where (x, y, z) is the Cartesian and-effector position.

(ii) *Minimum torque-change criterion*

$$V_T = \frac{1}{2}\int_0^{t_f} \sum_{i=1}^n (\frac{d\tau_i}{dt})^2 dt \tag{46}$$

where τ_i is the torque fed to to the ith actuator out of n actuators.

The two criteria V_J and V_T are closely related. The minimum torque-change NN model critically depends on the system dynamics and provides : (i) a computational model for the trajectory formation problem, (ii) understanding at the representation algorithmic and hardware levels of the problem. Trajectories derived from the minimum torque-change model are quite different from these of the minimum jerk model. The minimum torque-change model reproduces human-arm trajectories better. Kawato uses an iterative learning scheme which is of the Newton-line form in function space.

The internal robot dynamics model and the internal inverse dynamics model can be realised by a parallel distributed processing neural network of the form shown in Fig.11. For simplicity, consider a single-input single-output system. Hence only one neuron with synaptic plasticity is required. Extension to MIMO cases is straightforward. (Fig.14).

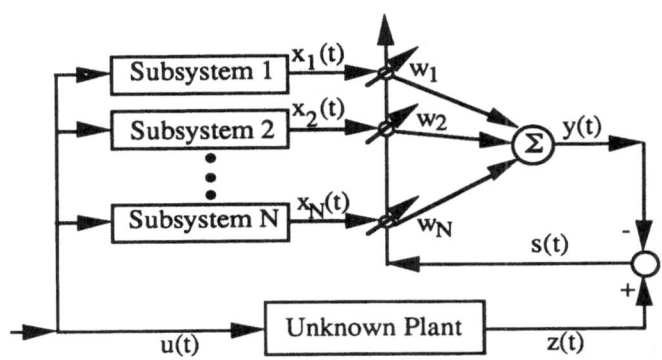

Fig.14 : NN nonlinear system identifier (involves many nonlinear subsystems and a neuron with heterosynaptic weights w_i).

The neuron approximates the output $z(t)$ by monitoring both the input $u(t)$ and the output $z(t)$ of this system. The input $u(t)$ is also fed to N subsystems, which nonlinearly transform it into the signals (variables) $x_i(t)$, i=1, 2, ..., N. If w_i denotes the synaptic weight of the ith input, then the output signal $y(t)$ is given by Eq. (39) , i.e

$$y = \mathbf{w}^T \mathbf{x}(t) \qquad (47a)$$

where

$$\mathbf{w}^T = [w_1, w_2, ... , w_N] \ , \ \mathbf{x}(t) = [x_1(t), x_2(t), ... , x_N(t)]^T \qquad (47b)$$

The second synaptic input to the neuron is the error $s(t)=z(t)-y(t)$ between the measured output of the unknown nonlinear system and the output $y(t)$ of the neuron. The ith synaptic weight w_i changes when the conjunction of the ith input $x_i(t)$ and the error signal $s(t)$ occurs, i.e

$$\tau \frac{d\mathbf{w}(t)}{dt} = \mathbf{x}(t)s(t) = \mathbf{x}(t)[z(t) - \mathbf{x}^T \mathbf{w}(t)] \qquad (48)$$

where τ is a time constant of change of the synaptic weight ($1/\tau=\mu$ is the learning rate). If the time constant τ is very long, then the synaptic weights converge to their optimal values. Actually, one may observe that Eq.(48) represents the steepest descent method, and the convergence is global.

The detailed structure of the Kawato's overall system employs the feedback-error-learning architecture (FELA) (Fig.4) and is shown in Fig.15. This system actually corresponds to a subsystem of the entire hierarchical architecture of Fig.11.

348

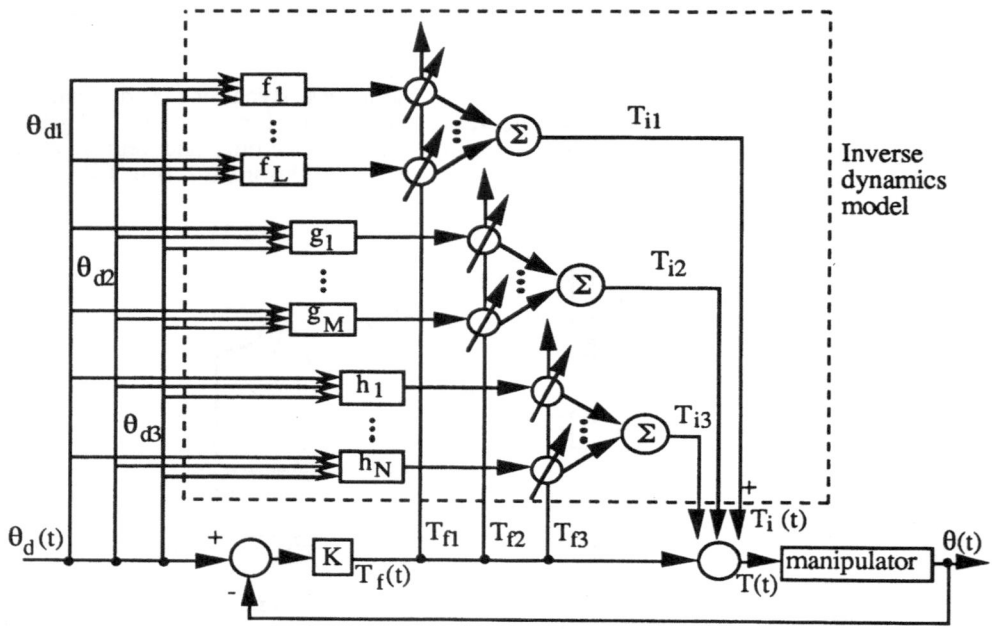

Fig.15. FELA scheme and inverse dynamics NN structure
for a 3-degrees-of-freedom robot

Here the control input u(t) is the total torque T(t) fed to the robot actuator and according to
Eq.(1) is the sum of the feedback torque $T_f(t)$ and the feedforward torque $T_i(t)$ provided
by the inverse dynamics model. The feeforward torque is required for smooth control of
fast movements and is learned from experience. The feedback torque is used for control at
the early stages of learning.

The learning equations for the synaptic weights w_{kj} (j=1,2,3) contained in the
inverse dynamics model are:

$$\tau\frac{dw_{k1}}{dt} = f_k(\theta_{d_1}(t),\theta_{d_2}(t),\theta_{d_3}(t))T_{f_1}(t) \qquad k = 1,2,\dots,L$$

$$\tau\frac{dw_{k2}}{dt} = g_k(\theta_{d_1}(t),\theta_{d_2}(t),\theta_{d_3}(t))T_{f_2}(t) \qquad k = 1,2,\dots,M \qquad (49)$$

$$\tau\frac{dw_{k3}}{dt} = h_k(\theta_{d_1}(t),\theta_{d_2}(t),\theta_{d_3}(t))T_{f_3}(t) \qquad k = 1,2,\dots,N$$

where f_k, g_k, h_k are the nonlinear transformations of the subsystems for the first, second
and third neurons respectively, and T_{f_j} (j=1,2,3) is the jth component of T_f which is fed to
the jth motor of the robot. The above FELA control scheme was applied to the Kawasaki-
Unimate PUMA 260 robot with great success.

3.5 Improved Iterative Learning Robot Neurocontroller

In this section the iterative learning robot neurocontroller described by Fu ans Sinha [58] will be briefly described. The main advantage of this iterative learning neurocontroller, which is based on a modified backpropagation algorithm, is its increased rate of convergence and the improved robustness over the common iterative controller. The learning process occurs between two consecutive operations of the robot and the neural learning control is implemented as feedforward control. On-line computation is needed only for the PD feedback controller, thus avoiding the intensive on-line computations required by other adaptive and robust controllers. An additional advantage of this learning controller is its cability to face the effects of flexibility, backlash and friction on the robot, not usually done by other controlers. The only problem is the lack of strictly theoretical verification of convergence. The neural learning controller needs retraining when the robotic trajectory has been changed.

A schematic presentation of the iterative learning neurocontroller is given in Fig.16,

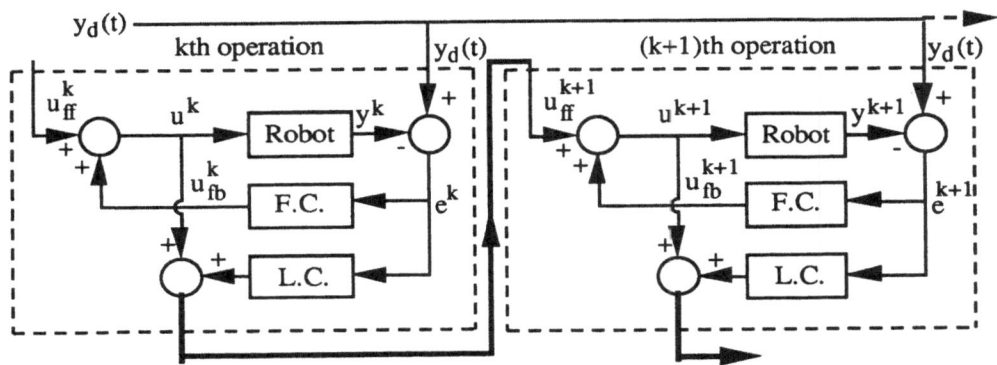

Fig.16. Pictorial representation of improved iterative learning
(F.C.=feedback controller, L.C. learning controller)

where $y_d(t)$ is the desired robot trajectory, and k is the robot operation index. The total control input u^{k+1} in the (k+1)th operation is the sum of the feedback control signal u_{fb}^{k+1} provided by F.C. (say a PID or an adaptive controller) and the feedforward control signal u_{ff}^{k+1} generated by adding the control signal u^k of the previous operation to the output of L.C.Thus this control belongs to the FELA type of control. An other piece of work on the iterative learning control of robots may be found in [59-60].

When the learning of the NN weights is based on the minimization of the error at a single-time instant, the rate of convergence is very small.However, since the functions in the output layer of the NN are linear, one can apply linear least squares learning which is much faster than the steepest-descent algorithm.

The training of the hidden layers is done through the gradient (steepest descent) scheme. The training aims at minimizing the criterion:

$$J(L) = \sum_{i=1}^{L} e^T(i)e(i) \qquad (50)$$

where L is the number of total training data, and $e(i) = y_s(i) - y(i)$ is the error between the real system Nx1 vector output and the NN Nx1 vector output (N is the number of the output modes). The index (50) differs from the one used in the original backpropagation algorithm of Rumelhart [61], since here the sum of the squared error over the total training samples is used. Thus, since the computation of the gradient needs an enormous computational effort for very high L (tens of thousands of training data) one has to resort to an iterative efficient method.

To this end, define J(k) by

$$J(k) = \sum_{i=1}^{k} e^T(i)e(i) \qquad (51)$$

and let w_m be the weights that connect the neurons in the lower hidden layer, or the input layer, to the mth neuron in the upper hidden layer. Then the original backpropagation learning rule is

$$w_m(k+1) = w_m(k) + \varepsilon \frac{\partial J(k)}{\partial w_m}]_{w_m = w_m(k)} \qquad (52)$$

where ε is the learning rate coefficient.

Now, the performance index (51) can be written as

$$J(k) = e^T(k)e(k) + J(k-1) \qquad (53)$$

and so

$$\frac{\partial J(k)}{\partial w_m}]_{w_m = w_m(k)} = F(k) + \frac{\partial J(k-1)}{\partial w_m}]_{w_m = w_m(k)} \qquad (54a)$$

where

$$F(k) = \frac{\partial[e^T(k)e(k)]}{\partial w_m}]_{w_m = w_m(k)} \qquad (54b)$$

The second term in (54a) is approximated by the Taylor's expansion

$$\frac{\partial J(k-1)}{\partial w_m}]_{w_m = w_m(k)} = \frac{\partial J(k-1)}{\partial w_m}]_{w_m = w_m(k-1)}$$

$$+ \frac{\partial^2 J(k-1)}{\partial w_m^2}]_{w_m = w_m(k-1)}[w_m(k) - w_m(k-1)] \qquad (55)$$

where the second order derivative is estimated by the relation

$$\frac{\partial^2 J(k-1)}{\partial w_m^2} = \frac{\partial J(k-1)/\partial w_m - \partial J(k-2)/\partial w_m}{w_m(k-1) - w_m(k-2)} \qquad (56)$$

Thus in view of (54a) the learning rule (52) becomes

$$w_m(k+1) = w_m(k) + \varepsilon F(k) + \varepsilon \frac{\partial J(k-1)}{\partial w_m}]_{w_m = w_m(k)} \qquad (57)$$

The improved gradient learning rule (57) is very much faster than the simple gradient rule. As for the learning coefficient ε, the smaller it is the smaller will the changes to the weights in the network be, which imply a better approximation at the cost of a slower learning rate. Making ε large so as to speed the learning rate there is the danger of oscillatory (unstable) performance. One way to increase ε and yet avoid the possibility of instability is to select it as

$$\varepsilon = (\alpha/k)/\|\partial J/\partial w_m\| \tag{58}$$

where α is a very small positive number and $\|\partial J(k)/\partial w_m\|$ denotes the Euclidean norm of the vector $\partial J(k)/\partial w_m$.

Now, let us study the training of the output NN layer. To this end consider the mth output neuron of Fig.17 and assume that there are M such output neurons. Defining $\mathbf{w}_m$ and $\mathbf{h}_m$ as:

$$\mathbf{w}_m = [w_{1m}, \dots, w_{im}, \dots w_{Nm}]^T , \quad \mathbf{h}_m = [x_1, \dots, x_i, \dots x_N]^T \tag{59a}$$

one obtains

$$y_m = \sum_{i=1}^{N} w_{im} x_i = \mathbf{h}_m^T \mathbf{w}_m , \quad m = 1,2,\dots,M \tag{59b}$$

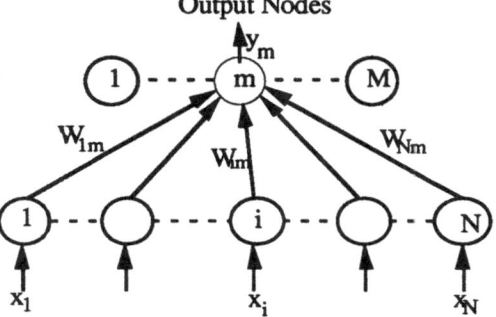

Fig.17. Output layer and one hidden layer of the NN

The training of the output weights $\mathbf{w}_m$ associated with the neuron m is performed by minimizing the criterion (m=1,2,...,M):

$$J_m(L) = \sum_{r=1}^{L} [y_{sm}(r) - y_m(r)]^2$$
$$= \sum_{r=1}^{L} [y_{sm}(r) - \mathbf{h}_m^T(r)\mathbf{w}_m(L)]^2 \tag{60}$$

where $y_{sm}(r)$ is the mth output of the real system.

It is an easy task to verify [62] that the resulting iterative learning algoritm is described by the equations

$$\mathbf{w}_m(k+1) = \mathbf{w}_m(k) + \mathbf{K}_m(k+1)[y_{sm} - \mathbf{h}_m^T(k+1)\mathbf{w}_m(k)] \tag{61a}$$

$$\mathbf{K}_m(k+1) = \mathbf{P}_m(k)\mathbf{h}_m(k+1)[\mathbf{h}_m^T(k+1)\mathbf{P}_m\mathbf{h}_m(k+1) + \mu]^{-1} \tag{61b}$$

$$\mathbf{P}_m(k+1) = \frac{1}{\mu}[\mathbf{I} - \mathbf{K}_m(k+1)\mathbf{h}_m^T(k+1)]\mathbf{P}_m(k) \tag{61c}$$

with initial conditions

$$\mathbf{w}_m(0) = \varepsilon \quad (\varepsilon \text{ small}) , \quad \mathbf{P}_m(0) = \gamma^2 \mathbf{I} \quad (\gamma \text{ large}).$$

Here, $\mathbf{K}_m(k+1)$ is the least squares gain matrix and μ is the forgetting factor.

The steps of the improved iterative back-propagation learning algorithm are summarized as follows:

Step 1: Choose the initial conditions $\mathbf{w}_m(0)$ and $\mathbf{P}_m(0)$.

Step 2: Feed the NN with input and output response vectors for iterations $k=1,2,...,K^*$.

Step 3: Train the output layer using the iterative least squares learning algorithm (61).

Step 4: Train the hidden layer using the improved gradient algorithm.

Step 5: Repeat Steps 2 to 4 until a desired accuracy is obtained.

The robot control is of the FELA type with a PD feedback contoller (Fig. 18). The NN is trained the inverse robot dynamics in the vicinity of the desired trajectory. At the initial phase the system is controlled as usually by the PD controller and stabilizes the system. As the learning procceds, the dominant part of the control signal is provided by the NN, while the PD controller simply supresses the disturbances. The weights of the NN are held fixed during the control period. When each operation of the control is finished, the NN is retrained with the operational data obtained from the last trial. The NN is trained the inverse robotic dynamic model by feeding the information of the robot trajectory (output) to the input layer of the NN.

The control is achieved exactly as described in Section 3.3 (Equations (34)-(38) and (44)). The result is a decoupled robot with each joint having damping ratio ζ and cyclic undamped frequency ω_n. Usually the damping ratio ζ is selected $\zeta=1$ (critically damped system).

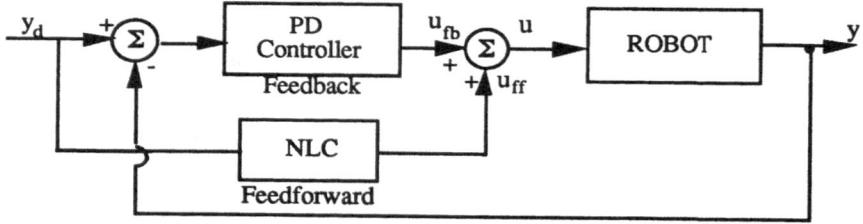

Fig. 18. FELA neural learning control scheme.

4. NUMERICAL EXAMPLES

In this section a set of numerical examples are presented which illustrate the type of results obtained by the majority of the techniques described in Sections 2 and 3.

4.1. Example 1: DIMA II Controller for Linear Systems

The neurocontroller described in Sec. 2.3 was applied by Gupta, Rao and Nikiforuk [20] to a class of linear plants with arbitrary dynamics (see Fig.6). The adjustable weights of the neurocontroller are initialized to some arbitrary values, and the system response is then obtained for a unit step input. In the following, the simulation results for a second-order plant are provided. The plant transfer function is

$$G_p(z) = \frac{\beta_0 + \beta_1 z^{-1} + \beta_2 z^{-2}}{\alpha_0 + \alpha_1 z^{-1} + \alpha_2 z^{-2}}$$

with $\beta_0=2.5$, $\beta_1=1.6$, $\beta_2=1$; $\alpha_0=1.8$, $\alpha_1=1$ and $\alpha_2=0.8$. This system has zeros at -0.32±j0.546 and poles at -0.278±j0.606. In this no-disturbance case the neurocontroller performance is shown in Fig.19. The performance of the neurocontroller for a second set of parameters: $\beta_0=1$, $\beta_1=0.8$, $\beta_2=1.2$; $\alpha_0=0.8$, $\alpha_1=1$ and $\alpha_2=0.6$, which corresponds to the zeros -0.4±j1.02 and the poles -0.625±j0.6, is shown in Fig. 20(a,b).

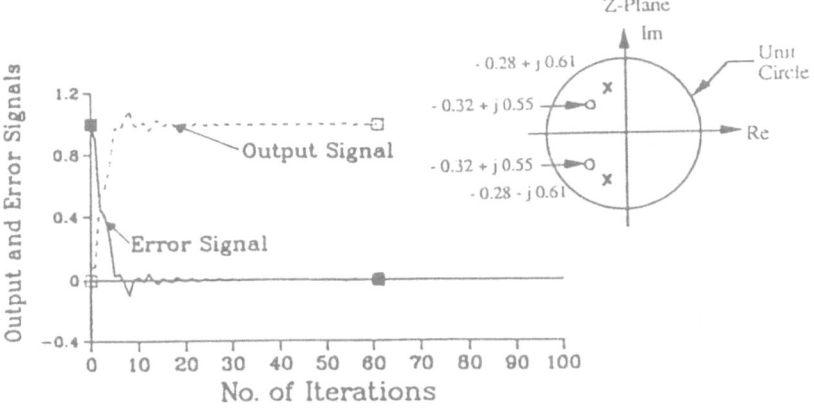

Fig. 19. Evolution of error and output signals for a 2nd order plant: No disturbance case

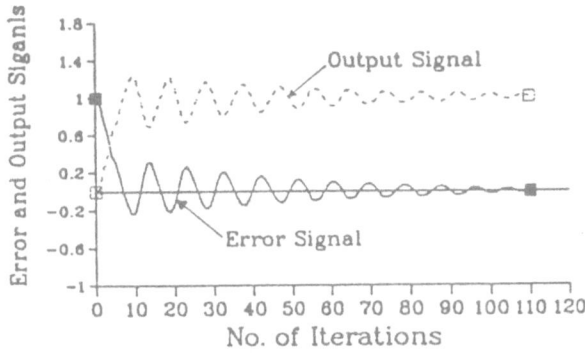

Fig. 20. Evolution of error and output signal for a different 2nd-order plant.

One observes from Figs. 19 and 20 that the NN controller is able to learn and adapt to different dynamic situations. However, the learning time increases as the plant poles and zeros move closer to the unit circle for the same initial values of the parameters.

Now consider a case where parameter disturbances are involved. The above system with the second set of parameters is again considered, where during the learning process the parameters α_1 and α_2 are perturbed by a step of magnitude 0.4 at an arbitrary time instant. The poles of the perturbed system lie at the positions $-0.875\pm j0.7$. The simulation results obtained in this case are shown in Fig.21. Similar results are obtained when the plant parameters are slowly time varying.

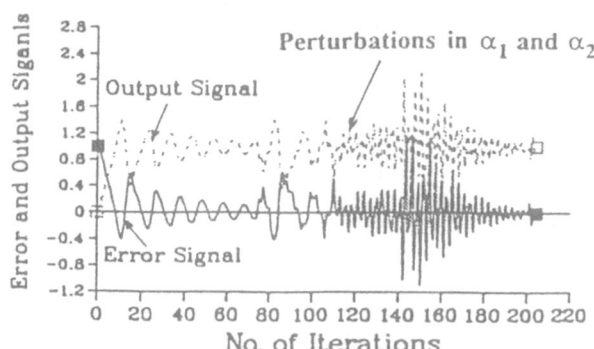

Fig. 21. Evolution of error and output signal for a system with parameter disturbances.

Finally the error and output signals obtained for a system corrupted by output random noise for a certain time are depicted in Fig. 22. One observes that the adaptive scheme remained causal and recovers to a desired performance level after the removal of the noise, although no filtering effect appears. The above results show that the system achieves the desired performance irrespective of the perturbations either in the structure or in the environment of the plant.

4.2. Example 2: Neurocontroller for CARMA Systems

The theory presented in Section 2.4 was applied to a number of CARMA systems including a robotic system that was modelled as CARMA System [29,33,34]. Here the results of a stable and an unstable system will be presented.

The stable system (see Eq.(17)) has the polynomials

$$A(z^{-1}) = 1+a_1z^{-1}+a_2z^{-2}, \qquad a_1=-1.6, \qquad a_2=0.63$$
$$B(z^{-1}) = 1+b_1z^{-1}, \qquad b_1=0.5$$
$$C(z^{-1}) = 1+c_1z^{-1}, \qquad c_1=0.5$$

with $\xi(k)\sim N(0,r)$. The set point $y_d(k)$ was selected to be a rectangular wave with period 100 and amplitudes ± 5. The initial values of the synaptic weights were set as [see (21) and (28)]
$w_1(0)=0.1, w_2(0)=-0.1, w_3(0)=0.1, w_4(0)=-0.1, w_5(0)=-0.1, w_6(0)=-0.1$ and
$s_1(0)=s_2(0)=s_3(0)=s_4(0)=s_5(0)=0.0$
The control results obtained by the adaptive (self-tuning) neurocontroller for $r=0.001$ and learning rates $\eta_1=0.001$ and $\varepsilon_1=0.01$ are shown in Fig.22.

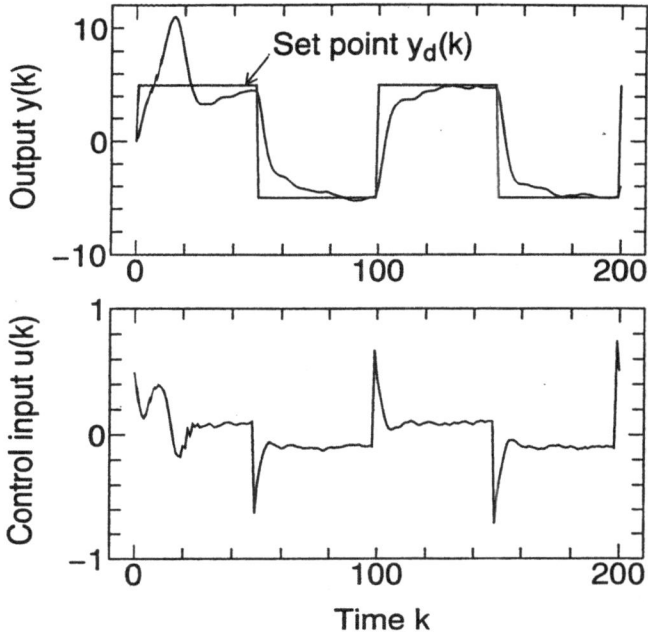

Fig. 22. Neurocontrol results for the stable case

One observes that the learning does not progress very much after about 50 samples, so that the control performance after that cannot be improved further. The variances for the output errors and the control signals for various values of the learning rate are depicted in Table 1.

TABLE 1
Control performance for the stable case

	Variance of e(t)	Variance of u(t)
$\eta_1 = 0.001, \varepsilon_1 = 0.01$	5.612	0.031
$\eta_1 = 0.002, \varepsilon_1 = 0.01$	5.662	0.037
$\eta_1 = 0.003, \varepsilon_1 = 0.01$	5.948	0.045
$\eta_1 = 0.004, \varepsilon_1 = 0.01$	6.176	0.052
$\eta_1 = 0.005, \varepsilon_1 = 0.01$	6.243	0.059

Now consider the unstable system with parameters $a_1=-2.1$ and $a_2=0.7$ (the parameters b_i and c_i are the same as before) and set the initial values of the weights as

356

$w_1(0)=0.5$, $w_2(0)=-1.0$, $w_3(0)=0.1$, $w_4(0)=-0.1$, $w_5(0)=-0.5$, $w_6(0)=-0.1$ and
$s_1(0)=1.0$, $s_2(0)=-0.5$, $s_3(0)=0.5$, $s_4(0)=0.1$, $s_5(0)=0.1$

The corresponding control results are shown in Fig.23 (where r=0.001, η_1=0.005 and ε_1=0.005) and the error and control variances are shown in Table 2. One observes that the generated control input stabilizes the plant very quickly at the transient stages. Of course in this case the initial weights were set closer to the true values. This means that the present controller is more effective when the normal plant is known but the plant parameters are varied during the operation.

TABLE 2
Control performance for the unstable plant

	Variance of e(t)	Variance of u(t)
$\eta_1 = 0.001$, $\varepsilon_1 = 0.001$	3.161	5.179
$\eta_1 = 0.002$, $\varepsilon_1 = 0.002$	3.066	4.912
$\eta_1 = 0.003$, $\varepsilon_1 = 0.003$	2.648	4.435
$\eta_1 = 0.004$, $\varepsilon_1 = 0.004$	2.278	4.114
$\eta_1 = 0.005$, $\varepsilon_1 = 0.005$	2.152	3.950

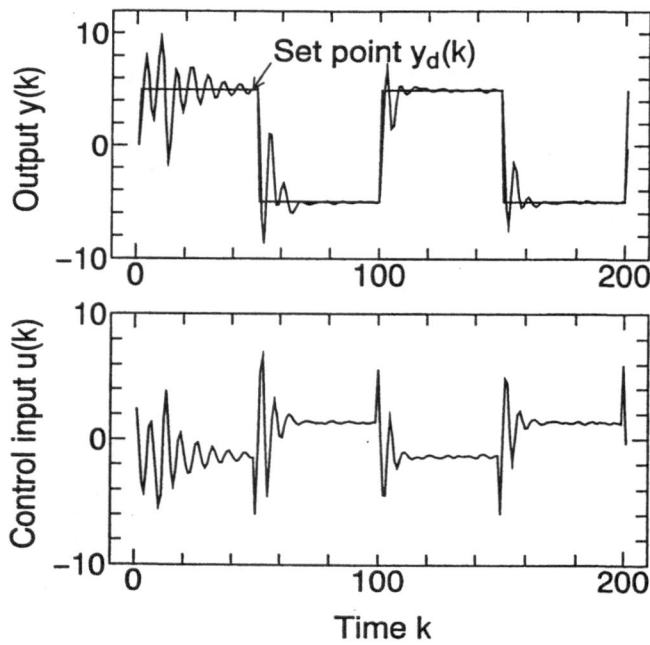

Fig. 23. Neuro control results for the unstable case.

4.3. Example 3: Supervised Neurocontrol of a Broom-balancing System

The broom-balancing (or inverted pendulum or cart-pole) system is a classic example of the application of NNs to control. The system consists of an inverted pendulum of length 2L and mass M mounted on a cart of mass M as shown in Fig.23A.

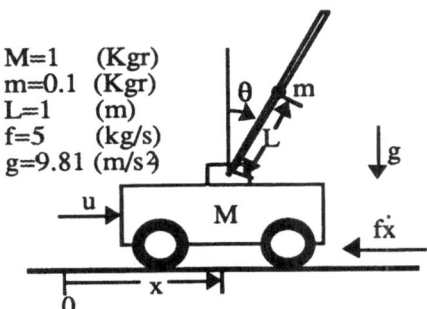

$$M=1 \quad (Kgr)$$
$$m=0.1 \quad (Kgr)$$
$$L=1 \quad (m)$$
$$f=5 \quad (kg/s)$$
$$g=9.81 \quad (m/s^2)$$

Fig. 23A. Broom-balancing system.

The goal of the controller is to keep the pole balanced (i.e. at the angle $\theta=0$) and maintain the cart at its origin (i.e. $x=0$) by applying a force u, in the horizontal direction. The original solution of the cart-pole problem was provided by Widrow and Smith [63-65] and Barto, Sutton and Anderson [12]. This was actually the first application of NNs to control, and involves all difficulties usually encountered in complex control problems, i.e. nonlinear, unstable and coupling features. Very often the cart-pole problem is used as a proof principle of NNs in control and robotics. Widrow and Smith used "bang-bang" control with a simple adaptive linear (ADALINE) NN. This device was trained by observing a human teacher operating the system manually. Guez and Selinsky [21,50] provide continuous rather than binary outputs.

The system dynamic equations are:

$$\ddot{\theta} = \frac{3}{4L} \left(g \sin \theta - \ddot{x} \cos \theta \right) \tag{62 a}$$

$$\ddot{x} = \frac{m \left[L (\sin \theta)\dot{\theta}^2 - \frac{3}{8} g \sin 2\theta \right] - f \dot{x} + u}{M + m \left(1 - \frac{3}{4} \cos^2 \theta \right)} \tag{62 b}$$

Clearly, the state vector z is $z = [x, \dot{x}, \theta, \dot{\theta}]^T$. The neurocontroller involves a feedforward NN with two hidden layers. The NN has four neurons in the input layer, sixteen neurons in the first hidden layer, four in the second hidden layer and one output. The activation functions of the input layer neurons are linear, and all others use sigmoidal functions. The state variables of the cart-pole system were used as inputs to the NN. As training data, periodic samples of the cart-pole system's state and the normalized control signal from the teacher were used. The NN is trained off-line via back error propagation. Training data is recorded over 1 to 2 minute period of operation. Guez and Selinski tried three types of neurocontrol, namely (i) supervised learning with a linear teacher, (ii) supervised learning with a nonlinear teacher, and (iii) supervised learning with a human as a teacher.

358

(i) Linear teacher

The cart-pole equations (62a,b) are linearized about $\theta=0$, i.e.

$$\ddot{x}=\frac{1}{M}(u-f\dot{x}), \qquad \ddot{\theta}=\frac{3}{4L}(g\theta-\ddot{x})$$

The teaching controller is

$$u=k_1x+k_2\dot{x}+k_3\theta+k_4\dot{\theta}$$

where $k_1=11.01$, $k_2=19.68$, $k_3=96.49$, and $k_4=35.57$.
The NN learned this linear control law after 20000 iterations of the back-error propagation
algorithm. The results are shown in Fig.24. The initial state used is $z=[2,0,-0.2,0]^T$.

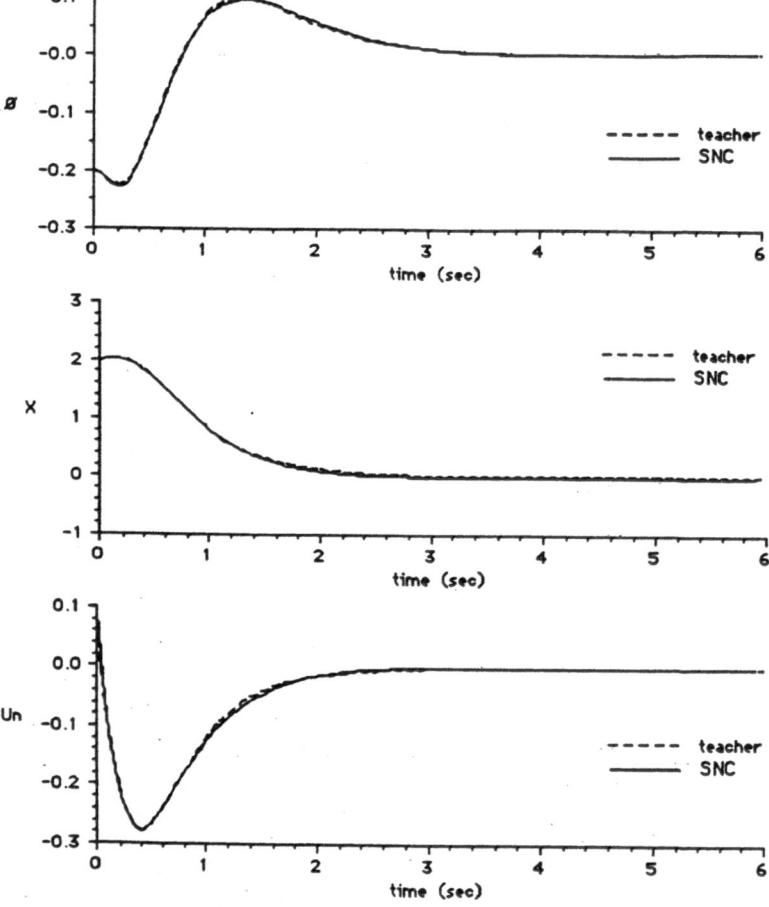

Fig. 24. Comparison of control results obtained by the linear teacher
and the supervised neurocontroller.

(ii) Nonlinear Teacher

The above neurocontroller, based on the linear teacher and the linearized model (about $\theta=0$), can stabilize the system only around the state space origin. To stabilize the system over the whole state space the feedback linearizing and decoupling technique is applied.

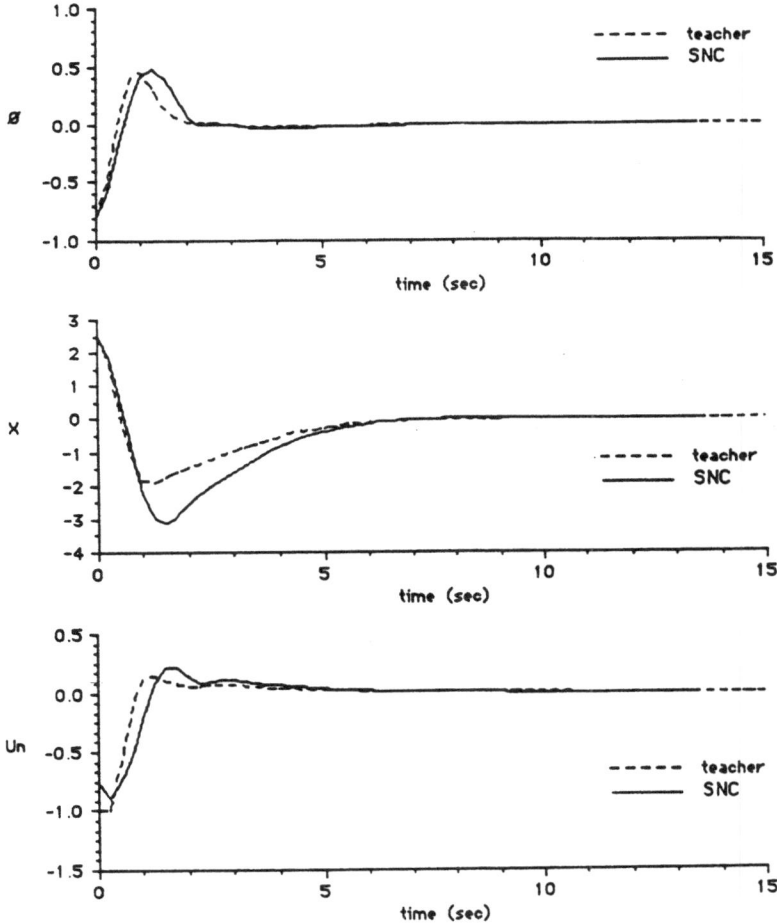

Fig. 25. Comparison of results obtained by the nonlinear teacher and neurocontroller.

Thus rewriting the dynamic equations (62a,b) as

$$\ddot{\theta} = \omega_1 - \omega_2 \ddot{x}, \qquad \ddot{x} = (\Omega_1 + u)/\Omega_2$$

360

where

$$\omega_1 = \frac{3}{4L} g \sin \theta, \qquad \omega_2 = \frac{3}{4L} \cos \theta$$
$$\Omega_1 = m\left[L \sin \theta \dot{\theta}^2 - \frac{3}{8} g \sin 2\theta \right] - f\dot{x}$$
$$\Omega_2 = M + m\left(1 - \frac{3}{4} \cos^2 \theta \right),$$

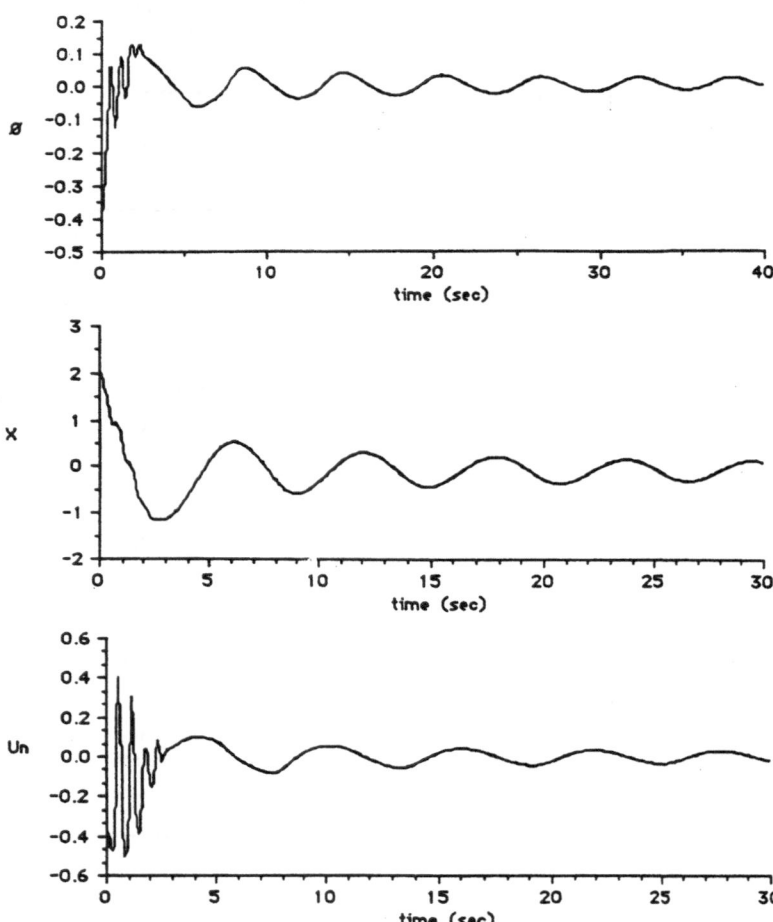

Fig. 26. Results of training a human supervised neurocontroller

the feedback linearizing and decoupling controller that must be used as automated teacher
is

$$u = \frac{\Omega_2}{\omega_1}\left[\omega_1 + k_1(\theta - \theta_d) + k_2\dot{\theta} + c_1(x - x_d) + c_2\dot{x}\right] - \Omega_1$$

with $k_1=25$, $k_2=10$, $c_1=1$ and $c_2=26$. Figure 25 shows the results obtained. The neurocontroller was fully trained in 80000 iterations of the back-error propagation algorithm. From Fig. 25 one observes that this nonlinear supervised neurocontroller has similar performance as the linear one (although not so much accurate).

(iii) Human Teacher

In this case examples generated by the responses of a human were used. Therefore the exact form of the control law is not known.
The human teacher was observing the cart-pole system on the computer screen and returning the system to the state space origin through a suitable control input device. Figure 26 shows the results obtained with learning length of 40000 iterations in the backward error propagation algorithm. One can see that the human supervised neurocontroller has learned to generate stable control from the examples of the human teacher responses.

4.4. Example 4: Feedback-Error Learning Robot Neurocontrol

The technique of Section 3.4. was applied to a 6-axis Kawasaki-Unimate PUMA 260 industrial robot, implementing the NN model in an HP 9000-300-320 microcomputer. Figure 27 shows a schematic of the three first axes of the robot. The dynamic model of the robot's three first axes is of the form of equation (34), where $D(q,t)$ and u(t) have the particular expressions:

$$D(q,t) = A+R(q), \qquad u(t) = CV(t) \tag{63}$$

with

$q = [\theta_1, \theta_2, \theta_3]^T$: the three first joint angles

$V = [V_1, V_2, V_3]$: voltage inputs to the three actuators

$A = diag[a_1, a_2, a_3]$: inertia of reduction gears and motors

R = inertia matrix of the robot

$C = diag[c_1, c_2, c_3]$: a voltage gain matrix

The inverse dynamics model of the robot is:
$$V = C^{-1}\{(A+R(q))\ddot{q} + h(q,\dot{q}) + B\dot{q} + g(q)\} \tag{64}$$

and represents a nonlinear mapping from q to V, i.e. it gives the required voltage input V(t) so that the desired joint angles' trajectory q(t) is realized. The nonlinear terms on the right-hand side of (64) are used as the subsystems appearing in Fig. 15.

362

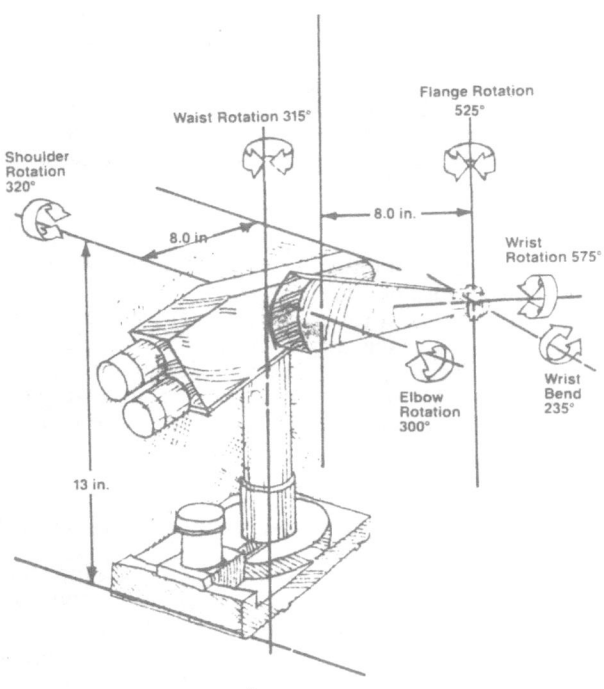

Fig. 27. Schematic of PUMA 260 robot

The number of subsystems for the first, second and third neurons are 17, 16 and 10 respectively. The synaptic weights w_{kj} in the inverse dynamics model of Fig.15 correspond to the coefficients of these nonlinear terms in the model (64). The voltage input to the motors (i.e. the feedback torque) is selected to be of the PD type {see (36)}, i.e.

$$V_{f_j} = K_{p_j} e_j + K_{v_j} \dot{e}_j, \qquad e_j = \theta_{d_j} - \theta_j \qquad (j = 1, 2, 3)$$

The overall experimental set-up (digital computer input-output data processor, control unit, and robot) is shown in Fig. 28. Photo encoders were used for measuring the joint angles at 10 msec time intervals, and the feedback and feedforward voltages (V_f, V_i) were computed with the same 10 msec inetrvals.

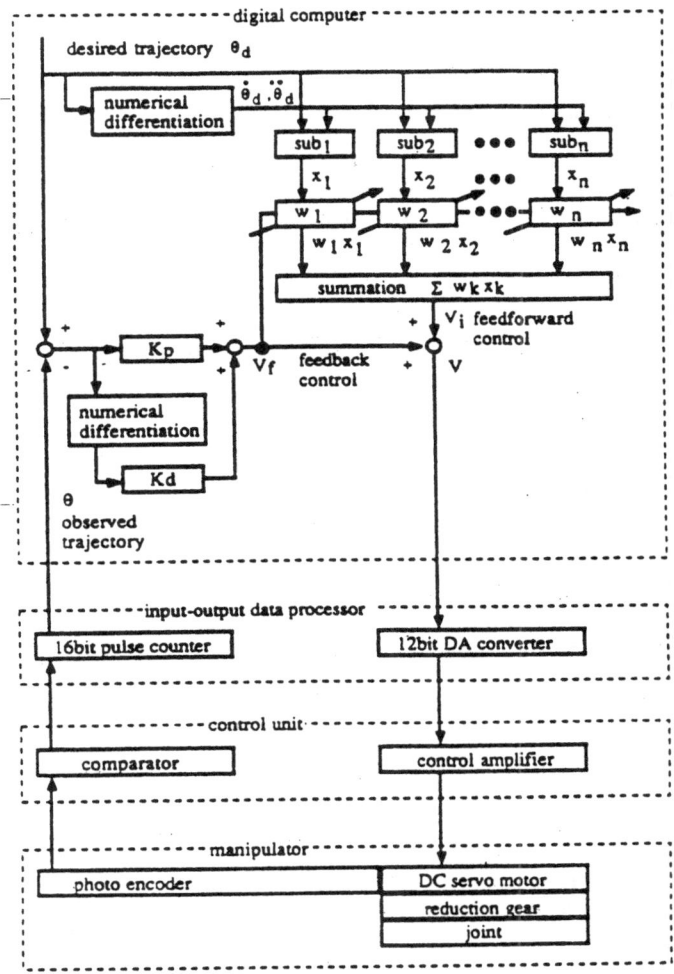

Fig. 28. Experimental set-up of feedback error learning control [35]

All joint angles derivatives required for computing the subsystems (i.e. $\dot{\theta}_d, \dot{\theta}$) were found numerically. The initial values for the synaptic weights were selected as zero, i.e. at the start of the learning experiment, the control of the robot was exerted only by the feedback voltage. In [35] the results of four experiments are reported, namely:

Experiment 1:
- Learning time constants: $\tau = 1000$ s (for all weights)
- Duration of desired trajectory: 6 sec (given 300 times to the control system repeatedly)
- Learning time: 30 min

- Feedback gains $K_p = 60$, $K_\upsilon = 1.2$

Experiment 2:
The time constants of the weights were selected such that their learning speeds to be more similar and thus obtained an improved overall learning speed (see Table 3).

Experiment 3:
A poorer feedback control (smaller feedback gains) was selected, i.e. $K_p = 20$, $K_\upsilon = 0.4$. All other conditions are the same as in Experiment 2.

TABLE 3

Time constants τ for the 43 weights in Experiment 2

	Joint 1	Joint 2	Joint 3
1	1000	1000	1000
2	1000	1000	1000
3	1000	1000	2000
4	3000	1000	1000
5	3000	1000	1000
6	3000	1000	1000
7	3000	1000	1000
8	700	1000	700
9	700	1000	500
10	700	1000	500
11	700	1000	
12	1000	1000	
13	2000	1000	
14	2000	700	
15	2000	100	
16	700	100	
17	700	100	

Experiment 4:
A payload of 315 gr weight was attached to the robot hand 15 min after the begining of the experiment during the 30 min learning period. All other conditions are the same as in Experiment 2.

Fig. 29 shows the evolution of the mean square error $\overline{(\theta_{d2} - \theta_2)^2}$ of the second joint obtained in Experiment 1 and Experiment 2. One can observe the substantial improvement of Experiment 2 over Experiment 1. This shows the beneficial effect of the right selection of the time constants for the modification of the synaptic weights. Actually, systematic ways for this selection are required.

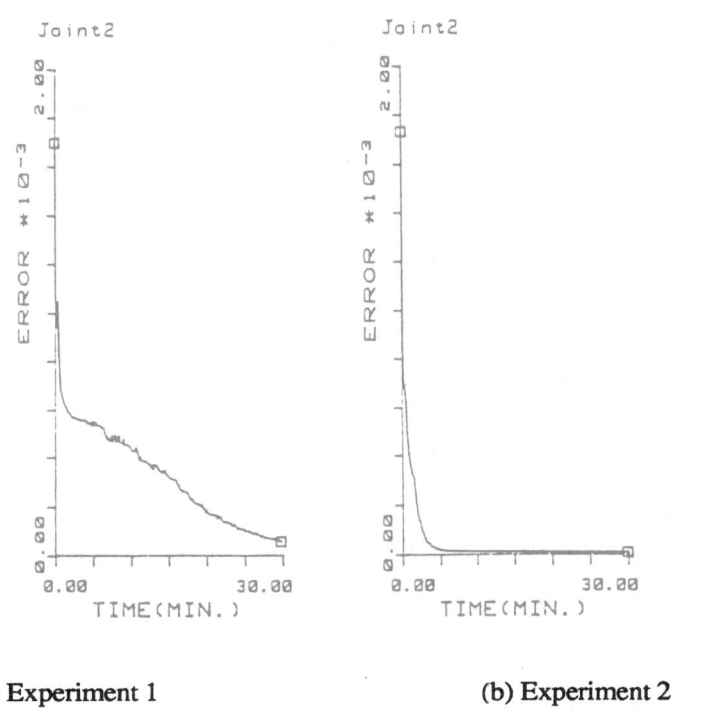

(a) Experiment 1   (b) Experiment 2

Fig. 29. Mean square error evolution for θ_2 (rad) during 30 minutes of learning

Figure 30 compares the control performance before (Fig.30a) and after (Fig.30b) 30 minutes of learning in Experiment 3. One can see that when the control depends only on feedback, overshoots and oscillation of the realized trajectory are observed (a), but after the 30 min of learning the desired and actual trajectory (almost) coincide. This experiment shows that the NN can learn even with a poor feedback signal.

Finally Fig. 31 shows the change of the mean square error $\overline{(\theta_{d2} - \theta_2)^2}$ of the second joint angle during a 30 min learning time. One observes a temporary increase of the error (rad) at the time of the payload attachment. This shows that the NN possesses very good adaptability to variations of the dynamic features of the controlled robot.

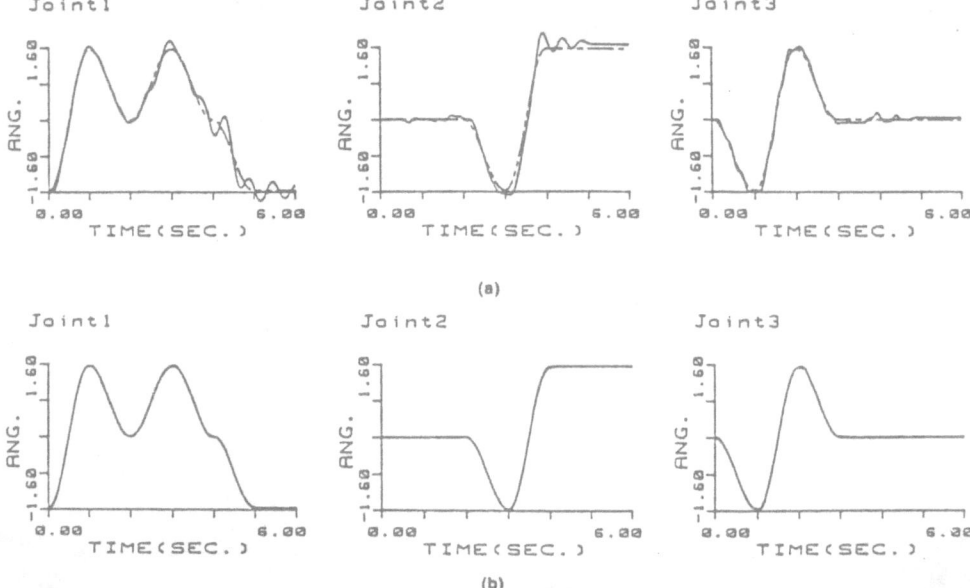

Fig.30. (a) Joint angles during 6 sec of a single repetition of the training movement
pattern before learning
(b) The same as (a) but after 30 minutes of learning

From these and other experiments one can draw the following general conclusions:

(i)   The mean square error of the trajectory decreases (i.e. the control behavior is improved) considerably during learning.
(ii)   As the learning proceeds, the inverse-dynamics NN model gradually takes the place of the feedback loop as a primary controller.
(iii) The NN controller has adaptibility to sudden changes in the robot dynamics.
(iv)   The NN has the ability to generalize learned motions. This means that, once the NN learns some motion, it can control quite different and faster motions.

4.5. Example 5: Iterative Robot Neurocontrol

The iterative neurocontroller of Section 3.5 has been applied to a simulated 2-link robotic model [58] (Fig.32). The full equations of this robotic model are:

$$\begin{bmatrix} \frac{1}{3}m_1L^2 + \frac{4}{3}m_2L^2 + m_2c_2L^2 & \frac{1}{3}m_2L^2 + \frac{1}{2}m_2c_2L^2 \\ \frac{1}{3}m_2L^2 + \frac{1}{2}m_2c_2L^2 & \frac{1}{3}m_2L^2 \end{bmatrix}\begin{bmatrix} \ddot{q}_1 \\ \ddot{q}_2 \end{bmatrix}$$

$$
+\begin{bmatrix} \frac{1}{2}m_2s_2L^2\dot{q}_2^2 - m_2s_2L^2\dot{q}_1\dot{q}_2 \\ \frac{1}{2}m_2s_2L^2\dot{q}_1^2 \end{bmatrix} + \begin{bmatrix} -\frac{1}{2}m_1gLc_1 - \frac{1}{2}m_2gLc_{12} - m_2gLc_1 \\ -\frac{1}{2}m_1gLc_{12} \end{bmatrix} = \begin{bmatrix} u_1 \\ u_2 \end{bmatrix}
$$

where $c_1 = \cos(q_1)$, $c_2 = \cos(q_2)$, $c_{12} = \cos(q_1 + q_2)$
       $s_1 = \sin(q_1)$, $s_2 = \sin(q_2)$.

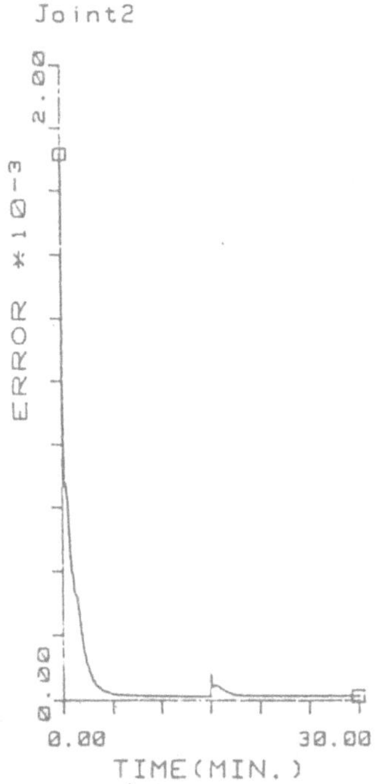

Joint2

Fig. 31. Evolution of mean square error θ_2 in Experiment 4

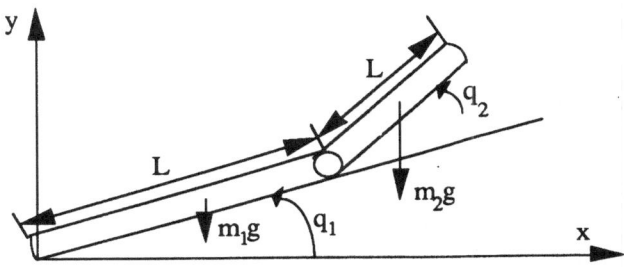

Fig. 32. 2-link planar robot

The desired joint-angle trajectory is described by q

$$q_{1d}(t) = 0.5t^2 + 2\sin(2.5t), \qquad q_{2d}(t) = t + \sin(5t)$$

The experiment was carried out using a NN with six neuron nodes at the input layer, 25 neuron nodes at the first hidden layer, 35 neuron nodes at the second hidden layer, and two neuron nodes at the output layer (symbolically N6,25,35,2). The robotic parameters in the model used were selected different than the true ones as follows: mass of link 1 = 1.7 Kg (true value = 2Kg), mass of link 2 = 3.1 Kg(true value = 2.5 Kg) and length of link L = 0.8m (true value = 0.6 m). Fig. 33 shows the trajectory (position, velocity) resulting from the control that is based on the approximate model which is used to bring the robot into the neighbourhood of the desired trajectory. The performance is of course not good, but it is good enough to start the learning process, out of which the neurocontroller will improve the performance by itself and will arrive at a satisfactory tracking precision after a few trials. The results of the training after two lessons are shown in Fig. 34. One observes that the control torque produced by the inverse dynamic controller and the torque learned by the NN are almost identical which shows that the training results are excellent.

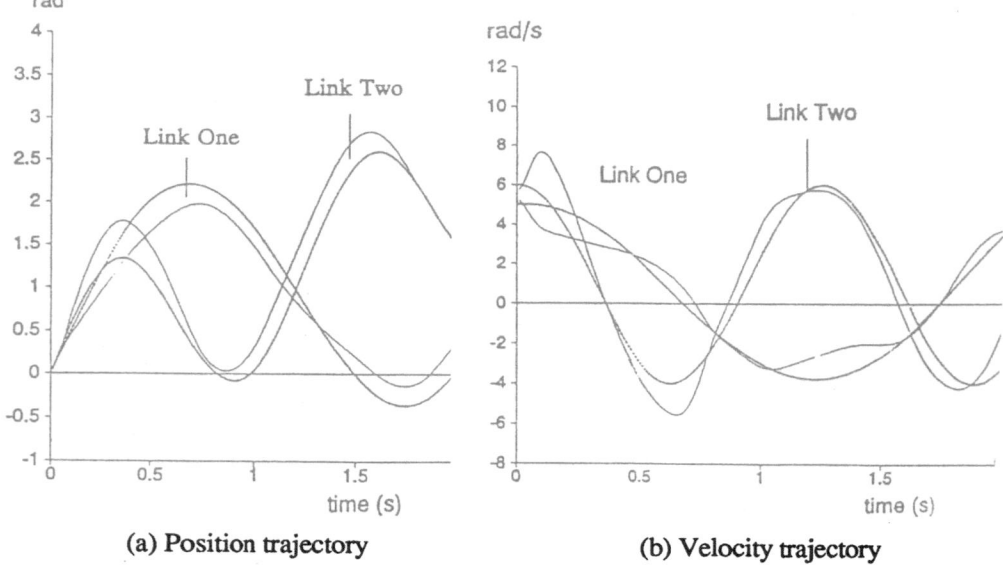

(a) Position trajectory   (b) Velocity trajectory

Fig.33. Trajectory resulting via the inverse dynamic scheme using the approximate robot model. (dotted line = desired trajectory, solid line = actual trajectory)

After the end of the NN training, the iterative learning controller was implemented as decscribed in Sec. 3.5. Here, only the feedback PD controller needs on-line computation, whereas the control signals produced by the NN are computed beforehanf off-line. The simulation results obtained at the first trial are shown in Fig. 35, whereas Fig. 36 shows the results of the third trial. One observes the improvement that takes place on the tracking precision as the number of trials increase.

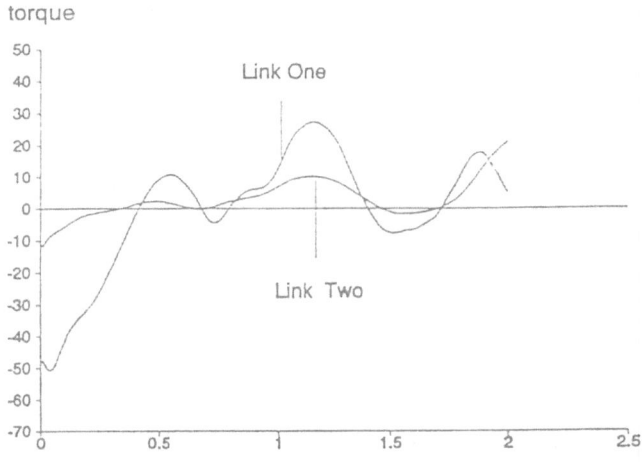

Fig. 34. Training results of the neurocontroller
(dotted lines and solid lines almost coincide)

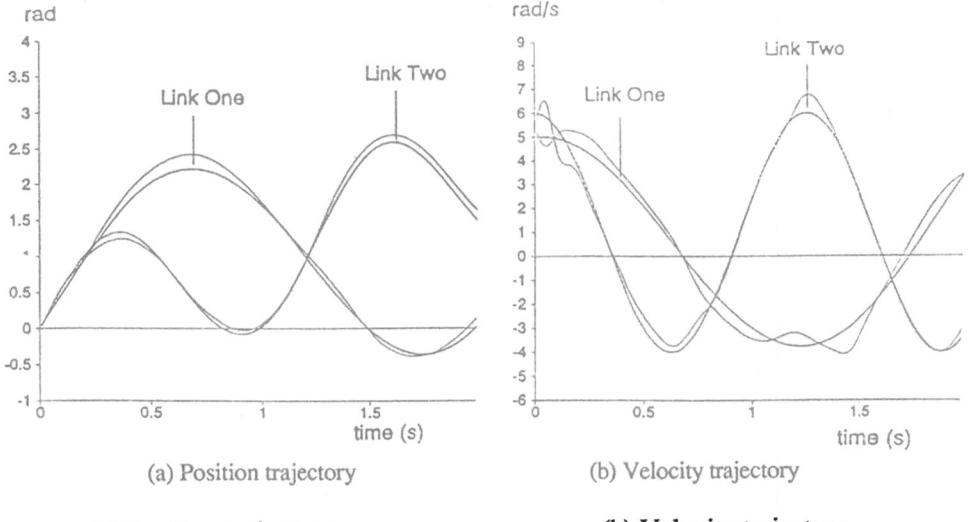

(a) Position trajectory

(b) Velocity trajectory

Fig. 35 Simulation results of the first trial

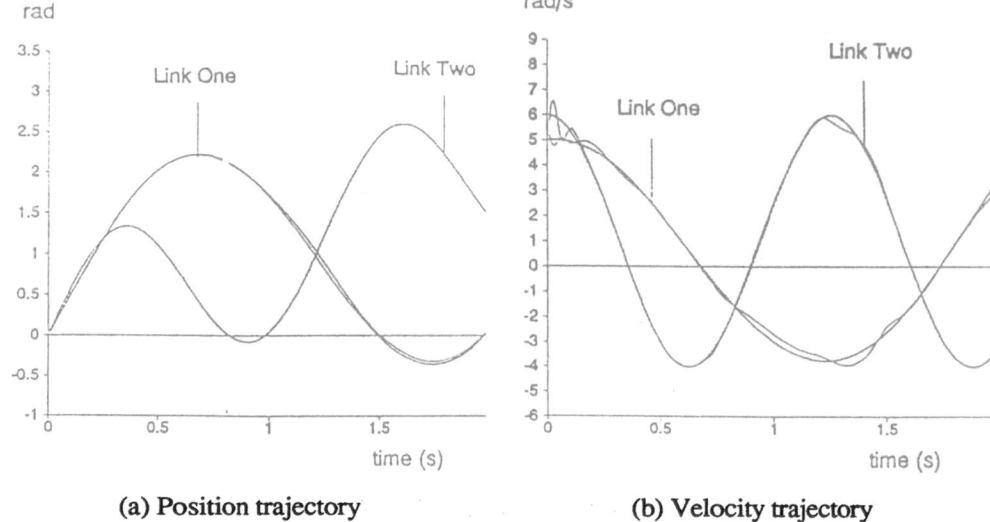

(a) Position trajectory
(b) Velocity trajectory

Fig. 36 Simulation results of the third trial

Fig. 37 depicts the absolute sums of the position tracking errors and the velocity tracking errors. Initially, the tracking performance improves very quickly, but the improvement slows down as time advances.

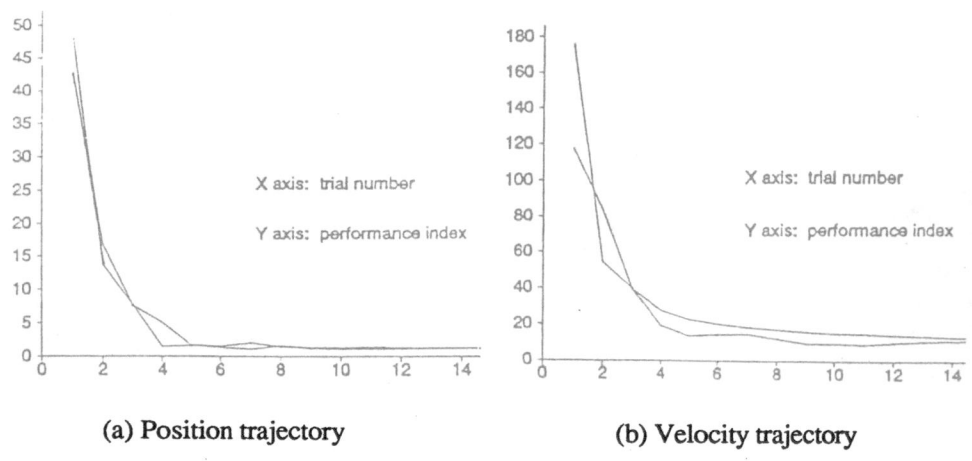

(a) Position trajectory
(b) Velocity trajectory

Fig. 37 Absolute sums of tracking errors versus trial number

As mentioned in sec.3.5 it is expected that the present iterative learning neurocontroller has better robustness and is less sensitive to noise than the traditional iterative learning controllers. This was really verified in the experiments. For example, Fig. 38 depicts the simulation results int he presence of 10% velocity measurement noise and 5% position measurement noise, while Fig.39 shows the measured position and velocity.

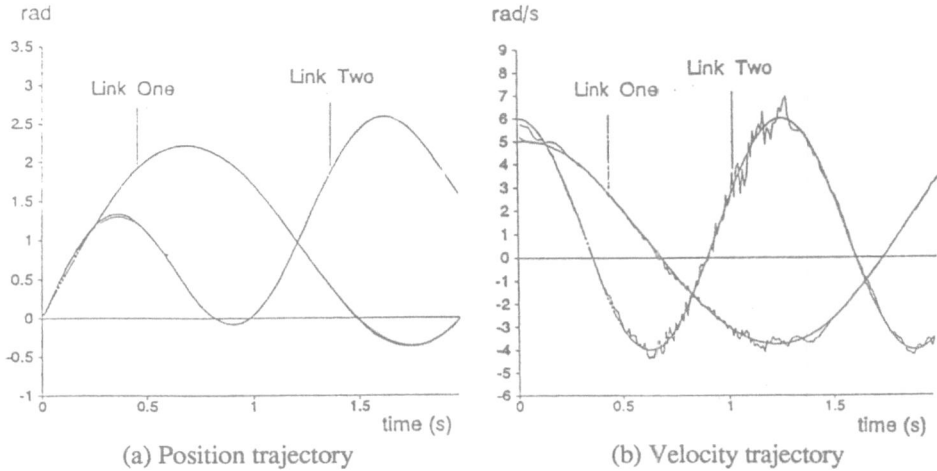

(a) Position trajectory   (b) Velocity trajectory

Fig. 38 Robot trajectories in the presence of noise.

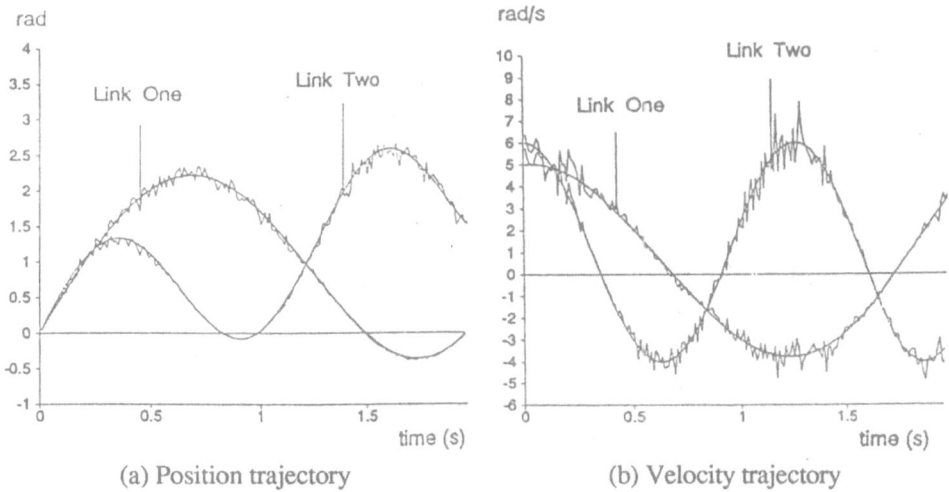

(a) Position trajectory   (b) Velocity trajectory

Fig. 39 Measured position and velocity.

Finally, Fig.40 shows the feedback control signal (u_{fb}) and the feedforward control signal versus time. It is easily seen that the control action is gradually shifted from the feedback controller to the feedforward as claimed in Sec. 3.5.

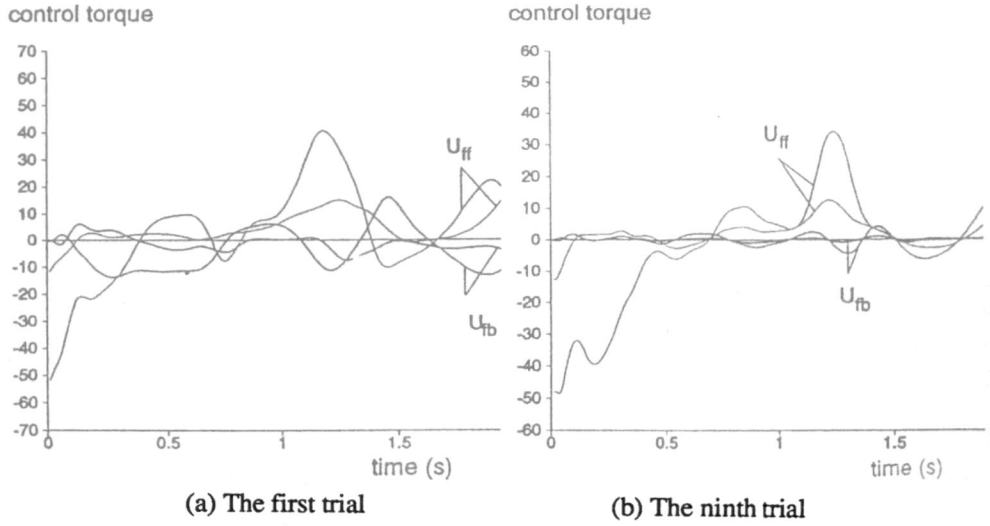

(a) The first trial   (b) The ninth trial

Fig. 40 Feedback and feedforward control signals versus learning time.

4.6 Unsupervised robot neurocontroller using hierarchical NN

The technique described in Sec. 3.3 was also applied to a simulated two-axis robot (Fig. 41) [21].

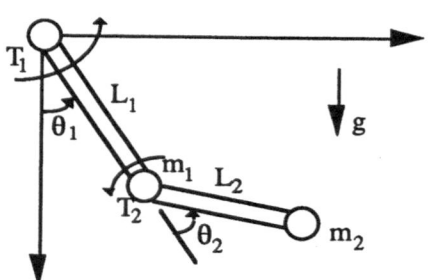

Fig. 41. A 2-link planar robot
($m_1=m_2=10$kg, $L_1=L_2=1$m, $b_1=b_2=5$kg/s).

The dynamic model of this robot, written so as to provide general forms of linearly separable nonlinear subsystems, is

$$w_{11}\ddot{\theta}_1 + w_{12}\cos(\theta_2)\ddot{\theta}_1 + w_{13}\ddot{\theta}_2 + w_{14}\cos(\theta_2)\ddot{\theta}_2$$
$$+ w_{15}\sin(\theta_2)\dot{\theta}_1\dot{\theta}_2 + w_{16}\sin(\theta_2)\dot{\theta}_2^2 + w_{17}\dot{\theta}_1$$
$$+ w_{18}\sin(\theta_2) + w_{19}\sin(\theta_1 + \theta_2) = T_1$$
$$w_{21}\ddot{\theta}_1 + w_{22}\cos(\theta_2)\ddot{\theta}_1 + w_{23}\ddot{\theta}_2 + w_{24}\sin(\theta_2)\dot{\theta}_1^2$$
$$+ w_{25}\dot{\theta}_2 + w_{26}\sin(\theta_1 + \theta_2) = T_2$$

The NN combines linearly the subsystems as in Example 4 (see Sections 3.4 and 4.4), but here use is made of a-priori knowledge of the robot dynamics to excite only specific subsystems of the dynamics. In this way the dimensionality of the synaptic weight space to be searched is reduced, a fact that reduces learning time and increases the likelihood of the weights, (w_i's) convergence to optimum values. The simulation results have actually supported this argument.

The optimum values of the weights as computed from the robot equations using the values shown in Fig. 41 are:

w_{11}	w_{12}	w_{13}	w_{14}	w_{15}	w_{16}	w_{17}	w_{18}
30	20	10	10	-20	-10	5	196.2

w_{19}	w_{21}	w_{22}	w_{13}	w_{24}	w_{25}	w_{26}
98.1	10	10	10	10	5	98.1

The training used an overall of 240390 iterations of the LMS algorithm and resulted in the following weight values

w_{11}	w_{12}	w_{13}	w_{14}	w_{15}	w_{16}	w_{17}	w_{18}
29.51	19.28	9.79	9.47	-19.28	-9.97	4.61	196.19

w_{19}	w_{21}	w_{22}	w_{13}	w_{24}	w_{25}	w_{26}
98.10	10.06	10.20	10.12	10.22	5.01	98

Figures 42 to 45 depict the controller performance as the identified compensation terms $C(\theta,\dot{\theta})$ are added to the linear controller

$$T = K_p e + K_v \dot{e} + C(\theta,\dot{\theta})$$

where $e = \theta - \theta_d$, $K_p = \text{diag}[400,100]$ and $K_v = \text{diag}[40,20]$.

Fig.42 gives the performance before any learning. Fig. 43 gives the performance with the gravitational terms learned and added ($C(\theta,\dot{\theta}) = \hat{g}(\theta)$ where $\hat{g}(\theta)$ is the output of the $g(\theta)$ NN). Fig.44 depicts the performance when the gravitational, centripetal, Coriolis and viscous friction compensation term is learned and added ($C(\theta,\dot{\theta}) = \hat{H}(\theta,\dot{\theta}) + \hat{g}(\theta)$, where $\hat{H}(\theta,\dot{\theta})$ is the output of the $H(\theta,\dot{\theta})$ NN). Finally Fig.45 shows the controller performance when all compensation terms are learned and applied. The overall control is now the following

$$T = \hat{D}(K_p e + K_\upsilon \dot{e}) + \hat{H}(\theta,\dot{\theta}) + \hat{g}(\theta)$$

where $\hat{D}(.)$ is the output of the $D(\theta,t)$ compensation NN (see Eq. 44).

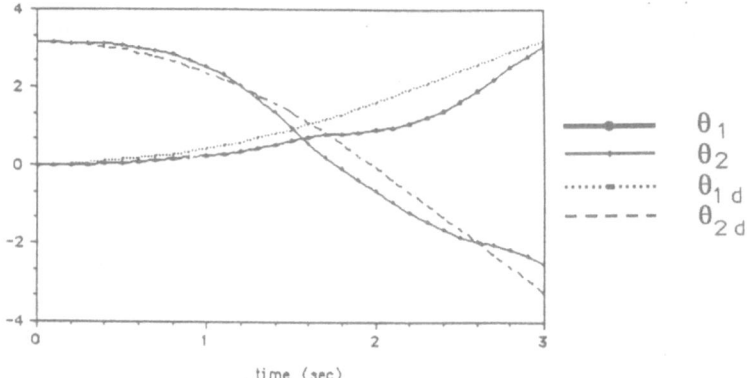

Fig. 42 Controller before learning: $C(\theta,\dot{\theta}) = 0$.

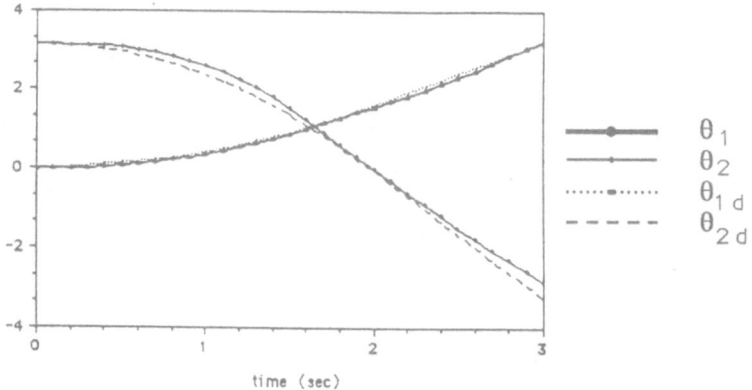

Fig. 43. Controller with $C(\theta,\dot{\theta}) = \hat{g}(\theta)$.

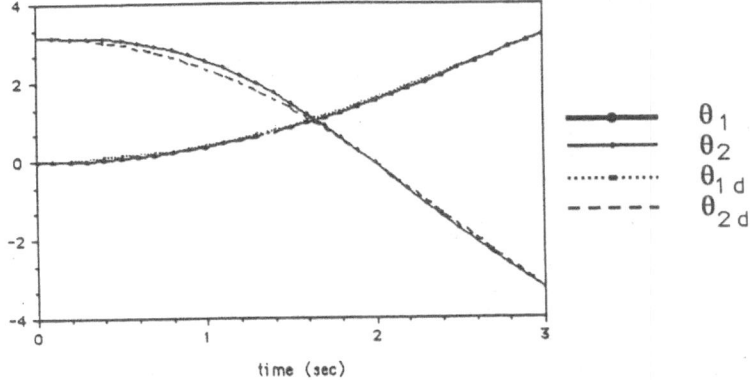

Fig.44. Controller with $C(\theta,\dot{\theta}) = \hat{H}(\theta,\dot{\theta}) + \hat{g}(\theta)$.

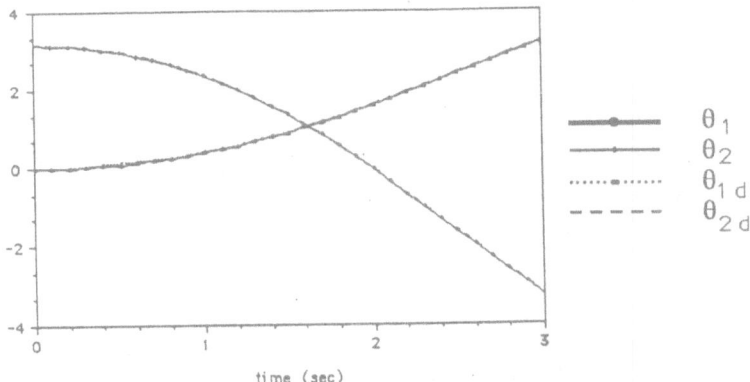

Fig. 45. Controller with full compensation.

5. CONCLUSIONS AND DISCUSSION

This chapter has presented the main issues of neural control and its application to robotic systems. Although the theory of NN's is at an advanced level of maturity, their application to control a robotic system is still at a rudimentary level, leaving much room for further improvements and developments. However as NN's are developed and understood further, these application problems will be solved more effectively. In most cases the MLP model is employed, and especially a NN with three layers. As we have seen in the appendix three layers are sufficient for representing any nonlinear mapping.

This chapter does not of course exhaust all results of NN approaches to robot control. However, the results are quite representative giving a very good picture of the present status of the field, and undoubtedly can be regarded as proof-of principle that NN's can solve robot control problems. Particularly, the chapter has started with a look at the general NN architectures for control, which have then been specialized for robot control (supervised and unsupervised). Robot control was treated through hierarchical (3-layers and 4-layers) NN's through the FELA technique, the minimum torque and the improved iterative learning approach. A set of numerical examples was included in the chapter in order to illustrate the kind of results that are obtained by neurocontroller. The improved iterative learning controller was shown to have a much faster rate of convergence and better robustness than the common iterative learning controller. A method was also outlined (Sec.3.3) where by use of a-priori knowledge converts a previously supervised learning technique to an unsupervised one. It was also shown that NN's can be taught, using a supervised learning paradigm, to produce stable control for a nonlinear dynamic system. The results of training with a human teacher show that stable control can be learned even if the exact form of the control law is unknown. Actually, the human trained adaptive controller proved to be more robust to parameter variations than a model based controller. The general result is that if some a priori knowledge of the process under control and its environment is available, no matter how little, can be effectively used to the structural design of neurocontrollers. A neural learning controller needs retraining when the robotic trajectory is changed. This is one drawback compared with conventional adaptive and robust controllers. In future, as NN's are improved and better understood, the solutions to the problems considered in this chapter will be much more competitive with conventional solutions. Open issues for investigation include stability, expert NN system design, and proper NN architectures.

6. APPENDIX: A BRIEF LOOK AT NEURAL NETWORKS

For the convenience of the reader who is unfamiliar with neural networks (NN) we give here a brief description of the mechanics of these networks [1-4,66].

The neural networks or connectionist models attempt to achieve high performance through an intensive interconnection of simple processing (computing) elements. Each node of the NN is a synapsis of inputs that passes the result through a nonlinear function (Fig.46).

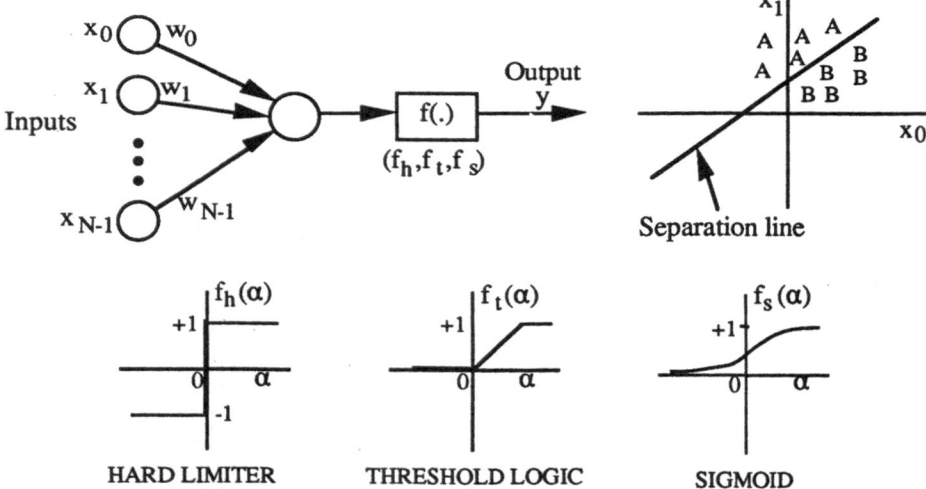

Fig. 46. Simple NN structure (perceptron) and usual nonlinear NN functions.

Here, by the term synapsis we mean summation of the N inputs x_i via suitable weights w_i. Fig.46 shows three standard non-linear functions that are used in NNs. More complex nodes can involve time integrations or other types of relations and more complex synapses than summation that may exhibit learning characteristics.

Neural nets are characterized by three major features, namely: their topology, the characteristics of their nodes, and their learning mechanism. The learning mechanism starts from an initial set of synaptic weights and suggests how these weights can adapt themselves during learning so as to achieve improved performance. Current research in NNs deals with the learning process and with the design procedures of NNs.

The six most important NNs are the following: single layer perceptron (SLP), multilayer perceptron (MLP), Kohonen network, Hopfield network, Hamming network and Carpenter / Grossberg network. A first classification of them is according to the type of their inputs (continuous, binary) and a second one is according to the form of learning (supervised or unsupervised). The classification of the above six NNs is as shown in Fig. 47.

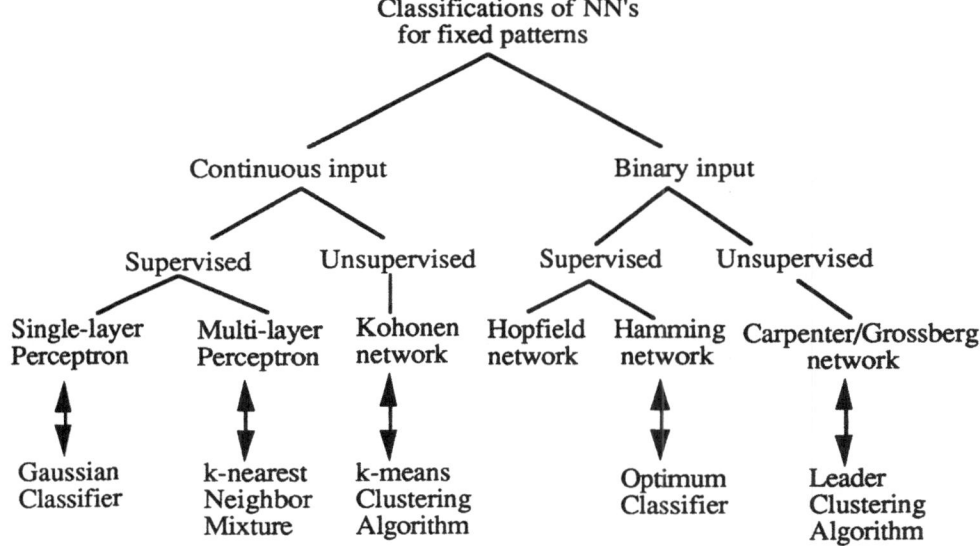

Fig. 47. Classification of the primary NNs

Due to space limitation we shall describe here the single-layer perceptron (SLP), the multi-layer perceptron (MLP), which is mostly used in control and robotics, and the Hopfield network that works with binary inputs.

6.1. Single Layer Perceptron (SLP)

The SLP has the ability to learn to recognize simple patterns, i.e. to decide if the input belongs to the A or the B class (Fig. 46). The network node computes a weighted sum, subtracts a threshold and passes the result through a hard limiting element so as the output y to obtain the values +1 or -1. If y=+1 then the decision is that the input belongs to class

A, and if y=-1 the input belongs to class B. The right hand side of Fig.46 shows the decision areas for a 2-input SLP. The operation of the SLP is exactly to form the separation line between the A and B input areas.

The learning mechanism of the SLP is the following.

Step 1:
Initialize the weights w_i and the threshold h, i.e. give to $w_i(0)$, i=0,1,...,N-1, and h small positive random values.

Step 2:
Present the inputs x_1, x_2,...,x_{N-1} and the value of the desired output $x_0=d(t)$.

Step 3:
Compute the actual output

$$y(t) = f_h[\sum_{i=0}^{N-1} w_i(t)x_i(t) - h]$$

Step 4:
Update the weights using the formula (δ-formula)
$w_i(t+1) = w_i(t) + \eta\ \delta(t)\ x_i(t)$, i = 0,1,...,N-1
where
$\delta(t) = d(t) - y(t)$,

$$d(t) = \begin{cases} +1 & \text{if } x_i \text{ belongs to class A} \\ -1 & \text{if } x_i \text{ belongs to class B} \end{cases}$$

and $0 < \eta \le 1$ is a gain fraction (learning rate gain).

Step 5:
Repeat from Step 2.

As one can see, initially we give arbitrary values to the weights and the threshold, and then the weights are updated so as to improve the separation line. Finally this line can separate the two classes A and B. The problem is that the separation line is not always a straight line. To overcome this difficulty we minimize the mean square error between the desired and the actual output (LMS method). The LMS algorithm of Widrow and Hoff replaces the *hard limiting function* by the *threshold logic* giving the value 1 to class A and 0 to class B. Therefore, in this case the NN classifies the input into class A if the output is greater than 0.5.

6.2. Multi-Layer Perceptron (MLP)

MLPs are extension of SLPs. They are feedforward NNs with continuous inputs and the layers that contain are distinguished in input layer, output layer and intermediate (hidden) layers. The main feature of MLPs is that the direct interconnection of nodes of the same layer is not allowed. Also the bypass of layers is not permitted. For example, when connecting an input node with an output node the bypass of hidden layers is prohibited (Fig. 48).

The superiority of MLPs over SLPs is illustrated in Fig. 49, where hard-limiting functions are assumed to be used in the nodes. The issue is to separate the decision areas A and B. The SLPs separate the decision areas by a straight line, and so they cannot approximate functions line the third XOR or like the functions required by general area schemes (last column of Fig. 49). With 2-layer NNs one can separate the decision space

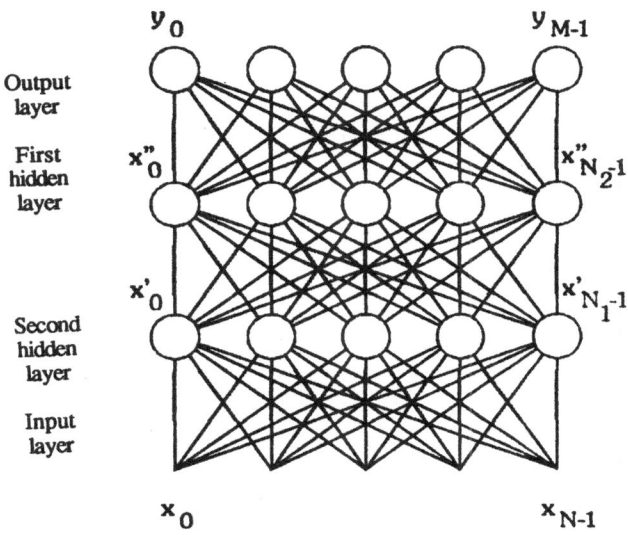

Fig. 48. An MLP with three layers.

by convex polygonal lines or schemes. The number of these lines depend on the number of nodes in the first layer.

The operation of MLPs is as follows. Up to the first layer they operate as many SLPs, the number of which is equal to the number of the nodes of the first layer. These separate the decision plane in semiplanes with straight lines. The section of these semiplanes is selected via the node of the second layer that performs an AND operation and thus the convex polygonal regions are formed. These sections however do not cover all cases, and so a third layer is needed which joints the resulting areas (by the second layer) through an OR operation. In this way any convex or nonconvex area can be obtained that covers the decision space. Thus, three layers are sufficient for the separation of any area within the decision space.

Any region in the decision space is bounded by a continuous function of N variables, where N is the dimension of the decision space. This function can be strongly nonlinear. The problem in NNs is to approximate these functions which in most cases are unknown. Kolmogorov has shown that *every continuous N-dimensional function can be computed by the linear summation of nonlinear continuously increasing functions* of a single variable. From this it follows that a NN with $N(2N+1)$ nodes and three layers can compute any separation funtion of the N-dimensional decision space.

The design of 3-layer NNs involves the selection of the number of nodes at each layer, the selection of the function of each node, and the selection of the learning algorithm for the proper tuning of the weights. If N is the dimension of the decision space, we need 2N nodes at the first layer. The number of nodes at the second layer depends on the form of the decision areas and in the worst case it is equal to the number of the unconnected areas of the input distribution. The number of nodes at the third layer depends on the number of the pattern classes at hand. As node function we usually use the sigmoid functions $f(x)=1/[1+e^{-(x-h)}]$ (see Fig. 46). Finally, as learning algorithm one can use the following back-propagation algorithm or its variations and improvements.

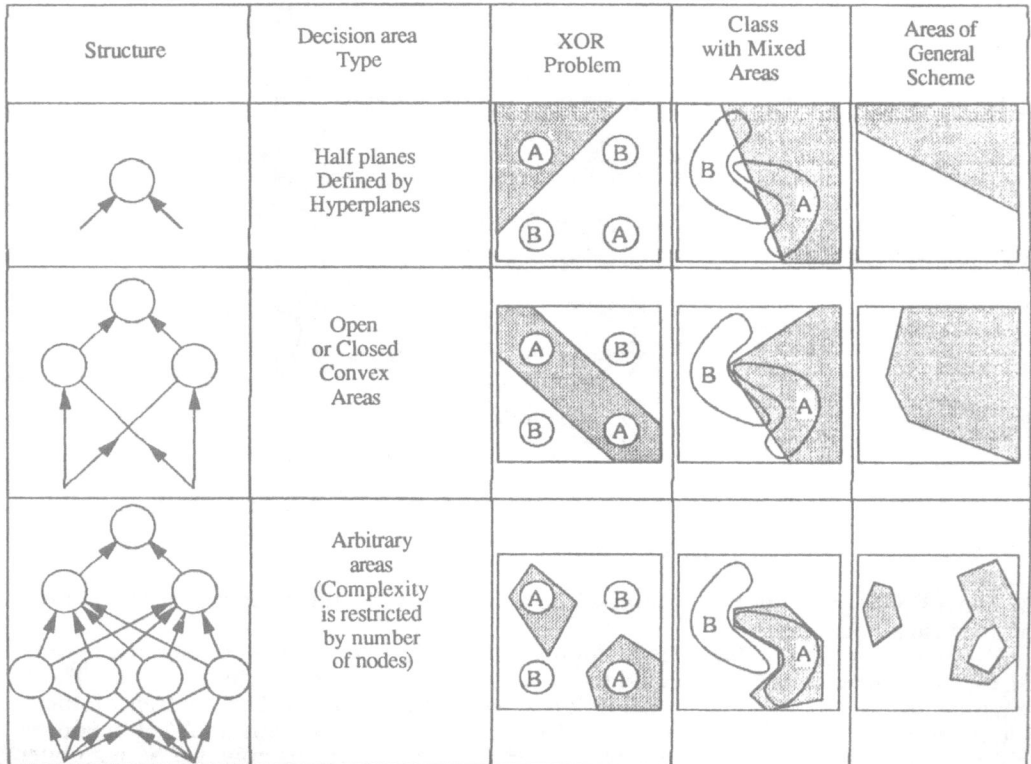

Structure	Decision area Type	XOR Problem	Class with Mixed Areas	Areas of General Scheme
	Half planes Defined by Hyperplanes			
	Open or Closed Convex Areas			
	Arbitrary areas (Complexity is restricted by number of nodes)			

Fig. 49. Capability of MLPs

Back-Propagation Algorithm

Step 1:
Initialize the weights and thresholds to small positive random values.
Step 2:
Present the input values $x_1, x_2,...,x_{N-1}$ and the desired output values $d_1, d_2,...,d_{M-1}$. If the NN is used as classifier, then put all d_i's to zero, except the one that expresses the class in which the input that must take the value 1 belongs.
Step 3:
Compute the actual outputs, using the sigmoid function and carrying out the computations that are shown in Fig. 48.
Step 4:
Update (improve) the weights starting from the output nodes and going backwards, using the formula

$$w_{ij}(t+1) = w_{ij}(t) + \eta \ \delta_j \ x_i'$$

where $w_{ij}(t)$ is the synaptic weight of the connection of node i with the node j of the next layer at time t, x_i' is the output of node i (or the input i of the NN), η is a gain factor (learning rate coefficient), and δ_j is an error term for node j defined as:

$$\delta_j = y_j (1-y_j) (d_j-y_j)$$

with y_j being the actual and d_j the desired output. If node j is a node of a hidden layer, then

$$\delta_j = x_j'(1 - x_j') \sum_k \delta_k w_{jk}$$

where k extends over all nodes of the previous layers. The thresholds of the internal nodes are adapted in a similar way, assuming that they are connection weights in subsidiary inputs of constant value. To speed up convergence one can add one more term as:

$$w_{ij}(t) = w_{ij} + \eta\, \delta_j x_i' + \alpha[w_{ij}(t) - w_{ij}(t-1)],\ 0 < \alpha < 1$$

Step 5:
Repeat from Step 2.

Actually, there are a number of problems with using MLPs. First, there is no guarantee of convergence to a local minimum, since the gradient used in the weight update is the gradient for a given training sample and may not represent the overall gradient. One can use a method called *batching* to compute an error based on many samples and so giving a more correct estimate of the gradient. This was done in Section 3.5 when forming the index (50).

6.3. Hopfield Network

The Hopfield network has many variations [2,67]. It accepts binary inputs (e.g. black and white images, ASCII character representation, etc) and is not suitable for applications where digitization is not advisable. Hopfield networks are used for associative memory and combinatorial optimization. The latter property has been the most frequently used for robotics applications. The Hopfield network is a 1-layer NN with feedback connected in a crossbar topology (Fig. 50). The hard-limiting nonlinearity is usually used, although other functions are also employed. The output equation of the network is:

$$y(t + 1) = f(Hy(t) + x)$$

The energy function under minimization is

$$J = -\frac{1}{2} y^T Hy - y^T x$$

where $x = [x_1, x_2,...,x_N]^T$ and $y = [y_1, y_2,...,y_N]$, H the matrix of interconnection weights, for the nonlinearity (hard-limiter), and J the performance index (energy function).

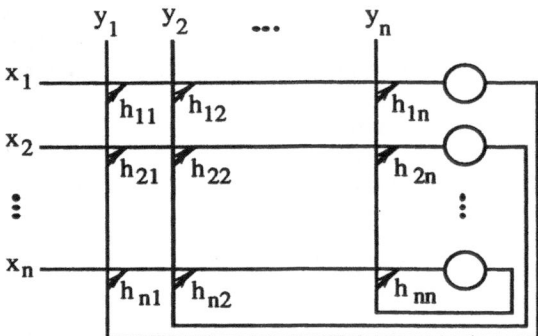

Fig. 50. The simplest variation of the Hopfield network

Conceptually, the output updating equation serves to perform a minimization of J. This property is actually used for solving combinatorial optimization problems [67,68]. Although at first glance the performance index J above appears to be quadratic, it is not since f(.) is nonlinear. Therefore, J can have multiple minima. As a combinatorial optimization example we mention here the work of Wilson and Pawley [68] where they treated the travelling salesman problem. In this problem a salesman must visit n cities so that to minimize the distance travelled over the entire tour. For solving this problem a performance index is defined so that the minimization of this function corresponds to a solution of the problem.

Let us now discuss the Hopfield network of Fig.51 which is used as an associative memory. This network has N nodes with function of the hard limiter type, with binary inputs and output values +1 and -1. The output of each node is applied to all other nodes through the weights w_{ij}.

The learning algorithm of this NN is the following.

Step 1:
Determine the weights at the interconnections

$$w_{ij} = \begin{cases} \sum_{s=0}^{M-1} x_i^s x_j^s & , \ i \neq j \\ 0 & , \ i=j \ , \ 0 \leq i \ , \ j \leq M-1 \end{cases}$$

where w_{ij} is the weight of the connection of node i with node j, and x_i^s is the element of the example pattern of class s.

Step 2:
Initialize with respect to the unknown input pattern, i.e. select

$$\mu_i(0)=0, \ 0 \leq i \leq N-1$$

where $\mu_i(t)$ is the output of node i at time t and x_i is the ith input element with value +1 or -1.

<u>Step 3:</u>
Update $\mu_i(t)$ up to convergence using the following updating (learning) rule:

$$\mu_i(t+1) = f_h [\sum_{i=0}^{N-1} w_{ij}(t)\mu_i(t)] \quad , 0 \leq j \leq M-1$$

where f_h is the hard-limiter function. When convergence is achieved, the output becomes unchanged at further iterations. Then, this output represents the example pattern that best matches the unknown input.

The main advantage of the Hopfield network is that there exists a proof for convergence for all variations. For example, the convergence proof of the network of Fig. 50 is based on Lyapunov analysis [67,68] and the convergence rate is independent of the number of nodes. However, although convergence is guaranteed, one can obtain sometimes a local minimum which is not desirable. When the network of Fig. 51 is used as associative memory, then the output obtained after convergence is used directly. When on the other hand is used as classifier, the output must be compared with each one of the M classes (prototype patterns). If it matches with someone of them, then this class is given as output. Otherwise as output the signal "disagreement" is given.

The main disadvantages of the Hopfield network are the following. (i) For satisfactory operation it needs a large number of nodes. For example, for 10 classes, 70 nodes and more than 10.000 connections are required, and (ii) a prototype pattern can be unstable (i.e. not to converge at step 3) if it has many common bits with some other prototype pattern. In this case the problem may be resolved through some suitable orthogonalization process.

Outputs (valid after convergence)

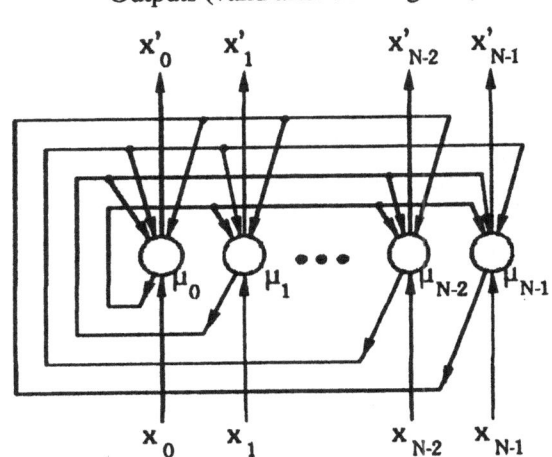

Inputs (applied at zero time instant)

Fig. 51. Hopfield network used as associative memory

384

REFERENCES

[1]   I. Aleksander and H. Morton, An introduction to neural computing, *Chapman and Hall*, London-N.Y., 1990.

[2]   J. Hopfield, Neural networks and physical systems with emergent collective computational abilities, *Proc. Nat. Acad. Sci. US*, Vol.79, 2554-2558, 1982.

[3]   T. Kohonen, Self-organization and associative memory, *Springer-Verlag*, Berlin,-N.Y., 1984.

[4]   D. Rumelhart and J. McClelland, Parallel distributed processing, *MIT Press*, Cambridge, MA, 1986.

[5]   M.M. Gupta, Introduction to adaptive control system design, In: *Adaptive Methods for Control System Design* (M.M.Gupta, ed.), *IEEE Press, Inc.*, N.Y., 1986.

[6]   K.J. Astrom, Theory and applications of adaptive control-A survey, *Automatica*, Vol.19, No.5, pp.471-481, 1983.

[7]   D. Bullock and S. Grossberg, A neural network architecture for automatic trajectory formation and coordination of multiple effectors during variable speed arm movements, *IEEE Intl. Conf. on Neural Networks*, 1987.

[8]   S. Grossberg and M. Kuperstein, Neural dynamics and adaptive sensory-motor control: Ballistic eye movements, *Elsevier/North Holland*, Amsterdam, 1986.

[9]   D.H. Nguyen and B. Widrow, Neural networks for self-learning control systems, *IEEE Control Syst. Magaz.*, pp.18-23, April, 1990.

[10] P.J. Antsaklis, Neural networks in control systems, *IEEE Control Syst. Magaz.* pp.3-5, April, 1991.

[11] MM. Gupta, D.H. Rao and H.C. Word, A neurocontroller with learning and adaptation, *IEEE Conf. Neural Networks*, Seatle, July 8-12, 1991.

[12] A.G. Barto, R.S. Sutton and C.W. Anderson, Neuronlike adaptive elements that can solve difficult learning control problems, *IEEE Trans. Syst. Man Cybern.*, Vol.SMC-13, pp.834-846, 1983.

[13] M. Kawato, K. Furukawa and R. Suzuki, A hierarchical neural network model for control and learning of voluntary movement, *Biol. Cybern.*, Vol.57, pp.169-185, 1987.

[14] M. Kawato, Y. Uno, M. Isobe and R. Suzuki, Hierarchical neural network model for voluntary movement with application to robotics, *IEEE Control Syst. Magaz.*, pp.8-16, April, 1988.

[15] D. Psaltis, A. Sideris and A. Yamamura, A multilayered neural network controller, *IEEE Control Syst. Magaz.*, pp.17-21, April, 1988.

[16] A. Guez, J.L. Eilbert and M. Kam, Neural network architecture for control, *IEEE Control Syst. Magaz.*, pp.22-25, April, 1988.

[17] A. Pellioniz, Sensorimotor operations: A ground for the co-evolution of brain theory with neurobotics and neurocomputers, *IEEE Intl. Conf. on Neural Networks*, 1987.

[18] A.G. Barto, R.S. Sutton and C.W. Anderson, Synthesis of nonlinear control surfaces by a layered associative search network, *Biol. Cybern.*, Vol.43, pp.175-185, 1982.

[19] K. Watanabe and S.G. Tzafestas, Learning algorithms for neural networks with the Kalman filters, *J. Intell. and Robotic Syst.*, Vol.3, No.4, pp.305-319, 1990.

[20] M.M. Gupta, D.H. Rao and P.N. Nikiforuk, Neuro-controller with dynamic learning and adaptation, *J. Intell. and Robotic Systems*, Vol.7, pp.151-173, 1993.

[21] A. Guez and J. Selinski, Neurocontroller design via supervised and unsupervised learning, *J. Intell. and Robotic Systems*, Vol.2, pp.307-335, 1989.

[22] K. Watanabe, K. Shiramizu T. Fukuda and S. Tzafestas, An iterative learning control for noisy systems by using linear neural networks, In: *Engineering Systems with Intelligence* (S.G.Tzafestas, ed.), *Kluwer*, pp.205-212, 1991.

[23] K. Watanabe, T. Fukuda and S. Tzafestas, An adaptive control for CARMA systems using linear neural networks, In: *Mathematics of the analysis and design of process control* (P.Borne, S.G.Tzafestas and N.E.Radhy, eds.), *North-Holland*, pp.525-534, 1992.

[24] S. Amari and M. Arbib, Competition and cooperation in neural nets, In: *Systems Neuroscience* (J.Meltzer, ed.), *Academic Press*, N.Y., pp.119-165, 1977.

[25] S.G. Tzafestas (ed.), Intelligent robotic systems, *Marcel Dekker*, N.Y., 1991.

[26] S.G. Tzafestas, Dynamic modelling and adaptive control of industrial robots: The state-of-art, *System. Anal. Modelling Simul.*, Vol.6, pp.243-266, 1989.

[27] C.Y. Ho and C. Cook, The application of spline functions to trajectory generation for computer controlled manipulators, *Digital Systems for Industrial Automation*, Vol.1, pp.325-353, 1982.

[28] J.P. Merlet, Force-feedback control in robotics tasks, In: *Intelligent Robotic Systems* (S.Tzafestas, ed.), Ch.9, pp.283-311, *Marcel Dekker*, (N.Y., 1991.

[29] S.G. Tzafestas, Adaptive, robust and fuzzy rule-based control of robotic manipulators, In: *Intelligent Robotic Systems* (S.Tzafestas, ed.), Ch.10, pp.313-419, *Marcel Dekker*, 1991.

[30] J.J. Craig, P. Hsu and S.S. Sasty, Adaptive control of mechanical manipulators, *Proc. IEEE Intl. Conf. on Robotics and Automation*, San Francisco, CA, pp.190-195, 1986.

[31] G. Ambrosino, G. Celentano and F. Garofalo, Robust model tracking control for a class of nonlinear plants, *IEEE Trans. Autom. Control*, Vol.AC-30, p.275, 1985.

[32] D. Whitney, Resolved motion rate control of manipulators and human prostheses, *IEEE Trans. Man-Maqchine Systems*, Vol.MMS-10, No.2, pp.47-53, 1969.

[33] H.N. Koivo and T.H. Guo, Adaptive linear controller for robotic manipulators, *IEEE Trans. Autom. Control*, Vol.AC-28, pp.162-171, 1983.

[34] S.G. Tzafestas et.al., Decentralized PID self tuning control of industrial robots based on multivariable identification, *Proc. 25th IEEE Conf. on Decision and Control*, Athens, Greece, pp.1888-1890, 1986.

[35] H. Miyamoto, M. Kawato, T. Setoyama and R. Suzuki, Feedback-error learning neural network for trajectory control of a robotic manipulator, *Neural Networks*, Vol.1, pp.251-265, 1988.

[36] M. Kawato, Y. Maeda, Y. Uno and R. Suzuki, Trajectory formation of arm movement by cascade neural network model based on minimum Torque-change criterion, *Biol. Cybern.*, Vol.62, pp.275-288, 1990.

[37] T. Iberall, A neural network for planning hand shapes in human prehension, *IEEE Conf. on Decision and Control*, pp.2288-2293, 1987.

[38] T. Iberall, A ballpark approach to modelling human prehension, *IEEE Conf. on Neural Networks*, Vol.4, pp.535-544, 1987.

[39] A. Guez and Z. Ahmad, Solution to the inverse kinematics problem in robotics by neural networks, *IEEE Conf. on Neural Networks*, Vol.2, pp.617-624, 1988.

[40] C.C. Jorgenson, Neural network representation of sensor graphs in autonomous robot path planning, *IEEE Conf. on Neural Networks*, Vol.4, pp.507-516, 1987.

[41] K. Tsutsumi and H. Matsumoto, Neural computation and learning strategy for manipulator position control, *IEEE Conf. on Neural Networks*, Vol.4, pp.525-534, 1987.

386

[42] K. Tsutsumi et.al., Neural computation for controlling the configuration of 2-dimensional truss structures, *IEEE Conf. on Neural Networks*, Vol.2, pp.575-586, 1988.

[43] H. Liu et.al., Building a generic architecture for robot hand control, *IEEE Conf. on Neural Networks*, Vol.2, pp.567-574, 1988.

[44] V. Seshadri, A neural network architecture for robot path planning, *Proc. 2nd Intl. Symp. on Robotics and Manufacturing: Research, Foundation and Applications*, ASME Press, pp.249-256, 1988.

[45] R. Eckmiller, Neural network mechanisms for generation and learning of motor programs, *Proc. IEEE Conf. on Neural Networks*, Vol.4, pp.545-550, 1987.

[46] J. Albus, A new approach to manipulator control: The cerebellar model articulation controller (CMAC), *J. Dyn. Syst. Meas. and Control*, pp.220-227, 1975.

[47] J. Albus, Data storage in the cerebellar model articulation controller (CMAC), *J. Dyn. Syst. Meas. and Control*, pp.228-233, Sept. 1975.

[48] J. Albus, Mechanisms of planning and problem solving in the brain, *Math. Biosci.*, Vol.45, pp.247-293, 1979.

[49] Y. Pao and D. Sobajic, Artificial neural-net based intelligent robotics control, *Proc. SPIE Conf. Intelligent Robots and Computer Vision*, Vol.848, pp.542-549, 1987.

[50] A. Guez and J. Selinski, A. Trainable neuromorphic controller, *J. Robotic Syst.*, Vol.5, No.4, pp.363-388, 1988.

[51] A. Guez et.al., Neuromorphic architecture for adaptive robot control: A preliminary analysis, *IEEE Conf. on Neural Networks*, Vol.4, pp.567-572, 1987.

[52] A. Guez et.al., Neuromorphic architectures for fast adaptive robot control, *Proc. IEEE Conf. on Robotics and Automation*, pp.145-149, 1988.

[53] W.T. Miller et.al., Application of a general learning algorithm to the control of robotic manipulators, *Intl. J. Robotics Research*, Vol.6, No.2, pp.84-98, 1987.

[54] R. Elsley, A learning architecture for control based on back-propagation neural networks, *IEEE Conf. on Neural Networks*, Vol.2, pp.584-587, 1988.

[55] Y. Pati et.al., Neural networks for tactile perception, *Proc. IEEE Conf. on Robotics and Automation*, pp.134-139, 1988.

[56] D. Yueng and G. Bekey, Adaptive load balancing between mobile robots through learning in an artificial neural system, *Proc. IEEE Conf. on Decision and Control*, pp.2299-2304, 1987.

[57] J.J. Slotine and W. Li, Adaptive manipulator control: A case study, *Proc. IEEE Conf. on Robotics and Automation*, Rayleigh, N.C., 1987.

[58] J. Fu and N.K. Sinha, An iterative learning scheme for motion control of robots using neural networks: A case study, *J. Intell. & Robotic Systems*, Vol.8, No.3, 1993.

[59] S. Arimoto, Bettering operation of dynamic systems by learning: A new control theory for servomechanism or mechanics systems, *Proc. IEEE Conf. on Decision and Control*, pp. 1064-1069, 1984.

[60] S. Arimoto, Robustness of learning control for robotic manipulators, *IEEE Conf. on Robotics and Automation*, pp. 1528-1533, 1990.

[61] D.F. Rumelhart, Parallel Distributed Processing, Vol.1: Foundations, *MIT Press* Cambridge, Mass, 1986.

[62] S.G. Tzafestas, Some computer aided estimation in stochastic control system identification, *Intl. J. Control*, Vol.12, No.3, pp.385-399, 1970.

[63] B. Widrow and F. Smith, Pattern recognition control systems, *Proc. Comp. and Info. Sciences (COINS) Symp.*, Spartan Books, Washington DC, 1963.

[64] F. Smith, A trainable nonlinear function generator, *IEEE Trans. Auto. Control*, Vol.AC-11, No.2, pp.212-218, 1966.

[65] B. Widrow, The original adaptive neural net broom-balancer, *Proc. IEEE Conf. on Circuits and Systems*, pp.351-357, 1987.

[66] R. Lippman, An introduction to computing with neural nets, *IEEE ASSP Magaz.*, pp.4-22, April 1987.

[67] D. Tank and J. Hopfield, Neural computation of decisions in optimization problems, *Biological Cybern.*, Vol.52, pp.141-152, 1985.

[68] G. Wilson and G. Pawley, On the stability of the travelling salesman problem of Hopfield and Tank, *Biological Cybern.*, Vol.58, pp.63-70, 1988.

12
CONTROL STRATEGY OF ROBOTIC MANIPULATOR BASED ON FLEXIBLE NEURAL NETWORK STRUCTURE

MOHAMMAD TESHNEHLAB[(*)], KEIGO WATANABE[(**)]
(*) Graduate School of Science and engineering,
Saga University, Honjomachi-1, Saga 840, Japan

(**) Department of Mechanical Engineering,
Faculty of Science and Engineering,
Saga University, Honjomachi-1, Saga 840, Japan

1. INTRODUCTION

The artificial neural networks (ANNs) application for computing has currently emerged as an important information processing technique. In some way, the ANNs are a parallel processing artichecture in which a large number of processing neurons are interconnected and the knowledge is represented by the connection weights between the neurons. The connection weights are adjusted through a learning process. The knowlegde is distributed over a large number of connection weights so that the operation of these networks degrade peacefully, even in some parts the connection weights are disconnected. But there is a big problem with this kind of structure. This kind of structure can be a good candidate for simple systems, not for large scale-systems and real applications.

Therefore, we attempt to utilize the ANN with new configurations to simplify the network's structure and learning processes to earn high-speed information processing and adaptivity. This means that the networks can learn and their performance may be improved with some basic new ideas from the traditional structure and learning process to make the ANN attractive for implementing robust and adaptive controllers. In our previous papers [1,2] we explored two kinds of sigmoid unit functions, bipolar and unipolar unit functions with parameters which give a changeable shape by changing the values of their parameters. Yamada and Yabuta [3] has already used a similar bipolar sigmoid unit function (only bipolar) to construct a direct neural network controller which is completely for a different purpose from ours. Their algorithms cause a steady state error in response and were used for only a single-input/single-output system. It is quite attractive to develop a new type of

S. G. Tzafestas and H. B. Verbruggen (eds.),
Artificial Intelligence in Industrial Decision Making, Control and Automation, 389–403.
© 1995 *Kluwer Academic Publishers.*

sigmoid unit function with the ability of adjusting its o wn shape, whenever necessary for improving the neural network performance. As a result, the sigmoid unit functions attain flexibility, in contrast to all of the former studies in which neural networks have fixed sigmoid unit functions at hidden- and output-layers [3-12].

In this chapter, the conventional feedback controller is configured in parallel to the ANN controller, in the same manner as Gomi and Kawato [4] and Miyamoto et al. [5], but with a different sigmoid unit function and learning algorithm. The main difference between our and their ANN structures is that the flexible bipolar neuron used here turns out to minimize the number of learning algorithms that is necessary for adjusting the connection weights and sigmoid unit function. Another difference of our and their structures is that not only the hidden-layer, but also the output-layer include the flexible bipolar neuron. Gomi and Kawato [4] used 13 nonlinear unit functions in hidden-layer and two linear unit functions at output-layer for learning impedance control. The difficulty of their structure technique is that the ANN needs a large number of learning algorithms to adjust the correction weights, because they used a large number of units at hidden-layer. As mentioned before, this kind of structure is not a good candidate for large scale-systems, and they also used linear functions at output-layer different from our function (flexible nonlinear activation function). Actually, our unit function is bounded according to desired trajectory or teaching signal. The proposed ANN is learned by the *feedback-error-learning* architecture suggested by Miyamoto et al. [5]. The learning approach is applied for learning both connection weights and sigmoid unit function parameters simultaneously.

In the following, we first introduce the bipolar unit function and its properties, and then provide the description on the fundamentals of proposed learning algorithms and ANNs scheme. We next give a brief explanation of manipulator and finally present the computer simulation results with a discussion of potential advantages of proposed approach over traditional ANN structure approaches.

2. THE REPRESENTATION OF BIPOLAR UNIT FUNCTION

The form of hyperbolic tangent as a general unit function is expressed as

$$f(x) = \frac{1 - e^{-x}}{(1 + e^{-x})} \tag{1}$$

This function is monotonically increasing with asymptotes to ± 1, as the $x \to \pm \infty$, which is useful when bipolar outputs are required. We define the bipolar sigmoid unit function (BSUF) as hyperbolic tangent function type with an additional parameter which is given by

$$f(x, a) = \frac{1 - e^{-2xa}}{a(1 + e^{-2xa})} \tag{2}$$

The proposed sigmoid unit function can have a different shape by changing the parameter a, as shown in Fig. 1. The another property of this function is given by

$$\lim_{a \to 0} f(x, a) = \frac{1 - e^{-2xa}}{a(1 + e^{-2xa})} = \frac{0}{0} \tag{3}$$

Thus, the l'Hospital rule can be applied to this function as

$$\lim_{a \to 0} f(x, a) = \lim_{a \to 0} \frac{(1 - e^{-2xa})'}{[a(1 + e^{-2xa})]'} = x \tag{4}$$

Then, it becomes linear when $a \to 0$, and becomes nonlinear for large values of a.

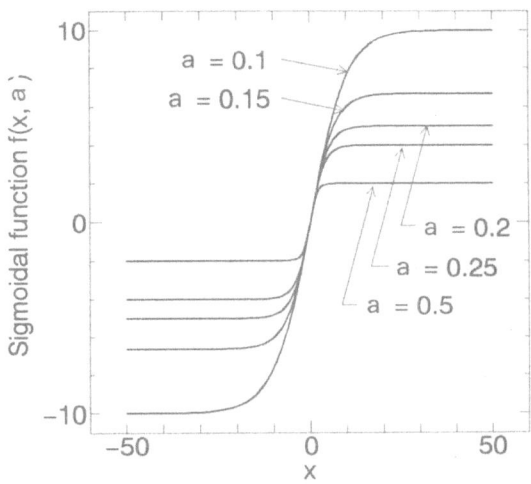

Fig. 1 The bipolar sigmoid unit function with changeable shape

3. LEARNING ARCHITECTURE

To train the neural networks to perform some task, we must adjust the connection weights in such a way that the error between the desired output and the actual output is reduced. This process requires that the neural network computes the derivative of the cost function with respect to the connection weight as an increment of the connection weight. The back propagation algorithm is the most widely used method for minimizing the error which proposed by Rumelhart et al. [8].It is here assumed that the neurons of the feedforward network utilize the flexible unit function discussed above. The object of looking at the derivative is to determine the rule for changing connection weights so as to minimize the descent of the cost function along the gradient. The minimization of cost function for a multi-layered network is well-known to be performed by the usual procedure of *back-propagation* (BP) algorithms.

The BP algorithms involve two phases. In the first phase, input signals are fed in a feedforward manner through the network to produce actual outputs. In the second phase, the output vector of a PID servo controller, $\tau_{PID} \in R^L$, is considered as the error to propagate backward through the ANN and used to adjust the sigmoid unit function parameters and connection weights. Thus, in a feedback error learning architecture, we consider the following cost function

$$J_g = 1/2 \sum_{i=1}^{L} (\tau_i - o_i^M)^2 \tag{5}$$

in which L denotes the number of control inputs and the relation of $\tau - \tau_n = \tau_{PID}$ has been used, where $\tau \triangleq [\tau_1, ..., \tau_L]$ is the actual input vector to the plant and $\tau_n \triangleq [o_1^M, ..., o_L^M]$ is the output vector from the ANN. The update equations for connection weights and sigmoid unit function parameters are given in following subsections.

3.1 The learning of connection weights

The J_g is minimized by adjusting the connection weights $w_{ij}^{k,k-1}$. Following the gradient descent algorithm, the increment $\Delta w_{ij}^{M-1,M}$ at output-layer M can be expressed as

$$\Delta w_{ij}^{M-1,M} = \eta_1 \delta_j^M o_i^{M-1} \tag{6}$$

where $\eta_1 > 0$ is a learning rate given by a small positive constant, and

$$\delta_j^M = (\tau_j - o_j^M) f'(i_j^M) \tag{7}$$

where $f'(.,.)$ denotes $\partial f(i_j^M, a_j^M)/\partial i_j^M$. Also, the increment of the connection weights at hidden-layer, $\Delta w_{ij}^{k,k-1}$, is expressed by

$$\Delta w_{ij}^{k-1,k} = \eta_1 \delta_i^k o_i^{k-1} \tag{8}$$

$$\delta_j^k = f'(i_j^k) \sum_l \delta_l^{k+1} w_{ji}^{k,k+1} \tag{9}$$

Usually, the update equations of connection weights can be expressed as

$$w_i^k(t+1) = w_i^k(t) + \eta_1 \delta_i^k o_i^{k-1} + \alpha_1 \Delta w_i^k(t) \tag{10}$$

$$w_i^M(t+1) = w_i^M(t) + \eta_1 \delta_i^M o_i^{M-1} + \alpha_1 \Delta w_i^M(t) \tag{11}$$

where α_1 represents the momentum coefficients of connection weights defined by $0 \leq \alpha_1 < 1$.

3.2 The learning of sigmoid unit function parameters

Now, find parameters a's in the BSUF to minimize the J_g, considering the same input-output relation between the k-th layer and the $(k+1)$-th layer. By employing the gradient descent algorithm, the increment of a_i^k, denoted by Δa_i^k, can be obtained as

$$\Delta a_i^k = -\eta_2 \frac{\partial J_g}{\partial a_i^k} \tag{12}$$

where $\eta_2 > 0$ is a learning rate given by a small positive constant.

At the output-layer M, the partial derivative of J_g with respect to a is described as

$$\partial J_g / \partial a_i^M = \frac{\partial J_g}{\partial o_i^M} \frac{\partial o_i^M}{\partial a_i^M} \tag{13}$$

Here, defining

$$\sigma_i^M \triangleq -\frac{\partial J_g}{\partial o_i^M} \tag{14}$$

gives

$$\sigma_i^M = -(\tau_i - o_i^M) \tag{15}$$

Therefore, the learning update equation for a at the output-layer neurons is obtained by

$$a_i^M(t+1) = a_i^M(t) + \eta_2 \sigma_i^M f^*(i_i^M, a_i^M) + \alpha_2 \Delta a_i^M(t) \tag{16}$$

where $f^*(.,.)$ is defined by $\partial f(., a_i^M)/\partial a_i^M$ and α_2 is a momentum coefficient defined by $0 \leq \alpha_2 < 1$.

The next step is to calculate a at the hidden-layer k. It is easily found that

$$\partial J_g / \partial a_i^k = \frac{\partial J_g}{\partial o_i^k} \frac{\partial o_i^k}{\partial a_i^k} \tag{17}$$

$$= \frac{\partial J_g}{\partial o_i^k} f^*(i_i^k, a_i^k) \tag{18}$$

where defining

$$\sigma_i^k \triangleq -\frac{\partial J_g}{\partial o_i^k} \tag{19}$$

we have

$$\partial J_g / \partial o_i^k = \sum_l \frac{\partial J_g}{\partial o_l^{k+1}} \frac{\partial o_l^{k+1}}{\partial o_i^k} \tag{20}$$

$$= -\sum_l \sigma_l^{k+1} \frac{\partial o_l^{k+1}}{\partial i_l^{k+1}} \frac{\partial i_l^{k+1}}{\partial o_i^k} \tag{21}$$

$$= -\sum_l \sigma_l^{k+1} f'(i_l^{k+1}, a_l^{k+1}) w_{i,l}^{k,k+1} \tag{22}$$

where $f'(.,.)$ denotes $\partial f(i_l^{k+1}, a_l^{k+1})/\partial i_l^{k+1}$. Gradually, it follows that

$$\sigma_i^k = \sum_l \sigma_l^{k+1} f'(i_l^{k+1}, a_l^{k+1}) w_{i,l}^{k,k+1} \tag{23}$$

Eventually, the learning update equations for the parameters a in hidden-layers is expressed as

$$a_i^M(t+1) = a_i^M(t) + \eta_2 \sigma_i^M f^*(i_i^M, a_i^M) + \alpha_2 \Delta a_i^M(t) \tag{24}$$

The entire process is repeated over the training input-output patterns and finally the network produces the desired responses.

4. NEURAL NETWORK-BASED ADAPTIVE CONTROLLER

We here apply the ANN to construct an adaptive controller for controlling a two-link manipulator. The manipulator model is shown in Fig.2. The equation of motion for the two-link manipulator can be described as

$$M(\theta)\ddot{\theta} + V(\theta, \dot{\theta}) = \tau \tag{25}$$

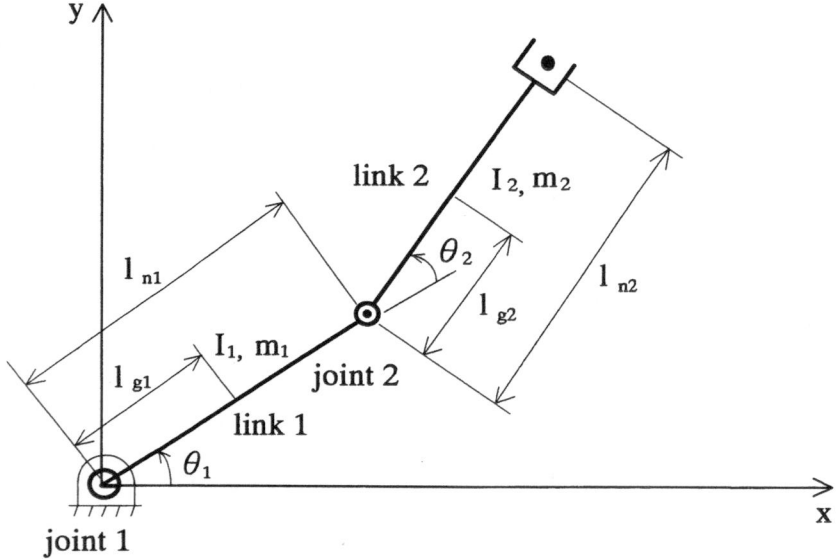

Fig. 2 Two-link planar manipulator

with

$$M_{11} = I_1 + I_2 + m_1 l_{g1}^2 + m_2[l_{n1}^2 + l_{g2}^2 + 2l_{n1}l_{g2}\cos(\theta_2)] \tag{26}$$

$$M_{12} = M_{21} = I_2 + m_2[l_{g2}^2 + l_{n1}l_{g2}\cos(\theta_2)] \tag{27}$$

$$M_{22} = I_2 + m_2 l_{g2}^2 \tag{28}$$

$$V_1 = -m_2 l_{n1} l_{g2} \sin(\theta_2)(2\dot{\theta}_1 + \dot{\theta}_2)\dot{\theta}_2 \tag{29}$$

$$V_2 = m_2 l_{n1} l_{g2} \sin(\theta_2)\dot{\theta}_1^2 \tag{30}$$

where $\tau \in R^2$ is the joint actuator torques with $\tau = [\tau_1 \quad \tau_2]^T$ and $\theta \in R^2$ is the generalized joint angles with $\theta = [\theta_1 \quad \theta_2]^T$, and $M(\theta) \in R^{2\times2}$ is a matrix, usually referred to as the manipulator mass matrix containing the kinetic energy functions of the manipulator. The vector $V(\theta, \dot{\theta}) \in R^2$ represents forces arising from Coriolis and centrifugal forces. Generally, the equation (25) is very complicated for all but the simplest manipulator configurations.

The block diagram of the control system is shown in Fig. 3. As can be seen from this figure, the summation of outputs of the ANN and feedback controller will be the actual input torque signals to the plant, which can be expressed as

$$\tau = \tau_{PID} + \tau_n \tag{31}$$

where τ_n is the output of ANN and τ_{PID} is the output of PID controller.

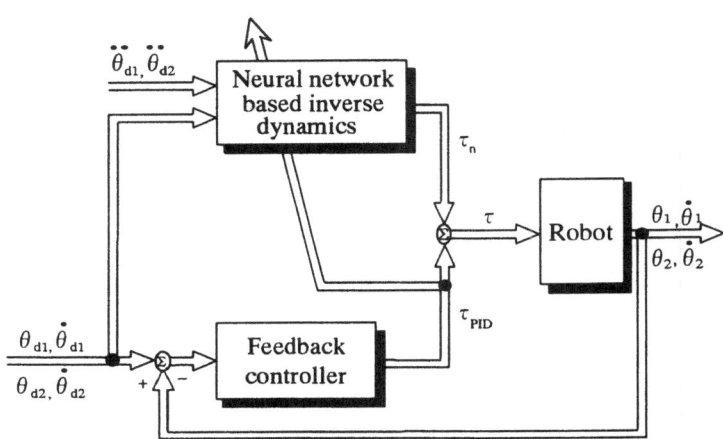

Fig. 3 The block diagram of neural network-based control system
using feedback-error-learning

4.1 The feedback-error learning rule

We applied the feedback-error-learning architecture to train the inverse dynamic model of two-link manipulator. In this learning rule, the PID feedback controller can be replaced by a general linear feedback controllers. The success of training depends directly on the proper choice of the feedback controller outputs, such as how select the gain values of conventional controller. In the feedback-error-learning scheme, the feedback gains are usually set as constant values. These values strongly affect on the output signal from the ANN, and the learning adjustment becomes fast if the feedback gain is selected as a large value. Thus, having stronger feedback error signals at the back-propagation to the ANN structure leads to faster convergence to minimize the cost function. The conventional PID controller is expressed as

$$\tau_{PID} = K_p e + K_i \int_0^t e \, dt + K_d \dot{e} \tag{32}$$

where $K_p \in R^{2 \times 2}$, $K_i \in R^{2 \times 2}$, and $K_d \in R^{2 \times 2}$ are diagonal matrices with positive K_{pj}, K_{ij}, and K_{dj} on the diagonals. The servo error, $e(t) = [e_1 \quad e_2]^T$, denotes the tracking error vector which is defined by

$$e(t) = \theta_d(t) - \theta(t) \tag{33}$$

As well known in the PID control, the servo gains K_i and K_d govern the influence of the past and future output error values and the K_p governs the influence of the present error on the controller. The plant outputs are $\theta_i, \dot{\theta}_i, i = 1, 2$, which may be different from the desired outputs $\theta_{di}, \dot{\theta}_{di}, i = 1, 2$.

4.2 Adaptation of neural network controller

The controller design objective is to find a controller such that the computed control commands, once applied to the plant, will give the plant output to follow some desired outputs. In the case of regulation, the plant states are required to approach the constant set-point state. In a tracking problem, the objective is to control the plant states to closely follow the time-varying desired states, which are often generated by a reference model. We adopt the ANN as a controller whose structure is the multilayered feedforward, which is most relevant to control applications. Here, we put our effort to use a very simple neural network while keeping the optimum performance. The learning process for the manipulator is done on-line learning in a certain time. During the training, the neural network learns the correct network output to obtain the desired trajectory output of the plant and decreases the feedback error signal simultaneously. Following this approach, the ANN may produces an approximate inverse dynamic model described as

$$\tau_n = \hat{M}(\theta_d)\ddot{\theta}_d + \hat{V}(\theta_d, \dot{\theta}_d) \tag{34}$$

where $\hat{M}(\theta)$ and $\hat{V}(\theta,\dot{\theta})$ are the estimates of $M(\theta)$ and $V(\theta,\dot{\theta})$ and $\theta_d, \dot{\theta}_d, \ddot{\theta}_d$ are the desired position, velocity and acceleration of the plant. Thus, the present ANN learns the following nonlinear relation,

$$\tau_n = f(\ddot{\theta}_d, \dot{\theta}_d, \theta_d, w, a) \tag{35}$$

where w and a are connection weights between any two layers and sigmoid unit function parameters at any layers.

5. SIMULATION EXAMPLE

In this section, we will make comparison between the traditional and our proposed ANN-based controllers. The inputs to the both ANNs are the desired system outputs $(\theta_{d1,2}, \dot{\theta}_{d1,2}, \ddot{\theta}_{d1,2})$ of two links in order to acquire the feedforward controller and to obtain desired responses. Our simulations of the adaptive control scheme for this system have the following assumptions:

1. A fourth-order Runge-Kutta-Gill subroutine was used to simulate the system dynamics.

2. The actual values of manipulator's parameters are given in Table 1.

3. The gains of conventional PID feedback controller are set as constants which are given in Table 2.

4. The initial values, learning rates, and momentum coefficients for the learning of connection weights and sigmoid unit function parameters of learning algorithms for different simulations are given in Tables 3 and 4.

It is assumed that the control sampling period is $T = 10$ [ms], the step width of the integration is 0.4 [ms], and the initial states are $\theta_1 = 0.5236$ [rad], $\theta_2 = 0.4363$ [rad], and $\dot{\theta}_1 = \dot{\theta}_2 = 0$. Also, the desired trajectories of the manipulator are assumed to be known as time functions of joint positions, velocities, and accelerations, that is, $\theta_{d2}, \dot{\theta}_{d2}$, and $\ddot{\theta}_{di}, i = 1,2$ are expressed as

$$\theta_{d1} = 0.5\cos(\pi t) \qquad \theta_{d2} = 0.5\sin(\pi t) + 1.0 \tag{36}$$

$$\dot{\theta}_{d1} = -0.5\pi\sin(\pi t) \qquad \dot{\theta}_{d2} = 0.5\pi\cos(\pi t) \tag{37}$$

$$\ddot{\theta}_{d1} = -0.5\pi^2\cos(\pi t) \qquad \ddot{\theta}_{d2} = -0.5\pi^2\sin(\pi t) \tag{38}$$

The simulations of traditional and proposed methods with three-layered configurations are explained. The traditional method adopts only the adjustment of connection weights with 6 linear units in the input-layer, 20 fixed bipolar units in the hidden-layer, and 2 linear units in the output-layer. Here, the simulation conditions are given in Tables 2 and 3. On the other hand, the proposed method consists of learning of both connection weights and sigmoid unit function parameters, whose structure has 6 linear units in the input-layer, 1 flexible bipolar unit in the hidden-layer, and 2 flexible bipolar units in the output-layer. Here, the simulation conditions are given in Tables 2 to 4. The proposed flexible neural network structure with three layers is shown in Fig. 4.

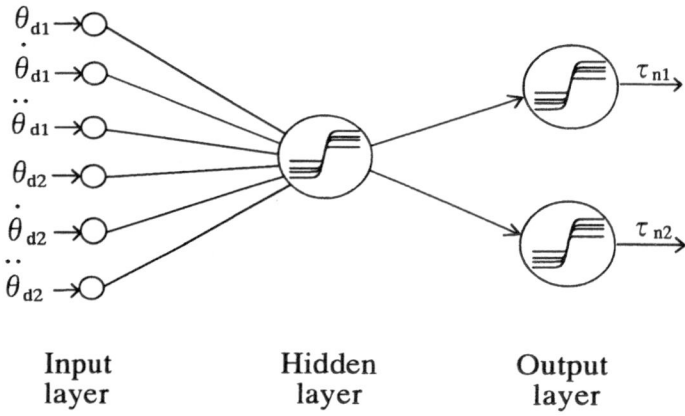

Input layer	Hidden layer	Output layer

Fig. 4 Three-layered neural network structure using flexible
bipolar sigmoid unit functions

The tracking responses of two links are shown in Figs. 5-a and 5-b. As a result of the sequential learning, the actual responses converge to the desired trajectories. This convergence in the proposed method is considerably faster than the traditional method, which means that the traditional learning process needs much more training time than the proposed one. It is impossible to learn a traditional neural network to have a desired performance unless the network has enough hidden units, or the network learning fails to find the optimal network connections. Thus, to find the optimal connections in the traditional method, the number of units at the hidden-layer must be chosen empirically, or might use two hidden-layers. Figure 6-a and 6-b show the resultant control inputs to the first and second links. The actual input signals are the summation of both ANN's output, τ_n, and PID's output, τ_{PID}. As can be recognized at the transient training, the PID controller works with

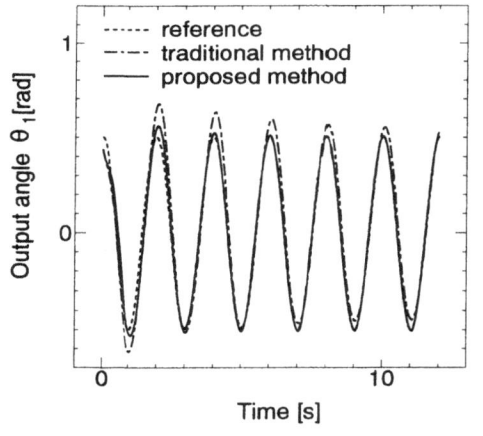

Fig. 5-a The trajectory control results of the first link

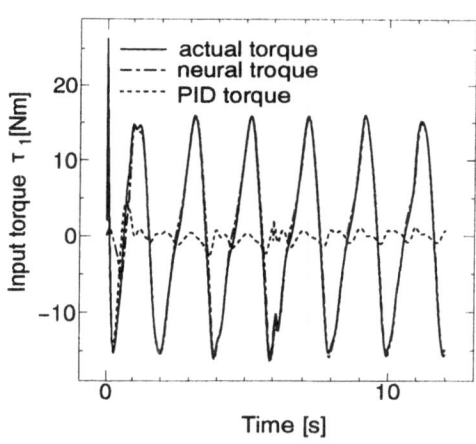

Fig. 6-a The actual input torques of the first link

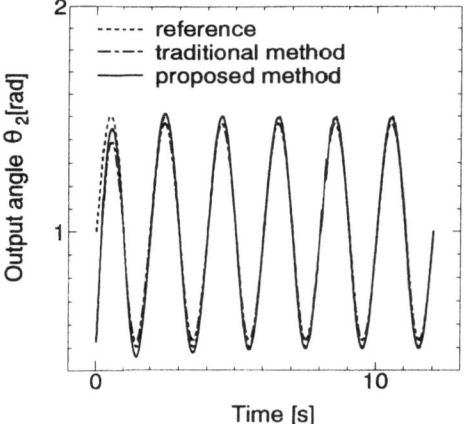

Fig. 5-b The trajectory control results of the second link

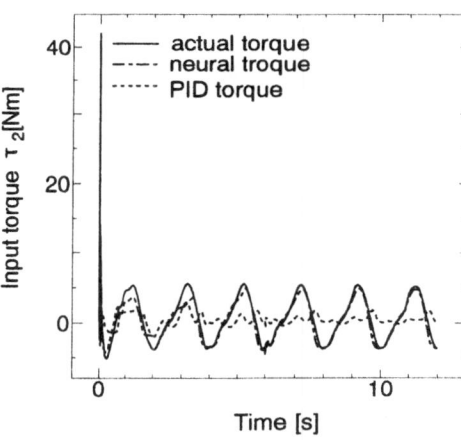

Fig. 6-b The actual input torques of the second link

400

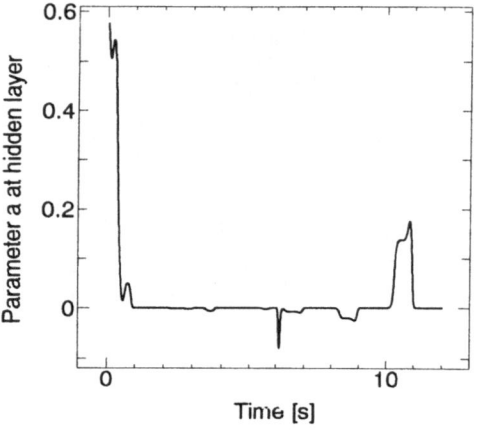

Fig. 7-a Time history of parameter *a* at hidden-layer

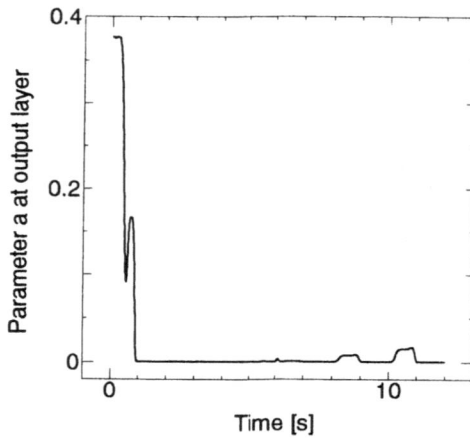

Fig. 7-c Time history of parameter *a* of second unit at output-layer

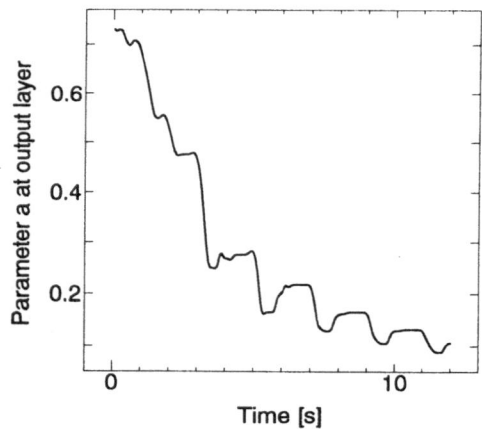

Fig. 7-b Time history of parameter *a* of first unit at output-layer

maximum input torques while the ANN works with minimum input torques, and by sequential learning the tracking errors have been minimized by adjusting the connection weights and sigmoid unit function parameters. In this state the output from the PID controller decreases and finally almost approaches to zero, and in case of some feedback error the feedback input torques again are fed back to the neural network. Figures 7-a, 7-b, and 7-c show the learning process of sigmoid unit function parameters at the hidden- and output-layers, respectively. The very interesting feature is that, especially in the hidden-layer unit and the first unit in the output-layer, the parameters show a variation around the zero, which means that these units worked as a linear function after some learning process.

Table 1 The physical parameters of two-link manipulator

	Link one ($i = 1$)	Link two ($i = 2$)
Mass m_i(kg)	5.0	5.0
Inertia I_i(kgm^2)	0.104	0.104
length l_{ni}(m)	0.5	0.5
length l_{gi}(m)	0.25	0.25

Table 2 The PID feedback gains

	K_{pi}	K_{ii}	K_{di}
$i = 1$	32	26	14
$i = 2$	88	25	5

Table 3 The initial uniform random number (URN), learning rates and momentum coefficients for the learning of connection weights

	between layers
initial values	URN[0.1.0]
learning rates	$\eta_1 = .00075$
momentum coefficients	$\alpha_1 = .005$

Table 4 The initial uniform random number (URN), learning rates and momentum coefficients for the learning of sigmoid unit function parameters

	for all units
initial values	URN[0,1.0]
learning rates	$\eta_2 = .0005$
momentum coefficients	$\alpha_2 = .00405$

6. CONCLUSION

We have demonstrated the feasibility of developing an adaptive controller based on neural networks for nonlinear dynamical plants. The principal aim of this chapter was to construct a different type of the ANN structure.

The adaptive inverse model with the proposed ANN structure was found to posses the good performance. The advantage of the proposed ANN over all other ANN structures is that it can reduce the amount of the training time, because the present ANN has a very simple structure. The feedback-error-learning was also used to train the ANN, in which the output torques from the PID controller were supplied to the ANN in order to adjust the connection weights and sigmoid unit function parameters. As a result, when the feedback control output is close to zero, the resulting manipulator control system can be considered to follow an ANN-based nonlinear inverse dynamic control rule, which means that the external feedback controller gradually could be replaced by the ANN inverse dynamics model. According to the traditional method under certain conditions, increasing the number of hidden neurons actually can improve the convergence speed of the learning. It should be noted, however, that beyond a certain limitation, it causes the training time becomes much longer.

REFERENCE

[1] M. Teshnehlab and K. Watanabe, "Self-tuning of computed torque gains by using neural networks with flexible structure, "*IEE Proceedings-D*, to be published, 1994.

[2] M. Teshnehlab and K. Watanabe, "The high flexibility and learning capability of neural networks with learning bipolar and unipolar sigmoid functions, "*Japan-U.S.A. symposium on flexible automation*, to be presented, Kobe, July 1994.

[3] T. Yamada and T. Yabuta, "Neural Network Controller Using Autotuning Method for Nonlinear Functions," *IEEE Trans. on Neural Networks*, Vol. 3, No.4, July, pp.595-601, 1992.

[4] H. Gomi and M. Kawato, "Nueral Network Control for a Closed-Loop System Using Feedback-Error-Learning," *Neural Networks*, Vol. 6, No. 7, pp.933-946, 1993.

[5] H. Miyamoto, M. Kawato, T. Setoyama, and R. Suzuki, "Feedback-Error-Learning Neural Network for Trajectory Control of a Robotic Manipulator," *Neural Networks*, Vol. 1, pp.251-265, 1988.

[6] R. T. Newton and Y. Xu, "Neural Network Control of a Space Manipulator," *IEEE Control Systems Magazine*, Vol. 13, No.6, Decem., pp.14-22, 1993.

[7] D.E. Rumelhart, G.E. Hiton, and J.L. McClelland, "A General Framework for Parallel Distributed Processing," in *Parallel Distributed Processing: Explorations in the Microstructure of Cognition*, vol.1, D. E. Rumelhart and J. L. McClelland, Eds., Cambridge, MA: MIT Press, pp.45-76, 1986.

[8] D.E. Rumelhart, G. E. Hiton, and R. J. Williams, "Learning Internal Representations by Error Propagation," in *Parallel Distributed Processing: Explorations in the Microstructure of Cognition*, vol.1, D. E. Rumelhart and J. L. McClelland, Eds., Cambridge, MA: MIT Press, pp.282-317, 1986.

[9] R. E. Nordgren and P. H. Meckl, "An Analytical Comparison of a Neural Network and a Model-Based Adaptive Controller," *IEEE Trans. on Neural Networks*, Vol. 4, No.4, July, pp.685-694, 1993.

[10] M.A. Sartori and P.J. Antsaklis, "Implementation of Learning Control Systems Using Neural Networks," *IEEE Control Systems Magazine*, Vol.12, No.2, April, pp.49-57, 1992.

[11] R.M. Sanner and D.L. Akin, "Neuromorphic Pitch Attitude Regulation of an Underwater Telerobot," *IEEE Control System Magazine*, Vol.10, No.2, April, pp.62-67, 1990.

[12] W.T. Miller, R.S. Sutton, and P.J. Werbos, *"Neural Networks for Control,"* The MIT Press, Cambridge, MA, 1990.

[13] S.-B. Yu and S.-R. Hu, "Neural Network For Ship Recognition, " *Proceedings of the International Conference on Fuzzy Logic and Neural Networks*, Iizuka, Japan, July 20-24, pp.325-328, 1990.

[14] D. Psaltis, A. Sideris, and A.A. Yamamura, "A Multilayered Neural Network Controller, " *IEEE Control Systems Magazine*, Vol.8, No.2, April, pp.17-20, 1988.

13
NEURO-FUZZY APPROACHES TO ANTICIPATORY CONTROL

L.H. TSOUKALAS, A. IKONOMOPOULOS, R.E. UHRIG
Center for Neural Engineering Applications
The University of Tennessee
Knoxville, TN 37996-2300

1. INTRODUCTION

Anticipatory systems are systems where change of state is based on information pertaining to present as well as future states. Cellular organisms, industrial processes, global markets, provide many examples of behavior where global output is the result of anticipated not only current state. In the global economy, for example, the anticipation of an oil shortage or of a significant default of foreign loans can have profound effects upon the course of the economy, whether or not the anticipated events come to pass [3]. Participants in the economy build up models of the rest of the economy and use them to make predictions. The models are more *prescriptive* (prescribing what should be done in a given situation) than *descriptive* (describing the options of a given situation) and involve strategies appropriately formulated in terms of *lookahead*, or anticipation of market conditions. In an industrial process, the prescriptions are typically Standard Operating Procedures (SOPs), dictating actions to be taken under specific conditions. The accumulated experience of various decision-makers at all levels of the process provides increasingly refined SOPs and progressively more sophisticated interactions amongst them and computer tools designed to assist them. As another example, consider a car driven on a busy highway. The driver and the car taken together are a simple, everyday example of an anticipatory system. An automobile driver makes decisions on the basis of predicting what may be happening in the future, not simply reacting to what happens at the present. Driving requires one to be aware of future system inputs by observing the curvature and grade of the road ahead, road conditions and the behavior of other drivers. Perceptual information received at the present, may be thought of as input to internal predictive models. Such a system, however, is very difficult to model using conventional approaches. In part the difficulty relates to the fact that conventional predictive models are unduly constrained by excessive precision. Generally, in situations like the driver-car system, it is important for a decision-maker (the driver) to use a *parsimonious description* of the overall situation, that is, a model at the appropriate level of precision. Predictions about the future are not very precise and of course they may be wrong. Yet, their efficacy does not rest on *precision* as much as on the more general issue of *accuracy* and their successful utilization. High levels of precision may not only be unnecessary for problems utilizing predicted values, they may very well be counterproductive. An over-precise driver may actually be a dangerous driver.

S. G. Tzafestas and H. B. Verbruggen (eds.),
Artificial Intelligence in Industrial Decision Making, Control and Automation, 405–419.
© 1995 *Kluwer Academic Publishers.*

Although anticipatory systems have been studied by a number of researches in the context of mathematical biology [10],[11], automata theory [15], preview control [14], and their epistemological roots may be traced back to Aristotle's views on causality, it is only recently that the advent of modern computing technologies makes it possible to employ them for complex system regulation and management [2],[16],[17],18]. In Japan, the Automatic Train Operator (ATO) used in Sendai's subway system employs anticipatory control strategies [20], and researchers at Tohoku University and Mitsubishi Research Institute have recently developed an innovative anticipatory guidance and control system for computer-assisted operation of nuclear power plants [19].

In preview control future information was considered as probabilistic in kind and the control problem seen as a problem of time-delay [14]. The situation is illustrated in figure 1 where a discrete control problem that lasts n time steps, presently at time i, is considered. Tomizuka postulated that up to certain time n_{la} past i, predictions can be made and utilized by the controller at i. Thus, the future is divided into *deterministic* and *probabilistic* parts as seen in the figure 1. The controller is assumed to make use of preview information with respect to a command signal (desired trajectory) from the present time i up to n_{la} time units into deterministic future. The quantity, n_{la} is the *preview time* (or *length of anticipation*) and is usually shorter than n, the problem duration, often one or two time steps. To make the solution applicable to a broader class of problems, measurement time delay, observation noise and driving noise were included in formulating the problem. The solution showed how to utilize the local future information obtained by finite preview (n_{la}) in order to minimize an optimality criterion evaluated over the problem duration n. It was found that preview dramatically improved the performance of a system relative to non-preview optimal performance, and a heuristic criterion about the preview time, n_{la}, was suggested, i.e., $n_{la} \approx 3 \times$ *(longest closed loop plant time constant).*

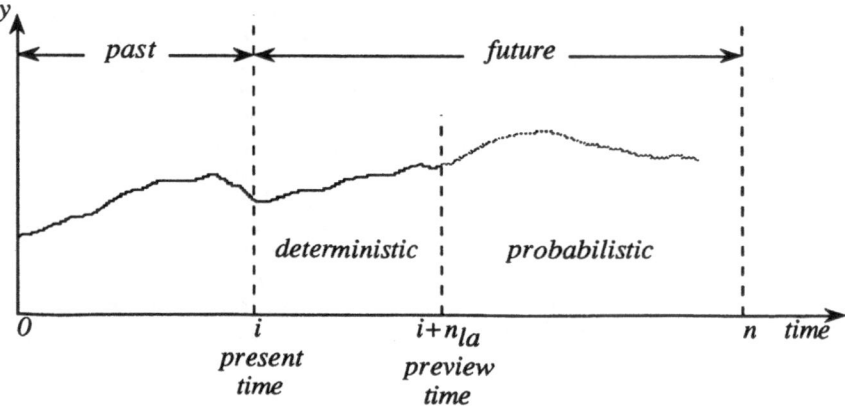

Figure 1. Prediction in finite preview problems.

The point of departure for our formulation is that future information is essentially *fuzzy* in nature, that is, predicted values are not imbued with stochastic or probabilistic type of uncertainty. Whatever can be said about the future does not come from measurements but models, hence, such predictions are fuzzy numbers, that is, linguistic categorizations of information pertaining to the future of the system. Generally, fuzziness is a property of language, whereas randomness is a property of observation and since there is no physical measurement pertaining to the future the mathematics of fuzzy sets may be more appropriate for anticipatory systems. Consider, for example, the process depicted in figure 2. At any time i we have available information from the present as well as information from the output of some predictive model. According to our formulation this is a fuzzy number (a fuzzy subset of the reals whose membership function is normal and convex [22], [6]). Therefore, the mathematical tools for utilizing it at time i ought to be fuzzy as well. The time Δt into the future, the anticipatory time step, depends on the nature of the problem and the predictive model used, and generally, need not be one or two time steps as is often the case in preview control. As is suggested in figure 2 the fuzziness of a prediction is postulated to depend on Δt in the sense that for greater Δt's we get fuzzier predictions.

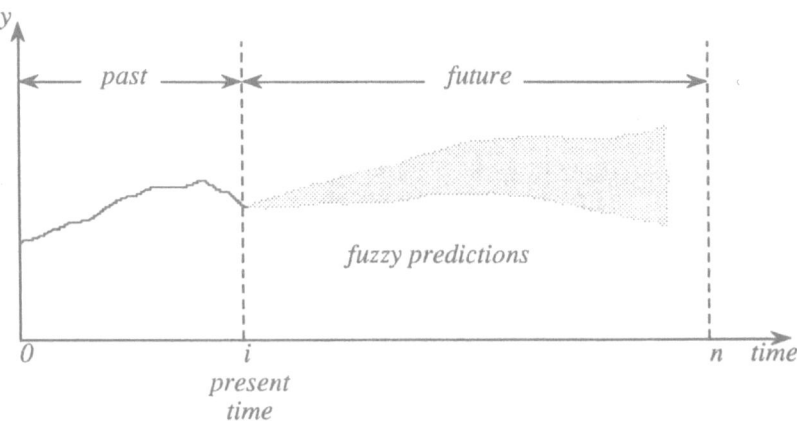

Figure 2. Predictions in anticipatory systems are seen as fuzzy numbers.

2. ISSUES OF FORMALISM IN ANTICIPATORY SYSTEMS

A system that makes decisions in the present on the basis of what may be happening in the future is thus envisioned to be different in two important respects: In the *language* used to formulate models of its behavior and in the method of *measurement* used to access future states. The first we call the *issue of formalism* and we address in this section while the latter we examine in the following section.

Consider the typical systems formulation in modern control theory. A system is described by a set of difference (differential) equations of the form

$$x(t + 1) = A\,x(t) + B\,u(t) + w(t)\,; \quad x(t_0) = x_0 \tag{1}$$
$$y(t) = C\,x(t) + v(t)$$

where, $\{u(t)\}$ is an $r \times 1$ input sequence, $\{y(t)\}$ is an $m \times 1$ output sequence, $\{x(t)\}$ is an $n \times 1$ state sequence, A, B, and C are appropriate transition matrices, x_0 some initial state, and, $w(t)$ and $v(t)$ are noise terms.

A system is called *anticipatory* if $x(t+1)$ and $y(t)$ are not uniquely determined by $x(t)$ and $u(t)$ alone, but use information pertaining to some future state $x(t + \Delta t)$, and/or input $u(t + \Delta t)$.

Looking at equations (1) we observe that it is rather difficult to include future information in (1) except by containing it within the noise terms as in the case of non-deterministic systems. In such a case one obtains sets of values $x(t+1)$ and $y(t)$ with each pair $[x(t), u(t)]$. Suppose that the values of x and y are subsets of some larger sets X and Y. If we denote these subsets of X and Y by X^{t+1} and Y^t we obtain mappings of the form

$$X^{t+1} = F[x(t), u(t)] \tag{2}$$
$$Y^t = G[x(t), u(t)]$$

Of course one could fuzzify this system as is done in the fuzzy literature [1],[7] by assuming that these X^{t+1} and Y^t are fuzzy subsets on X and Y respectively, and obtain a fuzzy system determined by conditional membership functions

$$\mu[x(t+1) \,|\, x(t), u(t)] \tag{3}$$
$$\mu[y(t) \,|\, x(t), u(t)]$$

Subsequently the compositional rule of inference may be used to calculate the fuzzy response of the fuzzy system to any fuzzy input. The problem however, of involving future information in the formulation of (1) still remains. Generally, if we do so the mappings in equations (2) cease to be *many-to-one* mappings, i.e., *functions*, but more general *many-to-many* mappings such as we have in *fuzzy relations*.

Consider again equations (1). Another point of view is to look at the equation signs " = " as assignment operators " := ", i.e.,

$$x(t + 1) := A\,x(t) + B\,u(t) + w(t)\,; \quad x(t_0) = x_0 \tag{4}$$
$$y(t) := C\,x(t) + v(t)$$

where, the assignment operator, ":=" is an *if/then* rule, which assigns the right hand side (*RHS*) of (4) to the left hand side (*LHS*) upon update. Now we are in the realm of logical implications and we can easily include terms such as $x(t + \Delta t)$, and $u(t + \Delta t)$ in our *if/then* rules. The calculus of fuzzy *if/then* rules is rather well known and provides an interesting alternative and enhancement of formulations such as (1) particularly for the purpose qualitative and complex system modeling [12],[22],[23]. Thus, an anticipatory system can be described by a collection of fuzzy *if/then* rules

$$R^N = \{R^1, R^2, \cdots, R^n\} \tag{5}$$

Each rule is a *situation/action* pair, denoted as $s \rightarrow a$, where both *present* and *anticipated situations* are considered in the *LHS* and *current action* in the *RHS*. The rules of (5) may be rewritten as

$$R^N = \{s^1 \rightarrow a^1, s^2 \rightarrow a^2, \cdots, s^n \rightarrow a^n\}$$
$$= \overset{n}{\underset{j=1}{f_\alpha}}(s^j \rightarrow a^j) \tag{6}$$

where f_α is an appropriate implication operator (Terano, 1992), (Lee, 1990). In many cases we can further partition the set of rules in (6) into rule-bases (*RB*) with each rule-base being responsible for one action, i.e.,

$$R^N = \bigcup_{p=1}^{r} [RB^p] \tag{7}$$

Rule-bases (7) can made to reflect temporal partitions, that is we can have rules that describe the state of the system at t, i.e., of the form

$$s(t) \rightarrow a(t) \tag{8}$$

as well as rules that describe the possible state of the system at some time latter, i.e.,

$$s(t + \Delta t) \rightarrow a(t). \tag{9}$$

Thus an anticipatory fuzzy algorithm can infer the current action, $a(t)$ on the basis of the present state $s(t)$ as well as anticipated ones $s(t + \Delta t)$.

Generally, the rules of (5) describe relations of a more general type than that of functions, i.e., *many-to-many* mappings. Such mappings have the linguistic form of fuzzy *if/then* rules, e.g.,

$$if \quad \mathbf{X} \quad is \; A \quad then \quad \mathbf{Y} \quad is \; B \tag{10}$$

where $\mathbf{X}$ *is* a fuzzy variable whose arguments are fuzzy sets denoted as A, and $\mathbf{Y}$ is a fuzzy variable whose arguments are the fuzzy sets B [24],[25]. Similar rules pertaining to future states are of the form

$$if \quad \mathbf{X} \quad will-be \; A \quad then \quad \mathbf{Y} \quad is \; B \tag{11}$$

where $\mathbf{X}$ is thought of as a situation variable and $\mathbf{Y}$ the corresponding action variable. Evaluating of formulations using rules such as (10) and (11) is the part of fuzzy inference called *generalized modus ponens*, and amounts to drawing conclusions on the basis of imprecise premises. In *generalized modus ponens* we are given a fuzzy implication, such as (10), and a new fact that *imperfectly* matches (i.e., matches to some degree) the antecedent part of (10) and we are asked to find the new consequent. This is formally stated as

$$\begin{array}{l} if \quad \mathbf{X} \quad is \; A \quad then \quad \mathbf{Y} \quad is \; B \\ \underline{\quad\quad \mathbf{X} \quad is \; A' \quad\quad\quad\quad\quad\quad} \\ \quad\quad \mathbf{Y} \quad is \; B' \end{array} \tag{12}$$

where $\mathbf{X}$ and $\mathbf{Y}$ are two fuzzy variables and A, and B are their respective fuzzy values, and A' and B' are the corresponding new values. Generally $\mathbf{Y}$ *is B'* is deduced by taking the *max-* composition* of $\mathbf{X}$ *is A'* with the implication relation of the rule. In terms of membership functions, we are given $\mu_A(x)$ and $\mu_B(x)$ and hence (through appropriate choice of implication operator [13]) the membership function of the fuzzy implication, $\mu_I(x,y)$. In addition, we know $\mu_{A'}(x)$ (through observation/measurement or prediction). The analytical problem involved in the *generalized modus ponens* of (15) may be simply stated as: given $\mu_A(x)$, $\mu_B(x)$ and $\mu_{A'}(x)$, compute $\mu_{B'}(y)$. We let the fuzzy implication of (10) be represented by the relation, $R(x,y)$, where

$$R(x,y) = \int_{(x,y)} \mu_I(x,y) / (x,y) \tag{13}$$

The membership function of B' is generally derived by the *max-* composition* of A' and $R(x,y)$, i.e.,

$$B' = A' \circ R \tag{14}$$

or in terms of membership functions

$$\mu_{B'}(y) = \bigvee_{x} [\,\mu_{A'}(x) * \mu_{I}(x,y)\,] \tag{15}$$

where, $*$ is a binary operation such us *max*, *min*, or *arithmetic product* and we take the maximum with respect to x of all the pairs inside the brackets. Figure 3 illustrates that generalized modus ponens works in a manner analogous to *evaluating* a function, only now we evaluate something more general than a function, that is, a fuzzy relation (a *many-to-many mapping*).

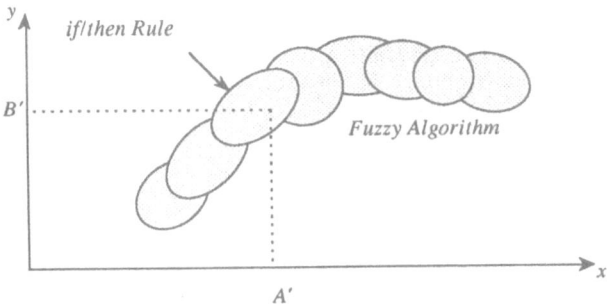

Figure 3. The compositional rule of inference is a mechanism for "evaluating," i.e., drawing inferences, from fuzzy linguistic descriptions.

Anticipatory control strategies may be based on global fuzzy variables such as *performance* where a decision at each time t is taken in order to maximize current as well as anticipated performance pertaining to $t + \Delta t$. *Performance* in this case is a fuzzy variable (with an appropriate set of fuzzy values) that summarizes information about the system allowing the system to make decisions about its change of state [9],[18],21. The observation/prediction of such variables can be addressed by the methodology presented in the next section.

Alternatively we may use fuzzy *if/then* rules to generate a decision from (7) and call a predictive routine to anticipate the effect of the proposed decision on the system output [20]. Additional rules may be called if the current decision will result in system behavior which is unacceptable. Consider for example the rule

*If the current decision (u_c) will cause the difference between the current and anticipated states to be **big**, then*

$$u = u_c(1 - \beta \cdot bigt) \tag{16}$$

where, β is a user-chosen parameter between 0 and 1 and *bigt* is the fulfillment function for the anticipated difference states. The parameter β may also be chosen by employing a predictive neural network [8].

3. Issues of Measurement and Prediction

Since the outputs of predictive models are fuzzy numbers we present here a method and an example of generating such numbers. It constitutes essentially the fuzzification (symbolization) of predictions generated by artificial neural networks in a process keen to measurement and thus we refer to it as *virtual measurement* [4],[5]. It should be remembered however that they are simply predictions involving fuzzy numbers.

Artificial neural networks consist of many simple processing units ("neurons") that can be globally programmed to perform a computation. They can be programmed to store, recognize and retrieve patterns, to filter noise from measurement data and may be used by themselves to control ill-defined processes. They are similar to fuzzy logic in that they both are *model-free-estimators*, i.e., they can estimate a function without an explicit analytical model of how outputs depend on inputs. A number of input-output pairs, called *examples*, are presented to the network and the connection weights are adjusted until the network has "learned" the underlying relationship that the examples represent. The process of presenting *examples* to the network is called *supervised learning* and the process of weight adjustment is referred to as *training*. The algorithm for training in the methodology presented is *back-propagation*. In virtual measurements neural networks are used to perform a mapping

$$f: M \to E \tag{17}$$

where, the domain M is the hyperspace of accessible variables such as temperatures and pressures in an engineering system and the output range E is a set of fuzzy numbers that constitute our predictions of fuzzy values referred to as *virtual measurement values (VMVs)* [4]. As discussed previously a fuzzy number is a normal and convex fuzzy set on the reals which models the value of a fuzzy variable at any given time, uniquely represented by a membership function [6],[25]. The fuzzy numbers used here have a trapezoidal shape. Trapezoidal membership functions are uniquely described by a set of four numbers, e.g., a given number $C = \{o_1, o_2, o_3, o_4\}$, where $0 \le o_1, o_2, o_3, o_4 \le 1$ and $\{o_1, o_2, o_3, o_4\}$ (from left to right) represents the universe of discourse components of the four corners of the trapezoid (from left to right). Such representation offers considerable advantage to computing speed.

The methodology for predicting fuzzy numbers has been presented elsewhere [4],[5],[17], and its main points may be summarized in the following steps:

1. Decide how many fuzzy values are necessary to adequately cover the range of the fuzzy variable to be predicted.

2. Determine the number of physically measurable variables that will be the basis (i.e., the input) of the virtual instrument.

3. Train one neural network per *VMV*, e.g., a program trained on five *VMVs* will require five trained networks.

4. Design an appropriate logic using the index of dissemblance (discussed below) to select which membership function will be the predicted value of the instrument at any given time.

The networks $N_1, N_2, \cdots, N_n$ comprising the virtual instrument are trained (in a process analogous to "calibration") with time series input vectors, and vectors $\{o_1, o_2, o_3, o_4\}$ representing fuzzy numbers as outputs. Each network learns each to map a constellation of input patterns to a particular linguistic label. The situation is illustrated in figure 4 where five inputs are used and n networks; hence this virtual instrument is calibrated with n fuzzy numbers. After training all networks $N_1, N_2, \cdots, N_n$ receive on-line time signals as inputs and produce a set of membership functions as outputs. Generally the outputs will be somewhat different from the membership functions the networks were trained for (the prototypes) and moreover one or at most two (if we allow overlap of membership functions) will represent correct values while the rest need to be ignored. It is thus important to identify the correct output. Since we consider each network's output to be a fuzzy number we use a *dissemblance index* [6] to estimate the outputs that are closest to the set of prototype membership functions on which we trained the networks and select one as the predicted fuzzy value [4].

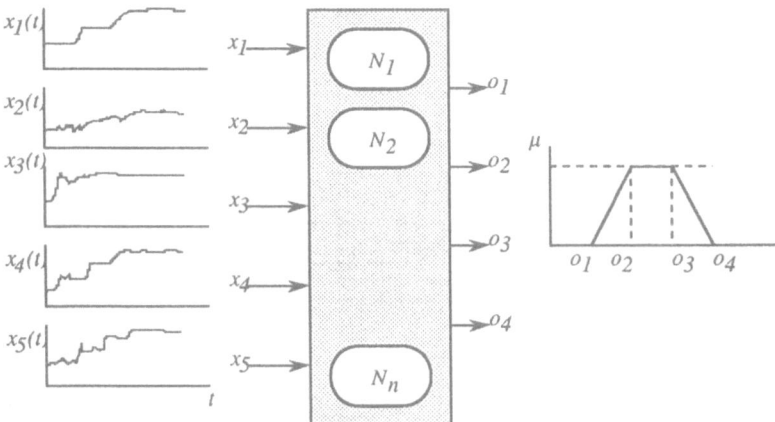

Figure 4. Each neural network in a virtual instrument maps a time-series input vector onto a vector $\{o_1, o_2, o_3, o_4\}$ representing a trapezoidal fuzzy number.

Using physically observable quantities to predict fuzzy values offers some unique advantages. A set of complicated time series is mapped to the universe of discourse of human linguistics through a neural network which acts as an interpreter of vital information supplied from the system. The information encoded in a time series is in the form of rate of increase/decrease, and maximum/minimum values attained over a period of time. The network is trained to represent this kind of "hidden" information in the form of membership functions which can be used for fuzzy inferencing such as shown in (12). The membership function provides sufficient information to predict the value of a fuzzy variable in the near future. Furthermore, a network trained to recognize a specific complicated time pattern (i.e., have a "crisp" value as output), will lose much of its ability to deal with noisy input signals since it will tend, for distorted inputs, to produce averaged forms of the desired output, missing therefore vital pieces of information.

As an example of the prediction method consider the following experiment. Actual data obtained during a start-up of the High Flux Isotope Reactor (HFIR) was used in order to test the methodology for predicting fuzzy values. HFIR is a three-loop pressurized water reactor operated at the Oak Ridge National Laboratory. A flow control valve on the secondary side of the system is used as the main mechanism for control (there is also a "trim flow control valve" for finer flow adjustments, as well as control rods). Although the signal sent to the motor of the valve is known, the actual position of the Secondary Flow Control Valve is not known and is rather hard to predict. The disk position is something that the operators of the plant "learn" how to estimate intuitively on the basis of experience. However, valve aging and varying plant operating condition as well as operator experience are major factors for substantial variations in the estimate of valve position.

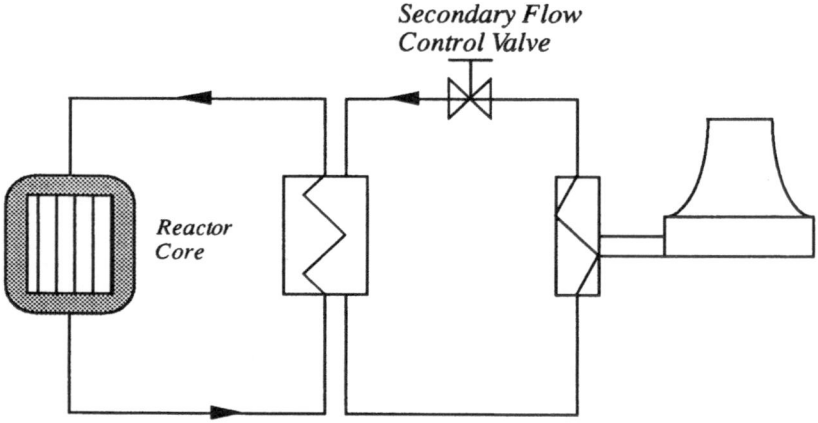

Figure 5. Schematic of the High Flux Isotope Reactor.

Five parameters were chosen as the basis for predicting the Secondary Flow Control Valve position: *neutron flux, primary flow pressure variation (DP), core inlet temperature, core outlet temperature* and *secondary flow*. All but the last one of the above mentioned time series, contain average values of the corresponding parameters of the three loop system. Figure 6 shows one of them, the secondary flow signal normalized in the range between zero and one. The parameters were selected in order to provide sufficient description of both the primary and secondary sides of HFIR during start-up. The time series of these five parameters are used to train five neural networks where each one of them has five nodes at the input layer and four nodes at the output layer (10 nodes in the hidden layer). In each network N (figure 4) there are five input nodes each receiving a time series from a physically measurable variable, and four output nodes representing the four corners of a trapezoidal membership function. The output is a membership function uniquely labeling a fuzzy value of the fuzzy variable describing the position of the Secondary Flow Control Valve, referred to as **VALVE_POSITION**. The data used for network training is normalized in the interval 0.1 to 0.9 and sampled every 16 seconds, with a total of 1240 samples available.

Designing a virtual instrument to predict **VALVE_POSITION**, requires first to partition its membership of discourse with the appropriate number of VMVs. We choose five vales, namely, *closed, partialy_closed, medium and partially_open* and *open*. Each value is represented by a membership function, i.e., $\mu_{closed}, \mu_{partially\_closed}, \mu_{medium}, \mu_{partially\_open},$ and μ_{open}. These five membership functions describe the position of the valve at every instant during the start-up period. The universe of discourse on which these membership functions are defined is the interval [0,1]. Thus, μ_{open} associates each point in the universe of discourse with the fuzzy value *open* at this point.

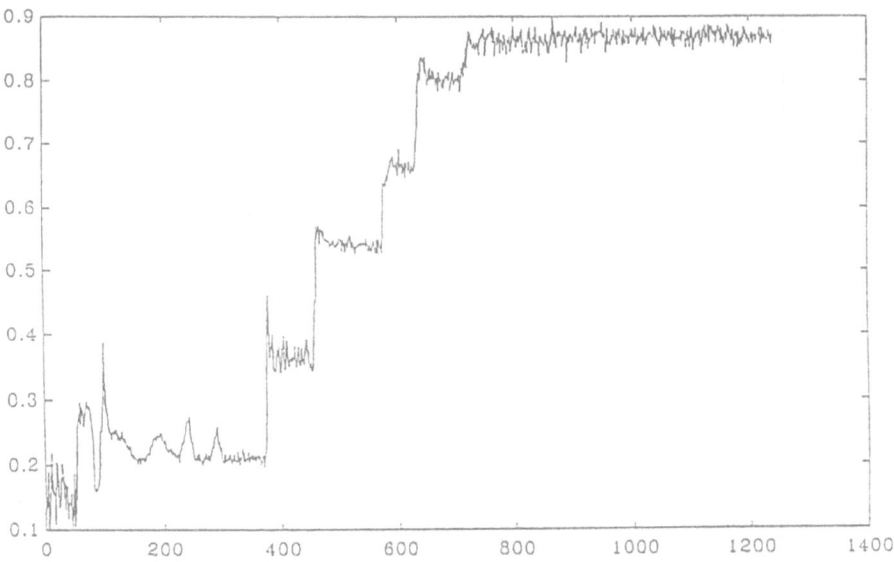

Figure 6. Secondary flow signal during start-up.

The membership functions representing the output of the predictive instrument in this particular study have trapezoidal shape or the degenerated (triangular) form of it, which is very useful for computations in the fuzzy control area. The membership function for *closed*, i.e., μ_{closed} is defined by a trapezoid with peak coordinates {(0.02, 0), (0.05, 1), (0.10, 1), (0.2, 0)}. Similarly, *partially_closed* , is represented by the trapezoid with coordinates {(0.15, 0), (0.2, 0), (0.30, 1), (0.4, 0)}, *medium* by {(0.35, 0), (0.4, 1), (0.50, 1), (0.6, 0)}, *partially_open* by {(0.5, 0), (0.6, 1), (0.7, 1), (0.75, 0)}, and *open* by {(0.7, 0), (0.82, 1), (0.85, 1), (0.90, 0)}. It is evident from the above geometrical schemes that there is an overlap between the membership functions used. The reason for the overlap is the fuzziness in the definition of the different states of valve position.

Figure 7 shows the prediction of the instrument during a startup of the reactor (1240 time steps). The valve is initially *closed* as seen by the membership function in the origin of the 3-d graph. It goes through the "medium" range rather quickly in the vicinity of 400-500 time steps and finally it becomes fully open after the 800 time step. Notice that this confirms rather well the trend shown in figure 6 where the secondary flow reaches its maximum value after about the 800 time step. To test the ability of each network to predict the valve position by calculating the right membership function at any particular time step, different levels of noise were introduced in the input signals. Initially up to 10% noise was introduced to all five input signals and the set of networks was tested with the "noisy" vectors. The appropriate networks fired at the corresponding time steps calculating the coordinates of the peaks of the corresponding membership functions with 98% accuracy. Henceforth there was an excellent prediction of the position of the disc valve during the whole time interval under consideration. In addition, 20% noise was introduced to all five input signals and the networks were tested again. The response of the system was indistinguishable form the previous case.

In addition, even when an input signal was dropped, actually substituted with random noise, the predictive instrument still predicted the valve position rather accurately. Figure 8 shows the output of the instrument when the secondary flow signal has been substituted with random noise. Comparing with figure 7 it can be observed that the instrument still indicates the valve position rather well. A series of statistical tests were conducted to confirm that the output of the instrument is actually within random error from the previous case. This is a significant tolerance to the informational hazard that the instrument was exposed to. Even with about 20% of its input information lost it still rather accurately measured the valve position. Similar results were obtained by dropping the other input signals one by one.

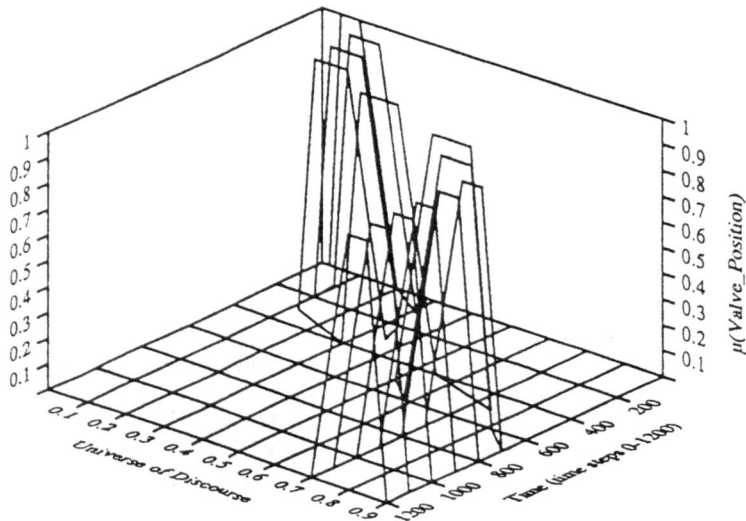

Figure 7. Virtual measurement for time steps 0-1240.

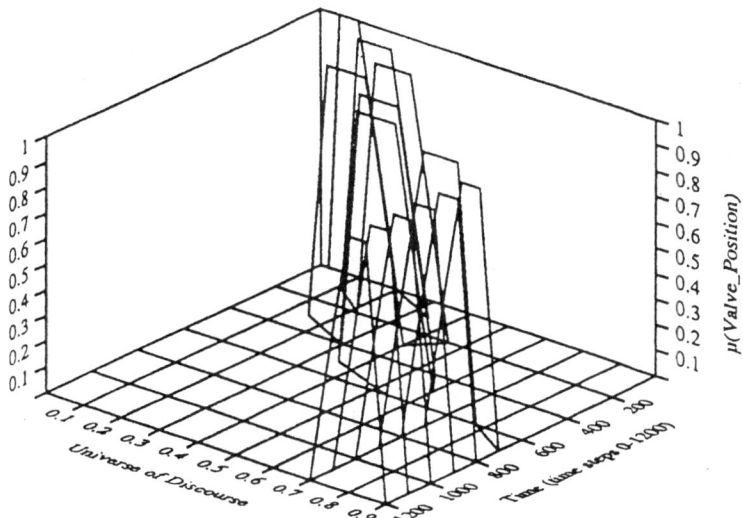

Figure 8. Virtual measurement for time steps 0-1240 when the secondary flow input signal has been substituted with 100% noise.

5. CONCLUSIONS

Anticipatory systems can utilize fuzzy predictions about the future in regulating their behavior through a new approach to observation that encodes system-specific knowledge at the level of the instrument called *virtual measurement*. It offers the possibility of timely and reliable estimation of variables not physically measurable (e.g., *performance* or *flow control disk position*) useful in the context for anticipatory systems whose change of states is based in present as well as future states. The virtual measurements contemplated for prediction are a form of *expert measurement*, i.e., measurement of system-specific fuzzy variables with functional significance. The values of such variables are fuzzy numbers representing, for example, the linguistic values of *performance*. Neural networks perform a mapping of physically measurable dynamic variables to fuzzy values each uniquely and unambiguously defined by a membership function. The failure-tolerance characteristics of such predictive instruments are demonstrated by performing rather accurate measurements and predictions of fuzzy values even with missing or partially distorted inputs. During the course of the lifetime of an anticipatory system both the physical characteristics of the system and the predictive instrument or model change. Calibration or fine-tuning of such predictive models may be achieved by additional training of the neural networks used. The development of an error monitoring algorithm on the basis of which further training may ensue (adaptively) is the subject of on-going research.

REFERENCES

[1]   Bellman, R.E., and Zadeh, L.A., *Decision-Making in a Fuzzy Environment*, Management Science, 17:4, pp. 141-164, December 1970.

[2]   Berkan, R. C., Upadhyaya, B.R., Tsoukalas, L.H., Kisner, R. A. and Bywater, R. L, *Advanced Automation Concepts for Large-Scale Systems*, IEEE Control Systems, Vol. 11, No. 6, pp. 4-12, October 1991.

[3]   Holland, J.H., *The Global Economy as an Adaptive Process*, The Economy as an Evolving Complex System, P. W. Anderson, K. J. Arrow, D. Pines, A Proceedings Volume in the Santa Fe Institute in the Sciences of Complexity, Addison-Wesley, 1988.

[4]   Ikonomopoulos, A., Tsoukalas, L.H., Uhrig, R.E., *Integration of Neural Networks with Fuzzy Reasoning for Measuring Operational Parameters in a Nuclear Reactor*, Journal of Nuclear Technology, Volume 104, pp 1-12, October 1993.

[5]   Ikonomopoulos, A., Uhrig, R.E., and Tsoukalas, L.H., *A Methodology for Performing Virtual Measurement in a Nuclear Reactor System*, Invited Contribution, Transactions of American Nuclear Society 1992 Winter Meeting, pp 106-109, Chicago, Illinois, November 15-20, 1992.

[6]   Kaufmann, A., and Gupta, M.M., Introduction to Fuzzy Arithmetic, Van Nostrand Reinhold, New York, 1991.

[7]   Kickert, W.J.M., Fuzzy Theories on Decision-Making, Kluwer, Boston, 1978.

[8]   McCullough, C. L., *Anticipatory Neuro-Fuzzy Control: A Powerful New Method for Real World Control*, Proceeding of IEEE International Workshop on Neuro Fuzzy Control, pp 267-272, Muroran, Japan, March 22-23, 1993.

[9]   Ragheb, M. and Tsoukalas, L., *Monitoring Performance of Devices Using a Coupled Probability-Possibility Method,* International Journal of Expert Systems, 1, pp 111-130, 1988.

[10]   Rosen, R., Anticipatory Systems, Pergamon Press, New York, 1985.

[11]   Rosen, R., Fundamentals of Measurements and Representation of Natural Systems, Elsevier North-Holland, New York, 1978.

[12]   Sugeno, M., and Yasukawa, T., *A Fuzzy-Logic-Based Approach to Qualitative Modelling*, IEEE Transactions on Fuzzy Systems, Vol. 1, No. 1, pp 7-31, February, 1993.

[13]   Terano, T., Asai, K., and Sugeno, M., Fuzzy Systems Theory and its Applications, Academic Press, Boston, 1992.

[14] Tomizuka, M., and Whitney, D.E., *Optimal Finite Preview Problems (Why and how is Future Information Important)*, Journal of Dynamic Systems, Measurement, and Control, pp 319-325, December 1975.

[15] Trakhtenbrot B. A. and Barzdin Y. M., Finite Automata Behavior and Synthesis, North-Holland Publishing Co., Amsterdam, 1973.

[16] Tsoukalas, L. H., Berkan, R. C., and Ikonomopoulos, A., *A Methodology for Uncertainty Management in Knowledge-Based Systems Employed within the Anticipatory Paradigm*, Proceedings of the Third International Conference of Information Processing and Management of Uncertainty in Knowledge-Based Systems, IPMU, pp. 77-80, Paris, July 2-6, 1990.

[17] Tsoukalas, L.H., Ikonomopoulos, A., and Uhrig, R.E., *Fuzzy Neural Control*, in Artificial Neural Networks for Intelligent Manufacturing, C.H Dagli, Ed., pp 413-434, Chapman & Hall, London, 1994.

[18] Tsoukalas, L.H., Ikonomopoulos, A., *Uncertainty Modeling in Anticipatory Systems*, Analysis and Management of Uncertainty , B.M. Ayyub, M.M. Gupta and L.N. Kanal Eds., *Machine Intelligence and Pattern Recognition Series*, pp. 79-91, Elsevier, North Holland. 1992.

[19] Washio, T., and Kitamura, M., *General Framework for Advance of Computer-Assisted Operation of Nuclear Plants - Anticipatory Guidance and Control for Plant Operation*, Proc. of Japan Atomic Energy Society, Oct. 1993, Kobe, Japan (in Japanese).

[20] Yasunobu, S., and Miyamoto, S., *Automatic Train Operation by Predictive Fuzzy Control,* in Industrial Applications of Fuzzy Control, M. Sugeno, Ed., pp 1-18, North Holland, 1985.

[21] Zadeh, L. A., *Fuzzy Sets and Information Granularity*, Advances in Fuzzy Set Theory and Applications, M. M. Gupta, R. K. Ragade, and R. R. Yager, Eds., pp. 3-18, North-Holland Publishing Company, 1979 .

[22] Zadeh, L.A., *Fuzzy Logic*, in IEEE Computer, pp 83-93, April 1988.

[23] Zadeh, L.A., *Outline of a New Approach to the Analysis of Complex Systems and Decision Processes*, IEEE Transactions on Systems, Man and Cybernetics, Vol. 1, pp. 28-44, 1973.

[24] Zadeh, L.A., *The Concept of a Linguistic Variable and its Application to Approximate Reasoning*, Information Sciences, Volume 8, pp 199-249, 1975.

[25] Zimmermann, H. J., Fuzzy Set Theory and its Applications, Kluwer-Nijhoff Publishing, Boston, 1985.

14

NEW APPROACHES TO LARGE-SCALE SCHEDULING PROBLEMS: CONSTRAINT DIRECTED PROGRAMMING AND NEURAL NETWORKS

YASUHIRO KOBAYASHI, HISANORI NONAKA
Energy Research Laboratory, Hitachi, Ltd.
7-2-1- Omika-cho, Hitachi-shi, Ibaraki-ken
Japan 312-19

1 INTRODUCTION

Industrial plant construction involves a large number of component installation tasks, such as setting up pumps, pipes, and other mechanical and electrical devices. During the construction, mutually related tasks make up a scheduling unit known as an activity, or a task. Computer aided methods have been employed to scheduling of activities, to improve construction efficiency and reliability.

Advanced scheduling methods have been developed on the basis of mathematical programming techniques [1] [2] or AI techniques [3] [4], but only a limited number have reached the stage of practical use [5] [6]. This is partly because highly automated methods are not necessarily applicable to ill-defined scheduling problems which planning engineers face in real world plant construction scheduling, and partly because highly interactive methods force planning engineers to make time-consuming scheduling refinements in a trial and error fashion.

The following points are commonly seen in scheduling problems for large-scale industrial plant construction.

(1) Various constraints on many interconnected activities must be taken into account.
(2) Schedules must be frequently revised to reach a satisfactory solution.
(3) Large numbers of activities are involved in the overall construction project.

Research and development efforts on advanced scheduling methods should be directed towards two goals: a new interactive method to build and revise schedules which will improve the revision efficiency; and a new automated method to facilitate schedule building through use of a more detailed scheduling model which will improve the solution quality.

The size of the scheduling problem characterizes the scheduling method. Hierarchical scheduling is a well established technique to deal with large-scale scheduling problems in the field of plant engineering related to design, manufacturing, construction and maintenance of industrial plants. The schedule of a large-scale system is decomposed

S. G. Tzafestas and H. B. Verbruggen (eds.),
Artificial Intelligence in Industrial Decision Making, Control and Automation, 421–445.
© 1995 *Kluwer Academic Publishers.*

into schedules of sub-systems, which are referred to as sub-schedules in this paper, according to the hierarchical structure of the system. These sub-schedules are planned as moderately sized sub-problems which are solved separately with the least interaction, and integrated to generate an overall schedule. The hierarchial scheduling technique includes the following steps: (i) scheduling of decomposed sections of a system; and (ii) integration of sub-schedules planned in step (i).

A knowledge-based interactive scheduling method has been applied to step (i) on the basis of constraint directed programming [7] - [9], since knowledge of scheduling experts is available and various kinds of domain-specific constraints should be incorporated at this level of scheduling. A similar interactive approach is applicable to step (ii), while the constraint propagation technique plays a less crucial role in constraint satisfaction because of the characteristics of constraints in this step.

For the second step a more automated method should also be pursued, since the problem is less ill-natured and can be straightforwardly formulated as an optimization problem with more tractable constraints. Neural networks [10] have been highlighted for their potential in dealing with combinatorial optimization problems since the Hopfield model of neural networks [11] was first successfully applied to a relatively small size traveling salesman problem [1] in the scheduling domain.

This paper covers both interactive and automated approaches to scheduling methods supporting a hierarchical scheduling system applicable to large-scale plant construction problems.

2 METHOD
2.1 Problem and Method Description
2.1.1 Problem

The problem is a combinatorial optimization problem to plan a feasible overall construction schedule with the most flattened resource usage. The objective function to be minimized is the daily resource usage variance integrated by time. Variables are starting dates of activities and sub-schedules. Constraints are imposed on activities in sub-schedules, and sub-schedules in the overall schedule.

As mentioned in the introduction, two kinds of scheduling problems are hierarchically solved in the domain of plant engineering; lower-level problems and upper-level problems. Figure 1 illustrates the conceptual scheme of hierarchical scheduling. Sub-schedules for all unit sections of a plant are planned as lower-level problem-solving. The overall schedule is integrated from sub-schedules as upper-level problem-solving.

423

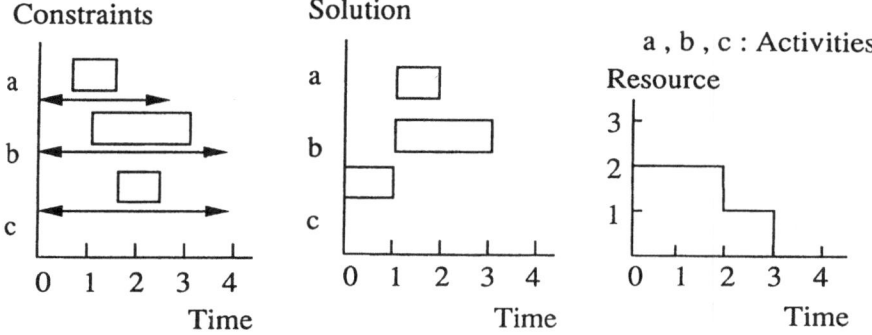

(a) Lower-level Scheduling (Sub-schedule A)

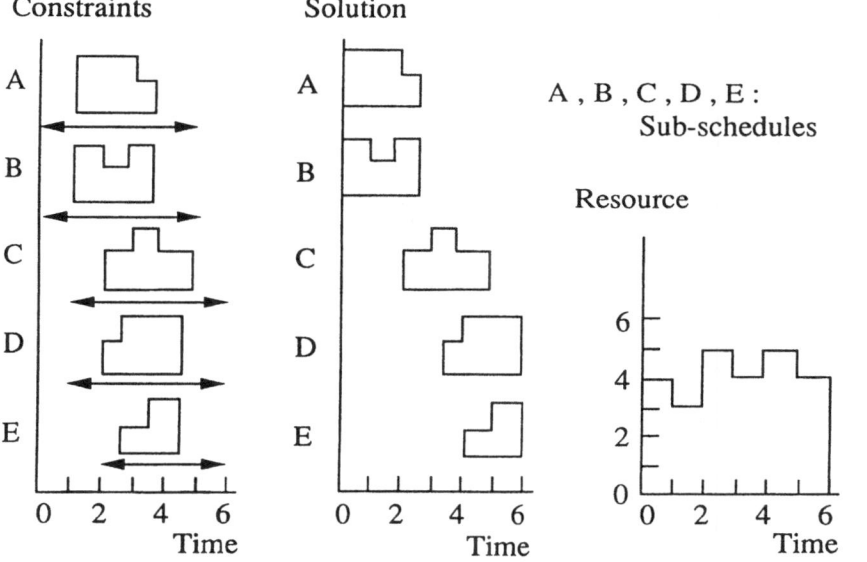

(b) Upper-level Scheduling (Overall Schedule)

Fig . 1   Hierarchical   Scheduling

2. 1. 2 Method

Scheduling methods are divided into two categories; interactive methods and automated methods. No cure—all methods are possible for large—scale and complicated scheduling problems in plant engineering. Both interactive and automated methods have features making them attractive to planning engineers. The former is advantageous for its high flexibility and wide range of applicability as a routine scheduling tool. The latter is advantageous for its high efficiency to reach the solution schedule based on a single criterion. Methods in both categories are developed and compared in this paper.

Scheduling problems are solved based on performance criteria. Typical performance criteria are completion time shortening and resource peak flattening in the overall schedule. Resource peak flattening is consistently pursued throughout this study, though the proposed methods are applicable to completion time shortening, with proper extension.

2. 2 Knowledge—Based Method for Lower—Level Problems
2. 2. 1 Interactive problem—solving

Problems are solved through interaction between a planning engineer and a sub—schedule generator in this approach. To support the planning engineer effectively, it is necessary to provide a sub—schedule generator as a scheduling tool in which the characteristics of the plant construction domain are properly reflected. The planning method consists of two processes: (a) an initial schedule is obtained under constraints based on a conventional mathematical programming technique for scheduling, and (b) the initial schedule is interactively improved, while a constraint propagation technique is utilized to resolve constraint violations automatically. Figure 2 shows the configuration of a scheduling system based on this method.

(1) Input data

(a) Construction task data: For activities to be scheduled, construction task data are prepared; they are a list of pre—tasks, an activity standard duration (i. e. the time to complete the activity), related resource types and their quantities required for the activity. Pre—tasks are activities which must be finished before starting the focused activity. The construction task data are used as work data on a datafile.

(b) Global constraints: Some constraints are imposed on most of the activities to be scheduled. The upper bound of the overall construction project length and that of the daily resource requirements are formulated by constraints of this type G.

(c) Local constraints: Some constraints are imposed on a limited number of activities. Most local constraints

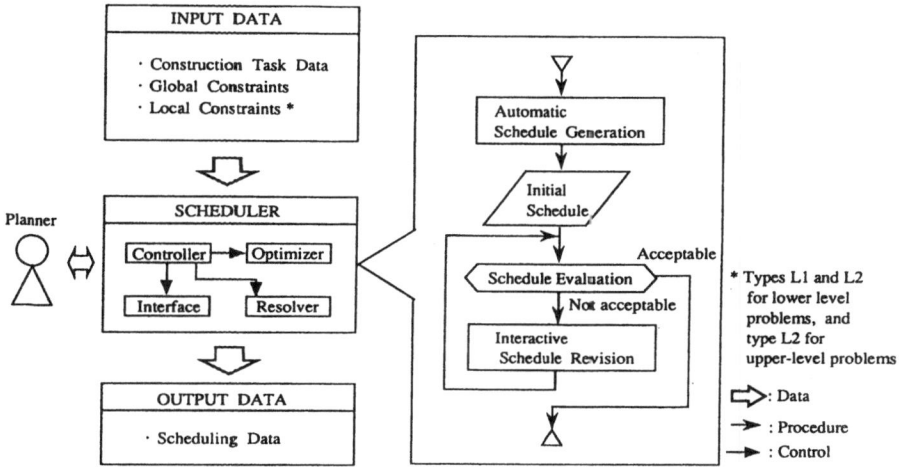

Fig. 2 Configuration of Interactive Scheduling Method

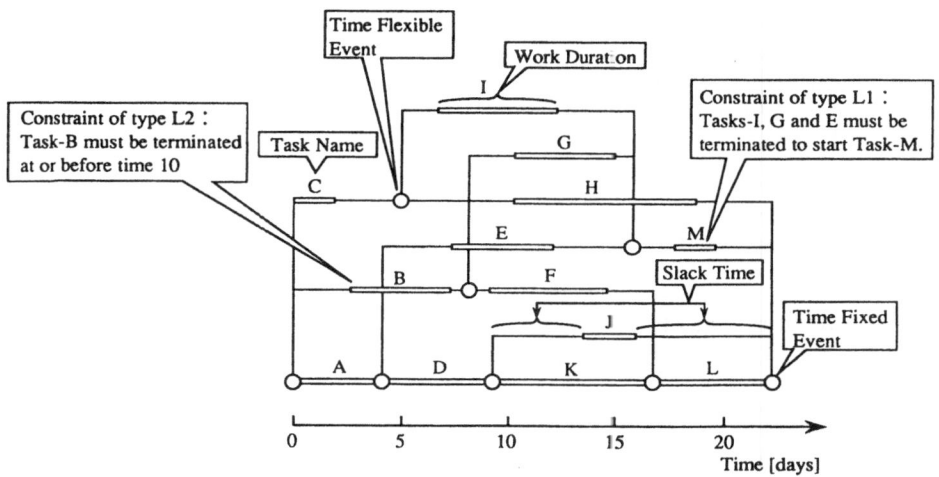

Fig. 3 Example of Unified Scheduling Chart

describe the relationship between two activities. Local constraints restrict start time or end time of tasks using a concrete date, like "Task–A must be terminated before 15 Nov. 1993", or in a relational way like "Task–B must be started within 2 days after Task–C is completed". Those constraints are described as declarative rules on a datafile in this knowledge–based method.

Constraints specifying an activity–to–activity relationship are referred to as type L1 constraints which are dominant in the lower level problem–solving step. Constraints specifying an activity–to–time relationship are refered to as type L2 constraints. Most of the latter are derived from conditions bounding an activity on the time axis; some simple examples are the ealiest starting date and the latest ending date of an activity. L2 constraints are dominant in the upper level problem–solving step.

(2) Automatic resource leveling

The distribution of the target resource requirement during the overall project term is leveled on the basis of a standard branch and bound method to prepare an initial sub–schedule. Details of this block are described later.

(3) Sub–schedule evaluation

The planning engineer evaluates the performance of the obtained sub–schedule from the sub–schedule generator. The sub–schedule is represented in a Unified Scheduling Chart (USC) as shown in Fig. 3. If the sub–schedule is an acceptable one, the scheduling process is terminated at this block.

(4) Interactive sub–schedule revision

If the subschedule is unsatisfactory, the planning engineer should revise it interactively, while any constraint violations caused in this block are automatically resolved. Details of this unique function are described in section 2. 2. 3. In this interactive process, the planning engineer can change the start time, end time and duration of activities, sequential order of activities, and global and local constraints, using the USC and a graphic interface.

(5) Output date

The final sub–schedule, represented in an USC, an advanced Gantt chart, and related attribute values make up the main output data.

2. 2. 2 Resource leveling function

The daily variation of required workers, machines and other resources is a kind of measure of scheduling performance and it should be minimized in the schedule for plant construction. Resource leveling is a major performance criterion in building an initial schedule in the first block. As the resource leveling problem is a combinatorial optimization problem, it generally takes considerable computational efforts to get the "best" (or the "optimal") solution. This method employs heuristics and the branch and bound method in the resource leveling procedure to get a "near best" solution in reasonable computational efforts. The process flow of the resource leveling function is shown in Fig. 4.

Scheduling heuristics are widely employed in selecting an activity to be scheduled, such as "Select a task that has the minimum slack time first" or "If some tasks have the same slack time, select a task which has the maximum resource requirement", to realize an efficient resource leveling procedure. One of the performance indices for resource leveling is the sum of the squares of the daily resource deviations during the overall project term.

There may be more than one schedule that satisfies the constraints on the sequence of activities, but some of them are not feasible because of timing constraints on individual activities. In this method, the feasible range of work time for the focused activity is checked so as to exclude futile trials which evaluate infeasible schedules. For example, candidate schedules are generated by modifying the work time of the selected task as long as the sequence of tasks is kept correctly (branch phase), then their feasibilities are checked to select the best schedule among the candidate schedules (bound phase).

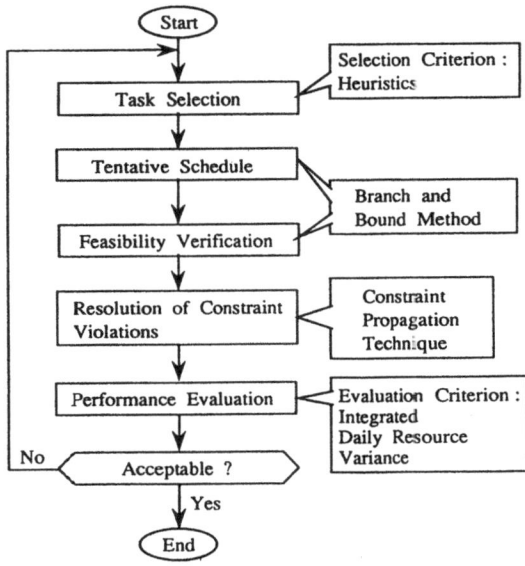

Fig. 4 Process Flow of Resource Leveling

2. 2. 3 Constraint satisfaction function

(1) Automatic constraint violation resolver

Constraints, as pieces of domain knowledge, play an essential role in schedule revision. As shown in Figure 5 the sequence of tasks and local constraints determines connections of tasks, which are represented by a constraint network. When constraint violations are caused by a schedule revision or other reasons, a constraint propagation technique is utilized to propagate the violations through the network, to adjust the start time or end time of some tasks, and finally to resolve the violations.

The configuration of the automatic constraint violation resolver is shown in Fig. 6. This function is composed of a monitoring program, modification program, and knowlege base. Each constraint is automatically translated into an LISP function, which determines whether the constraint is satisfied or not, and then it is recorded on the working memory.

The process of constraint violation resolution carried out by the modification program is as follows.

①Evaluate all the constraints. If all of them are satisfied, terminate the process. If there exist any constraint violations, select one of the violated constraints.

②Analyze the selected constraint to determine the task from which the constraint propagation is started. Here the focused task is known as a "master-task", and a task which is directly connected with the master-task by a constraint is a "slave-task".

③Modify start time or end time of the slave-task to resolve the focused-on constraint violation. Then regard the slave-task as a new master-task and evaluate the constraint which relates to it. If the constraint is not satisfied, return to step ②. And if all the constraints related to the master-task are satisfied, return to step ①.

Rescheduling is a conventional method to resolve the constraint violations, while the constraint propagation technique can resolve the violations efficiently by modifying a limited portion of the schedule.

(2) Monitoring program

The automatic constraint violation resolver solves constraint satisfaction problems by trial and error in the modification program. But this optimistic process does not ensure consistency of the schedule revision.

The monitoring program analyzes the history of constraint propagation, and detects abnormal phenomena caused by inconsistent constraints. There are two types of abnormal phenomena, LOOP and OSCILLATION. Table 1 summarizes some points regarding them. Suppose that r1, r2 and r3 are violated constraints. LOOP generates an endless repetition of the basic pattern, such as r1, r2, r3, r1, r2, r3, ⋯⋯, where the sequence r1, r2, r3 is the basic pattern. OSCILLATION

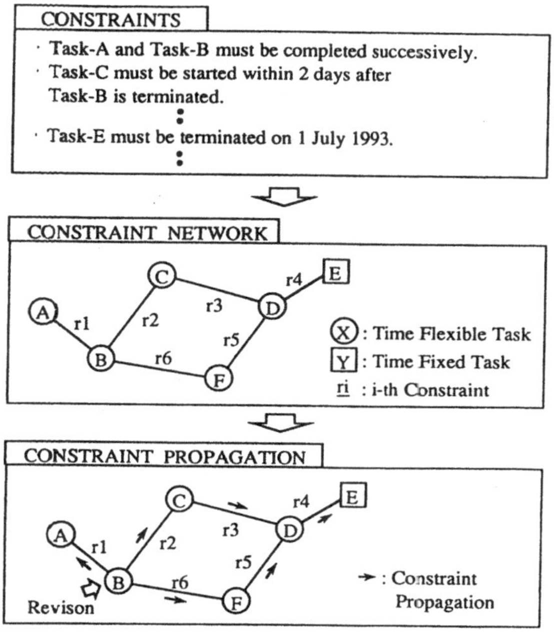

CONSTRAINTS

· Task-A and Task-B must be completed successively.
· Task-C must be started within 2 days after
  Task-B is terminated.
      ⋮
· Task-E must be terminated on 1 July 1993.
      ⋮

CONSTRAINT NETWORK

(C)   r4 [E]
(A) r1   r3 (D)
   r2   r5
(B)   r6   (F)

(X) : Time Flexible Task
[Y] : Time Fixed Task
ri : i-th Constraint

CONSTRAINT PROPAGATION

(C) → r4 [E]
(A) r1   r3 (D)
   r2   r5
(B) r6 (F)
Revison

→ : Constraint
    Propagation

Fig. 5 Generation and Utilization of Constraint Network

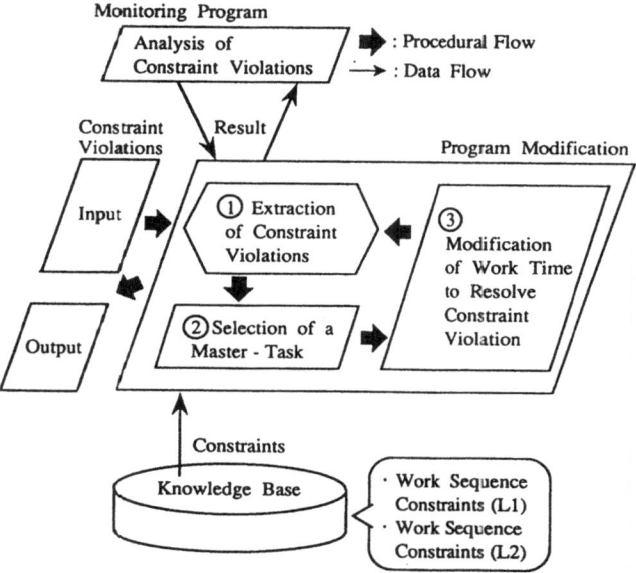

Monitoring Program

Analysis of
Constraint Violations

➡ : Procedural Flow
→ : Data Flow

Constraint
Violations   Result

Program Modification

Input

① Extraction
of Constraint
Violations

③ Modification
of Work Time
to Resolve
Constraint
Violation

Output

② Selection of a
Master - Task

Constraints

Knowledge Base

· Work Sequence
  Constraints (L1)
· Work Sequence
  Constraints (L2)

Fig. 6 Configuration of Automatic Constraint Violation Resolver

generates a repetition of the basic pattern and its reverse, such as r1, r2, r3, r3, r2, r1, r1, r2, r3, ⋯ ⋯. If the monitoring program detects those patterns in its rough analysis process, the history of the constraint propagation must be analyzed in detail, because those patterns do not necessarily denote the existence of a LOOP or OSCILLATION.

Here the amount to be modified is expressed for the focused data of task X to resolve a violated constraint rj in the i—th pattern Pi as $\Delta T(Pi, rj, X)$. If the equivalence relationship

$$\Delta T(Pi, rj, X) = \Delta T(Pi+1, rj, X)$$

is satisfied, it means that a LOOP exists. And

$$\Delta T(Pi, rj, X) = -\Delta T(Pi+1, rj, X)$$

signifies the existence of an OSCILLATION.
If the monitoring program detects a LOOP or OSCILLATION, it terminates the constraint propagation process, and informs the planning engineer of the abnormal phenomenon.

Automatic resolution of constraint violations assists planning engineers in revising the schedule without being hindered by troublesome constraint manipulation, and increases the planning efficiency.

Table 1   Abnormal Phenomena in
Constraint Propagation

	LOOP	OSCILLATION
Diagram	r1 C r3 A r2 B	r1 A r2 B r3
Rough Analysis	Repetition of Basic Pattern r1, r2, r3, r1, r2, r3 Basic Pattern Basic Pattern	Repetition of Basic Pattern and Its Reverse r1, r2, r3, r3, r2, r1 Basic Pattern Reversed Pattern
Detailed Analysis	State change in each BP is equivalent ?	State change in BP and that in RP are cancelled out ?

A, B, C   : Sample Tasks   BP : Basic Pattern
r1, r2, r3 : Violated Constraints   RP : Reversed Pattern

2. 3 Knowledge—Based Scheduling Method for Upper—Level Problems

The knowledge—based scheduling method is based on an interactive refinement process from an initial schedule. The initial schedule is generated by one of the automated scheduling methods with a simplified scheduling model, and improved interactively. Violations of constraints frequently caused in the interactive planning process are automatically resolved through the constraint propagation technique. This method is an extension of the approach for step (i). One of the major differences from the original one is that activities treated in this extended version have a timely variation in resource usage, and therefore this causes the heuristic man—machine interaction to be less tractable and less efficient.

2. 3. 1 Interactive problem—solving

Problems are solved through interaction between a planning engineer and an overall schedule generator. The overall schedule generator supports the planning engineeer in integrating sub—schedules, as the sub—schedule generator supports the planning engineer in integrating activities in the lower—level problems.

(1) Input data

(a) Sub—schedule data: Sub—schedules are prepared in the previous stage of lower—level problem—solving. Sub—schedules are defined by a date set of the relative starting date and duration of activities for a unit section in the plant construction project.

(b) Global constraints: In upper—level problem—solving, the overall construction project length is bounded and treated as a constraint. It is a global constraint, since all the sub—schedules are subject to this constraint.

(c) Local constraints: Most dominant constraints are imposed on the starting date in the upper—level problem—solving. This is because sub—schedules are handled independently of each other and integrated under the conditions of the earliest starting date and latest ending date of sub—schedules. These are constraints of type L2.

(2) Initial overall schedule generation

The initial overall schedule is prepared by a simple and quick procedure. Starting dates of all sub—schedures are set to the earliest date, which is explicitly obtained from the bounding constraints on the sub—schedule starting date. The initial schedule generated by this procedure does not supply a sufficiently flattened resource usage pattern in the time axis. This procedure assures that the starting point is an internal point in the feasible region, i. e. on the boundary of the feasible region. During the evaluation and revision cycle of the man—machine interaction, the search point is kept in the feasible region so that the all the sub—schedules

satisfy their bounding constraints in any overall schedule.
(3) Overall schedule evaluation
    The planning engineer evaluates the performance of the
overall schedule obtained from the schedule generator.   The
schedule generator has the same framework as the sub—schedule
generator.   The difference between these generators is
described in the subsequent section. If the overall schedule
is a satisfactory one, the scheduling process is terminated
at this block.
(4) Interactive overall schedule revision
    If the schedule is unsatisfactory, the schedule engineer
should revise the overall schedule interactively, by which
any constraint violations are automatically resolved. Since
most constraints are of type L2, correcting the violation of
bounding constraints is quite straightforward and efficiently
processed.   In this interactive process, the planning
engineer can change the start time of the sub—schedules,
using the USC and a graphic interface.
(5) Output data
    The final overall schedule, represented in the advanced
Gantt chart, and related attribute values are the main output
data.

2. 3. 2 Resource leveling function
    Major differences between lower—level problems and upper—
level ones are
(a) the profile of the resource usage pattern of activity, and
(b) the type of constraints to be treated.
In upper—level problems, a sub—schedule is regarded as an
activity.   The "activity" in this level has a resource usage
pattern which is not simple. Its complicated nature tends to
make the interactive process of resource leveling less
tractable.   The activity in the upper—level problem is mainly
imposed on by constraints of type L2. The straightforward
bounding constraints tend to make the resource leveling
process more tractable as suggested in the preceding section.
The tractability of this interactive problem—solving depends
on the balance of these two factors, and on the grain size of
the sub—schedules.

2. 4 Neural Networks for Upper—Level Problems
2. 4. 1 Neural network modeling for combinatorial
        optimization
    The scheduling problem discussed in this paper is a kind
of combinatorial optimization problem.   The Hopfield model
for neural network computation demonstrated the applicability
of neural networks to the traveling salesman problems (TSP's)
[11]. The TSP's are formulated with 0/1 integer variables
which are mapped to neural network unit outputs, and the
problem—solving is done through the energy minimization
process by the neural unit interaction.   The neural network

approach to TSP's has been extended and applied to the scheduling problem, since the interaction between neural units is more complicated in this problem domain than in the TSPs.

As modeled in the TSP's, an integer variable of the starting date of a sub-schedule is replaced by multiple zero-one variables in the present model. Its value is 1, if the sub-schedule starts on the corresponding date, and 0, if not. Potential starting dates of sub-schedules are represented by units of a neural network, and the outputs of units are converged so only one of the units for a sub-schedule is 1, and the others are 0 in the course of computation.

The resource usage of the overall schedule is calculated from outputs of units and the resource usage pattern of sub-schedules. The integrated variance due to daily resource usage variation is given by a quadratic form of the unit output vector. Explicit constraints come from the requirement that a sub-schedule appears only once in the overall schedule, though proper neural network modeling facilitates elimination of constraints related to the requirement that a sub-schedule is placed in the specified range of the overall schedule. These constraints are embedded into the objective function as a penalty term which is also given as a quadratic form of the unit output vector.

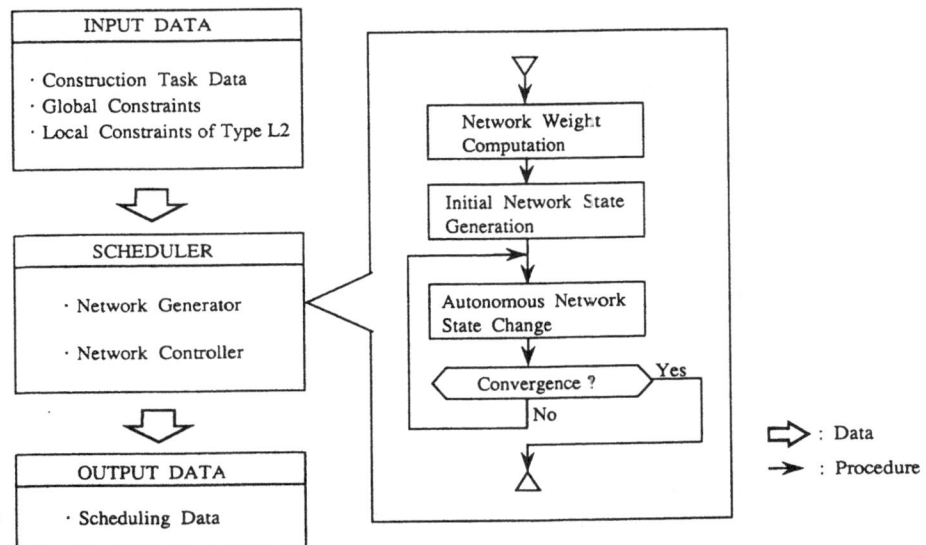

Fig.7 Configuration of Automated Scheduling Method

2. 4. 2 Automatic problem—solving in upper—level

Figure 7 shows the configuration of a scheduling system based on this method.

(1) Input data

(a) Construction task data: For sub—schedules to be scheduled, construction task data are prepared; (i) the feasible ranges for starting date of the sub—schedules, and (ii) the resource usage patterns in the sub—schedules. The former are given as boundary conditions of hierarchical scheduling, and the latter are given from the results of the lower—level problem—solving.

(b) Global constraints: The length of the overall schedule is bound to the maximum allowable project term. This is a typical example of constraints of type—G, when the resource peak reduction is the main scheduling target in this study. This constraint determines the deadline of the overall schedule, and it is finally masked by local constraints as suggested in the following.

(c) Local constraints: Sub—schedules are bound to the proper region in the overall schedule range. Each sub—schedule is limited to start at or after the earliest possible starting date, and to finish at or before the latest posibble ending date. The latest possible ending date is supposed to be the same as or before the deadline of the overall schedule. These straightforward constraints are examples of constraints of type L2.

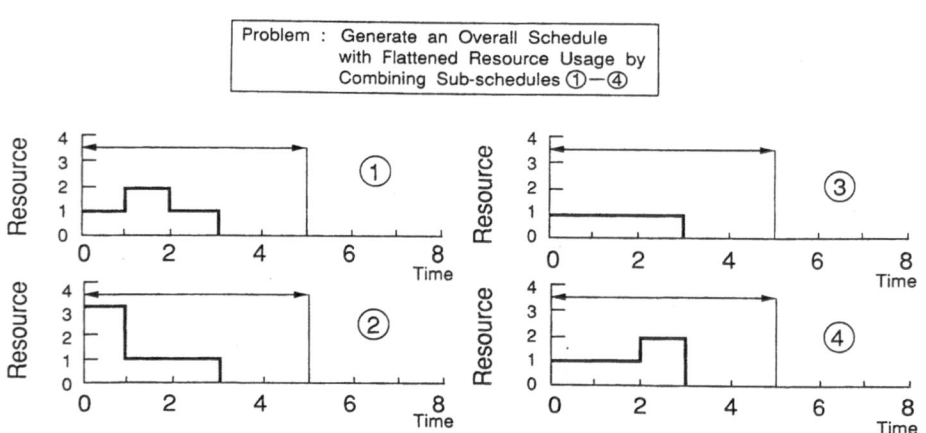

Fig. 8 Illustrative Scheduling Integration Problem

2. 4. 3 Formulation of upper—level problems

The formulation of the problem as a Hopfield model is summarized in the following. Figure 8 illustrates an example problem which determines the overall schedule with a flattened resource peak from four sub—schedules ①–④.

(1) Starting date of sub—schedule

The starting date of a sub—schedule is described by 0/1 integer variables for possible dates, not by a single date variable.

    u(i) : zero—one variable for date i,
           output value of neural unit i
    s(i) : sub—schedule index for date u(i)
    d(i) : day index for date u(i)

In the neural network for the example problem (Fig. 9) the starting date of sub—schedule ④ is formulated as a set of units 10 — 15. Sub—schedule ④ can start from 6 different times 0 — 5, or day indices 1 — 6. The output from neural unit 10 specifies if sub—schedule ④ starts from the first available day or not. But if it is 1, the sub—schedule starts from the day. If it is 0, the sub—schedule does not start from the day.

The resource chart shows that sub—schedule ④ starts from days 2 and 6, since units 11 and 15 give the non—zero value. This schedule obviously violates the constraint which requires that the sub—schedule should appear once in the overall schedule.

(2) Resource amount of overall schedule

The resource amount of the overall schedule at any date is represented by a linear function of 0/1 date variables for sub—schedules. This fact is derived from the variance of resource usage integrated through the overall project term being defined as a quadratic form of 0/1 date variables.

    p(k, m) : resource amount caused by sub—schedule k
              at m—th day from its starting date
    y(j) : resource amount of overall schedule at
           date j
           $y(j) = \sum_i \sum_m p(k, m) * u(i)$; $j = d(i) + m - 1$, $k = s(i)$

An overall schedule for the example problem is shown in Fig. 10. The dashed line in the resource chart gives the accumulated resource from sub—schedules ①–④. Sub—schedule ①–③ appear once within the schedule, but sub—schedule ④ appears twice. The second appearance of subschedule ④ contributes to the increase in resource amount in the latter part of the schedule $y(6)$ — $y(8)$.

    T: overall schedule length
    Y: average resource amount through overall schedule

       $\bar{y} = \sum_j y(j) / T$

    F: integrated variance of resource usage through

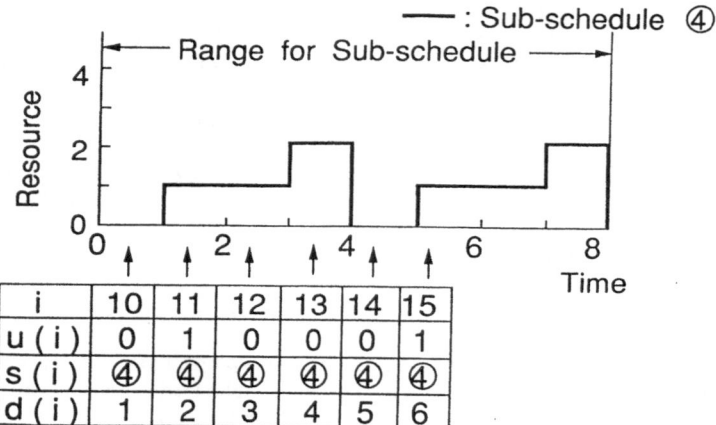

i	10	11	12	13	14	15
u (i)	0	1	0	0	0	1
s (i)	④	④	④	④	④	④
d (i)	1	2	3	4	5	6

Fig. 9 Optimization Problem Formulated with
Hopfield Model (1)
— Starting Date of Sub-schedule —

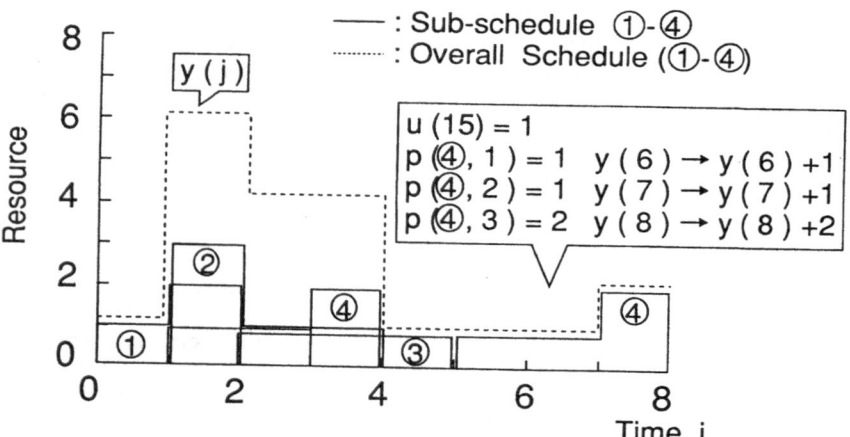

Fig. 10 Optimization Problem Formulated with
Hopfield Model (2)
— Resource Amount Overall Schedule —

overall schedule

$$F = \sum_j (y(j) - \bar{y}) **2$$

In this example case, $\bar{y}$ is 2.5 and F is equal to
(−1.5) + (3.5) + 2·(1.5) + 3·(−1.5) + (−0.5).

(3) Penalty term for constraints imposed on sub-schedules

The penalty term for constraints imposed on sub-schedules
is represented by a quadratic form of 0/1 date variables.

Mi(k) : index set of i satisfying k=s(i)
   G: penalty term for sub-schedule constraints

$$G = \sum_k \{ (\sum_{i \in Mi(k)} u(i) - 1) **2 \}$$

In the example case in Fig.10, G is equal to 1.0, since
sub-schedules ①-③ appear once, i.e. $\sum u(i) = 1$, and sub-schedule
④ appears twice, i.e. $\sum u(i) = 2$.

(4) Converted objective function

The energy function is, hence, the sum of the integrated
resource variance and the penalty term.

   E: energy function
alpha: weighting factor for penalty term
w(i, j) : connection weight between units i and j
         in neural network
t(i) : parameter for threshold characteristics of unit i

$$E = F + alpha*G$$
$$= (1/2) * \sum_i \sum_j w(i, j) *u(i) *u(j) + \sum_i t(i) *u(i)$$

(5) Network connection weights

The connection weights for a pair of units and the
threshold parameter for a unit can be obtained through the
comparing the coefficents of terms u(i)*u(j) and u(i) on the
right hand sides of the two equations above. The network
connection with a non-zero weight for the two units is
illustrated in Fig.11. Unless a pair of units has a
connection with a non-zero weight, there is no interaction
between them through the objective function and penalty term.

438

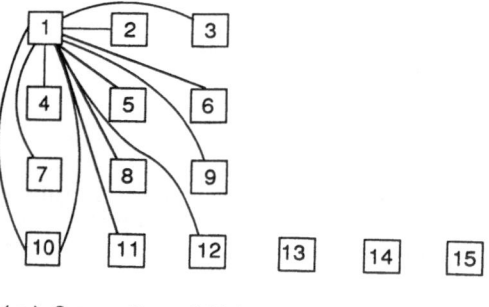

(a) Connection of Unit 1

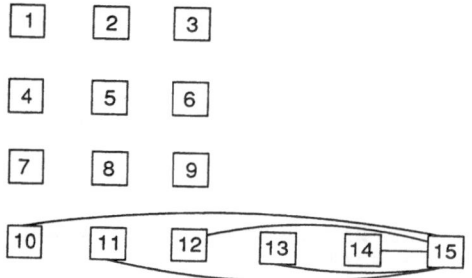

(b) Connection of Unit 15

Fig. 11 Nodal Connection for Sample Units
in Illustrative Problem Network

Table 2 Work Data of the Sample Section

No.	Task Name	Pre-tasks	Duration [days]	Resource [units/day]
1	RQI	——	30	4
2	TRI1	——	18	12
3	TRI2	TRI1	27	12
4	D1	TRI2	0	0
5	LPF1	TRI1	8	8
6	LPF2	LPF1	20	8
7	LPF3	LPF2, D1	22	8
8	LPW1	LPF1	10	4
9	LPW2	LPW1	80	4
10	LPW3	LPW2, LPF3	10	4
...	...	...	...	...
...	...	...	...	...
...	...	...	...	...

3 APPLICATION EXAMPLES

3.1 Scheduling systems

Scheduling systems have been developed through two different approaches. The first scheduling system based on the knowledge-based interactive method was written in COMMON-LISP on an engineering workstation HITAC-2050/32. The second scheduling system based on the Hopfield model neural network was written in FORTRAN and implemented on a mainframe computer HITAC-M280H.

3.2 Problem

3.2.1 Sub-schedule generation

The scheduling system was experimentally applied to a scheduling problem in power plant construction. The lower-level problems are solved to determine 80 sub-schedules for construction units which are called "areas" including 20 to 100 tasks, while the upper-level problem is solved to determine an overall schedule from those sub-schedules.

Table 2 shows work data format of a typical area. There were 30 tasks and 38 constraints for the order of tasks. Table 3 shows 13 local constraints for start time or end time of tasks. For each task, the standard time needed to complete it and the required daily resource amount were also prepared. One of the global constraints on sub-schedules is that the length of its critical path is equal to or less than its target project term. One sort of resource is taken in this problem, though multiple resources could be treated by the scheduling system after converting them into a single measure.

3.2.2 Sub-schedule integration

The scheduling systems were applied to the upper-level problem, i.e. the integration of sub-schedules from the lower-level problems. Sub-schedules for 80 areas from the first step of the hierarchical scheduling are combined to generate an overall schedule with flattened resource usage variation. In both interactive and automated methods, candidate starting dates are selected with intervals of 2-10 days, based on experience with the lower-level problems. In the neural network for the Hopfield-model-based method, about 1500 units are employed to represent candidate starting dates of sub-schedules.

3.3 Results

3.3.1 Sub-schedule generation results

Figure 12 shows a scheduling diagram in the user interface of the scheduling system. An initial schedule is supplied to the planning engineer before the interactive resource leveling process. Together with the USC, the time scale is given in the top region and the resource chart in the bottom region. All the tasks in the schedule were begun

Table 3 Local Constraints of Type L1
for the Sample Section

Constraint : Continuous Operation

No.	Constrained Tasks		
1	TRI1	and	TRI2
2	LPF1	and	LPF2
3	LPF2	and	LPF3
4	LPW1	and	LPW2
5	LPW2	and	LPW3
6	LSF1	and	LSF2
7	LSW1	and	LSW2
8	SPF1	and	SPF2
9	SPF2	and	SPF3
10	SPW1	and	SPW2
...	...	...	...
...	...	...	...
...	...	...	...

Table 4 Output Data of the Sample Section
after Resource Leveling

No.	Task Name	Starting Time	Completion Time	Earliest Starting Time	Latest Completion Time	Slack Time
...	...	...	...	...	...	...
...	...	...	...	...	...	...
...	...	...	...	...	...	...
21	SPF3	74	101	74	152	51
22	SPW1	56	63	52	77	18
23	SPW2	63	101	59	152	55
24	SPW3	101	106	101	157	51
25	SSI1	68	75	59	84	18
26	SSI2	75	111	66	157	55
27	SSI3	111	118	106	164	51
28	CPI	74	164	74	164	0
29	WC	84	164	66	164	18
30	BNI	0	10	0	164	154

at their earliest possible date, and the peak of the resource
chart was 61 units/day for the case shown.
    It took about 10 minutes to perform the resource leveling
using a 1.0MIPS class workstation.   The resource peak was
reduced to 44 units/day, with all the constraints being
satisfied.   The output data format of the schedule are shown
in Table 4.   Then an expert planner revised the schedule
interactively using the USC and a graphic interface to narrow
the resource peak.   After about 5 minutes work, the peak was
reduced further to 42 units/day, while the duration of one
task (SPF3) was prolonged and automatic resolution of
constraint violations was carried out.
    The transition of resources during the scheduling
procedure is shown in Fig. 13.   The planning engineer decided
to accept the schedule at this stage as a satisfactory one.
In this case, it took an expert planner more than 3 hours to
make a satisfactory schedule by hand.   Experience in cases
like this example suggested that with the use of this
intelligent interactive system, the scheduling efforts could
be reduced to approximately 10% of that by hand accompanied
by a conventional interactive scheduling system.
    Without using heuristics and functions for resource
leveling and constraint violation resolution, the efficiency
of the procedure deteriorates drastically, as does the
schedule quality.   As combinatorial optimization problems
often have more than one local optimum, some of which are far
worse than a near best solution, an unsophisticated method
easily leads to an insufficient solution.

3. 3. 2 Sub-schedule integration results
    Obtained overall schedules are shown in Figs. 14 (a) – (c)
for the reference solution with earliest starting dates, the
Hopfield-model-based solution, and the knowledge-based
interactive solution, respectively.   Results of (b) and (c)
are overall schedules with more flattened resource usage
variation than the reference solution.   This suggests that
these two methods are applicable to schedule integration to
obtain proper local optima in the optimization problem.
    Normalized peak values of resource usage are 2.35 for the
reference solution, and 1.67 for a solution by the Hopfield-
model-based method, and 1.53 for that by the knowledge-based
interactive method.   A comparison between these two results
regarding local fluctuation of resource usage variation, as
well as in normalized peak values, suggested that the less
automated and more demanding knowledge-based method gives a
solution of higher quality.
    These experimental results verified that two scheduling
systems are applicable as a practical tool for planning
engineers and that proposed scheduling methods, which are
based on the constraint directed programming and the neural
networks, are viable and promising for further extension.

442

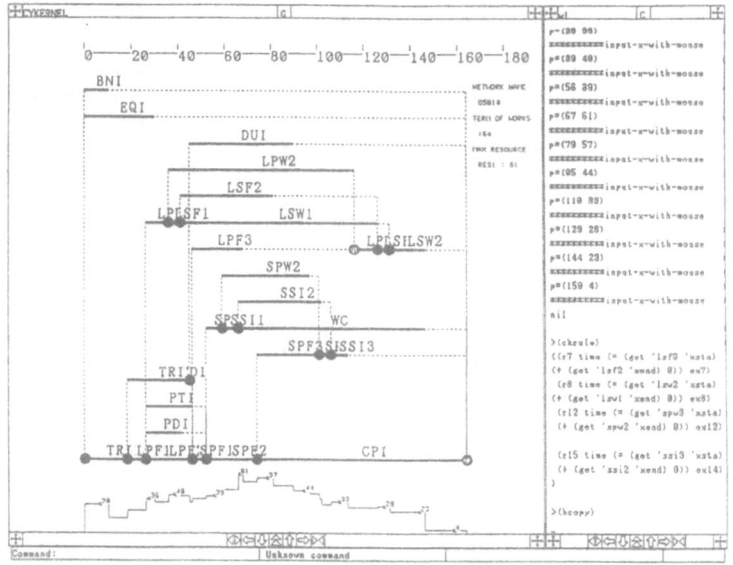

Fig. 12   User Interface of Scheduling System
(Sheduling Diagram Before Resource Leveling)

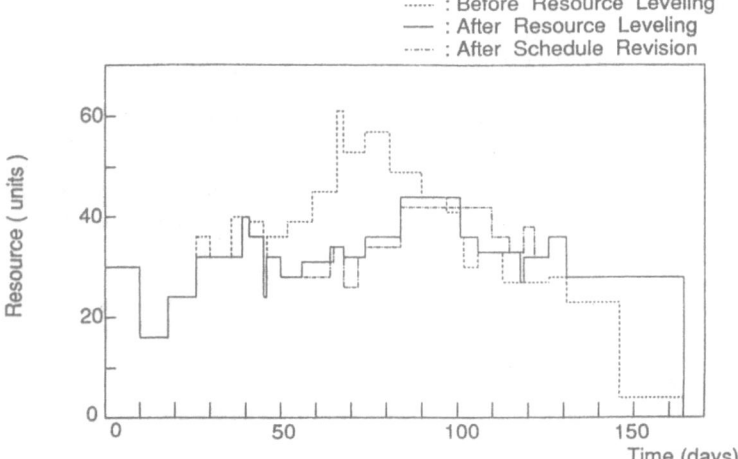

Fig. 13 Transition of Resources Diagram
Throughout the Scheduling Process

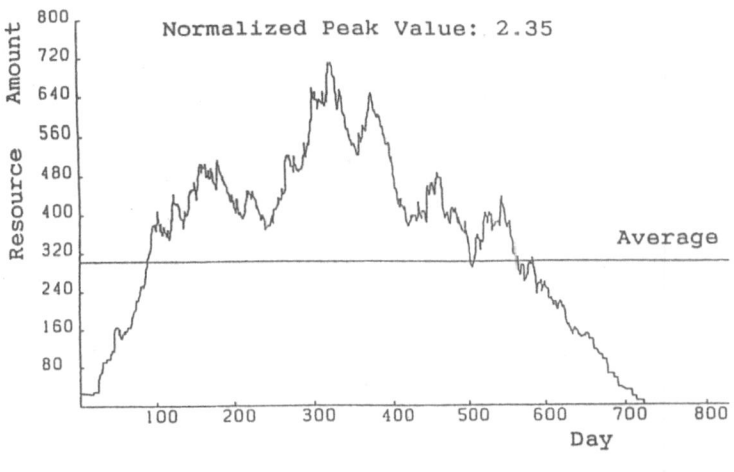

(a)   Reference Solution

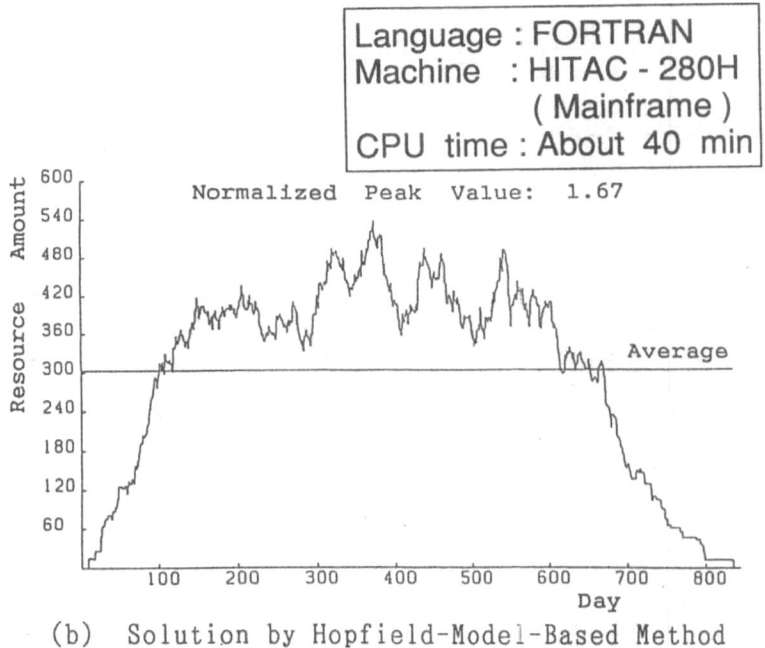

(b)   Solution by Hopfield-Model-Based Method

Fig. 14 Comparison of Schedule Integration Solutions (1)

444

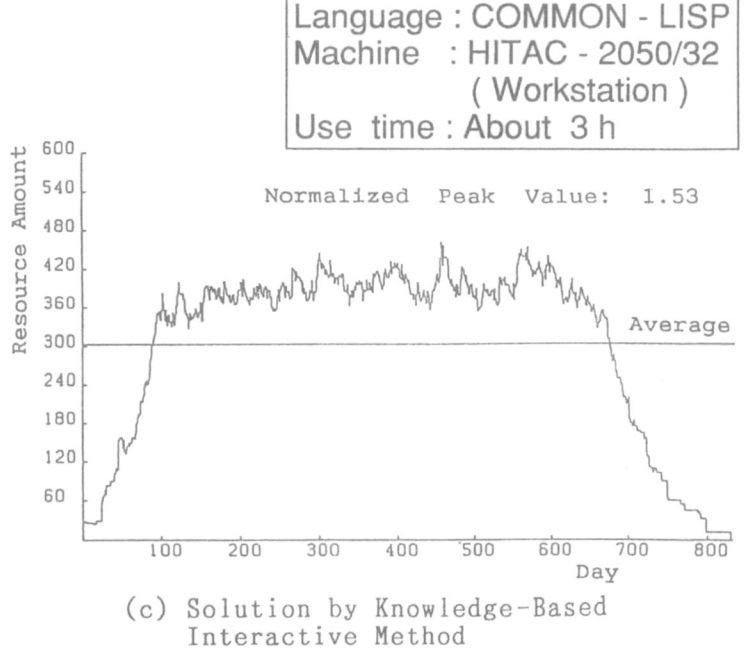

Language : COMMON - LISP
Machine : HITAC - 2050/32
(Workstation)
Use time : About 3 h

Normalized Peak Value: 1.53

(c) Solution by Knowledge-Based
Interactive Method

Fig. 14 Comparison of Schedule Integration Solutions (2)

4 CONCLUSIONS
Two kinds of AI methods has been developed and evaluated for large—scale scheduling problems in plant construction. A heiarchical scheduling process is employed to solve the large—scale scheduling problem in two steps: (i) generation of sub—schedules for construction units, and (ii) integration of the sub—schedules to generate the overall schedule.

The first method is based on constraint—directed programming and the second one is based on a neural network of the Hopfield model. The former has a unique function which identifies a constraint violation through an interactive schedule modification process for resource leveling, and automatically corrects the violation to ensure a feasible schedule. The latter realizes an automatic resource leveling function for the sub—schedule integration step through the network energy minimization process, using the neural network directed formulation.

Scheduling systems based on the proposed methods were implemented and verified as practical tools through their experimental application to a problem of generating a plant construction schedule with the leveled resource requirement under constraints. The results suggested that the first method is advantageous regarding resource leveling power, and that the second contributes more to reducing efforts of the planning engineer.

REFERENCES
[1] G. L. Nemhauser and L. A. Wolsey: Integer and Combinatorial Optimization, Wiley Interscience (1988)
[2] L. Davis, et al. : Schedule Optimization with Probabilistic Search, Proc. of 3rd Conference on Artificial Intelligence Applications, pp. 231—235 (1987)
[3] N. J. Nilsson: Principles of Artificial Intelligence, Springer—Verlag (1982)
[4] G. Bruno, et al. : A Rule—Based System to Schedule Production, IEEE Computer, pp. 32—39 (1986)
[5] H. Nonaka, et al. : A Support Method for Schedule Revision Using AI Techniques, Proc. of IAEA Specialists' Meeting on Artificial Intelligence, Helsinki, VTT Symposium 110, vol. II, pp. 219—231 (1990)
[6] M. Numao and S. Morishita: Scheplan — A Scheduling Expert for Steel—Making Process, Proc. of International Workshop in Artificial Intelligence for Industrial Applications, pp. 467—472 (1988)
[7] G. J. Sussman and G. L. Steele Jr. : CONSTRAINT — A Language for Expressing Almost Hierarchical Descriptions, Artificial Intelligence, vol. 14, no. 1, pp. 1—39 (1980)
[8] R. Dechter and J. Pearl: Network—Based Heuristics for Constraint—Satisfaction Problems, Artificial Intelligence, vol. 34, no. 1, pp. 1—38 (1987)
[9] M. S. Fox: Constraint—Directed Search: A Case Study of Job—Shop Scheduling, Research Notes in Artificial Intelligence, Pitman (1987)
[10] P. K. Simpson: Artificial Neural Systems, Pergamon Press (1990)
[11] J. J. Hopfield, et al. : Neural Computation of Decision Optimization Problems, Biolo. Cybern. , vol. 52, pp. 141—152 (1985)
[12] Y. Kobayashi, et al. : Application of Neural Network to Schedule Integration in Plant Engineering, Proc. Int. Neural Network Conf. (INNC—90), vol. 1, pp. 287—290 (1990)

PART 4
SYSTEM DIAGNOSTICS

15

KNOWLEDGE-BASED FAULT DIAGNOSIS OF TECHNOLOGICAL SYSTEMS

H. VERBRUGGEN(*), S. TZAFESTAS(**), E. ZANNI(**)
(*) Control Lab., T.U. Delft, Delft, The Netherlands
(**) IRCU, Comp. Engrg. Div., NTUA, Athens, Greece

1. INTRODUCTION

Fault diagnosis (FD) of man-made systems lies in the core of modern technology and attracts increasing attention by both theoreticians and practitioners. Actually, FD is one of the major concerns in industrial and other technological systems operation. In recent years a great deal of work has been done in the direction of designing systems (hardware and software) that are able to automatically diagnose the faults and malfunctions of an industrial process on the basis of observed data and symptoms. FD provides the prerequisites for fault tolerance, reliability and safety that are fundamental design features in any complex engineering system. Complex automatic industrial and other systems usually consist of hundreds of interdependent working parts which are individually subject to malfunction or failure. Total failure of these systems can present unacceptable economic loss or hazards to personnel or to the system itself. Hence, most modern systems involve: (i) a plan of maintenance which replaces worn parts before they malfunction or fail and (ii) a monitoring mechanism that detects a fault as it occurs, identifies the malfunction of a faulty component, and compensates for the fault of the component by substituting a configuration of redundant elements so that the system continues to operate satisfactorily. FD is actually this monitoring function and involves four subfunctions, namely *detection, prediction, identification,* and *correction of faults* during the on-line operation of the technological system at hand.

The two primary approaches to FD of industrial systems are:
- *Model-based* (MB) approach (using mathematical models and analytical redundancy)
- *Knowledge-based* (KB) (or artificial intelligence) approach (which tries to imitate the reasoning of human fault diagnosers and operators).

Of course several combinations of them can also be used.

A global view of MB-FD can be found in [1], and a wide set of issues on KB-FD are given in [2]. Our purporse in this chapter is to provide a survey study of the second approach. To draw up conclusions about the industrial plant, a diagnostic expert system is designed that uses a knowledge base where concepts are expressed in terms of inference rules, semantic nets, object oriented languages or a combination of these formalisms. The development of a KB-FD system for an industrial plant requires general specification tools that are suitable for different

S. G. Tzafestas and H. B. Verbruggen (eds.),
Artificial Intelligence in Industrial Decision Making, Control and Automation, 449–506.
© 1995 *Kluwer Academic Publishers.*

application areas. These tools must involve unambiguous concepts to acquire and express the various expertise levels, and a set of programs plus a suitable method to integrate and treat the successive knowledge layers.

A typical KB-FD Expert System consists of three components: an *inferece engine*, a *knowledge base* and a *working memory*. Declarative descriptions of expert-level information, necessary for problem solving, are stored in the knowledge base. The inference engine solves a problem by interpreting the domain knowledge stored in the knowledge base. The inference engine also records the facts about the current problem in a special purpose workspace, called working memory. The working memory may also include modules for a natural language communication with the user, a reasoning explanation interface as well as an automated knowledge acquisition tool.

The division between the knowledge base and inference engine has two important advantages. First, if all of the control structure information is kept in the inference engine, then one can engage the domain expert in a discussion of the knowledge base alone, rather than of questions of programming and control structures. Second, the versatility of the system is increased. If all of the task specific knowledge has been kept in the knowledge base, then it is possible to replace the current knowledge base by a new one and obtain a performance program for a new task. This does not mean that the inference engine and the knowledge base are completely independent. The knowledge base content is influenced by the inference engine, since the rules written for the knowledge base take into account the inference engine and its built-in control strategies. Expert Systems efficiency is highly affected by the structure of the knowledge base, and more specifically, by the way knowledge is represented.

A classification of fault types is the following:
- i) Solid-nonintermittent faults
- ii) Transient-intermittent faults
- iii) Critical faults
- iv) Single faults
- v) Multiple faults
- vi) Topologically active faults
- vii) Topologically inactive faults
- viii) Catastrophic/noncatastrophic faults

The consequences of *solid faults* can be reproduced, i.e. when a given sequence of input values produces a specific sequence of values at the output (that declare a given faulty behaviour with specific symptoms), then this input value sequence will always produce the same output sequence (symptoms). Thus these faults can always be detected with appropriate tests. The consequences of *transient faults* cannot be reproduced and so these faults are very difficult to be localized (detected). *Critical faults* affect directly the logic behaviour of the system (e.g. circuit) and so they always produce errors of the type stuck-at-1 (s-a-1) or stuck-at-0 (s-a-0) that can be easily detected. In the *single fault* case, the fault is located at only one component of the system. Thus the symptoms that are produced are due to only a single cause and can be detected. In the *multiple fault* case there are more than one fault and the corresponding errors are detectable with more difficulty. Sometimes a single fault appears to have the symptoms of a multiple fault. *Topologically active faults* are faults that produce new signal path sections, i.e. new internal variables. They actually change the schematic representation (topology) of the system (e.g. in the case where the system is a circuit this can be done via shortcircuits). The errors produced are in general nondetectable. *Topologically inactive faults* either don't affect the system topology or they affect it only indirectly by reducing the number of paths of the information flow (i.e. by reducing the number of internal variables). Catastrophic faults when they occur lead to a system catastrophy (e.g. explosion of a nuclear reactor) and every effort must be made to avoid them. Noncatastrophic faults (may be repairable or not) lead always to a safe failure of the system.

The structure of the present chapter is as follows. Section 2 presents the basic issues of knowledge representation and acquisition for fault diagnosis, summarizing nine tools for automated knowledge acquisition (TEIRESIAS, ETS, MORE, MOLE, SALT, KNACK, KADS, KEATS and KRITON). Section 3 provides a general discussion of first-and-second-generation diagnostic expert systems and in particular their relative merits and drawbacks. Shallow reasoning, deep reasoning and qualitative reasoning are discussed here in some detail. Section 4 discusses the fault-diagnosis methodologies and 2nd generation expert system architectures. Section 5 gives a survey of a number of digital systems diagnostic tools, namely the D-algorithm, Davis methodology, the IDM tool, the DART (I,II) tool, the IDT tool, the LES tool, and more briefly some other systems (Hamscher, Critter, Arby, NDS). Section 6 presents a general methodology for the development of fault diagnosis tools in the digital circuits domain, describing in full the constraint suspension technique and the candidate generation algorithm. Section 7 provides a brief exposition of a general methodology for the development of fault diagnosis tools in the process engineering domain which uses the digraph concept for the systems modelling. Finally, Section 8 discusses an implementation of the methodology of Section 6, including two simple examples of system-user dialogue that arrives at a diagnosis.

2. KNOWLEDGE REPRESENTATION AND ACQUISITION FOR FAULT DIAGNOSIS

2.1. Knowledge representation

Knowledge representation is the heart of knowledge engineering. To represent knowledge means to engineer knowledge: that is, to convert knowledge into an applicable form. An assortment of techniques is being used to represent knowledge including, first-order predicate calculus, rule-based systems, associative networks, frame systems, object-oriented systems and attribute grammar systems [3].

Rule-Based Systems: The methodology used in rule-based systems [4,5], originated from a framework called "production systems" proposed by Post [6] that was used in a wide spectrum of problems. A rule-based system nay be viewed as the assembly of three major parts: i) *working memory* that holds the facts, the goal and intermediate results, ii) *rule memory* which holds all the rules of the system, and iii) *rule* *intepreter* that decides what rules are applicable and in which order they should be executed or fired. Rule-based Expert Systems utilize both forward and backward chaining. However, due to their inability to represent structured knowledge, they combine some of the features of frame-based systems. Representative rule-based Expert Systems include the MYCIN [7], the DENDRAL [8], the Meta-DENDRAL [8] and the EMYCIN [9].

Frame-Based Systems: The concept of *frame*, first appeared in [10], is central in frame-based systems. Minsky describes frames as "data structures for representing stereotyped situations." In frame-based systems, knowledge is represented in data structures called frames. Frames may have a number of attribute descriptors, called slots. The slots may be either associated with a value or attached to a procedure written in a host programming language. Procedures that are activated when a slot is about to be updated and check the validity of the updates, are called *constraints*. Procedures triggered in response to assertions in slots, are called *demons*, and execute a series of commands. Procedures that are activated in response to queries on the values of the slots, are termed as *watchdogs*. Frames provide mechanisms to form some kind of hierarchical structure, and may be considered as a special kind of node in a taxonomy that represents objects or classes of objects. The frame at the top-most level of a hierarchical structure represents a so-called superclass, while frames at lower levels represent subclasses. Due to their structured representation, frames are very powerful in inherence-based inference and they perform adequately well in analogical reasoning and reasoning about events and procedures [11]. Although some frame-based systems combine rule-based features, the following systems can be basically considered frame-based: AM system [12] EURISCO [13] and PI [14].

Logic-Based Systems: Logic based systems are based on first-order predicate calculus and have as their main implementation instrument the PROLOG programming language. Knowledge is represented by simple and complex predicate statements. The fundamental proving (and problem solving) method employed in first-order predicate calculus is called *resolution refutation* [15]. Resolution refutation is the method by which, in order to prove proposition p, given a state description T, the negation of p (~p) is assumed, and conjugating ~p with T, a contradiction is attempted to be proved. If a contradiction exists, then p stands. Predicate logic, besides proving theorems, may also be used as a problem-solving method [16]. Logic-based systems, although capable of representing descriptive knowledge like frames [17,18], they cannot represent complex structures. For that reason, although they can support deductive methods, inductive methods, abduction and non-monotonic reasoning, among others, they are less suitable than frames for analogical and inheritance-based reasoning.

Object-Oriented Systems: Lately, much effort has been invested for the development of object-oriented systems for knowledge representation [19,20]. Objects are data structures, which besides being able to represent descriptive knowledge, they are also armed with private procedures, or methods, as they are also termed. The notion of a private procedures is that it can only be activated by the objects it has been designed for, and its scope is restricted entirely by those objects. Objects communicate with each other by messages. Messages may be viewed as triggering mechanisms for the execution of a method, but they are not considered as just another fancy name for procedure calling. Messages are not concerned with the information that is needed for the execution of the method, thus, they have no parameters. It is the responsibility of the recipient object to pass back a result, by activating whatever methods are appropriate. By enforcing this concept, it is obvious that good modularity is one of the features of the system.

In object-oriented systems, the inheritance mechanism has been expanded to include methods as inherited entities, as well. That is, methods defined in the superclasses are inherited by objects belonging in its subclasses.

Attribute Grammar Systems: Although this class could be discussed under the Logic-Based Systems, it is presented separately for its particular importance . Attribute Grammars (AG) were devised by Knuth as a tool for the formal specification of programming languages [21]. Recently, attribute grammars were also proposed as a tool for knowledge representation and logic programming [22]. The relation between attribute grammars and logic programming was theoretically investigated in [23]. In [24], it was shown that the parsing mechanism and the semantic notation of attribute grammars can be combined to represent the control knowledge and the knowledge base of logic programs, respectively. The practical implementation of this approach to situations where knowledge may be expressed in the form of logic rules was studied in [24]. A full account of the AG knowledge representation model along with solutions of the implementation problem of AG interpreters, as well as several extensions (to include "why" and "how" explanations, probabilistic and evidential inference etc.) and application examples may be found in [25].

Comparison of Knowledge Representation Methods: Knowledge representation is of critical importance to Expert Systems performance. A study comparing different methods of knowledge representation may be found in [26]. Comparisons are made on the efficiency and the degree of difficulty in implementation for rule-based systems, frame-based systems and logic-based systems.

In terms of implementation, the rule-based systems are the easiest to implement, while the most difficult are the frame-based systems. Due to the fact that pointers are used, the performance of frame-based systems remains almost constant, increasing the volume of knowledge representation. Rule-based systems have an overall acceptable performance, although their efficiency is deteriorating in the presence of high volume knowledge data. Logic systems are fairly difficult to be implemented and compared to the other two methods, less efficient; this is in accordance with intuition, because resolution is considered to be a time-consuming process.

Although the comparison provides a guideline for performance in knowledge representation, one must keep in mind that a specific representation scheme may be tightly coupled with the application domain of the Expert System. Each representation has its limitations and its strengths and it should be used in accordance to the specific implementation.

2.2. Knowledge Acqusition

Knowledge acquisition is the critical bottleneck in expert system development. It is the problem that many knowledge engineers consider to be the basic obstacle for the fast and correct development of expert systems. Without correct and complete knowledge, the expert system will lead to failure independently of how good is the design and implementation of its other components. Usually the knowledge acquisition process is very difficult and time consuming, whereas the incorporation of the acquired knowledge in a software tool is relatively easy.

Knowledge acquisition is distinguished in *human-aided* and *machine-aided* knowledge acquisition. Human-aided knowledge acquisition (elicitation) involves structured and unstructured interviews, focused meetings, questionnaires and rating scales, teachback interviews, ethnographic interviews, the technique of "20" questions, the laddered grid technique, and other formal techniques. Machine aided knowledge acquisition aims at eliminating the knowledge engineer and getting the expert working directly with a shell or tool. Looking at the literature one can observe that the current trend is towards machine-aided knowledge acquisition. One of the main reasons for this tendency, among others, is that human labour as a proportion of expert system development costs rises more quickly than other costs, although using a human intermediary appears to be less effective.

A survey of the human-aided knowledge acquisition techniques can be found in [27]. Here a survey of the knowledge-acquisition tools will be provided. The basic approaches for developing knowledge acquisition tools are: use of induction from examples (e.g. in EXPERT-EASE), use of knowledge base editors (e.g. TEIRESIAS), use of problem solving methods (e.g. ETS and MOLE), and use of language based approaches (e.g. KADS). Here the following knowledge acquisition tools, that reflect their historical development, will be reviewed: TEIRESIAS, Expertise Transfer System (ETS), MORE, MOLE, SALT, KNACK, KADS, KEATS and KRITON [79].

TEIRESIAS: The name of this knowledge acquisition tool comes from the ancient greek seer Teiresias. It contains knowledge about how MYCIN's knowledge is represented and used [28]. Its function is to help a domain expert add new knowledge to MYCIN. Actually it does not attempt to derive knowledge on its own, and so it is particularly applicable at the later stages of knowledge acquisition (viz refinements and expansion of the knowledge base). TEIRESIAS was developed at Stanford University [29]. Most system building aids are of the knowledge base editor type like TEIRESIAS.

ETS: This program is appropriate for supplying the initial knowledge base to TEIRESIAS, since it is designed to facilitate rapid prototyping "from scratch" [30,31]. ETS can interview experts, analyze the knowledge acquired, build knowledge bases and help the expert combine his knowledge with that of other experts. The expert's knowledge is expressed by ETS in the form of a repertory (rating) grid, and rules are generated accompanied by certainty factors. Actually, since very little training of the expert is required for using ETS, this tool can become an expert-driven tool, thus avoiding many of the human interviewing problems. ETS was followed by a larger tool, called AQUINAS, that makes the problem solving method more explicit to the expert using it.

MORE: It is similar in some ways to TEIRESIAS and ETS in providing a mechanism for interviewing domain experts [32,33]. MORE can detect gaps and errors in a knowledge base, and so it can drive the interaction process with the expert to lead to better diagnostic conclusions. It uses a kind of heuristic classification [34] and has been used for building fault diagnosis systems in domains such as computer networks.

MOLE: It is similar to TEIRESIAS and MORE since it can build expert systems for diagnosis and classification. It is also based on a kind of heuristic classification. Initially, it asks the expert to list all relevant events (symptoms and hypotheses) of the domain under study, and then asks for knowledge that explains the symptoms and helps to discriminate which hypothesis is the most likely explanation of a symptom. MOLE then employs this knowledge to construct a network of associations. Among the reported applications of MOLE are fault diagnosis of car engines and steel rolling mills [33].

SALT: Similar to MORE and MOLE this "single-mode" tool is suitable for building ES that use a propose-and-revise method for synthesizing complex tasks (such as construction, plan creation and design). It is easy to be learned by experts who do not have prior knowledge of programming. The interaction with the expert is conducted at the knowledge level by asking for a task description in terms of the types and relationships of the model within the tool. The applications of SALT include VT that configures elevator systems [35].

KNACK: Similar to MORE, MOLE and SALT this tool has been developed at Carnegie Mellon University [36]. It is concentrated at report generation, i.e. the expert is required to provide a skeletal report and the domain concepts and vocabulary. KNACK builds expert systems called WRINGERS which help the expert produce reports on plans and designs, and then to suitably integrate them. The WRINGER is then employed to use sample strategies provided by the expert in order to acquire values that instantiate the concepts in the generalized sample report.

KADS: This methodology and tool has been developed at Amsterdam University [37-39]. It focuses more broadly on the analysis of human problem solving and expert behaviour. The key idea of KADS is the interpretation model, where the generality of a language-based approach with the strength of a single-model system are combined. The KADS methodology was applied to printed circuit board manufacturing, credit guarrantee underwriting and commercial loans assessment [40].

KEATS: It consists of a tightly integrated set of tools operating in a bottom-up manner [41]. It actually incorporates four subsystems, namely (i) CREF - a cross-referencing editing facility, (ii) KDL - a knowledge description language, (iii) GIS - an intelligent graphical interface, and (iv) COPS - a context oriented production rule interpreter. KEATS overcomes the drawback of KADS methodology regarding the limited assistance provided in structuring the data [41]. However KEATS has the drawback that runs on the symbolics machine, whereas KADS runs on any system working under UNIX.

KRITON: This is a general purporse tool (named after a disciple of Socrates) similar to KADS and KEATS [42]. Its output is translated into an intermediate knowledge representation system that involves a description language in the form of a semantic net and a set of operator-argument structures derived from the protocol analysis. KRITON supports automated interviewing, forward scenario simulation, protocol analysis, combining repertory grid, semantic text analysis, and COLE -a machine learning-by examples system.

3. FIRST-AND SECOND-GENERATION DIAGNOSTIC EXPERT SYSTEMS

3.1. General Issues

The experience of a human expert, suitably coded, can form the knowledge base of an expert system. When the accumulated experience of an expert cannot help him (her) directly to face a diagnostic problem, he (she) has to resort to the fundamental principles that govern the system (problem) at hand. In situations of this type a good ES must have a similar reasoning ability. Although most of the existing ESs behave in a way similar to domain human experts, they do not usually have the more advanced human inference abilities (based on creativity, imagination and so on). This is due to the type of knowledge that is embodied in the knowledge base of the ES and to the representation and management style of the knowledge. The knowledge used in solving problems that need human experience is distinguished in two categories:
- Shallow or empirical or compiled knowledge
- Deep or functional knowledge

The shallow knowledge is usually drawn from human experts who have a personal experience on the particular knowledge domain, and is represented in rule form. These rules are rules of thumb that model the solution process in the way the expert proceeds in the problem solution. In other words, these rules constitute a processed form of the experience and perception gained by the expert after many years of work in that particular knowledge domain.

The deep knowledge describes the performance of the system (problem) under investigation in a cause-effect/deterministic way using the structure, the topology and the internal laws of the system operation. Deep knowledge is based on deep models that can derive their own behaviour for a given set of parameters and signals, and predict what should be the effects of changes in them.

3.2. First-Generation Expert Systems

The majority of first-generation expert systems (FGES) are designed and implemented on the basis of a shallow reasoning (i.e. reasoning with shallow knowledge). They do not possess essential knowledge for the way the hypotheses are related with the respective conclusions and the way in which the system at hand operates or should operate. Thus FGES possess some limitations, the principal ones being the following.

(I) *Knowledge-base limitations:* There is a difficulty in the initial construction of the KB. First, all possible relations must be enumerated via suitable look-up tables or dictionaries or complete decision trees. Second, much effort has to be given for ensuring internal consistency of the KB, i.e. the consistency of the rules, since the knowledge embodied in them is experiential and not the result of general principles governing the system at hand. The rules are arbitrarily provided and structured according to the subjective thoughts of an individual expert, and so there is an increased risk of contradiction.

(II) *Difficulties in the management and maintenance of the KBs* which are excessively large.

(III) *Difficulty in the extension of the KB* through the addition of new rules.

(IV) *Inference engine limitations:* The ES is unable to provide the reader with precise and sufficient justification on the reasoning way ("why", "how", "what-if", "when"). This is due to that the ES has not available causal replies on how the conclusions are drawn from the assumptions.

(V) *Difficulties in the overall operation of the ES.* The ES is unable to face problems that lie at the boundary or outside the area of its application, and to recognize when a problem goes beyond this boundary or concerns an incomplete part of the KB. In these cases, it is possible that the ES will proceed to solve the problem (by not taking into account the fact that it is not able to solve it) and so the user is uniformed on the fact that the results have doubtful validity. First-generation ESs are unable to use the experience gained in the course of their appication, and so to improve their performance with the time.

(VI) *Inadequacy of user interfaces:* The majority of first-generation systems have restricted ability to understand and accept information entered only in specific formats and representations/terminologies. They actually ignore any other information. Furthermore, the ESs using menu-driven interfaces are inflexible, although they save the user from having to learn these terminologies and formats. The lack of flexibility is mainly due to the fact that the system does not have an explicit data model and a user model. A data model helps the system to understand better the user information and abstract and draw conclusions in more intelligent manner. The user model helps in constructing the questioning for the elicitation of further knowledge by the user. These limitations of first-generation ESs have naturally led to the idea of developing second-generation ESs with common reference points the *deep knowledge* (in place of the shallow knowledge) and the use of new inference (reasoning) methods.

3.3. Deep Reasoning

As we have seen, the *deep knowledge* modelizes the system (problem) itself and the way it functions and performs. A large number of scientists have proposed several methods for reasoning with deep knowledge, the principal ones being the following [43-47]:
- Qualitative reasoning
- Causal reasoning

- Functional reasoning
- Reasoning from first principles
- Reasoning from structure and behaviour

All the above deep-level reasoning methods have a similarity in the way they examine the system at hand and differ only in their details. The fundamental principles of the deep reasoning methodology are the following:

(1) *The description of the behaviour of the system at hand must be compositional, i.e. it must follow from the structure and behaviour of its components.*

Here, by *structure* we mean the totality of the components that compose the system and their interconnections. By the term *behaviour* we mean the relation between the inputs and outputs of the system. Each component has a predefined performance, and the overall performance of the system is the result of the interaction of its components through the existing interconnections.

(2) *The changes that occur in the system are trasmitted locally from one component to another, via particular connections.*

In this way an ES that is based on this reasoning approach has the ability to follow, to control and to register (monitor) the way in which a change imposed or occuring in the system travels. Therefore, the ES is able to give causal explanations that are based on these local propagation rules.

(3) *The function of the system at hand expresses the relation between its behaviour and its purporse as it is seen by the user.*

For example, the time pointer of a clock is rotated about the centre, and its operation (purpose) is to show the time to the user. The operation of a component of the system at hand shows the way in which this component participates in the overall operation of the system. Thus, the examination of the operation of a system can facilitate the understanding of its behaviour and lead to important improvements in its design as, for example, in the replacement of a component by an other (with completely different structure) that performs the same operation.

3.4. Qualitative Reasoning

The qualitative consideration of a system provides useful conclusions about its performance without complex and time consuming numerical calculations which quite often are not able to reveal the real processes occuring in the system. For example, the fact: *"if the value of some parameter is positive and increasing, there is a possibility to have some hazardous state, whereas if it is negative and fixed there is not any reason for anxiety"*, constitutes a qualitative information from which one can draw some conclusions for the system state without knowing the exact value of this parameter. The qualitative reasoning approach involves the following three phases:

- system modelling
- model solution
- interpretation of the solution results

In the following, we briefly describe two particular qualitative methodologies: (i) Kuipers qualitative simulation and (ii) Dekleer and Brown envisioning technique.

(A) Kuipers qualitative simulation: According to Kuipers [47] the structural description of a physical system consists of a set of physical parameters and a set of constraints that determine the relations among the parameters. Each parameter is a real and continuously differentiable function of time and can take particular qualitative values that are called *landmark values*. These values represent the values at which some important changes of the system qualitative behaviour takes place. Thus, for each parameter, a time set is defined which consists of a completely ordered set of landmark values. Each parameter is also associated with its rate of variation in qualitative terms, i.e. positive (+), negative (-) and constant (0). The qualitative state of a parameter at some point of its time set, is represented by its position with respect to the landmark values (between two landmark values or equal to a landmark value) and the direction of change (+,0,-). The qualitative behaviour of the system is described by a differential equation that maps the relations existing among the parameters. The overall qualitative state of the physical system at some instant of time, is expressed by the set of the qualitative states of all the parameters at this time instant. Actually, two kinds of qualitative state change are considered, namely:

p-transitions which represent the transitions from some landmark value to an interval, and

l-transitions which represent the transitions from an interval to some landmark value.

Kuipers uses seven types of p-transitions and nine types of l-transitions.

For the modelling of the parameter relations five constraint types are used namely: *arithmetic, functional, derivative, inequality* and *conditional.* Kuipers uses these constraints to simulate the behaviour of a system with respect to time, although the time is not measured in time units (min or sec) but on the basis of the time instants at which there occur qualitatively discrete states of the physical system. Consequently, these time instants can have any time distance between them. The qualitative simulation method is as follows.

- At some time instant we are given a set of quantitative values for all physical parameters.
- Starting with this initial input state, the p- and l-transitions generate all possible successive states.
- These states are filtered through the constraints applied to the system in order to remove the states that are not realizable.
- Then a global pairwise consistency check is done which limits even more the number of possible successive states.
- If, finally, more than one successive states are produced, then the current state has multiple consecutive states and the simulation produces a tree.
- The above procedure is repeated for every possible successive state until a previous state appears again or a steady (equilibrium) state occurs with no further changes.

(B) Dekleer and Brown envisioning technique: This approach [45, 46, 48-50] is a kind of qualitative reasoning which is able to produce causal explanations for the behaviour of the system at hand. Its two principal characteristics are:
- It is a *kind of physics* that can be used to predict the qualitative behaviour of physical systems.
- It is a *causality theory* that can be used for the generation of explanations about the system behaviour which are satisfactory for the user.

This approach assumes that a physical system consists of discrete physical parts that are connected to a unique whole. The constituents of the system are:
- *The materials:* These are physical entities that are processed by the system, e.g. a fluid or an electric current.
- *The components:* These are physical elements (parts) of the system that can process and change the form of the materials. For each component type there exists a particular model that describes the possible behaviour with *qualitative equations* which are called *confluences* and are expressed in terms of the component parameters.
- *The conduits:* These are the parts of the system that transfer the material from one component to another, without affecting them in any way (conduits are for example the wires or tubes).

This approach to structural desciption ensures the capability to draw conclusions about the overall system performance that are solely based on the physical laws of the constituent parts and on the way by which these parts are communicated and interacted. Two basic principles must be adopted here namely the *"causality principle"* and the *"no function in structure principle"*. According to the causality principle everything that occurs in the physical world has a particular cause. Given that the various components interact only locally, i.e. that the neighbouring components share the value of the material that they receive via the conduits, one can causally identify the degree to which each component contributes to the behaviour of the overall structure. According to the "no function in structure principle" the rules that define the behaviour of the components are not directly or indirectly affected by the way the system operates.

The first principle helps in the understanding and explanation of the results of the method, whereas the second ensures that the behaviour of each component can be desribed and understood completely, independently of the system to which it participates. Thus, a given model of a componet can be used for the simulation of many different systems.

A component can be found in a number of different states, each one of which is described by a rule. Each rule is composed by a *definition part* and a *conditional part* (hypothesis) that controls whether some consequence holds. The definition part consists of qualitative equations (contributions) where the parameters of the component are involved. Specifically the general form of the model of a component is:

<Component Ci>:

<State 1>:<Def. Part Di1>

IF <Hypothesis Hij1> THEN <Consequence Rij1>

<State 2>:<Def. Part Di2>

IF <Hypothesis Hij2> THEN <Consequence Rij2>

The component models interact through links that are referred to the definition and control parts of the rules. Each variable can be involved in many qualitative equations and so it may be subject to many different influences. In each system each qualitative equation must be satisfied in isolation and the resulting qualitative values of a parameter must be compatible with all qualitative equations.

A single qualitative equation is insufficient for expressing all possible behaviours of a component over all the operation domain (spectrum) and so this domain should be divided in subdomains. Each subdomain represents a different state of the component, which is described by the corresponding qualitative equation.

The core of the approach is the qualitative differential equation, which is much different from the classical differential equations. This qualitative differential equation acts as a constraint upon the parameters and their derivatives and is related to the components restricting them to permissible combinations.

Dekleer and Brown technique works as follows:

- Starting from an initial state for each component, the definition parts are determined.
- Then an inference mechanism examines the validity of each condition in order to locate the consequences of the hypotheses that hold.
- The consequences represent the possible states to which the system can go if it continues to vary in the same way. From these consequences the next state arises.
- This procedure is repeated until a final steady state is reached.

The quantitative approach is quite often not satisfactory, mainly in cases where:

(i) There does not exist a quantitative model or the only available knowledge for the system is the general form of a relation.

(ii) There does not exist a general solution or it does exist but to find it an enormous computation effort is required.

(iii) There are involved non measurable variables as for example in medicine.

On the other hand, the qualitative approach, despite its advantages, has some problems such as:

(i) Inabiity to handle quantitative information.

(ii) Possibility to lead to fuzzy (equivocal) states, faked behaviours or to the omission of some states.

(iii) Inability to quantitatively represent the time.

The qualitative approach lacks over the quantitative reasoning in accuracy but it is better in discriminating fine qualitative differences of the system state that affect strongly the overall behaviour of the system. Also, the qualitative approach can be proved very useful in driving a quantitative modelization, as for example in constraining the search space (since a qualitative simulation can drive efficiently the search to particular paths), in identifying qualitative different regions within the search space, and in selecting the appropriate quantitative method.

3.5. Second-Generation Expert Systems

The first generation ESs are exclusively based on shallow knowledge and reasoning and are known as *rule-based systems*, since the knowledge is embodied in rule (or production systems) form. Deep knowledge could of course be coded in a simular rule-based fashion but little effort was devoted in this direction. The rule-based systems cannot easily permit extensions and modifications of the knowledge base, and if this can be done, management problems occur. Initially, these systems were considered quite satisfactory, since in most well selected domains were giving correct results in 90-95% of the cases. However, very soon the investigators observed their drawbacks such as reduced flexibility, reduced extendability etc. as discussed in Sec. 3.2. In this way, the class of second-generation ES (SGES) was born which makes extensive use of deep reasoning and through some other suitable care removes the limitations of first-generation expert systems.

Presently, it is quite common to combine the features of *shallow* and *deep* knowledge in unique inference engines and thus obtain very powerful ES with realistic, reliable and efficient expert behaviour that approaches very closely the human expert behaviour. The shallow part of the knowledge makes the ES faster and more effective in "routine" situations, while the deep part of the knowledge increases the effectiveness and reliability of the ES in facing more complex and difficult situations. A point of caution is that the deep reasoning may sometimes need a formidable amount of computation (computational explosion), and so special care is required to controllably avoid such cases. The deep knowledge is not based on the experience of a particular expert but on the general principles that govern the problem at hand. Therefore deep reasoning limits the possibility to get subjective solutions at a minimum. The above comments are sufficient to show why currently the effort of the ES developers is directed towards SGESs that employ both empirical and deep reasoning. In SGESs the reasoning and factual knowledge representation is explicit and this facilitates the process of knowledge acquisition, either manual or automatic. Also the functionality of all pieces of knowledge is explicitly represented via higher level knowledge. Second generation ESs should also have the ability to evolve on the basis of the gained experience. Finally, learning is not accomodated in first-generation systems, but it is considered as a fundamental characteristic of second-generation systems [51, 52].

4. A GENERAL LOOK AT THE FD METHODOLOGIES AND SECOND- GENERATION E.S. ARCHITECTURES

4.1. General Issues

The various applications of the E.S. technology that have so far been developed in the F.D. domain fall into a very wide spectrum of methodologies and architectures. The first systems were based on the coding of the empirical knowledge in the form of production rules. These rules associate directly the observed symptoms to their causes (faults). These systems which are known as *symptom-based expert systems* (SBES) belong clearly to the first-generation ES, and the diagnostic methods upon which were based use either *fault-dictionaries* or *diagnostic-trees*. The diagnostic tree is a special case of decision tree, where the search is restricted each time to particular paths. Both methods use look-up tables and so they suffer from the main problems of look-up tables; namely

- The larger the system at hand is, the larger the look up tables (the search spaces) are with the consequence of reducing dramatically the speed of the search algorithms.
- All the cause effect pairs must be known in advance, a fact that needs a time consuming processing for collecting these pairs. If this *a priori* knowledge is not complete, the diagnosis may be incomplete or impossible.

The more recent developments in the FD expert system design have led to the combination of the above shallow knowledge with more deep approaches. Thus now we have available ESs that have as a basic constituent system models that reflect the structure, the behaviour and the underlying physical principles of the system operation. These systems are called *model-based expert systems* (MBES) in contrast to the symptom-based ones.

Thus the newer approaches to expert system design support the distinction of two knowledge parts:

- *the shallow knowledge* which is usually represented by production rules, semantic nets, frames etc, and
- *the deep knowledge* which is usually composed by functional (operational) models and behavioural models of the system at hand.

A good deal of attention is given on how these two knowledge parts are represented, used and interacted in order to form a unified and consistent knowledge base , with the aid of a suitable inference engine which can drive the process of drawing conclusions and decide when and how the information of the two parts is to be used. Another characteristic of the recent development in the FD ESs is the *hierarchical structure*, i.e. the several models of the system at hand are structured in a hierarchical way, at various abstraction levels, a fact that improves the speed and resolution of the diagnostic tool, while simultaneously providing a good memory saving. The various sub-levels of a hierarchy are independent to each other, whereas at each level both shallow and deep knowledge can be embodied. A characteristic example of the above developments is the IDM (Integrated Diagnostic Model) tool (see Sec. 5.3) [53].

4.2. Diagnostic Modelling

In the diagnostic *theory formation* and *model construction* the basic goal is to design a model that expresses accurately the characteristics and the behaviour of the real systems. Thus any deviation between the performance of the real system and its model declares model insufficiency and the need for model improvement. The model is gradually optimised on the basis of empirical data. Fault diagnosis on the basis of structure and behaviour aims at detecting and locating faults, i.e. at determining the system components that are the cause of the observed deviations between the actual and the theoretical (ideal) performance of the system. The dominating efforts for system modelling for diagnostic purposes are concentrated on the modelling of *system structure* and *system behaviour.*

A *structural model* describes the *functional organisation* of the system at hand, i.e. the communication of its components via the input-output parameters, and the *physical organisation* i.e. the physical entities that make the system and its topology. Thus a structural model is obtained by the superposition of a *functional* and a *physical model.*

A *behavioural model* is actually a causal model that describes the system under study with the aid of logical cause-effect relations which concern the system and its components. The behaviour of each component is expressed (given) by the relations that connect the parameters of its inputs and outputs. Each logic variable contributes locally to the overall performance of the system, that follows deterministically from the superposition of the component behaviours.

Actually there exist many different ways for modelling a system, but for the purpose of FD the one described above through a *structural* and *behavioural* model seems to be the best in terms of analytical ability, effectiveness, flexibility and diagnostic reliability.

4.3. Second-Generation FD Expert System Architectures.

Here, a number of proposed second-generation architectures will be reviewed, explaining what desired requirements they meet and how this is done. These architectures are:
  i)   Structure and Behaviour (first principles) architecture.
  ii) Multi-level abstraction architecture
  iii) Model-based task-taxonomy architecture
  iv) Generic tasks architecture
The goal in designing and implementing second-generation E.S. was and still is to alleviate the limitations of first-generation E.S. (e.g. the flat knowledge representation and syntactic control). The development of the above architectures was done in parallel, i.e. with small mutual interaction or influence.

4.3.1 Structure and Behaviour Architecture

This architecture is due to Davis [44,54,55], and is based on a structural and functional model of the problem domain (i.e. on the "deep model"). Three basic representative deep-knowledge diagnosis techniques are:
  - Causal search/Hypothesis test technique
  - Constraint suspension technique
  - Governing equations technique

Causal search technique: The causal search technique is based on tracing of process malfunctions to their source. Causality is usually represented by signed directed graphs (digraphs), the nodes of which represent state variables, alarm conditions or fault origins, and the branches (edges) represent the influences between nodes. The digraph can include (besides the positive "+" or negative "-" influences on the branches) the intensity of the effect, the time delays and the probabilities of fault propagation along the branches. Digraph-based qualitative diagnostic techniques are popular since little information is required to construct the digraph and carry on the diagnosis (Kramer and Pawlowich, 1987).

Hypothesis formulation/testing: The causal search method is also called hypothesis formulation/hypothesis testing method since it follows the usual human diagnostics path, i.e. a cause for a system malfunction (upset) is postulated, the symptoms of the postulated fault are determined, and the result is compared with the system observables. Of course the search for the location of a fault can be narrowed by using appropriate heuristics or experiential knowledge and precedence rules in order to be able to resolve competing causal influences on the same process- variable. Hypothesis testing requires qualitative (non numerical) simulation of the effects of the postulated malfunctions, which needs predictions of the deviation of process measured variables as a result of faults. In simple cases the behaviour representation of the system under diagnosis can be the function, but in general, functional specifications include an account of the intentions for which the system or device is used (teleology). Moreover, very frequently, behaviour needs to be abstracted to a level higher than that at which the component is specified, e.g. in a electronic circuit.

Fault Detection by Interaction of Simulation and Inference: Having discussed the means for describing the structure (functional organization and physical organization) and behaviour, we now proceed to the main step of our work, i.e., to fault detection (troubleshooting). We first note that fault detection is equivalent to the detection of inconsistent signals or constraint violation. In the interaction of simulation and inference technique simulation produces expectations about correct behaviour on the basis of known inputs and components simulation (function) rules. Inference generates conclusions regarding the actual behaviour (function) on the basis of measured outputs and component inference rules. The comparison of the two, in particular the difference between them provides the basis for fault diagnosis.

Constrained suspension technique: This technique belongs to the approach of diagnostic reasoning from behaviour to structure or, more accurately, from misbehavior to structural defect. That is, given symptoms of misbehavior one wishes to determine the structural aberration. Although fault detection by constraint violation detection is good for simple systems, in a multicomponent system it may not be easy to locate the fault. That is facilitated by using the approach of suspending (disabling) constraints (i.e. rules of behaviour). The idea of constraint suspension is based on the observation that a faulty component can be regarded as a new component with different and uspecified properties. Then, constraint suspension means the removal of all the internal functional constraints imposed by the component at hand. If after the suspension of the constraints imposed by a component, the observed signals are consistent with the remaining constraints, then one concludes that a fault in that component justifies the observed data. In general, more than one component may exist, whose constraints when suspended, give a consistent set of signals. All these components are equivalent candidates. But a different set of observed data may rule out some of these candidates (simultaneous candidates are not considered here). Constraint suspension provides a mechanism for the careful treatment (management) of assumptions. More on constraint suspension will be presented in section 6.4.

4.3.2. Multi-levels of Abstraction Architecture

This is due to Patil and coworkers [56]. The architecture uses two levels of knowledge, a detailed pathophysiological level and a less detailed phenomenological level, and is able to reason at both levels. Both types of knowledge are modelled through causal networks involving "causes" links. The same knowledge is represented at multilpe levels of abstraction, i.e. the architecture supports parallel aspects of the same knowledge in increasing detail. Davis architecture also has this feature and the reasoning in both architectures is the same at each level of abstraction.

4.3.3. Model Based Task-Taxonomy Architecture

This architecture was proposed by Steels [57,58] and is based on a reasoning knowledge representation as a taxonomy of tasks (tasks structure). The control structure is actually a partial order on the task structure indicating the order in which tasks are to be executed. Strategies are formed by generic subtasks into which the tasks (that are generic) are decomposed. For a certain domain a generic primitive task is instantiated through an association of it with several domain procedures (problem-solving procedures) by which it is achieved. The problem-solving procedures operate on a domain model representing facts about the domain with no bias towards their usage.

4.3.4. Generic Tasks Architecture

This architecture is due to Chandrasekaran and Mittal [59] and has led to the design of a medical expert system called MDX/Patrec system [60,61]. This architecture employs both shallow (experiential) knowledge represented as a database of patterns, and deep knowledge allowing the application of first principles. Between these two extremes of knowledge there exist a knowledge and problem-solving structure which:
- has all the relevant dep knowledge "compiled" into it in such a way that it can treat all the diagnostic problems that the deep knowledge is supposed to handle if it is explicity represented and used in a problem solving
- can solve the diagnostic problems more efficiently (but no other types of problems i.e. of non diagnostic nature that the deep knowledge structure potentially could handle).

Compiled knowledge and shallow knowledge cannot be discriminated since both can be expressed in the same form (e.g. rules) and used in precisely the same way when making inferences. The difference of them lies in the way they are derived. Compiled knowledge is the summary result of a chain of reasoning based on a deep model that takes place during a consultation. The compilation of knowledge through a chain of reasoning captures the result for future use and the chain can be reproduced whenever required as an explanation or justification of the result. Eventually, one may expect that after the system has been employed on many occasions, it will have compiled sufficient knowledge to treat routinely occuring situations quickly and efficiently. See Van de Velde [62] for some results on this aspect. In the MDX/PATREC system the knowledge compiled from the deep knowledge is embodied into a taxonomy of concepts and the distribution of the reasoning across this taxonomy. The compound system MDX/PATREC employs the generic method of heuristic classification. The research on MDX/PATREC has produced the *generic task architecture* where a generic task is characterized by the domain knowledge structures it uses, its input and output information types and its control sructure. MDX and PATREC are instantiations of the generic tasks "hierarchical classification" and "knowledge-directed information passing" respectively.

Some other architectures of 2nd-generation systems are Clancey's Neomycin architecture [63] and the Oxford Medical System architecture (O'Neil, Glowinski anf Fox, [64] which due to space limitation are not discussed here (see [51]).

5. A SURVEY OF DIGITAL SYSTEMS DIAGNOSTIC TOOLS

Here a number of modern ES tools developed for the domain of digital systems will be briefly reviewed. Although non exhaustive, these systems provide a very good picture of the current status of this field. Before proceeding to our review the classical D- algorithm will be presented which is the base of many of these tools.

5.1 The D-Algorithm

The traditional approach to fault diagnosis in digital circuits was based on several variations of the *path sensitization* method. The most important of these variations is the so-called *D- algorithm* which is one of the very first attempts to diagnose faults in digital circuits. The D-algorithm is a method of generating and verifying checks, it is based on Boolean algebra and is constrained to the gate level. Because it performs an exhaustive search of the fault space, the computational cost increases exponentially with the number of gates in the circuit.

The fault model used in the D-algorithm is the so-called *stuck-at model*, i.e. two kinds of faults are considered;

s-a-1: stuck- at-1 and *s-a-0*: stuck-at-0, and indicate that the output value of a component is always 1 or 0, respectively, independently of the input value. As fault we mean the physical cause that produces an error, i.e. the faulty component of the circuit. The diagnosis method followed by the D-algorithm is the *path sensitization* technique. Here this method is applied to a simple circuit (Fig1) for illustrative reasons.

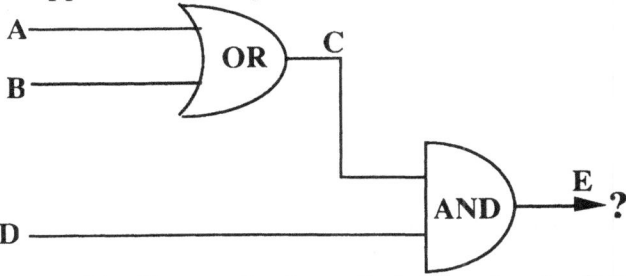

Fig 1. Path sensitization method applied to a simple digital circuit.

To examine if the single C is stuck-at-1, one must enter in the inputs the logical values that result in C=0, i,e. one must sensitize the error. Here this is done if we set the inputs A and B at the value 0. Then the result should be transmitted up to a measurable output (here up to E). Since this signal is the output of the AND gate, in order to have an error, the third input D must take the logical value 1. Thus if with A=B=0 and D=1 we have E=1, then we conclude that C is stuck-at-1. Of course, in order for the check to be complete, one must first ensure that E is not stuck-at-0.

The drawbacks of the above traditional diagnostic method are the following:

- The diagnosis starts with no any a-priori information and searches exhaustively a given fault space organized in the form of a diagnostic (decision)tree.
- The diagnosis cannot distinguish the fundamental difference between the diagnosis of faults in an operating system and the verification of a device when it is disconnected from the production line.
- The diagnosis is based on a given set of fault models that are directly enumerated and represent all the system faults that are up to now known. Since this theory is based on Boolean logic, it is directed towards faults that are modelled by a constant binary value (e.g. the stuck-at faults). The method is therefore unable to detect faults that appear for the first time. However the faulty behaviour is not necessary to match a specific a priori known fault model.

The above traditional diagnostic approach is mainly restricted to the generation and verification of checks. Newer approaches enlarge the definition of diagnostics so as to involve many other processes such as:

- Candidate generation
- Symptom generation
- Symptom interpretation
- Test design/generation
- Reasoning with functional models
- Fault location
- Fault diagnosis
- Fault repair

5.2. Davis Diagnostic Methodology

Davis and coworkers efforts [44,54,55,65-67] were devoted towards the development of a diagnostic tool which is able to carry out syllogisms and draw conclusions based on the structure and behaviour of a physical system. A large part of the work was devoted to the development of a language for the description of the structure of digital circuits on the basis of the VLSI simulation language DPL.

The basic idea of Davis methodology is that the diagnostic mechanism must be independent of the adopted fault model which must be varied dynamically without affecting the diagnostic methodology. Davis fault models are very much different from the models of other researchers. They are not predefined sets of particular system components that possibly are faulty, but sets of assumptions that the human expert is usually (tacitly) making when solving similar problems. Examples of such assumptions are:

(i)   the existence of a single not transmittable fault,

(ii) the assumption that the current has a predefined direction (although this may not physically possible),

(iii) the assumption that the diagram depicts the time topology and synthesis of the circuit and

(iv) the assumption that the given diagram satisfies the desired operation of the circuit (a fault that may not be true if these was an error in the design of the circuit).

If the system is not able to deal with a certain fault with the aid of some model, this means that this fault is outside the bounds of this model. The model is therefore enlarged, so as to embrace more classes of faults, and the system attempts again to determine the cause of the faulty behaviour.

Description language: Two kinds off hierarchy are used to describe the structure of a complex system, namely *functional* and *physical* hierarchy. The functional hierarchy describes the operational organization of the system, i.e. it modelizes the interactions among the various operational units in different abstraction levels. The physical hierarchy describes the physical organization of the system and usually has a less number of levels than the functional hierarchy. These two hierarchies are closely connected since they lead to the same generic components.

The basic elements of any functional hierarchy are:

(i)   the *module* (a classical black box),

(ii) the *ports* (the points at which the information enters and exits from a unit) and the *terminals.*

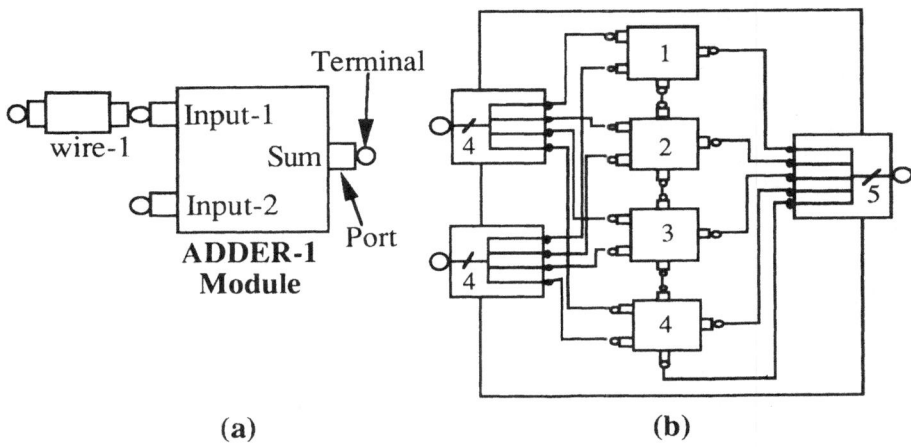

(a)   (b)

Fig.2. (a)   An adder module
       (b)   Next level structure of the adder showing that it is a ripple-carry adder.

Each unit has input and output ports. Each port has at least two terminals, one outside the black box and one or more inside. The ports provide a way of going from the higher level (the unit level) to the immediate lower (more detailed) level, i.e. the terminal level. The implementation of the physical hierarchical models is performed in a similar manner through the hierarchy of *boards, cabinets* and *chips*. The behaviour of each prototype unit is described by *simulation* and *reasoning* rules. The *simulation rules* correlate the inputs with the outputs (i.e. they modelize the real information flow through the system), and the *reasoning rules* correlate the outputs with the inputs. *The diagnostic method:* Starting from a deviation from the expected behaviour(i.e. from a symptom) the diagnostic inference engine tries to locate the structural units that are probably the cause of this misbehaviour, transmitting the erroneous behaviour backwards, i.e. from the outputs to the inputs using the reasoning rules. If more than one suspicious units exist, then further diagnostic checks are required to see which of them are really the reason for the erroneous behaviour. If this check isolates some basic (generic) component, then the problem is solved. Otherwise the above procedure is repeated to the immediately next lower functional level. If, however, the cause of the fault cannot be located, the fault model is extended, omitting the original hypothesis (hypotheses are ordered in decreasing limitation sequence). At the same time the model of the operational (functional) organization has to be enlarged to contain more types of interaction among the operational units in order to reflect the current fault model. If the current operational level is not sufficiently detailed, one has to move to the next lower level. If no fault model extension is feasible up to the level of the basic components, the immediately next hypothesis is diminished and the above procedure is repeated.

5.3 Integrated Diagnostic Model (IDM)

The IDM is a general tool for the design of diagnostic expert systems [53,68] that combines the :"shallow" and "deep" approaches and has been tested in several diagnostic applications in the areas of mechanics, chemistry and medicine. The two knowledge sources that compose the knowledge base of IDM are (Fig.3):

(1)   *The experiential knowledge base*
    This involves the *problem dictionary* and the *empirical network*. The problem dictionary is implemented in the form of frames, and is useful at the initial stages of the diagnosis for making worthy the information provided by the user. The empirical network is a 3-level semantic network. At each level of this network a different kind of knowledge is embodied which is necessary for the application of the diagnostic methodology.

(2)   *The physical knowledge base*
    This is coded in the form of a hierarchically structured functional model which is composed of functional units. This functional hierarchy is developed in two levels:
    - The higher level is called "system level" and simply characterizes the type of the system under examination.
    - The second level is the level of subsystems and describes the components of the system, their physical position in the system and the way of their communication.

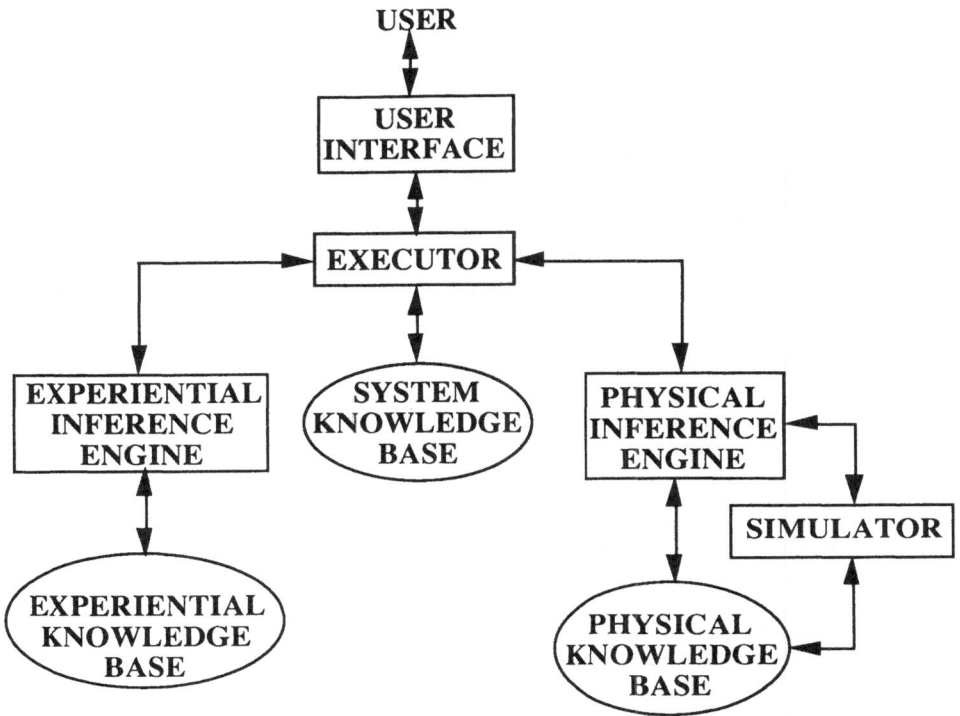

Fig.3. Architecture of the IDM tool.

Each subsystem of the original system can be considered as a new system with its own set of subsystems. In this way a recursive hierarchy is formed for the organization of the knowledge with reference to the system at hand, which allows its description with an arbitrary number of levels. Each subsystem level of this knowledge hierarchy, has a functional representation that simulates the way the subsystem at hand operates, with the aid of qualitative and quantitative information. This model is constructed by a set of basic functional units, which are able to compose a functionally complete representation, for applications in some particular knowledge domain. Each such basic (prototype) unit is represented by a node in a semantic net, which is correlated to a definite set of allowable values that indicates the acceptable values of the I/O parameters of the unit and their relation, and expresses the required qualitative information. The connections between the nodes represent logical or physical connections between the basic functional units. This type of representation is directed to make worthy the available qualitative information. Any available quantitative information is used only for the extension of the functional model.

The diagnostic method: One of the fundamental functions of the diagnostic inference engine is to decide *when, where* and *how* each knowledge source should be used. In the initial phase of the solution of a problem, the inference engine makes use of the problem dictionary, in order to constrain the spectrum of the possible cause of the erroneous performance. Hence, the IDM system starts by simulating the system at hand on the basis of the information it gets from the user, in order to see if the given symptoms really compose a compact problem and in these pieces of information are complete and compatible between them. As soon as the completeness and compatibility of the available pieces of information are ensured, the IDM tool tries to solve the problem using the experiential knowledge base. If a solution is not possible, then the IDM tool uses the physical knowledge base.

The utilization of the experiential knowledge base employs a *model driven technique*. The initial information is processed through a *bottom-up search* and as soon as a hypothesis is composed a *top-down search* is performed in order to determine possible checks and measurements that are needed for the examination of this hypothesis. Thus the IDM tool is first trying to translate the problem at hand into a state for which it has available some experiential knowledge. In parallel to this, it continuously informs the current functional model of the problem, which checks and ensures the validity and internal consistency of all the available information. The use of the experiential knowledge base helps to solve any problem of the standard (common) type for which there are available experiential rules. When the experiential reasoning is not able to solve the problem the IDM tool goes to the functional (deep) knowledge which is provided by the physical knowledge base. This inference mechanism makes use of this knowledge, using some very general inference rules that concern the evaluation of the I/O values of each functional unit, and decide which functional unit is to be examined each time.

The rules that drive the system to the detection/localization of a fault, examine the system I/O values only qualitatively and embody the classical shallow diagnostic process under very general terms. When the faulty functional units are localized, the IDM system can proceed to a more detailed diagnosis at a lower hierarchical level of the physical knowledge base, in the same manner. The IDM tool is one of the first attempts to develop a diagnostic tool that combines shallow and deep reasoning. The experiential part offers a general frame work for organizing the shallow knowledge of any domain. However, the physical knowledge base is directed towards the dynamic systems area encountered in engineering and electronics, and is not expected to be applicable in areas such as geology, voice recognition etc, where the functional processes are not represented by differential equations. An important characteristic of IDM is the way it combines the frame knowledge representations with the representation via distributed semantic nets. Most diagnostic ESs use production rules for the representation of shallow knowledge. The use of semantic nets offers several advantages. For example, it offers graphical (pictorial) representation (which is more convenient and understandable), it helps in the addition of new knowledge, and allows the use of natural language interfaces.

5.4 The Diagnostic Assistance Reference Tool (DART).

Under the name DART (Diagnostic Assistance Reference Tool) there have been presented an application (DART-1) and a diagnostic methodology (DART-2) for digital circuitry, within the framework of the so-called "Stanford Heuristic Programming Project"[69-71].

DART-1

This system constitutes an application of EMYCIN and its domain is the diagnosis of faults in the teleprocessing equipment (TE) of computers of IBM-370. The fault model assumed involves faults that appear either as inability of the users to enter the system from remote terminals or as inability of the operator to put in operation the TE network itself. These faults, that constitute a substantial part of the faults that may occur in the network appear as violation of the expected protocol sequence among the components of the network. The knowledge base of DART-1 consists of a set of production rules, that associate the possible violations in the execution sequence of the communication protocols of the higher levels, to specific components of the network that are responsible for this fault at the lower levels.

The method: According to the EMYCIN method the rule firing is controlled by a tree of the form shown in Fig. 4.

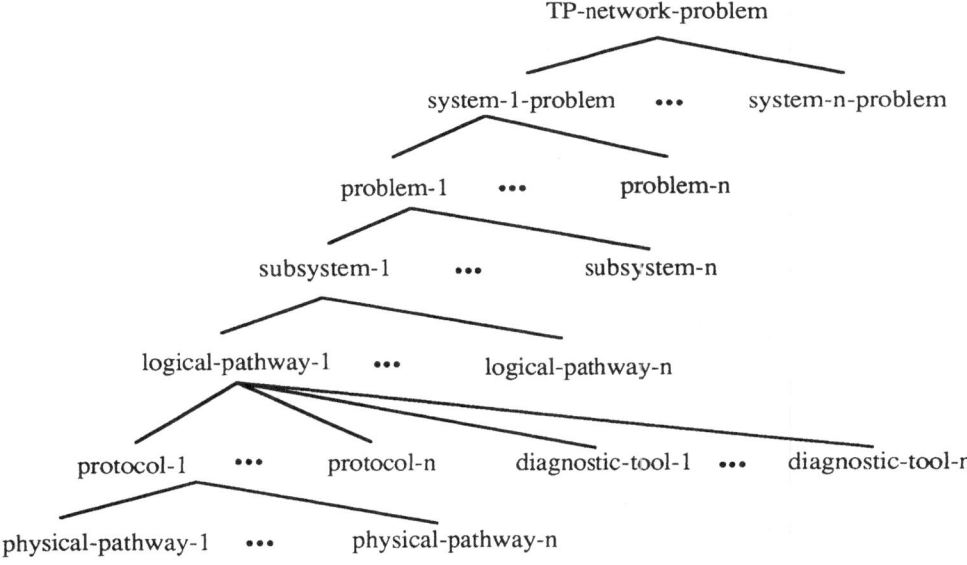

Fig.4. Diagnostic tree of TE network

This tree specializes the network problem to problems of particular systems, where a system consists of a CPU and its peripherals. Each problem of the system is decomposed in problems of the subsystems and each one of these problems in logic paths that sweep the system. A logic path is associated to a communication link between a peripheral and an application program. A problem in a certain logic path can be analyzed as a deviation from the normal operation sequence of a specific protocol in the given communication channel. Finally, this deviation can be reduced to a fault in a certain physical pathway that realizes the logic path. Each logic path is related to several diagnostic tools that can be used for the detection of the real exchanges that have been realized in this logical path in given time intervals.

These trees are generated dynamically for each particular TE network with the aid of production rules. During this process, the system requests from the user to select the diagnostic tools for the detection of the really occurred communication on these logic paths. The drawback of DART-1 is that the transfer of the protocols to the production rule formalism is a time-consuming process with doubtful results. Also, since the real protocol sequences are not directly represented with rules, the effectiveness of the explanation mechanism is not satisfactory.

DART-2

The basic characteristic of DART-2 is a general reasoning method, called resolution residue, which is a variation of the well known resolution principle. The method is used for the implementation of a theorem prover, where a number of heuristics are used to reject non useful conclusions. DART-2 was used for detecting faults in the TE part of the IBM-4331 system of a nuclear reactor. Theoretically, the methodology of Genesereth which is used in DART-2 constitutes a top down refinement strategy. Practically it can be coded as a two-step process that is repeated in each decision level.

Step 1: Attempt to form a proposition of the type
$$NOT(P_1) \text{ V } NOT(P_2) \text{ V } ... \text{ V } NOT(P_n)$$
where the proposition P_1 represents the fact that the component or connection 1 of the higher level, has fault.

Step 2: If there are more than one suspicious components or connections, the attempt to prove the proposition NOT(P) produces(generates) a number of propositions of the form
$$I_1 \text{ \& } I_2 \text{ \& } ... \text{ \& } I_n \text{ \& } 0 => \{NOT \text{ } (P_1) \text{ V } NOT(P_2) \text{ V } ... \text{ V } NOT(P_n)\}$$
Each one of the above propositions represents a diagnostic check and expresses the fact that if the input values $I_1, I_2, ..., I_n$ infer the value 0 to some measurable output, then some of the $P_1, P_2, ..., P_n$ has a fault. Such a diagnostic check is not required if one of the Os or ~Os follows directly from the values of the inputs $I_1, I_2, ...I_n$. The generation and execution of such diagnostic checks is repeated until a specific suspicious unit is isolated. If this is not possible, then the problem is characterized as undiagnosable.

The main drawback of DART-2 is that the heuristics employed do not reflect methods and practices of experts. Both Genesereth and Davis keep the assumptions that make the fault model independent of the inference engine. Davis method is based on the dynamic modification of the fault model since this is the way the experts work. On the contrary, Genesereth faces these assumptions as extra propositions, and these extra propositions are always useful in a system that is based on the resolution principle.

5.5 The Intelligent Diagnostic Tool (IDT).

The *Intelligent Diagnostic Tool* (IDT) constitutes a diagnostic methodology for hardware systems, which has been developed within the framework of the DEC system and has been presented by Shubin and Ubrich in 1982. This system was originally applied for fault diagnosis of the diskette RXO2 subsystem of a PDP 11/03 computer.

A *unit under test* (UUT) as for example the RXO2 diskette subsystem is modelled in the form of field replaceable units (FRUs) which are modelled by atomic units (AUs). Each AU can belong to only one FRU.

Contrary to the other systems that are described in this section the IDT system does not involve descriptions of the operational functional intedependencies among the FRUs and AUs, nor performance descriptions for the prototype unit forms (perhaps since here such descriptions are impossible). Due to this fact the diagnostic tests that concern the operation ability of the AUs cannot be generated dynamically but they are predetermined. Thus, if there exist such diagnostic tests for the particular FRU, the selection of the AUs must be based on them. In the opposite case, the tests must be designed so as to reflect the selected AUs. A diagnostic test is an action request, the result of which is characterized as "success" or "failure" and declares that some AUs have a fault or they are free from faults,respectively. The validity of this test depends on the validity of other structures that characterize the functional level of the AUs under examination. Thus the knowledge that forms a diagnostic test can be represented in the form of the following rule pair:

IF precondition (F)   IF precondition (T)
AND action (t) fails   AND action (t) succeeds
THEN conclusion (F)   THEN conclusion (T)

A test selection strategy determines the selection sequence of these rules. The *conjunction* between the hypotheses F and T puts the basis for the initiation of the action request for performing the operation t. Finally, although this is not directly declared, the diagnostic tests of IDT system are able to express empirical associations.

For the application of the IDT to the RXO2 diskette subsystem of the PDP11/03 the DECXXDP+ function/logical test sequences were used. For the representation of the relation of the structural description and the diagnostic tests the predicate calculus was used.

The method: The fault model adopted concerns an isolated fault of an AU. The goal of the diagnosis is to locate the FRU that contains the faulty AU (the AU is not necessary to be localised). The repair of the UUT of concern consists simply in replacing that particular FRU by a new one. The diagnostic procedure consists into the consecutive selection and execution of diagnostic tests and the translation of their results until the faulty FRU is localised (identified).

Each FRU, say FR(X), is associated with two general types, the OUTSIDE(X) and the INSIDE(X) that characterize the conjunctions of the AUs that lie *outside* or *inside* X, respectively. The diagnostic process is terminated when some FRU, say X, is localised for which either the INSIDE(X) is true, or the OUTSIDE(X) is false. In both cases, X is the FRU that has a fault. The results of the diagnostic tests are used for the simplification of these general types. To this end, the selection of these tests is made with criterion which test brings the diagnostic process nearer to such a simplification. Finally, for the tests selection one takes into account any information that is available to the user, about some FRUs or AUs which he (she) assumes that operate normally or are faulty.

The IDT system has improved the existing DEC package, by embodying to it some of the characteristics of the human diagnostic methodology, in particular the engineering habit to order diagnostic tests without having first ensured the validity of their hypotheses. One drawback is that the 2-level structure of the units cannot always reflect the complexity of many UUTs.

5.6 The Lockheed Expert System (LES)

LES is a general purpose expert system environment suitable for a variety of applications inclunding the diagnostic one. The information presented below concerns the fault diagnosis in a large signal switching network (Baseband Distribution Subsystem) and was presented by Perkins and Laffey in 1984 [72,73]. This network is referred to as BDS and contains a *Built-in Test Equipment* (BITE) which, however, is not considered to be very reliable.

The BDS is modelled in the form of a deterministic network that gives the characteristic signal paths of the system. The knowledge is represented with the aid of frames that describe the components of the BDS and their characteristics (e.g. type, position, connections with other components, the signal flow from them in a certain direction, etc). This description also involves a set of values which were collected from human experts, and describe the possible deviations from the general fault model.

The fault model adopted by the system, assumes that a component has a fault if the signal entering it does not appear at the output. The rules and the frames describing the components are implemented with the aid of a case grammar.

The Method: The diagnostic process adopts the assumption of a single (isolated) fault and the hypotheses made refer to the paths of the network that may possess a fault. Such a hypothesis is shown to be true when a component of the particular path has a fault, whereas it is rejected when it is seen that the signal can pass through this path way. The initial hypotheses set is derived by the BITE, via messages that it gives when the BDS presents a fault. The test of a hypothesis is made in a backward direction, from the outputs to the inputs. In the process of this test, the system may request from the user to make a check (test) through BITE, in order to obtain additional information.

The LES system has an extensive explanation facility through graphics and explanations. These explanations are rather *ad-hoc* and result mainly from the reversing of the AND/OR tree that is constructed from the application of the rules in each case.

5.7. Other Systems

Some other methodologies and tools include the *verification methodology* of Hamscher [74,75], the CRITTER-Redesign system of Kelly and Steinberg [76], the ARBY system of McDermott and Brooks [1982] and the *Network Diagnostic System* (NDS) of Williams (1983).

Hamscher's methodology focus on the verification and not on the fault diagnosis, and constitutes one more attempt to improve the D-

Algorithm by going far from the representation at the gate level. A device is modelled in the form of *primitive units* which communicate through information paths. With each primitive unit there are associated four rule categories, namely goal, behaviour, sensitization and conditional rules. The basic assumption made is that the primitive units are always fault-free (no fault can occur in any of the primitive units). This assumption makes the system nonrealistic.

The CRITTER methodology is focused on the verification of designs and examines their validity and robustness. In other words it examines if the behaviour of a design lies within the bounds of its specifications and how near it is in the violation of the specifications. The new issue in CRITTER is its ability to process symbolic expressions. The approach of the exhaustive transition of behaviours and specifications adopted makes this methodology inefficient for large and complex digital circuits.

The ARBY system is a diagnostic tool for large electronic devices and its design was based on the CADUCEUS medical diagnostic system [75,78]. The electronic devices are modelled in the form of hierarchically structured causal networks at several abstraction levels, sufficiently higher than the voltage level. The knowledge core that drives the diagnostic procedure is composed by "accounts for associations" and "evidence for associations". The part of the knowledge base that concerns the knowledge acquisition is represented in the form of interaction frames. The inference engine of ARBY consists of a HYPOthesis generator (HYPO) and an Interaction Frames Manager (IFM).

Finally, the NDS system is an application of ARBY for fault diagnosis in a national communication network. The assumed fault model covers single and multiple faults (transitive and nontransitive). The diagnostic strategy is a kind of top-down refinement strategy (ie. until a faulty FRU is detected). When a fault is detected it is repaired and a check follows to see if the network operates normally. If not, then the diagnostic procedure is repeated at the next higher level of the level where the repair has taken place. Such jumps towards the top are called "generalization jumps" in contrast to the "downwards refinement jumps".

6. A GENERAL METHODOLOGY FOR THE DEVELOPMENT OF FD TOOLS IN THE DIGITAL CIRCUITS DOMAIN.

Here, a general methodology for the development of diagnostic expert systems tools in the digital circuit domain is presented that posseses most of the features of second generation tools. This methodology is based on structure and behaviour and must make use of a suitable reasoning technique. This reasoning should be able to start from given symptoms, search for the structural deficiencies that are the causes of these symptoms, and explain the set of the given symptoms.

The implementation of a diagnostic tool using a reasoning mechanism that is based on structure and behaviour requires:
- A suitable description of the structure.
- A suitable desctiption of the behaviour.
- A flexible inference engine which is able to make effective use of the above two descriptions.

6.1. Description of the Structure

Actually there are many ways of structuring the information about the way the components of a system interact. These depend on the angle of attack in which the system can be described and analyzed. The most common ways are through the *physical* and *functional* organization.

Schematically the fault diagnosis from structure and behaviour works as shown in Fig. 5.

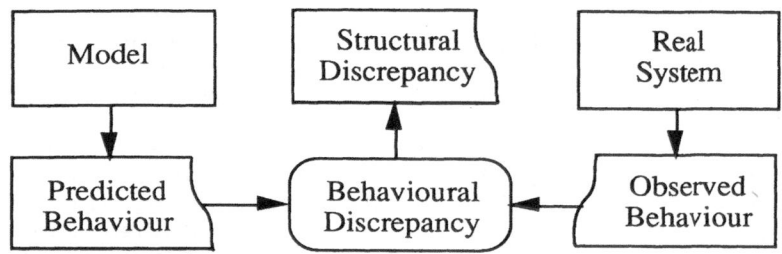

Fig. 5: Fault Diagnosis based on structure and behaviour

The *functional organization* describes a system in the form of a functional component hierarchy, at several abstraction levels, on the basis of the way the components interact, via their input-output parameters. The *physical organization* describes a system in the form of a physical unit hierarchy, which usually has less levels than the functional organization. The physical organization represents the way in which the system is connected, ie. the physical entities that compose the system and their relative position in the system.

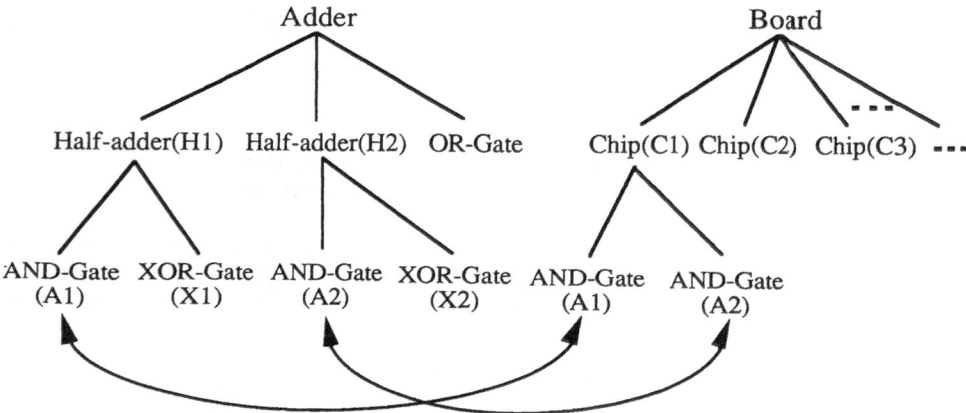

Fig. 6. Correlation of Physical and functional Organizations

Given that to each unit there corresponds a functional and physical description, the problem of relating them arises. This relation is easy to find since both descriptions end at the same primitive components. (Fig.6). Thus the physical position of some non primitive unit can be

found by superposition of the units it is constructed. The relation of these two descriptions gives the capability to answer to several important questions, as e.g:
- Which physical entities compose the section that operates as address register and at which point of the topology lie?
- What operations this four-AND-gate chip performs?

In determining the structural description of a unit it is very useful to define a prototype parametric sequence of instructions for the construction of its model, independently of the operation that the unit performs in the particular system at hand. The activaton of this instruction sequence with particular values produces a particular instance of the unit. Thus the structural description (model) of a unit is essentially a set of instructions for its construction. These instructions are executed by the system, and the result is the construction of data structures that modelize all the components of the system and their interconnection. To save the memory required, the concept of dynamic instantiation is invoked, ie. when some instance of prototype unit is made, only its shell and its ports are constructed (which is a kind of black box) whereas its interior is not constructed unless it is required.

6.2. Description of the behaviour.

Here, by the term behaviour it is meant the description of a unit in the form of a black box. This description must be able to reply in the question: How the information outgoing a certain unit is related with the information entered in it? This concerns the relations that connect the I/O parameters of a unit. These parameters contribute locally to the overall performance of the system at hand, which results from the deterministic behaviour of its components.

For the description of the behaviour there have been proposed several techniques such as:
- Single production rules that relate the inputs to the outputs (this is used when the system performance is simple)
- Petri-nets in the case of modelling parallel events
- Nonstructured code forms, in the case where more structured representation forms are shown to be insufficient or non representative.
- Several combinations of the above techniques.

Description language: The behaviour of a component of a complex system is described by one or more constraints that determine the existing relations among the appropriate I/O parameters. These constraints which may be qualitative or quantitative, describe the component behaviour under normal operation conditions (ie. all possible I/O signal combinations in the relevant pots). These constrains have double use. If some of the I/O values in the component ports are not known beforehand, they can be deduced from the known ones. If all I/O values are known a priori, one can test the validity of the given constraint. If the I/O values of a unit do not satisfy its constraint, then this unit does not show the expected behaviour and so it has a fault.

For example, the behaviour of an adder is described by the constraint:

$$Out = Inp1 + Inp2$$

which mathematically is equivalent to the following two constraints:

$$Inp1 = Out - Inp2$$
$$Inp2 = Out - Inp1$$

Logically for the description of the behaviour of the adder only one of the above relations suffices. However, for practical reasons in the simulation and application of the constraint suspension technique one must use all alternative expressions of the I/O parameter constraints of a given unit (or component). The set of all these relations, for all the units of the system at hand, constitutes a constraint network that describes the system behaviour in the normal behaviour state (ie. when all of its components function properly).

Such a network is compatible with the mathematical notion of constraints, according to which the information that is embodied in a constraint has not a directed character. However, in the modelling of general physical or technological systems this is not allways true. There is a clear distinction between the *simulation rules* (which represent the true information flow, eg. the flow of electric current) and the *inference rules* (which represent the flow of syllogisms, ie. of the conclusions that can be drawn about the system). This distinction is very important both for typical reasons of clarity of our reasoning schemes and for ensuring the correct and safe application of the simulation in cases of unexpectable events.

Thus we have to use two parallel but independent networks, namely:
- the *network of simulation rules* which models the system behaviour deterministically and causally, and
- the *network of inference rules* which models the way in which the human expert reasons in order to draw conclusions about the system behaviour.

Thus, one has to create two different instantiations of the system model. The first keeps the values of the I/O component parameters that result from the application of the simulation rules, and the second keeps the parameter values that result from the application of the inference rules. This mechanism of the independent propagation of the values in the two networks helps to produce an *interdependency network* which is used for the construction of the *explanation facility* (how-explanation facility, why-explanation facility, and what-if explanation facility).

6.3. The Diagnostic Mechanism

Having available descriptions for the structure and behaviour of a system we now need an efficient and flexible diagnostic mechanism that can lead the ES to a fast and reliable fault diagnosis. The diagnostic mechanism described here is based on the concept of *behaviour deviations* (discrepancies) and on the use of the *interdependency* network for the detection of the causes of these deviations.

More specifically, this diagnostic mechanism is developed in three stages:
1) Initially it simulates the system at hand, produces the expected output values and compares them with the observed ones. If there is a deviation then the system is faulty and the symptoms of this

faulty operation are the output parameters with different observed and simulated values.

2) Starting from these symptoms, the fault detection technique locates on the interdependency network all the system parameters that were involved in the calculation of the output parameters and for which the measurements give different values than the expected ones.

3) Then, the constraint suspension technique is applied which checks the overall consequence of the possible causes of this fault, with respect to the set of the observed symptoms and produces the final set of "suspicious" (candidate) components.

The diagnostic technique is based on the interaction of the simulation rules network and the inference rules network.

- The network of simulation rules (that models the system under normal operating conditions) generates the expected output values on the basis of the input data.
- The network of inference rules generates conclusions for the actual system performance, on the basis of the observed output values, i.e. it computes the I/O values which must appear at the parts of every component so as to obtain the specific values at the system outputs.

The interaction of these two networks constitutes a strong basis for fault diagnosis, as shown in the following algorithm for *generating suspicious components*. Before presenting this algorithm we first outline the fault detection technique and the constraint suspension technique.

Fault Detection Technique. As already said the simulation constructs an interdependency network which in the process of simulating the system registers the propagation of the values.

The idea behind the interdependency network is that *"any component belonging to some path that connects an erroneous output to some input, can possibly be the cause of the faulty behaviour of the system"*. In this way we can detect all possible starting points of this deviation and locate all possible causes of the problem, going backwards on the interdependency network. (Of course at this point it is tacitly assumed that the cause of the erroneous behaviour is a *solid-non-intermittent fault*).

In this way we generate interdependency chains which represent all possible paths that connect erroneous outputs with the inputs of the system. An interdependency chain is actually a sequence of units that belong to such a pathway.In the case of a single fault, the cause of the erroneous (annomalous) performance is one and only one component. Clearly, this component will apear in all the paths that are defined by the interdependency chains. Thus the component actually lies at the intersection of these paths.

In the case of multiple faults, there are more than one components that are the cause of the erroneous system performance. Each one of them can explain only part of the observed symptoms. In general, these components belong to different paths. Thus in this case, the set of all possible causes of the erroneous behaviour of the system, is defined by the union of the paths determined by the interdependency chains. However in both (single and multi fault) cases this set of possible causes results from local syllogisms, which may not be globally consistent with all observed symptoms. The global consistency check, is assured by the application of the constraint suspension technique to this restricted set of components.

6.4.The Constraint Suspension Technique

This technique is very powerful for condidate fault generation,i.e.for the generation/detection of the system components that are possibly responsible for the totality of the observed symptoms.
The constraint suspension comes from the method of *violated expectations* which, instead to assume some possible fault and explore its possible consequences (to check if they match with the observed symptoms), searches for deviations between the values resulting from the measurements and those which should result from the normal operation of the system at hand. Thus the method is able to detect a wide spectrum of faults,since as erroneous operation is considered any behaviour that differs from the theoretically expected one and not only those operations that result from already known and modelled fault types. Although the networks of simulation rules and inference rules are independent, for the application of the constraint suspension technique we consider their merging in a unified network, i.e. the so-called *network of constraints*, which involves the totality of rules that correlate the values of the I/O parameters of the system at hand. With the aid of the network of constraints, one can check the correctness of the operation of the system on the basis of the I/O values that appear at its ports.

When the system operates normally, all of its components operate correctly and so they satisfy all their constraints. Thus the propagation (transmission) of the I/O values of the system through the constraint network is possible.

However, when the system at hand suffers from a fault, i.e.some of its components do not function properly, the faulty components do not satisfy their constraints. So if we put the I/O values of the system into the constraint network and let them propagate, the procedure will go to a deadend, since some of the constraints will no longer be valid (i.e.they will have been violated). The case where all constraints are valid (non violated) and the particular input values produce erroneous output values is not possible.

This fact testifies the occurence of a fault, but in a complex system it is not clear which constraints are violated i.e. which components are faulty. The localization of the faulty components is based on the localization of the constraints, the removal (suspension) of which returns the internal consistency of the constraint network. The constraint suspension technique (as a fault diagnosis reasoning technique) in a hierarchically structured system works as follows.

Every component of the highest level is put successively out of operation by suspending its constraints (i.e.the equations that relate its input and output parameters). Then the input values and the measured output values of the system are put at the respective parts of the constraint network, which is called to propagate them. In this way one actually tries to answer the question: *"Is it possible for that particular component to have a fault that produces the observed symptoms?"* Or equivalently *"is it possible by the removal of this component from the system, to restore the normal operation of the system?"*.

Since the I/O parameters of this component enter the constraints of the neighbouring components, their values can be calculated from the other constraints of the system. If the suspension of the component's constraint (i.e.the deactivation of a component) restores the internal consistency of the constraint network, one can assume that this component is a possible cause of the erroneous function of the system. In the opposite case, this component is rejected, and the procedure is repeated with the next component of the higher level. In this way a set of candidate faulty components arises at the higher level, each one of which is considered as a new system and for which the constraint suspension method is repeated at the next lower level of analysis. If in the process of analysis of a component at the lower level, the method cannot detect (locate) any suspicious sub-component, the original component is rejected and the procedure returns to the higher level and continues by examining some other suspicious (candidate) component. This procedure is repeated as many times as they are required until finally some primitive faulty components are detected. The procedure assumes that the system might have more than one faults. If the cause of the erroneous system behaviour is a single fault, the method leads finally to the unique primitive component that has the fault. If there are more than one faults, their diagnosis can be made, successively, by deactivating at each level of the hierarchy each one component separately, then all the component pairs, next all the component triads, and so on. In this way the diagnosis of one, two, three e.t.c. faults can be performed.

The idea behind the constraint suspension method is that: *a component that operates erroneously (i.e. it does not present the expexted behaviour) can be considered as a new component, with different but till unknown properties* (i.e. with a new behaviour which is not expressed by the given constraints). This new behaviour is indicated by the signals that appear at the ports of the components. Thus, on the one hand, this method is able to locate the components that possibly operate erroneously, and on the other is able to offer information for the nature (symptoms) of this faulty behaviour. The constraint suspension technique possesses important advantages over the tranditional diagnosis methods (such as the D-algorithm of sec.5.1.) since it gives a systematic way for the location of all the components that possibly have contributed to the erroneous operation of the system. However, in its general form this technique is an exhaustive search technique, in the sense that at each abstraction level deactivates *all* the components (initially each component separately, then in pairs, then in triads and so on).

Given that our goal is the development of a technique as more general as possible, which will not be restricted by the complexity of the system at hand, we must modify the original constraint suspension technique so as to drive the search toward the *more suspicious* components in agreement with the symptoms each time. Indeed, the constraint suspension technique determines if some component might have a fault, by checking the existence of any set of values that can appear at its ports and can explain all the observed symptoms. However, the method itself is not able to know if some component really is the cause (totally or partially) of the erroneous behaviour of the system at hand (and so if there is reason to examine it or not).

At this point the FD technique proves to be very useful,since it helps in the detection of all the components that were involved in the computation of the output values that differ from the measured ones. In this way, all the available information about the erroneous bahaviour of the system is usefully employed and the procedure of locating suspicious (candidate) components is accelerated. The algorithm that follows presents a full description of the procedure for generating suspicious components, ignoring temporarily the distinction between the simulation rules and the inference rules, and working on a traditional network.

The Candidate Generation Algorithm

Step 1: Detection of Deviations
1.1. Introduce the input values of the device into the inputs of the constraint network (the simulation generates the expected values at the output of the faulty device).
1.2. Compare the theoretically expected values to the actually measured ones at the outputs of the device and detect the deviations.

Step 2: Determine the possible causes of the fault with the aid of the interdependency network
2.1. For each detected deviation in Step 1, traversing backwards the interdependence chain (i.e.starting from the expected output value and proceeding towards the inputs), obtain the whole set of components that were involved in this expected value.
2.2. The section of all the sets that are obtained in Step 2.1. constitutes the set of the components that belong to the paths of computing all the output values for which a deviation was observed, and, consequently, are possibly able to explain all the deviations.

Step 3: Global Consistency Check of the Candidate Components with the aid of Constraint Suspension
3.1. For each member of the components' candidate set arised in Step 2.2 remove the constraint that models its performance.
3.2. Put the values measured at the outputs of the device to the outputs of the constraint network.
3.3. If the remaining network is consistent with respect to the measured output values,
    Then   this suspicious component is globally consistent *and*
                its symptoms are declared by the I/O values at its ports *and*
                the component and its symptoms are added to the list of candidate (suspicious) components.
    Otherwise this component is removed (eliminated) from the set of candidate components.
3.4. Remove the output values from the constraint network.
3.5. Bring back the constraint that was deactivated (removed) in step 3.1.

3.6. Go to step 3.1.

This algorithm concerns the detection of a single fault. To detect multiple faults the following two modifications are sufficient:
1) In Step 2.2 we consider the *union* (instead of the section) of the sets of components which, according to the interdependency network, were involved in the computation of some expected output value.
2) In Step 3.1 we deactivate pairs, triads and so on, of members of the set of suspicious (candidate) components and check the overall consistency of the network.

6.5 Advantages of the Deviation Detection and Constraint Suspension Technique

The combination of these tools provides a very effective diagnostic mechanism with several advantages over the traditional approach. The main advantages are:

1. It consists an authentic diagnostic technique since it allows the detection of all the components that are possibly faulty (without having to resort to predetermined fault dictionaries, diagnostic trees, etc.).
2. It supports the use of hierarchical structures which help to obtain a complete and detailed description of the faulty system at hand.
3. Its reasoning is based on knowledge about the *structure* (as it is mapped in the schematic representation) and the *behaviour* of the system (which results deterministically from the behaviour of its components).
4. It defines the fault in terms of behaviour, i.e. in terms of deviation from the theoretically ideal behaviour.
5. It makes worthy all the available information regarding the nature of the erroneous functioning of the system at hand. The constraint suspension technique in its general form leads to an exhaustive search of all the system components, since in principle all of them are equally suspicious. However the deviation detection technique directs from the biggining the search to the more suspicious components thus leading more quickly to the fault detection
6. It gives information about the nature of the erroneous performance that shows some component which possesses a fault.
7. It provides a unified way for facing both the cases of single and multiple fault. Indeed the transition from the single fault to the multiple fault case needs only to change the number of components that are deactivated at each abstraction level.
8. It allows the use of multiple descriptions of the system, a fact that leads to an increase of the flexibility and the diagnostic power of the resulting tool. Simultaneously it ensures the possibility of making worthy the alternative descriptions in many different ways. Indeed these descriptions can be used as:
   - part of a computer code that can simulate the system operation
   - a basis for the execution of the diagnosis

- a knowledge base, that can provide information for the composition, the topology of the device, the communication and the interaction of its components, etc.
- a basis for an on line program for montitoring the operation of the device

9. It distinguishes different kinds of knowledge relative to the system under investigation and its organization in independent parts. The information regarding the composition and topology of the system is provided by the model of physical organization. The operational details are provided by the functional model, and the system behaviour is provided by the cause - effect (deterministic) relations. However the technique must be used with great care in practice since it tacitly assumes that the structural desrciption of the system is complete and provides exactly what is happening to the system, and that whatever is not mapped in this description it does not actually exist.

7. A GENERAL METHODOLOGY FOR THE DEVELOPMENT OF FD TOOLS IN THE PROCESS ENGINEERING DOMAIN

The methodology to be discussed here combines enhanced versions of classical failure models with inference techniques from AI [80]. The models used are fault propagation digraphs which represent the propagative aspects of faults and cause-effect knowledge bases based on augmented fault trees that represent the causal aspects of failures. Failure analysis processes that operate on these models utilize backtracking with constraint enforcment, implicit problem reduction, and bidirectional chaining of production rules.

Actually the failures of a system occur in two stages, the onset of faults in some units of the system (called failure sources) and the propagation of faults from these units to other units. Thus a process for failure analysis must initially try to locate the failure sources in the system by reasoning, based on the available alarm information and the knowledge it has about the failure characteristics of the system, and about the propagation of faults to units known to be faulty. Then it should try to infer the causes for the onset of failure at these sources. Thus the FD framework involves two phases:

(i) A failure source location phace,

(ii) A failure cause identification phase.

The framework consists of failure models and processes that operate on these models. Each phase is characterized by a failure model and a corresponding failure analysis process.

The methodology is pictorially illustrated in Fig. 7.

From a structural point of view the methodology is divided in two constituents; (i) *Failure knowledge*, i.e. a passive constituent that represents knowledge about the fault behaviour of the system at hand, and (ii) *Failure analysis*. i.e. an active constituent comprising processes that use the failure knowledge for diagnosis. From a functional point of view the methodology is divided into the failure source location phase (FSLP) and the failure cause identificaion phase (FCIP).

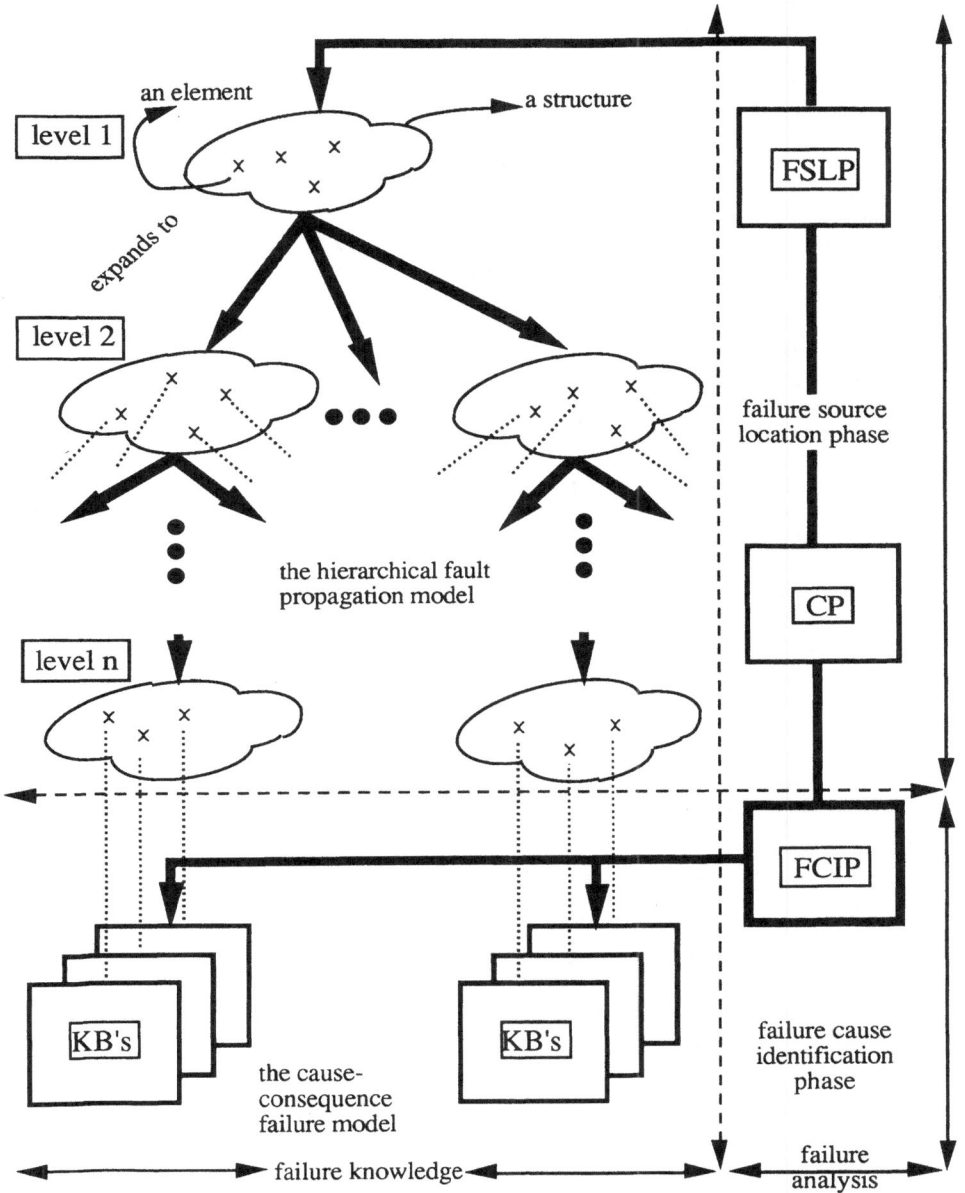

Fig. 7. Pictorial illustration of the FD methodology

FSLP: In this phase the failure origins are located (with maximum possible resolution) using the information available at the initiation of failure analysis. This phase is marked by the operation of a failure source location process on a hierarchical model of fault propagation. At the end of this phase a set containing subsystems that are likely sources of failure is generated from the structures at the lowest level of the fault propagation model hierarchy. The fault propagation hierarchical model contains a hierarchical representation of the available knowledge about the characteristics of fault propagation within the system. The granularity of system view increases with levels. Thus as one traverses down the levels of the model the resolution of view increases.

FCIP: This phase is marked by the operation of a failure ccause identification process on a cause-consequence failure model. The cause-consequence failure model contains knowledge about the causal aspects of failures. It involves a number of knowledge bases, each one corresponding to each subsystem (element) in the structures at the lowest level of the hierarchical fault propagation model for phase one. Each knowledge base contains heuristics as well as systematic knowledge about the temporal probabilistic, and causal aspects of various faults that can occur in the corresponding subsystem.
Initially, starting from failure events corresponding to all alarms that showed a normal state at the start of failure analysis, FCIP identifies all other faults that could not have occured. Then FCIP generates by forward reasoning, cause-effect chains of failure events that start from the failures indicated to have occured by alarms and possibly lead to the failure of the subsystem represented by the knowledge base. At the same time, FCIP selects some failure evevts to be used as starting points for the actual diagnosis. The diagnosis is carried out by backward reasoning from these points. FCIP returns a set of basic faults which are likely to have caused the onset of failure at the subsystem represented by the knowlwdge base. FCIP is repeatedly applied to the knowledge bases representing each subsystem in the failure source set generated by the previous phase.
Fault propagation hierarchical model: The model used for the fault propagation is a hierarchical set of level structured digraphs (directed graphs) [43]. The concepts used for this model are:

$S=\{s_1, s_2,...., s_n\}$ System with subsystems s_i

$R \subseteq (S_x S)$ Failure relation on S

$D=(V,E)$ Fault propagation digraph

$V=\{s_1,...., s_n\}$ Vertex set containing monitored and nonmonitored vertices

$E \subseteq (SxS)$ Edge set

$s_i R s_j$ A fault in subsystem s_i can propagate directly to subsystem s_j with probability $p(s_i, s_j) \neq 0$

$e_{ij}=(s_i, s_j)$ Directed edge in D from s_i to s_j and $e_{ij} \in E$ iff $s_i R s_j$

$A=[a_{ij}]$ Incidence matrix of D where $a_{ij}=1$ if $e_{ij} \in E$ and 0 otherwise.

The incidence matrix A is said to be a *full description* of R for S if the value of p_{ij} is known for all pairs s_i, $s_j \in S$. If p_{ij} is unknown for some pairs then A is called a partial description of R for S (for such pairs the value of a_{ij} is taken to be zero).

The reachability matrix $M=[m_{ij}]$ of S is given by

$$M=[I+A]^k=[I+A]^{k+1}$$

for any $k \geq k_0$ (and for no $k < k_0$) where k_0 is a positive integer, I is the identity matrix, and Boolean matrix multiplication is used to compute the powers.

The concept of level-structuring is defined through the *ancestors* and *descendants* functions of the vertex V [81]. An efficient level-structuring algorithm can be found in [82]. An example is given in [80].

Knowledge modelling and acquisition: The knowledge about cause-effect relations among faults is modelled by the cause-consequence knowledge bases which constitute the lowest level of the hierarchical knowledge representation structure. This knowledge is used in the failure cause identification phase of failure analysis.

Subsystems or units that are distinguishable under the finest granularity of the fault propagation model hierarchy correspond to vertices of digraphs at the final level of the fault propagation level. Failure source location (detection) is confined to these units. Once some of these units have been detected as fault sources, the basic faults that caused the failure of these units have to be determined. The knowledge required for this analysis is encoded in cause-consequence knowledge (rule) bases.

In [80] a systematic knowledge-acquisition procedure for developing the hierarchical failure model is also provided, which consists of fault propagation digraphs and cause-consequence knowledge bases of a given system. This procedure makes use of the augmented fault tree (AFT) concept as an intermediate representation from which the production rules for a knowledge base are derived. Due to space limitation the procedure is not described here, but it can be found in [80] together with a full description of the three fundamental processes for failure analysis, namely FSLP, FCIP and CP (control process). The fault diagnosis process based on FSLP, FCIP and CP can be suitably implemented in a distributed environment due to the inherent parallelism of FSL and FCIP [80]. One limitation of the methodology is due to the fact that the rule and path selection heuristics are based on measures that are fixed at design time and remain static. This restrict the flexibility of the selection process.

8. IMPLEMENTATION OF A DIGITAL CIRCUITS DIAGNOSTIC EXPERT SYSTEM (DICIDEX)

8.1 Introduction

The two most popular current languages for the implementation of expert systems are lisp (LISt Processing language) and PROLOG (PROgramming in LOGic). The first language has been developed in the USA at the beginning of 70s and was the first attempt for the development of AI tools. Many important tools such as KES and OPS5 were implemented in LISP. Prolog has been developed at the University of Marseilles (France) in the 70s and is still under further improvement and enhancement. In 1983 Prolog has been established in Japan as the basic programming language of fifth generation computer systems, although it is actually a logic programming language. This is due to its advanced properties such as the easy management of lists and symbols, and the direct representation and coding of concepts and relations. Our expert system DICIDEX (DIgital CIrcuits Diagnostic EXpert system) has been implemented using the LPA PROLOG PROFESSIONAL V3.0 language. This is a new Prolog implementation which is based on the Edinburgh (standard) syntax (known as DEC 10) and has been developed in U.K. (first version in 1981. V 3.0 version in 1989). The LPA Prolog Professional possesses a very fast incremental compiler which offers the possibility of on-line program development and quick prototype construction for larger areas. It also has an optimizing compiler that facilitates the construction of stand-alone programs. This Prolog version possesses many new important characteristics, namely:

(i)   smart I/O functions,
(ii)   possibility of declaring operators for the representation of logical concepts in a more natural language,
(iii)   easy implementation of menus,
(iv)   possibility of management of meta-variables,
(v)   a wide spectrum of built-in arithmetic and control predicates,
(vi)   possibility of on-line program tracing and debugging,
(vii)   extensive file management possibilities,
(viii) possibility of direct communication with the C and Assembly languages,
(ix)   two very powerful graphics packages, etc.

Two drawbacks are the limited evaluation space for some built-in predicates (e.g. findall, forall etc.), and the limited number of variables in predicates that are managed by dynamic data base. Therefore, this language is ideal for the development of prototype programs but is insufficient for the development of more complex systems.

8.2 DICIDEX Description

This FD-ES contains two files called DICIDEX and MODEL. The file DICIDEX constitutes the core of the FD tool and involves : (i) separate paths for the information acquisition and the interaction with user, (ii) basic domain rules that concern the area of digital circuits, and (iii) the inference mechanism. The file MODEL is actually a model base and contains the full structural description of the digital circuits that (presently) can diagnose. The clear distinction of these two parts of the knowledge base was selected in order to be able to modify and enhance the model base without the need to interfere the main body of the diagnostic tool. In this way the diagnostic mechanism is unaffected and independent from the system under study, each time. DICIDEX is applicable to combinatorial circuits that involve logical gates (AND, OR, XOR, NAND, 2-input NOR), 2-input half adders, 3-input adders, and 2-input (positive) multipliers. The gates are considered as primitive units, while all other units are composite units structurally defined in the file MODEL.

Structure description: All units and subsystems are described by predicates of the form:

$$\text{<unit-type>}(\text{<name>}, \text{inp}(I_1, I_2, ..., I_n), \text{out}(O_1, O_2, ..., O_n)) \qquad (1)$$

where the unit-type can be one of the following: "wire", "and gate" ,"or gate", "not gate", "nor gate", "xor gate", "nand gate", "half adder", "full adder" and "multiplier".

The *name* of a unit is a variable which, depending on the circuit where it participates, is instantiated to some Prolog name that declares the type of this circuit, (e.g. **m**(ultiplier) if the circuit at hand is a multiplier) and to the running index that characterizes it as a member of this circuit.

The arguments $inp(I_1, I_2, ..., I_n)$ and $out(O_1, O_2, ..., O_n)$ represent the I/O ports of the unit, whereas the variables I_i, O_i represent the respective signals. Each I/O signal is represented in the form of a list that contains the signal value in binary form.

For example an XOR gate is described by "xorgate"(h2, inp([A],[B]), out([S]))

The structural description of composite (non primitive) units is expressed by predicates of the form:

$$description (<unit\text{-}type>, (<unit\text{-}name>, inp(I_1, I_2, ..., I_n),$$

$$out(O_1, O_2, ..., O_n)), Components) \qquad (2)$$

The unit-type declares the type of the composite unit e.g. "half adder", "full adder" and "multiplier". The *name* of a complex unit is a variable which gives the capability to the user to give a particular name, code number, etc to the circuit at hand.

The structural description of a composite unit is represented in the form of a component list which is declared by the argument *Components*:

$$Components=[Comp_1, Comp_2, ..., Comp_n]$$

where each component Comp_i is of the form (1).

Behaviour description: As described in Section 6.3 the description of the behaviour of a complex (composite) system follows deterministically from the peformances of the totality of its components. On the other hand, the description of a component is done through the corresponding constraints that connect the I/O signals at its ports. More specifically, the simulation rules relate the outputs to the inputs, while the inference rules relate the inputs to the outputs. Thus for the description of the simulation we use the simulation rule network, but for the diagnosis (where both input and output signals are known) we use the constraint network, i.e. the totality of simulation and inference rules, and try to propagate values both towards the inputs and towards the outputs of the system.

Program structure: The main body of the diagnostic tool resides in the DICIDEX file, where the dicidex goal drives the diagnostic procedure. The dicidex goal consists of seven particular goals that perform the following operations:
- *display-banner:* It depicts on the screen some general information about the program.
- *initiate-system:* It initiates the system, by reading the MODEL file and loading the structural description of the system circuits to the dynamic data base.
- *get-information:* It collects all necessary knowledge from the user through the get-desired-function and get-device-type statements.

- *perform-task:* This statement guides to the dynamic data base to get the parametric model (G) of the structural description and store it to the file prototype. In this way later on it is able to present it to the user and give the command to start the simulation or the diagnosis {perform(simulation, (G), ...) or perform(diagnosis, (G), ...)}.

- *perform(simulation, (G), ...):* This statement realizes the simulation of the system. Prior to this it collects all the relevant information from the user with the aid of the statement *collect-data.* Particularly, it gets the name of the circuit under examination (via the *get-identifier* statement), the input signals (in the case of simulation), or the input and output signals (in the case of diagnosis). Then it simulates the system with the aid of simulate(system), instantiating the parameters of the parametric model (G) and storing them in a dictionary var.dic. It also stores that particular instance of the parametric model into the file *instance* and the analytic simulation results (i.e. the I/O signals for all components of the system) into the file *simfile* (with the aid of the statement *sim-out*)

- *perform(diagnosis, (G), ...)* This procedure carries out the diagnosis of the device selected by the user. Initially, it asks the ES to simulate the device in order to see if indeed there exist symptoms indicating faulty operations. Then it compares the measured and simulated output values, and if there is a discrepancy it calls the procedure *findfault* which localizes the possible causes of this deviation. These are the device components that are possibly faulty. Then the list of these probable faulty components is stored in the file *diagfile*, together with the signals at their ports, through the *diag-out* statement. The user can then read the diagfile at any desired time.

- *display-results:* This procedure presents the simulation or diagnosis results on the screen. Specifically, it presents the parametric model used for the simulation and diagnosis (present-model), the input data (present-inputs), the theoretically expected output values i.e. the simulated output values (*present-outputs*(simulation, ...)), the values of the model parameters during the simulation (present-bindings), the particular circuit model instance that results from the particular input signals (*present-instance*), and finally a detailed table with all the circuit components and their port values (*results - simulation*). If the user has asked for diagnosis, the presentation of the results continues with the observed output values that the user has given at the initiation of the system (*present-outputs(diagnosis, ...)*), the decision about the occurence of faulty behaviour, and the enumeration of the units that are possibly faulty together with the respective port signals (*present-candidates*). If the system finds that the device is faulty, but it cannot locate the faulty components, it delares its inability.

- *prepare print-out:* This procedure prepares a complete list with the information that appeared on the screen (this is done through the *display-results* predicate) and stores it in the file report from which the user can take a print-out.

- *end-system:* With this statement the system closes all the files that were constructed during the operation selected by the user. Then the ES asks the user if he (she) wishes to start a new dialogue (in the positive case the system empties the dynamic data base via the *restart* statement and calls again the procedure *dicidex*), or he (she) wishes to go back in the Prolog environment, or exit and return into the DOS.

Simulation mechanism: The simulation of the system at hand (System) is driven by the predicate *simulate* which has the argument [System | Components] that constitutes the system structural description. The *simulate* statement calls the recursive procedure *simul* which simulates successively all the system components (with the aid of the statement sml) and computes the signal values at their ports, as well as at the system output ports. These, theoretically expected output values of the device, are stored in the dynamic data base for two reasons, viz first to present them later to the user together with the other simulation results, and second to compare them with the measured values (in the case of diagnosis). Therefore we have:

```
simulate([System | Components]):-
                                simul(Components),!,
                                functor(System,D,3),
                                mem(System,[4,2],SO),
                                assertz(sim_out(D,[SO])).
    simul([]):-         !.
    simul([H | T]):-    sml(H), simul(T).
```

The various propositions sml (module), having ensured the correct application of the simulation (i.e. that the input values of the device at hand are already known), go to the respective truth tables and compute the expected output values of the device. For example, the simulation of a wire (the simplest type of device) is accomplished by:

```
    sml(wire(W, inp([WI], out([WO])))):-
                        nonvar(W1),!, wire_r(W1,W0).
```

and the corresponding truth tables which are expressed by the following prolog facts:

```
    wire_r  (1,1).
    wire_r  (0,0).
```

Similarly, in the case of a NAND gate we have the clauses:

```
    sml("nand gate"(N, inp([I1],[I2], out([NAND])))):-
                        evaluated([I1,I2]),!,
                        nand_r(I1,I2,NAND).
    nand_r(0,0,1).
    nand_r(0,1,1).
    nand_r(1,0,1).
    nand_r(1,1,0).
```

Finally, the simulation of a full adder (Fig. 8) is performed by:

```
    sml("full adder"(N, inp([I1],[I2],[I3]), out([C,S])))):-
                        evaluate([I1,I2,I3]),!,
                        full_add(I1,I2,I3,S,C).
    full_add(0,0,0,0,0).
    full_add(0,0,1,1,0).
    full_add(0,1,0,1,0).
    full_add(0,1,1,0,1).
    full_add(1,0,0,1,0).
    full_add(1,0,1,0,1).
    full_add(1,1,0,0,1).
    full_add(1,1,1,1,1)
```

494

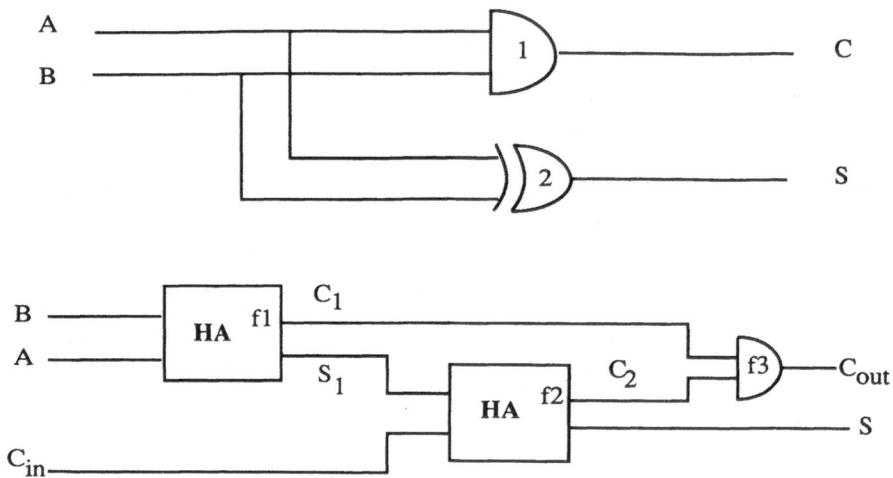

Fig. 8. Logic diagram of 1-bit full adder and half adder (HA)

Diagnostic mechanism: The diagnosis of the fault cause is based on the application of the constraint suspension technique. The realization of this technique is done through the statement *findfault(Comp, Structure)* which removes successively every component *Comp* of the system under examination from the totality of the components that make it (Structure). This is done via the statement *remove_item.* Then, it is checked if the removal of this component restores the internal consistency of the network that is formed by the remaining constraints. This is done via the statement *constraint_check.* In summary:

findfault(Comp, Structure):-

(remove_item(Comp, Structure, Rest),!,
constraint_check(Rest), !, fail.

constraint_check([ ]):- !.
constraint_check([H|T]):- validate(H)->constraint_check(T).

The statement *validate(Module)* realizes the union of the two simulation rules network and the inference rules network. This statement tries initially to simulate the particuular unit *Module* via the statement sml(Module). If this attempt fails this means that not all input values of the unit are known, and then an attempt is made to compute the signals at its ports with the aid of the inference rules which are generated by statements of the form *infer(Module),* i.e.

validate(Module):- (sml(Module),!); infer(Module).

As a simple example of the way the statements *infer(Module)* are generated we consider the gate NOT. In this trivial case if we don't know the input value of the gate, we can determine it solely by the output value. If however we also don't know the output value, it is impossible to draw any conclusion for the I/O values at the gate's ports and the clause fails. That is

```
infer("not gate"(N, inp[I], out([NOT]))):-
        nonvar(NOT), !,
        not_r(I,NOT).
infer("not gate"(_,_,_)):- !,fail.
```

The simple reasoning by which in the case of an OR gate one can find the signal values at its ports is the following:

1. If only the output value of the gate is known with value 0, then we conclude that both inputs have the value 0.
2. If only one input of the gate is known with value 1, then the output will also have the value 1.
3. If only one input of the gate is known with value 0, then the value of the other input will be equal to the output value of the gate.
4. In all other cases no conclusion can be drawn about the values of the I/O signals at the gate ports.

The above rules follow easily from the truth table of the OR gate:

```
or_r(0,0,0).
or_r(0,1,1).
or_r(1,0,1).
or_r(1,1,1).
```

Thus in the case of the OR gate the following three inference rules result. The first correesponds to the cases 1 and 3, the second corresponds to case 2 and the third to case 4:

```
R1) infer("or gate"(N, inp([I1], [I2], out([OR])))):-
                    ((evaluated([I1,OR]), I1 == 0,!);
                    (evaluated([I2,OR]), I2 == 0,!);
                    nonvar(OR), OR == 0,!)),
                    or_r(I1,I2,OR).
R2) infer("or gate"(N, inp([I1], [I2]), out([OR]))):-
                    (nonvar(I1), I1 == 1,!);
                    (nonvar(I2), I2 == 1,!),
                    OR=1.
R3) infer("or gate"(_,_,_)):- !, fail.
```

As a final example we present the clauses that correspond to the inference rules of the half adder:

```
R1) infer("half adder"(N, inp([I1], [I2]), out([C,S]))):-
                    ((evaluated([I1, C, S]),!);
                    (evaluated(I2, C, S]),!);
                    (evaluated(I1, S]),!);
                    (evaluated([I2, S]),!);
                    evaluated([I1, C], I1 == 1,!);
                    evaluated([I2, C], I2 == 1,!);
                    (nonvar(C), C == 1,!)),
                    half_add(I1, I2, S, C).
R2) infer("half adder"(N, inp([I1], [I2]), out([C, S]))):-
                    ((nonvar(I1), I1 == 0,!);
                    (novar(I2), I2 == 0,!),
                    C=0.
```

R3) infer("half adder"(_,_,___:-!, fail.

The above rules follow from the corresponding truth value table of the half-adder:

half_add(0,0,0,0).
half_add(0,1,1,0).
half_add(1,0,1,0).
half_add(1,1,0,1).

In all cases, the final rule is added so as to drive the whole set of rules to definite failure, via the predicate fail. In this way the system is informed that there does not exist any way for satisfying the goal and it has to direct its search to other components.

8.3. Examples of System-User Dialogues

Here we give two diagnosis dialogue examples which were obtained using particular data and printing the file *report* through the statement prepare_print_out.
Example 1
This example concerns the case of a 2-input half-adder with input values (1,1), expected output values (1 0), and measured output values (0,0).

OVERALL RESULTS
-----------------------*----------------------

The model employed for simulating the half adder's operation is :

PARAMETRISED SIMULATION MODEL
---------------------------------*-----------------------------------*-----------------
half adder(H,inp([A],[B]),out([C,S])).
and gate(h1,inp([A],[B]),out([C])).
xor gate(h2,inp([A],[B]),out([S])).

CURRENT SET OF INPUTS
---------------------*-----------*-------*--------------------
I(1) = 1
I(2) = 1

PREDICTED OUTPUTS
-----------------------*----------------------
O = 10

Under the current set of inputs, the model parameters were bound to the following values :

VARIABLE BINDINGS
-----------------------*----------------------

H = 0.
A = 1.
B = 1.
C = 1.
S = 0.

Hence, the instance of the model derived is :

INSTANTIATED SIMULATION MODEL
-------------------------------*-------------------------------*----------------

half adder(0,inp([1],[1]),out([1,0])).
and gate(h1,inp([1],[1]),out([1])).
xor gate(h2,inp([1],[1]),out([0])).

More specifically, the signals fed into, propagated through and output by the half adder model, are as follows :

SIMULATION RESULTS
-----------------------------*--------------------

COMPONENT	INPUT (S)	OUTPUT (S)
half adder 0	A = 1 B = 1	CS = 10
and gate h1	A = 1 B = 1	C = 1
xor gate h2	A = 1 B = 1	S = 0

The outputs actually obtained are :

OBSERVED OUTPUTS
------------------------*---------------------

O = 00

THE HALF ADDER UNDER CONSIDERATION, IS EVIDENTLY MALFUNCTIONING, i.e., **IT IS FUNCTIONING IN A WAY DIFFERENT FROM THE NORMALLY EXPECTED ! ! !**

The half adder components that are plausibly responsible for the device malfunctioning, along with their faulty behaviour, i.e., the (perhaps partially instantiated) signals at their ports, are :

DIAGNOSIS RESULTS
-----------------------------*--------------------

COMPONENT	INPUT (S)	OUTPUT (S)
and gate h1	A = 1 B = 1	C = 0

Example 2
Here a multiplier is considered with input values (1111, 1111), expected output values (1110, 0001), and measured output values (0110, 0001).

498

O V E R A L L　　R E S U L T S
--------------------*--------------------

The model employed for simulating the multiplier's operation is :

P A R A M E T R I S E D　　S I M U L A T I O N　　M O D E L
--------------------------------*--------------------------------*-----------------

multiplier(M,inp([A3,A2,A1,A0],[B3,B2,B1,B0]),out([R7,R6,R5,R4,R3,R2,R1,R0])).

and gate(m1,inp([A0],[B0]),out([R0])).

and gate(m2,inp([A0],[B1]),out([AND2])).

and gate(m3,inp([A1],[B0]),out([AND3])).

half adder(m4,inp([AND2],[AND3]),out([C4,R1])).

and gate(m5,inp([A0],[B2]),out([AND5])).

and gate(m6,inp([A1],[B1]),out([AND6])).

half adder(m7,inp([AND5],[AND6]),out([C7,S7])).

and gate(m8,inp([A2],[B0]),out([AND8])).

full adder(m9,inp([S7],[C4],[AND8]),out([C9,R2])).

and gate(m10,inp([A0],[B3]),out([AND10])).

and gate(m11,inp([A1],[B2]),out([AND11])).

half adder(m12,inp([AND10],[AND11]),out([C12,S12])).

and gate(m13,inp([A2],[B1]),out([AND13])).

full adder(m14,inp([S12],[C7],[AND13]),out([C14,S14])).

and gate(m15,inp([A3],[B0]),out([AND15])).

full adder(m16,inp([S14],[C9],[AND15]),out([C16,R3])).

and gate(m17,inp([A1],[B3]),out([AND17])).

and gate(m18,inp([A2],[B2]),out([AND18])).

full adder(m19,inp([AND17],[C12],[AND18]),out([C19,S19])).

and gate(m20,inp([A3],[B1]),out([AND20])).

full adder(m21,inp([S19],[C14],[AND20]),out([C21,S21])).

half adder(m22,inp([S21],[C16]),out([C22,R4])).

and gate(m23,inp([A2],[B3]),out([AND23])).

and gate(m24,inp([A3],[B2]),out([AND24])).

full adder(m25,inp([AND23],[C19],[AND24]),out([C25,S25])).

full adder(m26,inp([S25],[C21],[C22]),out([C26,R5])).

and gate(m27,inp([A3],[B3]),out([AND27])).

full adder(m28,inp([AND27],[C25],[C26]),out([R6,R7])).

C U R R E N T　　S E T　　O F　　I N P U T S
--------------------*-----------*------*--------------------

I(1) = 1111
I(2) = 1111

PREDICTED   OUTPUTS
-----------------------------*----------------------

O = 11100001

Under the current set of inputs, the model parameters were bound to the following values :

VARIABLE   BINDINGS
-----------------------------*----------------------

M = 0.
A3 = 1.
A2 = 1.
A1 = 1.
A0 = 1.
B3 = 1.
B2 = 1.
B1 = 1.
B0 = 1.
R7 = 1.
R6 = 1.
R5 = 1.
R4 = 0.
R3 = 0.
R2 = 0.
R1 = 0.
R0 = 1.
AND2 = 1.
AND3 = 1.
C4 = 1.
AND5 = 1.
AND6 = 1.
C7 = 1.
S7 = 0.
AND8 = 1.
C9 = 1.
AND10 = 1.
AND11 = 1.
C12 = 1.
S12 = 0.
AND13 = 1.
C14 = 1.
S14 = 0.
AND15 = 1.
C16 = 1.
AND17 = 1.
AND18 = 1.
C19 = 1.
S19 = 1.
AND20 = 1.
C21 = 1.
S21 = 1.
C22 = 1.
AND23 = 1.
AND24 = 1.
C25 = 1.
S25 = 1.
C26 = 1.
AND27 = 1.

Hence, the instance of the model derived is :

INSTANTIATED SIMULATION MODEL
--------------------------------*------------------------------*----------------

multiplier(0,inp([1,1,1,1],[1,1,1,1]),out([1,1,1,0,0,0,0,1])).
and gate(m1,inp([1],[1]),out([1])).
and gate(m2,inp([1],[1]),out([1])).
and gate(m3,inp([1],[1]),out([1])).
half adder(m4,inp([1],[1]),out([1,0])).
and gate(m5,inp([1],[1]),out([1])).
and gate(m6,inp([1],[1]),out([1])).
half adder(m7,inp([1],[1]),out([1,0])).
and gate(m8,inp([1],[1]),out([1])).
full adder(m9,inp([0],[1],[1]),out([1,0])).
and gate(m10,inp([1],[1]),out([1])).
and gate(m11,inp([1],[1]),out([1])).
half adder(m12,inp([1],[1]),out([1,0])).
and gate(m13,inp([1],[1]),out([1])).
full adder(m14,inp([0],[1],[1]),out([1,0])).
and gate(m15,inp([1],[1]),out([1])).
full adder(m16,inp([0],[1],[1]),out([1,0])).
and gate(m17,inp([1],[1]),out([1])).
and gate(m18,inp([1],[1]),out([1])).
full adder(m19,inp([1],[1],[1]),out([1,1])).
and gate(m20,inp([1],[1]),out([1])).
full adder(m21,inp([1],[1],[1]),out([1,1])).
half adder(m22,inp([1],[1]),out([1,0])).
and gate(m23,inp([1],[1]),out([1])).
and gate(m24,inp([1],[1]),out([1])).
full adder(m25,inp([1],[1],[1]),out([1,1])).
full adder(m26,inp([1],[1],[1]),out([1,1])).
and gate(m27,inp([1],[1]),out([1])).
full adder(m28,inp([1],[1],[1]),out([1,1])).

More specifically, the signals fed into, propagated through, and output by the multiplier model, are as follows :

SIMULATION RESULTS
-----------------------------*--------------------

COMPONENT		INPUT (S)	OUTPUT (S)
multiplier	0	A3A2A1A0 = 1111 B3B2B1B0 = 1111	R7R6R5R4R3R2R1R0=11100001
and gate	m1	A0 = 1 B0 = 1	R0 = 1
and gate	m2	A0 = 1 B1 = 1	AND2 = 1
and gate	m3	A1 = 1 B0 = 1	AND3 = 1
half adder	m4	AND2 = 1 AND3 = 1	C4R1 = 10
and gate	m5	A0 = 1 B2 = 1	AND5 = 1
and gate	m6	A1 = 1 B1 = 1	AND6 = 1
half adder	m7	AND5 = 1 AND6 = 1	C7S7 = 10
and gate	m8	A2 = 1 B0 = 1	AND8 = 1
full adder	m9	S7 = 0 C4 = 1 AND8 = 1	C9R2 = 10
and gate	m10	A0 = 1 B3 = 1	AND10 = 1
and gate	m11	A1 = 1 B2 = 1	AND11 = 1
half adder	m12	AND10 = 1 AND11 = 1	C12S12 = 10
and gate	m13	A2 = 1 B1 = 1	AND13 = 1
full adder	m14	S12 = 0 C7 = 1 AND13 = 1	C14S14 = 10
and gate	m15	A3 = 1 B0 = 1	AND15 = 1
full adder	m16	S14 = 0 C9 = 1 AND15 = 1	C16R3 = 10
and gate	m17	A1 = 1 B3 = 1	AND17 = 1

and gate	m18	A2 = 1 B2 = 1	AND18 = 1
full adder	m19	AND17 = 1 C12 = 1 AND18 = 1	C19S19 = 11
and gate	m20	A3 = 1 B1 = 1	AND20 = 1
full adder	m21	S19 = 1 C14 = 1 AND20 = 1	C21S21 = 11
half adder	m22	S21 = 1 C16 = 1	C22R4 = 10
and gate	m23	A2 = 1 B3 = 1	AND23 = 1
and gate	m24	A3 = 1 B2 = 1	AND24 = 1
full adder	m25	AND23 = 1 C19 = 1 AND24 = 1	C25S25 = 11
full adder	m26	S25 = 1 C21 = 1 C22 = 1	C26R5 = 11
and gate	m27	A3 = 1 B3 = 1	AND27 = 1
full adder	m28	AND27 = 1 C25 = 1 C26 = 1	R6R7 = 11

The outputs actually obtained are :

OBSERVED   OUTPUTS
-----------------------*----------------------

O = 01100001

THE MULTIPLIER UNDER CONSIDERATION, IS EVIDENTLY MALFUNCTIONING, i.e., IT IS FUNCTIONING IN A WAY DIFFERENT FROM THE NORMALLY EXPECTED ! ! !

The multiplier components that are plausibly responsible for the device malfunctioning, along with their faulty behaviour, i.e., the (perhaps partially instantiated) signals at their ports, are :

DIAGNOSIS   RESULTS
-----------------------*----------------------

COMPONENT		INPUT (S)	OUTPUT (S)
full adder	m25	AND23 = 1 C19 = 1 AND24 = 1	C25S25 = 01
full adder	m26	S25 = 1 C21 = 1 C22 = 1	C26R5 = 01
and gate	m27	A3 = 1 B3 = 1	AND27 = 0
full adder	m28	AND27 = 1 C25 = 1 C26 = 1	R6R7 = 10

9. CONCLUSIONS

This chapter has presented an overview of the knowledge-based system diagnosis. Starting with the knowledge representation and acquisition tasks which are a prerequisite for diagnosis, the classes of first-and-second-generation expert systems were discussed. Then, a look at the fault diagnosis methodologies and expert system architectures was made (general issues, diagnostic modelling, architectures). The chapter continued with a survey of diagnostic tools for digital systems presenting first the D-algorithm which is the base for many of them, Then, two general methodologies for the development of fault diagnosis tools were discussed, one for the digital circuits domain and one for the process engineering domain. Finally, an implementation of the digital circuits fault diagnosis methodology was presented with two particular examples.

First-generation fault diagnosis expert systems are at a mature state of advancement but they possess severe limitations in comparison

502

to second-generation expert systems. Second generation expert systems alleviate these limitations by beeing as deeper as possible also possessing automated knowledge acquisition/learning capabilities and generic task architecture. The current tendency is to develop and use knowledge-based diagnostic tools of the second-generation type. This area is still under further improvement and enhancement. Another area of current interest is the combination of knowledge-based fault diagnosis with model-based fault diagnosis (which is based on analytical redundancy and statistical methods)[1]. The resulting systems are hybrid systems which are known as *engineering-based expert systems* (EBES) for diagnosis and other engineering tasks [52, 83, 84].

REFERENCES

1. R. Patton, P. Frank and R. Clark, Fault Diagnosis in Dynamical Systems: Theory and Application, *Prentice-Hall Intl.*, New York-London, 1989.
2. S.G. Tzafestas, Knowledge-Based System Diagnosis, Supervision and Control, *Plenum Press*, New York-London, 1989.
3. S.G. Tzafestas, A.I. Kokkinaki and K.P. Valavanis, An Overview of Expert Systems, In: *Expert Systems in Engineering Applications* (S. Tzafestas, Ed.), <u>Springer Verlag</u>, pp. 25-51, 1993.
4. B.F. Hayes-Roth, Rule-Based Systems, *Communications ACM*, Vol. 26, No. 9, pp. 921-932, 1985.
5. R. Frost, Introduction to Knowledge Based Systems, *Collins Professional and Technical Books*, 1986.
6. E. Post, Formal Reductions of the General Combinatorial Problem, *Amer. J. Maths.*, Vol. 65, pp. 197-268, 1943.
7. E.H. Shortliffe, A Rule-Based Computer Program for Advancing Physicians Regarding Antimicrobial Therapy Selection, *Ph. D. Thesis*, Stanford Univ., 1974.
8. B.C. Buchanan and E.A. Feigenbaum, DENDRAL and META-DENDRAL: Their Applications Dimension, *Artificial Intelligence*, Vol. 11, pp. 5-24, 1978.
9. W.A. van Melle, A Domain-Interdependent System that Aids in Constructing Knowledge-Based Consultation Programs, *Ph. D. Thesis*, Stanford Univ., 1980.
10. M.L. Minsky, A Framework for Representing Knowledge, *The Psychology of Computer Vision*, pp. 211-277, *McGraw-Hill*, N.Y., 1975.
11. K.D. Forbus, The Qualitative Process Engine, In: *Readings in Qualitative Reasoning About Physical Systems* (D.S. Weld, J. deKleer, eds.), Morgan Kaufmann, 1990.
12. D.B. Lenat, On Automated Scientific Theory Formation: A Case Study Using the AM Program, *Machine Intelligence*, Vol. 9, pp. 251-283, 1979.
13. D.B. Lenat, EURISKO: A Program that Learns New Heuristics and Domain Concepts, *Artificial Intelligence*, Vol. 21, pp. 61-98, 1983.
14. P. Thagard, Computational Philosophy of Science, *MIT Press*, Cambridge, MA, 1988.
15. J.A. Robinson, Logic: Form and Function, *Edinburgh Univ. Press*, Edinburgh, 1979.

16. R.A. Kowalski, Logic for Problem Solving, North-Holland, Amsterdam, 1979.
17. R. Frost, Introduction to Knowledge Based Systems, *Collins Professional and Technical Books*, 1986.
18. G. Brewka, The Logic of Inheritance in Frame Systems, *Proc. 10th Intl. Joint Conf. on Artificial Intelligence*, pp. 238-488, 1987.
19. D.G. Bobrow and T. Winograd, An Overview of KRL: A Knowledge Representation Language, *Gognitive Science*, Vol. 1, No. 1, 1977.
20. A. Goldberg and D. Robson, Smalltalk 80: The Language and its Implementation, *Addison-Wesley*, Reading, MA, 1983.
21. D.E. Knuth, Semantics of Context-Free Languages, *Math. Syst. Theory*, Vol. 2, pp.127-145, 1968.
22. G. Papakonstantinou and J. Kontos, Knowledge Representation with Attribute Grammars, *The Computer Journal*, Vol. 29, No. 3, pp. 241-245, 1986.
23. P. Deransart and J. Maluszynski, Relating Logic Programs and Attribute Grammars, *J. Logic Programming*, Vol. 2, pp. 119-155, 1985.
24. G. Papakonstantinou, C. Moraitis and T. Panayiotopoulos, An Attribute Grammar Interpreter as a Knowledge Engineering Tool, *Angewandte Informatik*, Vol. 9/86, pp.282-288, 1986.
25. G. Papakonstantinou and S.G. Tzafestas, Attribute Grammar Approach to Knowledge-Based System Building: Application to Fault Diagnosis, In: *Knowledge Based System Diagnosis, Supervision and Control* (S.G. Tzafestas, ed.), Plenum, N.Y.-London(Ch. 7), 1989.
26. K. Niwa, K. Sasaki and H.Ihara, An Experimental Comparison of Knowledge Representation Schemes, *The AI Magazine*, Vol. 5, pp.29, 1984.
27. S.G. Tzafestas and A. Adrianopoulos, Knowledge Acquisition for Expert System Design, In: *Expert Systems in Engineering Applications* (S.G. Tzafestas, ed.), Springer-Verlag, Berlin-N.Y., Ch. 2, 1993.
28. R.J. Brachman, et. al., What Are Expert Systems?, In: *Building Expert Systems* (F. Hayes,-Roth, D.A. Waterman and D.B. Lenat, eds.), Addison-Wesley, Reading, MA, pp. 31-57, 1983.
29. D. Waterman, A Guide to Expert Systems, *Addison-Wesley*, Reading, MA, 1986.
30. J.H. Boose, Expertise Transfer for Expert System Design, *Elsevier*, Amsterdam, 1986.
31. J.H. Boose, Rapid Acquisition and Combination of Knowledge from Multiple Experts in the Same Domain, *Future Computing Systems*, Vol. 1, No. 2, pp. 191-216, 1986.
32. G. Kahn, S. Nowlan and J. McDermott, Strategies for Knowledge Acquisition, *IEEE Trans. on Pattern Analysis and Machine Intelligence*, Vol. PAMI-7, No 5, pp. 511-522, 1985.
33. J. McDermott, Making Expert Systems Explicit, *Proc. 10th IFIP Congress* (Dublin), Elsevier, Amsterdam, 1986.
34. W.J. Clancey, Heuristic Classification, *Artificial Intelligence*, Vol. 27, pp. 289-350, 1985.
35. S. Marcus, J. Stout and J. McDermott, VT: An Expert Elevator Designer that uses Knowledge-Based Reasoning, *AI MAgazine*, Vol. 9, No 1, pp. 95-112, 1988.

36. G. Klinker et al., KNACK:Report-Driven Knowledge Acquisition, *Int. J. Man Machine Studies*, Vol. 26, No 1, pp. 65-79, 1987.

37. J.A. Breuker and B.J. Wielinga, Use of Models in the Interpretation of Verbal Data, In: *Knowledge Acquisition for Expert Systems: A Practical Handbook* (A.L. Kidd, ed.), Plenum, N.Y., pp. 17-44, 1987.

38. J.A. Breuker and B.J. Wielinga, Knowledge Acquisition as Modeling Expertise: The KADS Methodology, *Proc. 1st Europ. Workshop on Knowledge Acquisition for KBS*, Reading Univ., Sept. 1987.

39. S.A. Hayward, Methodology: Analysis and Design for Knowledge Based Systems, *Esprit Project 1098-Report STC-Y-RR-001(3.0)*, STC, Harlow, 1987.

40. C. Hayball and D. Barlow, Skills Support in the ICL Kidsgrove Bonding Shop - A Case Study in the Application of CADS Methodology, Int. *Conf. on Human and Organizational Issues of Expert Systems*, Stratford-on-Avon, England, May, 1988.

41. E. Motta et al., Support for Knowledge Acquisition in the Knowledge Engineer's Assistant, *Expert Systems*, Vol. 5, No 1, pp. 6-28, 1988.

42. J. Diederich, I. Ruhman and M. May, KRITON: A Knowledge-Acquisition Tool for Expert Systems, *Int. J. Man-Machine Studies*, Vol. 26, pp. 29-40, 1987.

43. S.G. Tzafestas, System Fault Diagnosis Using the Knowledge-Based Methodology, In: *Fault Diagnosis in Dynamic Systems:Theory and Applications*(R. Patton, P. Frank and R. Clark, eds.), Prentice Hall Intl., Ch. 15 , 1989.

44. R. Davis, Diagnostic Reasoning Based on Structure and Behaviour, *Artificial Intelligence*, Vol. 24, pp. 247-410, 1984.

45. J. DeKleer, Local Methods for Localizing Faults in Electronic Circuits, *MIT AI Memo 394*, Cambridge, MA, 1976.

46. J. DeKleer, The Origin and Resolution of Ambiguities in Causal Arguments, *Proc. Int. Joint Conf. on Artificial Intelligence*, pp. 197-203, Tokyo, Japan, 1979.

47. B. Kuipers, Commonsense Reasoning About Causality:Deriving Behaviour from Structure, *Artificial Intelligence*, Vol. 24, pp. 169-203, 1984.

48. J. DeKleer and J.S. Brown, Assumptions and Ambiguities in Mechanical Mental Models, *Xerox PARC Report CIS-9*, Palo Alto, CA, 1982.

49. J. DeKleer and B.C. Williams, Diagnosing Multiple Faults, *Artificial Intelligence*, Vol. 32, No. 1, pp. 97-130, 1987.

50. J. DeKleer and B.C. Williams, Reasoning About Multiple Faults, *Proc. AAAI-86 Conf.*, pp. 132-139, 1986.

51. E.T. Keravnou and J. Washbrook, What is a Deep Expert System? Analysis of the Architectural Requirements of Second-Generation Expert Systems, *The Knowledge Engineering Review*, Vol. 4, No. 3, pp. 205-233, 1989.

52. S.G. Tzafestas, Second Generation Diagnostic Expert Systems: Requirements, Architectures and Prospects, *Proc. IFAC Symp. on Fault Detection and Safety for Technical Processes* (SAFEPROCESS' 91), Baden-Baden, Germany, Sept., 1991.

53. P.K. Fink, J.C. Lusth and J.W. Duran, A General Expert System Design for Diagnostic Problem Solving, *IEEE Trans. on Pattern*

Analysis and Machine Intelligence, Vol. PAMI-7, No 5, pp. 553-560, 1985.

54. R. Davis, Reasoning from First Principles in Electronic Troubleshooting, *Int. J. Man-Machine Studies*, Vol. 19, pp. 403-423, 1983.

55. R. Davis, Diagnosis via Causal Reasoning: Paths of Interactions and the Locality Principle, *Proc. AAAI-83*, Washington, D.C., pp. 83-94, 1983.

56. R. Patil, P. Szolovits and W. Schartz, Causal Understanding of Patient Illness in Medical Diagnosis, *Proc. Int. Joint. Conf. on Artificial Intelligence* (IJCAI-81), pp.893-899, 1981.

57. L. Steels, The Deepening of Expert Systems, *AI Memo 87-16*, AI Lab., VU Brussels, 1987.

58. L. Steels, Components of Expertise, *AI Memo 88-16*, AI LAB, VU Brussels, 1988.

59. B. Chandrasekaran and S. Mittal, Deep Versus Compiled Knowledge Approaches to Diagnostic Problem Solving, *Int. J. Man-Machine Studies*, Vol. 19, pp. 425-436, 1983.

60. B. Chandrasekaran, Generic Tasks in Knowledge-Based Reasoning: High Level Building Blocks for Expert System Design, *IEEE Expert*, Fall, pp. 23-30, 1986.

61. B. Chandrasekaran, Genetic Tasks as Building Blocks for Knowledge-Based Systems: The Diagnosis and Routine Design Examples, *The Knowledge Engineering Review*, Vol. 3, pp 183-210, 1988.

62. W. Van de Velde, Learning Heuristics in Second-Generation Expert Systems, *Proc. 6th Int. Workshop on Expert Systems and their Application*, Avignon, France, 1986.

63. W.J. Clancey and R. Letsinger, Neomycin: Reconfiguring a Rule-Based Expert System for Application to Teaching, *Proc. Int. Joint Conf. on AI-81*, pp. 829-836, 1981.

64. M. O'Neil, A. Glowinski and J. Fox, A Symbolic Theory of Decision Making Applied to Several Medical Tasks, *Proc. AIME-89*, pp. 62-71, 1989.

65. R. Davis, B.G. Buchanan and E.H. Shortliffe, Production Rules as a Representation in a Knowledge-Based Consultation System, *Artificial Intelligence*, Vol. 8, pp. 15-45, 1977.

66. R. Davis et. al., Diagnosis Based on Structure and Function, *Proc. AAAI Conf.*, pp. 137-142, August 1982.

67. R. Davis and H. E. Shrobe, Representing Structure and Behaviour of Digital Hardware, *IEEE Computer*, Vol. 16, No. 2, pp. 75-82, 1983.

68. P.K. Fink and J.C. Lusth, Expert Systems and Diagnostic Expertise in the Mechanical and Electrical Domains, *IEEE Trans. Systems, Man and Cybernetics*, Vol. SMC-17, p. 340, 1987.

69. M.R. Genesereth, Diagnosis Using Hierarchical Design Methods, *Proc. AAAI*, pp. 178-183, August, 1982.

70. M.R. Genesereth et al., A Meta-Level Representation System, *HPP-83-28* Stanford Univ. Heuristic Programming Project, Stanford, 1983.

71. M.R. Genesereth, The Use of Design Descriptions in Automated Diagnosis, *Artificial Intelligence*, Vol. 24, pp.411-436, 1984.

72. W.A. Perkins and T.J. Laffey, LES: A General Expert System and its Applications, *SPIE Proc.*, Vol. 485, Applications of Artificial Intelligence, p. 46, 1984.

73. T.J. Laffey, W.A. Perkins and T.A. Nguyen, Reasoning About Fault Diagnosis with LES, *IEEE Expert*, pp. 13-20, 1986.

74. W. Hamscher, Using Structural and Functional Information in Diagnostic Design, *Proc. AAAI-83*, pp. 152-156, 1983.

75. W. Hamscher and R. Davis, Issues in Diagnosis from First Principles, *AI Memo 394*, AI LAb. MIT, Cambridge, MA, USA, 1986.

76. V. Kelly and L. Steinberg, The CRITTER System: Analysing Digital Circuits by Propagating Behaviours and Specifications, *Proc. Natl. Conf. on AI*, Pittsburg, PA, pp. 284-289, 1982.

77. H. Pople, Heuristic Methods for Imposing Structure on Ill-Structured Problems, In: *Artificial Intelligence in Medicine* (P. Szolovits, ed.), AAAS Selected Symposium 51, 1982.

78. H. Pople, The Formation of Composite Hypotheses in Diagnostic Problem Solving: An Exercise in Synthetic Reasoning, *Proc. Intl. Joint Conf. on AI-77*, Cambridge, MA, 1977.

79. I.M. Neale, First Generation Expert Systems: A Review of Knowledge Acquisition Methodologies, *The Knowledge Engineering Review*, Vol. 3, No 2, pp. 105-145, 1988.

80. N. Hari Narayanan and N. Viswanadham, A Methodology for Knowledge Acquisition and Reasoning in Failure Analysis of Systems, *IEEE Trans. Systems, Man and Cybernetics*, Vol. SMC-17, No. 2, pp274-288, 1987.

81. J.N. Warfield, Structuring Complex Systems, *Batelle Memorial Inst.*, Columbus, Ohio, Monograph 4, April, 1974

82. S.V.N. Rao and N. Viswanadham, Graph Algorithms for Fault Diagnosis in Large Scale Systems, Tech. Rept. HIREL-SA-8, Indian Inst. of Sci., School of Automation, Bangalore, India, Nov. 1984.

83. A.I. Kokkinaki, K.P. Valavanis and S.G. Tzafestas, A Survey of Expert System Tools and Engineering-Based Expert Systems, In: *Expert Systems in Engineering Applications* (S. Tzafestas, Ed.), Springer-Verlag, pp. 367-378, 1993.

84. S.G. Tzafestas, AI Techniques in Computer-Aided Manufacturing Systems, In: *Knowledge Engineering* (H. Adeli, Ed.), Vol II: Applications, McGraw Hill, pp. 161-212, 1990.

16
MODEL-BASED DIAGNOSIS : STATE TRANSITION EVENTS AND CONSTRAINT EQUATIONS

KARL-ERIC ÅRZEN[(*)], ANDERS WALLÉN[(*)], THOMAS F. PETTI[(**)]
[(*)] Department of Automatic Control
Lund Institute of Technology
Box 118, S-221 00 Lund, Sweden

[(*)] W.R. Grace & Co.-Conn., Washington Research Center
7379 Route 32, Columbia, MD 21044, USA

1. Introduction

Computer assisted on-line monitoring and diagnosis is becoming an increasingly important part of modern process control systems. Diagnosis systems can be built from two different main principles: symptom-based diagnosis and model-based diagnosis. A symptom-based system aims at explicitly associate the symptoms of a fault, as indicated by the sensors, with the fault itself. The associations can be expressed as rules, tables, etc. The source of the diagnosis knowledge is experienced "expert" operators. Experience has shown that a purely heuristic approach like this has several drawbacks: knowledge acquisition difficulties, problems with completeness, unability to handle problems for which the experts have no solution, etc.

Model-based diagnosis systems are based on a model of the process. During diagnosis the model's predicted process outputs are compared with the real measured outputs. Discrepancies indicate faults. In model-based diagnosis the sources of the diagnosis knowledge are the process designers and engineers. Model-based diagnosis is nothing new in the control community. For long, diagnosis methods based on detailed process models, often on differential equation form, have been suggested and also in quite a few cases applied industrially. These include parity space approaches and observer-based approaches [Frank, 1990] and process identification based methods [Isermann, 1984]. However, a problem with these "classical" methods is the detailed dynamic models that they require.

S. G. Tzafestas and H. B. Verbruggen (eds.),
Artificial Intelligence in Industrial Decision Making, Control and Automation, 507–523.
© 1995 *Kluwer Academic Publishers.*

During the 1980s model-based on-line applications have received increased attention from the AI community. The difference from "classical" model-based methods is the nature of the models. The AI community focus on coarse models that often only give a qualitative description of the process behaviour. Models of these types include signed directed graphs and qualitative physics models represented, e.g., as confluence equations, constraint equations, etc. The diversity of models used and differences in how the models are actually used for on-line diagnosis, however, make it extremely difficult to evaluate different approaches against each other.

One part of the project "Knowledge-Based Real-Time Control Systems" [Årzén, 1990] has been to explore various approaches to model-based on-line diagnosis. The project has focused on methods that do not require complete dynamical models as these are not always available in the process industry. The project was a joint industrial project between ABB and the Department of Automatic Control, Lund Institute of Technology during 1988 to 1991. Within the project a common test process – Steritherm – has been used. Steritherm is a food engineering process for sterilization of liquid food products. A real-time simulation model of Steritherm has been implemented in the real-time expert system environment G2 from Gensym Corp [Moore et al., 1990]. Within the same environment a "knowledge-based" control system has been implemented that controls, monitors, and diagnoses the simulated Steritherm process. Within this control system various knowledge-based, as well as conventional, applications have been implemented. These include symptom-based monitoring, off-line troubleshooting based on fault trees, production scheduling, alarm analysis based on functional process models, and finally the topic of this paper, two different on-line model-based diagnosis schemes: the Diagnostic Model Processor (DMP) method and the MIDAS method.

A comparison between these two methods is especially interesting since the methods are so different. DMP uses quantitative equations as the basis for the diagnosis whereas MIDAS is based on qualitative signed directed graphs. DMP is based on taking periodic snapshots of the process variables and recalculating the equations whereas MIDAS is based on events and keeps an internal history of past events and may also predict future events.

In Section 2 an overview of the DMP method is given. MIDAS is presented in Section 3. The application of these methods to the Steritherm process is described in Section 4 and in Section 5 the results are compared. The material presented here is an extended version of [Nilsson et al., 1992].

2. DMP

The Diagnostic Model Processor method (DMP) [Petti and Dhurjati, 1991], [Petti, 1992] is based on quantitative constraint equations called model equations. By examining the direction and extent to which each model equation is violated and by considering the assumptions on which they depend, the most likely failed assumption (fault) can be deduced. Redundancy, sensor or analytic, which is available in the system leads to better performance because an assumption that is common to many violated equations is strongly suspect; whereas satisfaction of equations provides evidence that the associated assumptions are valid.

The process model consists of series of equations on residual form. The equations may contain anything that can be calculated at run-time including past sensor values, mean values, etc. Usually equations compare a sensor value to what it ideally should read, compare multiple sensors measuring the same physical value, compare a sensor reading a physical value with a calculation of of the physical value based on other indirect sensors, or express a general balance equation, e.g., a mass or energy balance.

Associated with each equation are tolerance limits which represent the expected (fault free) upper and lower values of the residual for which the equation is considered satisfied. Also associated with each equation is a set of assumptions which if satisfied guarantee the satisfaction of the equation. The j'th equation is denoted $e_j = c_j(P; a)$, where P denotes the process data and a indicates that each equation is dependent on the satisfaction of a vector of modeling assumptions.

A simple example of an equation and the associated assumptions is shown in Fig. 1. FT1 and DP1 are flow and differential pressure transmitters. The equation compares the measured and calculated pressure drops over the heat exchanger. Notice that some of the assumptions are explicit in the model equations, such as correct sensor readings, and some are implicit such as the fact that there are no piping leaks.

Since the residuals are not uniform in magnitude, they are transformed into a metric between -1 and 1 which indicates the degree to which the model equation is satisfied: 0 for perfectly satisfied, 1 for severely violated high, and -1 for severely violated low. These values constitute the satisfaction vector, sf, which is calculated using the model equation tolerances, τ. For the jth model equation,

$$sf_j = \frac{(e_j/\tau_j)^n}{1 + (e_j/\tau_j)^n} \tag{1}$$

The value of sf_j is given a positive value for a positive residual, e_j, and a negative value for a negative residual. The curve is a general sigmoidal function with

510

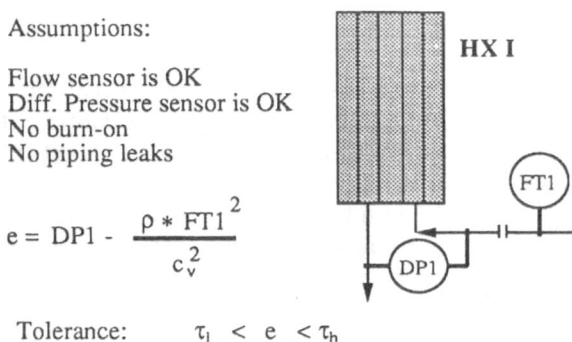

Assumptions:

Flow sensor is OK
Diff. Pressure sensor is OK
No burn-on
No piping leaks

$$e = DP1 - \frac{\rho * FT1^2}{c_v^2}$$

HX I

Tolerance: $\tau_l < e < \tau_h$

Figure 1. Example of the formulation of a model equation with its associated assumptions.

the steepness determined by the constant n. If the tolerances are not symmetric around the origin, the upper tolerance τ_h is used for a positive residual and the lower tolerance τ_l is used if the residual is negative.

A matrix of sensitivity values, S, which describes the relationship between each model equation and assumption is computed to weight the sf values as evidence. The ijth element of S, which represents the sensitivity of the jth model equation to the ith assumption is calculated as:

$$S_{ij} = \frac{\frac{\partial c_j}{\partial a_i}}{|\tau_j|} \tag{2}$$

The larger the partial derivative of an equation with respect to an assumption, the more sensitive that equation is to deviations of the assumption. Many model equations are non-linear in some assumptions; these partials are estimated by linear approximations. Assumptions which are implicit with respect to an equation (i.e., pump operation) are arbitrarily given a partial derivative equal to 1 or -1, unless experience suggests otherwise. Also, equations independent of an assumption have an associated sensitivity of zero. Explicit assumptions normally represent sensors which are used explicitly in the equations. In this case the sensitivity is expressed by the partial derivative of the equation with respect to the sensor according to Eq. 2. This expression is recalculated on-line. A sensor is assumed to either fail high or fail low, i.e., read either too high or too low.

Conclusions about the satisfaction of each assumption (fault state) is made by combining the evidence from the model equations, sf, with consideration to the sensitivity matrix S. This is done through the calculation of a vector of failure likelihoods, F, such that

$$F_i = \frac{\sum_{j=1}^{N}(S_{ij}\, sf_j)}{\sum_{j=1}^{N}|S_{ij}|} \tag{3}$$

where N is the number of model equations. It is evident that this method of combination allows the sf values of those equations which are most sensitive to deviations of assumption a_i to be weighted the most heavily in the calculation of F_i. The failure likelihood is interpreted as indicating a likely condition of assumption a_i failing high as the value of F_i approaches 1, while an F_i tending toward -1 indicates a likely failure low.

DMP allows the detection of non-competing multiple faults. This often leads to several assumptions' failure likelihood exceeding the likely limit that determines if they should be presented to the operator or not also when a single failure has occurred. To limit the number of faults presented and to direct attention to the most probable conditions a procedure is used to check the casual relationships between the assumptions. The procedure basically assumes a fault and checks the expected behaviour of the equations to see if other failure likelihoods should exceed the likely limit. If this is the case, the assumption is said to "explain" the appearance of the other assumptions. All likely assumptions are tested pair-wise and those assumptions that are not "explained" by other assumptions are considered top-level and get a special graphical indication.

G2 implementation

Equations and failure assumptions are represented as G2 objects. A dependency between an equation and an assumption is represented by a graphical connection, a dependence connection object, between them. Equations have attributes representing the residual, the tolerance limits, and the satisfaction value. The dependence objects have attributes characterizing the type of relationship (implicit or explicit), a formula to calculate the partial derivative of the equation with respect to the assumption, and a sensitive attribute. The calculation of satisfaction values and failure likelihoods are performed by generic formulas referring to the connections between equations and assumptions.

The objects implementing DMP in G2 are shown in Fig. 2. As can be seen in the figure the total DMP model makes up a network. From a computational point of view the DMP method is similar to an already trained feedforward neural network with one hidden layer. The input layer corresponds to the sensor values used in the equations. The hidden layer consists of the model equations and the output layer is the assumptions.

The DMP methods has also strong similarities with diagnosis methods based on parity equations or observer schemes. Since the model equations may contain any computable expression they may also contain dynamic equations. The

512

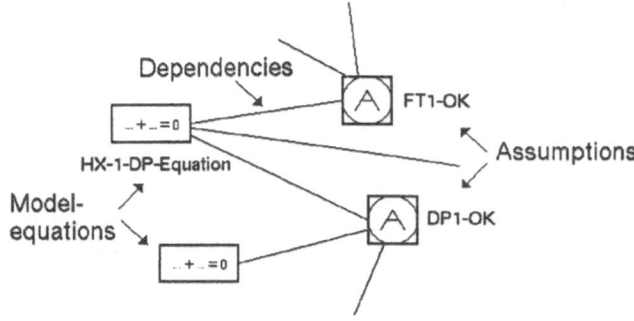

Figure 2. The objects implementing DMP in G2

satisfaction vector *sf* can be seen as a residual vector and the application of the sensitivity information in the dependencies can be seen as a way of structuring this residual vector so that only a fault-specific subsect (ideally with one element) of it becomes non-zero in case of a fault. The strong points of DMP are that it does not require a complete dynamic model, it may very well be used if, e.g., only static equations are known, and that the fault isolation is well integrated with the fault detection.

3. MIDAS

MIDAS (Model Integrated Diagnosis Analysis System) is an approach to model-based, on-line diagnosis developed at MIT by Kramer, Oyeleye, and Finch and described in [Oyeleye, 1989], [Finch, 1989] and [Finch *et al.*, 1990]. MIDAS belongs to the diagnosis methodologies based on qualitative causal reasoning about deviations from a nominal steady-state. During the diagnosis, events representing qualitative state transformations, e.g., that a sensor changes from NORMAL to HIGH, are clustered together into groups that each could be explained by a certain fault.

MIDAS models

MIDAS is based on a chain of different qualitative models. The final model in this chain is what is used on-line during the actual diagnosis.

SDG model: The first model is the Signed Directed Graph (SDG) that is derived from the physical equations of the process. A SDG describes the variables of the process as nodes and the qualitative relationships between the variables as arcs between the nodes. Nodes have the qualitative states 0, +, or − depending on if the process variable $x_i = 0, x_i > 0$, or $x_i < 0$. An arc between two nodes n_1 and n_2 has the sign $+(-)$ if an increase in n_1 causes an increase (decrease) in n_2. A zero-signed arc initiating and terminating at the same node indicates an integrator.

Included in the SDG is also the set of root causes that might affect the process. All the faults concerning the process that should be detectable must be incorporated in this set. For each fault there must exist a primary deviation variable, i.e., a variable included in the SDG which is first affected by the root cause, and a sign that tells in which direction this variable is deviated. An example of a SDG for a simple gravity flow tank is shown in Fig. 3

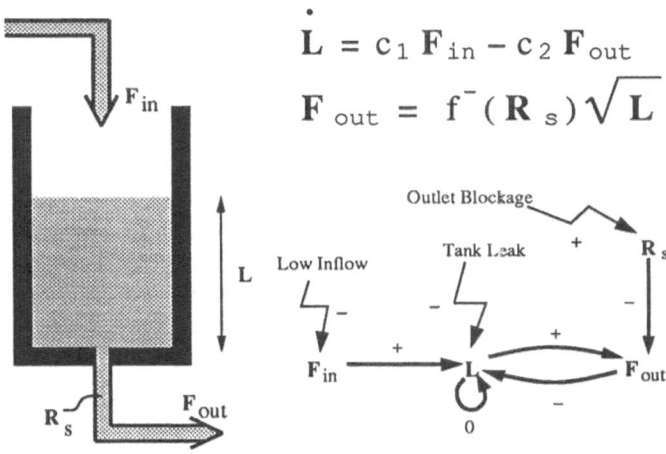

$$\dot{L} = c_1 F_{in} - c_2 F_{out}$$

$$F_{out} = f^-(R_s)\sqrt{L}$$

Figure 3. A gravity flow tank with its equations and SDG

The construction of a SDG for a whole process consists of combining together sub-SDGs representing the process components of the process. The sub-SDGs can be seen as generic SDG models that describe the qualitative models for classes of process components.

A problem with the SDG process models is that they only describe local, direct causalities between variables. Non-local causalities due to, e.g., feedback effects cannot be handled. Consider the tank in Fig. 3 once again. Assume that the pipe resistance R_s is somehow decreased. Following the arcs gives that F_{out} will increase, L decrease, F_{out} decrease, L increase, and so on. This leads to a chain of contradictions and the SDG does not tell whether the outflow and the level will ultimately be low or high, oscillate or return to their initial values.

ESDG model: To overcome the above problem MIDAS analyzes the loops in the SDG and insert additional non-physical arcs into the graphs. This leads to the Extended Signed Directed Graph (ESDG), which can explain the behaviour of the process through feed-forward paths only. The interpretation of the non-physical arcs and the algorithms for analyzing the graphs are found in [Oyeleye, 1989].

Event graph model: The model that MIDAS uses on-line during diagnosis is an event graph model. It consists of events, i.e., qualitative state transitions, root causes, and links connecting events with other events and with the root causes. A qualitative state of a variable is either high, normal or low, compared to the nominal steady-state value. That is, for every measured variable four events are possible: a transition from normal to high, from high to normal, from normal to low, and from low to normal.

The event graph model can be derived directly from the ESDG. Briefly, the transformation of an ESDG into an event graph is done by removing all unmeasured nodes in the ESDG, but for each measured node create four nodes in the event graph, each representing one of the possible state changes. The arcs in the ESDG are translated into arcs in the event graph. Each root cause is connected to the event that should be the first symptom of the fault. Further, there are links between the events in order to express possible fault propagation. These links may have conditions attached to them telling when the arcs are valid and what diagnostic conclusions that may be drawn when two events are linked together.

It is also possible to include other types of events in the event graph, e.g., operator actions, off-line test results, changes in trend, and quantitative constraint equations. However, the MIDAS methodology does not contain any structured way of building a model containing relations between such events. These have to be added on a purely heuristic basis.

An event graph for the single gravity flow tank in Fig. 3 where the level and the outflow are measured is shown in Fig. 4 Compared to the SDG in Fig 3, the set of root causes is somewhat extended. In the event graph the circles represent events. When an event has been detected. e.g., the flow gets low, either of the root causes pointing at this event has probably occurred, in this case outlet blockage or flow sensor low bias. The :NOT-conditions attached to most of the compiled links will be interpreted as follows: when the events at both ends of the link are detected, the root causes in the :NOT-conditions of the link are no longer likely. The :ONLY-IF-TRANSIENT means that the event at the termination of the link will not occur for any of the modeled root causes until the root cause has been corrected.

The model generation in MIDAS is summarized in the upper half of Fig. 5.

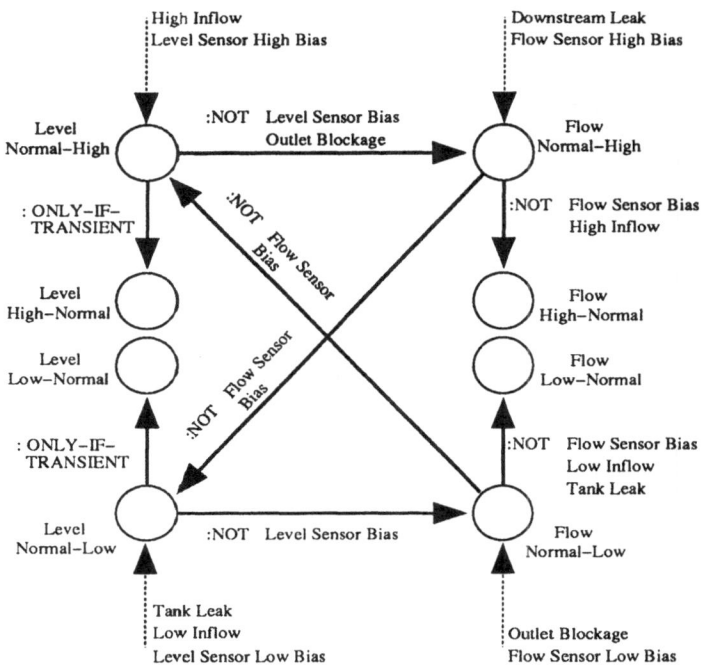

Figure 4. The event graph of a gravity flow tank. Adapted from [Finch *et al.*, 1990].

MIDAS diagnosis

The structure of the diagnostic function and the flow of information is shown inside the dashed line in Fig. 5.

Monitors: The diagnosis unit in MIDAS is linked to the process via monitor objects. Every sensor variable has an associated monitor which is responsible for detecting state changes, i.e. events, concerning the variable. The monitors in the G2 implementation contain information about the threshold levels between the qualitative states of the measured variables. When a monitor samples a sensor value, this value is first smoothed by a first-order filter.

In Fig. 5 there is an arrow labeled "Interrogation" from the event interpreter to the monitors. This indicates that the event interpreter may ask a monitor to predict a future event before this has actually occurred. This is done with polynomial, in its simplest case linear, prediction. If the monitor predicts an event to happen in the near future, a new predicted event is created and treated like a true recorded event, only with less probability of accurate detection.

The hypothesis model: The hypothesis model consists of clusters of dynamically created and interconnected objects representing previously detected events

516

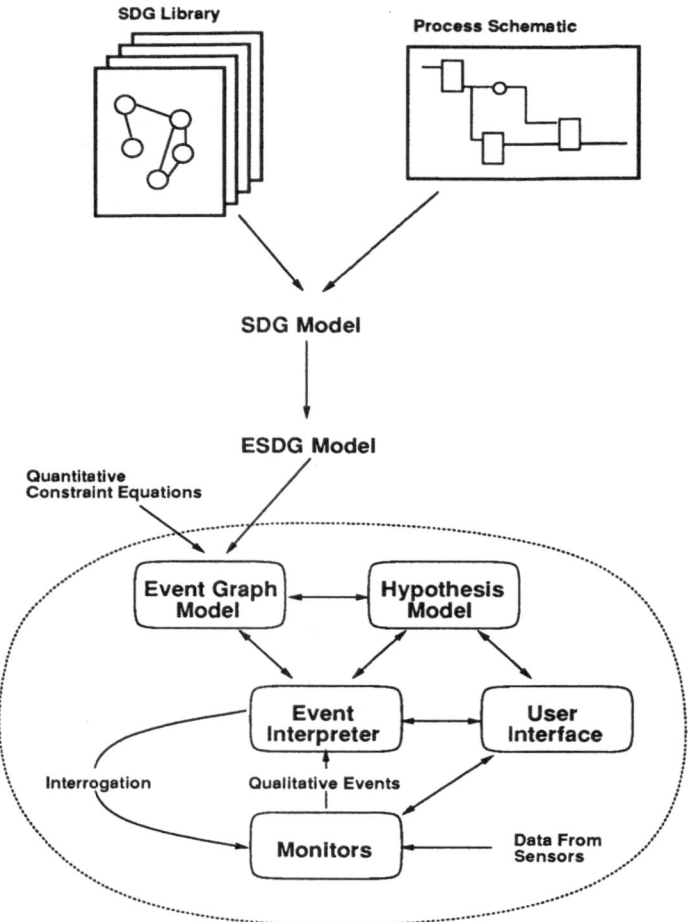

Figure 5. MIDAS model generation and the structure of the diagnosis.

along with the current fault hypotheses. Each cluster, occasionally called inferred malfunction (IM), contains recorded events (REs) that are causally linked in the event graph, and hypothesized root causes (HRCs) that may have caused the events. An IM can be seen as a subgraph of the event graph, since the recorded events and hypothesized root causes are copies of the corresponding components in the event graph. MIDAS assumes that exactly one HRC is the true fault in each IM. IMs are created and altered as new events are detected and the diagnosis evolve.

Fig. 6 shows a hardcopy in G2 of an IM. The icon of the IM can be seen on

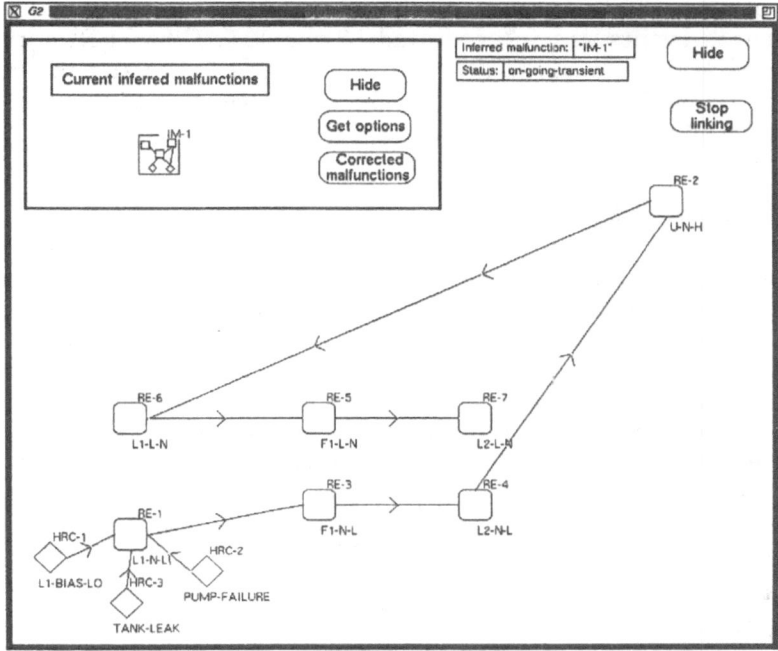

Figure 6. An example of an inferred malfunction

the small workspace at the upper left corner of the figure. The rest of the G2 window displays the REs and HRCs that are members of the IM. This IM has been created when MIDAS worked against a simulated tank process and a tank leak was simulated.

The event interpreter: When a monitor has detected a new event, this will be handled by the event interpreter, which is a set of procedures and algorithms used to link new events with old ones and to make a diagnosis from the observations.

The event interpretation consists of a procedure that is run for every recorded event. When a new event occurs a RE is created and catalogued in the hypothesis model. The interpreter tries to link the new RE together with clusters of old events. If the search for such a cluster, or IM, is successful, the RE will be incorporated in this cluster, otherwise the event interpreter creates a new IM containing the RE. The cluster containing the new event is examined to see which events in the cluster may be source events, i.e. events that can explain all other events in the cluster. Finally, the existing hypotheses are revised to include the information provided by the new event.

The basic strategy is to group related events together. Since the source events explain all other events in the cluster, all root causes in the event graph pointing at one of these events are incorporated in the IM. These HRCs are finally ranked, based both on a priori probabilities for the faults and the conditions associated with the links in the IM. Details about the event interpretation can be found in [Finch, 1989] and [Nilsson, 1991].

4. Steritherm diagnosis

Steritherm is a process for indirect UHT treatment of liquid food products. The product is heated up to the sterilization temperature and cooled down again in plate heat-exchangers. A water system with steam injection is used for heating and cooling. The process consists of six heat exchanger sections, three pumps, two PID controllers, one product balance tank, one water balance tank, one steam injector, eight temperature sensors, one flow sensor, two level sensors, and one differential pressure sensors. The process schematic for Steritherm is shown in Fig. 7.

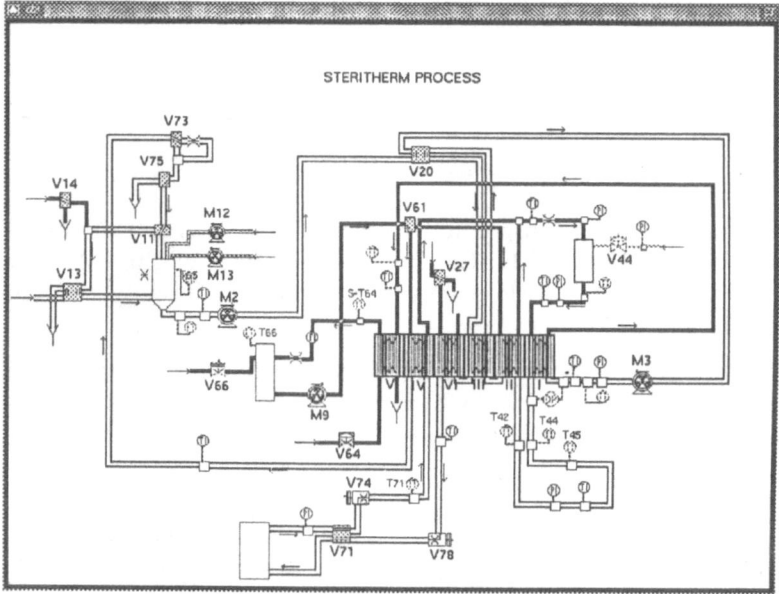

Figure 7. Steritherm process schematic

DMP Steritherm diagnosis

A total of 18 model equations and 17 assumptions are used in the DMP model. Of the 17 assumptions, 10 have to do with sensors and 4 concern the correct operation of pumps or valves. Additionally, 7 model equations which are activatable with supplied values coming from indicators that the operator manually has to read and enter are available. Of the total 25 model equations, 10 are mass or energy balance equations, 3 concern direct sensor redundancy, and 4 are simple limit checks on sensor values. The DMP network is shown in Fig. 8 with most of the dependencies hidden. The manually activatable model equations have an icon with a cross.

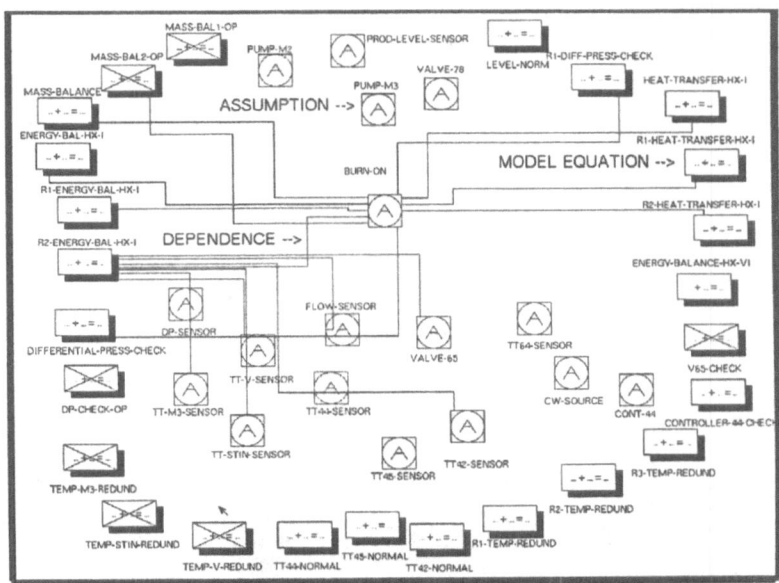

Figure 8. Steritherm DMP network

Many of the failure possibilities in the Steritherm process were examined to determine the effectiveness of DMP. The methods proved quite capable of identifying the correct fault situation. Only in a few cases was the correct fault accompanied by the possibility of failures which were not occurring. These cases, however, were as expected because the model equation relationships were identical for a few of the assumptions.

MIDAS Steritherm diagnosis

In the MIDAS application only about two thirds of the Steritherm was modeled. The SDG consisted of 68 nodes and 11 root causes. The root causes, i.e., faults,

520

were of the same type as in the DMP case. The corresponding event graph is shown in Fig. 9.

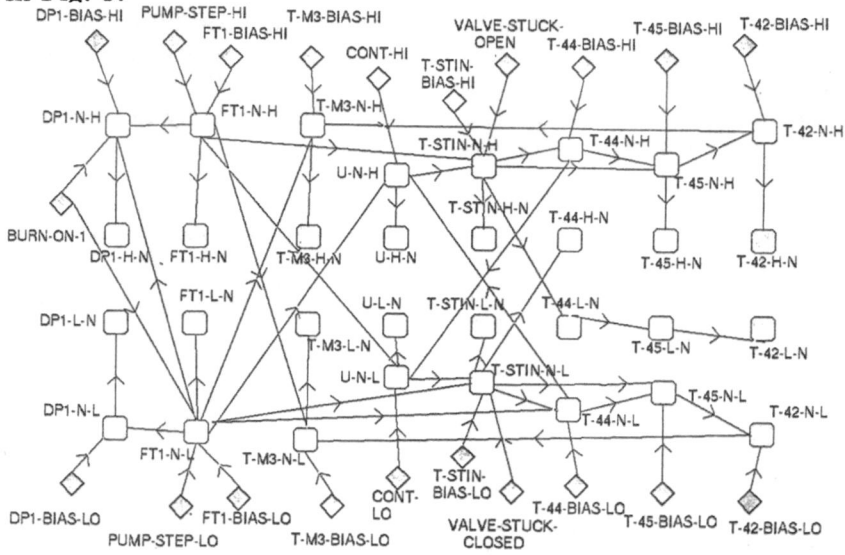

Figure 9. Steritherm event graph

The result of the fault simulations was that MIDAS in all cases succeeded in creating an accurate diagnosis, if not two malfunctions with interfering symptoms appeared simultaneously. However, a more thorough tuning of the monitors would have been needed to avoid drawing inaccurate conclusions during transient condition. This was not performed.

The DMP diagnosis was presented to the operator by highlighting the process components associated with the fault assumption. The MIDAS results were presented as text messages. In the process engineer interface both the DMP network and the dynamically created IMs were presented.

5. Comparisons

MIDAS and DMP are two quite different methods aimed at the same problem, on-line diagnosis. The results of the comparison showed that from a pure diagnostic point of view the methods were comparable. However, when comparing two methods also other aspects such as the tuning effort needed, the time needed to develop the model, the computational effort needed by the methods, etc., need to be taken into account.

DMP	Advantages	Disadvantages
	Non–crisp logic	Modelling effort
	Computationally simple	Snapshot
	Snapshot	
	Not restricted to steady–state	
	Relatively easy to tune	

MIDAS	Advantages	Disadvantages
	Semi–automatic model development	Computationally complex
		Difficult to tune
	Non–snapshot	Crisp logic
	Robust to out of order events	Non–snapshot
		Event order and timing not considered
		Restricted to steady–state operation

Figure 10. Advantages and disadvantages of DMP and MIDAS

The advantages and disadvantages of DMP and MIDAS are summarized in Fig. 10. DMP is simple from a computational point of view. The real expertise in DMP lies in the formulation of the equations something which the engineer must do manually. DMP is also memoryless. It is entirely based on snapshots of sensor values. This is both an advantage and a disadvantage. It makes the system computationally simple, but on the other hand probably quite poorly reflect how an experienced human operator would go about performing on-line diagnosis.

A strong advantage of DMP is that it is not based on crisp threshold values. Both the satisfaction values and failure likelihoods are real-valued numbers between −1 and 1. Due to this DMP is not particularly sensitive to the choice of tolerance limits. Since the equations may be dynamic, DMP is not restricted to diagnosis around a nominal steady-state operating point.

The implementation of MIDAS described here only makes use of the qualitative SDG models. As pointed out in [Kramer, 1990] the diagnosis accuracy of MIDAS is greatly enhanced if quantitative constraint equations also are used. From a computational point of view MIDAS is quite complex. The process knowledge behind MIDAS is the qualitative SDG models. The semi-automatic generation of an event graph from the SDGs is very attractive. In this project, however, this part of MIDAS was not used. Instead the SDG models derivation and the succeeding transformation to ESDG and event graph models were done with pen and paper.

The qualitative nature of MIDAS is attracting. However, somewhere in a diagnosis scheme the quantitative information must show up. In MIDAS the quantitative information shows up in the selection of threshold values. Since MIDAS is based on events the threshold levels are crisp. It was quite difficult to tune the threshold levels appropriately. MIDAS records past events. This memory type of function has both advantages and disadvantages. While it enhances the diagnosis it also gives problems with which information to discard and when to discard it. On purpose MIDAS does not take the order of events or time between events into account during diagnosis. The reason for this is the uncertainty about whether events arrive to the diagnosis system in the order they really occurred. However, in many types of processes the order and time of events can give useful information.

6. Conclusions

Two quite different model-based on-line diagnosis methods have been applied to the same simulated process. The diagnostic results of the methods were comparable. However, the methods proved to differ much with respect to tuning effort, model development time, and computational complexity.

Model-based diagnosis is currently an intensive research area both within the traditional control community and within the AI community. A problem with the area is that the methods used differ a lot both with respect to what types of models that are used and how these models are used on-line. Therefore, comparative studies of this kind are necessary in order to understand the relationships between the methods and their relative strengths and problems.

Acknowledgements

This work was supported by the IT4 project "Knowledge-Based Real-Time Control Systems" and by the TFR project "Integrated Control and Diagnosis", TFR-92-956.

References

ÅRZÉN, K.-E. (1990): "Knowledge-based control systems." In *American Control Conference (ACC '90)*, San Diego, California.

FINCH, F. (1989): *Automated Fault Diagnosis of Chemical Process Plants using Model-Based Reasoning*. PhD thesis.

FINCH, F. E., O. O. OYELEYE, and M. A. KRAMER (1990): "A robust event-oriented methodology for diagnosis of dynamic process systems." *Computers & Chemical Engineering*, **14:12**, pp. 1379–1396.

FRANK, P. M. (1990): "Fault diagnosis in dynamic systems using analytical and knowledge-based redundancy—A survey and some new results." *Automatica*, **26:3**, pp. 459–474.

ISERMANN, R. (1984): "Process fault detection based on modeling and estimation methods—A survey." *Automatica*, **20**, pp. 387–404.

KRAMER, M. (1990): "Process system diagnosis: Theory and practice." In *Proc. of the 1990 International Workshop on Principles of Diagno sis, Stanford, July 23–25*.

MOORE, R., H. ROSENOF, and G. STANLEY (1990): "Process control using a real time expert system." In *Preprints 11th IFAC World Congress*, Tallinn, Estonia.

NILSSON, A. (1991): "Qualitative model-based diagnosis—MIDAS in G2." Master thesis TFRT-5443, Department of Automatic Control, Lund Institute of Technology, Lund, Sweden.

NILSSON, A., K.-E. ÅRZÉN, and T. F. PETTI (1992): "Model-based diagnosis—State transition events and constraint equations." In *Preprints IFAC Symposium on AI in Real-Time Control*, Delft, The Netherlands. (Anders Nilsson has changed name to Anders Wallén).

OYELEYE, O. (1989): *Qualitative Modeling of Continuous Chemical Processes and Applications to Fault Diagnosis*. PhD thesis.

PETTI, T. F. (1992): *Using Mathematical Models in Knowledge-Based Control Systems*. PhD thesis, University of Delaware.

PETTI, T. F. and P. S. DHURJATI (1991): "Object-based automated fault diagnosis." *Chem. Eng. Comm.*, **102**, pp. 107–126.

17
DIAGNOSIS WITH EXPLICIT MODELS OF GOALS AND FUNCTIONS

JAN ERIC LARSSON
Department of Automatic Control
Lund Institute of Technology
Box 118, S-221 00 Lund, Sweden

1 INTRODUCTION

Industrial processes can be described and modeled in several ways, and the models obtained are used for many different tasks. However, most model types contain little or no *means-end* information, and thus provide no good support in diagnostic reasoning tasks.

This paper describes one type of explicit means-end models, *multilevel flow models,* (MFM), as developed by Lind [7]. Lind has suggested a syntax for a formal language and given general ideas on how to use the MFM representation. The contributions of this paper are descriptions of three methods or strategies for diagnostic reasoning using MFM:

- Measurement validation

- Alarm analysis

- Fault diagnosis

The measurement validation algorithm takes a set of measured flow values and uses any available redundancy to check consistency. A single erroneous flow measurement will be marked and corrected; if there are several conflicting values, the consistent subgroups of measurements will be marked but no flow value corrected.

The alarm analysis algorithm takes as input a set of alarm states such as *normal, low flow, high flow, low volume,* and *high volume.* Each alarm is associated with a corresponding MFM object, and the method can recognize the primary alarms, while the others are either primary or consequences of the primary alarms.

The fault diagnosis algorithm uses an MFM model to produce a "backward chaining" style of diagnosis. The input can come from questions answered by the user or from measured signals and triggering of rules. The system will trace down faults, provide explanations, and give remedies.

The main contributions within MFM has so far been made by Morten Lind and his group, [6], [7]. Other groups are also involved in projects, for example Sassen,

S. G. Tzafestas and H. B. Verbruggen (eds.),
Artificial Intelligence in Industrial Decision Making, Control and Automation, 525–534.
© 1995 *Kluwer Academic Publishers.*

[10], [11], [12], [13], and Walseth *et al* [14]. Previous articles of this project include [1], [3], [4], [5], and the Doctor's thesis [2].

2 BASIC IDEAS OF MFM

In multilevel flow modeling, a system is modeled as an artifact, i.e., a man-made system constructed with some specific *purpose* in mind. Thus, MFM contains some concepts from the natural sciences (mathematics and physics), and some from the human sciences (cognitive science and psychology).

The purposes of a system are modeled with the MFM concept of *goals,* i.e., objectives of running a system or using a process. The physical components of a system are used to provide one or several *functions.* These functions are the means with which the goals are achieved. The concepts of goals, functions, and components are explicitly represented in the MFM models.

Just as important are the different *relations* between goals, functions, and components. In MFM, these relations are explicitly described. A set of functions used to fulfill a goal are grouped together and connected to that goal via an *achieve* relation. If a subgoal is a necessary condition for a function to be working, it will be connected to the function via a *condition* relation. If a physical component is used for a certain function, the component object is connected to the function object via a *realize* relation. These relations connect the objects into a graph, i.e., the MFM model proper. Algorithms can then traverse this graph in order to perform different reasoning tasks.

It is important to observe that MFM models are *normative,* i.e., they describe how the system is supposed to work, not how it is actually working in the present state. Each difference between the model and the current state will be viewed as a fault. This is quite common in model-based approaches.

3 AN EXAMPLE OF A FLOW MODEL

The following example will hopefully help to explain the use of the basic concepts of MFM. The target process consists of a plate heat exchanger, see Fig. 1.

This is quite a small system, but it serves well to explain the concepts of MFM. The primary goal is to heat the product to a certain temperature, but a brief analysis shows that there are two subgoals also—having water and product available, i.e., bringing the media to the heat exchanger:

- G1: Heat product to certain temperature

- G2: Bring product to heat exchanger

- G3: Bring water to heat exchanger

The given example process is rather small, but there are many functions present:

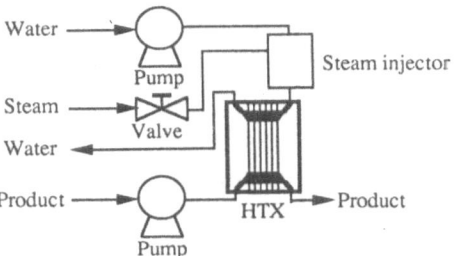

Figure 1: A heat exchanger system. The flowsheet shows how water is pumped through a steam injector, where it is heated with steam, and through a plate heat exchanger, where it heats the product.

- **F1**: Provide product

- **F2**: Transport product

- **F3**: Transfer thermal energy between media

- **F4**: Provide thermal energy

- Etc.

The third type of objects are the physical components. Note that the product and water tanks do not actually appear in Fig. 1:

- **C1**: Product tank

- **C2**: Product pump

- **C3**: Heat exchanger

- Etc.

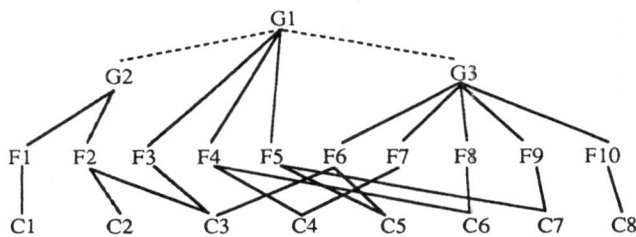

Figure 2: Goals, functions, components, and relations of the heat exchanger system.

These are the sets of goals, functions, and components. However, the relations between these objects are just as important as the objects themselves. First, the goal G1 is superior to G2 and G3, i.e., the latter are subgoals of G1. Thus, there is a *goal hierarchy*, formed by goal-subgoal relations. There are also relations between goals, functions, and components. For example, the heat exchanger component is used to realize the function of transferring thermal energy from water to product, and this function is used to achieve the goal of heating the product. In Fig. 2, both the goal hierarchy and the means-end relations are shown in a graph.

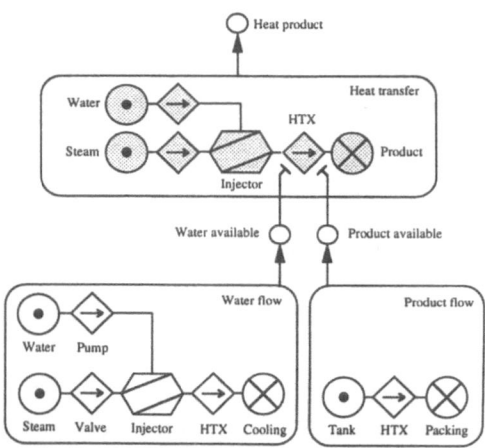

Figure 3: A flow model of the heat exchanger system. The goals and goal hierarchy is shown in the tree structure of the graph, while the functions are connected into flow paths in three networks. The topmost goal, to heat the product, is achieved by the heat transfer flow path, (upper network), while the subgoals, to bring water and product to the heat exchanger, are achieved by the water flow path, (lower left), and the product flow path, (lower right). The components and realize relations are not shown.

As can be seen, the graph of objects and relations is quite complex, even for a small process as the one in the example. In an MFM model, the goals, functions, and relations are represented in a graphical language. A model of the example process is found in Fig. 3.

4 THREE DIAGNOSTIC METHODS

The contributions of this paper are three diagnostic reasoning methods, which are easily and efficiently implemented with the help of MFM: measurement validation, alarm analysis, and fault diagnosis. These methods are thoroughly described and examples of their use are given in [2].

4.1 Measurement Validation

Industrial processes usually have many sensors which directly or indirectly measure the same variables. When mass and energy balance equations are taken into account, the set of measurements gives rise to redundancy. The proposed method uses some of this redundancy to check the measurements of the same mass or energy flow, and is thus called *measurement validation* or *data reconciliation.* A typical problem situation can be seen in Fig. 4.

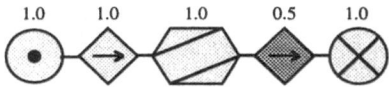

Figure 4: A flow path with flow values. One flow deviates from the rest. Thus, at least one of the measurements must be faulty; either the single one, the four, or all five are wrong. The method developed presents all these three hypotheses.

Here, the five flow measurements should agree, i.e., have more or less the same value. However, one of them clearly deviates from the rest. Three hypotheses could explain the situation:

- The single measurement is wrong, and the rest are correct.

- The single measurement is correct, and the rest are wrong.

- All measurements are wrong.

A failure to consider any of these possible explanations could be potentially dangerous, thus the method uses the sensor values to assign the different measurements to *consistent subgroups.* For each value there is a *validated value,* which can be set to a value different from the one actually measured. The method outputs the following information:

- Each consistent subgroup is presented. Thus the user can get an impression of the general agreement or disagreement of the measurements.

- A single deviating measurement is highlighted. Thus a probable fault can be detected and isolated quickly.

- If a single deviating measurement is surrounded by several consistent ones, the corresponding validated flow value will be set to that of the surrounding group.

The method has more features, for example, it uses a *flow propagation* algorithm to handle the case when no sensor is connected to a part of the model.

4.2 Alarm Analysis

Most processes are equipped with a large number of alarms, and in a failure state many of these will trigger. Some of them will be directly connected to the primary sources of faults, but other may be secondary, i.e., due only to consequential effects of the primary failures. The method helps to separate primary alarms from those that might be secondary, a task which can be vital in a fault situation.

Each flow function is associated with a working condition, which, if violated, will give rise to an alarm. Due to the semantic interpretations of the flow functions, certain faults may or will cause consequential faults in connected flow functions. An example is given in Fig. 5.

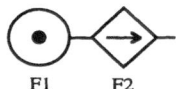

F1   F2

Figure 5: A source connected to a transport function. If the capacity of the source goes down, the transport flow will be forced out of its working interval, while if the desired flow of the transport increases, the source may or may not be able to supply it. If the desired flow through the transport decreases, however, the source is not affected. Thus, a low capacity in the source will cause a low flow through the transport; a high flow through the transport *may* cause a low capacity in the source; but a low flow through the transport will not cause any fault in the source.

The function F1 is a source that provides mass or energy for F2, a transport function. If F1 would loose its capacity of delivering the required amount of mass or energy, the transport would not be able to maintain a sufficiently large flow. If, on the other hand, F2 was to require too large a flow, F1 might not be able to provide it. Thus, two causation rules can be formulated for the connection of a source and a transport:

- A low capacity alarm in a source will cause a low flow alarm in a connected transport function.

- A high flow alarm in a transport function may cause a low capacity alarm in a connected source.

The method uses a set of causation rules like the ones above to analyze alarm situations and separate out those alarms that must be primary from those that might be secondary. However, no alarm is hidden from the operator, as there is always the possibility of multiple faults, i.e., a primary fault could look as if it was caused by another one.

The method uses a *consequence propagation* algorithm in order to handle unknown alarm states when parts of the process is not equipped with alarms.

4.3 Fault Diagnosis

This method uses the MFM model of a process to search for faults and gives explanations and remedies, much like a standard rule-based expert system would do. The MFM model contains information about the goals of a process, how these goals are achieved by networks of functions, how the functions depend on subgoals, and how they are realized by physical components. In a standard rule-based expert system, this information structure is implemented in rules, but in MFM it is explicitly described. Thus, a fault diagnosis can be easily implemented, as a search in the model graph. The strategy used is as follows:

- The user chooses a goal for diagnosis. If this is a top-level goal, the whole model, (and thus the entire process), will be investigated. However, the goal chosen could also be a subgoal, in which case only part of the process will be diagnosed.

- A search propagates downward from the goal, via the achieve relations, into the networks of flow functions, each of which is investigated.

- Each flow function may have a diagnostic question, which is asked in order to find out whether the corresponding physical component fulfills the function or not. Alternatively, there can be a rule or a realize relation to a physical component, whereby information about the working order of the function may be found.

- If a flow function conditioned by a subgoal is found to be at fault, or has no means of being checked, the connected subgoal is recursively investigated. If, however, a function is working, that part of the subtree can be skipped.

5 IMPLEMENTATION

The algorithms described above have all been implemented in G2, and tested on two target processes, a simple lab process with two cylindrical tanks, and Steritherm, a process for ultra-high temperature treatment of dairy products. The implementation is of moderate size, see Table 1, and is now commercially available as a toolbox for building and using MFM models in G2. The methods were successful in all test cases, but it should be noted that Steritherm is still a rather small-scale process. G2 was developed by Gensym Corporation, [8], [9].

The algorithms are local and incremental. They work in real-time, and propagate information along static links only. This makes them very efficient, and the effort as a function of model complexity increases at worst linearly with the size of the MFM models.

Measurement validation	*71 rules*
Alarm analysis	*63 rules*
Fault diagnosis	*21 rules*

Table 1: The three methods have been implemented as G2 knowledge databases.

6 COMPLEX SYSTEMS

The MFM modeling technology has several specific advantages when it comes to modeling complex systems. Some of these deserve to be further discussed:

- MFM gives a very strict top-down view of the process, with the most general and important goals appearing at the top level. These goals concern the operation and safety of entire plants. Diagnostic algorithms will first check these goals and then, in case there is more time, move downwards to more detailed goals. In this way, a graceful degradation is possible in case of hard real-time demands.

- The means-end view helps the human operators to handle large processes in an efficient way, as it makes it possible to reason about goals and functions on higher, plant-wide levels. This means that the operators may avoid the detail of the lower levels of representation.

- The diagnostic algorithms behave favorably when problems are scaled up. The computational effort for a diagnosis increases at worst linearly with model size, and often the increase is even lower, due to the local properties of the methods.

7 CONCLUSIONS

The paper has overviewed three newly invented and implemented diagnostic methods for use with multilevel flow models, MFM. The methods use MFM as a database and performs measurement validation, alarm analysis, and fault diagnosis. They have been implemented in G2 and successfully tested on two processes. The search algorithms are all very efficient and work in real-time, and their sensitivity to large scaling of models is at worst linear. Together with the implemented toolbox and demonstrations, the project provides a good example of the usefulness and power of means-end models.

8 ACKNOWLEDGEMENTS

I would like to thank my supervisor, Professor Karl Johan Åström, and the originator of MFM, Professor Morten Lind, and Doctor Karl-Erik Årzén for inspiration and support. This project has been an informal part of the Swedish IT4 project

"Knowledge-Based Real-Time Control Systems," and I also wish to thank the members of the project group.

References

[1] J. E. Larsson. Model-based alarm analysis using MFM. In *Proceedings of the 3rd IFAC International Workshop on Artificial Intelligence in Real-Time Control*, Sonoma, California, 1991.

[2] J. E. Larsson. *Knowledge-Based Methods for Control Systems*. Doctor's thesis, TFRT–1040. Department of Automatic Control, Lund Institute of Technology, Lund, 1992.

[3] J. E. Larsson. Model-based fault diagnosis using MFM. In *Proceedings of the IFAC Symposium on On-Line Fault Detection and Supervision in the Chemical Process Industries*, University of Delaware, Newark, Delaware, 1992.

[4] J. E. Larsson. Model-based measurement validation using MFM. In *Proceedings of the IFAC Symposium on On-Line Fault Detection and Supervision in the Chemical Process Industries*, University of Delaware, Newark, Delaware, 1992.

[5] J. E. Larsson. Diagnostic reasoning strategies for means-end models. *Automatica*, 30(5), 1994. To appear.

[6] M. Lind. Abstractions version 1.0—descriptions of classes and their use. Technical report, Institute of Automatic Control Systems, Technical University of Denmark, Lyngby, Denmark, 1990.

[7] M. Lind. Representing goals and functions of complex systems—an introduction to multilevel flow modeling. Technical report, Institute of Automatic Control Systems, Technical University of Denmark, Lyngby, Denmark, 1990.

[8] R. L. Moore, L. B. Hawkinson, M. Levin, A. G. Hoffmann, B. L. Matthews, and M. H. David. Expert system methodology for real-time process control. In *Proceedings of the 10th IFAC World Congress*, volume Vol 6, pages 274–281, Munich, 1987.

[9] R. L. Moore, H. Rosenof, and G. Stanley. Process control using a real-time expert system. In *Proceedings of the 11th Triennial IFAC World Congress 1990*, volume Vol IV, pages 241–245, Tallinn, Estonia, 1991.

[10] J. M. A. Sassen. *Design Issues of Human Operator Support Systems*. Doctor's thesis. Delft University of Technology, Faculty of Mechanical Engineering and Marine Technology, Delft, Nederland, 1993.

[11] J. M. A. Sassen and R. B. M. Jaspers. Designing real-time knowledge-based systems with PERFECT. In *Preprints of the 1992 IFAC/IFIP/IMACS International Symposium on Artificial Intelligence in Real-Time Control*, pages 625–630, Delft University of Technology, Delft, the Netherlands, 1992.

[12] J. M. A. Sassen, A. Ollongren, and R. B. M. Jaspers. Predicting and improving response-times of PERFECT models. In *Preprints of the 1992 IFAC/IFIP/IMACS International Symposium on Artificial Intelligence in Real-Time Control*, pages 709–714, Delft University of Technology, Delft, the Netherlands, 1992.

[13] J. M. A. Sassen, P. C. Riedijk, and R. B. M. Jaspers. Using multilevel flow models for fault diagnosis of industrial processes. In *Proceedings of the 3rd European Conference on Cognitive Science Approaches to Process Control*, pages 207–216, Cardiff, United Kingdom, 1991.

[14] J. A. Walseth, B. A. Foss, M. Lind, and O. Œgaard. Models for diagnosis—application to a fertilizer plant. In *Proceedings of the IFAC Symposium on On-Line Fault Detection and Supervision in the Chemical Process Industries*, University of Delaware, Newark, Delaware, 1992.

PART 5
INDUSTRIAL ROBOTIC, MANUFACTURING
AND ORGANIZATIONAL SYSTEMS

18
MULTI-SENSOR INTEGRATION FOR MOBILE ROBOT NAVIGATION

A. TRAÇA DE ALMEIDA, HELDER ARAÚJO, JORGE DIAS, URBANO NUNES
Electrical Engineering Department,
University of Coimbra,
3000 Coimbra, Portugal

1 INTRODUCTION

Available sensors for robot navigation are unreliable and noisy. Therefore there is a need to employ different types of sensors to acquire the information required for navigation. We discuss the different problems associated with the integration of several sensors in a mobile platform and present the approach we have developed to tackle these problems. We consider the particular problem of a mobile platform navigating in a 2D environment with a priori knowledge of its map. Unknown obstacles are allowed. In this chapter we are concerned with the integration of inertial sensors, odometry, sonars and active vision for navigation in a mobile robot.

2 SENSOR-BASED NAVIGATION

Navigation is a basic requirement of a mobile robot. A mobile robot is supposed to move between positions in an environment. To accomplish such a task it must be able to locate itself in the environment and must also be able to follow a path. These two functions, localization and path following, are the basic building blocks of what is usually called navigation. For that purpose sensors are essential for extracting information of some structures of the environment. The sensing process is essentially the acquisition of some physical measures. From that data the interesting features are extracted, constituting the perception of the environment. The role of extracting data from the environment is played by sensors. Therefore it is crucial that sensors be chosen according to the expected types of measures. In the case of navigation, sensors should provide information that enables the localization of the vehicle, obstacle detection and/or avoidance. Navigation by itself is well understood. Difficulties arise in getting the suitable

S. G. Tzafestas and H. B. Verbruggen (eds.),
Artificial Intelligence in Industrial Decision Making, Control and Automation, 537–554.
© 1995 *Kluwer Academic Publishers*.

data since there is a lack of adequate sensors and the existing sensors are unreliable and inaccurate. For navigation we need sensors that measure distances, velocities, accelerations and orientations. These physical quantities determine the type of sensors that have to be employed.

Two broad classes of navigation sensors can be considered, one of them used with dead-reckoning navigation (usually called internal sensors) and another one including all the sensors whose measures require external references. Dead-reckoning refers to navigation with respect to a coordinate frame that is an integral part of the guidance equipment. Dead-reckoning has the advantage that it is totally self-contained. However, dead reckoning suffers from several sources of inaccuracy. One of the main sources of inaccuracies is the cumulative nature of the errors whose sources include wheel slippage, and terrain irregularities. To reduce the problems with dead-reckoning, we can use landmarks or beacons along the trajectory. The localization of the mobile robot can be corrected by sensing these landmarks and beacons.

Reference guidance has the advantage that the position errors are bounded but the detection of external references or landmarks and real-time position fixing, may not always be possible. Reference guidance has the disadvantage of reducing the degree of autonomy of the vehicle by making it dependent of a certain environment. As a matter of fact in this type of navigation the vehicle must rely on a map to navigate from one beacon to the next and also to determine its location. To achieve a good accuracy and flexibility different types of sensors must be used. These sensors include vision, sonar, laser range finders and other sensors.

Dead-reckoning and external reference navigation are complementary and combination of the two approaches can provide very accurate positioning systems.

3 SENSORY SYSTEM

For mobile robots not all sensors are useful. Indeed a mobile robot puts a number of constraints on the usability of the sensors. These constraints result from the stringent requirements put by the autonomy of the mobile vehicle (namely power consumption, height, size and cost) and the type of information required for the vehicle operation.

Dead-reckoning is usually performed by odometry and/or inertial guidance sensors. Odometry consists on the measurement of distance by means of the integration of the number of turns of a wheel. Two types of odometry are currently employed in mobile robots, one that uses encoders fitted on the drive wheels, and another that uses a non-load bearing wheel.

On a mobile robot, the self motion, also called ego-motion, can be obtained from inertial information. This information can be used for different tasks such as navigation, trajectory generation or stabilization of the robot [Viéville 90]. The processes involved in these tasks have to be based on internal representations of the robot geometry and dynamics.

The theories of inertial navigation for airplanes or other vehicles and theirs related implementations cannot be directly used on robotics. In the case of airplanes the theories deal with long displacements around the earth, at high velocities, where quantities such as the radius of the earth or gravity field variations, are taken into account. In robot displacements, these considerations do not make sense because the trajectories are short and the gravity field can be considered constant along the path. Another aspect is the cost of an inertial system used in airplanes. These systems are considerably more expensive than other sensors used in a robot, like vision or laser range finders. Therefore, an inertial system for a mobile robot does not have exigencies similar to those of an airplane. For a mobile robot the trajectories are restricted and several realistic assumptions can be used. For a mobile robot the velocity and position can be calculated using the Newtonian laws of motion, and the gravity field can be assumed as constant during the robot displacements.

The available inertial sensors are of two kinds: linear *accelerometers* and *gyrometers* or *gyroscopes*. The linear accelerometers are sensors based on the second Newton's law that states that the acceleration of a particle is proportional to the resultant force acting on it. These sensors provide the acceleration by measuring the force required to induce the acceleration of a known mass. Linear accelerations induce inertial forces. These forces create displacements and these displacements can be measured. *Accelerometers* measure the acceleration along its eigen axis, but are also sensitive to transversal accelerations by a small factor (typically 2%) and, for that reason, a calibration procedure must be performed. The angular velocity can be measured by gyrometers. These systems are based on the measurement of Coriolis forces of a mobile submitted to angular motion. The elimination of other accelerations (gravity, angular acceleration, linear acceleration) is made by the appropriate combination of different elements. *Gyroscopes* are made of a solid body, with high angular velocity around an axis of symmetry, thus having an important kinetic momentum. This angular velocity induces angular stability in presence of perturbations. Such device is used to preserve a fixed inertial orientation, while the rest of the system is in motion. Both devices provide an information about angular velocity. The *gyrometer* output is a measure of the instantaneous angular velocity, and the *gyroscope* gives an output with the integration of the angular velocity during a period of time. The gyrometers have the advantage of being less costly and much easier to implement, than the gyroscopes. The theoretical results on gyrometers can be applied to gyroscopes because they provide the same physical information. In the case of the instantaneous motion the gyrometer is preferred because it directly gives the information of the angular velocity. The inertial information has the advantage of depending on the measurements carried out entirely within the vehicle, in accordance with the physical laws of motion and gravitation and not depending on information from somewhere outside the vehicle.

The displacements of the robot can be of two types: intrinsic or extrinsic to the movement commands. The intrinsic displacement is due to the action of robot actuators and the consequent displacement is partially predictable from the commands. In this case, the inertial sensors' information could be combined with odometric information to compute a better estimate of the actual displacement of the robot. This estimation process can also be useful on the calibration of the odometric and inertial devices. This combination is performed by using the redundancy of the different sources of information. The displacements of the robot not directly due to its actuators are referred to as extrinsic displacements. The inertial sensors, in cooperation with other sensors, can be used to estimate this type of displacements.

For reference guidance it is necessary to measure distances. For that purpose a wide variety of range sensors can be used. These sensors can be divided into two classes: active and passive ones. Active sensing computes the distance between the sensor and an object by observing the reflection of a reference signal produced by the object. Generally active sensing methods provide range data in a direct way without a significant computational cost (unlike what happens in most of the passive methods, e.g., stereo vision). In the first class ultrasonic and optical range sensors are the most commonly used. Distance can be measured using light in a number of different ways which include the following active methods: time-of-flight, phase modulation, frequency modulation, measure of the reflected signal intensity and active triangulation. The time-of-flight is the most commonly used principle in measuring distances by using ultrasonic transducers.

However, range does not have to be measured directly. An alternative approach is to infer range by indirect modes. Some of those approaches are explored in computer vision namely depth from stereo, depth from focus, depth from shading, and optical flow.

Considering our goal of implementing the navigation system for a mobile platform (Robuter) [Robosoft 91], we will concentrate on the issues related with sensor integration. Taking into account the basic features, availability, and cost of the sensors generally employed on robotics navigation systems, we decided to base our navigation system on odometry, sonar, vision and inertial sensors (accelerometers, gyrometers and inclinometers).

4 SENSOR INTEGRATION FOR LOCALIZATION: SOME METHODOLOGIES

The data obtained by a sensor are always corrupted by noise, sometimes with spurious or completely incorrect values. In practice, it is very important to explicitly manipulate the uncertainty of the measurements in order to effectively use the information provided by a sensor. For that purpose the sensors have to be modeled. From this point of view the sensor modeling can be considered as one of the important problems to be faced in robotics.

Sensors can be modeled quantitatively. The purpose of a sensor model is primarily to characterize the behaviour of the sensor concerning its measures. Methods for dealing with the uncertainty of the measures are necessary. Some of these methods are:

- Empirical methods, by giving some *a priori* error bound on the measures [Grimson 84];
- Fuzzy sets, by modeling uncertainty by fuzzy principles [Winston 84];
- Probabilistic modeling, by modeling a sensor as a statistical process [Durrant-Whyte 88].

The last method is widely used by the robotics research community. Some of the reasons are:

- The probabilistic theory is well developed in applied mathematics;
- This type of models provides a mechanism to specify the uncertainties in measurements;
- The probabilistic modeling can provide a consistent way to manipulate the uncertainties.

Most sensors in robotics give geometric or geometric related information about the environment. That is the case of a stereovision system, which can give a tridimensional description of the environment by processing the stereo images. The same case is observed for a set of ultrasonic sensors distributed around a mobile platform, whose information can be processed to build a local map of the environment. The information provided by odometers or by the inertial system is also related with the geometry of the environment. This implies that geometrical information is crucial for many domains of robotics, including navigation. In this sense probability theory can be applied to model the uncertainty in the geometrical information. By using this approach objects are considered geometric entities and these include locations, special primitives, and relations between primitives. These geometric entities can be considered as random variables with a probability density function. However, the geometric probability requires consistency of the data with the physical world, resulting in physical geometric constraints on functions and relations. The application of these principles originates two types of difficulties in the manipulation of the geometric uncertainties: the multi-dimensionality of some variables and the transformations of the geometric entities are normally nonlinear.

The correct modeling of the sensors' data is only part of the problem of sensor integration in mobile robots' navigation. Another aspect is the multisensor data fusion, which seeks to combine data from multiple sensors to perform inferences that cannot be possible from a single sensor alone. This includes the improvement of the accuracy of inferences such as the position of the mobile robot by using multiple sensors. In this point we propose an architecture where the data provided by odometry, inertial system, sonar and an active vision system are combined to be used by a mobile robot navigation system.

542

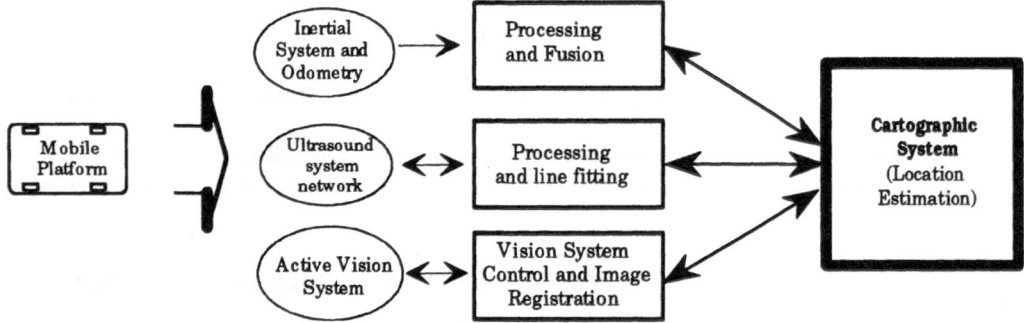

Figure 1 - Data integration system architecture

We will consider sensor integration within the framework of navigation in a known environment where new obstacles can show up and some obstacles can be removed. Therefore all data integration processes will be based on the previous knowledge of an environment map, even though this map may undergo a restricted set of changes (due to obstacle removal or addition). The methodologies for data integration will rely on the sensor models.

To integrate the information gathered by the sensors we considered a layered integration model schematically described in Figures 1 and 2.

4.1 Data integration - Intrinsic sensor level

At one level we pursue the integration of the data from the odometer, the accelerometers and gyrometer to estimate the location. From the accelerometers and gyrometer we extract estimates of the vehicle linear and angular velocities.

The inertial system forms a rigid body made up of two accelerometers and one gyrometer. A frame of reference $(O, \vec{x}, \vec{y})$ is attached to the inertial system as depicted in Figure 3.

An accelerometer measures the linear acceleration along its axis and the gyrometer measures the angular velocity about its axis. Since we consider that our vehicle moves on a plane these measures are complemented by the information given by a dual-axis inclinometer which measures the pitch and roll of the robot relative to gravity. That information enables the improvement of the results coming from the combination of accelerometers and gyrometer data. The linear acceleration has only two components whereas the angular velocity is defined by a one-dimensional vector.

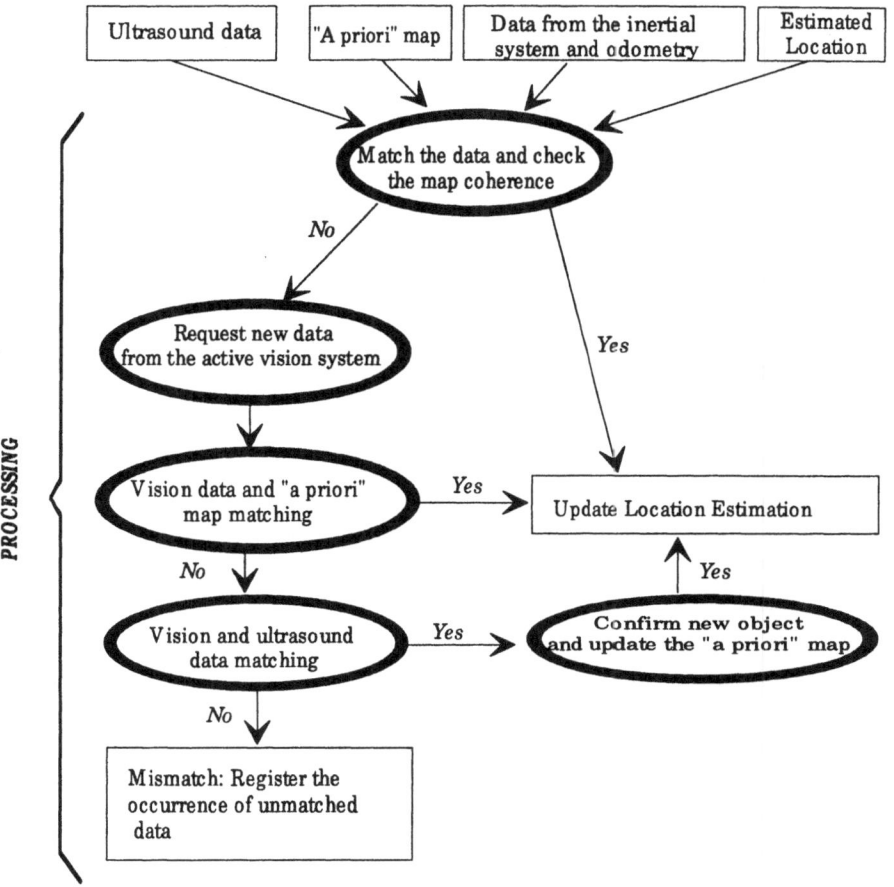

Figure 2 - Process of data Integration

Since it is not mechanically possible to put the two accelerometers and the gyrometer at the same point, the acceleration on each accelerometer is not the acceleration at the origin of the frame of reference. Moreover, the sensor model has to take into account that the exact location and orientation of each one of the sensors are not precisely known. The linear velocity is obtained by integrating the outputs of the accelerometers, and the orientation is obtained by integrating the angular velocity.

The odometry data and the results given by the inertial system data are continuously combined to obtain an estimate of the vehicle's position and orientation.

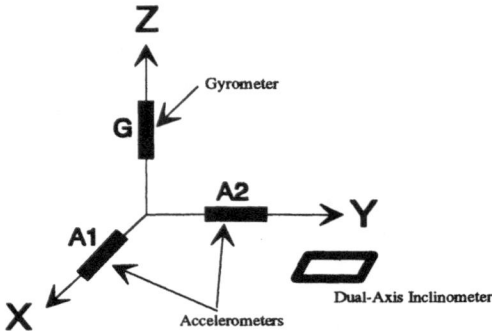

Figure 3 - Geometrical configuration of the inertial system

Two different approaches are being analysed to perform the data integration. One of them relies on using an Extended Kalman Filter (EKF) to estimate errors. The other relies on the combination of information based on the Bayes decision theory. The resulting information is passed to the part of the cartographic system responsible by the location estimation.

4.2 Data integration - Extrinsic sensor level

At another level we have to integrate the data from sonar with the estimate of the vehicle's position and orientation obtained from odometry and inertial system. There are several approaches for vehicle's location estimation. Next we describe two approaches that have proved to be applicable on actual platforms navigating in structured environments. One of them was even applied in a fully autonomous mobile platform that does not rely on an external computer.

The approach presented by John Leonard and Durrant-Whyte [Leonard 91], [Leonard 92] uses EKF to match beacon observations to a navigation map to maintain an estimate of vehicle's location. The other approach was presented by Ingemar Cox [Cox 90] and is used on the autonomous robot Blanche developed at the AT&T Bell Laboratories.

The J.Leonard and Durrant-Whyte approach denotes the position and orientation of the vehicle at any step k by a state vector $\vec{x}(k)$ containing the position and orientation of the vehicle on a Cartesian reference with respect to a global coordinate frame. Additionally J.Leonard and Durrant-Whyte define a system state vector composed by not only the previous location state vector but also an environment state vector containing the set of geometric beacon locations. The robot starts at a known location, and it has an a priori map of the locations of the geometric beacons. The map used in this approach is just a set of beacon's locations not an exhaustively detailed environment map.

Each beacon is assumed to be precisely known. At each time stop, observations of these beacons are taken, and the goal of the system is to associate measurements taken from the beacons with the corresponding map beacon to compute an updated estimate of the vehicle's position. The use of the EKF on this approach relies on two models: a plant model that describes how the vehicle's position changes with time in response to a control input, and a measurement model that expresses a sensor observation in terms of the vehicle location and the geometry of the beacon. The goal of the iterative computation used by J.Leonard and Durrant-Whyte is to produce an estimate of the robot location based on the previous estimate vehicle location, control input and new observations of the beacons. The algorithm employs the following steps: prediction, observation, matching and estimation. In this model the prediction of the location of the robot is obtained using the plant model and the knowledge of the command input, instead of odometric information used on the Blanche robot vehicle. After the prediction is performed, the system takes a number of observations of the different beacons and compares these with the predicted observations. Around each predicted measurement, the matching algorithm sets up a validation gate in whose result a decision for acceptance non-acceptance of beacon information is taken. When a single observation falls in the validation gate, a successful match is considered. Measurements that do not fall in any validation gate are simply ignored for localization. The final step is to use successfully matched predictions and observations to compute the updated vehicle location estimation.

In the other approach, presented by Ingemar Cox [Cox 90], for the mobile robot Blanche, the location estimation was developed to work in structured environments. A combination of odometry and optical range finder to sense the environment is used and a matching algorithm for the sensory data and a map was developed. The a priori map of the vehicle environment consists of a 2D representation based on a collection of line segments. In addition an algorithm is used to estimate the precision of the corresponding match/correction which allows the correction to be optimally combined with the odometric location to provide an improved estimate of the robot's location. Since the robot's location is to be constantly updated as the vehicle moves, there is a need to remove any motion distortion that may arise. This can be done by reading the odometric position at each range sample. The current position and range data are then used to convert the data point into to world coordinates for matching to the map.

The observations that motivated the derivation of the matching algorithm used on Blanche were [Cox 90]: 1) The displacement of the range image relative to the map is small and this assumption is almost always true, particularly if position updating occurs frequently; 2) Features extraction can be difficult and noisy. The use of a matcher algorithm that doesn't need feature extraction, but works directly on the range data, decreases considerably the computation time of the matching stage. Based on these observations, if the mismatch between the range image and the map is small, then for each point in the range image its corresponding line segment in the map is very likely to be its closest line segment in the map. This observation establishes that the determination of a partial correspondence between the range image and the map segment lines is reduced to a simple search to find the closest line to each point of the range image. The central feature of the approach was devising a method to use the approximately correct correspondence between range image points and map line segments to find a congruence that greatly reduces the mismatch. Iterating this method leads to the following algorithm: 1) For each point of the range image, find the line segment in the map which is nearest to the point; 2) Find the congruence that minimizes the total squared distance between the range image point and their corresponding line segments; 3) Move the points by the congruence found in 2); 4) Repeat steps 1-3 until the procedure converges. The composite of all the step 3 congruences is the desired total congruence. Any congruence can be described as a rotation by some angle θ followed by a translation $\vec{t}$.

For a structured environment the algorithm proposed by I.J.Cox has proven to be effective. Indeed the entire autonomous vehicle is self-contained, all the processing being performed on-board. The algorithms enable the range data to be collected while the vehicle is moving. For those reasons the mobile platform Blanche represents a high-level of performance at low-cost. However it is important to note that the range data is acquired by an optical sensor.

The J. Leonard and Durrant-Whyte algorithm requires much more computational effort and stops of the vehicle to acquire the range data. This algorithm uses dense sonar maps acquired by a servo-mounted sonar. We decided to use neither of these methods because we do not have a dense map of the environment (our environment is sensed by a set of 24 sensors).

For a specific instant k we first estimate the vehicle's position and orientation by combining odometry with inertial information (intrinsic location estimate).

To integrate the intrinsic location estimate with the matched position from sonar processing, we need to have estimates of standard deviation in the measurement of (X, Y, θ) for both the matcher and the intrinsic estimate. In our case the matcher uses the a priori knowledge of the environment map, the intrinsic location estimate of the vehicle and the sensor model to predict the data observed by the sensors. Based on the predicted points the features to be matched are determined. Also as a result of this process the sensors whose data will be considered for determining the environment features are selected. First of all the acceptance of the matching of each map feature with each measured feature is tested. This test is performed taking into account the uncertainties of the intrinsic location estimate and of the sensors' data. Based on the pairs predicted-measured features a new estimate of the robot's location is computed. The intrinsic estimate and the location estimate from the matcher are then combined. Given the locations and standard deviations from the matcher and the intrinsic estimate, the final value of the location and its corresponding standard deviation can be updated. If all the features were successfully matched this updated value if fed back to the odometry. This value will be considered the current position. Since the odometry is corrected after each matching, failure of the sonar or matching subsystem does not lead to immediate failure of the robot navigation system.

In the case where the matching of the sonar data with the a priori map fails, vision is used to verify the data. The mismatched data are used to determine which region of the environment should be imaged. Data from the vision system is used to check the non-matched features. Features are not matched due to two main reasons: a) erroneous, corrupted or missing sonar data; b) new objects in the environment.

The first non-matched features to be checked are those located on the robot's trajectory in order to avoid collision. Vision provides dense data, but its processing is computationally expensive. In order to reduce the computational cost of vision, we try to predict what should show up in the image. For that purpose we use the non-matched feature to select the region of the environment map that should be analyzed by the vision system. In addition the a priori map is used to generate a prediction of the features that should be visible in the image. As a result not all the image will have to be processed and the system will also be looking for specific features in the image. Therefore the high computational cost usually associated with the vision is significantly reduced. If the vision system confirms the prediction made based on the a priori environment map the non-matched sonar data is discarded. The location is then updated based on the intrinsic estimate and the sonar matched features. If the prediction is not confirmed by the vision system a hypothesis of a new object is generated. That hypothesis is then checked against the sonar data. If the match is confirmed the a priori map is updated with a new object. If the match is not confirmed the occurrence of unmatched data is registered. In this case the final location estimate will be the intrinsic estimate.

5 EXPERIMENTAL SETUP

The experimental setup is based on the Robuter mobile platform -- see Figure 4. This platform has two on-board microprocessors (MC-68020), for vehicle control. The sonar network is part of the on-board system and is controlled by a MC-68000. The communication with the mobile platform is assured by a wireless RS232 radio link. The sonar system is made up of 24 sensors distributed around the platform and the sensor data can be obtained at different spatial and temporal samplings. The platform has two rear wheels that can be actuated independently and two passive casters at the front.

The additional hardware for control and monitoring of the other sensors is on the top of the vehicle. All the additional hardware is based on PC-AT boards which communicate with the on-board computer through a VME/PC-AT bus interface.

5.1 Sensors' descriptions

Odometry: The rear wheels are equipped with optical encoders and each encoder has an associated counter. These two systems constitute the basis of the odometric system.

Figure 4 - Mobile robot with the active vision system

At each sampling period the signed integration of encoder pulses provides an estimate of angular displacement (respectively $\Delta\alpha_R$ and $\Delta\alpha_L$). The corresponding incremental translation at the nth time interval, ΔD_n, and incremental rotation, $\Delta\theta_n$, measured with respect to the mid-point of the back axis, are given by:

$$\begin{cases} \Delta D_n = R(\Delta\alpha_{R_n} + \Delta\alpha_{L_n})/2 \\ \Delta\theta_n = R(\Delta\alpha_{R_n} - \Delta\alpha_{L_n})/B \end{cases}$$

where R is the right and left wheel radius and B the distance between the two rear wheels. Using a first order approximation the updated location (X_n, Y_n, θ_n) is given by the following equations,

$$\begin{cases} X_n = X_{n-1} + \Delta D_n \cos(\theta_{n-1} + \Delta\theta_n/2) \\ Y_n = Y_{n-1} + \Delta D_n \sin(\theta_{n-1} + \Delta\theta_n/2) \\ \theta_n = \theta_{n-1} + \Delta\theta_n \end{cases}$$

These equations approximate the trajectory as a sequence of constant curvature segments (see [Wang 88]). The problem of modeling odometry is also treated in [Crowley 89], and [Chenavier 92]

Sonar system: The sonar system uses 24 ultrasonic Polaroid sensors [Robosoft 92]. Sensors are grouped in nodes, each node being activated separately. Each sensor, belonging to an activated node, emits an ultrasonic pulse of 16 cycles of a 50 kHz square wave. The emitted pulse has a beamwidth of about 20° including the side lobes. The echoes, resulting from the reflection of the ultrasonic pulse on the surrounding objects, are measured by electronic circuitry that gives an estimate of the distance in number of pulses. These sensors operate within a range between 12 cm and 12 m. The network of ultrasonic sensors is supervised by a VME board, based on a MC-68000 as described before.

A range modelling of the sonars was performed [Moita 93]. Figure 5 shows the measured distance versus orientation for several ranges from 12 cm to 6 m. The same data is depicted in 3D on Figure 6. Three main conclusions were reached: a) the measured distance is a linear function of the real distance; b) the upper bound for the measured distance standard deviation is a linear function of the range with a minimum value of 2 mm and maximum value of 5 mm; c) the angular uncertainty decreases with range as represented in Figure 7.

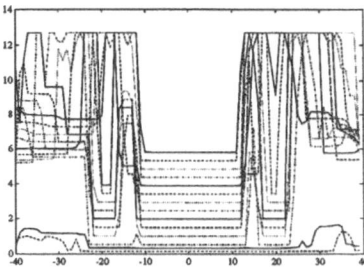

Figure 5 Overlapped sonar responses (range vs orientation)

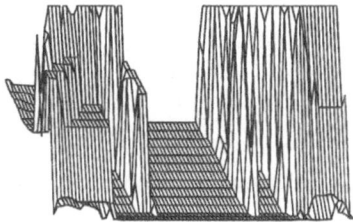

Figure 6 Tridimensional display of the overlapped sonar responses
(range vs orientation)

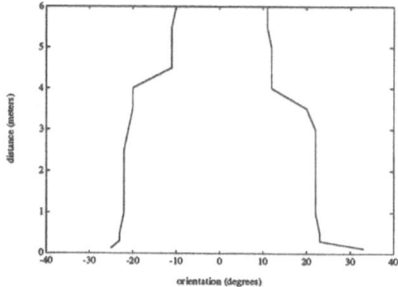

Figure 7 Visible cone of the sonar sensors (range vs angle)

Vision system: The vision system is based on an active head. The mechanical structure of the this system is illustrated in Figure 8. The system measures 295mm x 260 mm x 315 mm, weighting 8.9 Kg. The cameras´ support platform has three rotational degrees of freedom (see Figure 9). Each degree of freedom has two limit switches and an incremental optical encoder. The optical encoder is used as a feedback sensor. There is also a reference position, which is required for calibration. The mechanical structure is made up of two blocks: the cameras' platform and the block of the rotational degrees of freedom.

Figure 8- Mechanical structure of the active vision system

The cameras' platform enables the vergence motion of the cameras. The block on which the cameras' platform is mounted has three degrees of freedom whose rotation axes are mutually orthogonal. One of them has a vertical axis of rotation enabling movements in azimuth, whereas the other two have horizontal rotation axes, enabling movements in elevation. The azimuth degree of freedom has a range of 340° ($\pm 170^{\circ}$) whereas the other two have a range of 90° ($\pm 45^{\circ}$). Each one of these degrees of freedom has a precision of 0.005°. Stepper motors are used as actuators, and have speeds that can vary between 11.35°/sec and 129°/sec.

The zoom, focus and aperture of each camera are actuated by variable speed DC motors. Feedback on each of these degrees of freedom is provided by an absolute position potentiometer.

The stepper motors and the DC motors are controlled by dedicated hardware based on the AT-BUS.

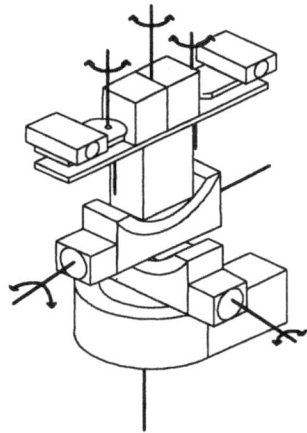

Figure 9 - Degrees of freedom of the mechanical system

Inertial system: The inertial system is constituted by two low-cost linear accelerometers, a piezo-electric gyrometer and a dual-axis inclinometer. The two linear accelerometers are piezo-resistive and they measure the acceleration along two orthogonal axes (Figure 3). This type of accelerometers has the advantages of small size, low output impedance and wide frequency response - from 0.1 Hz to 1 KHz - see Figure 10. These sensors have typically a sensitivity of 1V/g at 100 Hz. Such sensors measure the acceleration along a given axis, but are also sensitive to transversal accelerations. Due to this fact, a calibration of the sensor is needed.

The gyrometer (gyrostar from *Murata*) has a bandwidth of 7 Hz, without hysteresis and a relative precision of 3%. This sensor can measure angular velocities up to $\pm 90^{\circ}$/sec.

552

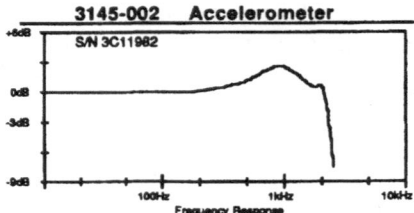

Figure 10 - Typical frequency response of the linear accelerometer
(from *ICSensors* data sheets)

The inclinometer (AccuStar II from *Lucas Sensing Systems Inc.*) gives information of the rotation angle about two orthogonal axes. The range of this sensor is $\pm 20^{\circ}$ with a resolution of 0.01°. The information provided by the inclinometer enables the direct determination of the vehicle attitude providing useful information for navigation on uneven terrain. This sensor gives the vehicle attitude even when the vehicle is not moving. The inclinometer is a dual-axis structured sensor. The Figure 11 illustrates the distribution of the sensors on the mobile platform.

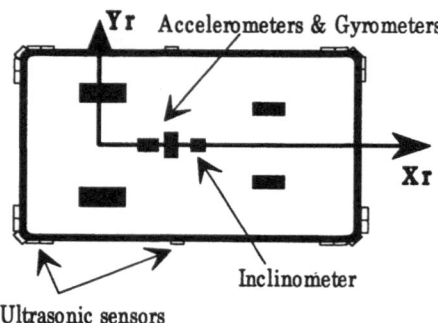

Figure 11- Robuter's sensor locations

6 CONCLUSIONS

Mobile robot navigation depends heavily on the quality of sensor information. Therefore, when designing a navigation system one has to carefully consider the types of sensors to be used as well as the related issue of combining and integrating the information. Even for the simple case of robot localization in a known and structured environment there still are open issues when considering that the robot navigates in a real environment. The problems are much more complex and difficult when one considers navigation in an unknown and unstructured environment. In these cases both the problem of sensor integration and the problem of dynamic map building are more demanding.

These considerations led us to restrict ourselves to the problem of navigating in a structured environment with a priori knowledge of its map. As a result of these considerations we decided to build our navigation system around a set of sensors that includes an inertial system, odometry, a sonar network and an active vision system. Integration of information from the odometry and the sonar network is currently dealt with by our research group within the framework of the project PO-ROBOT, sponsored by NATO [PO-ROBOT 93]. These sensors permit us not only to develop a localization system but also to implement an obstacle detection and avoidance system to cope with changes in the environment. The focus of our research lies on sensory integration and its relationships with the issues of obstacle detection, localization and map representations.

Data integration and fusion itselves have many issues that require deeper consideration and study. From our point a view some of these issues are sensor modeling, data association and process estimation. We are currently tackling these subjects within the framework of the development of a sensor system for a mobile platform.

We are also planning to add a structured lighting system to our platform that will enable low cost range computation based on the vision system.

REFERENCES

[Chenavier 92] F.Chenavier and J.L.Crowley, "Position Estimation for a Mobile Robot Using Vision and Odometry", *IEEE Int.Conf. Rob.Automation*, pp.2588-2593, 1992.

[Cox 90] Ingemar J.Cox, "Blanche: Position Estimation For an Autonomous Robot Vehicle", in *Autonomous Robot Vehicles*, I.J.Cox and G.T.Hilfong (Eds.), Springer-Verlag, pp. 221-228, 1990.

[Crowley 89] J.L.Crowley, "Asynchronous Control of Orientation and Displacement in a Robot Vehicle", *IEEE Int.Conf. Rob.Automation*, pp.1277-1282, 1989.

[Durrant-Whyte 88] H. Durrant-Whyte, "Uncertainty Geometry in Robotics", *IEEE Journal Robotics and Automation* - 4 (1), pp. 23-31, 1988.

[Grimson 84] W. Grimson, T. Lozano-Perez, "Model-Based Recognition and Localization From Sparse Range or Tactile Data", *Int.Journ. of Robotics Research*, 5 (3), 1984.

[Leonard 91] John J.Leonard and H. Durrant-Whyte, "Mobile Robot Localization by Tracking Geometric Beacons", *IEEE Trans.Rob.Automation*, 7 (3), June 1991, pp. 376-382.

[Leonard 92] John J. Leonard and H. Durrant-Whyte, *"Directed Sonar Sensing for Mobile Robot Navigation"*, Kluwer Academic Publishers, 1992.

[Moita 93] F.Moita, A.Feijão, and U.Nunes, "Polaroid Ultrasonic Sensor Modelling", Po-Robot Document ISR-C/1/Nov.93/ST1/T2.

[PO-ROBOT 93] Full plan of the project *"PO-ROBOT - Multi-purpose Portuguese Flexible Mobile Robot"*, prepared by Isabel Ribeiro, Institute of Systems and Robotics, March 1993.

[Robosoft 91] Robosoft SA, *"Robuter^{TM} User's Manual"*, August 1991.

[Robosoft 92] Robosoft SA, *"Local Area Ultrasonic Network: User's Manual"*, 1992.

[Viéville 90] T. Viéville, O. Faugeras, "Computational of Inertial Information on a Robot", *Robotics Research, The Fifth International Symposium*, Hirofumi Miura and Suguru Arimoto (Eds.), MIT Press, 1989, pp.57-65, ISBN: 0-262-13253-2.

[Wang 88] C.Ming Wang, "Location Estimation and Uncertainty Analysis for Mobile Robots", *IEEE Int.Conf. Rob. Automation*, pp. 1230-1235, 1988.

[Winston 84] P. Winston, *Artificial Intelligence*, Addison Wesley, Reading, MA, 1984.

19

INCREMENTAL DESIGN OF A FLEXIBLE ROBOTIC ASSEMBLY CELL USING REACTIVE ROBOTS

ELPIDA S. TZAFESTAS[(*)], SPYROS G. TZAFESTAS[(**)]

[(*)] LAFORIA-IBP
Université Pierre et Marie Curie
4, Place Jussieu
75252 Paris, Cedex 05, France

[(**)] Intelligent Robotics and Control Unit,
National Technical University of Athens
Zographou 15773, Athens, Greece

In this paper we advocate the use of reactive robots for incremental design of flexible robotic assembly cells. To this end, we propose a layered reactive architecture for assembly robots which is shown to have many desirable properties such as robustness, reactivity and incrementality. We then describe a series of successive simulation experiments to validate our incremental design choices for the assembly cell. Those choices are shown to translate to smooth modifications to the basic architecture, and is each an attempt to improve the overall cell's performance. This gives rise to a number of interesting observations, namely the existence of non multiplicative factors, and the stabilising impact of the use of simple adaptive components. We thus proceed naturally from the more primitive robot to the adaptive one, hoping to gain some insight into the parameters that determine the performance of a cell and must be used to drive its design.

1. Introduction

The design of robotic assembly cells has up to now centered on the design and construction of the manipulation and assembly components which presented a significant numeric complexity (Ranky 1991, Richard et al. 1991). The development of novel behavioural techniques (Connell 1990, Malcolm & Smithers 1988) together with the advances in sensor technologies (Tzafestas 1988, Van Eck & Mathys 1993), allow both the easy incorporation of a variety of physical/mechanical possibilities into a single robot, as well as the shift of the research focus to the exploration of appropriate integration methods in order to meet the enhanced performance requirements set by the modern production flexibility and

S. G. Tzafestas and H. B. Verbruggen (eds.),
Artificial Intelligence in Industrial Decision Making, Control and Automation, 555–571.
© 1995 *Kluwer Academic Publishers.*

adaptivity objectives (Almgren 1989, Tzafestas 1990, Tzafestas & Tzafestas 1991b, Nof & Drezner 1993).

In this direction, we have studied the option of a reactive layered control system for the robots that participate in such a flexible assembly cell as a powerful tool that will help us design incrementally more complex and efficient assembly cells with the minimum effort. **Incrementality** is considered as a desirable property of both the robotic control architecture and the design process : For the robotic control architecture, it stands for the possibility to easily add features to the robots without rebuilding everything from scratch, but rather by reusing existing components, whereas for the design process it stands for the possibility to add constraints and goals to the operation of the cell. The long term goal of the project is both to proceed to autonomous rather than just flexible manufacturing cells, as well as to automate their design. Incrementality is then the vehicle that will allow us to pass forth and back from robot specifications to cell operation specifications, evaluating design options for the robots, modifying operational needs and so on. In contrast with previous reactive assembly robots work (Malcolm & Smithers 1988), our approach focuses heavily on evaluation of design alternatives based on global metrics such as completion times. We believe that the results presented here are an indication that reactive architectures are good candidates for our long term goals.

The plan of the paper is as follows : In section 2, we describe the example assembly cell, enumerate the fundamental assumptions behind our design experiments and present the knowledge representation scheme adopted, while in section 3 we present the basic architecture of the robots and state its most prominent features. In section 4 we present the minimal cell, whereas in sections 5 to 8, we describe a series of successive simulation experiments to validate our incremental design choices for the assembly cell. Each choice is shown to translate to smooth modifications of the basic architecture, and is an attempt to improve the overall cell's performance. We also discuss the obtained results and a number of interesting phenomena observed : the existence of non multiplicative design factors and the stabilising impact of the use of simple adaptive components. Finally, in section 9, we conclude our work and mention directions of future research.

2. Description of the assembly cell

The role of an assembly cell is to receive a number of parts that arrive on a conveyor belt and assemble them to products according to predefined assembly specifications.

Our example assembly cell consists of a transfer robot, an assembly robot, a part storage pallet and two conveyor belts (fig. 1).

The **transfer robot** senses the arrival of parts on the input conveyor belt, performs a recognition operation on each and loads those parts that have been recognised as requested ones. These parts are subsequently stored to the pallet for the assembly robot to retrieve.

The **assembly robot** retrieves from the pallet pairs of parts that match and assemble them according to the production specifications. Intermediate parts representing incompletely assembled products are returned temporarily to the pallet, while completely assembled products are output on the second belt.

There is also a **global database** maintained by a computer stand-alone in the cell or remote. This database does not contain any information on the assembly process, but records a symbolic state of all the products involved in the process. The robots are equipped with a repertoire of behavioural components -as in (Malcolm & Smithers 1988)- to manipulate and assemble those known parts. Those components are indexed by symbolic names, so that all the reasoning takes part in this symbolic level, while we let the behavioural components work out for themselves the details of the motions. In fact, as will be described in the next section, this reasoning is minimised, so that it can be best regarded as a "sensing" operation on the database data.

Product representation : An oriented Petri net representation has been adopted for the symbolic assembly knowledge (Tzafestas & Tzafestas 1991a). In such a net, places represent subassemblies and transitions represent potential assembly operations. Places may participate in more than one transitions and all transitions have exactly two input and one output places. Two products have been modelled using this representation scheme : the gear box (Fox 1986) and a virtual product called box1 and defined by the authors on the basis of sharing parts with the gear box and showing symmetries.

Basic process assumptions

i) The **rate of arrival** is compatible with the rate of assembly. The requested parts do not arrive all at a time, rather they arrive at regular intervals which permit our transfer agent to process (recognise and eventually load) arrived parts uniformly. This means that no deadlock can occur in the cell, i.e. we don't expect the pallet to get filled in a rate much greater than the assembly rate.

ii) Adopting a grid-shaped pallet, open from both sides, we can safely assume that there is **no conflict** between the two robots as far as the access of the pallet is concerned : all storage operations will be carried out on places denoted as free in the database, while all retrieval operations will occur on places denoted as occupied in the database.

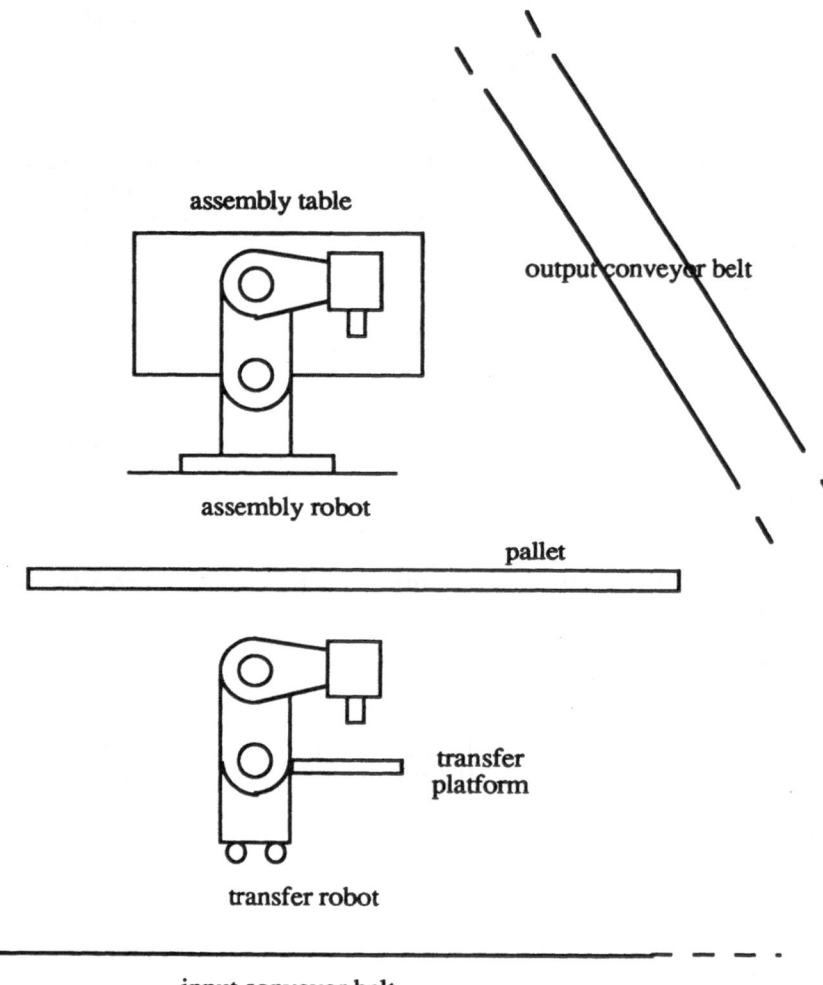

Fig. 1 : The assembly cell layout

iii) The manipulation tasks involved have **variable durations** so that the cell can be regarded as a network of asynchronously operating agents. For the sake of our simulations those durations have been set to 1, 2, 3 and 5 for recognition, loading, pallet access and assembly respectively. Those chosen durations are qualitatively realistic, i.e. the relations between them are realistic. Reasoning and database access durations have been neglected, since they are usually several orders of magnitude less than the actual manipulation durations.

iv) All simulated experiments described in subsequent sections were performed in contexts where no abnormalities were allowed to occur, that is, although error handlers were in place, no abnormal events were prescribed during simulation. The reason behind this choice is that **design options have to be evaluated in the case of normal operation,** while error handlers are there to assure the overall system converges back into normal operation in case abnormalities occur. Of course, evaluating the adequacy of error handlers is a parallel design problem which will not be further discussed here. We stress however the point that the architecture presented below is by nature very fault-tolerant to high-level (process) abnormalities, so that the only errors left to lower-level handlers are the mechanical ones, like a part falling off a manipulator, and it is shown in (Malcolm & Smithers 1988) that a reactive approach may be used to handle those situations as well.

3. Basic architecture of the robots

Each robot consists of **a repertory of behavioural components** that are preprogrammed and **a "reasoning" control system** that chooses the best component to execute, according to the state of the production at any given moment.

In this work we have simulated the operation of the control system, assuming all behavioural components are in place. We will demonstrate in the following sections how slight changes to just the control system without any changes to the manipulation components lead to significant improvement of the overall performance of the cell.

The control systems of the transfer and the assembly robots are given in fig. 2. Each consists of a number of layers laid in a decreasing priority order (our approach bears some resemblance with the other subsumption and layered approaches found in the literature, such as (Brooks 1986) and (Steels 1990). Each layer corresponds to a generic indexed task and contains its own local constant and/or variable data and an evaluation check which returns true or false according to some robot and/or database data. Among all tasks that are active -i.e. whose evaluation checks return true- the highest priority one is selected for execution, which translates to some local index of this task passed as a global index to the actuation level, i.e. the behavioural components. The term indexing is borrowed from (Malcolm & Smithers 1988) and stands for the translation process between real-world data, such as part positions, and symbolic representations used for on-line processing during decision-making. Indexing also assumes that relevant data are only relevant due to certain conditions that are locally true in the component that uses them, so that a chosen_part index won't be global to an assembly robot, but local to the task that uses it and hence prone to dynamic modification. This assumption enforces distributedness of data processing and focusing on locally relevant data, which has the side-effect of significant acceleration of the processing, since no global world model construction, monitoring and updating is anymore necessary.

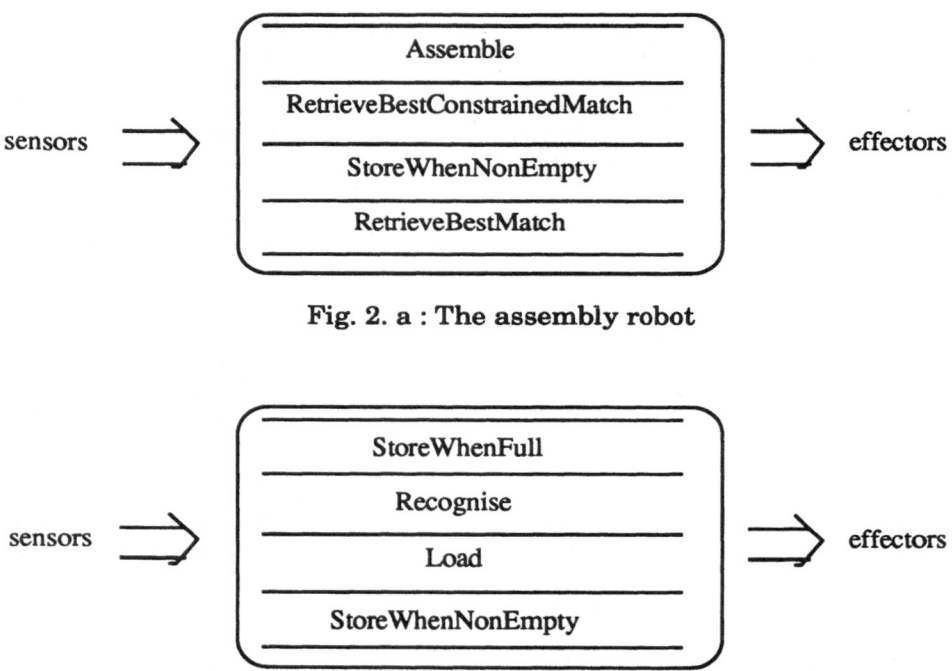

Fig. 2. a : The assembly robot

Fig. 2. b : The transfer robot

As an example, the pseudocode for the evaluation check of the RetrieveBestConstrainedMatch task is given below :

```
if robot is empty return(false);
part := search-database(best-match-for, robot's carried part, pallet);
if part exists { index part; return(true) };
return(false);
```

Note here that the check involves only local data of the robot or data of the global database. It does not involve any knowledge on other robots, on the pallet or on process metrics and objectives.

Summarising the features of this architecture :

i) **Narrow-mindedness** : Each layer is autonomous from the other ones and involves only local knowledge and knowledge of the (global) state of products that can be directly retrieved from the global database.

ii) **Selfishness** : Each robot having no knowledge on the other robot(s) or on process metrics and objectives, we can therefore regard it as an agent "committed" to act as much as possible and this pursuit of local goals leads to the achievement of the global goal.

iii) **Robustness** and **reactivity** : Again, having no process knowledge implies that each robot is reactive to any changes that might occur, so that no process event can be regarded as unexpected and the overall operation of the cell will be consistent even in abnormal cases. Imagine, for instance, the case of some malevolent human that moves arbitrary parts from the pallet. The assembly robot detecting the inconsistency, it will report it to the database and the latter will update itself. From this point, the process will continue as normal, without processing this event as an abnormality. Hence, the robots are reactive to opportunities and robust to faults.

iv) Last but not least, **modularity** and **incrementality** : The tasks using symbolic indexes to parts, we need not develop every new task from scratch. Instead we re-use existing tasks for refinement and tailoring, especially when the indexed manipulation components remain the same, as in the case of the various access pallet tasks that have been developed during this project.

In one word, the power of the reactive approach lies in its simplicity: Reactive agents are powerful because they make little or no assumptions on the kind of "knowledge" they can process. In fact, they use crude representations of the world they are in, rather than elaborate precise models of it. This leads to improved fault-tolerance conterbalanced by loss of optimality: **reactivity and representation precision are orthogonal properties.**

4. Case 1 : The minimal assembly cell

We have simulated the operation of the minimal cell described in section 2 for the task of producing 5 gear boxes and 3 boxes of type box1 and obtained results for 40 complete runs. The simulation testbed was developed on a Sun workstation under the operating system Unix 4.2BSD and it was written in Smalltalk version 4.0.The results are summarised in table 1.

Assembly Robot :		
	Retrievals from pallet : min 91, max 104, avg 98.05	
	Storage to pallet : min 8, max 21, avg 15.05	
Completion time : min 794, max 904, avg 850.025		

Table 1 : Results in 40 runs for the minimal cell

Performance of the system : Given that for each product the number of assembly operations required is constant and does not depend on the order of the assemblies, a measure of the performance of the overall cell is the number of access pallet operations performed by the assembly robot.

5. Case 2 : Extending the robots architecture

To improve the performance of our cell we observe that the number of retrievals can be decreased by largening the capacity of the worktable on which the assembly robot is docked. Not surprisingly, the modifications that we had to incur to the basic robotic architecture were very small :

i) We had to add an extra counter for the number of occupied places of the worktable and an extra constant for its capacity.

ii) We had to replace the StoreWhenNonEmpty task by a StoreWhenAlmostFull task whose evaluation check is *load* >= *capacity - 1* and which interestingly translates to StoreWhenNonEmpty when *capacity = 2*.

We have simulated the operation of the new cell for a range of values for the worktable capacity. The results that we have obtained after 40 runs for each case are summarised in table 2.

As one has probably been able to expect qualitatively, we observe a **stabilisation of the performance** after *capacity = 13* where the number of access pallet operations of the assembly robot equals the number of parts. Any further increase in the size of the worktable does not affect the performance of the cell.

Table capacity	Access pallet	Time
3	106.45	838.7
4	99.3	818.88
5	93.95	814.63
6	92.35	816.78
7	89.7	805.48
8	87.55	793.55
9	84.6	795.98
10	83.9	794.83
11	84.15	798.9
12	83.1	794.38
13	83	785.625
14	83	782.6
15	83	778.9
16	83	786.98
17	83	784.05
18	83	791.65
19	83	785.78
20	83	797.03

Table 2 : Average in 40 runs for a range of table capacities

6. Case 3 : Using more than one assembly robots

An alternative extension has been to introduce more than one assembly robots built on the same basic architecture. Now we expect an improvement on the completion time of the production task since more than one assembly operations will be feasible in the same time.

The only modification we had to apply to the basic architecture concerned the access of the pallet by the assembly robots, since now conflicts may arise in manipulators paths as well as in physical pallet positions. To solve this problem we introduced a simple locking technique: the assembly robots access the pallet only when it is not locked, in which case they lock it for as long as their access lasts.

Again, we have simulated the operation of the new cell for a range of values for the worktable capacity. The results that we have obtained after 40 runs for each case are summarised in table 3.

Number of assembly robots	Time
2	733.8
3	712.9
4	699.53
5	704.1
6	702.5
7	693.3
8	689.8
9	691.5
10	704.78
11	696.45
12	693.15
13	696.9
14	707.08
15	701.98
16	698.38
17	693.95
18	697.1
19	698.98

Table 3 : Average in 40 runs for a range of assembly robot populations

This time, we observe a **stabilisation of the performance** after *number of robots = 8*, which could again have been qualitatively expected since at any moment during the process only a small number of assembly operations are feasible in the same time and introducing more robots beyond that point won't affect the performance of the cell. Clearly, the number of assembly robots appears to be a more drastic performance parameter than the size of the worktable. This feature may be attributed to the fact that the presence of an assembly robot triggers an assembly operation which is the primary performance-related operation, while a greater worktable triggers an access-pallet which is a secondary -supporting- operation.

7. Case 4 : Combining cases 2 and 3 - Interacting factors

Then, what about combining the previous two extensions ? We simulated the operation of the assembly cell for a range of pairs of values for the table capacity and the number of assembly robots. The results are summarised in table 4.

Table capacity

Number of assembly robots	3	4	5	6	7
2	651.23(14)	639.42(28)	630.833(34)	642.25(36)	616(39)
3	623.65(17)	612(37)	-	-	-
4	631(17)	572(39)	-	-	-
5	622(20)	-	-	-	-
6	620.8(30)	-	-	-	-

Table 4 : Average in 40 runs for a range of values for table capacity against robot population (number of deadlocks in parentheses)

The first observation is the **occurrence of deadlocks** whose probability increases dramatically with table size. Deadlock occurs whenever the assembly robots carry parts that match only with parts carried by others and not stored to the common pallet. The deadlock arises because the assembly robots do never return parts to the pallet when they are not almost full. To solve the deadlock problem we have only added as the lowest priority task a time-out storage task (fig. 3) that becomes active whenever the assembly robot has stayed idle continuously for a given period and already carries a part, which implies a deadlock or potential for a deadlock.

Our design problem then shifted to choosing the time-out parameter. We expect that there must be an optimal value for the time-out parameter, small enough to not waste much idle time repetitively and big one to not be activated too often and lead to useless pallet access operations. To this end, we performed a whole series of simulations for several values of the time-out parameter and we were surprised to find the optimal value around 14 which was explained as being exactly the average cycle-period of a part when used without any delay : *14 = 1 (Recognise) + 2 (Load) + 2 x 3 (2 x AccessPallet) + 5 (Assemble)*. Furthermore, we observed a **plateau of optimality** around this value (14) rather than a clear peak. We don't know whether this applies more generally than our example task, so we have planned to perform in the near future more extensive simulations to better understand this issue.

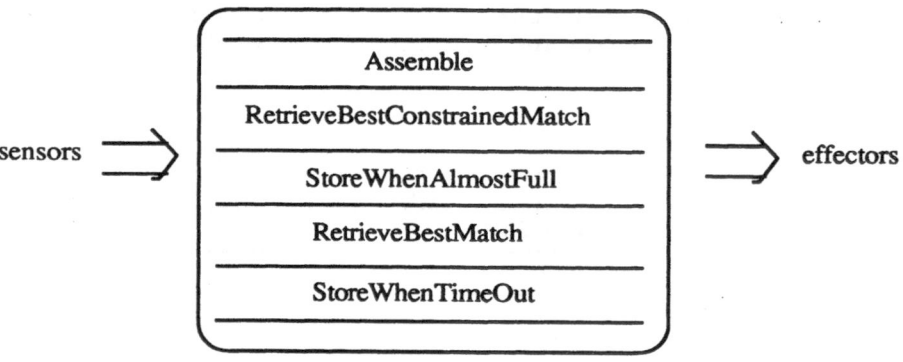

Fig. 3 : The time-out assembly robot

Of course, the introduction of this time-out task, while solving the deadlock problem, has the undesirable side-effect of degrading the performance of the system and comparative results for table sizes 3 to 7 against 2 to 6 assembly robots are given in table 5.

Returning to the deadlocks observed, we now understand that the two performance enhancing factors, worktable size and number of assembly robots, are not multiplicative, but rather they interact in an undesirable way. The *important engineering achievement* here is that we have managed to get rid of this undesirable interaction by introducing a simple time-out task. But, **however simple this time-out task might appear, it changes the shape of the curves** : rows and columns in table 5 don't demonstrate the same decrease tendency that appears in table 4. Instead, the L-shaped curves become U-shaped, that is there is some region which exhibits optimality and the values below or above that region lead to degraded performance. The time-out task is then far more drastic than it appears at first glance.

Table capacity

Number of assembly robots	3	4	5	6	7
2	642.23	643.3	650.6	654.28	656.43
3	635.15	637.65	632.33	641.45	639.88
4	621.45	621.98	625.08	626	626.25
5	648.18	641.6	641.98	638	637.75
6	657.4	660.55	649.98	641.98	646.1

Table 5 : Average in 40 runs for a range of values for table capacity
against robot population (time-out parameter = 14)

8. Case 5 : The adaptive robot - Commitment to product

Still we expect to improve the overall performance if we observe that there is interference between the robots : often a product's parts float between different robots and assembly operations that regard the same final product are distributed through the robots which may lead to an unnecessarily large number of pallet access operations - travel of parts between robots. We would like to somehow resist to that drawback and to do so we examined what do human workers do in similar cases. Humans avoid this kind of conflict in parts by "committing" themselves to products so as to minimise interferences. However, this commitment is not predefined and not static. A human does not stay idle if (s)he feels of help, but rather (s)he participates in the process even if this is only local and temporary. A human is then adaptive to the state of the process, even if (s)he does not have any explicit knowledge about it.

We implemented a kind of commitment in the above sense by introducing an additional variable for each assembly robot, which we have explicitly called *commitments* and which has as its value a list of pairs (product_id, commitment) sorted according to the second value. We have also refined all the pallet access tasks by replacing random examination of the available parts by *examination-highest-commitment-first*. Hence, a robot accesses more often parts that belong to products that it is more committed to and finally all processing concerning a given product is localised to only one or -in the worst case- two robots. This commitment-based technique is implemented as follows :

- First all robots start with equal *commitment = 1* for every product.

- Access pallet tasks examine available parts by decreasing order of commitment to the corresponding products, and

- Each time an assembly operation has taken place, the corresponding robot updates locally its commitment by adding a small constant to it.

commitment (t+1) = commitment (t) + reinforcement_value

This simple reinforcement of the commitment value was designed so as to achieve two goals :

i) **Commitments should be dynamic** : a robot should be more committed to a product it has already processed (because we assume that a part involved may already be carried by that robot).

ii) **Commitments should be relative** : a robot should be more committed to a product which is closer to its final state (complete assembly), because then a final product may be output earlier.

To comply with these two goals the reinforcement constant was defined as *1/depth* of the product type, where depth is the maximum number of sequential transitions from start to complete assembly as found from the petri net representation of the product.

Comparative results for table sizes 3 to 7 against 2 to 6 assembly robots are given in table 6.

By comparing tables 5 and 6 we can observe that the introduction of the commitment technique has a **stabilising and smoothening effect** on the performance of the system :

i) Variations between different settings are smaller, since this technique leads to fair **load balancing** between robots, and

ii) The larger the table size, the larger the gain in performance, and inversely there may be loss in performance for table sizes close to 2, since then the load balancing between robots is less important than the greedy state-driven operation.

Overall, this adaptation mechanism doesn't distort the shape of the curves (compare table 5 to table 6), although it shifts the optimal region further to the right of the rows. As a result, this mechanism, in conjunction with the time-out task discussed, manages to restrict this optimal operation region around small values for both the number of the assembly robots and the table capacity (3 to 5 assembly robots and 4 to 6 table capacity, in the present example). It is an issue of current research to identify and compare other adaptation mechanisms with the hope to find one that yields optimal results in the same value region for a set of production specifications.

Table capacity

Number of assembly robots		3	4	5	6	7
	2	648.8	640.83	643.18	648	648.35
	3	630.43	627.53	630.23	633.98	631.56
	4	624.18	623.78	622.63	617.88	623.9
	5	651.6	630.05	645.7	634.83	638.4
	6	669.1	664.65	643.55	645.4	646.48

Table 6 : Average in 40 runs for a range of values for table capacity against robot population (time-out parameter = 14, adaptive)

9. Conclusions and further work

This series of simulations is an attempt to gain some insight into the **operational design of robotic cells**, and the above results allow us to draw a number of early conclusions and identify initial sets of design parameters and techniques.

As cell design should be driven by optimisation considerations, namely by our struggle toward optimising the performance of our cell, it is important to unravel those parameters that would lead to a more drastic improvement or those that are negative and should be counterbalanced by others.

So far and for the example class of assembly cells, we have identified three important parameters :

i) **Number of robots** : This is the more expensive one. We would like to be able to drastically improve performance of very low numbers of robots.

ii) **Size of storage sites** : This is a secondary parameter, because it does not explicitly trigger the operation of the robots. We would like to be able to better exploit larger sizes of sites.

iii) **Robot architecture** : This is our primary design field. We are bound to find an optimal architecture for our robots given specific values or value intervals for the two previous parameters. Our choice of a layered reactive architecture allowed us to explore many options with very little reprogramming effort, but rather by integration of previous work. Two different directions have been identified :

- **Introducing additional tasks**, for example the time-out task. This task was found to have drastic impact on the behaviour of our robots.

- **Locally refining tasks**, for example introducing the adaptive commitments technique. This was found to stabilise and smoothen the performance of our robots.

We believe that there are many architectural options that deserve further study, such as incidental rather than time-out representations, adaptive tasks rather than adaptive variables etc. There are also a number of simplifying assumptions that have been adopted at this stage, such as the uniform durations of different tasks, the absence of tools etc., and that need to be relaxed in a more in-depth study.

We plan to address some of these issues in our future work and refine our initial conclusions regarding the design of robotic assembly cells. Furthermore, we plan to introduce different types of robots, such as catalyst robots whose role is to trigger other robots' operation, so as to allow emergence of social phenomena that we would like to study and exploit. Finally, work is underway toward complete definition of a manufacturing cycle based on locally reactive cells, i.e. toward a factory that could be seen as a society of interacting and cooperating autonomous cells.

References

R. Almgren (1989) : On knowledge-based planning and programming systems for Flexible Automatic Assembly, University of Linköping, Studies in Science and Technology, *Thesis no. 176, LiU-Tek-Lic-1989:16*, 1989.

R. A. Brooks (1986) : A Robust Layered Control System for a Mobile Robot, *IEEE Journal of Robotics and Automation*, Volume RA-2, Number 1, 1986, pp. 14-23.

J. Connell (1990) : *Minimalist mobile robotics : a colony-style architecture for an artificial creature*, MIT Press, Cambridge, Massachusetts, 1990.

B. R. Fox (1986) : The implementation of opportunistic scheduling, *1986 Intelligent Autonomous Systems Conference*, Amsterdam, Dec. 1986.

C. A. Malcolm, T. Smithers (1988) : Programming assembly robots in terms of task achieving behavioural modules : First experimental results, *Proc. Intern. Advanced Robotics Programme, 2nd Workshop on Manipulators, Sensors and Steps Towards Mobility*, Manchester 1988, also *DAI Research Paper 410*, Edinburgh University.

S. Y. Nof, Z. Drezner (1993) : The multiple robot assembly plan problem, *Journal of Intelligent and Robotic Systems*, Vol. 7, No. 1,pp. 57-71, Feb. 1993.

P. G. Ranky (1991) : Flexible robot work cell design by simulation, in *Intelligent Robotic Systems*, by S. G. Tzafestas (Ed.), Marcel Dekker, 1991.

J. Richard, F. LePage, G. Morel (1991) : Computer/Programmable Control of a Flexible Manufacturing Cell, in *Microprocessors in Robotic and Manufacturing Systems*, by S. G. Tzafestas (Ed.), Kluwer, 1991.

L. Steels (1990) : Cooperation Between Distributed Agents through Self-Organisation, in *Proceedings IEEE International Workshop on Intelligent Robots and Systems (IROS) '90*, Japan, 1990.

S. G. Tzafestas (1988) : Integrated sensor based intelligent robot systems, *IEEE Control Magazine*, April 1988.

S. G. Tzafestas (1990) : Artificial Intelligence Techniques in Computer-Aided Manufacturing Systems, in *Knowledge Engineering*, by H. Adeli (Ed.), McGraw-Hill, 1990.

S. Tzafestas, E. S. Tzafestas (1991) : The Blackboard Architecture in Knowledge-Based Robotic Systems, *Advanced Study Institute on Expert Systems and Robotics*, Corfu, juillet 1990, Springer Verlag, 1991.

S. G. Tzafestas, E. S. Tzafestas (1991) : Intelligent FMS Control using the Blackboard Model, *European Control Conference*, Grenoble, juillet 1991.

J. L. Van Eck, P. Mathys (1993) : Sensors, in *Applied Control : Current Trends and Modern Methodologies*, by S. G. Tzafestas (Ed.), Marcel Dekker, 1993, pp. 149-192.

20

ON THE COMPARISON OF AI AND DAI BASED PLANNING TECHNIQUES FOR AUTOMATED MANUFACTURING SYSTEMS

A.I. KOKKINAKI, K.P. VALAVANIS
Robotics and Automation Laboratory
The Center for Advanced Computer Studies and Apparel-CIM Center
The University of Southwestern Louisiana
Lafayette, LA 70504-4330, USA

1 Introduction

This research has been motivated by the challenge to survey, summarize and compare traditional and distributed AI planning techniques derived for automated manufacturing systems, including CIM, FMS, robotic assemblies and intelligent robotic systems. All these methods are examined to identify their advantages and limitations when applied to solve a specific problem in an automated manufacturing environment.

Planning is an integral function of all systems, where the term "system" is considered in its most general form. Development/derivation of planning techniques related to a specific type of systems, requires knowledge of the different disciplines of AI, Operations Research, Management Science, Mathematics, Engineering and Psychology. Understanding of a human decision-making process may be useful in the development of automated planning methods. Furthermore, research in automated planning aims to understand, explain, predict and improve human decision-making by enhancing it, or even replacing it in restricted domains, with automated planning methods; a list of expert systems suitable for planning and scheduling may be found in [103].

Planning may be defined as the sequence of compatible actions/tasks necessary to achieve one or more explicitly stated goals by satisfying a set of system function constraints and priorities. Scheduling is the process of assigning specific resources and time periods to primitive actions. Planning characteristics depend heavily on the specific application domain. Although many of the basic issues have been well understood empirically, they have not been adequately formalized from the mathematical point of view [163].

S. G. Tzafestas and H. B. Verbruggen (eds.),
Artificial Intelligence in Industrial Decision Making, Control and Automation, 573–629.
© 1995 *Kluwer Academic Publishers.*

From the traditional AI point of view, planning is tightly related to problem solving; problem solving programs often incorporate a plan construction part whose execution will transform their current state into the desired goal state. As noted in [223], the problem solving domain may vary significantly, i.e., route planning [63, 180], puzzle-like type of problems [154], VLSI design [129], etc. Therefore, *planning from the traditional AI perspective requires the derivation of efficient general purpose procedures that can accommodate at least one solution for a wide variety of situations, given a set of initial data and constraints.* For example, in computer automated manufacturing systems, valid constraints include execution time and wasted resources. One of the major drawbacks of traditional AI approaches that reflects upon planning also, is the fact that systems operate effectively only in a very limited domain of expertise and the apparent solution of injecting "more knowledge" (common sense knowledge) into the system does not lead to more powerful systems.

As justified in [56], Distributed AI (DAI) techniques attempt to overcome the brittleness associated with traditional AI techniques by orchestrating effectively a society of systems so that a diverse collection of expertise is achieved. Moreover, DAI methods are particularly useful for tasks that are of inherently distributed nature [45, 201], or in domains where the size and other system characteristics depict that a single computer system solution is not effective. *Planning from the DAI perspective, or multi-agent planning, involves a network of nodes or agents that construct a partial plan depending on their prospective of the situation. Agents are able to communicate/cooperate with a group of other agents to interchange information and to produce a more extended plan.* Current research in multi-agent planning is focused on agents coordination and synchronization as well as theoretical models of belief.

To evaluate (somehow) the potential/performance of the surveyed (AI and DAI based) methods when applied to an automated manufacturing environment, it is important to establish some criteria. First, in an automated manufacturing environment, it is imperative to have a well-structured, robust planning method. It is also essential to have a good utilization of time and resources; waste of time as well as resources that remain idle are highly undesirable. Thus, it is important for a planning method to assign the tasks as uniformly as possible among the resources, in order to minimize their idle time and to minimize the time required for the completion of the plan. In an automated manufacturing environment, there exist and cooperate several robotic systems. Therefore, distributed planning techniques seem to be more of a natural choice. However, in accordance with the former two criteria, a distributed planning method needs to be carefully designed with respect to cooperation configuration and synchronization techniques, to avoid redundant communication and an unwise distribution of incompatible plans among different agents (robotic systems), yet achieve an effective distribution load. Moreover, in an automated manufacturing environment, the planning method must be capable of replanning in case an unexpected situation arises, and to plan opportunistically, that is, to have the flexibility to reach the same goal state in different ways, depending on the availability of system resources. Finally, a planning method with an incorporated learning subsystem should be able to monitor the execution of its plans, learn from unsuccessful plans and replan the process [225]; evolution in generating more effective plans is critical to improve system performance with time.

Planning in the manufacturing environment is challenging, because production conditions may vary with time [84]. Two alternative solutions have been proposed:

- Planning ahead of time (off-line planning) may lead to a plan execution that is inefficient under the specific operational conditions. Planning ahead of time with learning is a method able to enhance its capability to choose the most efficient way of planning and execution of complex jobs [183, 217].
- Planning in real time guarantees an efficient plan. However, plan generation requires long computation time and this may prevent agents from operating under their full capabilities.

Real-time planning is preferable in unfamiliar and/or unstructured environments. However, planning ahead of time with learning may be also used in these environments.

Figure 1 presents a classification of the examined planning methods and demonstrates the structure of this research. Section 2 summarizes traditional AI planning methods; short evaluations of them are also incorporated. In cases where it is applicable, a comparative study is attempted. Section 3 presents the DAI approach to planning; an overview of current multi-agent research topics is also included. Section 4 includes representative planning systems developed in accordance with DAI methods and techniques. Section 5 presents representative synchronization approaches to distributed planning techniques. Since learning and planning are tightly coupled in planning systems that act in a dynamic environment, their interrelation is studied in Section 6. Finally, some general remarks, guidelines for further research and conclusions are presented in Section 7.

2 Traditional Artificial Intelligence Planning Systems

Traditional AI planning techniques may be classified into:

- Theorem proving based planning systems,
- Blackboard based architectures, and,
- Planning methods suggested for assembly.

576

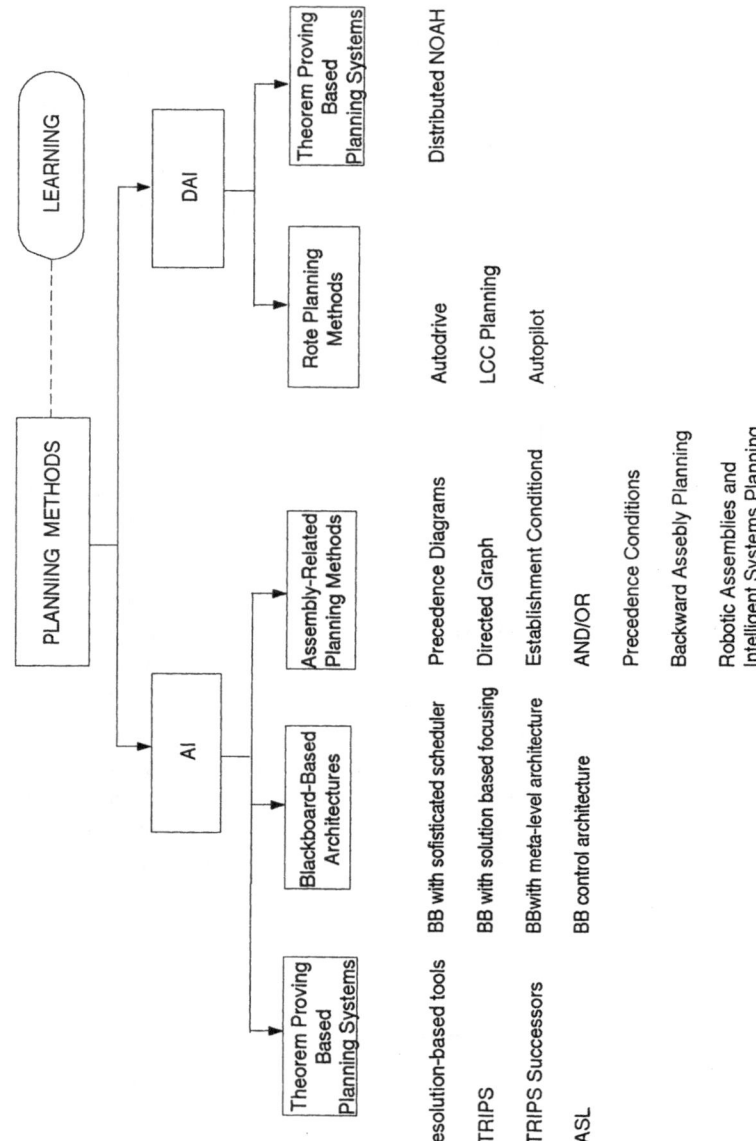

Figure 1. Classification of existing planning methods

2.1 Theorem Proving Based Planning Systems

Early planning systems have been influenced by AI theory and have focused towards a general domain independent implementation. Such planning systems are mostly concerned with important AI issues, like the halting problem (the system is not guaranteed to terminate, unless a solution configuration has been reached), the frame problem (there is a need to represent even the relations that are not affected by the execution of the system), and, the combinatorial explosion problem (the vast search space which abstracts the solution of not trivial problems).

2.1.1 Resolution-based tools

Green in [75] suggested a resolution-based tool in which operators are represented as implications and a situation variable is included in well-formed formulas that describe a specific state. The system may indicate if a specific goal-state is attainable by constructing series of acts that transform the initial state to the goal state. This planning tool suffers from the combinatorial explosion problem, the halting problem, and the frame problem. However, this has been a pioneering planning tool.

2.1.2 STRIPS

STRIPS (Stanford Research Institute Problem Solver) [63] has been implemented to plan robot movements of objects in a set of interconnected rooms. STRIPS consists of many procedures that perform various tasks including maintaining a list of goals, selecting a goal to work next, searching the operators that can be used towards the current goal, etc. States and objects form the world model and are represented by predicate argument expressions. Permissible actions, when activated, update (insert and/or delete) facts in the world model. A simple example is presented in the sequel. The initial set, W, indicates that a robot exists in roomA, and two objects obj1 and obj2 exist in roomA and roomC respectively. The action push(X, Y, Z) means that object X in roomY should be transferred to roomZ. In this example the action push(obj1, roomA, roomB) results in the transformation of the W into W'.

```
W = {at(robot, roomA), at(obj1, roomA), at(obj2, roomC)}

push(X, Y, Z)
      preconditions: at(robot, Y), at(X, Y)
      delete list:   at(robot, Y), at(X, Y)
      add list:      at(robot, Z), at(X, Z)
```

```
W' = {at(robot, roomB), at(obj1, roomB), at(obj2, roomC)}
```

STRIPS uses the means-ends-analysis method for the search space. This method, which is based on some heuristic estimation of the distance between the current and the goal state, tries to minimize that distance in every step it takes. STRIPS concentrates in solving the weak points of the resolution-based tool presented by Green. It is interesting because it introduces a number of fundamental ideas related to the AI planning perspective. However, its search strategy is not adequate for non trivial planning problems and it can not attain the goal state in problems that require the world model to become more disordered than it already is, before the goal state can be approached.

2.1.3 STRIPS Successors

STRIPS successors include triangle tables [62], NOAH [180] and NONLIN [210]. Triangle tables improved the capability to recover from errors but only within one fixed sequence [84]. NOAH and NONLIN suggest a hierarchical approach to planning. In the higher level of abstraction, large units of actions are viewed as series of primitive actions. Planning is organized as a partial order of these units. Successive refinement is performed at the lower levels until the plan is fully specified. It is not unusual to revise decisions within a level of abstraction in most planning problems. However, it might be the case that decisions at a higher level of abstraction need to be revised, and this is totally undesirable.

2.1.4 NASL

NASL is the computer program (language interpreter) developed based on a theory of problem solving suggested by McDermott in [127, 128, 130, 131]. As noted in [131], this theory was developed as an attempt to satisfy both *analytical adequacy* and *additivity*. The language, whose interpreter is NASL, is capable of expressing rules for selecting and scheduling plans. A problem is treated as a non-primitive task; a task is defined as an arbitrary activity. A basic design decision is that every task should have a unique reduction. Thus, instead of an AND/OR graph representation as is the classic approach [157], a plan is represented as a pure AND graph. A plan for a problem is a network of subtasks. Planning and executing are tightly interleaved. Instead of building a complete plan before attempting plan execution (as done in STRIPS and NOAH), this attempt allows tasks to be executed as soon as scheduling rules permit it. The error correction issue is raised and some primitive tools to deal with it are provided. The main weakness of this approach is that it does not allow a failing task (a task with an error-correcting subtask that has not been completed) to be abandoned. Progress in this domain requires a flexible logic of time and action.

2.2 Blackboard-Based Architectures

The blackboard architecture is a problem-solving mechanism initially developed for the HEARSAY speech-understanding system [60], but was also used for multiple task planning [79]. Problem-solving is an incremental and opportunistic process in blackboard architectures. Problem-solving is achieved by combining effectively solution elements. All solution elements that are generated during the problem-solving process are recorded in a global database called the blackboard by independent processes called knowledge sources. Knowledge sources have a condition action format. A general description that specifies when a knowledge source may contribute to the solution is encapsulated in the condition part of a knowledge source. The action part specifies the knowledge source behavior. Any triggering of a knowledge source is recorded in a Knowledge Source Activation Record (KSAR). Knowledge sources are modular and act independently. However, by modifying information on the blackboard they may trigger another knowledge source execution. In that sense, they are also interactive. The scheduling mechanism of a blackboard architecture selects a single KSAR to execute its action. Thus, although blackboard systems exhibit some behavioral innovations, they are very similar to the basic von Neuman processing cycle.

Variations of the basic blackboard architecture emerge from the way the control problem is handled. In solving the control problem, a system decides either implicitly or explicitly, a series of fundamental alternatives, including: the selection of the problem it will attempt to solve, the selection of problem solving methods and strategies, the criteria to apply for evaluation of distinct problem solutions, etc. Summarizing, in solving the control problem, a system determines its own cognitive behavior. According to the control problem proposed solutions, blackboard architectures may be classified in the following four categories:

- The blackboard with sophisticated scheduler,
- The blackboard with solution based focusing,
- The blackboard with meta-level architecture, and,
- The blackboard control architecture.

2.2.1 The blackboard with sophisticated scheduler

Hearsay-II [41, 60, 119] is considered a blackboard architecture with sophisticated scheduler. Hearsay-II selects the next KSAR to be executed based on the evaluation provided by control heuristics that are incorporated in the scheduling mechanism. This is a powerful mechanism to achieve considerable flexibility and opportunism, to make explicit decisions that solve the control problem, and, to adopt heuristics that focus on useful action attributes. However, the blackboard with sophisticated scheduler cannot pursue recursive subgoals of its control heuristics, it cannot generate new heuristics favoring actions that have become feasible recently, it does not provide variable grain-size control heuristics, it does handle control heuristics dynamically and does not plan strategic action sequences dynamically.

2.2.2 The blackboard with solution based focusing

HASP [155, 156], which is classified as a blackboard with solution based focusing, relies on a complex program to solve the control problem. Instead of selecting the KSAR with the highest ratings as the blackboard with the sophisticated scheduler, it selects sequentially specific blackboard events and executes knowledge sources triggered by each one. The order in which the triggered knowledge sources are executed is implementation depended. The blackboard with solution based focusing provides less flexibility than the blackboard with sophisticated scheduler. Thus, in addition to the disadvantages listed above, this implementation does not adopt heuristics that focus on useful action attribute. However, it is considered to be more efficient.

2.2.3 The blackboard with meta-level architecture

MOLGEN [116, 204, 203, 222] is a blackboard with meta-level architecture and it is a multiple layered system, in which each layer acts as a level of control for the layer below it. Actions may be classified either as domain actions, or as meta-level actions that operate based on domain actions. Meta-level actions may also operate based on meta-level actions. At any given meta-level, all actions are executed to identify the ideal action to execute at the lower level. The first level represents the domain problem. Higher levels represent such things as possible actions on domain objects and criteria for selecting and combining those actions. However, although all control knowledge sources and decisions operate at some metalevel(s), there is no fixed mapping of metalevels onto control knowledge sources.

MOLGEN has three layers of control, also known as planning spaces. Each level contains its own objects and operators. Figure 2 presents MOLGEN's planning spaces with operators shown on the left hand side.

The blackboard with meta-level architecture is capable of making explicit decisions that solve the control problem. Moreover, it is capable to adopt heuristics that

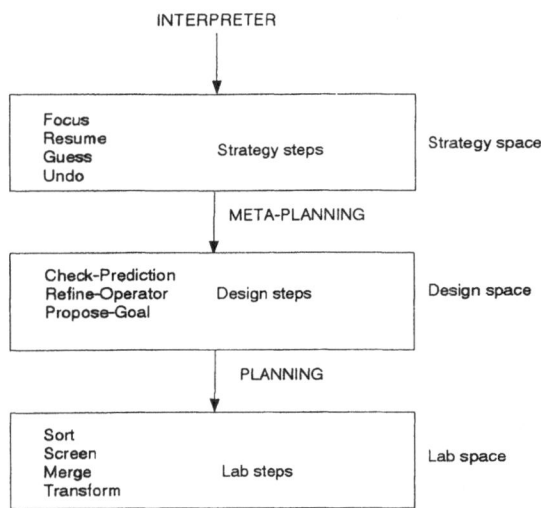

Figure 2. Planning spaces in MOLGEN (from [203]).

focus on useful action attributes. It suffers from the same weaknesses as the blackboard with a sophisticated scheduler.

2.2.4 The blackboard control architecture

OPM [79] extends and elaborates the basic blackboard architecture. The blackboard control architecture defines explicit domain and control blackboards. The domain blackboard records solution elements for the current domain problem. The control blackboard records decisions about the systems's own desirable, feasible and actual behavior, thus, solution elements for the control problem. Accordingly, domain and control knowledge sources are also defined. Domain knowledge sources operate primarily on the domain blackboard. They are domain-specific and are determined by the application system designer. Control knowledge sources operate primarily on the control blackboard. They articulate, interpret, and modify representations of the system's own knowledge and behavior. A simple, adaptive scheduling mechanism has been designed to manage both domain and control KSARs. The blackboard control architecture exhibits some very desirable behavioral goals. It is capable of making explicit decisions that solve the control problem, it can reconcile desirability and feasibility of actions, it may adopt variable grain-size control heuristics and it can also adopt, retain and discard heuristics that focus on useful action attributes dynamically. Moreover, it can integrate multiple control heuristics, plan dynamically strategic action sequences and reason about control

582

Characteristics	BSS	BSBF	BMLA	BCA
Make explicit decisions that solve the control problem.	T	T	T	T
Reconcile desirability and feasibility of actions	F	F	F	T
Variable grain control heuristics	F	F	F	T
Heuristics focusing on useful action attributes	T	F	T	T
Dynamic manipulation of control heuristics	F	F	F	T
Deside how to integrate multiple heuristics	F	F	F	T
Dynamic plan strategic action sequences	F	F	F	T
Reason about control vs domain actions	F	F	F	T

BSS: Blackboard with Sofisticated Scheduler   T: TRUE, F: FALSE
BSBF: Blackboard with Solution Based Focusing
BMLA: Blackboard with Meta-Level Architecture
BCA: Blackboard with Control Architecture

Table 1. Summary of basic characteristics of various blackboard systems (taken from [79]).

versus domain actions. This system is computationally intensive. Therefore, its use may be prohibited for some applications.

Table 1 summarizes the characteristics of the presented alternative black-board architectures. Time efficiency and computational complexity have not explicitly been presented in the reviewed literature and for that reason they are not included.

2.3 Assembly Planning and Assembly Sequences Representations

There has been considerable research in computer aided assembly planning [112], because it provides a systematic way to search for an optimal solution and it may also provide additional information pertinent to the assembly of a product. This information is essential since it may be used for more effective modification of the product and linked to the design level, the assembly floor layout and the workcell level for programming instructions.

Assembly sequence representations affect both the design of the planning system and the design of the intelligent control system. Thus, the selection of an assembly sequence representation is critical. Some of the criteria for the selection of a representation may be summarized based on:

- Storage requirements,
- Whether the representation of the assembly may be derived easily from the description of the assembly so that it might be automated,
- How difficult it is to map the representation into another representation, and,
- Whether the time dependence and time independence between assembly tasks can be represented.

An algorithm that generates an assembly sequence representation must also be correct (only feasible sequences are generated) and complete (all feasible sequences are generated). A prerequisite for a proof of correctness and completeness of the algorithm is a proof that the representation of assembly sequences used by the algorithm is correct and complete. In the sequel, some background terminology (based on [86, 112]) is provided, six different assembly sequence representations are presented and their interrelation is examined. The correctness and completeness of these representations are also addressed [86].

An *assembly*, A, is a cluster of parts assembled together by a certain assembly sequence, which maintains a particular geometric relationship among parts [112]. Formally, an assembly may be defined as $A = \{P(A), G_L(A), G_P(A), \Pi(A)\}$. $P(A)$ is the set of parts of A. $G_L(A)$ is the attributed liaison representation of A and contains information on part configurations, the geometry and relative pose of the assembly parts, the interconnection mechanisms of part connections, and the local freedom of motion in part mating. $G_P(A)$ is the special process and constraint forest associated with assembly A. Processes associated with an assembly may include tasks as testing, painting, etc. Special processes may impose certain constraints (possibly different) during each phase of an assembly process. $\Pi(A)$ is the set of all feasible assembly sequences for A.

A *subassembly* of an assembly A, $S_i|A$, is an assembly that has a non-empty proper subset of A. A subassembly has either only one part or is such that every part has at least one surface contact with another part in the subset (it is assumed that there exists a unique assembly geometry for each pair of parts). A *direct subassembly* of an assembly A is denoted as $S_i^d|A$. $S_i^d|A$ can be assembled with A - $S_i^d|A$ at the last step of assembly in one possible assembly sequence.

Assume two subassemblies that are characterized by their sets of parts θ_i and θ_j; if θ_k, where $\theta_k = \theta_i \cup \theta_j$, is also a subassembly, then the process of joining θ_i and θ_j is called an *assembly task*. An assembly task is *geometrically feasible* if there is a collision-free path to bring the two subassemblies, presumably far apart, into contact. An assembly path is called *mechanically feasible*, if it is possible to establish the attachments on the contacts between the two subassemblies.

The ordered set of assembly tasks $\tau_1 \tau_2 ... \tau_{n-1}$ is called an *assembly sequence* if there are no two input tasks that have a common input subassembly, the input subassemblies to any task τ_i is either a one-part subassembly or the output subassembly of a task that precedes τ_i and the output subassembly of the last task in the sequence is the complete assembly. An assembly sequence is said to be feasible if all its assembly tasks are geometrically and mechanically feasible and the input subassemblies of all tasks are stable.

The *assembly process* consists of a succession of assembly sequences.

The assembly planning, AP(A) of an assembly A is the process of generating a set of assembly sequences $\Pi_a(A) \subseteq \Pi(A)$ based on the given $P(A)$, $G_L(A)$, $G_P(A)$ and the criteria for selecting desirable assembly sequences.

The *Assembly Graph of Connections* [P, C] is an undirected graph with P nodes that correspond to a part in the assembly and C connections; there is one edge in C connecting each pair of nodes whose corresponding parts have at least one surface contact. An assembly may be represented by an Assembly Graph of Connections, under the assumption that whenever two parts are joined, all contacts between them are established.

Assembly state is called the configuration of the parts instantiated at the beginning (or at the end) of an assembly task. An assembly state may be represented by a n-ary binary vector $\underline{x} = [x_1, x_2, ..., x_n]$ (n connections in the product), and x_i is T (true) whether the ith connection is established, or F (false) otherwise. With this representation the initial state of a five-connection assembly is [FFFFF] and the final is [TTTTT]. An assembly state may be also represented by the set of formed subassemblies. It is easy to see that for an assembly with four parts A through D, the initial state is represented as {(A) (B) (C) (D)} and the final state as {(A B C D)}. The two representations of assembly states are dual, i.e., given one representation and the connections graph, the other representation may be easily obtained.

A list of different assembly representations follows.

- Precedence Diagrams,
- The Directed Graph representation,

- Establishment Conditions representation,
- AND/OR representation,
- Precedence Relations between the establishment of one connection and states of the assembly process, and,
- Precedence Relations between the establishment of one connection and the establishment of another connection.

Planning methods for Robotic Assemblies which do not use assembly representations include:

- Backward Assembly Planning, and,
- Robotic Assemblies and Intelligent Robotic Systems planning method.

2.3.1 Precedence Diagrams

Precedence diagrams, an off-line plan generation method, have been proposed in [67]. Plans are given to the robot as a set of operations with minimal ordering constrains. Plans are represented as precedence diagrams and actually encompass several possible sequences of operations that lead to the assembly of a specific product. This technique has limited flexibility but its main problem is that for most products, no single partial order can encompass every possible assembly sequence.

2.3.2 The Directed Graph Representation

The directed graph representation of assembly states can explicitly encompass the set of all assembly sequences. The directed graph representation is a directed graph with nodes that represent the stable state partitions of the set of parts in an assembly. The nodes in this representation may be either a partition of the set of parts [83], or a subset of the connections of pairs of parts [23, 48, 49]. A directed edge exists between node k and l if there exist a feasible assembly task to transform state k to state l. The directed graph of feasible assembly sequences of an assembly whose set of parts is P, is the directed graph $[X_p, T_p]$ in which:

$$X_p = \{\Theta \mid [\Theta \in \Delta(P)] \wedge [\forall \theta \, (\theta \in \Theta) \Rightarrow (sa(\theta) \wedge st(\theta))\,]\} \tag{1}$$

$\Delta(P)$ is the set of all partitions of P, $sa(\cdot)$ is a predicate that denotes whether a subset of parts forms a subassembly, and, $st(\cdot)$ is a predicate that denotes whether a subassembly

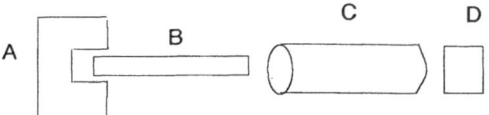

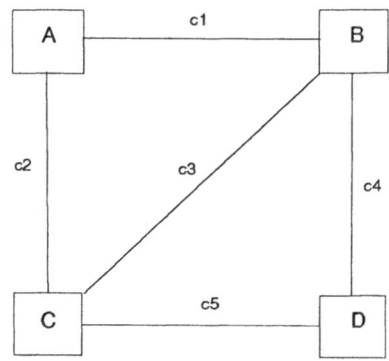

Figure 3. The graph connections for a four part assembly.

is stable. T_p is formally described as follows:

$$
\begin{aligned}
T_p = \{ \ & (\Theta_i, \Theta_j) \mid [(\Theta_i, \Theta_j) \in X_p \times X_p] \\
& \wedge [\mid \Theta_j - (\Theta_i \cap \Theta_j) \mid = 1] \\
& \wedge [\mid \Theta_i - (\Theta_i \cap \Theta_j) \mid = 2] \\
& \wedge [\mathbf{U}(\Theta_i - (\Theta_i \cap \Theta_j)) \in \Theta_j - (\Theta_i \cap \Theta_j)] \\
& \wedge [mf(\Theta_i - (\Theta_i \cap \Theta_j))] \\
& \wedge [gf(\Theta_i - (\Theta_i \cap \Theta_j))] \ \}
\end{aligned}
\tag{2}
$$

$U(A, B, ..., Z)$ represents the union of all these sets and the two predicates gf(•) and mf(•) denote whether the assembly task they take as input is geometrically and mechanically feasible, respectively.

To demonstrate this representation, an example is presented in the sequel. For that example, we assume that an assembly consists of four subparts A, B, C and D. Figure 3 presents an assembly and its graph of connections.

Figure 4 represents the directed graph of feasible assembly sequences. The subassembly of A connected to B may be represented as (A B), or else, using the array

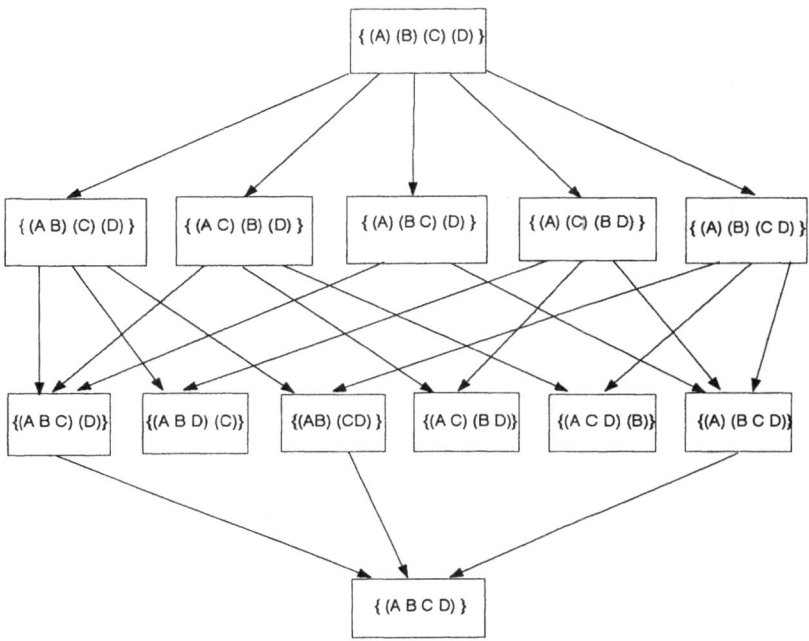

Figure 4. Directed graph of feasible assembly sequences

representation as (TFFFF). A path of feasible sequences whose initial node is { (A) (B) (C) (D) } and final node is {(A B C D)} represents a sequence of tasks to form the specified assembly. Some nodes in the graph have no ancestors because the assembly sequence they represent is infeasible.

2.3.3 Establishment Conditions

Establishment conditions of assembly sequences have been introduced in [23]. If the states of the assembly process are represented by n-ary binary vectors as described above, then a set of logical expressions can be used to encode the directed graph of feasible assembly sequences. Let $\Xi_i = \{ \underline{x}_1, \underline{x}_2, ..., \underline{x}_m \}$ be the set of states from which the ith connection can be established without precluding the completion of the assembly. The *establishment condition* for the ith connection is the logical function:

$$F'_i(\underline{x}) = F'_i(x_1, x_2, ..., x_n) = \sum_{i=1}^{m} \prod_{j=1}^{n} \gamma_{ij} \tag{3}$$

That is, the sum and the product operations are the logical operations OR and AND, respectively; γ_{ij} is either x_j if the jth component of $\underline{x}_i$ is true or $\overline{x_j}$, otherwise. The detailed example for establishment conditions for the assembly presented in Figure 3, is provided in the sequel.

$$
\begin{aligned}
F_1(\underline{x}) = &\ \overline{x_1} \cdot \overline{x_2} \cdot \overline{x_3} \cdot \overline{x_4} \cdot \overline{x_5} + \overline{x_1} \cdot x_2 \cdot \overline{x_3} \cdot \overline{x_4} \cdot \overline{x_5} + \\
&\ \overline{x_1} \cdot \overline{x_2} \cdot x_3 \cdot \overline{x_4} \cdot \overline{x_5} + \overline{x_1} \cdot \overline{x_2} \cdot \overline{x_3} \cdot \overline{x_4} \cdot x_5 + \\
&\ \overline{x_1} \cdot x_2 \cdot \overline{x_3} \cdot x_4 \cdot \overline{x_5} + \overline{x_1} \cdot \overline{x_2} \cdot x_3 \cdot x_4 \cdot x_5
\end{aligned} \tag{4}
$$

$$
\begin{aligned}
F_2(\underline{x}) = &\ \overline{x_1} \cdot \overline{x_2} \cdot \overline{x_3} \cdot \overline{x_4} \cdot \overline{x_5} + x_1 \cdot \overline{x_2} \cdot \overline{x_3} \cdot \overline{x_4} \cdot \overline{x_5} + \\
&\ \overline{x_1} \cdot \overline{x_2} \cdot x_3 \cdot \overline{x_4} \cdot \overline{x_5} + \overline{x_1} \cdot \overline{x_2} \cdot \overline{x_3} \cdot x_4 \cdot \overline{x_5} + \\
&\ x_1 \cdot \overline{x_2} \cdot \overline{x_3} \cdot \overline{x_4} \cdot x_5 + \overline{x_1} \cdot \overline{x_2} \cdot x_3 \cdot x_4 \cdot x_5
\end{aligned} \tag{5}
$$

$$
\begin{aligned}
F_3(\underline{x}) = &\ \overline{x_1} \cdot \overline{x_2} \cdot \overline{x_3} \cdot \overline{x_4} \cdot \overline{x_5} + x_1 \cdot \overline{x_2} \cdot \overline{x_3} \cdot \overline{x_4} \cdot \overline{x_5} + \\
&\ \overline{x_1} \cdot x_2 \cdot \overline{x_3} \cdot x_4 \cdot \overline{x_5} + \overline{x_1} \cdot \overline{x_2} \cdot \overline{x_3} \cdot x_4 \cdot \overline{x_5} + \\
&\ \overline{x_1} \cdot \overline{x_2} \cdot \overline{x_3} \cdot \overline{x_4} \cdot x_5 + \overline{x_1} \cdot x_2 \cdot \overline{x_3} \cdot x_4 \cdot \overline{x_5} + \\
&\ x_1 \cdot \overline{x_2} \cdot \overline{x_3} \cdot \overline{x_4} \cdot x_5
\end{aligned} \tag{6}
$$

$$
\begin{aligned}
F_4(\underline{x}) = &\ \overline{x_1} \cdot \overline{x_2} \cdot \overline{x_3} \cdot \overline{x_4} \cdot \overline{x_5} + \overline{x_1} \cdot x_2 \cdot \overline{x_3} \cdot \overline{x_4} \cdot \overline{x_5} + \\
&\ \overline{x_1} \cdot \overline{x_2} \cdot x_3 \cdot \overline{x_4} \cdot \overline{x_5} + \overline{x_1} \cdot \overline{x_2} \cdot \overline{x_3} \cdot \overline{x_4} \cdot x_5 + \\
&\ x_1 \cdot \overline{x_2} \cdot \overline{x_3} \cdot \overline{x_4} \cdot x_5 + x_1 \cdot x_2 \cdot x_3 \cdot \overline{x_4} \cdot \overline{x_5}
\end{aligned} \tag{7}
$$

$$
\begin{aligned}
F_5(\underline{x}) = &\ \overline{x_1} \cdot \overline{x_2} \cdot \overline{x_3} \cdot \overline{x_4} \cdot \overline{x_5} + x_1 \cdot \overline{x_2} \cdot \overline{x_3} \cdot \overline{x_4} \cdot \overline{x_5} + \\
&\ \overline{x_1} \cdot \overline{x_2} \cdot x_3 \cdot \overline{x_4} \cdot \overline{x_5} + \overline{x_1} \cdot \overline{x_2} \cdot \overline{x_3} \cdot x_4 \cdot \overline{x_5} + \\
&\ \overline{x_1} \cdot x_2 \cdot \overline{x_3} \cdot x_4 \cdot \overline{x_5} + x_1 \cdot x_2 \cdot x_3 \cdot \overline{x_4} \cdot \overline{x_5}
\end{aligned} \tag{8}
$$

To provide an insight to the establishment conditions, the first establishment condition, $(F_1(x))$, is analyzed. $(F_1(x))$ corresponds to the fact that connection c_1 can be established without precluding the completion of the assembly in the following cases: no connection has been established, or only connection c_2 is established, or only connection c_3 is established, or only connection c_5 is established, or only connections c_2 and c_4 are established, or only connections c_1 and c_2 are not established.

It should be noted that in $(F_1(x))$ there is no term corresponding to the fact that c_4 is established, because this is a dead end path.

As it is apparent, the establishment conditions can be complex. Simplification may be available using Boolean algebra rules and do-not-care conditions. Establishment conditions may be obtained from both directed graph and AND/OR representation [86].

2.3.4 The AND/OR Representation

The AND/OR representation has been presented in [82, 84, 85, 86]. In the AND/OR representation the nodes are the subsets of P, where P is a set of parts, that characterize stable subassemblies, and the hyperarcs correspond to the geometrically and mechanically feasible assembly tasks. The formal definition follows.

Consider an assembly that contains the following set of parts: $P = \{p_1, p_2, \cdots, p_n\}$. The AND/OR graph of feasible assembly sequences is the $[S_p, D_p]$ graph, where S_p is the set of all stable subassemblies, $\Pi(P)$ is the set of all possible subsets of P and D_p is the set of all feasible assembly tasks:

$$
\begin{aligned}
D_p = \{ \ & (\theta_k, \{\theta_i, \theta_j\}) \mid [\theta_i, \theta_j, \theta_k \in S_p] \\
& \wedge [U(\{\theta_i, \theta_j\}) = \theta_k] \\
& \wedge [mf(\{\theta_i, \theta_j\})] \\
& \wedge [gf(\{\theta_i, \theta_j\})] \}
\end{aligned}
\tag{9}
$$

$$
S_p = \{\theta \in \Pi(P) \mid sa(\theta) \wedge st(\theta)\}
\tag{10}
$$

To illustrate this representation, the same example as before, is presented in Figure 5.

The useful feature of the AND/OR graph representation is that it encompasses all possible assembly sequences. In cases where the assembly has more than five parts, the AND/OR graph representation requires less space for representation storage than other methods [83].

2.3.5 Precedence Relations between the establishment of one connection and states of the assembly process

This method has been introduced in [48], and is used for the generation of all assembly sequences. Some necessary definitions and conventions are presented in the sequel. To denote that the establishment of the ith connection must precede a state, s, whose binary representation, $\underline{x}$, validates a logical function $S(\underline{x})$, we use the following notation: $c_i \rightarrow S(\underline{x})$. For compactness reasons, $c_i + c_j \rightarrow S(\underline{x})$ is just another notation for $[c_i \rightarrow S(\underline{x})] \vee [c_j \rightarrow S(\underline{x})]$.

If $S(\underline{x_k}) \Rightarrow \exists l[(l < k) \wedge (c_i \in \gamma_l)]$, for $k = 1, 2, \cdots, n$, where $(\underline{x_1}, \underline{x_2}, \cdots, \underline{x_n})$ is the binary vector representation of the assembly sequence and $(\gamma_1, \gamma_2, \cdots, \gamma_{n-1})$ is the representation of the assembly sequence as an ordered sequence

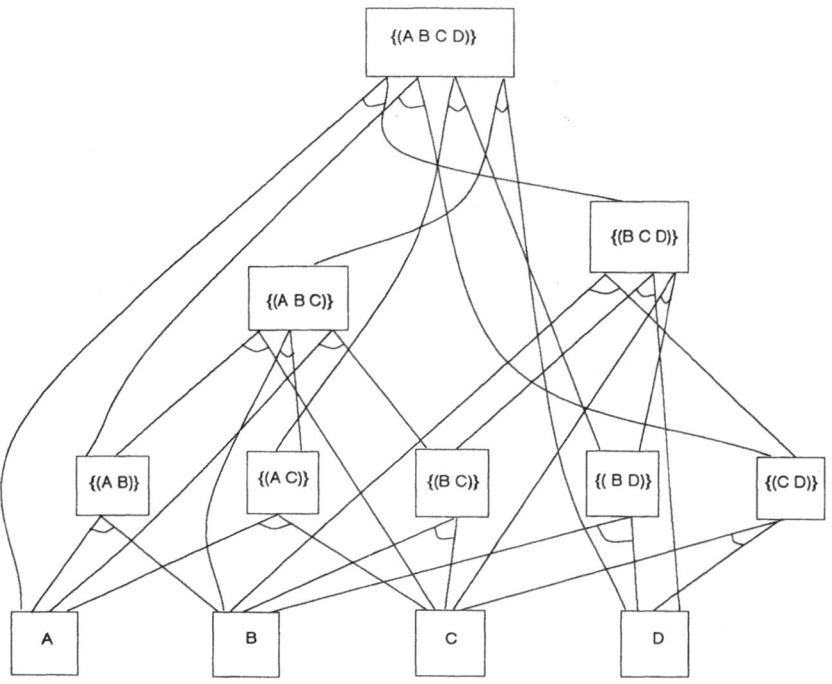

Figure 5. AND/OR representation of feasible assembly sequences

of subsets of connections, the assembly sequence satisfies the precedence relationship $c_i \rightarrow S(\underline{x})$.

To better illustrate this point, an example is presented. The assembly in Figure 1 is used again. Assume an assembly sequence; its representation as an ordered sequence of binary vectors is: $([FFFFF][TFFFF][TTTFF][TTTTT])$, or the equivalent representation as an ordered sequence of subsets of connections is: $(\{c_1\}\{c_2, c_3\}\{c_4, c_5\})$. This assembly sequence satisfies the precedence relationship $c_1 \rightarrow x_2 \cdot x_3$. This is so, because the logical function $S(\underline{x}) = x_2 \cdot x_3$ is true only in the third and the fourth state and the connection c_1 is established on the first assembly task. The same assembly sequence, however, does not satisfy the precedence relationship $c_4 \rightarrow x_1 \cdot x_2 \cdot x_3$, because the establishment of the connection c_4 occurs on the third assembly task, i.e., after the third state when the value of the logical function $S(\underline{x}) = x_1 \cdot x_2 \cdot x_3$ becomes true.

2.3.6 Precedence Relationships between the establishment of one connection and the establishment of another connection

This representation is focused on the *precedence relationships*, not between a connection and states as above, but between a connection and another connection.

Precedence relationships between the establishment of two connections have been presented in [48]. This representation is actually a formalization of the answers to the following two questions:

- What connections must be undone when the ith connection is established?
- What connections must not be left to be done after the ith connection is established?

To denote that the establishment of the connection c_i must precede the establishment of the connection c_j, the following notation is used: $c_i < cj$. In cases where the establishment of the connection c_i must precede, or be simultaneous to the establishment of the connection c_j, the following notation is used: $c_i \leq c_j$. Similarly, as above, $c_i + c_j < c_k$ is equivalent to $(c_i < c_k) \vee (c_j < c_k)$, and, $c_i < c_j \cdot c_k$ is equivalent to $(c_i < c_j) \wedge (c_j < c_k)$.

An assembly sequence whose representation as an ordered sequence of binary vectors is $(x_1, x_2, \cdots, x_n)$ and as an ordered sequence of subsets of connections is $(\gamma_1, \gamma_2, \cdots, \gamma_{n-1})$, satisfies the precedence relationship $c_i < cj$ if $c_i \in \gamma_a$, $c_j \in \gamma_b$ and $a < b$. Similarly, an assembly sequence satisfies the precedence relationship $c_i \leq c_j$, if $c_i \in \gamma_a$, $c_j \in \gamma_b$ and $a \leq b$.

An example of precedence relations between connections based on the assembly described in Figure 1, is presented in the sequel. An assembly sequence whose representation as an ordered sequence of binary vectors is $([FFFFF][TFFFF][TTTFF][TTTTT])$, or, as an ordered sequence of subsets of connections $(\{c_1\}\{c_2, c_3\}\{c_4, c_5\})$ satisfies the precedence relationship $c_2 < c_4$ and $c_2 \leq c_3$, but does not satisfy the precedence relationships $c_2 < c_3$ and $c_2 \leq c_1$.

Each feasible assembly sequence of a given assembly can be uniquely characterized by a logical expression consisting of the conjunction of precedence relationships between the establishment of two connections. For example, for the assembly sequence presented above, the following logical expression characterizes it uniquely:

$$(c_1 < c_2) \wedge (c_2 < c_4) \wedge (c_2 \leq c_3) \wedge$$
$$(c_3 \leq c_2) \wedge (c_4 \leq c_5) \wedge (c_5 \leq c_4)$$

2.3.7 Backward Assembly Planning

The backward assembly planning, BAP, is usually implemented in the form of backward search of assembly sequences [111, 112, 113]. The main idea is that an assembly may be obtained by reversing the order of actions in the corresponding disassembly. Although this is not always applicable, this approach has the advantage that by disassembling an assembly into subassemblies, the *precedence relationships* that are satisfied are immediately known, whereas in other methods an exhaustive search may be necessary.

The backward assembly planning of an assembly A, BAP(A), recursively identifies and selects direct subassemblies $S_i^d|A$ and decomposes A into $S_i^d|A$ and A - $S_i^d|A$ until all remaining subassemblies consist of a single part. In order to select an optimal assembly plan, it is critical to identify the best way to decompose the assembly. However, as noted in [112], if local optimization methods are employed in the search space every time direct subassemblies need to be selected, then a globally optimal solution may not be reached. To overcome this burden, AO^* is used. Moreover, BAP has incorporated mechanisms to handle cases that the reverse of the disassembly sequence does not produce an assembly sequence. Finally the BAP may be easily parallelized to shorten the required assembly time.

2.3.8 The Robotic Assemblies and Intelligent Robotic Systems planning method

This is a general method that may be applied to Robotic Assemblies (RA), as well as Intelligent Robotic Systems (IRS). It has been presented in [217] and enhanced in [218, 221]. Planning for RAs is viewed similarly to planning for IRSs, modeled in terms of a hierarchical three level (organization, coordination, and execution) system based on the theory of intelligent controls [183, 184, 186]. In accordance with the mathematically proven principle of increasing intelligence with decreasing precision [189], all planning and decision making actions are performed at the highest level, the organizer. Initial knowledge, which includes a set of definitions establishing the organizer workspace and a set of operational procedures, is provided to the system to build its knowledge base and self-organization procedures. The organization level algorithm consists of five stages namely: input classification, machine reasoning, machine planning, machine decision making, machine feedback and machine memory exchange.

This is a very general opportunistic planning technique applicable to a wide range of cases that may be modeled hierarchically. The main advantage is that the learning algorithm that is incorporated allows maximum flexibility in the sense that plans may not be judged in advance, but rather they may be evaluated during training sessions.

However, this method since it is so structured, requires the specification of compatible pairs of events that might be difficult to pinpoint in cases where the number of events is large.

2.3.9 Conclusive Remarks for the AI planning methods

In this section, several different planning techniques of assembly processes have been presented. Depending on the application, one method of assembly sequences may be preferable over another. It is widely accepted that the directed graph representation is easy to understand and to implement. Thus, it may be selected when in each assembly task only one part is added. The useful feature of the AND/OR graph representation is that it encompasses all possible assembly sequences. In cases that the assembly has more than five parts less space is required for the AND/OR graph representation. The precedence relations representations are also very compact, therefore useful in applications where storage space is critical. The planning method for robotic assemblies and intelligent robotic systems is very general and suitable for opportunistic planning, but it may require significant analysis performed prior to its execution.

3 Distributed Artificial Intelligence Planning systems

Distributed Planning (DP) may be viewed as a subtopic of the Distributed Problem Solving (DPS) research area; The DPS attempt to problem solving involves a collection of decentralized and loosely coupled problem solvers each of which may reside on its own distinct processor [196]. A distributed planning system or multiagent planning consists of a network of semi-autonomous processing nodes, or agents, each cooperating with a selected group of others to achieve a common goal. The agents, based on their local perception of the network solving activities are capable of producing sub-plans and select an appropriate strategy to communicate, cooperate, or negotiate with the other agents.

In general, there are two main classes of distributed planning systems:

- Those in which a single intelligent agent formulates a plan and "distributes the solution" [44] into all participating agents that execute parts of the plan, and,
- Those in which a group of agents formulate the plan and some (or all) of them execute it.

In the former case, a planner must be able to represent and reason about the belief and goals of other agents. In the later case, it is important to study which are the appropriate communication patterns among the agents. In both cases, however, synchronization issues are of major concern. Research in distributed planning may be classified into three distinct categories:

- Coordination methods suitable for specific classes of problems,
- Theories of belief or persuasion, and,
- Synchronization issues of multiple agents that act and interact in a given environment.

Since emphasis is put on issues related to automated manufacturing systems, synchronization mechanisms and coordination patterns are elaborated in more detail than the theoretical issues of belief. In the sequel, a brief overview of these research issues is presented, suggested solutions are summarized and annotated bibliography is provided.

3.1 Coordination in multi-agent planning

One of the main issues in multiagent planning is the communication pattern being used among the agents, because this relates to the selected coordination process. Depending on the characteristics of the problem and the communication links, four major categories of configurations may be distinguished:

- The agents communicate all their information to the coordinating nodes which generate and distribute the plans [27, 39, 69, 98, 202],
- The agents perform their task semi-autonomously, then selectively communicate their sub-plans to other nodes which make them converge into a global plan in a functionally-accurate/cooperative manner [40, 119],
- The Contract Net protocol which depicts that agents are grouped into small groups within which they have extensive communication connections to permit them to negotiate and to contract out tasks in the network [46, 197]. An extension of this approach is suggested in [38], and,
- The Partial Global Planning suggested in [55, 53, 57], which is a unified framework capable of supporting the preceding three different configuration styles through the use of partial global plans; it allows the agents to coordinate their major steps for achieving partial results of their plans using asynchronous protocol.

Another major issue associated with multiple agents cooperation is the preservability of *global coherence* [198]. Global coherence of a network aims at increasing performance by diminishing idle processing times of the agents, identical or conflicting activities between two or more agents and misallocation of assignments [121, 118]. Global coherence is attainable, only if all the agents have a complete history of activities and future intentions of all other nodes; however, because of system reliability and communication limitations, this may be either impractical, or infeasible as pointed out in [58]. A presentation of different suggested architectures for the organization of such an environment may be found in [201]. As noted in [120], it may also be desired that distributed systems are allowed to have incorrect and inconsistent intermediate results provided that they end up with acceptable final results (within a range). This approach relaxes the requirement of global coherence and introduces the concept of a functionally accurate system. Functionally accurate systems are complex, may have many intermediate solutions, may derive a correct solution in many different ways. Their main advantage is the reduced amount of required intercommunication, a self correcting behavior and their inherited parallelism.

3.2 Theories of Belief

Theories or models of belief are necessary for the formulation of a global plan from an agent that will be executed in a distributed environment. A plan generator in a distributed planning environment often formulates plans based on incomplete knowledge. It usually elaborates on the initially incomplete plan based on incoming information and/or accumulated knowledge. Therefore, agents need a model of what other agents know, plan, or indent to do, to form their own plans under time constraints. AI models of belief and ascription as well as criticism about them may be found in [9, 14, 15, 91, 98, 99, 122, 148, 149, 172, 190, 194]. Two problems related to belief theories are:

- The Knowledge Precondition Problem for Actions (how does an agent reasons that it knows how to perform an action), and,
- The Knowledge Preconditions Problem for Plans (how can one say that an agent is capable of successfully executing a plan) [151].

Different approaches suggested to address these problems include [126, 149, 150, 151].

3.3 Synchronization of Multi-Agents

Research in multi-agent synchronization may borrow techniques from theories and languages developed for concurrent programming (in association with the Operating Systems and Data Bases fields). However, there is a significant difference between the multi-agent synchronization and concurrent programming [182]. In concurrent programming, there exists a number of processors and a number of processes that need to be synchronized to use these processors. In multi-agent synchronization, however, the notion of processors becomes obsolete; rather, multi-agent synchronization is action-oriented; it is of major importance to be able to represent actions, to reason about the satisfaction of their prerequisites and the order (timing) that these actions were executed.

Traditional synchronization techniques like cooperating sequential processes [50], CSP [80] and path expressions [28] have influenced multi-agent synchronization. Methods that may be used for multi-agent synchronization besides concurrent programming include Modal Logic [102], Dynamic Logic [174], approaches based on Temporal Logic [7, 8, 52, 77, 110, 125, 132, 170, 177, 182] and Procedural Logic [71].

Although these methods have not yet been applied to automated manufacturing systems, they are potentially useful, because of the similarities that exist between the two domains. For that reason, representative examples of plan synchronization/coordination methods are included in Section 5.

4 Distributed Planning Systems

4.1 Rote planning using distributed techniques

Distributed planning for finding a path from an initial state to the final destination may be applied to a variety of cases as auto-piloting and auto-driving in a dynamic environment. Several existing methods are summarized in the sequel.

4.1.1 Autodrive

Autodrive [226] is a planning system for multiple agents. It is currently operating in the simulation mode. The planning system achieves its goal by instructing the appropriate actions to take place. All data concerning road information are divided into data segments and kept in the data base. Each segment has a segment controller that decides whether the location of the automobile justifies the activation of that particular segment. Actions taken by other agents or other events may result in the update of the stored data. One of the major problem in the coherence of the data base, is the representation of different actions through the medium of time. By monitoring the changes concerning a particular data segment, segment-controllers exercise control over the information filtered to the planning system.

4.1.2 Location-Centered, Cooperative planning

Location Centered, Cooperative (LCC) planning has been suggested by Findler in [65] as a method to control air traffic in order to avoid incidents (conflicts or violations). The obvious way to avoid incidents is for some aircraft (one or more) to change flight parameters. The following four rules formalize priorities and chronological orders:

1. The aircraft with the highest *urgency factor*, U, i.e., the one involved in a incident within the shortest period of time, is selected to replan its route first,
2. An aircraft status, S, which is a sum of the emergency level and the fuel availability, is also used as a selection criterion,
3. To compensate for those aircraft heavily involved into many incidents, a measure that is the sum of incidents and the number of all other aircraft involved in those incidents, is also used as a priority rule, and,
4. The lower priority rule indicates that in a presence of tie among the other ones, the aircraft with the least number of planes surrounding it, is selected to replan its route.

After the selection of the appropriate airplane is performed, the planning algorithm has to indicate what navigation commands have to be issued (i.e. CLIMB(x), DESCEND(y), TURNRIGHT), when they have to be issued, and what parameters should be applied. Each processing node participating in the LCC mode of planning (see Figure 6) obtains information either through communication with other nodes or through radars. The *Plan Generation Unit* is responsible for the plan development. Given a restricted knowledge of the world both in spatial terms (only the aircraft in its neighboring space are taken into consideration) and in temporal terms (only a relatively small time interval is examined), the plan generator produces the initial plan. The *Look Ahead Unit* is responsible for the simulation of the current plan. It is consulted by the *Incident Detector*, which identifies and specifies any possible incidents based on the information stored in the *Knowledge Base*. If an incident is predicted, or if the updates from the *Sensing Unit* vary from the representation of the world in the Plan Generation Unit, the Plan Generation Unit modifies the current plan. The *Negotiation Unit* is responsible for the work distribution among the airplanes involved and may perform conflict resolution among them, if necessary. The *Control Unit* is responsible for the coordination and control of the above procedures.

4.1.3 Autopilot

The Autopilot planning method presented in [213], is used for the air traffic control (ATC) in a simulation mode. Every aircraft entering the specified airspace gets a clone (an identical copy) of the Autopilot function that performs all necessary planning for its navigation. In that sense, Autopilot is not a truly distributed method, it is better characterized as an object-centered planning technique. However, in the same article

598

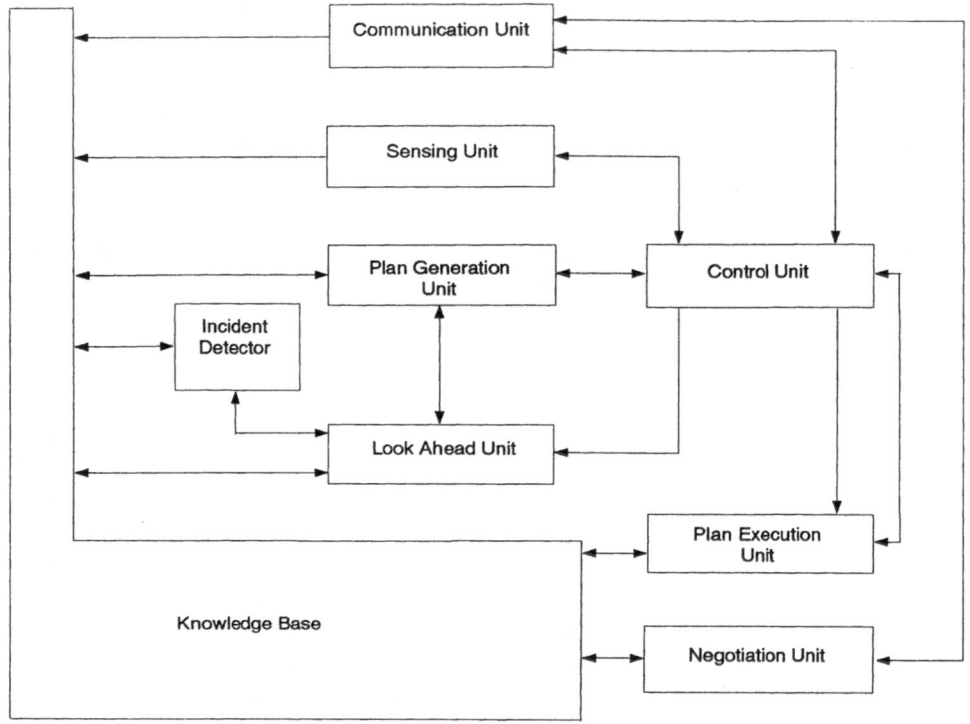

Figure 6. Design of a processing unit in LCC architecture (taken from [65]).

distributed planning extensions to Autopilot are also discussed. The structure of the Autopilot planner is shown in Figure 7.

The *World Model* contains information about aircraft locations, flight plans, destinations, etc. Location updates are performed by the *Sensor* unit. The *Plan Generator* produces candidate plans. The *Evaluator* checks for possible conflicts among the newly released plan and other aircraft plans. If conflicts are detected, then the Plan Generator refines the initial plan. Possible reactions include timing actions, delaying and course alternation. The *Communicator* may either request necessary data, or request from other aircraft to modify their routes.

The plan is represented as a schema that has various slots that need to be filled. Planning generation is incremental. Parameters may be determined during plan generation, plan evaluation, or even plan refining. Distributed planning architectures are

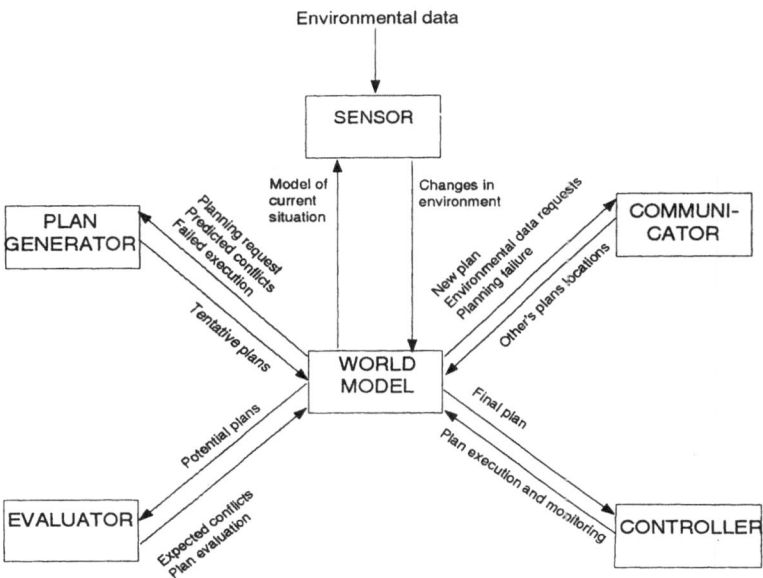

Figure 7. Internal Design of Autopilot (taken from [213]).

studied as alternatives, or extensions in the object-centered architecture. Four different schemes are presented in the sequel:

- The *Object-Centered Autonomous, No Communication* schema assumes the intentions/plans of other aircraft based on their current altitudes, bearings and nearest exit fixes, or airports along their current flight paths. This schema eliminates the communication requirements, but it is associated with an inherited uncertainty on other aircraft plans. For that reason, the sensor unit has to constantly monitor and detect changes in the presumed model. If conflicts are detected, each planner attempts to patch its current plan. The existence of global rules of road, precedence rules ensures the effective cooperation of the planning systems.

- The *Object-Centered Autonomous, Limited Communication* schema requires posting of plans and route maps from all aircraft; thus, uncertainty in plans of other planes is avoided on the expense of communication requirements. Each plane entering the airspace has to plot a conflict-free plan with respect to other aircraft already on flight. It is worth noting that the plan generation may still be incremental. However, when a conflict-free or the best available plan is confirmed, it is posted as final and the Controller monitors execution of the plan.

- The *Object Centered Hierarchically Cooperative* method is effectively the reverse of the strategy described above. Each plane entering the airspace determines its optimum path and suggests to the aircraft with projected conflicts to modify their plans, one at a time. Each modified plan is posted so that update attempts from other planes will be valid. In case that another aircraft is unable to find a successful patch in its plan, the initial aircraft abandons its best plan and proceeds with the second best.
- The *Object Centered Asynchronously Cooperative* schema is a decentralized attempt to find a route for aircraft entering the airspace. Instead of proceeding sequentially as above, from the most favorite plan to the ones lower in the hierarchy, a plane broadcast its set of potential plans to all aircraft that come in conflict with it. Partial solutions are found on other planes planning systems and sent along with the set of assumptions under which the solution is generated. These partial solutions are maintained in the initial aircraft planning system. A plan is established, if all conflicts from a path have been eliminated through the proposed patches and the proposed refined plans do not conflict with each other.

These different proposed schemes may be selected depending on the circumstances of a particular case: when the communication cost is higher than the cost of locally generating plans, or it is fairly easy to generate an effective plan, then the former two methods should be preferred. If the time required for the complete plan generation is critical, then the asynchronous cooperative model method should be selected.

4.2 Distributed NOAH

Sacerdoti's NOAH method suggests a hierarchical approach to planning that uses successive refinement of series of actions to achieve the specified goal. Its structure makes it suitable for parallelism [39]. The expansion of every abstract action may be assigned to a different processing unit. The reduction in the search space, which is achieved by the coupling of non-linear and hierarchical planning, implies less inter-node communication in the network. Finally, the generated plans are formulated in a way that is directly applicable towards parallel execution.

To illustrate how the distributed version of NOAH works, an example is presented in the sequel. The initial state and the desired final state of the problem are shown in Figure 8.

In this example, two processing units are needed. Both of them are given the initial state. However only the first one, P1, is provided with the goal state. P1 assigns the second part of the goal to a different processing unit, P2. The way the nodes interact and expand their actions is shown in Figures 9 10 and 11.

Figure 8. Initial and Goal state for the Distributed NOAH example.

5 Distributed Planning Synchronization Examples

5.1 CSP Influenced Synchronization Method

The multi-agent plan synchronizer [206, 207] provides a formal framework for plan synchronization among different agents heavily influenced by the parallel programming language CSP [80].

A formal theory of actions, environment, agents and plans is fully described in [207]. Theoretical models of multiagent activity may be also found in [70, 105, 211, 212]. For the purposes of this paper, it is sufficient to provide the following definitions.

An event is an indivisible, discrete transformation of the world, possibly associated with some preconditions that must hold at the time it is executed.

An action is a set of possible finite sequences of events that has a beginning and an end. Multiple agents operating in parallel may execute actions simultaneously.

The state of the environment in which actions are executed consists of a world state and a set of actions currently being executed.

Agents interact with the environment using operators. Let A denote the set of operators, then strings over the alphabet $A \times \{begin, end\}$ are interchanged between agent and the environment.

An agent is defined over a set of nodes (agent states) and a set of arcs defining allowed state transitions with associated messages.

Plans are defined recursively. Let A denote the set of operators, M denote the set of memory states and S denote the set of signals. Then:

- $\forall \alpha \in A :$ α is a plan for executing a single action,
- $\forall m \in M, s \in S :$ $(set\,m)$, $(set\,s)$, $(guard\,m\,s)$ are synchronizing primitives, and,
- If p_1 and p_2 are two plans, then $p_1; p_2$ is the plan to execute them in sequence, $p_1 \| p_2$ is to execute them in parallel, $p_1 \mid p_2$ is to execute one of the two in a non-deterministic way and p_1^* executes p_1 arbitrary number of times.

602

　　　　　　　　　　　　PLANNING REFINEMENT PROCESS

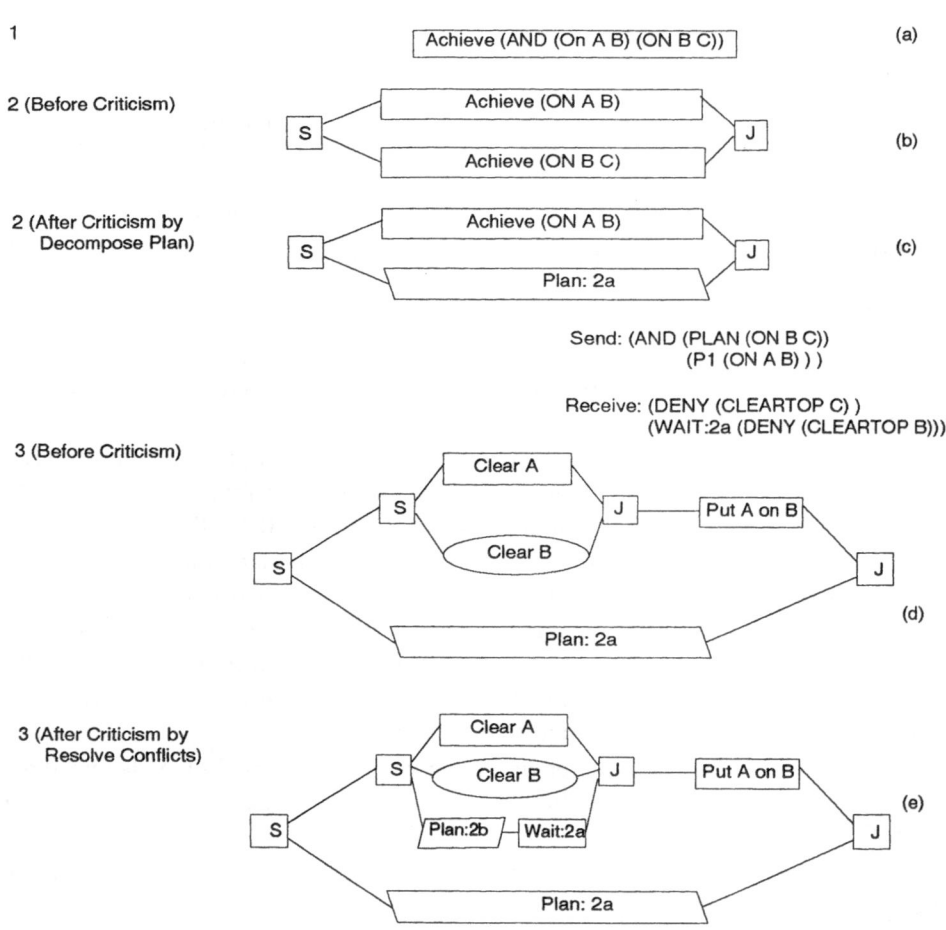

Figure 9. Distributed NOAH: P1 planning processing (taken from [39]).

To illustrate the use of the synchronization primitives consider the following example. Three robot arms R1, R2, R3 are used to transfer three objects A, B, C respectively from the initial state to the goal state as shown in Figure 12.

The following program controls the correct execution of the synchronized parallel plan [206]:

LEVELS   PLANNING REFINEMENT PROCESS

4 (Before Criticism)

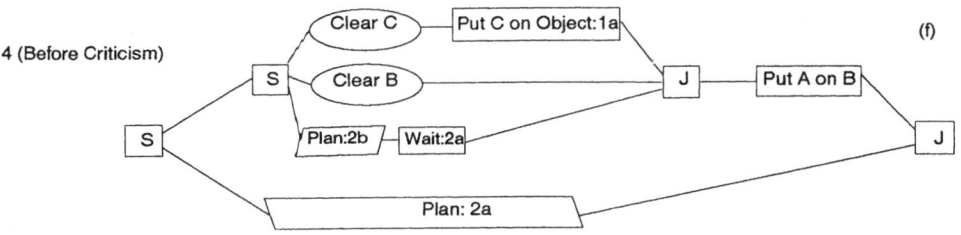

(f)

4 (After Criticism by
    Resolve Conflicts)

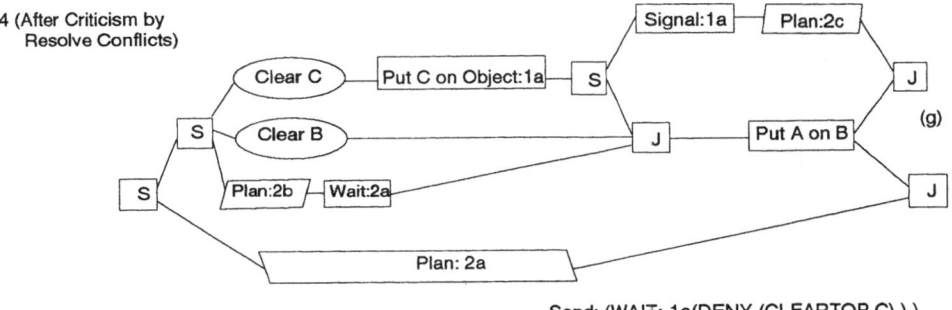

(g)

Send: (WAIT: 1a(DENY (CLEARTOP C)))

KEY

GOAL:

PHANTOM:

MODEL:

Figure 10. Distributed NOAH: P1 planning processing (continued).

```
((PARALLEL
 ((SEND (BEGIN 1))
  (START (R1 R2 R3) (A X) (B Y) (C Z)))
  (PARALLEL
   ((PICKUP R1 A X)(SEND 2 1 1))
    (SEND (BEGIN 2 1 2))(PUTDOWN R1 A Y))
   ((PICKUP R2 B Y)(SEND (END 2 2 1))
```

LEVELS   PLANNING REFINEMENT PROCESS

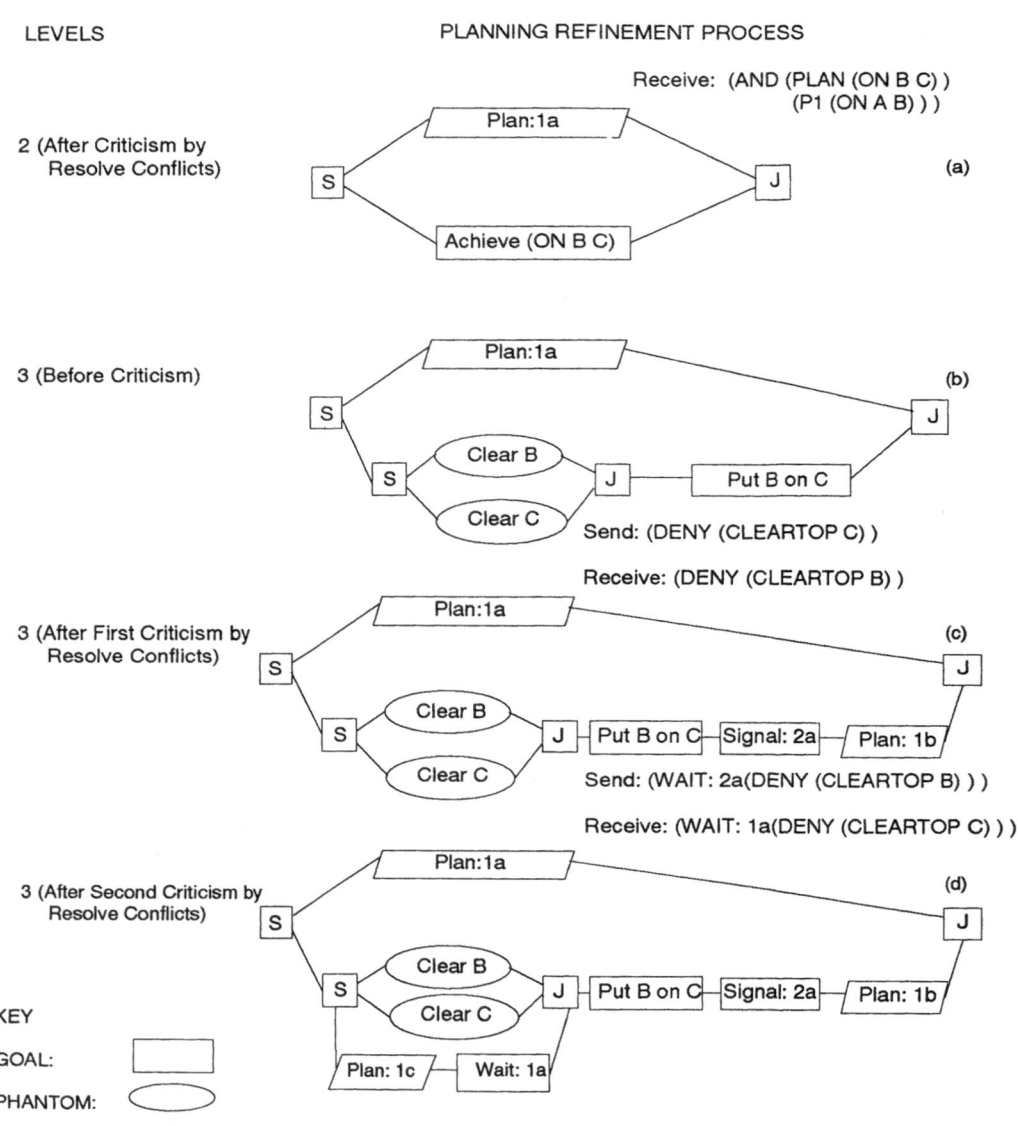

Figure 11. Distributed NOAH: P2 Planning Processing.

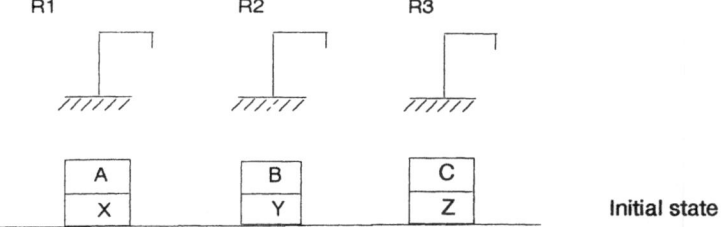

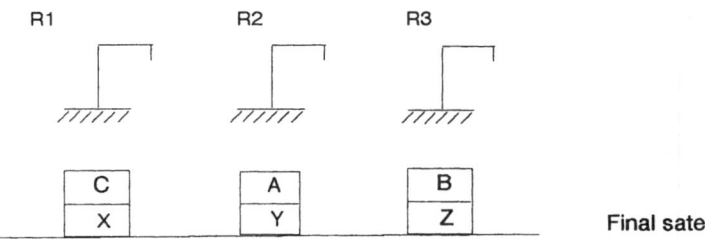

Figure 12. The initial and the goal state of multi-agent plan synchronizer example (taken from [206]).

```
(SEND (BEGIN 2 2 2)(PUTDOWN R2 B Z))
((PICKUP R3 C Z) (SEND (END 2 3 1))
 (SEND (BEGIN 2 3 2)) (PUTDOWN R3 C X))))


((RECV (BEGIN 1)
 (SETQ N 2)
 (WHILE (NOT(EQ N 13))
   (SELECT-ONE-OF
     (IF (AND (EQ N 2)(RECV (END 2 3 1)))
         THEN (SETQ N 5))
     (IF (AND (EQ N 2)(RECV (END 2 2 1)))
         THEN (SETQ N 4))
. . .
     (IF (AND (EQ N 5)(RECV (END 2 1 1)))
         THEN (SETQ N 6))))))))
```

5.2 Partial Plan Synchronization

A framework that uses partial global plans to promote different styles of agent cooperations has been suggested in [53, 55, 57]. The style the agents should cooperate depends on many different factors. However, there might be cases that call for several styles simultaneously and this is supported by the partial global plans framework.

In this model, three different abstractions of plans, depending on the number of agents that are collaborating to generate a plan, are distinguished:

- A Local Plan is the representation of a plan that is maintained by the agent that is pursuing the plan. The plan's objective, the order of major plan steps, expected time required for the plan's completion and detailed list of primitive actions involved in the accomplishment of the plan, are recorded at this level,

- A Node Plan is the representation of a plan that agents use in their intercommunication. It contains the same information as the local plan with the exception of the list of primitive actions. Instead, it contains the long-term order of the planned activities and an estimate of how long each activity will take. The planner uses this information to generate the node plan's *activity-map*, a sequence of activities where each activity has a predicted starting and ending time and result track, and,

- A Partial Global Plan, PGP, is the representation of how nodes are working towards the accomplishment of a plan at a higher level of abstraction. It contains information about that goal, the major concurrent planning activities and intentions of the nodes and how partial solutions formed by agents may be integrated to accomplish the goal at the higher level. PGP formation is highly dynamic; an agent's PGP is the same as its local plans, initially; however as information flows from other agents, it involves. A *PGP's activity map* contains information about interleaved activities, their estimated start and finish time and what task or part of task it is working on. This information may be used by the planner to produce a more efficient (but not necessarily optimal) reordering.

The planner uses the activity map to construct the solution-construction-graph which facilitates communication decisions [58]. A more detailed view of the agents planning activity is presented in Figure 13.

5.3 Logic Based Plan Synchronization

Rosenschein [179], has proposed another method for coordinating actions of multiple-agents based on the belief-representation suggested in [100].

Every agent keeps a FACT-list that contains its beliefs about the world and plans of other agents. Moreover, every agent maintains a GOAL-list that contains its current goals, too. The predicate HASCAP(agent, operator) denotes the capability of

1. receive network information;

2. find the next problem solving action using network model:
   (a) update local abstract view with new data;

   (b) update network problem, including PGPs, using changed local and received information (factoring in credibility based on source of information);

   (c) map through the PGPs whose local plans are active, for each:
       i.   construct the activity-map, considering other PGPs;
       ii. find the best reordered activity-map for the PGP;
       iii. if permitted, update the PGP and its solution-construction-graph;
       iv. update the affected node-plans

   (d) find the current-PGP (this node's current activity);

   (e) find next action for node based on local plan of current-PGP;

   (f) if no text action (local plan inactive) then go to 2b (since local plans may have changed), else schedule the next action;

3. transmit any new and modified network information

Figure 13. Planning Activities of an Agent (taken from [55]).

the agent to execute the operator, while the predicate WILL-PERFORM(agent, operator) represents that the agent will execute the operator. The following two axioms appear in agents knowledge bases:

$$HASCAP(agent, operator) \wedge GOAL(agent, operator) \supset$$
$$WILL - PERFORM(agent, operator)$$
$$FACT(x, {}'HASCAP(agent, operator) \wedge GOAL(agent, operator) \supset$$
$$WILL - PERFORM(agent, operator)')$$

Four primitives are supplied for communication purposes: REQUEST, CAUSE-TO-WANT, INFORM and CONVINCE with the following interpretations:

REQUEST(x, y, act): x requests y to adopt act as a goal,

CAUSE-TO-WANT(x, y, act): x causes y to adopt act as a goal,

INFORM(x, y, prop): x informs y of prop, and,

CONVINCE(x, y, prop): x convinces y to believe prop.

The following two axioms specify conditions that depict the acceptance of a proposition from the receiver:

$$MASTER(x, y) \supset ACCEPT(x, y, act)$$
$$MASTER(x, y) \supset BE - SWAYED(x, y, prop)$$

Finally the operators WHEN-GET and PAUSE may be used by agents to declare when(until when) the agent will adopt a specified goal:

PAUSE(agent, precondition, aim): agent decides to wait until precondition is satisfied before adopting aim.

WHEN-GET(agent, precondition, aim): agent adopts aim when he knows that precondition is satisfied.

6 Application of Learning to Planning

As it has been noted repeatedly [24, 101, 178], machine learning is an essential element of any intelligent system. It is interesting to examine how machine learning can be incorporated to improve planning systems. Improvements may include enlargement of the planning system data base by acquiring new facts or rules, or reduction of the data base by eliminating irrelevant data, or acquiring new problem solving methods and strategies, or even improving currently existing techniques or strategies in terms of speed and accuracy.

Although many definitions of learning have been proposed [117, 137, 195], in this paper, machine learning is defined according to the definition given in [30], as *"the computational methods with which intelligent systems acquire new knowledge, new skills and new ways to organize existing knowledge"*. Knowledge acquisition is a conscious process that allows the agent to expand common sense, theoretical, methodological, and technical knowledge. Skill improvement involves a gradual refinement of intellectual or manual skills. Conceptual learning is a subclass of machine learning [138] and it may be distinguished into cognitive modeling (concerned with the modeling of concept learning in live beings) and the engineering approach (oriented towards the exploration of all existing learning mechanisms). Several basic concept learning methods have been identified, however, classification of the learning techniques may be performed in many different ways [96, 139, 140]. In the sequel, a classification of learning methods is presented in accordance with the study given in [95].

Some of the existing learning methods may be used in conjunction with planning techniques to form a more efficient planning system. Learning can contribute to i) knowledge acquisition for creating and executing plans, ii) acquisition of plan refinement knowledge, and, iii) detection of the plan failure and learning from this event [61]. The existing learning methods that may be used for planning enhancement are specially marked (double border) in the included classification.

Learning methods that may be used for planning enhancement include:

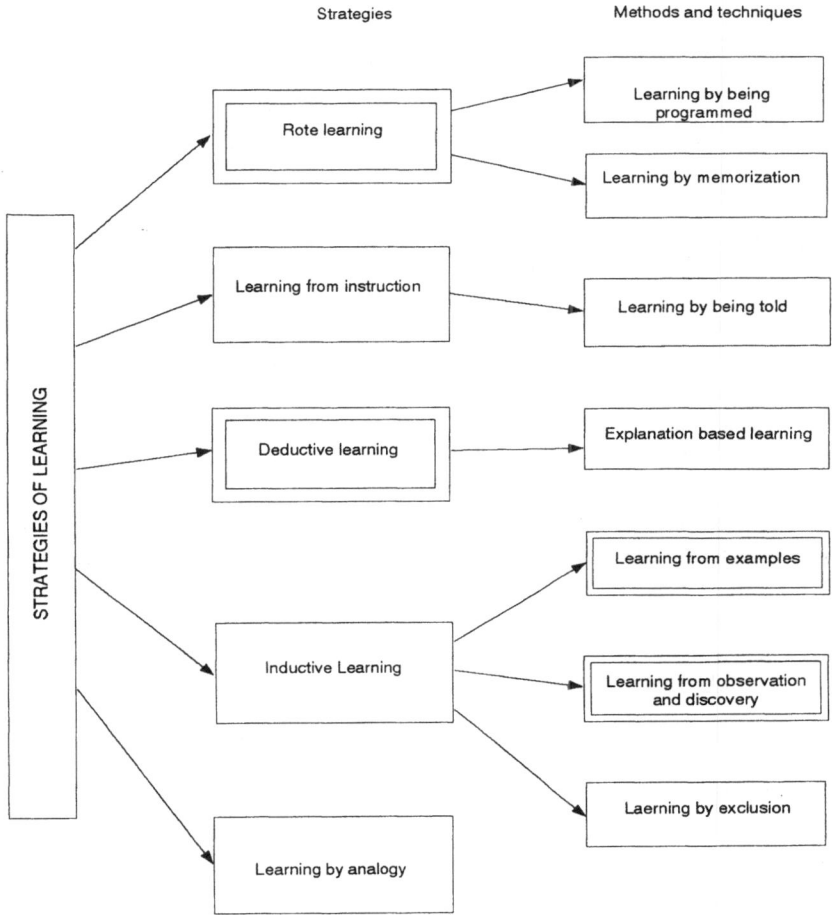

Strategies   Methods and techniques

Double Bordered Rectangulars denote learning methods/strategies that may be used with planning systems also.

Figure 14. Classification of learning methods according to the type of inference (from [95]).

- Rote Learning is a very primitive method of learning; knowledge considered to be useful for the specific application is provided at the performance task level, it is accepted and used whenever needed. Rote learning may be viewed as learning by being programmed and as learning by memorization. An example of rote learning is Samuel's checkers player [36].

- Deductive Learning that involves knowledge reformulation and transformation, knowledge compilation, creation of macro operators, etc [95]. Explanation-based learning, which is a deductive learning strategy, may be applied to planning and decision making process. In order to better illustrate the way explanation-based learning works an example from [61] is provided. Examples of systems that use deductive learning methods include SOAR [106, 107], LEX [146] and PET [173].

- Two of the inductive learning techniques are: learning from examples and learning from observation. The main difference between these two techniques is the way the input is provided to the system.

  In learning from examples, a set of examples and counter examples relative to the concept is provided to the system and the system induces a general concept description. In this category, one may classify the following systems: decision tree induction [29], learning with genetic algorithms [81] similarity-based learning [145, 139], CLUSTER/2 [139], ID3 [175] and the classification approach for IC manufacturing control [229].

  In learning from observation (unsupervised learning) the input is in the form of facts that are classified and organized by the learning system itself. Learning of observation may further classified as passive observation and active experimentation. Application systems that use learning from observation include AM [114], EURISKO [115], BACON [108], STAHL [238], GLAUBER [109] and IDS [158].

  Planning systems with embodied learning methods include STRIPS [63], HACKER [208], PRODIGY [144], GERRY [237] and CHEF [78]. A systematic approach is given in [143]

7 Conclusions

An assortment of planning techniques both from the traditional and distributed AI point of view, has been presented. From the review of the current literature it is worth noting that there is a lack of general or generic planning tools that may be customized for specialized areas. Most planning methods that have been developed, depend heavily on the particular domain. As a consequence, it is difficult to compare these planning methods in detail. It might be of interest to examine whether the expert systems successful paradigm can be repeated with respect to planning systems. It is also interesting to note that there is not adequate mathematical formulization in the area, although there exist several proposed methods.

Section # of planning techniques	R	TME	RME	D	F	L
2.1.1	F	F	N/A	F	F	F
2.1.2	T	F	N/A	F	F	T
2.1.3	T	F	N/A	F	F	F
2.1.4	T	T	N/A	F	F	F
2.2.1	T	F	N/A	F	T	T
2.2.2	T	F	N/A	F	T	F
2.2.3	T	F	N/A	F	T	F
2.2.4	T	F	N/A	F	T	T
2.3.7	T	T	N/A	F	F	F
2.3.8	T	T	N/A	F	F	F
4.1.1	T	T	T	T	F	F
4.1.2	T	T	T	T	T	F
4.1.3	T	T	T	T	T	F
4.2	T	T	T	T	F	F

Figure 15. Evaluation of the examined planning methods.

As it was noted in the introduction of this chapter, the planning methods were examined with respect to their potential contribution to an automated manufacturing system environment. Summarizing, it is of our interest to evaluate a planning technique based on robustness (R), time management effectiveness (TME), resource management effectiveness (RME), distributed nature (D), flexibility (F), and, learning capabilities (L). The following table in Figure 15 provides a general evaluation for the methods presented based on the criteria stated above.

References

[1]   Acar, L., and Ozguner, U., "Design of Knowledge-Rich Hierarchical Controllers for Large Functional Systems", *IEEE Transactions on Systems, Man and Cybernetics, Vol. 20*, No. 4, 1990, 791–803.

[2]   Agre, P. E., and Chapman, D., "Pengi: An Implementation of a Theory of Activity", In *Proceedings of the National Conference on AI* (Seattle, WA, 1987), pp. 268–272.

[3]   Albus, J. S., "Theory and Practice of Intelligent Control", In *Proceedings of the 23rd IEEE COMPCON* (1981), pp. 19–39.

[4]   Albus, J. S., "Outline for a Theory of Intelligence", *IEEE Transactions on Systems, Man and Cybernetics, Vol. 21*, No. 3, 1991.

[5]   Albus, J. S., "A Reference Model Architecture for Intelligent Systems Design", In *An Introduction to Intelligent and Autonomous Control*, Kluwer Academic Publishers, 1993.

[6]   Albus, J. S., McLean, C., Barbera, A. J., and Fitzerald, M. L., "An Architecture for Real-Time Sensory-Interactive Control of Robots in a Manufacturing Environment", In *4th AC/FIP Symposium on Information Control Problems in Manufacturing Technology* (Gaithesburg, MD, 1982).

[7]   Allen, J. F., "An Internal Based Representation of Temporal Knowledge", In *Proceedings Seventh International Joint Conference in Artificial Intelligence* (1981), pp. 221–226.

[8]   Allen, J. F., "Towards a General Theory of Action and Time", *Artificial Intelligence, Vol. 23*, No. 2, 1984, 123–154.

[9]   Allen, J. F., "Recognizing Intentions from Natural Language Utterances", In *Computational Models of Discourse*, M. Brady and R. C. Berwick, Eds., MIT Press, Cambridge, MA, 1993.

[10]   Ammons, J. C., Lofren, C. B., and McGinnis, L. F., "A Large Scale Workstation Loading Problem", *Annuals of Operations Research, Vol. 3*, 1985, 319–332.

[11]   Antsaklis, P. J., M., P. K., and J., W. S., "Towards Intelligent Autonomous Control Systems: Architecture and Fundamental Issues", *Journal of Intelligent and Robotic Systems, Vol. 1*, No. 4, 1989, 315–342.

[12]   Antsaklis, P. J., M., P. K., and J., W. S., "An Introduction to Autonomous Control Systems", *IEEE Control Systems Magazine, Vol. 11*, No. 4, 1991, 5–13.

[13]   Antsaklis, P. J., and Passino, K. M., "Introduction to Intelligent Control Systems with High Degrees of Autonomy", In *An Introduction to Intelligent and Autonomous Control*, Kluwer Academic Publishers, 1993.

[14] Appelt, D. E., "A Planner for Reasoning about Knowledge and Action", In *Proceedings of the First Annual Conference of the American Association for Artificial Intelligence* (Stanford, CA, 1980), pp. 131–133.

[15] Appelt, D. E., "Planning Natural Language Utterances to Satisfy Multiple Goals", PhD thesis, Stanford Universityly, 1981.

[16] Astrom, K. J., "Expert Control", *Automatica, Vol. 22*, No. 3, 1986, 277–286.

[17] Barber, S., Harbison-Briggs, K., Mitchell, R., and Tiernan, J. C. M., "Symbolic Representation and Planning for Robot Control Systems in Manufacturing", In *AI Applications in Manufacturing*, A. Famili, D. S. Nau, and S. H. Kim, Eds., AAAI Press, 1992.

[18] Barbera, A. J., Albus, M. L., and Fitzerald, M. L., "Hierarchical Control of Robots Using Microcomputers", In *Proceedings of the 9th International Symposium on Industrial Robots* (Washington, DC, 1979).

[19] Bastos, J. M., "Batching and Routing: Two Functions in the Operational Planning of Flexible Manufacturing Systems", *European Journal of Operational Research, Vol. 33*, 1988, 230–244.

[20] Berrada, M., and Stecke, K. E., "A Branch and Bound Approach for Machine Loading in Flexible Manufacturing Systems", *Management Science, Vol. 32*, 1986, 1316–1335.

[21] Bertolotti, E., "Interactive Problem Solving for Production Planning", In *AI Applications in Manufacturing*, A. Famili, D. S. Nau, and S. H. Kim, Eds., AAAI Press, 1992.

[22] Bose, P., "An Abstraction-Based Search and Learning Approach for Effective Scheduling", In *AI Applications in Manufacturing*, A. Famili, D. S. Nau, and S. H. Kim, Eds., AAAI Press, 1992.

[23] Bourjault, A., *"Contribution a une approche methodologique de l' assemblage automatise, elaboration automatique des sequences operatoires"*, These d'etat, Universite de France-Comte, Besancon, France, 1984.

[24] Bratko, I., "AI Tools and Techniques for Manufacturing Systems", *Journal of Robotics and Computer Integrated Manufacturing, Vol. 4*, No. 1/2, 1988, 27–31.

[25] Brooks, R. A., *"Planning is just a way of avoiding figuring out what to do next"*, Working Paper 303, MIT Artificial Intelligence Laboratory, Cambridge, MA, 1987.

[26] Butler, J., and Otsubo, H., "ADDYMS: Architecture for Distributed Dynamic Manufacturing Scheduling", In *AI Applications in Manufacturing*, A. Famili, D. S. Nau, and S. H. Kim, Eds., AAAI Press, 1992.

[27] Cammarata, D., McArthur, D., and Steeb, R., "Strategies of Cooperation in Distributed Problem Solving", In *Proceedings of the Eighth International Joint Conference in Artificial Intelligence* (1983), pp. 767–770.

[28] Campbell, R. H., and Habermann, A. N., "The Specification of Process Synchronization by Path Expressions", In *Lecture Notes in Computer Science, vol. 16*, Springer Verlag, 1974.

[29] Carbonell, J. G., "Introduction Paradigms for Machine Learning", *Artificial Intelligence, Vol. 40*, No. (1-3), 1989, 1–9.

[30] Carbonell, J. G., and Langley, P., "Machine Learning", In *Encyclopedia of Artificial Intelligence, Vol. 1*, John Wiley, 1987, pp. 464–488.

[31] Carbonell, J. G., Michalski, R. S., and Mitchell, T. M., "An Overview of Machine Learning", In *Machine Learning: An artificial Intelligence Approach, vol I*, J. G. Carbonell, R. S. Michalski, and T. M. Mitchell, Eds., Morgan Kaufmann, 1983.

[32] Chang, T. C., "TIPPS: a Totally Intergrated Process Planning System", PhD thesis, Virginia Polytechnic Institute, 1982.

[33] Chang, Y. L., Sullivan, R. S., Bagchi, U., and Wilson, J. R., "Experimantal Investigation of Real-Time Scheduling in Flexible Manufacturing Systems", *Annuals of operations Resource, Vol. 3*, 1985, 355–378.

[34] Chapman, D., *"Planning for Conjuctive Goals"*, Technical Report 802, MIT Artificial Intelligence Laboratory, Cambridge, MA, 1985.

[35] Coad, P., and Yourdon, E., *"Object-Oriented Analysis"*, Yourdon Press, Englewood Cliffs, NJ, 1991.

[36] Cohen, P. R., and Feigenbaum, E. A., Eds., *"The Handbook of Artificial Intelligence Volume III"*, Pitman, 1982.

[37] Computer, I., "Special Issue on Autonomous Intelligent Machines", *IEEE Computer, Vol. 22*, No. 6, 1989.

[38] Conry, S. E., Meyer, R. A., and Lesser, V., "Multistage Negotiations in Distributed Planning", In *Readings in Distributed Artificial Intelligence*, A. H. Bond and L. Gasser, Eds., Morgan Kaufmann, San Mateo, CA, 1988.

[39] Corkill, D. D., "Hierarchical Plannning in a Distributed Environment", In *Proceedings Sixth International Joint Conference in Artificial Intelligence* (1979), pp. 168–175.

[40] Corkill, D. D., "A Framework for Organizational Self-Design in Distributed Problem Solving Networks.", PhD thesis, University of Massachusetts, Amherst, MA, 1983.

[41] Corkill, D. D., and Lesser, V. R., "A Goal-Directed HearsayII Architecture, Unifying Data-Directed and Goal-Directed Control", In *Proceedings National Conference on Artificial Intelligence* (Pittsburg, PA, 1982).

[42] Darbyshire, I., and Davies, B. J., "EXCAP: An Expert Generative Process Planning System", In *Proceedings of the IFIP WG 5.2 Working Conference on Knowledge Engineering in Computer Aided Design*, Amsterdam: North-Holland, 1985, pp. 291–303.

[43] Dasgupta, S., *"Testing the Hypothesis Law of Design: The Case of Britannia Bridge"*, In Press, 1994.

[44] Davis, R., *"A Model for Planning in a Multi-Agent Environment, Steps Towards Principles of Teamwork"*, AI Working Paper, MIT, Cambridge, MA, 1981.

[45] Davis, R., and Smith, R. G., *"Negotiation as a Metaphor for Distributed Problem Solving"*, Artificial Intelligence Laboratory Memo No. 624, MIT, Cambridge, MA, 1981.

[46] Davis, R., and Smith, R. G., "Negotiation as a Metaphor for Distributed Problem Solving", *Artificial Intelligence, Vol. 20*, 1983, 63–109.

[47] Davis, W. J., and Jones, A. T., "A Real-Time Production Scheduler for a Stochastic Manufacturing Environment", *International Journal of Computer Integrated Manufacturing, Vol. 1*, 1988, 101–112.

[48] De Fazio, T. L., and Whitney, D. E., "Simplified Generation of all Mechanical Assembly Sequences", *IEEE Journal Robotics and Automation, Vol. RA-3*, No. 6, 1987, 640–658.

[49] De Fazio, T. L., and Whitney, D. E., "Corrections on Simplified Generation of all Mechanical Assembly Sequences", *IEEE Journal Robotics and Automation, Vol. RA-4*, No. 6, 1988, 705–708.

[50] Dijkstra, E. W., "Cooperating Sequential Processes", In *Programming Languages*, F. Genuys, Ed., Academic Press, New York, 1968.

[51] Dougherty, E. R., and Giardina, C. R., *"Mathematical Methods for Artificial Intelligence and Autonomous Systems"*, Prentice Hall, 1988.

[52] Drummond, M. E., "A Representation of Action and Belief for Automatic Planning Systems", In *Reasoning About Actions and Plans, Proceedings of 1986 Workshop at Timberline, Oregon*, M. P. Georgeff and A. L. Lansky, Eds., Morgan Kaufman, Los Altos, CA, 1987, pp. 189–211.

[53] Durfee, E. H., *"Coordination of Distributed Problem Solvers"*, Kluwer Academic, Boston, MA, 1988.

[54] Durfee, E. H., and Lesser, V. R., "Coordination through Communication in Distributed Problem Solving Network", In *Distributed Artificial Intelligence*, M. N. Huhns, Ed., Pitman, London, UK, 1987, pp. 29–58.

[55] Durfee, E. H., and Lesser, V. R., "Using Partial Global Plans to Coordinate Distributed Problem Solvers", In *Proceedings Tenth International Joint Conference Artificial Intelligence* (1987), pp. 875–883.

[56] Durfee, E. H., and Lesser, V. R., "The Distributed Artificial Intelligence Melting Pet", *IEEE Transactions on System, Man and Cybernetics, Vol. 21*, No. 6, 1991, 1301–1306.

[57] Durfee, E. H., and Lesser, V. R., "Partial Global Planning, A Coordination Framework for Distributed Hypothesis Formation", *IEEE Transactions on System, Man and Cybernetics, Vol. 21*, No. 5, 1991, 1167–1183.

[58] Durfee, E. H., Lesser, V. R., and Corkill, D. D., "Increased Coherence in a Distributed Problem Solving Network", In *Proceedings Nineth International Joint Conference Artificial Intelligence* (1985), pp. 1025–1030.

[59] Egilmez, K., and Kim, S. H., "Teamwork among Intelligent Agents: Framework and Case Study in Robotic Service", In *AI Applications in Manufacturing*, A. Famili, D. S. Nau, and S. H. Kim, Eds., AAAI Press, 1992.

[60] Erman, L. D., Hayes-Roth, F., Lesser, V. R., and Reddy, D. R., "The HearsayIII Speech Understanding System: Integrating Knowledge to Resolve Uncertainty", *Computing Surveys, Vol. 12*, 1980, 213–253.

[61] Famili, A., and Turney, P., "Application of Machine Learning to Industrial Planning and Decision Making", In *AI Applications in Manufacturing*, A. Famili, D. S. Nau, and S. H. Kim, Eds., AAAI Press, 1992.

[62] Fikes, R. E., Hart, P. E., and Nilsson, N. J., "Learning and Executing Generalized Robot Plans", *Artificial Intelligence, Vol. 3*, 1972, 251–288.

[63] Fikes, R. E., and Nilsson, N. J., *Artificial Intelligence, Vol. 2*, 1971, 198–208.

[64] Findeisen, W., *"Control and Coordination in Hierarchical Systems"*, Wiley, New York, 1980.

[65] Findler, N. V., and Lo, R., "An Examination of Distributed Planning in the World of Air Traffic Control", *Journal of Parallel and Distributed Computing, Vol. 3*, 1986, 411–431.

[66] Firschein, O., *"Artificial Intelligence for Space Station Automation"*, Noyes, NJ, 1986.

[67] Fox, B. R., and Kempf, K. G., "Opportunistic Scheduling for Robotics Assembly", In *Proceedings IEEE International Conference in Robotics and Automation* (1985), pp. 880–889.

[68] Freedman, R. S., and Sylvester, W. A., "OPGEN: The Evolution of an Expert System for Process Planning", *AI Magazine, Vol. 7*, No. 5, 1986, 58–70.

[69] Georgeff, M., "Communication and Interaction in Multiagent Planning", In *Proceedings Eighth International Joint Conference in Artificial Intelligence* (1983), pp. 125–129.

[70] Georgeff, M. P., "Actions, Processes and Causality Reasoning About Actions and Plans", Morgan Kaufmann, 1986.

[71] Georgeff, M. P., Lansky, A. L., and Bessiere, P., "Procedural Logic", In *Proceedings of the Nineth International Joint Conference in Artificial Intelligence* (1985), pp. 516–523.

[72] Gevarter, W. B., *"Artificial Intelligence"*, Noyes, NJ, 1992.

[73] Ginsberg, M. L., *"Universal Planning: An Almost Universally Bad Idea"*, Technical Report, Department of Computer Science, Stanford University, Stanford CA, 1989.

[74] Glorioso, R. M., and Colon Osorio, F., *"Engineering Intelligent Systems"*, Digital Press, 1980.

[75] Green, C., "Theorem-Proving by Resolution as a Basis for Question-Answering Systems", In *Machine Intelligence, vol. 4*, American Elsevier, New York, NY, 1969.

[76] Hadavi, K., Hsu, W. L., Chen, T., and Lee, C. N., "An Architecture for Real Time Scheduling", In *AI Applications in Manufacturing*, A. Famili, D. S. Nau, and S. H. Kim, Eds., AAAI Press, 1992.

[77] Halpern, J., Manna, Z., and Moszkowski, B., "A Hardware Semantics Based on Temporal Intervals", In *Proceedings Nineteenth ICALP*, Springer Lecture Notes in Computer Science, Vol. 54, 1983, pp. 278–292.

[78] Hammond, K. J., "CHEF: A Model of Case-Based Planning", In *Proceedings Fifth National Conference on Artificial Intelligence* (1986), pp. 267–271.

[79] Hayes-Roth, B., "A Blackboard Architecture for Control", *Artificial Intelligence, Vol. 26*, 1985, 251–321.

[80] Hoare, C., "Communicating Sequencial Processes", *Communications of the ACM, Vol. 21*, No. 8, 1978, 666–677.

[81] Holland, J. H., "Escaping Brittleness: The Possibilities of General-Purpose Learning Algorithms Applied to Parallel Rule-Based Systems.", In *Machine Learning*, R. S. Michalski, J. G. Carbonell, and T. M. Mitchell, Eds., Morgan Kaufmann, 1986.

[82] Homem de Mello, L. S., and Sanderson, A. C., "Planning Repair Sequences Using the AND/OR Graph Representation of Assembly Plans", In *Proceedings IEEE International Conference on Robotics and Automation* (1988), pp. 1861–1862.

[83] Homem de Mello, L. S., and Sanderson, A. C., "Task Sequence Planning for Assembly", In *Proceedings Twelfth World Congress Scientific Computation* (1988), pp. 390–392.

[84] Homem de Mello, L. S., and Sanderson, A. C., "AND/OR Graph Representation of Assembly Plans", *IEEE Transactions on Robotics and Automation, Vol. 6*, No. 2, 1990, 188–199.

[85] Homem de Mello, L. S., and Sanderson, A. C., "A Correct and Complete Algorithm for the Generation of Mechanical Assemply Sequences", *IEEE Transactions on Robotics and Automation, Vol. 7*, No. 2, 1991, 228–240.

[86] Homem de Mello, L. S., and Sanderson, A. C., "Representations of Mechanical Assemply Sequences", *IEEE Transactions on Robotics and Automation, Vol. 7*, No. 2, 1991, 211–227.

[87] Homem de Mello, L. S., and Sanderson, A. C., "Two Criteria for the Selection of Assembly Plans, Maximizing the Flexibility of Sequencing the Assembly Tasks and Minimizing the Assembly Time Through Parallel Execution of Assembly Tasks", *IEEE Transactions on Robotics and Automation, Vol. 7*, No. 5, 1991, 626–633.

[88] Horrowitz, E., and Sahni, S., "Computing Partitions with Application to the Knapsack Problem", *Journal of the ACM, Vol. 21*, 1974, 277–292.

[89] Hunt, E. B., Marin, J., and Stone, P. J., *"Experiments in Induction"*, Academic Press, 1966.

[90] Jin, V. Y., and Levis, A. H., "Compensatory Behavior in Team Decision Making", In *Proceedings of the IEEE International Symposium on Intelligent Control* (Philadelphia, PA, 1990), pp. 107–112.

[91] Kautz, H. A., "A Circumscriptive Theory of Plan Recognition", In *Intentions in Communication*, P. R. Cohen, J. Morgan, and M. E. Pollack, Eds., MIT Press, Cambridge, MA, 1990.

[92] Keller, R., "The Role of Explicit Control Knowledge in Learning Concepts to Improve Performance", In *Machine-Learning Technical-Report-7*, R. University, Ed., New Brunswick, NJ, 1987.

[93] Kimemia, J. G., and Gershwin, S. B., "An Algorithm fot the Computer Control of a Flexible Manufacturing System", *IIE Transactions, Vol. 15*, 1983, 353–363.

[94] King, J. R., "Machine-Component Group Formulation in Production Flow Analysis: An Approach Using a Rank Order Clustering Algorithm", *Int. J. Prod. Res., Vol. 18*, No. 2, 1980, 213–232.

[95] Kocabas, S., "A Review of Learning", *The Knowledge Engineering Review, Vol. 6*, No. 3, 1991, 195–222.

[96] Kodratoff, Y., and Michalski, R. S., Eds., *"Machine Learning: An Artificial Intelligence Approach, Volume III"*, Morgan Kaufman, 1990.

[97] Kokkinaki, A. I., and Valavanis, K., "On the Comparison of AI and DAI Based Planning Techniques for Automated Manufacturing Systems", *In Print*, 1993.

[98] Konolige, K., "A Deductive Model of Belief", In *Proceedings of the Eighth International Joint Conference in Artificial Intelligence* (1983), pp. 377–381.

[99] Konolige, K., "Defeasible Argumentation in Reasoning about Events", In *Proceedings of the International Symposium on Machine Intelligence and Systems* (Torino, Italy, 1989).

[100] Konolige, K., and Nilsson, N. J., "Multiple-agent planning systems", In *Proceedings of the First Annual Conference of the Anerican Association for Artificial Intelligence* (1980), pp. 138–141.

[101] Koren, Y., *"Robotics for Engineers"*, McGraw-Hill Book Company, New York, NY, 1985.

[102] Kripke, S., "Semantical Considerations on Modal Logic", *Acta Philosophica Fennica, Vol. 16*, 1963, 83–94.

[103] Kusiak, A., *"Intelligent Manufacturing Systems"*, Prentice Hall, 1990.

[104] Kusiak, A., and Chow, W. S., "Efficient solving of the group technology problem", *Journal of Manufacturing Systems, Vol. 6*, No. 2, 1987, 117–124.

[105] Ladner, R. E., "The complexity of Problems in Systems of Communicating Sequential Processes", In *Proceedings of the Eleventh ACM Symposium on Theory of Computing* (1979), pp. 214–223.

[106] Laird, J. E., Newell, A., and P., R., "Chunking in SOAR", *Machine Learning, Vol. 1*, 1986, 11–46.

[107] Laird, J. E., Newell, A., and P., R., "SOAR: An Architecture for General Intelligence", *Artificial Intelligence, Vol. 33*, 1987, 1–64.

620

[108] Langley, P., "BACON1 A General Discovery System", In *Proceedings of the Second National Conference of the Canadian Society For Computational Studies* (1978).

[109] Langley, P., Simon, H. A., Bradshaw, G. L., and Zutkow, J. M., *"Scientific Discovery: Computational Explorations of the Creative Processes"*, The MIT Press, 1987.

[110] Lansky, A. L., and Folgesong, D., "Localized Representation and Planning Methods for Parallel Domain", In *Proceedings of the Sixth National Conference on Artificial Intelligence* (1987), pp. 240–245.

[111] Lee, S., "Disassemply Planning by Subassembly Extraction", In *Proceedings of the Third ORSA/TIMS Conference on Flexible Manufacturing Systems* (MIT, MA, 1989), Elsevier Science.

[112] Lee, S., "Backward Assemply Planning", In *AI Applications in Manufacturing*, A. Famili, D. S. Nau, and S. H. Kim, Eds., AAAI Press, 1992.

[113] Lee, S., and Shin, Y. G., "Automatic Construction of Assembly Partial-Order Graph", In *Proceedings of the 1988 International Conference on Computer Intergrated Manufacturing* (Troy, New York, 1988), RPI.

[114] Lenat, D. B., "On Automated Scientific Theory Formation: A Case Study Using the AM Program", *Machine Intelligence, Vol. 9*, 1979, 251–238.

[115] Lenat, D. B., "EURISKO: A Program that Learns New Heuristics and Domain Concepts", *Artificial Intelligence, Vol. 21*, No. 1, 1983, 61–98.

[116] Lenat, D. B., Davis, R., Doyle, J., Genesereth, M., Goldstein, I., and H., S., "Reasoning about Reasoning", In *Building Expert Systems*, F. Hayes-Roth, D. A. Waterman, and D. B. Lenat, Eds., Addison-Wesley, Reading, MA, 1983.

[117] Lenat, D. B., and Feigenbaum, E. A., "On the Threshold of Knowledge", In *Proceedings of the Tenth International Joint Conference in Artificial Intelligence* (1987), pp. 1173–1182.

[118] Lesser, V. R., "A High Level Simulation Testbed for Cooperative Distributed Problem Solving", In *Proceedings of the Third International Conference in Distributed Computer Systems* (1982), pp. 341–349.

[119] Lesser, V. R., and Corkill, D., "Functionally Accurate Cooperative Distributed Systems", *IEEE Transactions on Systems, Man and Cybernetics, Vol. 11*, No. 1, 1981, 81–96.

[120] Lesser, V. R., and Corkill, D. D., "The Application of Artificial Intelligence Techniques to Cooperative Distributed Processing", In *Proceedings of the Sixth International Joint Conference in Artificial Intelligence* (1979), pp. 537–540.

[121] Lesser, V. R., and Erman, L. D., "Distributed Interpretation: A Model and Experiment.", *IEEE Transactions on Computers, Vol. 29*, No. 12, 1980, 1144–1163.

[122] Litman, D. J., and Allen, J. F., "A Plan Recognition Model for Subdialogues in Conversations", *Cognitive science, Vol. 11*, No. 2, 1987.

[123] Liu, D., "Intelligent Manufacturing Planning Systems", In *Proceedings IEEE COMPINT-Computer Aided Technologies* (1985), pp. 552–554.

[124] Lum, H., Ed., *"Machine Intelligence and Autonomy for Aerospace Systems"*, AIAA, Washington DC, 1988.

[125] Manna, Z., and Wolper, P., "Synthesis of Communicating Processes from Temporal Logic Specifications", In *Proceedings of the Workshop on Logics of Programs, Lecture Notes in Computer Science*, Springer Verlag, 1981.

[126] McCarthy, J., and Hayes, P., "Some Philosophical Problems from the Standpoint of Artficial Intelligence", In *Machine Intelligence 4*, Meltzer, B., 1969.

[127] McDermott, D., "A Deductive Model of Control of a Problem Solver", In *Proceedings of the workshop on pattern-directed inference systems*, SIGART Newsletter no. 63, 1977, pp. 2–7.

[128] McDermott, D., "A Deductive Model of Control of a Problem Solver", In *Proceedings of the Fifth International Joint Conference in Artificial Intelligence* (1977), pp. 229–234.

[129] McDermott, D., *"Flexibility and Efficiency in a Computer Program for Designing Circuits"*, MIT AI Laboratory Technical Report 402, 1977.

[130] McDermott, D., "Circuit Design as Problem Solving", In *Proceedings of the IFIP workshop in Artificial Intelligence and Pattern Recognition in Computer-Aided Design*, Springel Verlag, New York, NY, 1978.

[131] McDermott, D., "Planning and Acting", *Cognitive Science, Vol. 2*, 1978, 71–109.

[132] McDermott, D., "A Temporal Logic for Reasoning about Actions and Plans", *Cognitive Science, Vol. 6*, 1982, 101–155.

[133] Mendel, J., and Zapalac, J., "The Application of Techniques of Artficial Intelligent to Control System Design", In *Advances in Control Systems*, C. T. Leondes, Ed., Academic Press, NY, 1968.

[134] Mesarovic, M., Macko, D., and Takahara, Y., *"Theory of Hierarchical, Multilevel Systems"*, Academic Press, 1970.

[135] Meystel, A., "Intelligent Control: Issues and Perspectives", In *Proceedings IEEE Workshop on Intelligent Control* (1985), pp. 1–15.

622

[136] Michalski, R. S., "Pattern Recognition as Rule-Guided Inductive Inference", *IEEE Transactions on Patern Analysis and Machine Intelligence, Vol. 2*, No. 4 July, 1980, 349–361.

[137] Michalski, R. S., "Understanding the Nature of Learning: Issues and Research Directions", In *Machine Learning*, R. S. Michalski, J. G. Carbonell, and T. M. Mitchell, Eds., Morgan Kaufmann, San Mateo, CA, 1986.

[138] Michalski, R. S., "Concept Learning", In *Encyclopedia of Artificial Intelligence*, S. C. Shapiro, Ed., John Wiley, 1987, pp. 185–194.

[139] Michalski, R. S., Carbonell, J. G., and Mitchell, T. M., Eds., *"Machine Learning: An Artificial Intelligence Approach"*, Morgan Kaufmann, San Mateo, CA, 1983.

[140] Michalski, R. S., Carbonell, J. G., and Mitchell, T. M., Eds., *"Machine Learning: An Artificial Intelligence Approach, Volume II"*, Morgan Kaufmann, San Mateo, CA, 1986.

[141] Michalski, R. S., and Kodratoff, Y., "Research in Machine Learning: Recent Progress, Classification of Methods and Future Directions", In *Machine Learning: An Artificial Intelligence Approach, vol. 3*, Y. Kodratoff and R. S. Michalski, Eds., Morgan Kaufmann, San Mateo, CA, 1990.

[142] Minton, S., *"Learning Effective Search Control Knowledge: An Explanation Based Approach"*, 1988.

[143] Minton, S., Ed., *"Machine Learning Methods for Planning"*, Morgan Kaufmann, 1993.

[144] Minton, S., Johnston, M. D., Philips, A. B., and Laird, P., "Solving large-scale constraint satisfaction and scheduling problems using a heuristic repair method", In *Proceedings of the Eightth National Conference in Artificial Intelligence* (Menlo Park, CA, 1990), pp. 17–24.

[145] Mitchell, T., "Learning and Problem Solving", In *Proceedings of the Eightth International Joint Conference in Artificial Intelligence* (Karlsruhe, Germany, 1983), pp. 1139–1151.

[146] Mitchell, T., Keller, R. M., and Kedar-Cabelli, S. T., "Explanation Based Generalization: A Unifying View", *Machine Learning, Vol. 1*, No. 1, 1986, 47–80.

[147] Mooney, R., "A General Explanation-Based Learning Mechanism and its Application to Narrative Understanding", *Technical Report UILU-ENG-87-2269*, 1988.

[148] Moore, R., "Reasoning about Knowledge and Action", In *Proceedings of the Fifth International Joint Conference in Artificial Intelligence* (Cambridge, MA, 1977), pp. 223–227.

[149] Moore, R., "Reasoning about Knowledge and Action", *SRI Technical Note 191*, 1980.

[150] Morgenstern, L., "A First Order Theory of Planning, Knowledge and Action", In *Proceedings of the Conference on Theoretical Aspects of Reasoning About Knowledge* (Los Altos, 1986), M. Kaufmann, Ed.

[151] Morgenstern, L., "Knowledge Preconditions for Actions and Plans", In *Proceedings of the Tenth International Joint Conference in Artificial Intelligence* (1987), pp. 867–874.

[152] Morin, T. L., and Marsten, R. E., "An Algorithm for Non-Linear Knapsack Problems", *Management Science, Vol. 22*, 1976, 1147–1158.

[153] Nau, D. S., and Gray, M., "SIPS: An Application of Hierarchical Knowledge Clustering to Process Planning", In *Integrated and Intelligent Manufacturing*, C. R. Liu and T. C. Chang, Eds., The American Society of Mechanical Engineers, 1986, pp. 219–225.

[154] Newell, A., and Simon, H. A., *"Human Problem Solving"*, Prentice Hall, Englewood Cliffs, NJ, 1972.

[155] Nii, H. P., and Aiello, N., "AGE (attempt to generalize): A Knowledge-Based Program for Building Knowldge-Based Programs", In *Proceedings Sixth International Joint Conference in Artificial Intelligence* (Tokyo, Japan, 1979), pp. 645–665.

[156] Nii, H. P., Feigenbaum, E. A., Anton, J., and Rockmore, A., "Signal-to-Symbol Transformaton: HASP/SIAP Case Study", *AI Magazine, Vol. 3*, 1982, 23–35.

[157] Nilsson, N. J., *"Problem Solving Methods in Artificial Intelligence"*, McGraw Hill, New York, NY, 1971.

[158] Nordhausen, B., and Langley, P., "Towards an Integrated Discovery System", In *Proceedings of the Tenth International Joint Conference in Artificial Intelligence* (1987), pp. 198–200.

[159] Olsder, G. J., and Suri, R., "Time Optimal Control of Parts from Routing in a Manufacturing System with Failure Prone Machines", In *Proceedings of the nineteenth IEEE Conference on Design and Control*, 1980.

[160] O'Rorke, P. V., "LT Revisited: Experimental Results of Applying Explanation Based Learning to the Logic of Principia Mathematica", In *Proceedings of the Fourth International Machine Learning Workshop* (Irvine, CA, 1987), pp. 148–159.

[161] Ozguner, U., "Decentralized and Distributed Control Approaches and Algorithms", In *Proceedings of the 28th IEEE Conference on Decision and Control* (Tampa, FL, 1989), pp. 1289–1294.

[162] Passino, K. M., "Restructurable Controls and Artificial Intelligence", *McDonnell Aircraft Internal Report IR-0392*, 1986.

[163] Passino, K. M., and Antsaklis, P. J., "Artificial Intelligence Planning Problems in a Petri Net Framework", *Dept. of Electrical and Computer Engineering, University of Notre Dame, Technical Report 880*, 1988.

[164] Passino, K. M., and Antsaklis, P. J., "Fault Detection and Identification in an Intelligent Restructurable Controller", *Journal of Intelligent and Robotic Systems, Vol. 1*, 1988, 145–161.

[165] Passino, K. M., and Antsaklis, P. J., "On the Optimal Control of Discreete Event Systems", In *Proceedings of the 28th IEEE Conference on Decision and Control* (Tampa, FL, 1989), pp. 2713–2718.

[166] Passino, K. M., and Antsaklis, P. J., "A System and Control Theoretic Perspective on Artificial Intelligence Planning Systems", *Journal of Applied Artificial Intelligence, Vol. 3*, 1989, 1–32.

[167] Passino, K. M., and Antsaklis, P. J., "Event Rates and Aggregation in Hierarchical Discrete Event Systems", *Journal of Discrete Event Dynamic Systems: Theory and Applications, Vol. 1*, No. 3, 1992, 271–287.

[168] Passino, K. M., and Ozguner, U., "Modeling and Analysis of Hybrid Systems: Examples", In *Proceedings of the 1991 IEEE International Symposium on Intelligent Control* (Arlington, VA, 1991), pp. 251–256.

[169] Peek, M. D., and Antsaklis, P. J., "Parameter Learning for Performance Adaptation", *IEEE Control Systems Magazine, Vol. 10*, 1990, 3–11.

[170] Pelavin, R., and Allen, J. F., "A Formal Logic of Plans in Temporally Rich Domains", *Proceedings of the IEEE, Special Issue on Knowledge Representation, Vol. 74*, No. 10, 1986, 1364–1382.

[171] Peterson, J. L., *"Petri Net Theory and the Modeling of Systems"*, Prentice Hall, 1981.

[172] Pollack, P. R., "Plans as Complex Mental Attitudes", In *Intentions in Communication*, P. R. Cohen, J. Morgan, and M. E. Pollack, Eds., MIT Press, Cambridge, MA, 1990.

[173] Porter, B. W., and Kibler, D. F., "Experimental Goal Regression: A Method for Learning Problem-Solving Heuristics", *Machine Learning, Vol. 1*, No. 3, 1986, 249–286.

[174] Pratt, V. R., "Semantical Considerations on Floyd-Hoare Logic", In *Proceedings Senentheenth IEEE Symposium on Foundations of Computer Science* (1976), pp. 108–121.

[175] Quilan, J. R., "Learning Efficient Classification Procedures and their Application to Chess and Games", In *Machine Learning: An Artificial Intelligent Approach*, R. S. Michalski, J. G. Carbonell, and T. M. Mitchell, Eds., Morgan Kaufmann, 1983.

[176] Rajagopalan, S., "Formulation and Heuristic Solutions for Parts Grouping and Tool Loading in a Flexible Manufacturing System", In *Proceedings Second ORSA/TIMS Conference on Flexible Manufacturing Systems* (1986).

[177] Rescher, J., and Urquhart, A., *"Temporal Logic"*, Springer Verlag, 1971.

[178] Rivest, R. L., and Remmele, W., "Machine Learning: the Human Connection", In *Siemens Review* (1988), vol. 2, pp. 33–37.

[179] Rosenschein, J. S., "Synchronization of Multi-Agent Plans", In *Proceedings Second National Conference in Artificial Intelligence* (1982), pp. 115–119.

[180] Sacerdoti, E. D., "Planning in a Hierarchy of Abstraction Spaces", *Artificial Intelligence*, No. 5, 1974, 115–135.

[181] Sacerdoti, E. D., *"A Structure for Plans and Behavior"*, Elsevier North-Holland, Amsterdam, 1977.

[182] Sandewall, H., and Ronnquist, R., "A Representation of Action Structures", In *Proceedings Fifth National Conference in Artificial Intelligence* (1986), pp. 89–97.

[183] Saridis, G. N., *"Self-Organizing Control of Stochastic Systems"*, Marcel Dekker, New-York, NY, 1977.

[184] Saridis, G. N., "Toward the Realization of Intelligent Controls", *Proceedings of the IEEE, Vol. 67*, No. 8, 1979, 1115–1133.

[185] Saridis, G. N., "Intelligent Controls for Advanced Automated Processes", In *Proceedings of the Automated Decision Making and Problem Solving Conference* (NASA CP-2180, 1980).

[186] Saridis, G. N., "Intelligent Robotic Control", *IEEE Tranactions on Automatic Control, Vol. 28*, No. 5, 1983, 547–556.

[187] Saridis, G. N., "Foundations of the Theory of Intelligent Controls", In *Proceedings IEEE Workshop on Intelligent Control* (1985), pp. 23–28.

[188] Saridis, G. N., "Knowledge Implementation: Structures of Intelligent Control Systems", In *Proceedings IEEE International Symposium on Intelligent Control* (1987), pp. 9–17.

[189] Saridis, G. N., "Analytic Formulation of the Principle of Increasing Precision with Decreasing Intelligence for Intelligent Machines", *Automatica, Vol. 25*, No. 3, 1989, 461–467.

[190] Schmidt, C. F., Sridharan, N. S., and Goodson, J. L., "The Plan Recognition Problem: An Intersection of Artificial Intelligence and Psychology", *Artificial Intelligence, Vol. 10*, No. 1, 1978.

[191] Schoppers, M. J., *"Representation and automatic synthesis of reaction plans"*, UIUCDCS-R-89-1546, Department of Computer Science, University of Illinois at Urbana Champaign, Urbana, IL, 1989.

[192] Shavlik, J. W., *"Generalizing the Structure of Explanation in Explanation-Based Learning"*, Technical Report UILU-ENG-87-2276, CSL, University of Illinois at Urbana-Champaign, Urbana, IL, 1987.

[193] Shavlik, J. W., and Dietterich, T. G., *"Readings in Machine Learning"*, Morgan Kaufmann, San Mateo, CA, 1990.

[194] Sidner, C. L., "Plan Parsing for Intended Response Recognition in Discourse", *Computational Intelligence, Vol. 1*, No. 1, 1985.

[195] Simon, H. A., "Why Should Machines Learn?", In *Machine Learning*, R. S. Michalski, J. G. Carbonell, and T. M. Mitchell, Eds., Morgan Kaufmann, 1983.

[196] Smith, R. G., "A Framework for Distributed Problem Solving", In *Proceedings of the Sixth International Joint Conference in Artificial Intelligence* (1979), pp. 836–841.

[197] Smith, R. G., "The Contract-Net Protocol: High Level Communication and Control in a Distributed Problem Solver", *IEEE Transactions on Computers, Vol. 29*, No. 12, 1980, 1104–1113.

[198] Smith, R. G., and Davis, R., "Frameworks for Cooperation in Distributed Problem Solving", *IEEE Transactions on Systems, Man and Cybernetics, Vol. 11*, No. 1, 1981, 61–70.

[199] Smith, S. F., Kempf, K. G., and Kempf, N. K., "Exploiting Local Flexibility during Execution of Pre-Computed Schedules", In *AI Applications in Manufacturing*, A. Famili, D. S. Nau, and S. H. Kim, Eds., AAAI Press, 1992.

[200] Stecke, K. E., and Talbot, F. B., "Heuristics for Loading Flexible Manufacturing Systems", In *Flexible Manufacturing Systems: Recent Developments in FMS, CAD/CAM, CIM*, A. Raouf and S. I. Ahmad, Eds., 1985.

[201] Steeb, R., Cammarata, S., Hayes-Roth, F. A., Thorndyke, P. W., and Wesson, R. B., "Distributed Intelligence for Fleet Control", *RAND Corp. Rep. R-2728-ARPA*, 1981.

[202] Steeb, R., Cammarata, S., Narain, S., Rothenberg, J., and Giarla, W., "Cooperative Intelligence for Remotely Piloted Vehicle Fleet Control", *RAND Corp. Rep. R-3408-ARPA*, 1986.

[203] Stefik, M., "Planning and Meta-Planning MOLGEN: Part II", *Artificial Intelligence, Vol. 16*, 1981, 141–169.

[204] Stefik, M., "Planning with Constrains", *Artificial Intelligence, Vol. 16*, 1981, 111–140.

[205] Stiver, J. M., and Antsaklis, P. J., "A Novel Discrete Event System Approach to Modeling and Analysis of Hybrid Control Systems", In *Proceedings 29th Allerton Conference on Communication, Control and Computing* (University of Illinois at Urbana-Champaign, 1991).

[206] Stuart, C., "An Implementation of a Multi-Agent Plan Synchronizer", In *Proceedings of the Nineth International Joint Conference in Artificial Intelligence* (1985), pp. 1031–1033.

[207] Stuart, C., "An Implementation of a Multi-Agent Plan Synchronizer Using a Temporal Logic Theorem Prover", *SRI AI Center*, 1985.

[208] Sussman, G. J., *"A Computational Model of Skill Acquisition"*, Technical Report 297 MIT AI Lab, 1973.

[209] Tate, A., *"Project Planning Using a Hierarchic Nonlinear Planner"*, Technical Report 25, Department of Artificial Intelligence, University of Edinburgh, Edinburgh, UK, 1976.

[210] Tate, A., "Generating Projects Networks", In *Proceedings of the Fifth International Joint Conferenceon in Artificial Intelligence* (1977), pp. 888–893.

[211] Tennenholtz, M., and Moses, Y., *"On Cooperation in a Multi-Entity Model"*, Technical Report, Weismann Institute, 1989.

[212] Tennenholtz, M., and Moses, Y., "On Cooperation in a Multi-Entity Model (Preliminary Report)", In *Proceedings of the Eleventh International Joint Conferenceon Artificial Intelligence* (1989), pp. 918–923.

[213] Thorndyke, P. W., McArthur, D., and C., C., "AUTOPILOT A Distributed Planner for Air Fleet Control", In *Proceedings of the Nineth International Joint Conferenceon Artificial Intelligence* (1983), pp. 171–177.

[214] Tsatsoulis, C., "Using Dynamic Memories in Planning and its Application to Manufacturing", PhD thesis, School of Electrical Engineering, Purdue University, 1987.

[215] Tsatsoulis, C., and Kashyap, R. L., "Case-Based Reasoning and Learning in Manufacturing with the TOLTEC Planner", *IEEE Transactions on Systems, Man and Cybernetics, Vol. 23*, No. 4, 1993, 1010–1023.

[216] Turner, P. R., *"Autonomous Systems: Architecture and Implementation"*, Jet Propulsion Laboratories, Report No. JPLD-1656, 1984.

[217] Valavanis, K. P., "The Analytical Design of Intelligent Machines", PhD thesis, Electrical and Computer Engineering Department, Rensselaer Polytechnic Institute, Troy, NY, 1986.

[218] Valavanis, K. P., and Carello, S. J., "An Efficient Planning Technique for Robotic Assemblies and Intelligent Robotic Systems", *Journal of Intelligent and Robotic Systems, Vol. 3*, No. 4, 1990, 321–347.

[219] Valavanis, K. P., and Saridis, G. N., "Information Theoretic Modeling of Intelligent Robotic Systems, Part I: The Organization Level", In *Proceedings of the 26th Conference on Decision and Control* (Los Angeles, LA, 1987), pp. 619–626.

[220] Valavanis, K. P., and Saridis, G. N., "Information Theoretic Modeling of Intelligent Robotic Systems, Part II: The Coordination and Execution Levels", In *Proceedings of the 26th Conference on Decision and Control* (Los Angeles, LA, 1987), pp. 627–633.

[221] Valavanis, K. P., and Stellakis, H. M., "A General Organizer Model for Robotic Assemblies and Intelligent Robotic Systems", *IEEE Transactions on System, Man and Cybernetics, Vol. 21*, No. 2, 1991, 302–316.

[222] Weyrauch, R. W., "Prolegomena to a Theory of Mechanized Formal Reasoning", *Artificial Intelligence, Vol. 13*, 1980, 133–170.

[223] Wilenskey, R., "Meta-Planning: Representing and Using Knowledge About Planning in Problem Solving and Natural Language Understanding", *Cognitive Science, Vol. 5*, 1981, 197–233.

[224] Wilkins, D. E., *"Monitoring the Execution of Plans in SIPE"*, Technical Report, SRI International, Menlo Park, CA, 1984.

[225] Wilkins, D. E., *"Practical Planning: Extending the Classical AI Planning Paradigm"*, Morgan Kaufmann, San Mateo, CA, 1988.

[226] Woods, S., "Dynamic World Simulation for Planning with Multiple Agents", In *Proceedings of the Eightth International Joint Conferenceon Artificial Intelligence* (1983), pp. 69–71.

[227] Wos, L., *"Automated Reasoning: 33 Basic Research Problems"*, Prentrice Hall, NJ, 1988.

[228] Wu, S. D., and Wysk, R. A., "An Application of Discrete Event Simulation to On-Line Control and Scheduling in Flexible Manufacturing", *International Journal of Production Resources, Vol. 27*, 1989, 1603–1623.

[229] Zaghoul, M. E., "A Machine-Learning Classification Approach for IC Manufacturing Control Based on Test Structure Measurements", *IEEE Transactions on Semiconductor Manufacturing, Vol. 2*, No. 2, 1989, 47–53.

[230] Zeigler, B. P., "Knowledge Representation from Newton to Minsky and Beyond", *Journal of Applied Artificial Intelligence, Vol. 1*, 1987, 87–107.

[231] Zeigler, B. P., "DEVS Representation of Dynamical Systems: Event Based Intelligent Control", *Proceedings of the IEEE, Vol. 77*, No. 1, 1989, 72–80.

[232] Zeigler, B. P., and Chi, S. D., "Model-Based Concepts for Autonomous Systems", In *Proceedings of the IEEE International Symposium on Intelligent Control* (Philadelphia, PA, 1990), pp. 27–32.

[233] Zeigler, B. P., and Sungdo, C., "Model-Based Architecture Concepts for Autonomous Systems Design and Simulation", In *An Introduction to Intelligent and Autonomous Control*, Kluwer Academic Publishers, 1993.

[234] Zhong, H., and Wonham, W. M., "On the Consistency of Hierarchical Supervision in Discrete-Event Systems", *IEEE Tranactions on Automatic Control, Vol. 35*, No. 10, 1990, 1125–1134.

[235] Zhu, D., and Latombe, J. C., "New Heuristics Algorithms for Efficient Hierarchical Path Planning", *IEEE Tranactions on Robotics and Automation, Vol. 7*, No. 1, 1991, 9–20.

[236] Zrimec, T., and Mowforth, P., "Learning by an Autonomous Agent in the Pushing Domain", In *Toward Learning Robots*, W. Van de Velde, Ed., MIT Press, 1993.

[237] Zweben, M., Deale, M., and Gargan, R., "Anyttime Rescheduling", In *Proceedings of the DARPA Workshop on Innovative Approaches to Planning and Scheduling* (1990).

[238] Zytkow, J. M., and Simon, H. A., "A Theory of Historical Discovery: The Construction of Computential Models", *Machine Learning, Vol. 1*, 1986, 107–137.

21
KNOWLEDGE-BASED SUPERVISION OF FLEXIBLE MANUFACTURING SYSTEMS

A.K.A. TOGUYENI, E. CRAYE, J.C. GENTINA
Laboratoire d' Automatique et d' Informatique
Industrielle de Lille (L.A.I.I.L.)
(URA_CNRS D 1440), Ecole Centrale de Lille,
B.P. 48, 59651 Villeneuve d' Ascq, Cedex, France

1.   SUPERVISION AND AI-TECHNIQUES

Supervision of Flexible Manufacturing Systems (FMS) covers different kinds of activities (Fig. 1).

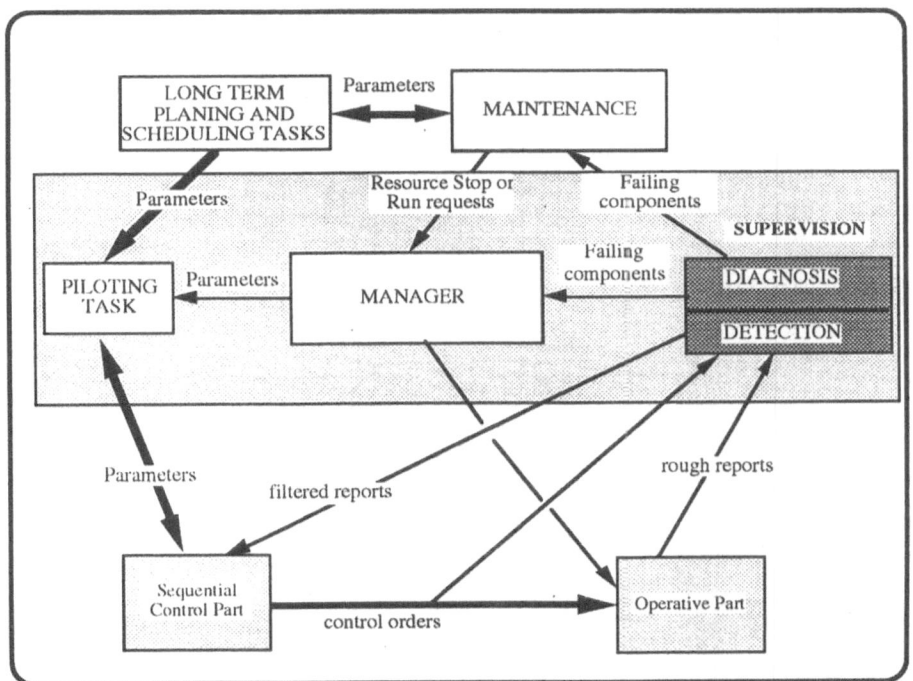

Figure 1 : Integration of the main functions of the supervision level

S. G. Tzafestas and H. B. Verbruggen (eds.),
Artificial Intelligence in Industrial Decision Making, Control and Automation, 631–662.
© 1995 Kluwer Academic Publishers.

The first one we present in this chapter is related with strategy of piloting. More precisely, due to the flexibility of the process, different choice can be effective for the control of production flow. Critical resources can also be shared and the solving of such a problem is generally executed at the supervision level.

The second task of the supervision concerns the management of the working modes. We propose in that way to use an expert system to deduce from the real state of the machines the consistent state of the whole process.

In the third section of this chapter we present the monitoring function which is attached to the watching of the process. We propose to use AI techniques in order to design and to implement a reactive diagnosis in the context of on-line monitoring.

2. PILOTING FUNCTION.

2.1. Introduction.

To solve the problem of industrial productivity, the complexity of Flexible Manufacturing Systems has grown considerably. Their use implies a close cooperation between many distributed devices such as Computerised Numerically Controlled machines, Programmable Logic Controllers and Central Host site which computes the task planification, supervises the production and assumes the degraded modes and transitory states which have been decided by the recovery part.

In order to give an operational answer to the problem of a distributed implementation, the description of the control of the sequential automatisms has been dissociated from the general plan which regulates and synchronizes these automatisms. In this way, a multi-level structuration is presented figure 2.

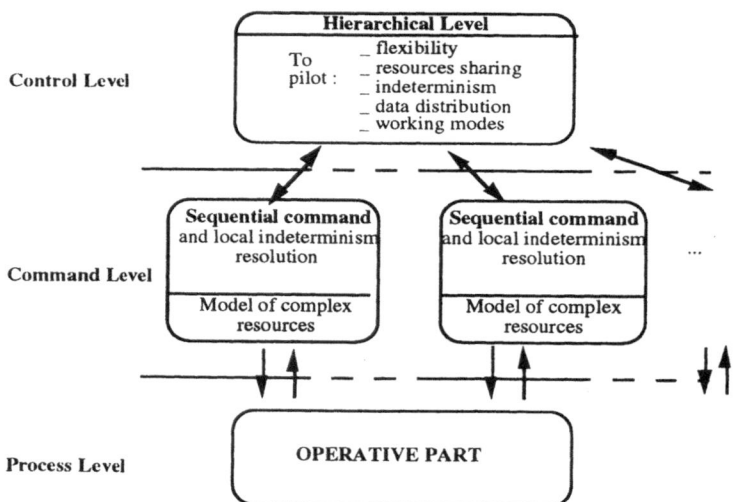

Figure 2 : The multi-level structuration of the Control Part

It is on a such decomposition that a knowledge-based system using artificial intelligence techniques has been retained. The differences of behaviour between the design and implementation models will be introduced in the first part. These differences will justify our structured approach and the definition of the hierarchical level. In the second part, the use of the hierarchical level and how it was designed with artificial intelligence will be shown.

2.2.   Problems met from design to implementation.

At the time of the design of the control system, the programmer attends first to the normal working mode of his automatisms. He obtains control graphs and models of complex resources in Petri nets or Grafcet [1] languages for example. At this step the simulation has to validate the dynamic behaviour of the design model. Two studies are necessary : a qualitative valuation and a quantitative valuation. The first one verifies that the dynamic behaviour of the installation does not involve any failure in its evolution. With this valuation, it must be possible to detect, to analyse and to solve the "dead-locks" and the indeterminations misled by a wrong conception or by an incomplete definition of the control. The quantitative valuation enables the dimension of the operative part to be estimated. By temporisation, it verifies the performances of the production corresponding to the specification requirements.

At this level, the simulation consists of a central way to validate the control system. The simulator manages a set of records in which, both the static description of the command and control levels, together with the dynamic data which characterize the behaviour of the marking of the graphs, are modelled. A single program, the simulator, manipulates these information and the evolution of events. This approach differs from the reality of the implementation on two main points. The hierarchical level on one hand, the sequential control on the other hand are distributed over several controllers; their sequences are necessarily not synchronized and cannot be correctly simulated by a single program.

In the same way, the distribution of the data raises another problem. The set of the distributed information is the same as the centralised structure : each piece of information is effectively present. However, the centralisation conceals the network communication between the controllers in order to exchange the data on their own states. Moreover, the consistency of the information is also concealed by the simulation.

As an example, let us consider two process sharing out a critical resource. The retained solution, during the design stage, decides to model the state of the resource (free or taken) at the command level. This solution is based on the use of adaptive Petri Nets [2]. On the other hand, the rules to allocate it, are modeled at the control level (the hierarchical level) in order to keep the best flexibility and to solve the indeterminism at the latest time. This approach is shown in figure 3. The marking of the places Alloc/P1 and Alloc/P2 depends on the rules of the hierarchical level. Such a model can easily be simulated. The simulator disposes of a set of records, unique and fully accessible. Therefore, the indetermination of functioning is solved by the global view of the whole system. However, what happens if the two process are distributed between two controllers ?
    - How can the state of the critical resource be represented ?
    - Must we manage the robot from a single controller and implement communication protocol ? Such a method obviously leads to an asymmetry.

634

- Must we duplicate the virtual image of the resource on every programmable controller ?

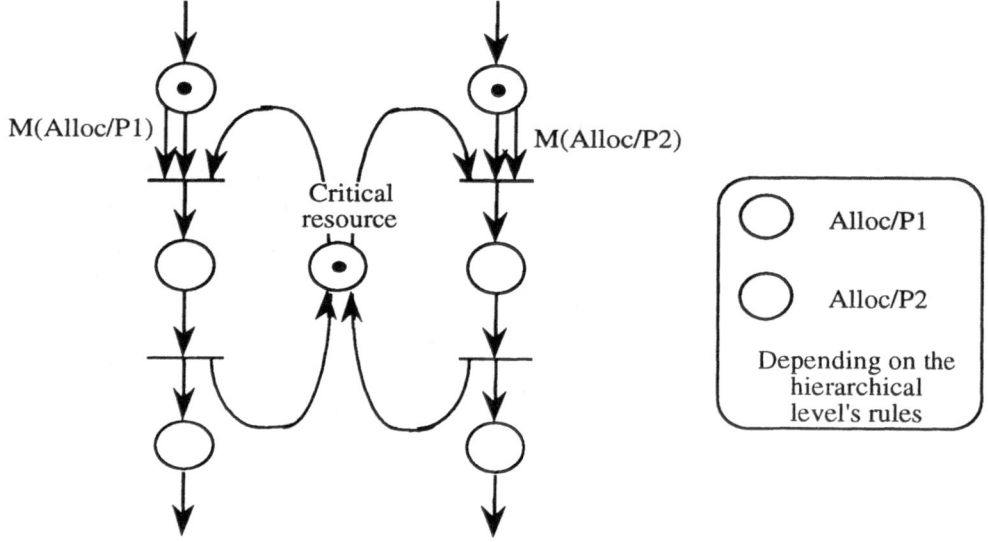

Critical
resource

M(Alloc/P1)

M(Alloc/P2)

Alloc/P1

Alloc/P2

Depending on the
hierarchical
level's rules

Figure 3 : Modelling of the sharing of a critical resource with adaptive Petri Nets [2].

The data are no longer global, nor indivisible. In particular, the resource symbolized by the place R cannot be implemented as modeled. The implementation on industrial devices differs from the initial model.

In order to solve the problem of a distributed implementation, the description of the sequential automatisms has been dissociated from the general plan which regulates these automatisms. It allows to structure the approach of design because one problem at once is solved.

2.3.   The knowledge-based system.

It consists of a set of production rules driven by an inference engine in forward chaining. A functional language has been voluntarily chosen as description tool for the hierarchical level to the detriment of the more conventional use of an imperative language. This choice can be justified for several reasons :
- homogeneity of the retained language (Le_Lisp from I.N.R.I.A [21] in relation to other complementary works developed by the L.A.I.L concerning a progressive and modular design methodology for the control of flexible systems [3] [4],
- important flexibility and facility to develop the hierarchical level itself.
But, especially from the user's point of view, the use of such a language offers an interactivity unequalled by any imperative compiled language. Indeed, the definition of control structures to drive a flexible unit neither need compilation at any time, nor link. The user converses with the system in a "pseudo-natural" way ; he can accede to the

whole of his work : it means both the data of his model and the way to use them. This interactivity is kept up during the whole life of the hierarchical level : design, final adjustment, exploitation. It is a fundamental requirement for the viability of the system. The designer corrects progressively and often afterwards the borderline cases in the working of his installation. Thus, the interactivity of the system never implies the unity of the hierarchical level. Each modification is easily made at little cost.

The hierarchical level consists of a set of production rules [5] driven by an inference engine in forward-chaining. Let us now define its different components.

2.3.1.   The fact bases

• *Presentation* ,
An internal fact designates a type-variable of which the initial value is given at the system configuration, and which can later change depending on the results of the inference; such a variable is essentially local at the hierarchical level. Conversely, an external fact is a type-variable whose value is given by the state of the operative or command parts. To know or change the value of an external fact, a local network in order to communicate with the installation has to be used. Thus, the hierarchical level has two separate fact bases; the first one is named internal, the second one external.

• *Description of type-variables.*
Our purpose is not to present here an exhaustive list of type-variables defined in the fact bases. The difference between an internal variable and an external one is just detailed on the example of a boolean. To define an internal boolean, the user must give its name, its type (boolean) and its initial value. To define an external boolean, he must also give its name and its type; he does not give the initial value because it depends on the state of the command part. However, he has to inform the fact base about the location of this boolean in the command, on which programmable logic controller this boolean can be found and what is the boolean variable of the controller which is linked to it. With these information, the hierarchical level is able to associate with the boolean, the communication protocol which is used to know or modify its value. In that way, the data processing is completely transparent for the designer. He never has to be concerned with the network primitives and thus he can more easily stay at the level 7 of the O.S.I. (Open System Interconnection) model.
The user disposes of boolean, integer variables, character strings but also variables of timer-type or inputs/outputs that he uses as he wishes with the production rules to define the hierarchical level.

2.3.2.   The rule bases

The syntax of a production rule is the following :
i) a rule identifier,
ii) a statement like [condition] -> [action].
The [condition] part is made up of a set of predicates whose value depends on the state of the two fact bases. The [action] part modifies these latter and the operative and command part with pre-defined operators.

- *Detail of a condition*,

[condition] <==> {(internal premises)} {(external premises)}
i) A premise consists of :
- a pre-defined reserved word in the system,
- a set of data (variable or not), parameters of the concerned word.
ii) An internal premise is a premise which only uses internal data in the micro-computer; that means these data are not subordinate to the external process.
iii) An external premise is a premise which has at least one external variable. In order to know its value, the hierarchical level must do a request communication to a controller through the local network.

- *Detail of an action* ,

[action] <==> {(internal consequent)} {(external consequent)}
i) A consequent has a same syntax as a premise.
ii) An internal consequent is a consequent which changes the value of an internal variable. There is sometimes a request to the network but the external process is in no way modified.
iii) An external consequent is a consequent which modifies at least the value of an external variable, consequently the state of the extern process.

- *Rule bases and meta-rule bases* ,

We have in the hierarchical level two separate rule bases. The rule base, the structure of which has been defined in the previous paragraph, effectively drives the distributed devices and the operative part. It supervises the installation and, with the gathered data, modifies the process. Nevertheless, a base of meta-rules, which is used as a first filter during the inference, has been defined. It saves times because there is no need to examine the whole rule base. The meta-rule base, with only internal data, chooses the rules which will be candidate for the entire inference. In this way, the number of rules which work with the local network, the slowest task of the system, are minimized.

- *The inference engine* .

The principle of the inference engine consists in a width forward-chaining adjusted to the hierarchical level (Fig. 4).

It works in two successive phases; firstly, an inference on the meta-rule base; secondly, an inference on the rule base. The next choices have been retained for more efficiency :
i) selection of rules whose internal premises have been verified,
ii) selection of rules whose external premises have also been verified,
iii) parallel starting of internal consequent and memorizing of external consequent,
iv) repetition of the three previous points until there is no more candidate rule,
v) starting of external consequents.
The selection according to the internal premises is important because it does not spend times. It is also for this reason that we start the external consequents only at the end of the inference. In that way, we can consider that the external process are frozen during a whole cycle of inference; for this reason, the hierarchical level

never has to accede more than once to an external data during a same cycle of inference.

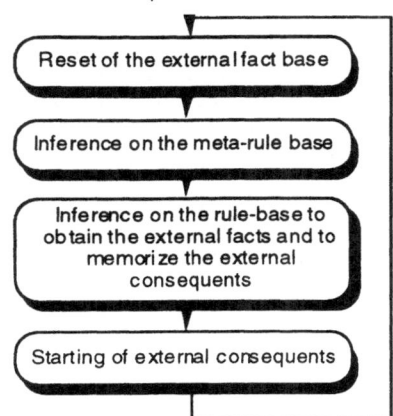

<u>Figure 4 : The principle of the inference engine</u>

2.4. Conclusion.

The hierarchical level has been developed to solve the indeterminisms of routing, the resources sharing, the data distribution over distributed devices. Due to the complexity of the whole system and the important amount of different facts, the inference engine uses a forward chaining. It allows an optimal strategy to be defined : the objective consists in the saturation of the fact bases without any "a priori" knowledge of which goals are to be proved. Such a mechanism is data driven, makes the definition of the rules easier and facilitates the trace of the inference engine [6].

Last but not least, reliability is a major interest of the hierarchical level. Modelling the changes of working modes, taking the failures into account, adding degraded modes or priorities, parametrizing a flexible production, all these notions can be supported by the hierarchical level. Next section will show how recovery has been studied; the modification of the working modes are assumed to be done by the hierarchical level previously defined.

3. MANAGER OF WORKING MODES.

3.1. Introduction

Management of working modes consists in the effective command of the machines by integrating changes of their states. These states can be modified by the sending of orders or by the occurrence of a failure independent of the orders. The management is complex because the machine must be considered not only as an independent entity but also as a member of a set of machines.

The management of working modes also consists in taking into account the strategic choices of production defined in the planning level. It implies that both the Control Part and the Operative Part must be adapted to the new load of the workshop.

Such complexity of management is difficult to be ensured by an operator, also we have designed a manager of working modes. Its aim is to integrate the various working modes of a production unit by taking into account behaviours constraints. It must detect the behavioural inconsistencies of the process by analysing the resources states and it must take actions for recovery purposes.

The manager of working modes is an expert system which implements two operating modes : a manual mode which allows the operator to enter orders to stop or to start a resource, and an automatic mode integrating the functioning described previously in this part.

The manager is built from a knowledge obtained by the modelling of the process to be surveyed.

3.2.   Representation and modelling of the process.

The aim of the modelling is to obtain the best possible description of the process to be commanded. The quality of the representation is important in that a detailed survey of the process is offered and also the certitude of giving proper controls. Nevertheless, it would be an error to try to model the process in too much detail because it would be too heavy to use in operational time. The idea of this modelling is to have a representation of the process where the cooperation and the interactions between machines are explicit.

3.2.1.   Modelling of machine behaviour.

The modelling approach proposed here distinguishes two kind of machines : real or **effective machines** and **virtual machines**.

Definition : An effective machine [7]
It is a machine which can have many working modes and moreover can be controlled : its states can be modified by the sending of orders.

For example, a robot, a conveyor, a machine-tool, a turnout ... are effective machines in a workshop.
In the most general case the behaviour of an effective machine can be represented by a **State diagram** with six states (Fig. 5) :

- State 0 called "**live_off**" corresponds to a machine that is switched off,
- State 1 called "**live_on**" corresponds to a machine that is switched on,
- State 2 called "**initialised**" corresponds to a configured state to resume work,
- State 4 called "**running**" corresponds to the normal working mode,
- State 5 called "**degraded_running**" corresponds to a functioning under a degraded working mode,
- State 3 called "**stop_under_failure**" corresponds to a stop after the diagnosis of a failure.

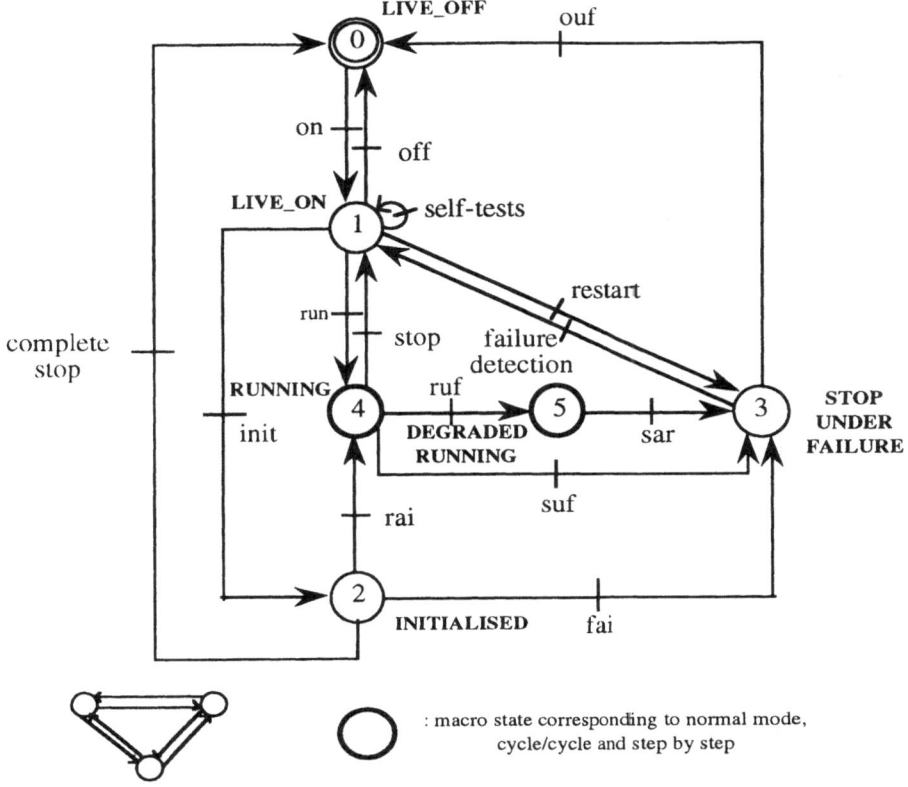

Figure 5 : The state diagram of an effective machine

These states are bounded by constraints which correspond to "actions" such as working orders or detection of an event which may modify the behaviour of the machine.

The basic idea of this modelling is to group together, machines whose behaviour are tied up from the point of view of the behaviour of the whole system. This idea leads us to introduce the concept of a virtual machine.

Definition : A virtual machine [7].
It is the association of two or more effective or virtual machines, whose behaviour is the same as that of an effective machine. These behaviours are tied up by operational constraints which do not permit to analyse them individually.

The last definition implies that a virtual machine is made up of sub-machines whose behaviours are inherited by it. It means that the state of the virtual machine is deduced from the states of it sub-machines. When the state of a sub-machine has been changed, the virtual machine reaches a new state (e.g. one of the states enumerated in the behaviour of an effective machine) through transitory-states.

640

Definition : A transitory-state [7].
A virtual machine is in a transitory state if the states of some of it sub-machines are inconsistent with respect to the states of the other sub-machines of the association.

3.2.2.   Structural description of the process.

Let us apply the structural description to the flexible cell of EC Lille (Fig. 6).

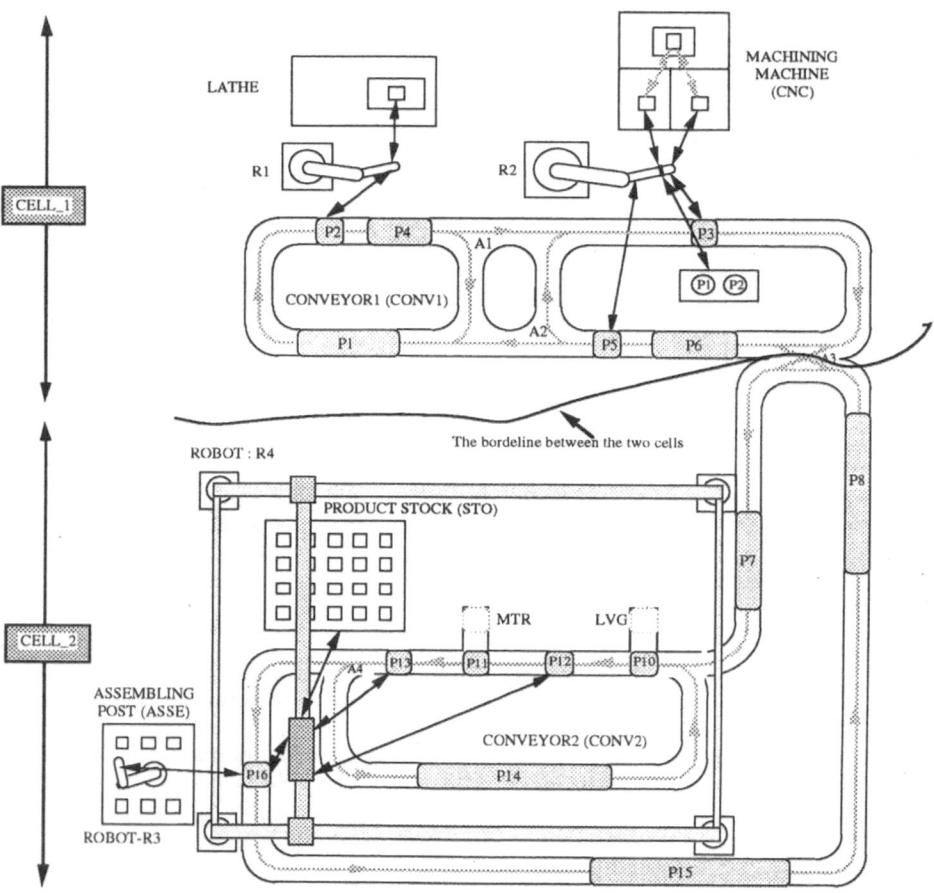

Figure 6 : The flexible cell of EC Lille

In order to manage the working modes of the process, the modelling principle consists of grouping the machines from an operational point of view. The method is based on a bottom-up analysis where the operational constraints are exploited. These constraints are :

● **Constraints on states** ,

- cooperation constraint **CC** : two ore more machines cooperate closely and are strongly dependent on each other,
- divided constraint of cooperation **CCP** : two or more machines share a resource.
- exclusion constraint CE : this is the opposite case of cooperation,
- Structural constraint : the machines are serially ordered or in parallel.

● **Constraints on changes of states,**

- proceeding constraint **CPRO** : protocols of running among machines are taken into account . For example, it can be precedence relation between "on" orders ;
- observability constraint **CO** : it is the possibility of accession to the information required for the control of the machines.

During a phase of specifications, the programmer indexes the whole state of constraints in the manufacturing system. A table defines the different bonds between the machines by marking them. From these information, the designer can make groups or form virtual machines. The analysis starts from elementary machines whose operational constraints are expressed in a table. The result of this first phase is illustrated by figure 7.

	R1	LATHE	R2	CNC	LPC1	CONV1	LPC2	R3	ASSE	STO	R4	CONV2
R1		CC1			CO1	CC5						
LATHE	CC1				CO1	CC5						
R2				CC2	CO1	CC5						
CNC			CC2		CO1	CC5						
LPC1												
CONV1	CC5	CC5	CC5	CC5	CO1							
LPC2												
R3							CO2		CC3			CC6
ASSE							CO2	CC3				CC6
STO							CO2				CC4	CC6
R4							CO2			CC4		CC6
CONV2							CO2	CC6	CC6	CC6	CC6	

M1=R1+LATHE; M2=R2+CNC ; M3=R3+ASSE ; M4=R4+STO

Figure 7 : First step of machine aggregation in virtual machines

The same analysis is then achieved step by step with the virtual machines obtained at each level. At the end the process structure is represented by a tree as shown in Figure 8.

642

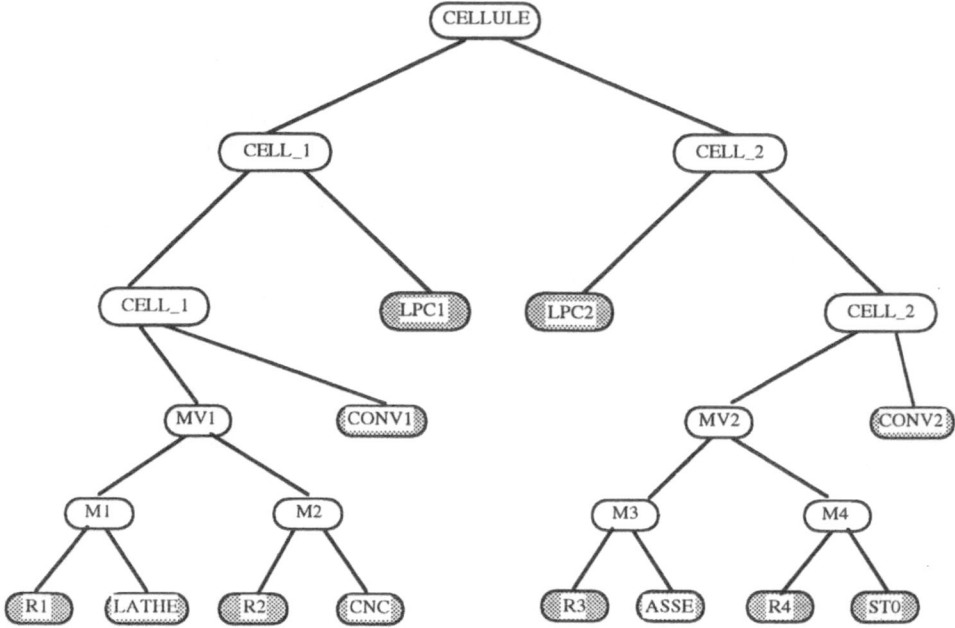

Figure 8 : The structural composition of machines : a bottom-up analysis

In that figure the different structural levels of the process show clearly the importance of each machine within the process.

3.3.   The manager framework

3.3.1.   The basic concepts.

The manager functioning is divided in three steps (Fig. 9) :
- a step where the goals to be solved are generated ,
- a step of goal solving,
- a step of correctness of the solving.

To each of these steps correspond basic concepts whose understanding is important in the view of this problem solving implementation. First of all, let us introduce the concept of a goal.

Definition : A goal [7].
It is the modification of the working mode of a machine

According to the environment of the manager, a goal can be generated directly from an order sent by the operator (through the interface operator) or by a superior level of the hierarchical level. The goal is then analysed by the manager in order to ensure that it is coherent with respect to the state of the system. For example, the order "on" entered by an operator will be interpreted as a goal to be solved.

Goals can also be generated, indirectly, after the modification of the state of one or more machines. More particularly, a goal can be bond with each transition of the state diagram of each effective machine :

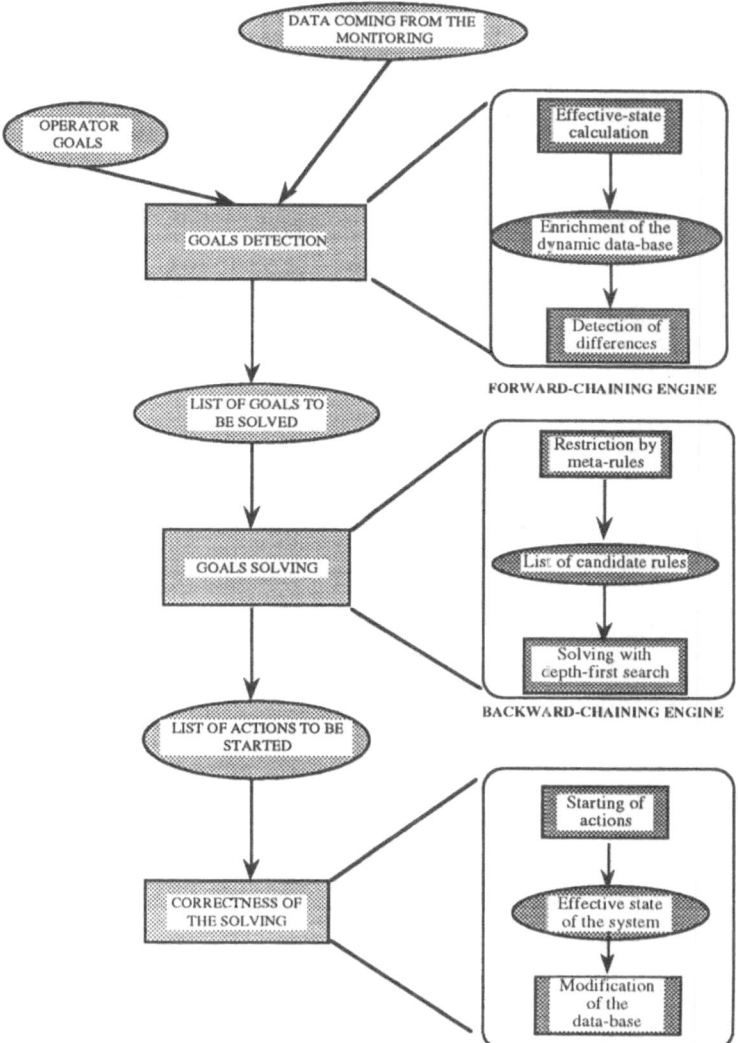

Figure 9 : The manager organisation-chart

TRANSITION	DERIVED ACTION	DERIVED GOAL
0->1	on	?on
1->0	off	?off
1->1	self-tests	?self-tests
1->2	init	?init
1->3	failure-detection	?failure-detection
1->4	running	?running
2->0	complete-stop	?complete-stop
2->3	failure-after-init (fai)	?failure-after-init (fai)
2->4	run-after-init (rai)	?run-after-init (rai)
3->0	off-under-failure (ouf)	?off-under-failure (ouf)
3->1	restart	?restart
4->1	stop	?stop
4->3	stop-under-failure (suf)	?stop-under-failure (suf)
4->5	run-under-failure (ruf)	?run-under-failure (ruf)
5->3	stop-after-ruf (sur)	?stop-after-ruf (sur)

Table 1 : Actions and goals derived from the state diagram of a machine

The second concept that we introduce here is the "state" concept. We distinguish three kinds of state :

- an **effective state** : It is the real state of a machine obtained directly from data issue from the process,
- a **computed state** : It is a state of a machine computed by the manager,
- an **anterior state** : It is the last state of a machine which has been validated by the manager.

The last basic concept is the action concept.

Definition : The actions [7]
They are all the operations that permit to go from a stable state to another stable state. They are triggered by the success of a goal solving.

Actions can be of different nature : effective actions (see Table 1), the modification of the manager data base in order to take into account the changes into the process, the modification of the piloting parameters ... The transition of a machine between two states can require several actions. In this case, all of them must be generated in the cycle of inference.

3.3.2. Specification of the inference engines and the associated knowledge.

The manager is structured around a data base that contains two kinds of knowledge : a static knowledge obtained from the description of machines and a dynamic knowledge corresponding to the current state of each machine. Schematically, the manager is invoked in two cases by the modification of its data base :

- for recovery purposes, when the internal structure of the process has changed ; in this case we have a bottom-up flow of control.

- for strategic control required by the higher levels of control (including the operator interventions on the system) ; in this case we have a top-down flow of control.

The structuring proposed here allows us to take into account by the same way, these two flows of data.

3.3.2.1.   The generation of goals.

This step is achieved by a rule-based system, operating in a forward chaining.

3.3.2.1.1.   Description of rules.

These rules are written according to the traditional formalism of production rule. To write the premises of the rules, two predicates are used :

- *state_test*

This predicate is useful to test if the state of a sub-machine of the considered machine has changed.

- *state_evaluation*

This predicate permits to compare the anterior_state of a machine with its effective_state.

Corresponding respectively to each of the previous predicates, state_computing and goal_computing are used in the writing of the consequents of rules :

- *state_computing*

This predicate permits to compute the new effective state of a machine knowing that at least one of its sub-machines state has changed. It is worth understanding here that this computing depends on the constraint type hat link the components of the considered machine. As an example, in the cell of EC Lille, let us consider M1 made up of the lathe and the Afma robot. If the lathe is stopped, the following rule must be applied :
R1 : (*state_test* M1)   -->   (*state_computing* M1)
It implies that the effective state of the machine M1 is computed to be "live_on" according to the CC constraint between the lathe and the robot (Fig. 10).

- *goal_computing*

To change effectively the state of a machine after the change of the state of one of its component, some actions must be achieved. To know what kind of actions are to be achieved, the manager begins to generate as many goals to be solved than transitions met between the anterior state and the effective state. This generation is started by using the predicate goal_computing in the consequent part of some rules

In consequence, two kinds of rules are applied in this step :
type1 : (*state_test* <machine>)   -->   (*state_computing* <machine>)
type2 : (*state_evaluation* <machine>)   -->   (*goal_computing* <machine>)

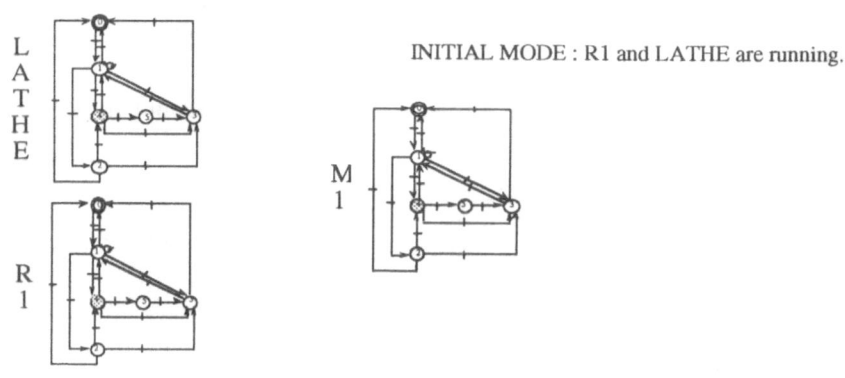

LATHE
R1
INITIAL MODE : R1 and LATHE are running.
M1

Figure 10a : Modification of the state of a virtual machine

LATHE
R1
DEGRADED MODE : LATHE has been stopped for an external reason.
M1
Current State
Anterior State
LATHE
R1
FINAL MODE : R1 has been stopped by the manager.
M1

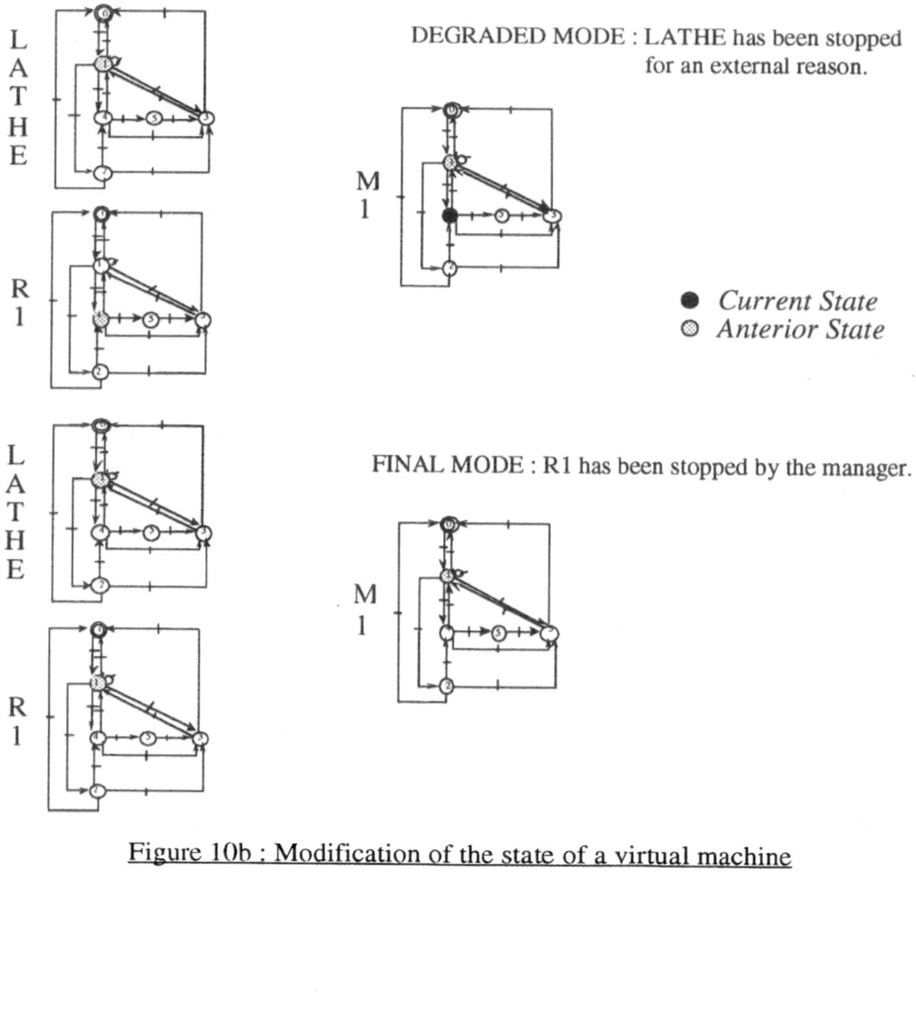

Figure 10b : Modification of the state of a virtual machine

3.3.2.1.2. Description of the inference engine

The main characteristics of this engine are the following :

- *a width forward-chaining* ;

This choice has be done because this mechanism corresponds to a bottom-up propagation of machines data through the tree modelling the process structure. Indeed, as a machine of level i+1 inherits the states of its sub-machines of level i, before computing its effective state, we must have established the state of its sub-machines.

- *an irrevocable functioning*,

- *a monotonous functioning*.

The two last characteristics explain that the engine does not make any attempt and never refutes its previous conclusions. It only infers new facts according to its fact base.

3.3.2.2.   Goal Solving and action triggering
It is the second step of the manager functioning. It starts when goals to be solved have been generated in the first step.

3.3.2.2.1.   Description of rules

These rules are used to translate a goal to be solved in sub-goals or actions in typical contexts during the exploitation of the process. Their syntax is the following (given in BACKUS-NAUR FORM) :

```
<rule>          ::=     <triggers>              <-      <consequents>
<triggers>      ::=     <goal> / <fact>
<consequents>   ::=     <goal> / <actions> / <constraint>
<goal>          ::=     <goal-operator><parameter>
<fact>          ::=     <fact-operator><parameter>
<action>        ::=     <action-operator><parameter>
<constraint>    ::=     <constraint-operator><type-of-constraint>
```

In this formalism, <parameter> always refers to a machine.

Facts are useful to reduce the rules started in each inference cycle. They are used like constraints in constraint programming languages (PROLOG III, CHARME, PECOS). They are introduced by the reserved-word "**machine_state**" which enable to take into account the state of machines in the solving.

Goal-operators and action-operators are listed in table 1. Constraint-operator is the reserved word "**constraint**".

These rules are off-line generated from research trees corresponding to the structural model of the process presented previously in part 3.2.2.. For each goal is assigned a research tree. Its root is the initial goal and its leaves are sub-goals associated with some parts of effective machines. For example, let us considered the virtual machine M1 made up with the AFMA robot (R1) and the LATHE. If M1 is in running and the goal is (?stop M1), the solving will use the following rule :

rule1 : (**?stop** M1) (**machine_state** running) <- (**?stop** AFMA) (?stop LATHE)
(**constraint** CC)
rule2 : (?stop AFMA) (**machine_state** running) <- (**stop** AFMA)
rule3 : (?stop LATHE) (**machine_state** running) <- (**stop** LATHE)

3.3.2.2.2.   Description of the inference engine

The engine is characterized by a backward chaining with backtracking and non-monotonic control (Fig. 11). The solving is started when goals are put in the GOAL_BASE (the base of goals to be solved). First of all, the goals are ordered to favour the highest machines of the hierarchy concerned by actual problems. This enables the manager to have a global vision of problems to be solved. In this case. the following steps are executed :

Step 1 : Choice of candidate rules
Depending on the machines concerned by the goal to be solved, the initial rule base is reduced in order to improve the filtering step. This is performed by a forward engine using meta-rules based on the structural description of the process.

Step 2 : The sort of the candidate rules.
By using the QuickSort algorithm, the BASE_OF_CANDIDATE_RULES is sorted depending on the priority of each rule. These priorities are heuristically fixed by the operator according to its troubleshooting experience. The sort is achieved only once in a cycle of inferences because the priorities are static.

Step 3 : The filtering step.
Starting with the first goal of the GOAL_BASE, this step consist of searching the rule whose triggers are instantiated by the FACT_BASE. Thus, its consequents are stacked in the GOAL_BASE (for sub-goals and constraints) and in an ACTION_BASE (for the actions). In the GOAL_BASE, the added sub-goals are evaluated before the other goals : it is the mechanism of the **depth-first search.**
When whole the evaluation of the consequents is successful, the initial goal is solved. The actions of the ACTION_BASE are then executed and the solving restarts with another goal. Else, the current rule is removed from the BASE_OF_CANDIDATE_RULE and this step is repeated with another rule. In this case, the actions of ACTION_BASE are unstacked.

The solving is completed when the GOAL_BASE or the BASE_OF_CANDIDATE_RULES is empty.

3.4.   Conclusion

The manager software has been completely developed in Le_Lisp on VAX. As it only uses standard features of Le_Lisp as "*plist*", it can be easily carry on other computers running Le_Lisp package. Concerning its integration in CASPAIM tools, it is already connected to the piloting software through the concept of actions. In order to manage the working modes, the actions of the manager can modified the data-base of the meta-rules of the piloting function.

As the piloting software, the manager has been developed according to a centralised architecture. In order to be more reactive to the modification of the surveyed system, we are studying a **hierarchical architecture** today. This architecture will be defined in the context of the global supervision of a manufacturing system. The following section presents our specifications to design a monitoring function in this context.

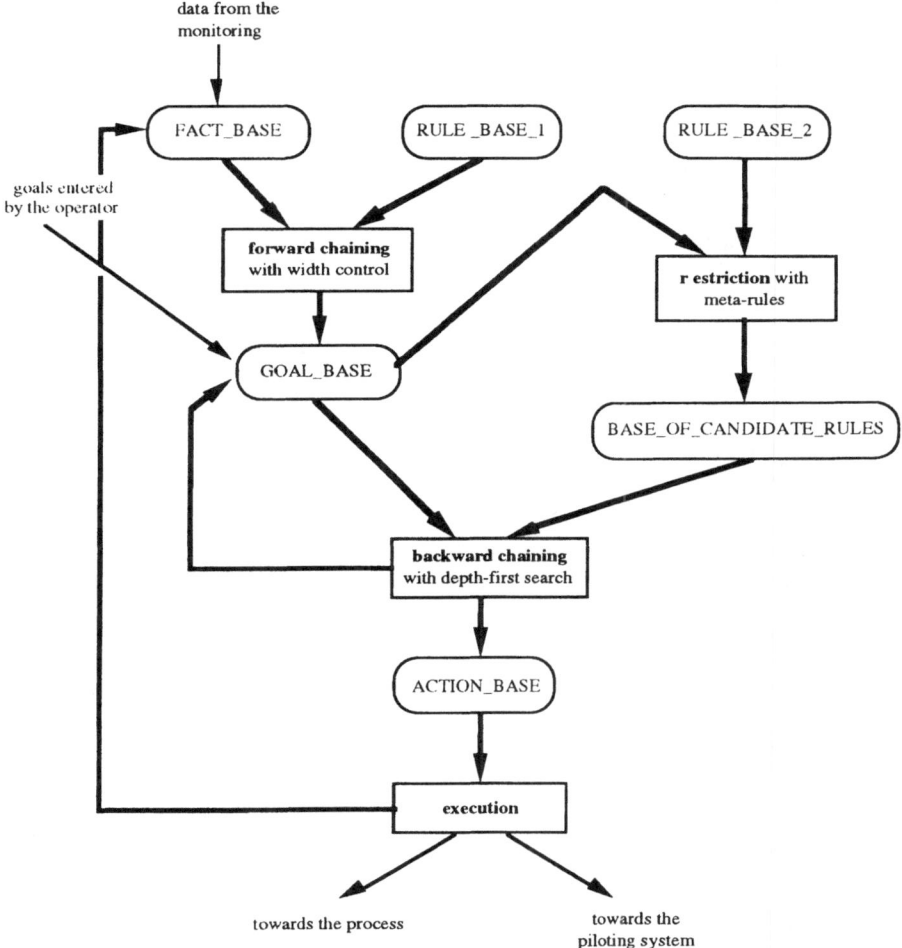

Figure 11 : The backward-engine organisation-chart

4. A MODEL-BASED DIAGNOSTIC SYSTEM FOR ON-LINE MONITORING.

4.1. Introduction

In past few years some authors as A.E.K. Sahraoui [8], [9] have introduced the on-line monitoring concept to be used in order to survey flexible workshop. The main functions of such a system are the detection , the diagnosis and the failure-recovery. In this section we are going to present the main features of our approach to implement an on-line diagnostic system.

4.2. The modelling method.

In the last decade a lot of systems based on artificial-intelligence techniques have been developed for diagnostic purpose. However most of them are Expert Systems of first generation [10]. These systems are characterised by a shallow knowledge which consists in an "associative representation of heuristically determined cause-effect or symptom-fault relations" [11]. This technique presents some limits specially with respect to the problem of how to ensure the completeness of the encoded knowledge. In order to deal with this problem, the authors propose to base the on-line diagnosis on a functional modelling of the process to be surveyed.

The basic idea of this modelling is to extract a causal model from the stucture of a system and from its global behaviour. Consequently, the method is divided in three main steps.

Step 1 : It is structural decomposition where the topology of the system is gradually described.

Step 2 : This step consists of a functional decomposition of each component described in the structural model. For each of them, the designer only retains the generic functions [12] which could be failed during the system life.

Step 3 : In this step, the designer has to establish the **causal links** between the functions of the components. This is achieved from the functions of the system down to the elementary functions of basic components. The "structuro-functional" model obtained in previous step is used as support to drive the process. For each function, the designer has to answer about the following questions :
1 - Which sub-functions enable to create the function ?
2 - Which higher level functions use this one to achieve their services ?
Contextual functions are added in this step. Figure 12 shows an extract of the functional model established for a machining-machine.

In [12], we proved that the obtained model called **Functional Graph** (FG), is a causal model useful to study the **observability** of the system to be surveyed. This model is also used to implement a distributed-diagnostic system whose main features will be presented in the following parts.

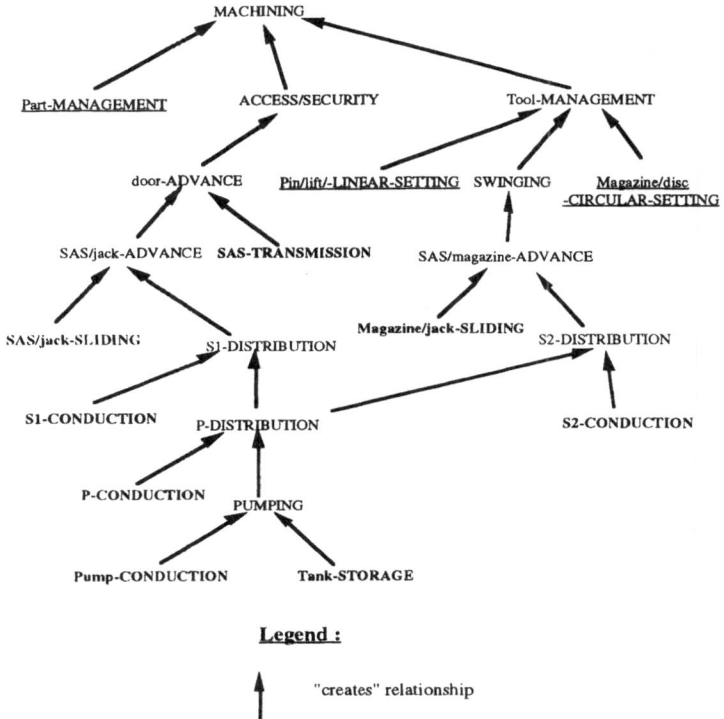

Figure 12 : An extract of the Functional Graph of a machining-machine

4.3.   The Causal Temporal Signature or CTS.

Our approach of the diagnosis in FMS context is based on two main hypotheses :
1) the process observations are dynamic,
2) the multiple-failures hypothesis.

In order to take into account these two hypotheses, we introduced in [20], [9] and [13] the concept of Causal Temporal Signature (CTS).

4.3.1.   Definition of symptoms

Following De Kleer's definition [14], [15], a symptom is "any discrepancy between the predicted behaviour of a system and its observed behaviour". To detect symptoms, a temporal model specifying the local behaviour of the process is defined for each sensor (Fig. 13). This temporal model links a Start-Event (it is generally a control order) to a sensor report which must occur within a temporal window called **validation interval** and noted I_{CRi} (CRi is a sensor).

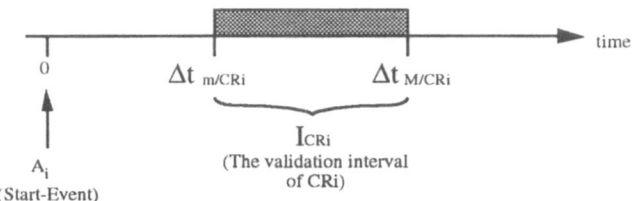

Figure 13 : The temporal model used for process reports interpretation.

From this generic model two kinds of symptoms have been introduced in [20] :

Definition : Symptoms of type I noted S^1_i.
These events characterise reports not occurred in the validation interval in which they were expected. These symptoms are always generated at the end of their validation intervals.

Definition : Symptoms of type II noted S^2_i.
They characterise unexpected reports. They are reports which are not occurred within a validation interval.

In symptom notation, the exponent always defines the type of the symptom and the suffix specifies the "sensor" which emits the report (Notice here that at an abstraction level, a machine can be considered as a sensor).

4.3.1.1. CTS formalism.

A CTS is a kind of rule based on the concept of **temporal redundancy** [9]. Formally a CTS is defined as follows :

Definition 5 : A Temporal Constraint Signature or CTS.
We call CTS all relations linking a ordered sequence of characteristic events to one or several causes.

The most general form of a CTS is :

$$CTS_i : (occ\ Start\_Event_j) * (tc_1\ S_1) * (tc_2\ S_2) * \bullet\bullet\bullet * (tc_k\ S_k) ===> causes$$

It is worth noting here that **a CTS is not a production rule**. The premises of a CTS are verified progressively according to the occurred events. In this formalism :
- **occ** is a predicate whose function is to fix the temporal reference concerning a type of event called Start-Event ; this temporal reference is used by the inference engine to estimate the temporal constraint (tc) of the first symptom of the characteristic event sequence.
- **Start_Event$_j$** represents an order sent by the Control Part to the Operative Part ; this event occurrence serves as temporal reference both for the detection of symptoms and for the interpretation of symptoms at diagnostic level.
- **tc_k** is a temporal constraint which links symptom S_k to the previous event in the sequence.

- "*" is the sequential operator ; this temporal operator may be translated by "**will-be-followed-by**".

In CTS formalism, the causes can be of two types :

■ **certain,**

In this case, the inference leads to definitive conclusion :
- the failure of a function of the system and the cause is then noted (**fail** <functions>),
-or the failure of a sensor and he cause is then noted (**lock** <sensor> <value>).

In this formalism "*fail*" and "*lock*" are predicates of our modelling language and <value> parameter identifies the type of a sensor locking. This information is useful for recovery task.

■ **uncertain.**

In this case the local diagnostic process can only infer hypotheses on the faulty state of :
- the function and the assumption is written (**H1-fail** <function>),
- a sensor ; the assumption is written (**H1-fail** (lock <sensor> <value>)).

"*H1-fail*" is also a predicate. It means that it is "strongly possible" that the considered fact is true. Furthermore, we will introduce another predicate for hypothetical fact which will give another kind of modality useful for the Overall Diagnosis efficiency.

To express the temporal constraints in a CTS, designers can use three types of temporal entities :

- **Instant** noted $\Delta t,$
- **Period** noted **p** and defined by an interval,

- **Duration** noted **d** and defined by a doublet (Δt, v) where "Δt" is a date defined relatively to the date of previous event (event-1) occurrence ; "v" gives a delay where the second event (event-2) must be true ; $v \in \mathfrak{R}$.
For more details about the meaning of temporal constraints and the rules to write CTS, readers can refer to [20] and [13].

4.3.1.2.   CTS recognition scheme

The solving principle consists in extracting symptoms belonging to each CTS, from a sequence of symptoms at the entry of the diagnostic task. Its main features are : the **closed-world hypothesis**, a **breadth-first search** and the **one-symptom-for-one-CTS hypothesis**. Thus, the mechanism consists of the building of a solving tree whose each node represents a specific solving context called **world** (it is a set of assumptions). Basically, the solving is controlled with three kinds of events :
- the start-events,
- the symptoms,
- the expiration of a watching-dog that corresponds to the rejection of an assumption.

At "i" a given level of the solving tree, the jth world $M_{i,j}$ is defined by the following sets :

- $R_{i,j}$: it is the set of CTS which are not in processing,
- $L_{i,j}$: it is the set of CTS in processing,
- $V_{i,j}$: it is the set of the inferred causes.

$M_{0,0}$ is the initial world defined by $R_{0,0} = \{CTS_1, CTS_2, ...,CTS_n\}$ (e.g. the base of CTS), $L_{0,0} = \{ \ \}$, $V_{0,0} = \{ \ \}$.

The resolution is completed only one world $M_{i,j}$ ($i \neq 0$ and $j=0$) with $V_{i,j} \neq \{ \ \}$ and $L_{i,j} = \{ \ \}$ remains. If $V_{i,j} = \{ \ \}$, the entry sequence of symptoms must be sent to the operator. This feature is useful for debugging steps, because it allows the designer to complete the knowledge base. If $V_{i,j} \neq \{ \ \}$, the inferred causes can be exploited for recovery purpose.

With all these specifications, the CTS recognition algorithm is the following :

```
(1) (* Initiation *)
i <- 0; (* level of the solving tree *)
j <- 0; (* number of world - 1 *)
new-j <- 0 ; (* new number of world at level i+1 *)
(* Definition of M0,0)
Ri,j <- {CTS1, CTS2, ...,CTSn};
Li,j <- { };
Vi,j <- { };
(2) (*Interpretation of Event *)
For k <- 0 to j do begin
        If not (watching-dog(Mi,k)) then begin
                If Symptom then
                        (* "aux" identifies the number of CTS of a world matched
                        by the current Symptom *)
                        If match(Symptom, Mi,k, aux) then begin
                                new-j <- new-j +aux ;
                                build-world (Symptom, Mi,k, aux ); (* this function
                                builds "aux" son-worlds from the matching of
                                Symptom with the CTS of Mi,k *)
                        end
                else begin
                new-j <- new-j+1
                M(i+1),new-j <- Mi,k
                end
        else erase (Mi,k) ;
end ;
j <- new-j -1;
i <- i+1 ;
(3) (* Interpretation of worlds *)
If (j=0) and (i≠0) and (Vi,j ≠ { }) then begin
        send (Vi,j) ; (* Starts the Overall Diagnosis *)
        If Li,j = {} then goto (1) else goto (2) ;
        end
else begin
```

> **If** $(V_{i,j} = \{\})$ **then** *alert-operator* **else** goto (2).
> end ;
> end.

In order to validate the CTS recognition algorithm, we have developed a prototype in Le_Lisp with a centralised approach of the diagnosis.

4.4. The multi-agent framework of the diagnostic system.

To deal with the real-time constraint applied to the diagnosis, we are developing a multi-agent framework based on the functional model of the process to be surveyed. The idea is to use FG as a network (called FG_network) to transfer the assumptions generated by local **agents** [16], [17]. From the model-based approach point of view, two steps can be formally distinguished in the diagnostic process : the **local diagnosis** and the **Overall Diagnosis**. However, from problem solving point of view, the local diagnosis implements a **localisation** task and the Overall Diagnosis is divided in a **identification** task and in a **prognostic** task [18], [13].

4.4.1. The local diagnosis or localisation.

The localisation is using the CTS formalism that has been presented previously. This solving task is local to some of the FG_network called **Triggering Node**. This type of node must be both controllable and observable [12]. The activity of a Triggering Node begin when it receives symptoms. Depending on the quality of its conclusions it will send different types of assumptions through FG_network. This is why it is called Triggering Node.

4.4.2. The Overall Diagnosis.

4.4.2.1. The management of assumptions.
This second step of the diagnosis is based on the propagation of assumptions via FG_network. Two types of assumptions are managed :

■ *Identification assumptions typed by H1-fail.*

They are managed by an identification mechanism. They are associated to one of the following contexts :

- The local inference of uncertain conclusions (refer to § 4.2.4)
- The propagation of certain conclusions from a Triggering Node for identification purpose. Let us consider the example of figure 14. Assume that door-advance Node is a Triggering Node and that local diagnosis has inferred that the function is failing : **(fail door-advance)**. From this data, the identification process will propagate the assumptions **(H1-fail SAS-TRANSMISSION)** and **(H1-fail SAS/jack-advance)** to the respective nodes. This type of assumption means that at least one of these nodes is sure to be failing.

■ *Validation assumptions typed by H2-fail.*

They are generated when we have to confirm or reject an uncertain conclusion produced by a Triggering Node. As regards the example of figure 14, the data **(H1-**

fail door-advance) leads to propagate **(H2-fail SAS-TRANSMISSION)** and **(H2-fail SAS/jack-advance)** to the respective nodes.

This type of assumption introduces another form of modality. It means that the node "may-be-failed". In other terms, as regards to our previous example, it means that the two propagated assumptions could be both rejected. Furthermore, we will see that the distinction of these two types of assumptions brings more efficiency to the reasoning mechanism.

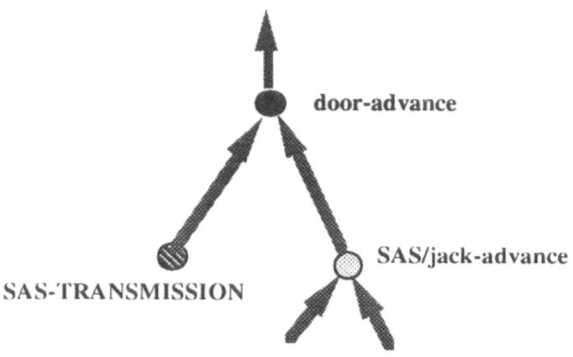

door-advance

SAS-TRANSMISSION

SAS/jack-advance

Figure 14 :

The assumptions are managed using two types of strategies [13] :

■ *Strategy of competitive analyses.*

This is used with multiple failure assumption. All the hypotheses have the same probability. Consequently all the investigation ways are examined in parallel.

■ *Strategy of discriminating analyses.*

The basic idea of this strategy is that if a function fails then all functions of the superior level using its services will also be failed in a specific context. Thus, to reject the hypothesis of the failure of a function, it can be sufficient to verify that one of its dependent function works well.

4.4.2.2.   The main mechanisms

The identification and the prognosis are implemented by using three propagation processes :

■ *Identification propagation.*

This is a top-down mechanism of refinement of causes generated by a father-node. The refinement requests are implicitly answered by the prognostic propagation.

■ *Validation propagation.*

This is mechanism similar to the previous but based on the propagation of H2-fail assumptions. In this mechanism, validation requests are explicitly answered by a backtrack propagation called **H2-answer**.

■ *Prognostic propagation.*

This is a bottom-up mechanism of diffusion started after a localisation whose issues are certain, or after the identification of the origin of a failure (Fig. 15).

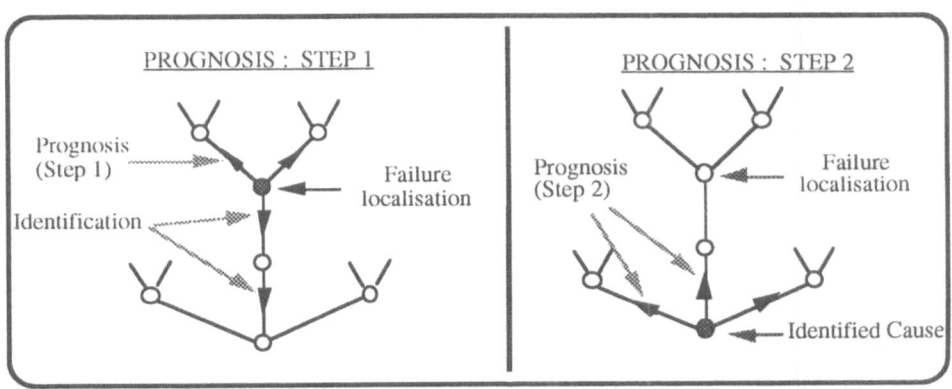

Figure 15 : The different cases of prognosis.

Since one of the prognosis goals is to freeze some of the control functions (for safety purpose), its conditions of stopping must be specified, in order to avoid a systematic propagation up to the Principal Node of FG-network. These conditions are established depending on the degree of redundancy of each function. To simplify, they can be expressed by the following rule :

If a function can be found in the system which can give the same service as the failing function then the prognosis must be stopped.

The propagation of assumptions between the agents is controlled by using the message mechanism of oriented-object languages. In the agents, the assumptions are confirmed or rejected by using two types of treatment depending on the observability of the agent : the **direct-validation** treatment and the **indirect-validation** treatment. These treatments are implemented by using the **demon** mechanism of frame languages.

4.4.2.3.   The classification of the agents.

According to their structural or dynamic properties, four classes from agents have been stated (Table 2) :

658

Agent Class	Structural property	Dynamic properties	Mechanisms used
Triggering Node	at least one son	controllable directly observable	all excepted H2-answer and direct-validation
Direct-validation Node	at least one father	directly observable	identification prognosis direct-validation
Indirect-validation Node	several fathers or sons	indirectly observable	validation-propagation identification prognosis H2-answer indirect-validation
Simple Node	one father and one son	indirectly observable	validation-propagation identification prognosis H2-answer

Table 2 : The classification of agents.

We are developing a tool of prototyping of diagnostic system in KOOL [19] which is an hybrid language. It implements objects and frames specifications. We have stated an hierarchy of agent classes as illustrated by figure 16.

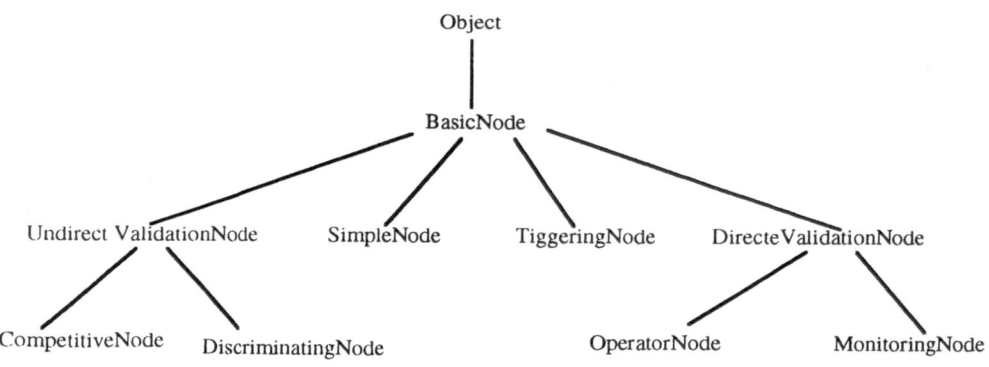

Figure 16 : Agents hierarchy under KOOL.

As an example, let us present the specification of the BasicNode class :

```
{ Class : NoeudBase
Super : Object
<-- Nom-fonction     : () ; Gives the name of the function modeled by the node
      Mode           : Mono
      Type           : String
-- Genre             : ()
      Mode           : Mono
      Type           : [Triggering, D_validation, I_validation, Simple]
-- Fils              : [ ] ; list of its sons
      Mode           : Multi
      Type           : BasicNode
-- Pères             : [ ] ; list of its fathers
      Mode           : Multi
      Type           : BasicNode
-- Etat              : () ; It is the state of the function
      Mode           : Mono
      Type           : [normal, degraded, unavailable]
      Init           : normal ; Initial value of Etat
      WhenFilled : [prognosis] ; The prognosis is started when Etat=unavailable
-- Nb-alternatives : () ; Specifies the degree of redundancy. If Nb-alternative=0 then
                     Etat := unavailable
      Mode           : Mono
      Type           : Number
      Default        : 1 ; Specifies the default value
-- Composant         : () ; give the name of the current component associated to the
                       function.
      Mode           : Mono
      Type           : String
-- H1-fail           : [ ] ; List of H1-fail assumptions
      Mode           : MultiH1
      WhenFilled     : [identification]
-- H2-fail           : [ ] ; List of H2-fail assumptions
      Mode           : MultiH2
      WhenFilled     : [propagation-H2]
      WhenRemoved : [reponse-H2]
-- Diag-externe      : [ ] ; list of external data send by other agents
      Mode           : Multi
      WhenFilled     : [evaluer1] ; Sorts the data of Diag-externe

Methods              :
      AddData        : AjoutDonnee > }
```

The function called "*AjoutDonnee*" enables the system to add data in Diag-externe attribute. It is activated by the messages send by other agents.

4.5. Conclusion

In this section, we have presented the main specifications of CASPAIM project to implement a multi-agent system for one-line diagnosis system. At the moment, we are developing an ergonomic and friendly tool based on these specifications. The tool is using KOOL package whose interest is to enable the designer to built a prototype supported by Le_Lisp, and after the validation step, allows him to compile is specification in C code.

In future, we are interested by the problem of the sharing of agents on real computers. This implies to solve the problem of how to synchronise the clocks associated with each of these computers in order to have the same temporal reference.

5. GENERAL CONCLUSION

We have illustrated in this chapter three applications of AI-techniques at the supervision of Flexible Manufacturing Systems. Historically, we have first developed the piloting function [6] and the manager software [7], and thereafter the monitoring function [13]. This explains that the techniques are different. The piloting function and the manager software use both a traditional programming with production rules. Instead the diagnostic function of the monitoring has been developed using oriented-object programming.

The right definition of supervision tasks and the design methodology of these tasks are not currently be well defined and stabilised. The presented works have to be considered as a first attempt in that way. A major difficulty concerns the impossibility of using a sufficient powerful model of the control of the system able to mix the different point of views related with :
- low level control,
- decision and strategy to drive in a optimal the flexible flow,
- the discrete evolution of the working modes due both to the reactive adaptation of the process for giving the constraints of production and also to unpredictable failures.

Different kinds of mathematical models have been proposed to model the process according to one of these purposes : Petri Nets, queuing network, finite automatons ... They are not sufficient to cover the real complexity of the control of FMS. We believe that AI-techniques are then an alternative solution to help the design and the implementation of the supervision level.

6. REFERENCES

[1] Advances in Grafcet, APII - Volume 27 - n° 1/1993

[2] D. Corbeel, J.C. Gentina, C. Vercauter, *"Adaptive Petri-nets for real-time applications"*, IMACS Digiteh'84, Patras (GRECE), 1984.

[3] M. Kapusta, Génération assistée d'un graphe fonctionnel destiné à l'élaboration structurée du modèle de la partie commande pour les cellules de production flexibles dans l'industrie manufacturière, Doctorate thesis of the university of Lille (France) in 1988

[4] D. Cruette, Méthodologie de conception des systèmes à événements discrets : application à la conception et la validation hiérarchisée de la commande de cellules flexibles de production dans l'industrie manufacturière, Doctorate thesis of the university of Lille (France) Février 1991.

[5] J.L. Laurière, *"Représentation et utilisation des connaissances."*, T.S.I., vol 5, n°1, pp.25-42,n°2,pp. 109-133, 1982.

[6] E. Craye, De la Modélisation à l'Implantation automatisée de la commande hierarchisée de cellules de production flexibles dans l'industrie manufacturière, Doctorate thesis of the university of Lille (France) in 1989

[7] S. Bois, Intégration de la gestion des modes de marche dans le pilotage d'un système automatisé de production, Doctorate thesis of the university of Lille (France), Novembre the 28th, 1991.

[8] A.E.K. Saharaoui, Contribution à la surveillance et à la commande d'atelier, Doctorate thesis of the university of Toulouse (France) in 1987.

[9] A.K.A. Toguyeni, E. Craye, E. Castelain, *"From the traitement of failures to the management of working modes "*, Proceedings of IMACS MCTS 91, Vol. 1, pp. 595-601, Lille, may 1991.

[10] B. Chandrasekaran, J.W. Smith, J. Sticklen,*"Deep Models and Their Relation to Diagnosis""*Artificial Intelligence in Medecine, Vol. 1, 1989, pp. 29-40.

[11] G. Spur, D. Specht, S. Weiss, *"Integration of learning approaches for maintenance tasks"*, in Computers in Industry, Elsevier Editions, N° 17, pp 269-277.

[12] A.K.A. Toguyeni, E. Craye, J.C. Gentina, *"An approach for the placement of sensors for on-line diagnostic purposes"*, in the Procedings of IFAC'93, Sydney, Australia, 1993.

[13] A.K.A. Toguyeni, Surveillance et Diagnostic en ligne dans les ateliers flexibles de l'industrie manufacturière, Doctorate thesis of the university of Lille (France) in 1992.

[14] J. de Kleer, J.S. Brown, *"A Qualitative Physics Based on Confluences"*, Artificial Intelligence , Vol. 24, 1984, pp.7-83.

[15] J. de Kleer, B.C. Williams, "*Diagnosing Multiple Faults*", Artificial Intelligence, 1987, vol. 32, n°1, pp 97-130.

[16] B. Chandrasekaran, "*Towards a taxonomy of problem solving types*", AI Magazine, 4 (1), 1983, pp. 9-17.

[17] J.P. Haton, N. Bouzid, F. Charpillet, M.C. Haton, B. Lâasri, H. Lâasri, P. Marquis, T. Mondot, A. Napoli, Le raisonnement en intelligence artificielle, chapitre 10, 1991, pp. 387-429.

[18] A. Rault, "*Detection and diagnosis system and model-based approach*", in the Proceedings of AIPAC'89, Nancy, 1989, pp. 34-43.

[19] S.A. Bull, "*KOOLv2 : Guide d'utilisation (1)* ", 95 F2 89SM, 1990
S.A. Bull, "*KOOLv2 : Guide d'utilisation (2)* ", 95 F2 24SR, 1990
S.A. Bull, "*KOOLv2 : Manuel de référence*", 95 F2 98SM, 1990

[20] A.K.A. Toguyeni, E. Craye, J.C. Gentina, "*A method of temporal analysis to perform on-line diagnosis in the context of Flexible Manufacturing System*", Proceedings of IECON' 90, Vol. 1, pp. 445-450, Pacific Grove-California, november 90.

[21] I.N.R.I.A., Le Lisp version 15.21, 25 décembre 1987.

22

A SURVEY OF KNOWLEDGE-BASED INDUSTRIAL SCHEDULING

K.S. HINDI, M.G. SINGH
Department of Computation
UMIST
Manchester, U.K.

1   Introduction

Industrial scheduling problems are invariably very complex; they require taking into account such different and conflicting factors as due date requirements, cost restrictions, production levels, machine capabilities, alternative production processes, order characteristics, resource characteristics and availability. Computationally, even the simplest of such problems are very hard.

Traditional approaches to industrial scheduling, based mainly on operational research (OR), have been concerned primarily with generating off-line schedules. But, in fact, the requirement is for real-time, on-line, dynamic scheduling, based on the actual state of the system. Unfortunately, traditional scheduling approaches fail to meet this requirement. Moreover, they are not altogether satisfactory even for generating off-line schedules, for the following reasons:

- due to computational considerations, OR-based models are usually drastically simplified, so much so that the problem they deal with may have only a tenuous relation to the real one.

- algorithmic approaches usually have the objective of optimising a goal function, based on a single parameter and find it difficult to cater for multiple objectives.

- the scheduling process often entails decisions based on heuristic considerations that cannot be integrated into a mathematical model.

The deficiencies of classical scheduling approaches aroused interest in knowledge-based scheduling systems. Research in this area has tended to reflect the mainstream AI research of the time during which it was conducted [63]. Thus the first work on applying AI to job-shop scheduling was Gere's dissertation [33], which incorporates

663

S. G. Tzafestas and H. B. Verbruggen (eds.),
Artificial Intelligence in Industrial Decision Making, Control and Automation, 663–685.
© 1995 *Kluwer Academic Publishers.*

some of the Newell and Simon's ideas on 'complex information processing' of the late 50s and early 60s. Iskander's dissertation [38] applies some of the findings of pattern-matching research from the late 60s and early 70s to the job-shop problem. When logic programming began to arouse interest during the late 70s, Bullers, Nof and Whinston [16] used predicate calculus to model and solve manufacturing problems. During the past few years, increasing effort has been expended on attempting to apply AI production systems and results from the area of AI planning and search.

However, most knowledge-based scheduling systems reported in the literature are still in the research or prototypical phase and only a limited number is actually in use. Among the reported systems, job-shop scheduling systems are the most common subject for AI research, possibly because of the difficulty of job-shop scheduling. Flexible Manufacturing Systems are the second most common subject, probably because the dynamic nature of these systems frequently precludes effective scheduling by manual or static methods. Recently, project scheduling, under resource and / or with time windows constraints has been receiving attention.

In the following overview [17], particular attention is paid to the knowledge representation formalisms and search techniques employed by some of the most promising systems. In the interest of clarity, an attempt has been made to examine knowledge acquisition and representation issues and plan generation techniques separately, even though they are in practice almost inseparable. Occasionally, reference will be made to recent findings in AI that could prove beneficial scheduling.

2 Knowledge Acquisition

Knowledge-based systems are successful to the extent that the knowledge embodied in the system is a complete and accurate representation of the problem domain [25]. Traditional knowledge acquisition techniques include interviews, protocol analysis and machine learning techniques. The common feature of these techniques is that they all depend to various degrees on the expert's cooperation.

However, in industrial scheduling the more complex and novel the problems are, the less likely it is that experts on them exist [68]. This is reflected in current knowledge-based scheduling systems. The literature on well known systems reveals very little, if anything, about knowledge acquisition. Most scheduling rules are obtained from OR theory and only the characteristics and features of the manufacturing system concerned and the organisational constraints are extracted through interaction with managers.

Simulation-assisted knowledge acquisition could be a practical response to the scarcity of expertise [66]. This approach is suitable in situations where it is possible to establish a model to predict the effect of changes in input parameters on system performance. The system designer becomes the domain expert through simulation experiments with the model.

Another problem faced by system designers is the inability of production schedulers to articulate fully the mental processes they use to devise a schedule. For this reason, the knowledge acquisition method adopted in [40] was to analyse actual schedules in order to derive a set of rules which appeared to produce the same net result and to be intuitively acceptable to the production scheduler himself.

Alternatively, trace-driven knowledge acquisition by an automated, protocol-assisted method [66] may be used. The method is only applicable when it is possible to develop an unambiguous model of the underlying system such that it is possible to enumerate all the choices that a decision maker could have made at any point in time. Experts operate a simulated version of the system of interest and the analyser keeps a sequential record of all their interventions as well as a record of the state of the system at the time of the intervention. The task of the analyser is then to identify decision rules that yield these interventions.

A similar approach was followed in the Furnace Scheduling Advisory System (FSAS) [45], which produces schedules for a multi-pass glassing and furnacing operation for glass-lined vessels. As no direct interaction with the future users of the system was possible, the designers inferred basic scheduling rules from the process description. These rules were subsequently confirmed by the users.

In a recently developed prototype system for production scheduling, the Learning-Aided Dynamic Scheduler (LADS) [58], machine learning (learning by experimentation) is adopted to enhance the knowledge acquisition capability. The most prominent feature of this system is that it can formulate operators containing both quality and quantity type attributes.

It is probably true to say that many knowledge-based scheduling systems acquire their knowledge mainly from their designers, as they themselves become, during the design process, acquainted with both classical scheduling theory and the characteristics of the manufacturing system they model. Moreover, it is extremely difficult to capture information in highly dynamic environments [40, 58]. Thus, dynamic acquisition, through machine learning, may prove a realistic alternative.

3 Knowledge Representation

Widely used knowledge representation schemes fall into the following categories: logic, rules and slot and filler representation systems. This section examines some scheduling systems from a knowledge representation perspective. Not many systems rely solely on one scheme; most use a combination. However, in the following discussion, systems are classified according to the primary formalism they adopt.

3.1 Logic-based systems

Most systems under this category are based on first order logic; they use predicates to describe the problem area and functions to assist in decision making.

Parrello and Kabat [54] developed a scheduler for car assembly line sequencing, based on an automated reasoning program. Predicates, called 'demodulators', are used to represent the attributes of each car, the difficulty factor of each attribute and a set of penalty functions which evaluate alternative sequences. The system relies heavily on functions and predicates evaluated over sets of objects in the domain rather than individual objects. Thus, it uses a second-order, rather than a first-order, knowledge base. However, it appears that the resulting system was too computationally intensive to be of practical use.

Similar logical expressions are adopted in SONIA [23, 24], a knowledge-based scheduler which supports both predictive and reactive scheduling. Within SONIA, a shop schedule is represented as a set of resources, manufacturing orders and operations to which various kinds of constraints are attached. Individual resources (e.g., machines) are grouped into compound resources (e.g., work areas). Time tables composed of reservation constraints are associated with resources. A reservation constraint is logically expressed by a formula (reserve res t_1 t_2 n list-of-motives). The meaning conveyed by such a formula is that n individual resources from the group 'res' are unavailable throughout the time interval (t_1 t_2). The list of motives explains why they are unavailable. New constraints can be deduced from existing ones by constraint propagation. Collinot and Pape in previous work [22] advocate the use of a flexible propagation system for temporal constraints in which the amount of computational effort spent is not fixed once and for all. They demonstrate how pieces of knowledge within the domain of shop scheduling can be used to dynamically adjust the amount of propagation.

Like SONIA, the Shop Activity Manager (SAM) [14], which is a multicell job shop scheduler, uses both predictive and dynamic scheduling allowing for both well planned schedules and real time response to system feedback. Information relating to the shop and the scheduling task is represented by PROLOG predicates. Job descriptions, and alternative routings are also described in a similar manner.

Logic-based scheduling systems treat the epistemological part of the AI problem [46] successfully, but they fail to address its heuristic part [5] satisfactorily, due to the separation of representation and processing. Moreover, most systems based on first order logic are unable to maintain consistency between basic and derived facts under updates.

3.2 Rule-based systems

Some systems depend strongly on rules, even though knowledge might be partly represented differently. The system of Kerr and Ebsary [40] uses a relational database to store information relating to orders, operations, alternative routes and work centre status. Six categories of rules are identified: order priority rules, operation precedence rules, work centre loading rules, job dispatch rules, general contingency rules and time conversion rules.

Similarly, FSAS [45], which is written in a rule-based programming language, OPS5, uses a database to record information about the objects which are processed, the order in which they should be processed and temporal information. A set of rules guides the scheduling activity.

A system, SCHEDULE [43], is claimed to be able to solve several scheduling problems. The knowledge base consists of an algorithm base and reduction rules. The algorithm base contains procedures for solving thirty seven well-known types of scheduling problems. All thirty seven procedures are of polynomial-time complexity. The reduction rules of the knowledge base show possible reductions among scheduling problems which can be represented by reduction digraphs. The reduction rules are expressed by adjacency lists.

Over a hundred rules relating to the scheduling problem, a list of many references that analyse them, and a classification scheme can be found in [52]. Motivated by the over-abundance of scheduling rules reported in the literature and their complexity, [4] presents a framework for an expert system to assist the user in selecting the appropriate rules to meet objectives.

3.3   Frame-based systems

Most well known systems, including ISIS [30, 31, 59, 53, 48, 42] and OPIS [51, 59, 60, 61], which were developed at Carnegie Mellon University, belong to this category.

The ISIS modelling system is based on the Schema Representation Language (SRL) [30, 31]. SRL has its origins in Bartlett's 'schemata' [6], which have come to be known as 'frames' [47, 69], and 'units' [13]. In fact, ISIS can be considered as a multi-layer system for modelling manufacturing organisations in SRL. The basic concepts are those of states, objects and acts. Acts transform states and objects. Time and causality are primitive concepts in the language. Time relations provide time ordering among states and acts. Causal relations define how states enable acts, and acts cause states. A manufacturing operation is defined as an act, and time and causality relations link it to other states and acts. Operations may also be defined in multiple levels of abstractions. Resources are defined as objects, and attributes and physical structure may be defined for an object. Allocation of resources is defined as a state of possession by some operation or resource with a specified time relation (e.g., duration). Orders are also represented with these primitives. An order is simply a goal state to be achieved by scheduling the appropriate operations.

The hierarchy of abstractions which is made possible by the frame-based nature of ISIS provides a good model of the problem while decomposing the domain knowledge into reasonable, manageable pieces. Furthermore, since frames describe general or expected properties of entities belonging to some entity type, the absence of a value for a slot or the presence of an inappropriate value may signify an abnormal situation. In this case the system might deal with the anomaly using demons.

This type of inference enables the definition and generation of constraints within the ISIS architecture, by providing the capability to attach constraints to a schema (frame) via its slots and values. Constraints may be generated dynamically by attaching constraint generators (demons) to relations in the model. ISIS also supports a number of high level interfaces for communicating constraints to the system. The constraint editor is used to formulate constraints. Driven by knowledge of the underlying constraint representation, the editor provides guidance to the user in specifying or revising information relating to alternatives, relevance and importance. However, constraints may not always be satisfiable. Therefore, the representation of alternatives is important. ISIS adds the specification of relaxations to the representation of constraints. Relaxations may be defined either as predicates or choice sets, which can be discrete or continuous. Associated with a relation is a preference measure which determines the preferred relaxations among those available.

OPIS like ISIS uses frames to represent domain knowledge. It aims to remove a weakness of ISIS in respect of conflict oriented scheduling situations (especially where there are bottleneck machines). Simultaneously two scheduling perspectives are considered: resource-based and order-based subproblems are opportunistically generated in an attempt to directly address as many important constraint conflicts as possible. The solution components known as Knowledge Sources (KSs) communicate through a global structure called the blackboard [60, 61]; a system configuration derived from the HEARSAY-II blackboard architecture [28].

Another system which models the factory environment using frames is the Reinforcement Scheduling System (RESS-I) [11, 12]. The modelling system of RESS-I is not as sophisticated as that of ISIS and OPIS, but it attempts to generate better schedules using a different mechanism for resource allocation.

3.4   Multi knowledge representation systems

Some systems adopt nets and frames in modelling the manufacturing environment and a rule-based approach to expressing constraints and dispatching rules. For example, in OPAL [8, 9, 10], a KBS for industrial scheduling, the description of the workshop, of processing sequences of parts and production requirements is carried out by means of a set of structured objects related by inheritance links within a semantic net. The description of the schedule in progress is done by means of a precedence graph. Empirical knowledge about priority rules and their influence on production objectives as well as practical knowledge (provided by shop floor managers) about technological constraints are represented by 'if ... then ...', rules, stored in a rule base. The 'if' part of the rules is defined using Fuzzy sets.

The theory of Fuzzy sets [72, 73] provides a means for dealing with vagueness. Other methods for deriving inferences from domain and problem knowledge, where both the knowledge and its implications are less than certain are Bayesian reasoning [27], measures of belief and disbelief [62] and Dempster/Schafer theory of evidence [70], where a distinction is made between uncertainty and ignorance. Recently,

INFERNO [55], a new calculus for dealing with uncertainty in expert systems, which uses two values to represent the uncertainty of a statement, has been developed. These developments may prove useful for knowledge-based scheduling.

Another system which uses nets and frames to represent the manufacturing environment is SOJA [44]. A semantic net is used to represent partial schedules and the system examines this graph and resource requirements to decide which scheduling rule to activate next. Frames are used to represent both shop details and scheduling rules.

A prototype designed by Arthur Andersen [15], and named Dymamic Rescheduler (DR), is presented as an automated assistant for managing change and adjusting an active schedule to accommodate new events in the production environment. Its data base contains schemata representing schedule information, work order requirements, cell information, and product details that have been input by the user. Its knowledge base consists of four sets of rules: hypothesising operators, constraint rules, solution evaluators and transitional operators. Hypothesising operators generate alternative hypothetical schedules. Given an active schedule and a new situation in the production environment, they generate all possible schedules that accommodate the change. This of course implies that the system can, in some cases, be grossly inefficient due to the combinatorial nature of the problem.

It is clear from existing systems, that 'if ... then ...' rule bases and frames are the most widely used representation paradigms. However, most of the work that has been carried out on knowledge representation indicates the various schemes are in fact implementable, to an extent, in one another [34]. The decisive factor in choosing a representation becomes then a question of economy, i.e., of choosing a scheme that will represent particular forms of knowledge with the least amount of effort, measured by the amount of knowledge that has to be explicitly encoded to generate the appropriate behaviour. Thus, it seems likely that hybrid representations, consisting of rules, frames and other schemes will prove to be most effective in knowledge-based scheduling systems.

4 Temporal Issues

Reasoning about time is an area of growing concern in knowledge-based scheduling research, since scheduling systems take decisions with reference to time.

The constraint-based analysis (CBA) module of OPAL [8, 9], deals with the partial ordering of operations and time constraints. By propagating due-dates and earliest starting times along the directed graph induced by precedence relationships among operations, a window is assigned to each operation. This window defines the time span in which processing must take place. Comparing the respective locations of the windows of two not yet ordered operations on the same resource, it is possible sometimes to prove that one operation should precede the other or that the windows do not allow for the processing of the operation.

ISIS and OPIS represent time as intervals and relations between intervals [1, 2]. The time during which a state or act may occur is defined by a time interval, as opposed to a single point, or by reference to known time intervals. A time-line schema is defined to specify the units of time, the time scale and the functions to manipulate them. A time interval is represented by an instance of a time-interval schema, with the following slots: BEGIN-TIME, END-TIME, DURATION and DATED-BY. The DATED-BY slot points out the time-line in which the interval is defined. Ordinal (relational) specification of time is represented by slots in a schema. If act A is to occur before act B, then a BEFORE slot will exist in A with a value of B. Other time relations specified are: during, meet, overlap, time-equal, and the inverse relations: includes, met-by, overlapped-by.

It is probably true to say that temporal issues will have to receive a great deal of further attention in future knowledge-based scheduling systems, for temporal reasoning to become a viable operational tool. A detailed analysis of temporal intervals and their theoretical background is given in [1, 2, 3]. Vilain and Kautz [67] show how a fragment of an interval-based representation language can be expressed using a point-based language and benefit from the tractability of the latter. Representation of temporal intervals in predicate logic is discussed in [16]. Kahn [39] presents a time specialist; a program knowledgeable about time in general, which can be used by a higher level program to deal with the temporal aspects of its problem-solving.

5 Control Mechanisms

Most knowledge-based scheduling systems view the scheduling problem as a search process involving the selection and sequencing of resources over time to achieve the production objectives, subject to a set of constraints. As a result, the control mechanisms of such systems comprise search strategies which are applied at different stages of the problem solving activity.

5.1 Forward-reasoning systems

Rule-based systems which support either meta-rules or domain-independent inference procedures for rule selection belong to this category.

An example of this control mechanism is that of SCHEDULE, which uses a forward-reasoning procedure to scan a reduction digraph in a depth-first fashion to decide which rule to activate next. A more sophisticated forward-reasoning mechanism is that in [40]. In this system whenever any type of event occurs (e.g., a particular operation is completed, a customer order arrives, a machine breaks down), a search is performed within the relevant subset of the rule base to collect all the rules whose conditional parts are satisfied. A conflict resolution strategy based on specificity is then applied to eliminate all rules but one. The remaining rule is then fired by taking action on the particular set of conditions that are satisfied.

In FSAS [45], before actual scheduling, an expert system selects the preferred combinations of items which will utilise the furnace fully. Different types of rules are applied during this phase (e.g., readiness check rules, priority rules, component determination rules, etc.). OPS5 produces a set of possible instantiated rules and decides which rule to fire next by applying one of two alternative conflict resolution strategies: lexical-ordering(LEX) or means ends analysis (MEA) [26]. Both recency and specificity filters are used in these strategies. Once the preferred combination of items is determined, another set of rules is used to perform short term and long term scheduling.

5.2 Constraint-directed and opportunistic systems

Some scheduling systems use a hierarchical search method adopted from planning systems such as MOLGEN [64, 65] and expert systems such as CRYSALIS [71]. A hierarchical search method generates a hierarchy of representations of a schedule in which the highest is an abstraction of the schedule and the lowest is a detailed schedule. The rationale behind a hierarchy of representations is that it allows the distinction to be made between problem solving actions critical to the success of a schedule and non-critical actions which are implementation details.

The control mechanism of ISIS [30, 31, 59] is a hierarchical constraint-directed reasoning system. The scheduling method used is a single order method. The main feature is that it formalises various scheduling parameters as constraints (e.g., schedule objectives, resources, preferences) on the system's knowledge base, uses these constraints to guide the search and heuristically generates the schedules. In each scheduling cycle an order is selected based on its category and due date. The system then proceeds through a level of analysis of existing reservations (capacity analysis) to determine the earliest start time and latest finish time for each operation of the selected order. The time intervals generated at this level are codified as constraints and passed to the next level (resource analysis) where a constraint-directed search is performed to obtain reservation time bounds for each resource required for the operations on the selected order. The search can be either forward from the order's requested start date or backward from the order's due date according to the nature of the order. The search operators are defined in a rule-based approach focusing on order priority. The search space is composed of states representing partial schedules and a beam search is performed to sequence the application of search operators. Based on the resource reservation time bounds, a detailed assignment of resources and time intervals for each operation are derived in order to minimise the work-in-progress time. ISIS has a rudimentary reactive component as well; the invalidation of reservations by unforeseen events (e.g., machine breakdowns) results in a minimal rescheduling of only the affected orders while attempting to maintain previous reservations.

A constrained directed-search is also performed by SOJA [44]. The main difference between ISIS and SOJA is that the former represents some qualities of a schedule, such as shop stability, number of tool changes, etc., as additional but relaxable constraints, whereas the latter translates them into scheduling rules that guide the satisfaction of constraints. The actual scheduling of the resources is performed by a specially built forward chaining inference mechanism.

A development of the ISIS searching mechanism is that of OPIS [51, 49]. The latter belongs to the category of planning systems which use opportunistic search methods to derive a plan. As mentioned before, the OPIS framework is a variation of the HEARSAY-II blackboard model. However, unlike the original model that treated Knowledge Sources (KSs) as more or less equal in that they were all directly under the control of a single heuristic scheduler, OPIS KSs are divided into two categories: co-operative and conflicting. This results in a distribution of system expertise that allows distinct types of planning issues to be addressed in isolation [51]. The scheduling process commences by the *resource scheduler* KS generating at the first stage a schedule for the processes of the bottleneck machines followed by a second stage whereby the scheduling is completed by the *order scheduler* KS which deals with all outstanding processes and machines. The most costly limitation, in terms of the quality of the solution, is the requirement that the *order scheduler* KS and *resource scheduler* KS must each return solutions that are feasible with respect to the existing state of the partial schedule developed [51]. This implies that the solution returned by the resource scheduler KS may not be a good one but the rest of the problem solving effort must conform and extend this solution.

OPIS 1 [61], contrary to its predecessor recognises the fact that bottleneck resources often 'float' over time and attempts, using heuristics, to predict resource contention, the time span of concern and the degree of criticality.

Another opportunistic scheduling system is SONIA [23, 24]. Both its predictive and reactive components are co-ordinated by a dual blackboard architecture [37]. The domain blackboard contains information relating to the evolution of the scheduling process and consists of three areas: Results, Capacity and Conflicts. The control blackboard gathers pieces of information which concern the control activity. Several areas are distinguished: Problem, Sub-problems, Strategies, Agenda, Policies and Chosen action. SONIA's predictive component is based on SOJA's constraint-directed system. The reactive component is primitive and its best reaction to an unexpected event is a limited shift to the right of some operations.

In OPAL [8, 9, 10], the search mechanism applied is depth-first, but in conjunction with a Constraint-Based Analysis module (CBA) [29], which aims at characterising feasible solutions of the scheduling problem by taking advantage of the interaction between the resolution of conflicts for resource utilisation and the limit times associated with jobs. This is achieved by using knowledge which consists of facts relating to operations and resources, and of rules which allow inferring new facts from the facts which are already known. When no feasible schedule exists, the data should be modified to recover feasibility. For this to be possible a constraint relaxation mechanism is required. The ISIS approach is adopted for this purpose.

Two levels of reasoning activity in job shop scheduling are identified by [11, 12]: a choice generation level and a decision making level. It is claimed that the key problem in scheduling is the resource-constrained decision making. It is argued that although systems like ISIS do not provide a strong method for the problem of resource allocation. Instead, they satisfy the constraints opportunistically and base the scheduling decisions only on local analysis. To deal with this problem, a 're-inforcement planning' strategy is proposed and embedded in RESS-I. The essence of this is that the system chooses a critical resource and builds a rough utilisation plan (reinforcement plan) as a basis for generating decision making guidelines for constructing detailed schedules. Currently the system builds the reinforcement plan assuming that every order will finish on time and work centres have infinite capacity. This method shows the loading condition on each work centre over time. Detailed scheduling is then performed using the SLACK rule, preference constraints and the information provided by the reinforcement plan. After each schedule the reinforcement plan is updated to reflect the current detailed scheduling decisions.

However, although this method is intuitively appealing, problems can arise due to the difficulty of keeping the reinforcement plan valid during the detailed scheduling activity.

5.3 Mixed-control systems

SAM [14], a prototype system for scheduling of multiple manufacturing cells, uses both forward and backward reasoning strategies. It consists of four modules: loader, dispatching scheduler, monitor and exception handler. The loader whose function is to detect and avoid resource bottlenecks is implemented in PROLOG, as a state-space search process, in which pruning heuristics are used to restrict the size of the search tree. Job routings are chosen to best satisfy capacity constraints in the system. Detailed information is not provided about the other three modules. However, both the dispatching scheduler which performs low level scheduling and the monitor which provides feedback from cells to the other three modules are implemented in PROLOG. The exception handler, because of its ill structured task which involves looking out for the occurrence of any of a large number of different types of critical situation and responding to these when they happen, is implemented in OPS5.

The job-shop scheduling system of Charalambous and Hindi [18, 19], which is discussed in the following section, also uses mixed control.

An interesting approach to industrial scheduling would be an adaptation of Bech-ler's artificial memory [7]. Bechler argues that a simulation package can be used to record the behaviour of a production system when different production rules are applied to alternative initial states of the system. By clustering similar alternative initial states and similar target states derived from the simulation, triplets of initial clusters, rules and target clusters can be recorded. These clusters can then be used at a later stage, possibly by a scheduling system, to decide which rule should be applied to a given state to achieve a specific goal.

6 KBSS

The above overview of current knowledge-based industrial scheduling systems reveals that they appear to suffer a major drawback, most evident in the case of knowledge-based job-shop scheduling systems. Although most provide sophisticated modelling environments, they fail to offer powerful and efficient control mechanisms. The search procedures they employ have heavy memory requirements and the time complexity of the control algorithms is rarely taken into consideration. Thus, they fail to produce good schedules when the problem is large and the constraints tight.

The Knowledge Based Scheduling System (KBSS) [18, 19] is a job-shop scheduling system, which has been designed taking account of the above consideration. In addition, KBSS seeks to address at the same time two scheduling concerns, namely attempting to minimise the makespan and satisfy the due dates as nearly as possible. It is worth noting that few existing systems attempt to satisfy due dates, even though this is the performance criterion considered paramount by scheduling practitioners.

The underlying algorithms have been designed to solve general job-shop problems, where operator constraints and dynamic features are not considered. However, the system can handle such features as the dynamic arrival of jobs and machine breakdowns. Such perturbations are dealt with in a deterministic manner, by partial re-scheduling from the point of interruption onwards.

The system assumes an open job-shop, where no provisions for inventory are made. Technological constraints impose precedence relations among operations and the user can specify machine preferences. These constraints are used to delineate the search space and guide the search process. The optimisation criteria are to minimise the makespan and to meet the due dates.

To achieve the above objectives, a flexible control structure has been constructed which consists of three modules: the Primary Scheduler (PS), the Heuristic Scheduler (HS) and the Controlled Backtracking Scheduler (BS). PS is a rule-based scheduler which attempts to minimise the makespan and meet the due dates. To this end, a single pass procedure is applied which yields solutions fast with minimum memory requirements. If PS fails to meet the due dates, then HS is activated in order to consider more than one possible solution. As a result, both memory requirements are higher and speed is lower in comparison with PS. If the user is not satisfied with the outcome, BS can be invoked. The latter is a scheduler based on a specially designed backtracking strategy, which accepts from the user a target makespan and attempts to find a solution which will meet the due dates within the target makespan. The three schedulers can be accessed either in sequence as above or independently one from another. The schedules produced are 'active' schedules [32].

6.1 The Primary Scheduler (PS)

In general, when the makespan is considered, better schedules tend to be produced when jobs with heavy unprocessed workloads are selected first, by using, for example, the MWKR (Most Work Remaining Rule) [36]. However, of the various measures of performance that have been considered in research on scheduling, the measure that is most important for practical applications is the satisfaction of pre-assigned job due dates [20]. The work in [21] suggests that three factors are important in meeting due dates:

1. A function of job due date, to pace the progress of individual jobs and reduce the variance of the lateness distribution.

2. Consideration of processing time, to reduce congestion and to get jobs through the shop as quickly as possible.

3. Some foresight, to avoid selecting a job from a queue which, when the imminent operation is completed, will move on into a queue which is already congested.

The above observations guided the design of the time-transcending heuristic which lies at the heart of the scheduling algorithm of PS. Such heuristics perform three basic functions [52]:

1. determine a priority rating for each job

2. schedule the next operation of the job with top priority

3. re-evaluate the priority ratings and repeat, always scheduling the operation of the job with top priority

In PS, the priority for each job is considered to be the ratio of remaining work over due date, with ties resolved arbitrarily. The remaining work for a job is defined as the sum of the processing times of its operations that have not been scheduled yet. The processing time of an operation is independent of the machine on which it will be executed, since all alternative machines are assumed to have the same processing rate.

Once a job is selected, PS selects and schedules an operation from it, using a two-level hierarchy of IF THEN rules. The first level of the hierarchy is responsible for the selection of the operation. A procedure then calculates the earliest possible starting time for the selected operation, before the second level of the hierarchy of rules is entered to find an appropriate machine, and complete the scheduling process for the selected operation.

First level

1st rule: IF there is a starting operation not already scheduled THEN select it and proceed to level 2.

2nd rule: IF causal constraints dictate that an operation has to follow THEN select that operation and proceed to level 2.

3rd rule: IF temporal relationships give a set of operations eligible for scheduling THEN select the member of the set with the longest processing time and proceed to level 2.

Employing the longest processing time criterion in the third rule, so that operations with heavy processing requirements are scheduled as early as possible, is in line with the general strategy of selecting jobs with heavy unprocessed workloads.

In addition to specifying the operation to be scheduled, the first level finds its earliest possible starting time. If the operation is determined by rule 1, then the earliest starting time is the start point of the schedule. Otherwise, it is determined by the completion times of the operations which must precede the chosen operation.

Second level

1st rule: Find the set of idle intervals on machines of required type such that the interval can accommodate the operation.

2nd rule: IF the set is not empty THEN select interval providing tightest fit; schedule operation and exit.

3rd rule: Select earliest available machine of type required.

These rules seek to produce an active schedule by performing left shifts whenever possible. The choice of the interval which gives the tightest fit is motivated by a desire to leave as much room as possible for later placements. If no suitable idle gaps are found, then the selected operation is scheduled as early as possible.

6.2   The Heuristic Scheduler (HS)

The Heuristic Scheduler is based on a combination of A^* and beam search techniques and employs the scheduling rules of both levels of the primary scheduler to schedule an operation.

The A^* algorithm is a variation of the best-first search developed for additive cost measures [35]. Thus, it employs an additive evaluation function $f(n) = g(n) + h(n)$, where $g(n)$ is the cost of the currently evaluated path from a start node s to a node n and $h(n)$ is a heuristic estimate of the cost of the path remaining between n and some goal node. A^* constructs a tree T by generating all successors of a given node.

Starting with the root node s, A^* selects for expansion that leaf node of T which has the lowest f value, and only maintains the lowest f path to any node.

However, A^* suffers, albeit less drastically, from the same memory limitations as breadth first search [41]. To reduce memory requirements, A^* was modified by limiting the size of the search tree in a manner similar to that employed in beam search. Whereas in the latter, the α best nodes at each level are retained, where α is a small pre-defined number; in HS, the best α active nodes are kept and the rest are pruned away without generating any successors. The value for α can either be specified by the user, or adjusted by HS according to the size of available memory. This approach on the one hand does not require all the nodes in the OPEN list to be expanded before choosing which children nodes to discard, and on the other hand retains the flexibility and heuristic power of A^*, even though it reduces the amount of expanded nodes to suit memory capacity.

The procedure terminates when a schedule is completed, i.e., a solution is returned as soon as the last operation is scheduled. At each node of the search tree, the rules of the primary scheduler are applied for each incomplete job. As a result one operation from each incomplete job is scheduled.

The f value of a node is calculated by the following heuristic:

1. add the processing times of the remaining operations.

2. divide the total obtained from 1 by the number of machines available.

3. Add the result to the endpoint of the rightmost operation on all machines.

6.3 The Backtracking Scheduler (BS)

In ordinary backtracking, search is focused on the lower levels of the search tree, by virtue of the fact that the overwhelming majority of the nodes of the search tree are at these lower levels. However, practical experience gained from attempting to solve a large number of job-shop problems shows that the quality of a schedule is determined, to a large extent, by early decisions during the initial stages of the formation of the schedule. Moreover, the combinatorial nature of the job-shop scheduling problem means that any practical algorithm will have to truncate the search in some way, if solutions are to be arrived at in reasonable computation time.

These considerations argue for developing a backtracking strategy with twin objectives. First, that it leads to considering at most a truncated search tree representing a limited number of alternative schedules, specified by the user. Secondly, it takes, at the same time, account of the significance of early placements by retaining a larger proportion of the decision nodes at the higher levels of the search tree.

Consider a node n that appears in both the complete enumeration tree and its truncated counterpart. Let $b(n)$ be the number of successors of n in the former and $c(n)$ be the number of its successors in the latter. If all $b(n)$ values were known, it would be possible to estimate the $c(n)$ values such that they are proportionately smaller and such that the total number of the implied schedules in the truncated tree is as desired. Since the $b(n)$ values form a non-increasing sequence from the start node at level 0 to the nodes at the last level, proportionality between the $c(n)$ and $b(n)$ values will ensure that the $c(n)$ values will also be in non-increasing order. This, in addition to the fact that deleting a node in the search tree leads also to deleting all its successors, ensures that proportionately more nodes are retained at the upper levels of the truncated search tree, leading to more emphasis on early decisions.

Two questions remain. The first is how to calculate the $b(n)$ values. The answer to this question lies in observing that there is no need to calculate these values a priori. As the search tree is developed, the $b(n)$ value for a node can be calculated the first time it is visited simply by counting all possible moves from it.

The second question is how to estimate the value of a constant $0 < h < 1$ such that $c(n) = h * b(n)$ $\forall n$, when it is desired to consider at most S alternative schedules. Estimating h is discussed in detail in [19].

7 Reactive and Real-Time Scheduling

knowledge-based scheduling systems have been criticised as having a smart planning phase, followed by a blind execution phase, which fails in real situations characterised by complexity, uncertainty and immediacy. This certainly is the major difficulty faced by attempts to build knowledge-based scheduling systems for automated manufacturing systems. In this domain, the need is for reactive scheduling and real-time scheduling. Having a reactive scheduling ability can be defined as having the ability to revise a schedule effectively, quickly and minimally, in response to unexpected events, taking into consideration what has been implemented so far. On the other hand, having a real-time scheduling ability can be defined as having the ability to take scheduling decision on-line, under sever time constraints.

However, there seems to be within the AI community some confusion regarding reactive and real-time scheduling, with some doubting the feasibility of the whole enterprise of reactive scheduling or 'situated action' . There is also some doubt about how reactive scheduling differs from non-reactive scheduling, since changed situations can be similar to the original ones, so much so that the techniques used to solve the original problem can also be used to solve the revised one.

Nevertheless, in real-time applications, rescheduling is usually subject to the constraint that there is a time limit for the selection of appropriate action. In some contexts, it may be possible to identify in advance various events, compute the appropriate plans and store them efficiently, so that the resulting 'plan cache' can be used on-line. As a special case, it may be feasible to compute an appropriate move or action, rather than a plan, corresponding to each event, in which case the resulting system is essentially a pattern classifier transforming inputs to primitive output actions. It is interesting to note that this is reminiscent of both the 'artificial memory' of Bechler and also of pattern classification by neural networks.

However, if reactive and real-time scheduling are based, as described above, on associating reactions with events, then it is very likely that the resulting system is myopic, in the sense of being based on the immediate situation. One possible remedy is to invoke a planner of a specific kind, rather than a plan, in response to each specific event. This approach is reminiscent of current knowledge-based systems which have rule-selector modules to select appropriate rules in response to various situations.

8 Conclusions

The field of knowledge-based industrial scheduling is, perhaps, still in infancy. A large number of the studies reported in the literature offer mere ideas. The majority of the systems actually built are still either research prototypes developed by a university or research group to investigate the application of AI methods to industrial scheduling, prototypes for industry, i.e., systems developed by a university or research group for an industrial partner, or industrial prototypes, developed within industry for an application. Very few scheduling systems appear to be operational, and it is debatable how much knowledge is contained in them [14].

Knowledge-based scheduling systems are often equated with attempts to mimic human schedulers. But this does not appear to be the view adopted by most researchers and developers in the field, who seem to be aware of the weaknesses of this approach. These weaknesses stem from the fact that human schedulers [63] are unable to deal with many variables at the same time, and therefore resort to locally greedy strategies. To cope with complexity, they also add artificial constraints, adopt arbitrary planning horizons and adopt inappropriate levels of abstraction and resort to scheduling by crisis management. In addition, they have to adapt to lack of information which might well be available to a computerised system.

To assess the stage of development of Knowledge-based scheduling systems, an analogy from the blocks world, popular in AI research, may be appropriate. This is a world of blocks of different sizes, which are numbered with the letters of the alphabet. In it, there is a robot that can perform various primitive actions, like picking a block up, putting it on the table or on another, etc. The objective is to devise a set of primitive actions to reach a goal configuration, such as a stack of the blocks arranged alphabetically. The following is a hierarchy of planning tasks aimed at achieving the objective:

1. from a pre-defined starting configuration

2. from any starting configuration

3. from any starting configuration, in the presence of a baby who wanders in the room, occasionally knocking the stack off and moving the blocks around.

4. from any starting configuration, in the presence of a mischievous baby who occasionally tries deliberately to knock the stack off and perhaps hides the blocks or even throws them at the robot.

It is debatable whether there is a need in the realm of industrial scheduling to consider problems of the last degree of difficulty, but there is plainly a case for considering problems of the third degree. Yet, current efforts seem to be capable of dealing with problems at a level of difficulty somewhere between the first and second degrees.

In conclusion, it could be said that this survey has revealed that the way to be traversed before knowledge-based scheduling becomes an effective operational tool is still long indeed. It may even be said that Artificial Intelligence workers, like control theorists and operations researchers before them, underestimated the difficulty of industrial scheduling problems. Hybrid solutions combining the points of strength of both approaches may be the right way forward.

References

[1] J. F. Allen, An interval-based representation of temporal knowledge, 7th Int. Conf. on AI, Vancouver 1981.

[2] J. F. Allen, Maintaining knowledge about temporal intervals, Dept. of Computer science, Univ. of Rochester, Technical report TR–86.

[3] J. F. Allen, Planning using a temporal world model, Proc. 8th Int. Joint Conf. on AI (IJCAI–83), Karlsruhe, Germany 1983.

[4] S. M. Alexander, An expert system for the selection of scheduling rules in a job shop, Computers and Industrial Engineering, vol. 12, no. 3, pp. 167–171, 1987.

[5] A. Barr and E. A. Feigenbaum, The Handbook of Artificial Intelligence, Vol. 1, Pitman 1983.

[6] F. C. Bartlett, Remembering, Cambridge University Press, 1932.

[7] E. Bechler, J. M. Proth and K. Voyiatzis, Artificial memory in production management, report no. 336, Institute National de Recherche en Informatique et en Automatique, 1984.

[8] E. Bensana, G. Bel and D. Dubois, OPAL: A knowledge based system for industrial job-shop scheduling. Laboratoire LSI, Universite Paul Sabatier 31062, Toulouse Cedex (France).

[9] E. Bensana, G. Bel and D. Dubois, OPAL: A multi-knowledge based system for industrial job-shop scheduling, Int. Jour. Prod. res.,1988, vol. 26, no. 5, pp. 795-819.

[10] E. Bensana, M. Correge, G. Bel and D. Dubois, An expert system approach to industrial job-shop scheduling, IEEE International Conference on Robotics and Automation, San Francisco, Apr. 1986.

[11] Bing Liu, Scheduling via reinforcement, Int. jour. for artificial intelligence in engineering, vol. 3, no. 2, April 1988, pp. 76-85.

[12] Bing Liu, A reinforcement approach to scheduling, ECAI 88: Proc. 8th European Conf. on Artificial Intelligence, 1-5 Aug. 1988, Munich, W. Germany, pp. 580-585.

[13] D. Bobrow and T. Winograd, KRL: Knowledge representation language, Cognitive Science, Vol. 1, No. 1, 1977.

[14] J. Bowen, P. O'Grady, H. Nuttle, M. Terribile, An artificial intelligence approach to loading workstation resources in a distributed job shop controller, Computer-Integrated Manufacturing Systems, vol.2, no.1, February 1989, pp. 21-28.

[15] M. C. Brown, The Dynamic Rescheduler: Conquering the Changing Production Environment, Proceedings of the fourth conf. on AI applications, San Diego, CA, USA, 14-18 March 1988, IEEE Comp. Soc. Press, pp. 175-80.

[16] W. I. Bullers, S. Y. Nof and A. B. Whinston, Artificial intelligence in manufacturing planning control, AIIE Transactions, 12, 4 (1980), pp. 351–363.

[17] O. Charalambous and K. S. Hindi, 'A Review of Artificial Intelligence-Based Job-Shop Scheduling Systems', Information and Decision Technologies, Vol. 17, pp. 189202, 1991.

[18] O. Charalambous and K. S. Hindi, KBSS: A Knowledge Based Job-Shop Scheduling System, Production Planning and Control, Vol.4, No.4, pp. 304310, 1993.

[19] O. Charalambous and K. S. Hindi, A Knowledge Based Job-Shop Scheduling System with Controlled Backtracking, Computers and Industrial Engineering,Vol. 24, No. 3, pp. 391400, 1993.

[20] R. W. Conway, W. L. Maxwell and L. W. Miller, *Theory of scheduling*, Addison-Wesley, (1967).

[21] R. W. Conway, 'An experimental investigation of priority assignment in a job shop', *RAND corporation memorandum RM–3789–PR*, Febr., (1964).

[22] A. Collinot and C. le Pape, Controlling constraint propagation, 10th Int. Conf. on AI, Milan 1987.

[23] A. Collinot, C. le Pape and G. Pinoteau, SONIA: a knowledge-based scheduling system, Int. journal for artificial intelligence in engineering, vol.3, no.2, 1988, pp. 86-94.

[24] A. Collinot and C. le Pape, Controlling the behavior of knowledge sources within SONIA, Proc. of the 22nd annual Hawaii Int. Conf. on System Sciences, vol. 3: Decision Support and Knowledge Based Systems, Kailua-Kona, HI, USA, 3-6 Jan. 1989.

[25] N. J. Cooke, The elicitation of domain-related ideas: stage one of the knowledge acquisition process, Expert Knowledge and Explanation the Knowledge-language interface, Charlie Ellis, Ellis Horwood Limited, 1989.

[26] T. A. Cooper, N. Wogrin, Rule-based programming with OPS5, 1988, Morgan Kaufman Publishers.

[27] R. D. Duda, P. E. Hart and N. J. Nilsson, Subjective Bayesian methods for rule-based inference engines, In Nilsson (ed.), Readings in Artificial Intelligence, Webber, 1981.

[28] L. Erman, F. Hayes-Roth, V. Lesser and D. Reddy, The HEARSAY–II speech understanding system: integrating knowledge to resolve uncertainty, Computing Surveys 12(2), June 1980.

[29] J. Erschler and P. Esquirol, Decision aid in job-shop scheduling: a Knowledge Based Approach, 1986 IEEE Int. Conf. on Robotics and Automation, San Francisco, April 1986.

[30] M. S. Fox, Constraint directed search: a case study of job-shop scheduling, Morgan Kaufman, 1987.

[31] M. S. Fox and S. F. Smith, ISIS — a knowledge based system for factory scheduling, Expert Systems Journal, vol.1, no.1, 1984.

[32] S. French, *Sequencing and scheduling, An introduction to the mathematics of the job shop*, Ellis Horwood, (1982).

[33] W. S. Jr. Gere, A heuristic approach to job-shop scheduling, PhD. thesis, School of industrial administration, Garnegie Institute of technology, 1962.

[34] J. Harhen, M. Ketcham and J. Browne, Artificial Intelligence and Simulation of manuf. systems. In IFIP 86, H. J. Kugler (ed.), Elsevier Science.

[35] P. E. Hart, N. J. Nilsson and B. Raphael, 'A formal basis for the heuristic determination of minimum cost paths', *IEEE Transactions on Systems, Man and Cybernetics*, SMC–4, no. 2, pp. 100–107, (1968).

[36] A. C. Hax and D. Candea, Production and Inventory Management, Prentice-Hall, 1984.

[37] B. Hayes-Roth, A blackboard architecture for control, Artificial Intelligence, vol. 26, no. 3, 1985.

[38] W. H. Iskander, An investigation of the use of AI in solving the job-shop sequencing problem. PhD. thesis, Dept. of industrial engineering, Texas Tech. University, August 1975.

[39] K. Kahn and G. A. Gory, Mechanizing temporal knowledge, Artificial Intelligence, vol. 9, 1977, pp. 87–108.

[40] R. M. Kerr, R. V. Ebsary, Implementation of an expert system for production scheduling, European Journal of Operational Research, vol. 33, no. 1, pp. 17–29, January 1988.

[41] R. E. Korf, 'Optimal path-finding algorithms', in V. Kanal, and V. Kumar (eds.), *Search in artificial intelligence*, Springer Verlag, (1988).

[42] A. Kusiak, M. Chen, Expert systems for planning and scheduling manufacturing systems, Eur. jour. of operational research, vol. 34, no. 2, March 1988, pp. 113-130.

[43] A. Lamatsch, M. Morlock, K. Neumann, T. Rubach, SCHEDULE-An Expert-like system for machine scheduling, Annals of Operations Research, vol. 16, 1988, pp.425-438.

[44] C. Lepape, SOJA: A daily workshop scheduling system, Expert systems 85, Proc. of the 5th technical conf. of the BCS specialist group on expert systems, Univ. of Warwick, 17–19 Dec. 1985.

[45] M. Litt, J.C.H. Chung, D. C. Bond, G. G. Leininger, J. Hall A scheduling and planning system for multiple furnaces, Engineering applications of artificial intelligence, vol. 1, pp. 16–21, March 1988.

[46] J. McCarthy, J. P. Hayes, Some philosophical problems from the standpoint of artificial intelligence. In D. Michie and B. Meltzer (Eds.), Machine intelligence 4. Edinburgh, pp. 463-502.

[47] M. Minsky, A framework for representing knowledge, in P. Winston (ed.), The psychology of computer vision, , McGraw-Hill.

[48] B. Nardi, Westinghouse unveils ISIS-II factory order systems, Intellinews vol. 1, no.4, 1985.

[49] N. Muscettola, S. F. Smith, A probabilistic framework for resource constrained multi-agent planning, IJCAI-87, pp. 1063-1066.

[50] N. J. Nilsson, Principles of Artificial Intelligence, Springer-Verlag, 1982.

[51] P. S. Ow and S. F. Smith, Towards an opportunistic scheduling system, 9th Annual Hawaii International Conference on System Sciences, 1986.

[52] S. S. Panwalkar and W. Iskander, A survey of scheduling rules, Operations Research, vol. 25, no. 1, Jan-Feb 1977.

[53] P. Papas, ISIS: A project in review, Symposium real time optimization in automated manufacturing facilities, National Bureau of Standards, Gaithersburg MD, January 1986.

[54] B. D. Parrello, W. C. Kabat, Job-Shop Scheduling using automated reasoning: a case study of the car-sequencing problem, Journal of Automated Reasoning, vol. 2, 1986, pp. 1-42.

[55] J. R. Quinlan, INFERNO: A cautious approach to uncertain inference, The Rand Corporation, 1700 Main str., Santa Monica, California.

[56] B. Raphael, The frame problem in problem-solving systems, Artificial Intelligence and Heuristic programming, 1971, American Elsevier.

[57] M. S. Salvador, Scheduling and sequencing. Handbook of operations research: Models and applications, J.J Moder, S.E. Elmaghraby (editors), Van Nostrand Reinhold, New York, 1978.

[58] Shinichi Nakasuka, Taketoshi Yoshida, New framework for dynamic scheduling of production systems, Int. Workshop on Industrial Applications of Machine Intelligence and Vision (MIV-89), Tokyo, pp. 253–259, April 10–12, 1989.

[59] S. F. Smith and M. S. Fox, Constructing and maintaining detailed production plans. Investigations into the development of knowledge-based factory scheduling systems, AI Magazine, Fall 1986.

[60] S. F. Smith, Peng Si Ow. The use of multiple problem decompositions in time constrained planning tasks, 9th Int. Joint Conf. on AI, Aug. 1985, LA.

[61] S. F. Smith and Peng Si Ow, Integrating multiple perspectives to generate detailed production plans, ULTRATECH, AI in Manuf. Conf., Long Beach, Sept. 1986.

[62] P. Snow, Bayesian inference without point estimates, AAAI–86, Fifth National conf. on AI, Philadelphia, August 1986.

[63] M. S. Steffen, A survey of artificial intelligence-based scheduling systems. Fall Industrial Engineering Conference, Dec. 7–10, 1986.

[64] M. Stefik, Planning with constraints (MOLGEN:Part1), Artificial Intelligence, vol. 16, 1981.

[65] M. Stefik, Planning and Meta-Planning (MOLGEN:Part2), Artificial Intelligence, vol. 16, 1981.

[66] A. Thesen, L. Lei, Knowledge Acquisition Methods for Expert Scheduling Systems, Proc. of the 1987 Winter Simulation Conf., Atlanta, GA, USA, pp. 709–714, 14–16 Dec. 1987.

[67] M. Vilain and H. Kautz, Constraint propagation algorithm for temporal reasoning, AAAI–86, Fifth National Conf. on AI, August 1986, Philadelphia, Morgan Kaufman.

[68] C. K. Whitney, Building expert systems when no experts exist (FMS application), Proc. 1986 IEEE Int. Conf. on Robotics and Automation, Washington, DC, USA, 1986, pp. 478-85.

[69] T. Winogard, Representation and understanding, in Bobrow Collins (ed.), Studies in cognitive science, 1975.

[70] J. Yen, A reasoning model based on an extended Dempster-Shafer theory, AAAI–86, Fifth national conf. on AI, August 1986, Philadelphia, Morgan Kaufman.

[71] R. E Young and M. A. Rossi, Toward knowledge-based control of flexible manufacturing systems, vol. 20, no. 1, IEEE trans., March 1988.

[72] L. A. Zadeh, Commonsense knowledge representation based on fuzzy logic, Computer 16 (10), pp. 61–65.

[73] L. A. Zadeh, Fuzzy sets, Information and control, 8, pp. 338–353, 1965.

23

REACTIVE BATCH SCHEDULING

VICTOR J. TERPSTRA, HENK B. VERBRUGGEN
Delft University of Technology
Department of Electrical Engineering, Control Laboratory
P.O. Box 5031, 2600 GA Delft, The Netherlands

A design for a robust, reactive, on-line, scheduler is presented. It makes a **prediction** of the effects of the schedule and tries to **optimize** the global plant performance within a constrained environment. It is able to deal with constraints on **state variables**. Next to routing and sequencing choices, it can also optimize real-valued control variables (e.g. batch sizes and throughputs). These two features make the scheduler is especially useful for mixed-batch/continuous plants. It adapts the schedule on-line (reactive) to handle disturbances and failures. Robustness analysis can be made for a schedule by which guaranteed non-failing schedules can be generated. The scheduling technique is model based and generically applicable to a wide class of plants. The model is built up of independent units which are represented by objects from a library.

The scheduler consists of a planner which formulates the scheduling problem, an integer scheduler which finds production paths and sequences for the batches through the plant and a non-integer scheduler which optimizes timing and flows within a path by transforming this schedule into an NLP problem. The scheduler can also be used for cyclically operated batch plants. The scheduler has been partly completed and is implemented in G2. The results show an increase in the average production rate when compared with non-reactive schedules.

S. G. Tzafestas and H. B. Verbruggen (eds.),
Artificial Intelligence in Industrial Decision Making, Control and Automation, 687–722.
© 1995 *Kluwer Academic Publishers.*

1 Introduction

1.1 Project

This article reports some of the work within the scope of the SCWERE Project (Supervisory Control With Embedded Real-time Expert systems). This is a joint project of the Departments of Informatics, Electrical, Mechanical and Chemical Engineering of the Delft University of Technology; it is linked to several industrial organizations.

The aim of this project is to develop plant-wide control systems on a supervisory level by applying artificial intelligence techniques [Terpstra, 1991]. The supervisory control systems are envisioned as being applied in the areas of fault detection and diagnosis [Terpstra,1992], planning in response to faults, scheduling, optimization, on-line model updating, prevention of alarm inflation, etc. Emphasis is given to the on-line and real-time aspects of these tasks.

1.2 Scheduling

Scheduling in chemical batch plants deals with the allocation of resources and timing of events (e.g. starting batches on units). By using a prediction of the plant behaviour it tries to accomplish a certain production. Scheduling can be a major problem in the operation of batch plants. It is of major interest if the quality of the schedule (the timing of events) greatly influences the production rate or the production costs and thereby the profitability of the plant. This is the case when there are problems like parallel units which must be synchronized, production deadlines, constraints on state variables (e.g. minimum volumes), shared resources (e.g. steam, electric power, available manpower), etc. An optimal schedule should maximize production at minimal costs. Scheduling is even necessary if (bad) schedules can be constructed which could violate essential constraints.

In practice, scheduling is mostly done by hand. The human scheduler searches for a feasible schedule which does not violate any constraints, and he is likely to generate the schedule which can be found most easily. This will probably not be the schedule which performs best. The problem starts if an optimal schedule is required. Complexities which arise are caused by: a large number of degrees of freedom, a large time horizon, severe interaction between units (which therefore cannot be solved by decomposition) and a non-trivial optimizing function. The human scheduler has great difficulties in handling this complexity as a whole. The foregoing considerations were the motive for the research being performed on automatic schedulers.

To date, most of the proposed automatic schedulers operate off-line. Given the nominal values of the process parameters (like processing times), an optimal schedule is constructed [Wellons, 1989a+b], [Patsidou, 1991], [Pantelides, 1993]. This schedule is applied to the plant without feedback. If the behaviour of the plant can be predicted accurately, the scheduling will yield an optimal and non-failing production process.

However, in practice, and especially in chemical plants, this is not the case. The processing times of chemical reactions can vary widely. Because a schedule is based mainly on these processing times, these variations can have a major influence on its performance. Recently, off-line schedules which can handle a certain amount of variation in the processing times have been introduced [Djavdan,1992]. This schedule must be flexible enough to cope with the disturbances. In other words it is made 'compliant', which has two disadvantages. First, it cannot change its schedule in-line, which urges the compliant schedule to be based on pessimistic estimates of processing times. Therefore, in practice, this compliant, off-line calculated schedule often introduces unnecessary waiting times, which results in a lower production rate. Second, compliant schedules can only handle small disturbances. Large disturbances can cause the schedule to fail and in a worst-case situation causes a plant to shut down. This is the major problem in our case study (see § 1.3 "Example case"). One solution is to adapt the schedule on-line by means of a reactive scheduler. Cott [1989] suggests a method for on-line schedule modifications by which small changes are added to the starting times in the off-line schedule.

This paper proposes a more rigorous and flexible way to deal with disturbances and failures. Instead of adapting the off-line schedule, our scheduler is able to calculate a completely new schedule at each 'sampling time'. With this method, all possible degrees of freedom can be used (including flows), in contrast to Cott [1989] who only changes starting times. Thus bigger disturbances can be handled faster. The scheduler cannot only cope with disturbances but also with expected and non-expected unit failures. The proposed scheduler always attempts to optimize the schedule.

Assuming knowledge of the process variability in terms of a statistical probability distribution, it is possible to generate a schedule (both off-line and on-line) with a given probability of failure. Or, given absolute boundaries on the process variability, even a guarantee of no failure. The acceptability of failures can be given as a constraint and/or as a part of the optimizing function of the scheduler (see § 8 "Robustness analysis").

The scheduling method is model-based and therefore independent of this specific case study. There is a clear separation between the scheduling part and the model of the plant. Therefore, the same scheduling algorithm can be reused for a wide class of mixed batch/continuous plants (see § 6 "Non-integer scheduler").

1.3 Example case

To illustrate the scheduling problem, we applied the proposed scheduling techniques to an example case study of a combined-batch/continuous plant. The plant consists of several reactors, separator, stock vessels, measuring vessels, buffer tanks and a distillation column. The reactors' processing times vary. The degrees of freedom for the scheduler are: starting times, processing times (if they can be manipulated), batch sizes and the flow into the distillation column.

The most essential operational constraint for this plant is the volume of the buffer tank. An empty tank is not allowed, because if so, the flow into the distillation column becomes zero and the column will trip. Because the half product degrades over time, the buffer tank must be as small as possible.

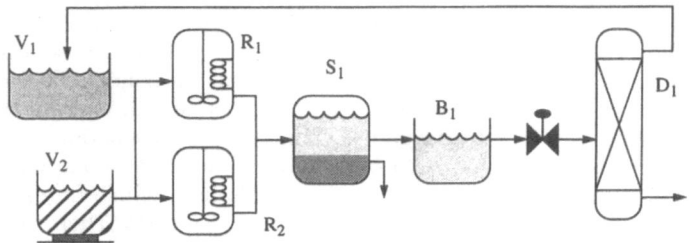

Figure 1   Simplified version of the case study, which consists of two stock vessels (V_1 and V_2), two batch reactors (R_1 and R_2), a batch separator (S_1), a buffer tank (B_1) and a continuous distillation column (D_1).

A simplified version of the case study is used as an example in this paper (see Figure 1).

1.4   Definitions

In this research an attempt is made to conform to the standards laid down in the draft Batch Control System standards [ISA,1990]. Within the scope of this paper the ISA configuration has been simplified and is represented as a 3-layer hierarchy.

At the lowest (unit) level, a unit is viewed as a collection of equipment modules (e.g. tanks, valves, heaters, etc.). At this layer, a unit is controlled by the batch management. It executes one operation.

One layer higher is the train/line layer. It consists of a collection of units arranged in serial or parallel paths which is used to make a complete batch. This layer is controlled by the scheduler. It executes a procedure.

The highest layer (in this hierarchy) is the area or corporation layer which consists of a collection of trains/lines and higher organizational structures. This (production) planning layer develops, given customers orders, a production plan which consists of production location, product type, amounts and deadlines. At this level MRP tools generate master schedules which are the input for the scheduling layer.

The scheduler, as described in this paper, aims at the train/line level in which the units are viewed as black boxes. Because the interaction between units takes place at the level of phases, the scheduling controls the timing of the phases. The execution and control is left to the batch management.

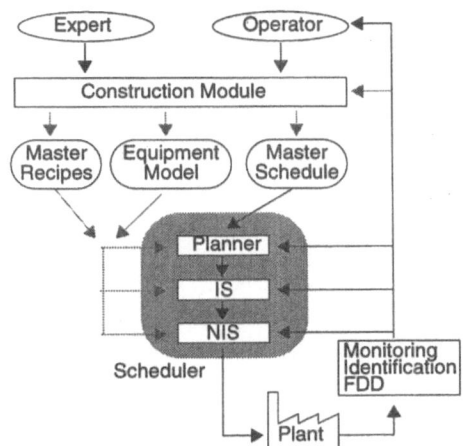

Figure 2   Architecture of the scheduler. A construction module helps the users
to build the three models. The scheduler consists of three modules:
Planner, Integer Scheduler (IS) and Non-Integer Scheduler (NIS).
They are able to operate reactive: state feedback from Monitoring is
used to update the schedule. Identification may (automatically)
update system parameters, Fault Detection and Diagnosis (FDD)
causes a change in the equipment model

2   Scheduling strategy

The scheduling problem can be defined as followed: to fulfil the *objective*, given a *model of the process* by controlling *control variables*.

The objective is defined as a set of orders for products, possibly with deadlines and an optimization function.

The model of the process contains: a set of units, each having its own properties, a topology (network) by which the units are interconnected, a set of recipes which define how (in generic terms) a product must be fabricated and a set of orders which tell what has to be produced.

The control variables are the setpoints for the DCS of the batch plant. Or, more precisely, the start and stop times of the phases of units and the amplitude of the flows through the interconnecting pipes between the units.

Figure 2 shows the overview of the architecture.

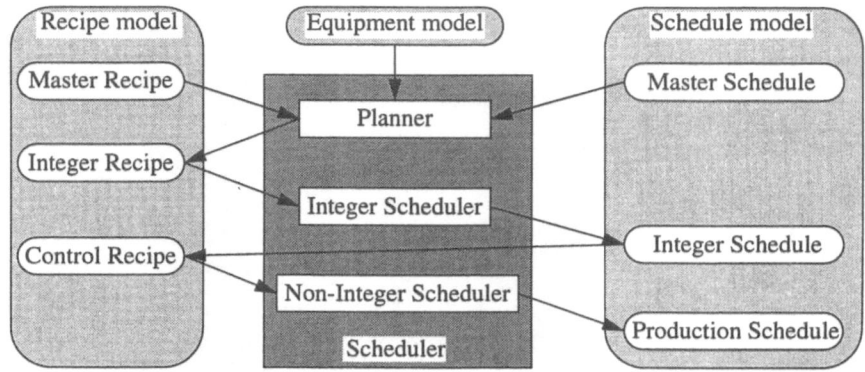

Figure 3   The scheduler consists of three modules which communicate
              through intermediate models

2.1 Modelling

An important aspect of the scheduler is the clear separation between the procedures of
the scheduler and the models which give the domain specific information. This has two
advantages. First, the scheduler can be used for a wide class of plants within the process
industry. Second, the scheduler can automatically, and even on-line, accept changes in
the models. This last feature is especially useful for the reactiveness of the scheduler.

First the interactive *construction module* is used. An *expert-user*, who knows the
construction of the plant and the subtleties of the production process, constructs two
models. The *equipment model*, which represents the physical constitution of the plant,
and the *master recipes*, which describes the procedures and formulas, used for the manufacturing of the products in the plant. The models are hierarchically structured. Basic
building blocks are grouped by the expert scheduler into 'higher' building blocks. Each
'higher' building block is put in a library and can be used repeatedly during the construction process, making this process less time consuming and more consistent. The *operator-
user* defines in the construction module the *master schedule*, a list of product orders with
deadlines which he receives from plant management. The models are the input for the
scheduling modules.

2.2 Modularity

The scheduler is split up into three modules which each solve a part of the problem: a
planner, an *integer scheduler* and a *non-integer scheduler* (see Figure 2). The modules
exchange data using intermediate models (see Figure 3).

First, the planner generates a *plan* how to manufacture the product(s). It formulates the scheduling problem by selecting all the degrees of freedom. Using qualitative

reasoning and rough estimates it excludes a lot of impossible combinations. The degrees of freedom can be *integer* or *non-integer* (real-valued) variables.

Second, the integer choices within the degrees of freedom are made by the integer scheduler. One integer choice basically represents one *path* for each batch through the plant, and the *sequencing* between these batches.

Third, the non-integer scheduler assigns quantitative values to the remaining real-valued control variables (i.e. exact timing and magnitudes of flows). The essence of the non-integer scheduler is a transformation of the optimization problem into an explicit NLP formulation. From that point on, a general NLP solver can be used to optimize the control variables.

A more detailed discussion on these modules will follow in the next paragraphs.

2.3   Prediction and cycles

As previously stated, the objective of the planning and scheduling is formulated as a list of orders and deadlines. The scheduler acts as a supervisory model based predictive controller. It uses a prediction horizon which includes the production of all orders.

However, if the plant should produce one product for a relatively long time (or always, in a single product plant), the plant is used in a *cyclic* way, i.e. it constantly repeats the same schedule each cycle. Mixed-batch/continuous plants especially are used in this way. The batch part of such a plant will try to emulate a continuous production to ensure a smooth interface with the continuous parts. The case study is a specific example of this type of plant.

The objective of such a plant is to optimize the production rate. Therefore, the scheduler must calculate a cyclic schedule which optimizes the production over one cycle. This means that the scheduler can use a prediction horizon of one cycle. The schedule will result in a sequence of cyclically repeated states of the plant. This is called a *cyclic state*, which is equivalent to a steady state in a pure continuously operated plant. The optimization of this cyclic state schedule is done by the *cyclic state scheduler*. The calculation of this schedule can be done off-line.

The cyclic state scheduler is almost the same as the 'normal' scheduler. The only differences are some simple changes in the order list and the objective function.

2.4   Reactive behaviour

In on-line mode, the off-line calculated schedule is used as an initial solution and is executed. If the plant would behave without any disturbances, this would lead to the perfect execution of the schedule. However, this will never be the case, especially not when dealing with chemical reactions which introduce uncertainties in the models. These disturbances are handled in two ways.

First, by making the scheduler *reactive*. The scheduler is designed as a *predictive, plant wide controller*. The formulation of the optimization problems as they are being used in the scheduler modules, include the current state of the plant. At each 'sample' the current state is updated. Depending on the type of disturbance compared to the 'setpoint' (the previous calculated schedule), the different modules are restarted.

Second, by introducing *robust schedules* (see § 2.5 "Robustness").

The scheduler shows different types of reactive behaviour according to the type of disturbance.

A disturbance can temporally and randomly act on a system parameter (e.g. a duration of a reaction). In the case of a relatively small disturbance, the reactive scheduler has only to adapt the non-integer control variables.

If a disturbance is temporal but relatively large, the schedule must be changed more rigorously by selecting a different integer schedule.

In the case of non-temporal, long-term disturbances a new cyclic state schedule must be calculated. This is useful in cases where constant but small deviations of a system parameter (like a reaction time) occur, or where long-term, large deviations in the model (like a failure of a unit) occur.

2.5 Robustness

Knowledge about the process variability is used to calculate the effects of these variabilities on the possible violation of *critical constraints*. A critical constraint is a constraint which can be violated by a bad schedule and which violation results in high costs. An example is a constraint on the minimum volume of a buffer tank which is a buffer between the batch part of the plant and the continuous part. If the volume becomes zero, the feed to the continuous section dries up and it will cause an expensive trip of the plant. To prevent such a situation, *robustness constraints* are added to the optimization problem which will guarantee that, whatever (pre-defined) disturbance takes place, the scheduler will be able to adapt the schedule to prevent a constraint violation.

3   Modelling

As the scheduler is model based, it follows that the (graphical) representation of the information about the plant is important. The models must contain all the information needed, but each piece of information may occur just once to prevent inconsistencies. Furthermore the models should have a structure that resembles the reality, so that it is easy to build, validate and maintain. Finally, for the sake of clarity and compatibility the models and the terminology should correspond with the standards for batch control systems as they are laid down by the ISA working group SP88 [ISA, 1992].

Three types of models are used: the *equipment model*, the *recipe model* and the *scheduling model*. The equipment model provides information about the physical structure of the plant. The recipe model contains knowledge about the process. The scheduling model gives for each level of the scheduling process the goals that are set by the superior level. All the three models are hierarchically structured.

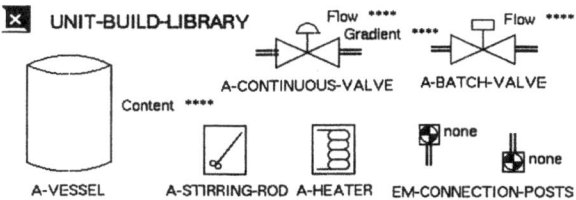

Figure 4   Equipment library. It contains the standard equipment modules from which a unit can be constructed.
(Stirring-rod and heater are special attributes which do not involve scheduling).

3.1   The equipment model

The equipment model describes the physical constitution of the production plant. It is a hierarchical model. The lowest level is formed by *elements*, which are simple actuators and sensors. On top of the hierarchy stands the (business) *corporation*, an organization that coordinates the operation of one or more plants.

Three levels of the equipment model are of interest to our scheduler: *production cells (trains)*, *units* and *equipment modules*. An equipment module is a functional group of equipment. Examples of equipment modules are vessels, filling systems, etc. (see Figure 4). A unit is a group of interrelated equipment modules, used to perform an operation on the batches that enter the unit. A unit operates relatively independently and it can not operate on more than one batch simultaneously. Examples of units are distillation columns, reaction vessels, buffer tanks, etc. (see Figure 5). A production cell or train is a collection of several units, arranged in serial and/or parallel paths, used to make a complete batch. The scope of the scheduling activity of our scheduler is restricted to one production cell (see Figure 6).

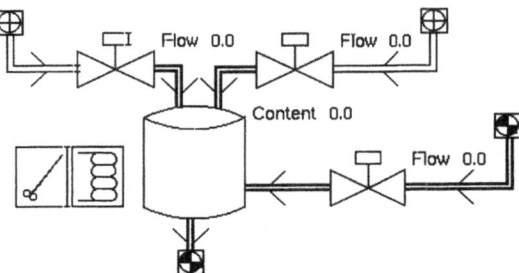

Figure 5   Illustration of a unit. This example consists of a reactor vessel, three controlled input batch valves and one uncontrolled output (which flow will be determined by the input of the unit which is connected to the output).

696

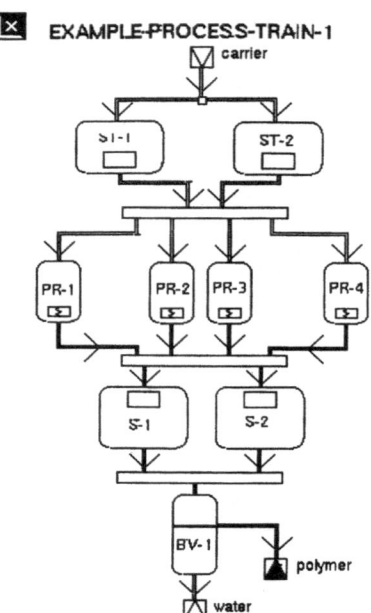

Figure 6   Example of a production cell (or train). It consists of several units
            connected by a piping topology.

Also, units can be defined which represent *shared resources* (e.g. steam and electricity generators, personnel, etc.). They are modelled as a source of flow (steam, electricity). The flow is distributed and arrives at inputs (batch or continuous) in the consuming
units. Phases in the recipe define how much will be used. On the one hand, the flow can
be constrained to a maximum, which is useful if a shared resource has an limited instantaneous capacity (e.g. a steam generator). On the other hand, also the integral of this flow
can be constrained to model a limited storage.

The attributes of a unit can be split into two groups.

The first group are attributes for 'planning' purposes. The planner compares these
attributes with the equipment requirements of the master recipe in order to decide which
operations are allowed to be (possibly) scheduled on the unit. Examples are material
constraints, equipment facilities like heaters, stirring-rods, etc.

The second group of attributes define the control and state variables plus the
equipment constraints on them. A vessel introduces a state variable (a volume) and possibly two constraints (a min. and max. volume). A batch valve is assumed to be a controlled valve which results in a pulse-shaped flow. The start- and stop-times of this pulse and
the magnitude of the flow are real-valued control variables. A continuous valve is also
assumed to be a controlled valve which results in a ramp-shaped flow. The start- and
stop-times of the ramp and gradient of the flow are the control variables. This means that

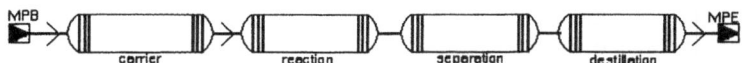

Figure 7   Illustration of the procedure of the master recipe consisting of a sequence of operations. This example shows a single sequence. However, also splitting and merging of product streams and alternative sequences are possible.

the flow is a state variable. By constraining the gradient, the limited dynamics of a continuous unit is modelled.

3.2   The master recipe

The master recipe is the highest hierarchical level of the recipe model that is used by the scheduler. It contains the complete set of information that specifies the control requirements for manufacturing a batch of a particular product. The master recipe for a product consists of a procedure, a formula and some equipment requirements. The procedure defines and orders the *operations* to be performed for making a general class of products (see Figure 7). The formula is a set of parameters that distinguishes this particular product from the other products with the same procedure. These are modelled as absolute values of or *aspect ratios* between the amounts of feed material and the timing of (reaction) phases. The equipment requirements specify the type and size of the equipment needed for every operation in the procedure.

An operation is executed in one unit. It consists of some sequences of *phases*. Each sequence of phases corresponds with a port of the unit (normally a valve). A phase describes a state of the port. For example, a series of phases could be: 'charging' - 'reaction' - 'discharging' - 'waiting' (see Figure 8). The different series of phases are synchronized to order the actions of different valves.

If all phases within an operation on a unit are synchronized, defining a sequence for all ports separately results in a single sequence of phases for the whole operation. This will be the 'normal' situation for a 'normal' unit. However, some types of units may

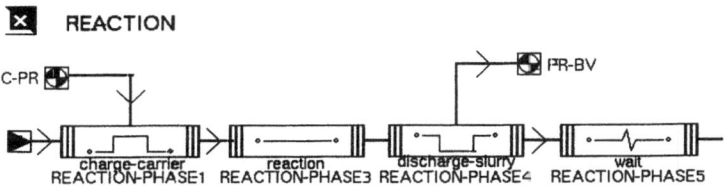

Figure 9   Illustration of one operation consisting of a sequence of phases.

have ports which behave independently (non-synchronized). In that case, for those ports separate sequences of phases must be defined. An example of the last is a buffer tank with one or more inputs and one or more outputs. Filling and emptying phases for these ports need not to be synchronized.

3.3   Master schedule

The master schedule is the highest hierarchical level of the scheduling model that is used by the scheduler. In the master schedule the goals are expressed, that are set by the master production scheduling, a management activity. It contains all the product orders (customer orders) that are assigned to the production cell. The following characteristics are given for each production order: the product, the quantity, the deadline or due date. Also here the criterion function is given which defines what is optimal. Examples are simple criteria as to minimize the production time of the full set of orders, which is equivalent to maximizing the production rate. It becomes more complex if penalty functions are defined for not meeting deadlines or if change-over costs are included. In theory, any criterion function can be defined as long as it is expressed as a function of the control and state variables as defined in the equipment and recipe models. However, it is obvious that the more complex the criterion function becomes, the more complex (i.e. time consuming) the optimization will be.

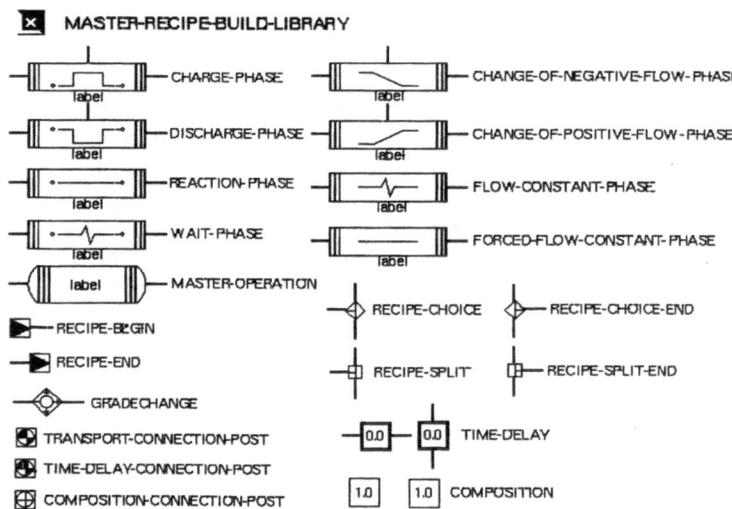

Figure 8   Master Recipe library. It contains the basic classes of phases from which an operation can be build.

3.4   The degrees of freedom of the scheduler

This paragraph gives an summary of way a plant is modelled and of the degrees of free-dom of this model which can be optimized. Thereby, an impression is given of what type of batch scheduling problems can be modelled.

The plant is modelled as a topology of units. The topology defines the possible connec-tions (pipes) between the units. A unit is viewed as a black box, only its interactions with the other units are modelled. Since the interactions change per phase, its output behav-iour is modelled up to the level of phases. A phase defines what the ports of the unit do. A port is modelled by a batch or a continuous in- or output flow. The flow is assumed to be controlled, batch ports assume a pulse-shaped flow, continuous ports a ramp-shaped flow. Each unit may contain a state variable in terms of a volume. A reaction phase is rep-resented as an 'elapsed time' model. The batch size and the number of batches an order is split, are also control variables.

Up to this point, the plant is seen as a 'water model'. However, also some other control variables concerning the recipes can be included. One is the aspect ratios (in the recipe) between the amounts of materials (input or output). Changing these ratios will lead to *flexible recipes*. Second, the time a reaction takes can be an algebraric function of some control variables (e.g. a batch size). Third, the procedure of operations may contain alternative paths.

Constraints can be defined on, of course, the control variables, but also on state variables (volumes, flows) and timing (deadlines, maximum storage time).

The state of a plant is defined as: the volume of every vessel, the flow through every con-tinuous valve, the type phase and operation every ports currently executes, the start-time of that current phase and the elapsed time within that phase up to the current point in time.

4   Planner

The task of the planner[1] is to generate a plan of how to achieve the scheduling objective. The planner should refine the objective and formulate the actual scheduling problem at a more detailed abstraction level. In other words, it is a *refinement* of the objective. The objective is formulated in abstract terms, i.e. in terms of input/output of the overall plant (a set of orders- and deadlines), viewing the plant as a black box. The 'plan' should be formulated in more detail, i.e. in terms of the interaction between units as the compo-nents of the plant, viewing the units as black boxes. The plan is a list of batches and per batch all possible paths through the plant, it states all possible ways to produce the prod-uct at this specific plant. In fact, it *formulates* the scheduling problem by selecting all the

---

1. The term "planner" is used in the Artificial Intelligence connotation of the word.

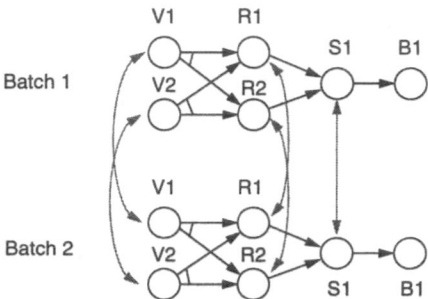

Figure 10   Integer recipe of the case study. It represents all path choices for two batches. The solid arrows represent the sequencing of operations within one batch. Two arrows leaving an operation is one binary control variable. The shaded arrows represent sequencing choices between operations of different batches which are fighting for a limited resource (one unit). This is also one binary control variable.

degrees of freedom (the control variables) which are necessary to fulfil the objective. These must be given a quantitative value in order to achieve a feasible and optimal schedule.

The output of the planner, the 'plan', is called the 'integer recipe'. It is specifically aimed to model all integer choices and constraints on them. For every batch which is to be scheduled, it contains all possible paths for that batch to traverse through the plant. This is called a 'path-network'. To prevent two batches claiming the same resource (e.g. a reactor, steam supply, personnel, etc.) at the same time, resource constraints between the path-networks are added. These resource constraints introduce sequencing choices between the operations of different batches (see Figure 10). shows a screen dump of a different integer recipe as modelled in the implementation.

5   Integer scheduler

The objective of the integer scheduler is to find the best integer schedule. It makes the integer choices within the degrees of freedom. An integer schedule defines a *path* for each batch and the *sequencing* between those paths. As input it receives the integer recipe from the planner. The output is the integer schedule, which is sent to the non-integer scheduler.

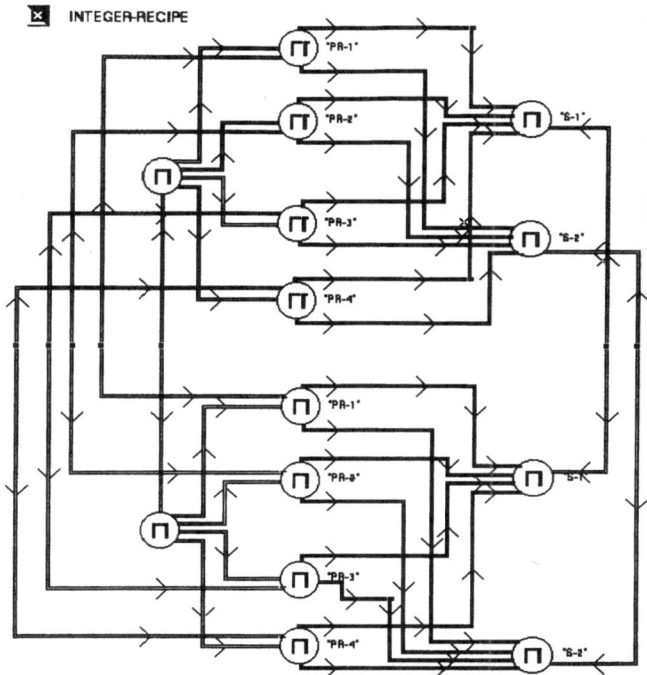

Figure 11   Integer recipe of the example plant of Figure 6. The solid lines represent the path choices, the hollow lines the sequencing choices. Each circle is a step. A step is a contraction of several phases which independently do not introduce an extra integer choice (e.g. 'reaction' and 'wait').

This problem is, in fact, the usual scheduling problem as it appears in practice in many situations in which integer choices have to be made. This is the step in which the notorious combinatorial explosion takes place. The main difference between the stated scheduling problem and other scheduling problems is that, even if all integer control variables are chosen, the effect of the schedule is still not known exactly. This has two causes:

1.   There remains a set of non-integer control variables which are still not optimized.
2.   The system parameters may have an uncertainty.

The above means that, initially, the value of the optimization function is uncertain. Therefore, it is not known whether the integer solution is optimal and further the feasibility is uncertain. To obtain the best possible values for both, the non-integer scheduler must be started to calculate the best fully quantitative schedule.

Therefore, in order to find the best integer schedule, the integer scheduler must generate all possible integer schedules, it must ask the non-integer scheduler to calculate

702

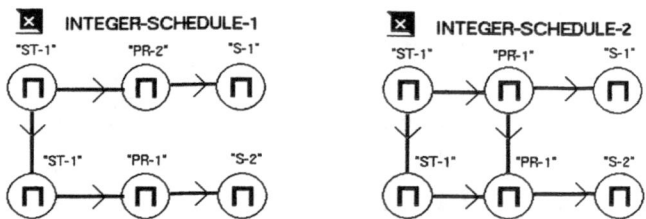

Figure 12   Illustrations of two integer schedules generated from the integer recipe of Figure 11.

the value of the optimization function for all these schedules to select the best one. Theoretically this is possible, but practically (probably) not, because of the combinatorial explosion of the number of integer schedules.

The above defines the essential problem for all schedulers which is: how to find the optimal solution without enumerating all integer combinations. In other words, is it possible to exclude sets of integer combinations while searching in the search tree? The solution for the problem is based on two methods: consistency maintenance and bounding of solutions. (An elaborate discussion on these techniques can be found in [Terpstra, 1994b])

Consistency maintenance

    When an integer choice has been made, a lot of other integer choices become irrelevant or determined. For instance, if a path through a unit of a certain stage is selected, all other paths through the alternative units must become "no path" and all sequencing choices with respect to these units become irrelevant.

    An integer consistency maintenance system ensures that these qualitative infeasible choices will never be tried while searching. This system uses constraint propagation which is a very important artificial intelligence technique to limit the search [Kumar, 1992].

Bounding of solutions

    If a partial solution is "integer" feasible it doesn't mean that the solution is feasible with respect to the real-valued variables. An 'estimator' is used which is able to give an estimation of either the *feasibility* or *optimality* of a partly defined schedule. This estimator defines a bound and, therefore, whole branches of the search tree can be pruned by using a Branch-and-Bound method.

The output of the integer scheduler is the integer schedule (see Figure 12). This sequence of operations defines a partly ordered set of qualitative actions which must be given a quantitative value. This integer schedule is transformed into a control recipe (see Figure 14 and Figure 13). These actions are the starting times of phases in the operation, flows through controlled valves, batch sizes, etc.

The complete scheduling problem, including the integer scheduling, is, in fact, a kind of Mixed Integer NonLinear Programming (MINLP) problem. The integer scheduler takes care of the integer part which is responsible for exponential numbers of solutions. Intelligently limiting this number is very important. Wellons [1989b] presents some interesting ideas.

Others [Patsidou, 1991], [Pantelides, 1993] end up with a MILP model. The reason our scheduler results in an nonlinear model is because the flows are also taken as a control variable.

The proposed method is based on a *continuous time model* (just as Patsidou [1991]). In contrast to the *discretized time model* which limits the operation to use only an integer number of discrete time slots [Pantelides, 1993]. Pekny [1993] shows that the use of a continuous time model may result in an enormous reduction in time to find a solution.

Note that the proposed method doesn't generate an explicit MI(N)LP problem, like [Patsidou, 1991], [Pantelides, 1993] and [Pekny, 1993] who use separate, generic MILP solvers. Not using 'standard' MI(N)LP solvers allows a dedicated search strategy using consistency checking and specialized bounding algorithms.

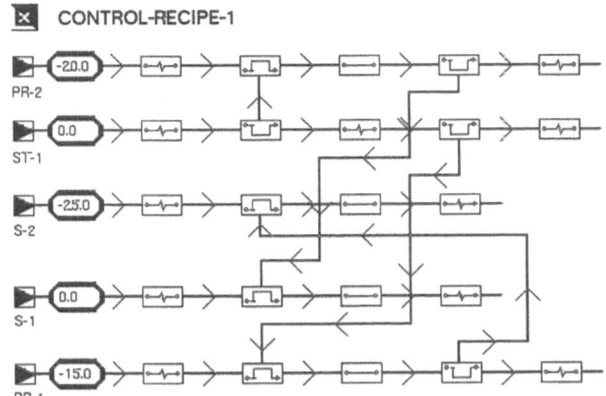

Figure 13   Control recipe of the integer schedule-1. This is a 'qualitative Gannt chart. The numbers on the left are a part of the current state of the plant: the time at which the current phase was started.

704

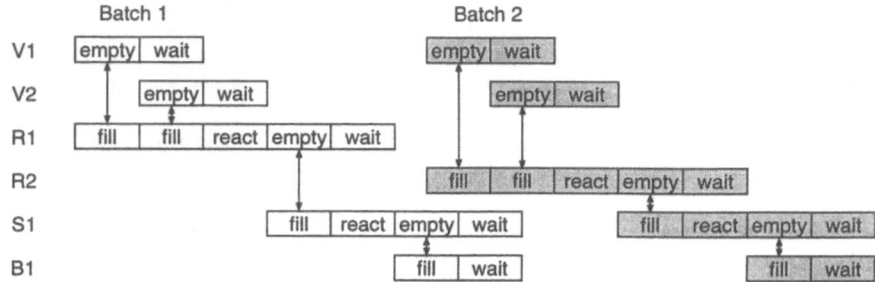

Figure 14   Part of the control recipe of the case study. Each unit performs several operations in which each consists of phases. The integer schedule determines the path of a batch through the plant (which is represented by the synchronization between the phases of the different units) and the sequence between the batches.

6   Non-integer scheduler

The objective of the non-integer scheduler is to optimize the non-integer control variables of one integer schedule. The actual input model is the control recipe. The difference between the integer schedule and the control recipe is only a matter of ordering the same data.

The optimization is done by transforming the optimization problem into an explicit NLP model[1]. For the optimization, a general NLP solver can be used, or a dedicated NLP solver (see § 6.2 "Dedicated NLP solver").

6.1   Generation of NLP model

Two techniques are essential in the generation of the NLP model:
1.   Solving differential equations.
2.   Dealing with constraints on state variables.

The input for the optimization problem consists of the equipment model, the master recipe and the integer schedule (see § 6 "Non-integer scheduler"). The output is an NLP model.

---

1. In this research, an NLP model is defined as a non-linear version of the LP-model. That is, a set of, possibly, non-linear algebraic inequalities (constraints) and a non-linear algebraic optimization function. Both constraints and optimization function are purely expressed as functions of the control variables. So, differential equations and/or state variables are not used in the NLP model.

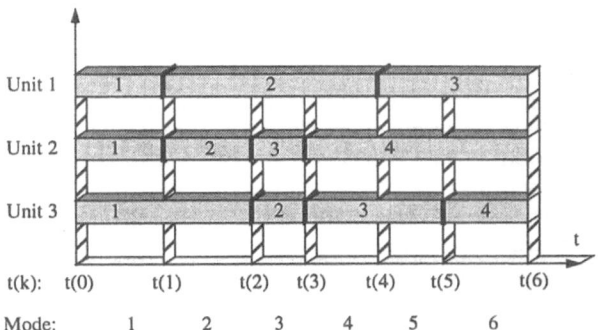

Figure 15   Each unit performs an operation which consists of phases. Within one phase the unit acts as a continuous system. On a global level, in between two phase transitions (one *mode*), the whole plant can be modelled as a continuous system.
The variable *t(k)* represents the time instance at which the *k*-th global discrete event takes place.

The algorithm consists of several steps:

Step 1 aggregation of local models into a global model.
Up to this point, all knowledge is represented as locally as possible. This step collects all this knowledge and makes it global. This involves:
   a.   The generation of a combined discrete event/continuous model.
   b.   The enumeration and globalization of all constraints.

Step 1a: generation of the discrete-event/continuous model.
A batch plant is a combination of a discrete event and a continuous system. Some of the manipulable variables are discrete events and act as switches (e.g. on/off, open/closed). Between such discrete events the process is said to be in one *mode* and can be described as a continuous system (see Figure 15).
To model such a hybrid system, a new modelling method is proposed. It consists of two sets of equations. One for the discrete event system (Eq. 1) and one for the continuous system representation (Eq. 2)

$$z(k+1) = z(k) + G(z(k), u_{de}) u_{de} \qquad \text{(Eq. 1)}$$

$$\dot{x}(t) = Ax(t) + B(z(k)) u_c \qquad \text{(Eq. 2)}$$

The model for the continuous system is a classical linear state space representation. However, with one exception: the *B*-matrix, which is a function of the discrete event state vector $z(k)$. The *B*-matrix can be seen as a representation of the discrete valve-positions in the plant. It switches the in- and output flows. The input vector u_c contains the continuous input variables (e.g. the flows or the derivatives of the flows). The flows in the system are assumed to be controlled by primary control loops. Therefore, the discrete event state $z(k)$ does not influence the A-matrix.

The discrete event system determines the sequence and timing of the discrete events, i.e. the changes in the discrete event state $z(k)$. These changes depend only on the timing variables of the input vector. These are stored in u_{de}. The matrix G is a relatively complicated function of $z(k)$ and u_{de}. It determines the next global discrete event to happen and, as a result, it adds the proper parts of u_{de} to the discrete event state vector $z(k)$. It acts as a switching matrix: its elements are 0 or 1.

The input vectors u_{de} and u_c contain all scheduling variables up to the scheduling horizon.

The evolution of $z(k)$ depends only on u_{de}. The changes of the discrete event state vector $z(k)$ do not depend on the continuous state vector x, as they may be in other mixed discrete event/continuous systems [Visser, 1994]. Therefore, first, the evolution of $z(k)$ can be determined and, second, the evolution of x.

Step 1b: The enumeration and globalization of all constraints

This mixed discrete event/continuous model is generated from the object-oriented model in which all knowledge was represented locally. With this step, the system representation is made global. The following step is to make the constraints global. This is nothing more than an administrative job: all variables must be given a unique global name and the in- and output variables which are linked according to the topology are put into equality equations. Thereafter, one of them can be eliminated in the optimization problem (e.g. the output-flow of unit 1 is the input-flow of unit 2, which results in only one global variable).

Step 2: solving the discrete event/continuous model.

A part of the constraints are the constraints on state variables, for instance, on the maximum or minimum level of a buffer tank. These constraints must hold also in the future, during the scheduling horizon, therefore, they contain expressions on future states. These constraints must be transformed into algebraic constraints on control variables. Therefore, the future states are expressed as functions of the initial state and the control variables during the complete scheduling horizon. This function follows from the symbolic evolution of the discrete event/continuous model. In other words, the differential equations are solved symbolically.

Step 3: reduction of the number of constraints on state variables.

Another problem with constraints on state variables is their testing and, in particular, *when* to test them. These constraints must hold at every time instant. However testing all of them continuously would result in a very large NLP model. This problem is solved by analysing the qualitative shape of the function of the state variable (and thus of the expression using the state variable) (see Figure 16).

Within one mode, all inputs are monotonic functions. Therefore, the extremes appear at the start or end of a mode, which is the time instance when a discrete event takes place. In a sequence of modes, the extremes are the time instances at which there is risk of constraint violation. Only at these points in time will the constraint be tested. If the

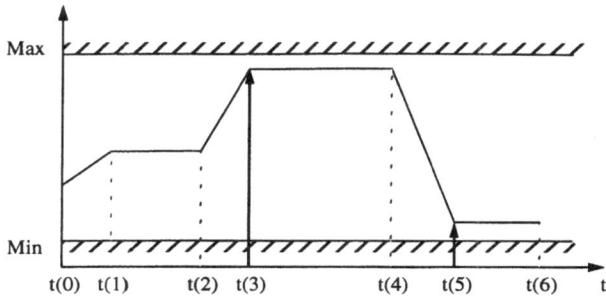

Figure 16   The analysis of t qualitative trend of a state variable gives a minimal number of time instances at which constraints on this variable must be tested.

schedule satisfies the constraint at this time instance, at all other time instances the constraint will automatically be satisfied.

6.2   Dedicated NLP solver

The non-integer scheduler produces an NLP problem which is nonlinear in a specific way. It is possible to use this knowledge and solve the problem in a more efficient way than general NLP solvers.

The basic technique is *constraint propagation*, this is how the feasible area is determined. If the optimization function is monotonically dependent on a set of control variables, the optimum can be easily found by setting the control variables in the extremes of their feasible area while continuing the constraint propagation. This will be the case if, for instance, solely the production rate must be optimized. The output is a fully quantified production schedule (see Figure 17).

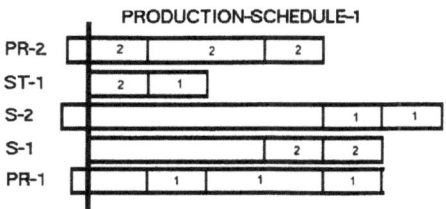

Figure 17   Gannt chart representing the quantified production schedules belonging to the integer schedule 1 (see Figure 12).

7   Reactiveness

Reactiveness is defined as the ability of the scheduler to *adapt* the schedule in an *on-line* situation in response to *disturbances*. The reactive scheduler is a kind of *predictive controller*. Predictive controllers are based on prediction of the future behaviour of the process to be controlled. In order to predict the future, it uses a model of the process, the current state of the process and proposed trajectories for the control variables. The prediction is used to optimize the future behaviour by using the future control variables as the degrees of freedom.

7.1   Horizons

The scheduler will predict and optimize the behaviour over a certain, limited horizon. This horizon is called the *prediction horizon* (H_p). When output of the scheduler is needed, the first element of the scheduler output sequence, determined by the last optimization, is used to control the process. In on-line, reactive use of the scheduler, at the next 'sample' this procedure is repeated using the latest measured information. This is called the *receding horizon* principle.

Another type of horizon is the *control horizon* (H_c). This horizon includes all control variables which are used as a degree of freedom in the optimization process. The *remaining prediction horizon* (H_{rp}) includes the control variables of H_p in the time interval after H_c (see Figure 18). In formula:

$$H_c + H_{rp} = H_p \qquad \text{(Eq. 3)}$$

In scheduling, H_c is normally equal to H_p and therefore H_{rp} is zero. In 'conventional' predictive control, the control variables in H_{rp} are all given the value of the last control variable in H_c: $u(t_0+H_c)$. This can't be done in scheduling because the value of $u(t_0+H_c)$ is meaningless at later points in time $(t > t_0 + H_c)$. However, in the process of the determination of specific types of robustness constraints, H_c is smaller because the control variables in H_{rp} are given a preset value [Terpstra, 1994a].

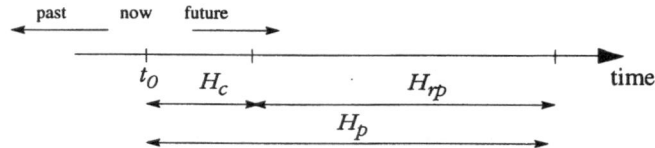

Figure 18   Illustration of the definition of the horizons as used by the scheduler:
a. H_p, the prediction horizon.
b. H_c, the control horizon.
c. H_{rp}, the remaining prediction horizon.

In case of cyclic state scheduling, the scheduling horizon will normally be one (qualitative) cycle. A shorter horizon produces bad results: the schedule attempts to optimize the first part of the cycle, without considering the effects on the second part. A longer horizon (e.g. 2 cycles) contains the same information twice. Because it does not contain new information, the schedule for the second cycle will not be different from the first one.

7.2   Sample Rate

Normally when used as a direct controller, predictive control is a discrete-time controller concept, using a fixed sample time in between the discrete points in time at which the input is updated and the output is executed. However, the scheduler is a supervisory controller at a higher level. Information is coming from all kinds of information sources, probably running on different kinds of computers. The output of the scheduler consists of setpoints for other controllers, again probably running on other computers. That means that there exists no synchronized and fixed sampling rate. Therefore, the scheduler must be able to deal with asynchronous *events* occurring in a continuous, real-valued time scale.

7.3   Three Control Loops in Scheduler

Because of the modular approach in the scheduler (dividing the scheduler in the planner, IS and NIS), there exist three control loops for each of the three modules. Each control loop adapts the schedule in a different way. This NIS re-optimizes the NLP model and thereby changing only the real-valued control variables. The IS control loop is able to change path and sequencing choices. The planner must adapt the degrees of freedom in the integer recipe in cases of failures or rush orders.

It must be decided under which conditions these control loops must be (re)started. For each control loop, an error signal is defined (see § 7.4 "Error Signal"). When one or more of these errors occur (they become non-zero), the relevant control loop is (re)started.

(Re)starting the IS also implies eventually a (re)start of the non-integer scheduler. First, it will use the NIS to evaluate candidate integer schedules. Concurrently (possibly in parallel, see § 7.8 "Parallelism"), the NIS will continue executing the old integer schedule. Only after a final choice for a new integer schedule is made, the NIS will change over to execute the new integer schedule. The same holds for the planner with respect to the integer and non-integer scheduler. It generates one integer recipe which must be evaluated by the IS, which itself uses the NIS. This runs concurrently with the IS-control loop. In theory, it might even generate several integer recipes. This means that at the same time, several 'instantiations' of the IS and especially the NIS are active (see Figure 19). (At which points in time they must be started, is discussed in § 7.5 "Timing".)

710

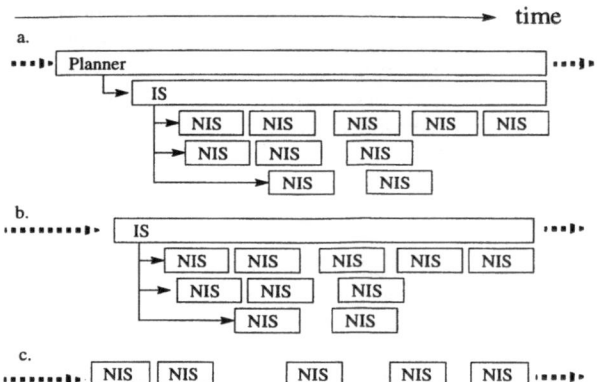

Figure 19   Concurrency and/or parallelism in the three control loops within the scheduler:
a. The planner control loop.
b. The integer scheduler control loop.
c. The non-integer scheduler control loop.
The loops repeat and can (re)start lower loops.

7.4   Error Signal

The input signal for the scheduler is not an explicit error signal, being the difference between the setpoint trajectory and measured output. In fact, the output of the scheduler is a setpoint trajectory in itself. The input is the current state of the plant, the objective (the master schedule) <u>and</u> the models of the plant (flowsheet and recipe's) because they may change also.

Nevertheless, an error signal is defined and used as an indication to (re)start the scheduler. The error signal is defined on the basis of the 'state' of the plant in a broad sense:

- x_{NI} = all non-integer state variables, i.e. volumes, flows, starting times and durations of current phases.
- x_I = all integer state variables, i.e. the current sequencing and path choices as mentioned in the integer recipe.
- m_{quant} = the quantitative, real-valued parameters of the equipment model of the plant, the master recipe's and the master schedule.
- m_{qual} = the qualitative aspects of the equipment model (the flowsheet, qualitative attributes of units), the master recipe and the master schedule.

Based on these variables, three error signals are defined on the basis of $\hat{x}$ and $\hat{m}$ (the predicted/expected state and model) and x, and m (the actual, measured state and model):

$$e_{NI} = (\hat{x}_{NI} - x_{NI}) + (\hat{m}_{quant} - m_{quant}) \qquad \text{(Eq. 4)}$$

$$e_I = \hat{x}_I - x_I \qquad \text{(Eq. 5)}$$

$$e_{qual} = \hat{m}_{qual} - m_{qual} \qquad \text{(Eq. 6)}$$

These three definitions are used in the three control loops of the scheduler.

7.5 Timing

Because there is no fixed sample rate which determines all timing, several decisions concerning the timing of the scheduler have to be made explicitly:
1. When (re)start the three control loops.
2. When to change from executing an old schedule, to executing a new schedule.

Events may occur while the scheduler is reasoning. The ability to use the new information and to change the reasoning, is called *non-monotonic reasoning*.

In general, one can say that no new schedule is needed as long as the process behaves as predicted.

Restart of the non-integer scheduler.

(Re)start of the NIS control loop is only needed if the e_{NI} is non-zero (Eq. 4). The NIS is relatively fast. When, for the first time, a new candidate integer schedule has to be evaluated, a new NLP model must be generated and optimized. In the control loop however, the integer schedule, and therefore also the NLP model, remains the same and it needs only to do the optimization. However, the error is real-valued and therefore will probably be non-zero after each update of the state variables, resulting the NIS to restart often. Two solutions for this problem exist:
1. Introduction of an explicit, absolute sample rate which depends on the time constants of the process.
2. Keeping the event driven approach, but introducing a threshold to prevent the control loop from being too 'nervous':

$$\text{Restart NIS if} \quad e_{NI} > e_{NIthreshold} \qquad \text{(Eq. 7)}$$

The value of the threshold may be different for each element of the vector e_{NI} and may depend on how critical an error on the state variable is. The individual thresholds must be 'balanced' in such a way that each has the same influence on the critical constraints.

These two options result in a conflict of interest: how to optimize the 'scheduler response time'. A high sample rate or low threshold will consume a lot of processing time which cannot be used by the planner or integer scheduler. One can imagine an advanced optimization algorithm that computes the 'optimal' sample rate or thresholds. However, future experiment have to show whether such (time consuming) algorithm is useful.

Restart of the integer scheduler.

(Re)start of the IS control loop is only needed if the e_I is non-zero (Eq. 5). The IS is relatively slow due to the *combinatorial explosion*. e_I deals with integer variables which do not change 'fast' compared to real-valued variables. They inherently act as thresholds. Therefore there is no need for a sample rate and the (re)start of the NIS can be event driven.

An error $e_I > 0$ will definitely cause a restart of the IS. However, if $e_I > 0$ the NIS may ask for a new integer schedule in two cases:
1. The current integer schedule becomes <u>infeasible</u>. Detection is straightforward: NLP solver will fail.
2. The current integer schedule becomes no longer <u>optimal</u>. The detection of non-optimality is more difficult. Alternative integer schedules are needed which have been evaluated and updated(!) by the NIS. Thereby, the value of the criterion function of the alternatives can be compared with the current schedule.

If these alternative integer schedules are not available, the IS must be restarted.

Restart of the planner.

(Re)start of the planner control loop is only needed if the e_{qual} is non-zero (Eq. 6). The planner control loop is of course the slowest of the three. This is not directly due to the processing time of the planner itself, but is mainly determined by the calculation times of the (theoretically several) calls to instances of the IS (see Figure 19). e_{qual} deals with qualitative changes in the process model, recipes and master schedule which do not change fast. Therefore the (re)start of the planner can be an event driven activity.

Start executing a new schedule.

When to start executing a newly calculated schedule. The scheduler uses *progressive reasoning* or an *any time algorithm* to generate a feasible schedule as soon as possible. After that, it starts generating better schedules. This means, that the time at which a new schedule is chosen and started to execute, can be chosen.

The largest part of the schedule consists of control variables which are future events. It is useless to choose and start executing a new schedule in which the next to be executed event takes place in the future. A new schedule needs only to be selected if a direct schedule action is needed. Therefore, the, up to that moment, best schedule is potentially selected. At the time the first schedule action in this schedule is needed, only <u>this</u> action is executed. The scheduler can still continue to generate alternative schedules, however only those which include the same first action as just taken.

A small part of the schedule consists of actions in terms of setpoint changes of continuously variable control variables such as the gradient of the throughput of a continuous unit in the plant. The problem with taking this control action immediately, is that alternative, possibly better integer schedules which do not use this action, will be excluded.

A balance must be made between the cost of delaying this action and the possible profit of a better schedule. This can be done heuristically or mathematically:

1. Heuristically. If the restart of the scheduler was caused by the occurrence of a large disturbance, a fast respond is probably preferable.

2. Mathematically. A *sensitivity analysis* can be made on the effect of a delay on the optimality and feasibility. To every candidate integer schedule, an extra waiting time is added before every first schedule action. As a result, the NLP model contains an extra variable. While 'normal' optimizing, this extra wait time will be set to zero and has no influence. For the use of the sensitivity analysis, it can be used for a 'what-if' analysis.

Mathematically determining the possible profit of a better schedule, is very difficult without evaluating all alternatives. And this is just what this whole analysis is trying to prevent. (Note that this is one of the key questions in scheduling.) Therefore, heuristic will be needed.

7.6 Progressive Reasoning

In 'conventional' direct predictive controllers, the calculation time of the control loop is fixed and known. This is due to the fact that the algorithm is programmed in a straightforward procedure using always the same computational operations. In more complex and especially in AI-based 'controllers' such as direct intelligent controllers [Krijgsman, 1993], and also the scheduler, the calculation or 'reasoning' time varies. It also might take a very long time before a final 'conclusion' is drawn, especially in unusual, complex situations (e.g. the occurrence of a disturbance or failure). However, any type of controller must produce an answer within a limited calculation time. This limited time is forced by the process, either by a fixed sample rate or by events as in the case of the scheduler (see § 7.2 "Sample Rate"). Therefore, for controllers which don't have a fixed calculation time, a *progressive reasoning* [Krijgsman, 1993] or *any time algorithm* is needed.

The idea behind progressive reasoning is, that the algorithm will produce a (feasible) solution as fast as possible. The remaining time up to the moment a solution is needed, is used to improve the previous solution. If a better one is found, it replaces the previous. In this way, (almost) at 'any time' a solution is available.

Within the scheduler, progressive reasoning plays a crucial role. Due to the possible very large computation times (as a consequence of the *combinatorial explosion*), progressive reasoning is inevitable. The possibilities to achieve progressive reasoning are discussed in the following paragraphs. The general idea is to first skip computations which are not needed to find a feasible solution.

The NIS

Progressive reasoning hardly plays a role within the NIS. When generating of the NLP model, it might be acceptable to skip the, time consuming, generation of some of the advanced robustness constraints [Terpstra, 1994a].

The IS

The IS searches through a tree, possibly using heuristics to guide the search. At each node a constraint propagation procedure is executed to estimate a bound on the criterion function, which is used in a branch-and-bound algorithm. The first time, a depth-first search is done to a feasible (and hopefully reasonable good) integer schedule. This first solution is used to bound the search. This first solution is send to the NIS and used as the first schedule in the progressive reasoning. Each time during the search, an end node is reached, a better candidate integer schedule is generated and could be send to the NIS and used as the next schedule in the progressive reasoning. An alternative method is mentioned in the next point.

General experiences with the optimization of combinatorial problems show that the optimal solution is often found very early in the search (it even might be the first one). The remaining search is only needed to prove that this is indeed the optimum. Therefore, this type of progressive reasoning is a useful strategy.

The interface between the IS and NIS

Without using the progressive reasoning approach of the previous point, the IS generates a list of all candidate integer schedules. This list is sorted using heuristic guidelines, concerning the:

1. Criterion function. The constraint propagation algorithm results in an interval for the criterion function. The list can be sorted using criteria as 'min-min' (the integer schedule with the minimal value of the lower bound of its interval), 'min-max', 'min-average', etc.
2. Robustness. It is likely that, a schedule out of a group alternatives which are similar, will be more robust than an alternative which is very different.
3. Reactiveness. By selecting a schedule out of a group with the same first schedule action, all members of this group remain candidates after this action should be executed. Using such a schedule, calculation time can be won to use for continued evaluation of the alternatives.

In theory, all these candidates must be evaluated by the NIS to guarantee that the optimal schedule will be found. This sequence of evaluated schedules can be used as the progressive reasoning output of the scheduler. Because the list is sorted (in contrast to the sequence of schedules in the previous point), the first schedule has the highest probability of being the best one. Concluding, this option will take more time to generate the first schedule, but that schedule will be of a higher quality.

An alternative option is to mix both strategies by using them concurrently: the IS generates an, in time increasing, list of candidate schedules.

7.7 Anticipatory Schedules

As described previously (see § 7.5 "Timing"), the control loops are mainly *event driven*. The need for starting to execute a new schedule is also event driven. If both events are the same, the calculation time of the scheduler causes a delay in the response time. Especially the IS control loop is time consuming. The response time would be highly

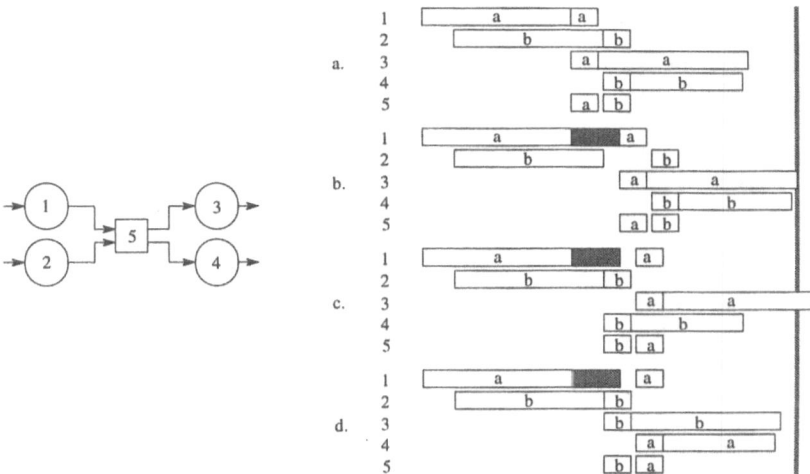

Figure 20   Example of the generation of anticipatory schedules. A disturbance
in the reaction phase of batch "a" in unit "1" is assumed. (Note:
reactor 4 is faster than reactor 3.)
a. The nominal schedule progress as predicted.
b. The progress of the current integer schedule when the disturbance
takes place.
c. The first anticipatory schedule: the sequence of the operations on
the piping network "5" is changed.
d. The second anticipatory schedule: both the sequence of the
operations on the piping network "5" is changed and the paths for
the batches "a" and "b" are swapped.

decreased, if pre-computed (integer) schedules would be available 'off shelf' to deal with
disturbances. This is only feasible if the kind and timing of a possible disturbance can be
predicted. These type of schedules anticipate to future events and are therefore called
anticipatory schedules.

With respect to the IS, the kind and timing of possible future (integer) errors e_I can be
predicted while the number of these different possible errors is limited. This can be done
by a *sensitivity analysis* on the execution of the current schedule. Given the current state
of the plant, the future integer states x_I are predicted while applying disturbances on the
parameters of the model. An example is give in Figure 20.

Using this technique in the planner is only feasible if failures can be predicted. For
instance, when there is a phase in the operation of a unit that is especially critical.

7.8   Parallelism

Parallelism is not a specific feature of a predictive controller, but it improves the computational performance of the scheduler and thereby it influences the timing. The type of parallelism, as used here, is that on the level of high level tasks. The execution of one task is done on one processor. This section does not discuss possibilities of using parallelism within one task, as for example parallel constraint propagation algorithms or parallel NLP solvers, etc.

The scheduler is very suited to take advantage of parallelism:
1. The three control loops can run in parallel.
2. Each instance of the NIS can run in parallel. This will especially speed up the IS control loop because it has to start instances of the NIS a lot to evaluate candidate integer schedules.
3. The tree-search algorithm as used in the IS calls at every node a constraint propagation procedure. Each different node can be evaluated in parallel. The algorithm of the search strategy (e.g. depth-first or breath-first) should be changed to make efficiently use of the parallelism.

8   Robustness analysis

Scheduling of batch processes in the chemical industry has some specific characteristics. One is the *uncertain nature* of chemical processes (e.g. wide variations in processing times). The other is that some plants have constraints which have very *costly effects* when violated. For instance, if the buffer tank from which a continuous distillation column is fed dries up, the column will trip.

The proposed scheduler deals with the disturbances in two ways. First, the scheduler is *reactive*. Second, the amount of compliancy is calculated explicitly out of statistical knowledge of the disturbances. This is called robustness analysis [Terpstra, 1994a].

All parameters which have variations are represented by a statistical probability distribution. In our case study we used an uniform distribution U(min,max), in other words, assuming a *bounded* disturbance. The cyclic state scheduler generates a schedule on the basis of nominal values.

Because of the unknown variations, after one cycle the actual state x ends up in an area $x_{cs} \pm \Delta x_f$ ($\Delta x_f > 0$). This will result in a 'cone' within which the state will end if all parameters vary independently within their distribution given the execution of a fixed, cyclic state schedule u_{cs} (see Figure 21).

The reactive scheduler can adapt the schedule during the cycle. However, only within certain limits because of the constraints on its manipulable variables. Given the maximal adapted schedule u_{dmax} instead of a fixed schedule u_{cs} this will also result in a cone (see Figure 22). If Δx_a is positive the process is not guaranteed to be controllable.

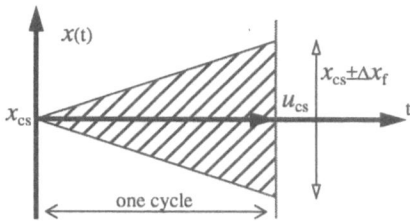

Figure 21   Area in which the state $x(t)$ can end up after one cycle given a fixed schedule u_{cs} and all parameters varying maximally within their limits.

However, Δx_a will normally be negative because of the maximum opposite steering by the scheduler. The more negative, the more robust the reactive scheduler is.

Another indication of robustness is the time needed to recover from a disturbance. Given a certain size of the error vector $x - x_{cs}$, it will take the scheduler a certain number of cycles to guarantee a return to the cyclic state x_{cs}. The larger the error vector, the more cycles it may take. Until the error vector reaches a limit value after which no guaranteed recovery is possible (see Figure 23)

Using these techniques, the robustness of a given schedule can be analysed. It is also possible to use this information while scheduling, to produce a robust schedule. The error is expressed as a symbolic function of the control variables. This expression can be used in two in two ways.

First, by including extra robustness <u>constraints</u> in the NLP problem. A constraint can be placed on the maximum error of x. For instance, "it must be possible to bring the plant back to the cyclic state x_{cs} within two cycles".

Second, by including robustness <u>criteria in the optimizing function</u> of the NLP problem. Now the schedule will be optimized for robustness.

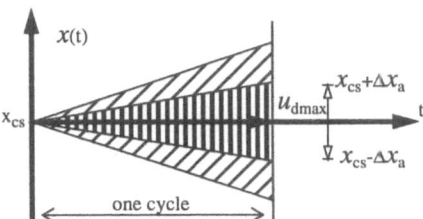

Figure 22   Area in which the state $x(t)$ can end up after one cycle given a maximal adjusted schedule u_{dmax} and all parameters varying maximally within their limits.

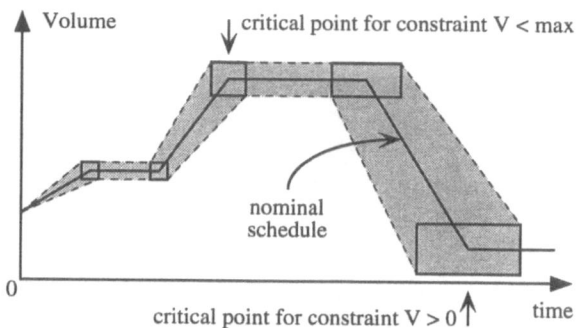

critical point for constraint V < max

Figure 24   Illustration of how accumulative uncertainties of bounded disturbances (the gray area and boxes) constrain the nominal schedule.

In addition to the use of the expression of the error, also the probability of violating *critical constraints* is used. A critical constraint is a constraint which, when violated, will have very serious and unwanted effects. This probability is also expressed as a symbolic expression of the control variables and can thereby be used as a constraint or a criteria in the optimization problem. The result can be, for instance, a schedule which guarantees a certain volume in the buffer tank. In other words, the scheduler can guarantee a non-interrupted production despite (limited) disturbances (see Figure 24).

It is also possible to add constraints to guarantee a safe operation in case a unit fails unexpectedly, which is interesting if the effects of a schedule failing are costly and/or if a unit has a high failure probability.

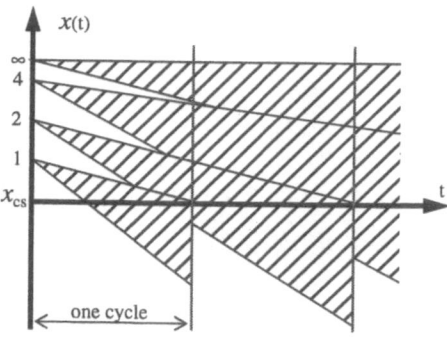

Figure 23   The error $x - x_{cs}$ can be classified by the number of cycles needed to guarantee a return to the cyclic state given worst conditions.

9   Implementation and results

At the moment, the implementation of the scheduler is proceeding. It is being implemented in the real-time expert system shell G2. The first versions of the modelling environment (see Figure 26), the integer scheduler and the non-integer scheduler (see Figure 27) are running.

Calculations show an increase in the average production rate of the case study (see Figure 25). Further, the size of the disturbances the proposed scheduler can handle is much larger than that of a conventional schedule. The conventional scheduler cannot cope with failures of units, while the proposed scheduler can. What is more, the probability of a shut down is much lower, or, with the same failure probability, a much smaller buffer tank can be chosen.

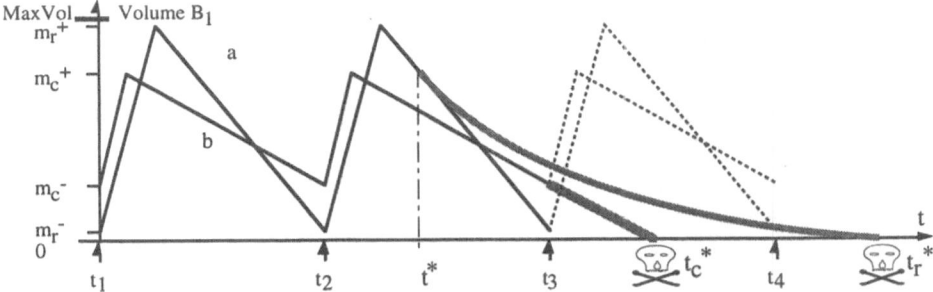

Figure 25   This example shows the improvements in steady state production and robustness of the new reactive scheduler (a) compared with the old compliant schedule (b) applied to the case study. Shown is the most critical state variable: the volume of the buffer tank B_1 (see § 1.3 "Example case"). At times t_n the buffer tank is filled with a batch coming from the batch unit S_1. The steepness of the ramp downwards is determined by the flow out, and is therefore proportional to the production of the plant. The compliant schedule uses wide safety margins m_c^+ and m_c^-.

Suppose the reaction in one of the reactors takes longer than expected, B_1 is not filled at t_3, but later. In the case of the compliant schedule, if the batch does not arrive before t_c^*, the buffer will be empty and the distillation column will trip. In the case of the reactive scheduler, the delay is noticed at t^* and the schedule is adapted immediately, which means that the throughput to the distillation column is decreased (slowly because of a constraint). The arrival of a new batch may be delayed until t_r^* to prevent a trip. This permitted delay is long enough to deal with even a complete failure of the reactor.

10 Conclusions

A robust, reactive scheduler for mixed batch/continuous plants has been developed and implemented. The use of the scheduler will result in optimal schedules, which guarantees stability and gives a better performance than conventional schedules. Because of its reactive nature, it can cope with larger disturbances. The scheduler is model based, therefore it can be easily applied to a wide class of batch plants.

11 References

Cott, B.J., S. Macchietto (1989). Minimizing the effects of batch process variability using on-line schedule modification. *Computers in Chemical Engineering*, Vol. **13**, No. **1/2**, 105-113.

Djavdan, P. (1992). Design of on-line scheduling strategy for a combined batch/continuous plant using simulation. Proceedings of ESCAPE 1, 24-28 May 1992, Elsinore, Denmark. Supplement of Computers and Chemical Engineering. 281-288.

Krijgsman, A.J. (1993). *Artificial intelligence in real-time control*. Ph.D. thesis, Control Laboratory, Delft University of Technology, Delft, The Netherlands.

Kumar, V. (1992). Algorithms for constraint-satisfaction problems: a survey. *AI Magazine*, Spring 1992, 32-44.

ISA-dS88.01 (1992). Batch Control Systems Models and Terminology, Draft 5, December 1992.

Pantelides, C. C. (1993). Unified frameworks for optimal process planning and scheduling. *Foundations of Computer Aided Process Operations (FOCAPO) II*, 18-23 July 1993, Crested Butte, Colorado, USA.

Patsidou, E.P., J.C. Kantor (1991). Scheduling of a Multipurpose batch plant using a graphically derived mixed-integer linear program model. *Ind. Eng. Chem. Res.* Vol. **30**, No. **7**, 1548-1561.

Pekny, J. F. and M. G. Zentner (1993). Learning to solve process scheduling problems: the role of rigorous knowledge acquisition frameworks. *Foundations of Computer Aided Process Operations (FOCAPO) II*, 18-23 July 1993, Crested Butte, Colorado, USA.

Terpstra, V. J., H. B. Verbruggen and P. M. Bruijn (1991). Integrating information processing and knowledge representation in an object-oriented way. *IFAC Workshop on computer software structures integrating AI/KBS systems in process control*, Bergen, Norway. IFAC. 19-29.

Terpstra, V. J., H. B. Verbruggen, M. W. Hoogland and R. A. E. Ficke (1992). A real-time, fuzzy, deep-knowledge based fault-diagnosis system for a CSTR. *Proceedings of the IFAC Symposium On-line fault detection and supervision in the chemical process industries*, Newark, Delaware, USA. IFAC. 26-31.

Terpstra, V.J., R.M. de Bruijckere, H.B. Verbruggen (1994a). Robustness in reactive batch scheduling. *Process Systems Engineering (PSE) '94*, 30 May - 3 June, 1994, Kyongju, Korea.

Terpstra, V.J. (1994b). *Intelligent supervisory process control*. Ph.D. thesis (to appear). Control Laboratory, Delft University of Technology, Delft, The Netherlands.

Visser, H.R. (1994). *Dynamic modelling of switching systems*. Ph.D. thesis (to appear). Control Laboratory, Delft University of Technology, Delft, The Netherlands.

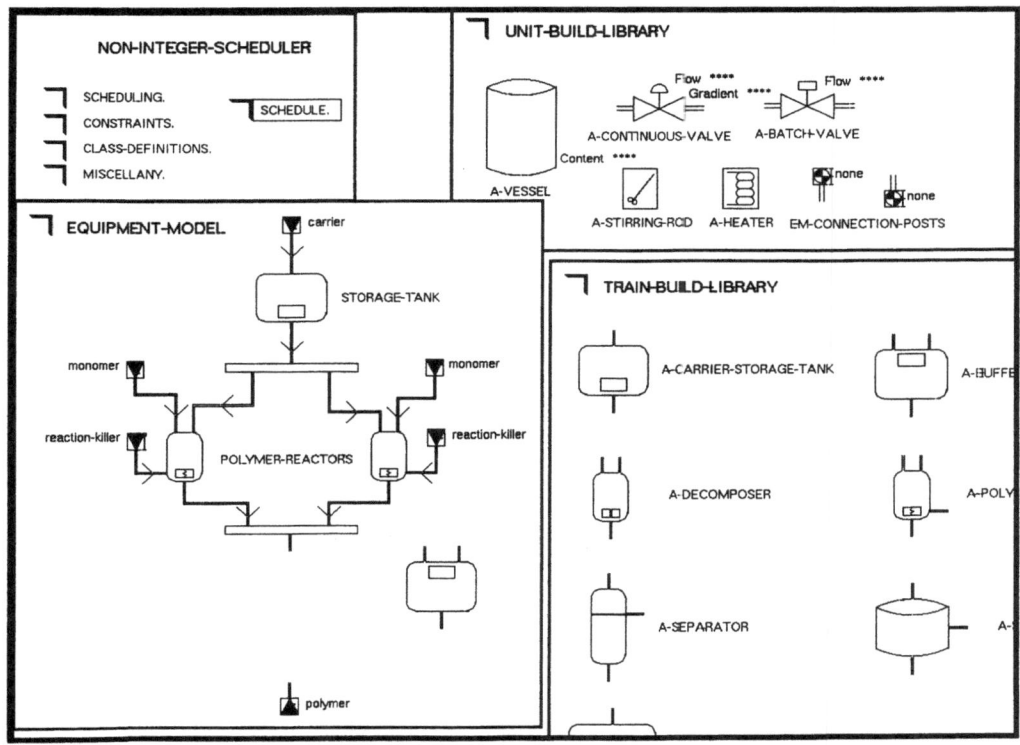

Figure 26   Screen dump of the plant modelling environment of the implementation in G2. The upper right window contains the basic equipment items from which the plant-type dependent unit-classes in the lower right window are built up. In the left window, a specific plant is being modelled by defining a topology between instances of the unit classes.

722

Wellons, M.C., G.V. Reklaitis (1989a). Optimal schedule generation for a single-product production line - I. Problem formulation. *Computers in Chemical Engineering*, Vol. **13**, No. **1/2**, 201-212.

Wellons, M.C., G.V. Reklaitis (1989b). Optimal schedule generation for a single-product production line - II. Identification of dominant unique path sequences. *Computers in Chemical Engineering*, Vol. **13**, No. **1/2**, 213-227.

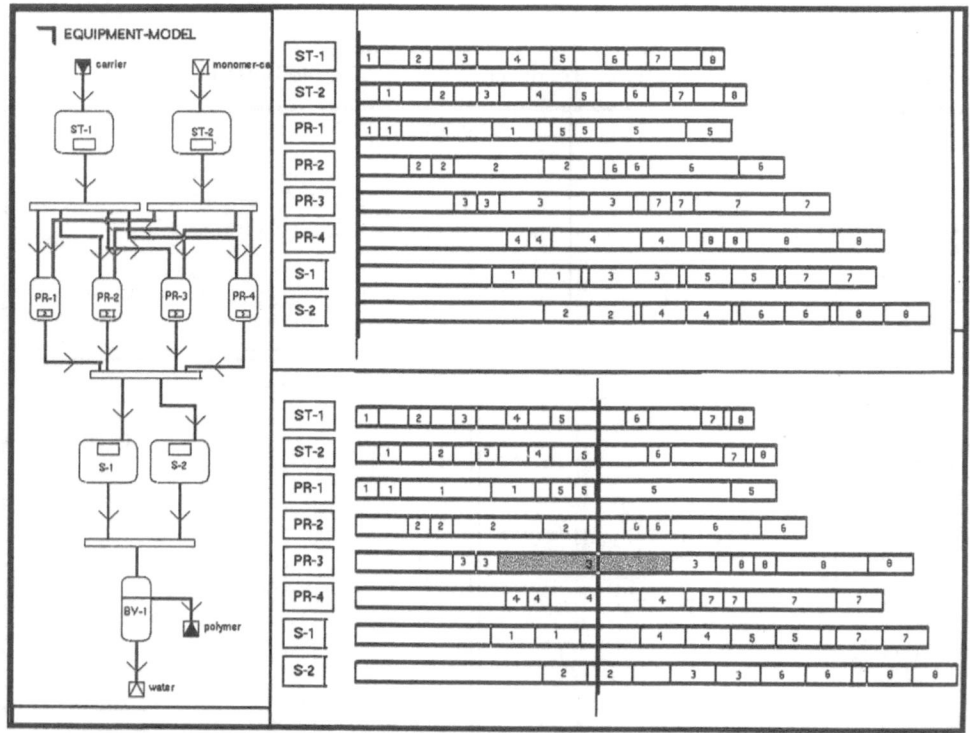

Figure 27   Screen dump of the output of the non-integer scheduler: final schedule. It is visualized using a classical Gannt-chart (upper right window).

The lower right window shows an on-line situation. At current time, there is a set of batches being processed in the plant. Due to the disturbance in the processing time in one of the units (PR-3), the schedule is changed with respect to the original cyclic state schedule. Note that apart from changes in starting times, also paths of batch have been changed.

24

APPLYING GROUPWARE TECHNOLOGIES TO SUPPORT MANAGEMENT IN ORGANIZATIONS

ANTONIOS MICHAILIDIS, PELAGIA-IRENE GOUMA,
ROY RADA
Department of Computer Science
University of Liverpool
Liverpool L69 3BX, UK

"Collaboration technology may in fact be the necessary precursor to real gains in industrial automation." Lynn Conway

1 Introduction

When group interaction takes place via computers (e.g. in organizations), then it is worth considering the effect of the computers on managers and on the environment in which they work. It is expected that proper information systems can be used to communicate effectively to the manager (as well as to the user generally) the information that is generated. Studies performed by members of our group have indicated that distributed group work requires new technologies that complement existing ones while maintaining the support that face-to-face provides by means of group meetings. Groupware is such a technology as it aims to support communication (information sharing and exchange) so that decision-making can be more informed and to assist with the coordination of collaborative group work to enhance cooperation. The provision of systems that make the management processes easier for organizations is one of the goals of groupware.

Our large research group has developed a sequence of collaborative hypermedia systems under the name of MUCH (which stands for Many Using and Creating Hypermedia). The latest version is based on networked UNIX workstations, has its own database system, and an X-window interface. This system has been used to introduce a groupware tool used for on-line planning and reporting documents so that it could be used as a management support tool for our organization. This groupware tool appeared to offer a potentially valuable contribution to assisting management by acting as a self-regulating source and repository of group information.

2 Groupware

The term Groupware is used to describe the technology that has been developed in the last few years to enhance group working. It is designed to harness multi-user computers

723

S. G. Tzafestas and H. B. Verbruggen (eds.),
Artificial Intelligence in Industrial Decision Making, Control and Automation, 723–755.
© 1995 *Kluwer Academic Publishers.*

and networks to give groups of workers the kind of productivity gains individual users already enjoy from single-user systems. As a tool, the computer is valuable in providing a user with massive information processing capacity, and enhancing the potential for productive communication. Groupware technology seeks to provide an optimal configuration of these capacities for people working collaboratively. It seeks to support information-rich person-to-person communication. This communication may be synchronized in time or not, and it may occur across great or small physical distances.

2.1 Groups And Computer-Supported Cooperative Work

The beginning of person-to-person communication using computers is closely connected with the appearance of the first extensive computer network, the ARPANET, in 1969. Its main feature was electronic mail [85]. Since then, computerized networks have become very popular communication technologies. It is now believed that computers might offer the potential required in order to improve the interaction of the human users.

When people work in groups, (e.g. in organizations), they need to interact and cooperate in order to be productive. Cooperative work usually needs to be supported by proper means, to bring about an effective performance of the group task. Computer-Supported Cooperative Work is the interdisciplinary research field that focuses attention "on issues such as interpersonal communication and coordination, and how these are mediated by, and may be assisted by, computer technology" [45, 2, 38, 36, 37, 80]. Among the technologies being developed to support work in groups is groupware.

Groupware is a novel computer-based technology representing a group's common task and shared information space. It aims to help actively groups of collaborative users by improving the group processes and performance. Groupware is primarily concerned with facilitating interactions among the collaborative parties. The artificial intelligence perspective in groupware aims at designing groupware systems that facilitate the group processes by becoming active participants themselves in an interaction process, thus providing information to users based on anticipated needs. The evolution of artificial intelligence might have a significant contribution to groupware by transforming technology from a passive agent in the collaboration process in an active agent that enhances interaction [25]. In any case, it is the expertise of the human participants that guides most of the interaction. The study of the relationship between the groups of users and the technology falls into the area of social cybernetics [52]. Understanding the fit of technology into the workflow would be critical in ensuring the successful role of technology as an active agent in group interaction [25].

2.2 Groupware Taxonomy

Groupware is a multidimensional concept, whether it is given emphasis on its group-related perspectives or it is viewed from the -ware's point. Two kinds of taxonomy for groupware systems have been proposed. The first taxonomy focuses on issues such as time and space, while the second taxonomy is based on the functionality offered by groupware applications.

2.2.1 Dimensions of Groupware

Of major significance for the characterization and evaluation of a groupware system are the following parameters [52]:

- *time:* the duration of the task-focussed interactions; These can occur either in real-time (synchronously) or asynchronously (at different times);
- *place:* whether the participants in the interaction process are co-located or distributed (in distant geographical locations);
- *task:* indicates the nature of the tasks that need to be performed for the attainment of the group's goals;
- *group context:* that is the group's environment and its features, social protocols and policies, etc.;
- *group composition:* this includes the biographical and personality characteristics of the individual members of the group, their abilities and attitudes, and finally their position in the group [59].
- *human machine allocation:* refers to the extent of the automation of the tasks to be performed; and
- *artifact or process focus:* this determines the kind of support the groupware application provides, that is either focussed to the product of a work process(e.g. document production) or to the process itself(e.g. management).

2.2.2 Classes of Groupware

Six general classes of groupware systems have emerged during the last decade: Message systems, Co-authoring and Argumentation Systems, Group Decision Support systems and Electronic Meeting Rooms, Computer Conferencing, Intelligent Agents, and Coordination Systems [25, 77]. These classes are overlapping and particular applications may fall into more that one class.

Message Systems. This is the most mature and widely used category of groupware applications. Message systems include electronic mail, text-based electronic conferencing, and electronic bulletin board systems. Advances in computer networking and the wide use of message systems have created the information overload problem. To overcome this problem the need for increased message structure that would allow the processing of some semantic information conveyed through the message has been arisen. For instance, some systems like Information Lens [56] allow the processing and filtering of messages based on their content.

Co-authoring and Argumentation Systems. Co-authoring and argumentation systems aim at supporting group problem-solving that involves negotiation and argumentation. Co-authoring systems provide support to a group of authors during the phases of writing (e.g. brainstorming, information gathering, planning, drafting and revising). They allow users to work either asynchronously or at the same time. They offer mechanisms for separating comments from text, defining roles, and notification of the group members' actions. Argumentation systems support the structured development of arguments and negotiation in a group setting. Argumentation systems, such as gIBIS [12], are based on the IBIS method [76]. This method attempts to model the development of arguments and multiparty negotiation on the basis of the conversations that take place during this process.

Group Decision Support systems(GDSS) and Electronic Meeting Systems (EMS).
Computer-based support for group decision making emerged before the CSCW paradigm. It has grown from research in social organization and particularly group decision-making, computer support for decision making, and technological advances in computer-based communications and computer-based information services [51]. Group decisions occur as the result of the exchange of information among members [22]. Group behavior is directed towards the convergence of members on a final decision. The types of information exchange that take place during a decision-making session include proposal exploration, analysis, expressions of preference, argumentation, socializing, information seeking, information giving, proposal development and proposal negotiation. The goal of GDSS is to alter the communication process within groups. The greater the degree of change in communication introduced by the technology, the more dramatic the impact on the decision process and, presumably, on the decision outcomes. The need for the use of computer tools for improving group performance in decision making was paralleled by developments in computer conferencing and electronic mail. Initially GDSS was viewed as an expansion of existing computer-based DSS without realizing the potential of the new technology on the group decision making process. GDSS promises to increase performance of group decision making by reducing time-consuming meetings that distract decision-makers from other critical activities while retaining or even increasing the quality of the resulting decisions [51].

Electronic Meeting Systems promise to take advantage of information technology to enhance the meeting process and outcomes. EMS target a wide range of meetings as, a meeting - includes any activity where people come together, whether at the same place at the same time, or in different places at different times [65]. EMS emerged from early developments in Group-Decision Support Systems (GDSS) and CSCW systems. GDSS are task-oriented in the sense that they support a group of people making decisions. CSCW systems are a more general class of systems that provide support for communication and coordination in working groups. Decision-making, communication and coordination are integral parts of EMS and both GDSS and CSCW systems are expected to inform the design of an EMS. An EMS has been defined as [21]:

An information technology-based environment that supports group meetings, which may be distributed geographically and temporally. The information technology environment includes, but is not limited to, distributed facilities, computer hardware and software, audio and video technology, procedures, methodologies, facilitation, and applicable group data. Group tasks include, but are not limited to, communication, planning, idea generation, problem solving, issue discussion, negotiation, conflict resolution, systems analysis and design, and collaborative group activities such as document preparation and sharing.

EMS should have a positive effect on meeting process and outcomes. That is, ideally, EMS should increase process gains and decrease process losses that occur during a meeting. To achieve this goal one needs to have an adequate understanding of both the meeting process and how EMS affects it. Meetings, like any kind of group work that involves group interaction, are highly dynamic. The introduction of EMS technology may affect group meetings in three dimensions: (i) meeting process and outcomes, (ii) methods that groups use, and (iii) the environment where the meeting is held.

Computer Conferencing Systems. These systems provide a shared information space. Users interact through the shared information space either synchronously or in real-time using various types of information such as text, audio, video, etc. Depending form of the interaction and the type of information communicated and stored in the shared information space the following categories of conferencing systems can be identified: text-based conferencing, collaborative hypermedia, real-time conferencing, desktop conferencing, and multimedia conferencing [78]. Text-based conferencing systems consist of a number of groups or conferences. Each group focuses on a particular topic and consists of a sequence of messages. Users subscribe to one or more groups and post messages or replies to messages. In contrast to text-based conferencing systems, collaborative hypermedia systems allow non-linear structuring of information. Multiple users create independently a graph or network structure where chunks of different types of information are connected by links [13]. Real-time conferencing systems allow a group of co-located or distributed users to interact synchronously through their terminals. Desktop conferencing has emerged from the integration workstation technology and real-time conferencing technology. Desktop conferencing systems allow the screen and the contents of windows to be manipulated by any workstation while all the users view the whole screen. Multimedia conferencing systems provide, in addition to the shared screen facilities offered by real-time conferencing and desktop conferencing, audio and video links [1, 79, 90].

Intelligent Agents. Systems that incorporate AI techniques aim to transform systems from passive agents that process and present information into active participants in interactions (e.g. virtual conversants) [25, 64]. Intelligent agents are used to perform certain set of tasks and so that their behavior resembles that of the other users of the system. Intelligent agents may support the group interaction process by reducing the amount of the control information that is processed by the participants. Liza [33] is an example of a groupware toolkit that includes intelligent agents.

Coordination Systems. Coordination systems aim at helping group members to coordinate and adjust individual activities in a harmonious way towards the accomplishment of a shared goal [82] cited in [25]. These systems allow users to be aware of their own and others' actions and they also provide automatic notification mechanisms. There are four models that underlie the development of coordination systems. Form-oriented models support coordination by modeling organizational procedures and activities as fixed processes. Procedure-oriented models attempt to program organizational procedures and implement them as software. Conversation-oriented models are based on speech-act theory [81] which holds that language (either written or oral) may affect the action of the person who originates it and its recipients [91]. When people are involved in conversations, in addition to exchanging messages that convey information, they also perform speech-acts. Communication structure-oriented models support coordination by modeling organizational relationships in terms of the roles [9] that people assume and their relationships.

It has been claimed that group support and group pressure can make individuals to adapt to the collaborative technology much quicker and easier than it has been expected [15]. It should be noted though, that groupware systems should not impose any constraints on the use of existing technological applications. Moreover, the new technology should complement rather than replace the existing technologies. The users should decide themselves which means suits best to their communicative needs and make their technological transitions accordingly [62].

2.3 Review of Groupware Systems

Following are briefly described some characteristic groupware systems that have been used so far in organizational environments:

Windows for Workgroups [42] have facilities for email (to which files can be attached), scheduling (appointments diary and time planner) and prioritized task lists with due dates. Individual appointment books can be consulted by other group members or by the system, which can make group appointments automatically. It also supports 'Chat'. Further workgroup applications are expected to be released in the near future.

Lotus Notes is known as "an information manager for groups" [54]. It mainly supports "workflow management", by routing material to the correct people to work with it and also ensuring that these individuals perform the required task. It was designed for commercial use, aiming at assisting group communications in a manner that could change the way people work. Notes connects work groups, co-located and/or distributed. People in such groups can share information and collaborate across physical barriers. However, as it is not an interactive system, its contribution in assisting communication between the members of an organization does not yet extend to simulating face-to-face, personal interaction. Another feature that this software lacks is the ability to allow "users to access it all at once to share their ideas" [84]. Lotus Notes works with a standard desktop scanner to make paper-based information accessible to the groups of people who need to act on it. Therefore, it becomes easy to find and share this information across the office, or around the world. Among the benefits it provides are: capture and store electronically a wide variety of document images; index, display, print or route images simultaneously to multiple recipients; helps keep record current; creates a single point of access for related information - any Lotus Notes desktop; and integrates imaging capacity into the collaborative, flexible desktop environment.

The Coordinator [57, 30] was a PC program, "an early form of group productivity software, or groupware", which mainly used an email system to coordinate the work of people within the work groups. It was based on speech-act theory [91]. Speech-act theory's developers, focusing on conversation-based communications, have argued about the existence of recurrent patterns of conversational structures in organizations. The identification and support of these patterns might probably be the key point to discover the ways for providing support to organizational work [92]. Coordinator provoked extreme emotions, both positive and negative, depending on the character of the organization in which it was applied. Users who preferred a rather flexible way of communicating found it inconvenient to work under the pressure of the task-oriented e-mail messages they were receiving. Consequently, the designers of the system were forced to modify it in order to become less oppressive [40].

The *CHAOS* prototype was built after a linguistic model of the office, based on speech-act theory. Its aim was to support coordination and increase the cooperation of the office activity [19].

GroupSystems is a software from Ventana which provides anonymous, real-time, text-based interaction for workgroups. The GroupSystems architecture consists of three major components, an EMS meeting room, a meeting facilitator, and a software toolkit [86]. Usually the leader of the group wishing to use the system meets the GroupSystems facilitator and determines the tools to be used in the meeting and develops the meetings agenda. The idea generation phase of the meeting occurs in anonymous mode. At the end of this phase the ideas are organized into a set of key ideas followed by a prioritization process which results in a short list of ideas. Then the participants make plans of how to realize the ideas. The process is repeated until consensus is reached. Usually at the end of the meeting a large volume of ideas and plans for actions is produced.

3 Management

When group interaction takes place via computers (e.g. in organizations), then it is worth considering the effect of the computers on managers and on the environment in which they work. It is expected that proper information systems can be used to communicate effectively to the manager (as well as to the user generally) the information that is generated. Studies performed by members of our group have indicated that distributed group work requires new technologies that complement existing ones while maintaining the support that face-to-face provides by means of group meetings [62]. Groupware aims to support communication (information sharing and exchange) so that decision-making can be more informed and to assist with the coordination of collaborative group work to enhance cooperation. The provision of systems that make the management processes easier for organizations is one of the goals of groupware. Before talking about the groupware applications for management, it is essential to discuss the activities that take place in an organizational environment.

3.1 Organizations

Organizations can be studied using a diversity of approaches, depending on which perspective one is looking at this concept (e.g. social sciences, economics, business studies) [89]. Each approach highlights particular aspects of the organization. Different definitions for the organization have been provided in different contexts. An integrated approach, though, considers human organizations as extended groups consisting of elements who cooperate or compete to achieve the organizational goals [20, 6]. The term organizational intelligence indicates the achievement of organizational goals, using appropriate means (media or tools), and getting positive results. Organizations may become more intelligent via appropriate design and use of information technology applications [10].

The element or fundamental unit of work in the modern organization is the group, not the individual [85]. These groups, having complementary expertise, are aggregated within the organization performing various activities. Collaboration is a goal-oriented process and constitutes the most essential part of group work, as it involves the sharing of the effort by the group members to achieve the common goal. Collaborative work requires communication links among the collaborators, novel ways to solve problems that might arise, planning of activities and monitoring the plan's development [23]. The dynamic nature of collaborative work implies the changing of the structure of the group's operations or even of the physical layout of an organization [66, 27]. This continuous change is compatible to the dynamic nature of any organization. In addition, it is an expression of the level of organizational intelligence.

According to organizational information processing theory, organizations are seen as "social structures constituted to gather and interpret information about the environment and use it to convert other resources into outputs such as products and actions" [75] but have limited information capacity [16]. A key objective of the organizational strategy is to obtain control over information [35]. The information processing that takes place in organizations aims at the effective handling and reduction of task uncertainty and equivocality. Task uncertainty refers to the lack of information in the specification of the tasks to be carried out, the skills and resources required, and the sequencing, timing and duration of the task, whereas equivocality indicates ambiguity and confusion in the task performance, thus requiring proper organization. Organizational information processing is also characterized by sharing, as it is performed mainly by groups of individuals [16]. The level of task uncertainty suggests the amount of information that is necessary to be processed by the collaborators for coordination purposes [11].

To assure that meaningful information flows continuously, through timely and relevant communication, is one of the most critical processes in organizations [18]. Thus, organizations need to develop appropriate internal structures for information processing. That is, to allocate tasks and responsibilities to individuals and groups within the organization, and to design systems capable of providing effective communication and integration of the efforts [16]. Such structural design, should communicate the required amount of information to the appropriate members of the organization in order to improve the coordination and control activities associated with the management of the organization and minimize uncertainty.

Communication can determine such intraorganizational behavior. The communication patterns which are developed in an organization might influence the relations among its members and also become indicators for the various organizational structures. Every organization should be capable of providing a listing of reporting relations-known as the authority structure [14]. In developed organizations, information flows along communication channels in any direction throughout the hierarchical structure of the organization, as illustrated in Figure 1. This communication flow in organizational hierarchies is determined as follows [14]:

- the communication needs of the managers and the operating personnel (employees), as these needs are induced by the organizational structure; and
- the opportunities for communication this structure allows between them.

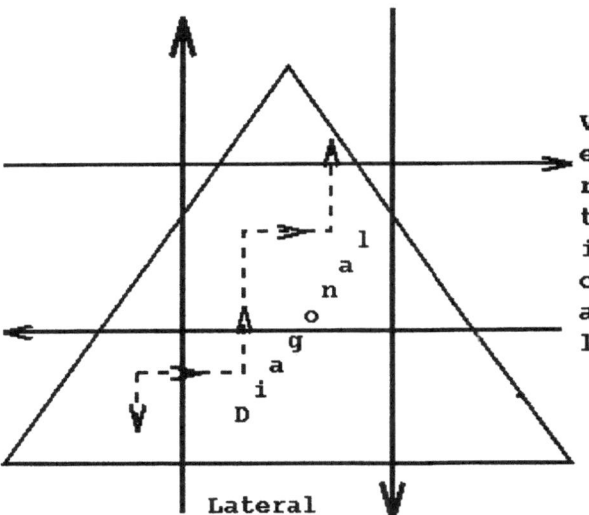

Figure 1: Routes of communication in developed (hierarchical) organizations.

Information technology[1] applications in organizations surely impose some impact on the way people communicate [48, 28](Kling, 1991). This belief, along with the reduced cost of computation and communication, allowed computerized networks to emerge and become powerful human resource tools. Computerized networks provide channels of communication among the communicating parties, a function especially advantageous when they are widely dispersed. In modern organizations people use different means to communicate, including face-to-face, memos, telephone, email, and facsimile. Each of them offers different possibilities to the user to reach a thoroughly reliable communication.

According to Ellis, Gibbs, & Rein [25], it is not so hard to imagine "...the electronic workplace-an organization-wide system that integrates information processing and communication activities". The new collaborative technology exists separately from the conventional tools (e.g. telephone, face-to-face conversations). Groupware in particular is believed that might help in filling the gaps in communication between distributed groups [5]. By providing a shared work space, groupware systems allow people to engage in multidirectional, long-term, conversation-based communications on-line [52].

The communication structure of a group is the key to better coordination of the activities performed by the group members [11]. Coordination has been defined as "the act of working together harmoniously" [57] and involves a variety of processes which are important for the group in order to perform efficiently its tasks. Specifically, coordination of a group's activities is necessary to enhance the effectiveness of communication and cooperation among the members of the group. It 's role is to adjust individual work efforts in an integrated and harmonious way towards the attainment of the group's ultimate goal [25].

Computerized networks are used to connect cooperating groups and Group Support Systems (GSS) are implemented to facilitate coordination of the work in organizations of any size. The close relation between coordination tasks and the way information is acquired and processed, probably indicates a dependence of coordination-related activities upon the application of information technology in a cooperative work environment. Information technology has been expected to have considerable impact on the organizational structure and the management performance in organizations [69, 85].

Furthermore, as coordination of an organization's operations is mainly the job of their managers, groups become essential management tools and the group management becomes important [32]. Contingency theory suggests that organizations should follow sensitive management strategies to allow for their internal needs to balance with the external environmental changes [45]. Therefore, organizations process information in order to accomplish internal tasks, to coordinate the various activities performed within

---

[1]The term information technology is used to denote the full scale of computer technologies and tools, including hardware, software, and communication systems.

their environment, and to interpret the external environment. Information requirements are in close connection with particular structural characteristics and behavior in the organization [17]. Coordination processes can determine, in a way, organizational structure and thus, they are possibly directly affected by the information technology.

In the next two sections, emphasis will be given to the management processes and how the applications of IT in general and groupware technologies in particular may affect management of organizations.

3.2 Managing Organizations

Management, as a concept, describes the wide range of activities carried out by those members of organizations who are formally responsible for the work of at least one other person in the organization [4]. "Manager's working processes have not been understood in depth yet" [61]. According to traditional perspectives, managers are considered to build their actions on strictly rational models. That implies that managers are expected to follow the sequence of setting and evaluating goals, assessing the strategy to be followed in order to reach these goals, and finally making the best choice. The conventional approach to management functions, tools, and systems is presenting them in terms of planning, controlling, staffing, organizing, and directing [50]. However, many studies have shown that in practice managers are rarely bound to rational principles alone. Moreover, they rely heavily on intuition [43]. As a result, it has been claimed that management is closer to art rather than science [61].

It was Mintzberg [61] that first discovered some of the actual characteristics of the manager's job. Since then it is known that managerial work is varied, brief, fragmented and highly interpersonal. Managers are considered to work long hours, and spend much time in oral communication [50]. The extent to which the above characteristics are presented has been found to depend upon the rate of change of the organization concerned, whereas the role of the manager varies with the size of the organization [58], and on the level of the hierarchy that the manager represents. In general, the nature of the managerial work is characterized by uncertainty, diversity, and a great amount of information [50] and thus the managers have to deal with "ambiguity, inconsistency, novelty, and surprise" [43]. Therefore, the managers try to find decision rules, information sources, and structural designs that could provide them with adequate understanding to cope with uncertainty. It is critical for the actual managerial processes to be modeled completely and successfully [67], but that still has not happened.

Modeling the managerial processes is a way towards investigating what part of these processes could be supported by the technology. It has been claimed that there are certain activities, that the manager performs, which allow for them to be partially or fully automated, such as planning, information gathering (collecting), and resource allocating activities [61]. Investigating which activities are programmable might facilitate the design of rational systems that could provide support to the manager in handling non-recurrent situations that are left undetected by systematic approaches [43]. As reported before, the recent advances in networked computing have allowed the use of information technology as part of the group-interaction process [69]. More generally, information technology is expected to affect significantly both the organizational structure and the management functions.

Recognizing that the traditional division of the management functions into categories can roughly represent the actual situation, then it might be worth focusing in particular categories (as performed by managers and facilitators) and the support that computers can offer in these areas. For instance, managers appear to perform planning and organizing activities in a less systematic way than it was expected. Planning is a very difficult task and managers should spend a lot of their time and attention to perform it. It has been claimed that the more difficult the situations that the managers face are, the more they avoid analytical planning. Even when analysis is used, this is done in conjunction with intuition [43]. An information system could contribute in making the planning tasks easier by sorting out the unnecessary information (noise) the managers receive and by processing the essential data provided [50].

Organizing, as well as staffing, require many other activities to take place so that they can be done properly. Thus, organizing involves the attainment of effective organizational processes and the management of general goals [43]. These, in turn, can be achieved by proper tasks' allocation, roles' distribution (mapping to people), decisions' or changes' implementation, etc. A tool that could facilitate the task-role assignment process and offer flexibility in the presentation of the organization structure might be valuable complement for the manager.

Since managers are in charge of the attainment of the goals of an organization, they should be capable of taking control over all the aspects of the organization's production. In addition, as managers target in increasing the overall performance of the organization for which they are responsible, it is important to have the ability to monitor their groups' work and assess the progress of the various projects in the most accurate way [44]. Directing and controlling require a strong set of cooperative relationships [50]. Managers seek support to fulfil the above requirements. Computers, on the other hand, may support the evaluation of a given job. As a consequence, the use of computers and other electronic devices [41] (such as video cameras, microphones) to monitor employee performance at work is increasing.

A major problem of today's manager is what is called: "the dilemma of delegation [60]. The term "delegation of authority" implies that someone authorizes in her / his behalf the right to control. It is one of the most critical parts of a manager's job to decide what, when, and to whom to delegate. About the decision-making activities, it has also been found that managers "devote most of their attention to the tactics of implementation rather than the formulation of strategy" [43]. It has been claimed that decision-making should be located where the relevant information is immediately available. In a decentralized organization, where employees may have better access to the continuously changing information, decision-making powers could be delegated to the employees [67]. Studies of how organizations are evolving have indicated that the closer to where work is being done (i.e. at lower levels of management) the related decision-making process is moved, the more productive the organization becomes. On the other hand, the personnel in the various positions of authority, as defined by the organization, retain the responsibilities for the activities of their subordinates. The delegator thus requires information on the decisions taken by subordinates and their outcome. There are therefore cases where the decision will be delegated, and other cases where the manager alone wishes to be involved in the decision process.

Concluding, it should be emphasized that harmonising the efforts towards the accomplishment of a job in workplaces involves diverse strategies of social control [48]. The manager's job is in great part to achieve this harmonious result and for that reason s/he spends a significant proportion of her/his time working in groups. The nature of managerial job makes it difficult to be modeled or predicted following logical analysis. However, some of the managerial activities can be automated and on this belief is based the design of computer-based systems to support management. D. Engelbart, whose contribution in changing the idea people had about the computer from being a simple machine to an active communication system is priceless, claimed that: "Conventional project-management operations can be augmented through the use of computer-based project-management tools with the enriching services of dialogue support, document development, and plans, commitments, schedules and specifications" [26].

3.3 IT Systems for Management-Support in Organizations

Organizations' information capacity is limited, despite the fact that they have to deal with information complexity. Therefore, organizations need to develop information processing mechanisms that can cope efficiently with variety, uncertainty, coordination, and an unclear environment [16]. Moreover, it has been found that the most frequent and time consuming activities performed by the managers of the organizations involve the giving and receiving of information [58]. Thus, information systems are introduced in organizations as regulators of the necessary information for the operations performed in such contexts.

Information systems are impersonal and thus incapable of detecting, measuring, and communicating equivocal issues. As such, they cannot cope with equivocality in organizational environments [16]. There can be two obvious suggestions: either to incorporate artificial intelligent agents in the systems to make them more "intelligent", or insist on the human factor, as human beings can interpret and respond to ambiguous situations. The latter approach introduces people as facilitators of the organizational processes, in both their technical and social aspects. In addition, there are potential dangers accompanying the computer assistance [8], such as:

- information overload, when the system provides the users with greater amount of information than the ones they can handle. In this case, the presence of a system's facilitator might be necessary so as to assure that computer assistance allows groups to deal with large amounts of information by presenting it at a rate that is appropriate for processing by the group.
- provision of less contextual embedding for the participants.

When an information system designed to support work group coordination is introduced in an organization, a contingency view is suggested by the transaction costs model. From this perspective, if the system satisfies the needs of the work group, then it will facilitate changes in the coordination processes by acting on the fundamental transaction costs themselves. The design of such systems aims at [11]:

- standardizing tasks, i.e. reducing task uncertainty;
- standardizing interfaces between execution of granular tasks ;
- facilitating control and monitoring of performance of the actors; and
- enhancing communication by reducing hierarchical barriers and opening new communication channels.

There are three main categories of advanced information technology applications that are used to support managerial work in organizations and these are: expert systems, groupware systems and Computer-Integrated Manufacturing systems.

Expert Systems: Expert systems are advanced applications built by knowledge engineers and experts who together translate expertise into a knowledge base plus inferencing mechanism. While the expert system may directly solve problems, it also communicates knowledge from an expert to a user [73], (using different representations and reasoning methods) (see MAIL-MAN, an expert system for management) [63].

Groupware systems' effect on organizational structure depends on the degree of social structure they impose, and vice versa. For instance, if a groupware tool alters the nature of participation in a group process, impacts on the decision quality, and the outcomes of that process. On the other hand, the reactions of organizations with strict hierarchical tradition to the application of such a technology are different than those of more flexible environments (see Coordinator [91, 40]).

Computer-Integrated Manufacturing (CIM) systems, i.e. systems designed to facilitate coordination of manufacturing activities. "CIM involves networking the entire manufacturing enterprise into a single, integrated information system" [39]. The design of a CIM system takes into account mainly the difference in the data flow at the various levels of management [71, 49]. So, this integrated system usually provides monitoring and performance evaluation mechanisms for the employees (agents).

Focusing on groupware applications, it should be emphasized the claim made by Ellis,

Gibbs, & Rein [25] that groupware systems and groups are "intimately interacting entities". In order for a task process to be augmented technologically in a successful and efficient way, there should be equilibrium between the quality of the social processes and procedures and the appropriately structured technology. There are two broad classes of Group Support Systems [21]:

- those driven by communication needs (e.g. Email)
- those driven by problem solving, planning decision needs (e.g. group decision support systems).

Electronic mail services have provided "fast, asynchronous group communication, as well as one-to-one communication", reducing the need for face-to-face meetings, decreasing coordination costs consequently [85, 55]. However, as Email is designed with minimal functionality it is perceived as a low level, routine information disseminator, thus has not been used to support managerial work in organizations to a larger extend [63]. Computer communication systems also allow for conferences to be held via computers with participants distributed at many remote locations [14]. Such conferencing systems have been widely adopted and are considered to contribute in the elimination of the powers of status and dominance, thus promoting the organizational strengthening (although they might also result in organizational instability) [68]. The tendency of such systems to eliminate hierarchy or conflict implies the risk of not providing the appropriate corrective mechanisms required for the coordination processes to become efficient.

Kraemer and King [51] have stated that most tools for support of group work facilitate discrete aspects of decision processes, such as display of data, communications among members of the decision group, etc. The provision of systems that make decision-making easier for organizations is one of the goals of groupware. A number of group decision support systems (GDSSs) that appear to support collaboration and coordination in distributed organizations have already been developed and proposed. Some of these systems were designed in accordance with the speech-act theory [92]. GDSSs hold out the promise of improving group decision-making by reducing time-consuming meetings that distract decision-makers from other critical activities, while retaining or even increasing the quality of the resulting decisions. GDS systems usually provide tools such as electronic brainstorming tools, presentation tools and electronic notepads. The technology itself would not be able to adequately support meetings unless it encompassed the procedures and methodologies that people use in meetings.

3.4 Comparing R&D Department with Organizations

Our study focuses on the implementation of a groupware tool to support management of an R&D organization, which is a University's department. Universities can be considered as organizations in that they exist to achieve concrete ends which are capable of rational analysis and that they face many problems common to most modern organizations [53]. However, there are differences between universities and organizations as the former are permanent multipurpose organizations, undertaking teaching, research and public services. We will consider here that universities are organizations which have corporate responsibilities, and which possess powers to manage the activities of their members in order to carry out these responsibilities. Departments are not autonomous units but interdependent parts of a unitary organization. Nevertheless, the management policy that the university follows is considered to delegate authority and initiative for most activities to the constituent groups and individuals.

It will not be an extremity to state that the Department under study, although existing within the university, operates as a private enterprise, thus it is an R&D organization. Some of the characteristics of this organization we emphasize are: the development of a systematic organizational structure for the co-ordination of all the departmental resources (human and others) in order to achieve the stated purposes, the simulation and the facilitation of the various projects, problems of containing costs, and of maintaining the capacity to innovate, etc. The affairs of this R&D organization require a clear framework of management within which to be conducted. This framework should be based upon the groups within which members operate, it should establish a structure for these groups and their interactions, it should specify the terms and roles of those people who hold responsibilities within the structure, and it should incorporate the channels through which decisions are reached.

4 Case Study

In order to gain insight on how the above considerations might work in practice, a groupware tool, namely the Plan-Document (into a hypermedia system-Many Using & Creating Hypermedia), has been applied to support the management of an R&D organization. This organization operates within a University. The structure of this organization has been modeled using a modified object-oriented group model language, called Activity Model Environment. Since the roles of the members of the organization and the required activities have been determined and the rules that govern the functions to be performed have been established, then the model has been implemented to the tool design.

The Plan-Document design was based on the claim that among the activities, that the managers perform, that could become in some degree automated are those of scheduling, information collection, and resource allocation. The objectives and the activities, the roles and the rules are kept in the Plan-Document. Thus, this groupware tool is designed so as to allow for planning of the activities to be performed, to keep a record of the progress of the various projects by storing the reports of each group's members, and to facilitate the allocation of resources. In a sense, it works as a kind of organizational memory. An additional feature is the Plan-Facilitator, which is one of the managers of the organization, that supervises (and coordinates at times) the overall use of the system and mediates the interactions among the users.

4.1 Modeling the Organizational Structure

Human organizations act in ways that could be characterized as intelligent. Thus, it has been claimed that human organizations' intelligence results from a number of factors including: (i) its structure; (ii) its codified knowledge (procedures manuals, memos, reports, etc.); (iii) its culture (myths, stories, jokes, etc.) and determines in a way the effectiveness and survival of the organization [7].

Organizational research and theory building aim to understand and predict the appropriate structures for the specific organizational situations. These structures and the associated with them internal systems facilitate the interactions and communications to reach the favorable levels of coordination and control of organizational activities [16].

One way to apply AI to organizational design is to model the organization as an object-based system [7]. In this study a model of our R&D organization is proposed, based on the AME model.

4.2 The AME Model

The Activity Model Environment (AME) is a prototype object-oriented tool for exploring models of organizations [83]. The AME prototype consists of a database and an associated rule-based formalism for representing activities and organizational states. Users interact with the model by creating and playing roles. Activity related communication proceeds via the exchange of messages between roles. Eight components of the framework can be identified:

- *Activities* - are sets of tasks performed by groups of role instances for achieving a set goal.
- *People* - are placeholders for actual individuals.
- *Roles* - which specify the responsibilities and duties of the people playing the roles.
- *Workspaces* - the conceptual spaces in which work takes place and which contain resources associated with roles.
- *Messages* - are objects that flow between the role instances associated with an activity.
- *Information Units (iunits)* - elementary ("atomic") units of information used in building messages.
- *Rules* - regulations used to constrain the behavior of components.
- *Functions* - operations are carried out by roles and messages as part of an activity.

In AME, roles, people, workspaces, iunits and messages are represented as objects and are stored in the Organizational Manual which is a database acting as reference both for users of the AME, and for the AME itself. To describe an activity one should present the input/output states as well as roles and messages involved in its execution. The people in the organization have object entries in the Organizational Manual associated with them. Each entry specifies the roles that each person is authorized to play. A person interacts with AME through specified role instances. Each person may hold several role instances at any one time.

Roles define responsibilities that are taken by one or more people. A role instance consists of the person instance undertaking the role, the set of role rules, and a role agent. The role agent is executed by the system and might undertake some of the person's responsibilities. The role agent uses the role rules in the performance of the role. A workspace is a conceptual work area that contains resources associated with a particular role. Multiple role instances may be associated with a workspace. The workspace also contains message handling resources.

Messages are used for role instances communication. Messages collect and transfer information associated with activities between roles. They exist for the lifetime of an activity. There are different types of messages (e.g. memos, notices, forms). Messages are composed from groups of information units (iunits) which are atomic information objects. An iunit has a name, fields and a set of completion rules associated with it. Rules define and constrain the behavior of roles, messages, and iunits under specified conditions. Finally, functions are atomic operations performed within group communication (e.g. instantiate-message, fill-field). They must be executed entirely by one role instance or role agent.

4.3 The Modified Version of AME

In this section the AME model of the R&D organization is presented. The database or Organizational Manual in this modified version of the AME model is the Plan-Document, a groupware tool of the MUCH system. The Plan-Document acts as a mechanism for enhancing participative management. The integration and flow of the data through a central point makes access to the required information much easier. The partial-mediator is known as the Plan-Facilitator in this specific model.

4.3.1 Activities

The various groups in this organization conduct research in connection with groupware and hypermedia, collaborative authoring, and reuse and courseware development. Research is also applied by means of participating in externally funded projects, developing computer tools, teaching and publishing media. An external relations group cooperates with organizations outside the university (see Table 1). Frequently, members of the organization are involved in several activities at once. As such, these groups have a matrix organization [34]. The various activities of the constituent groups are inter-related. There is usually a dependence on each others work and in the pursuit of one's interests.

4.3.2 People

Each person in the R&D organization has an entry in the Plan-Document. This entry specifies the role that the person has been assigned to play.

Courseware Development	Tool Development	Publishing Book/CD-ROM	Management	Externally Funded Projects
Analysis	Requirements	Inform. gathering	Research	Proposal Develop.
Design	Specification	Planning	Human Resources	Socializing
Development	Design	Writing	Budgets	Politicking
Implementation	Implementation	Revising	Monitoring	
Evaluation	Testing	Evaluation	Coordination	
	Maintenance		Evaluation	

Table 1: The activities performed by the elements of the organization under study.

4.3.3 Roles

The identification of a number of different roles does not imply that each role is filled by a different person (the fewer the number of people, the better), but a person could be charged with more than one role or a role could be filled by more than one person. It is important to assign precise roles to people, since without a clear definition and sharing of responsibilities, at a given moment some team members could be repeating other people's work; the effects of this on the development time and costs are easy to imagine. It is important to note that to the extent that the roles are not filled by people with appropriate skills, the product will suffer. In such a case, it is essential for people to be flexible to exchange roles. The roles and their features that have been assigned to the members of the organization are presented below:

- *Managers*: There are several layers in the management structure of the organization, ranging from the president and the top management group to the middle managers and the line managers. The managers at all levels perform similar tasks, such as planning the development of a particular project, make decisions about the work to be done, supervise, monitor and evaluate their subordinates. However, authority and responsibility increase vertically, from the bottom layer to the top.
- *Researcher*: The role of the researcher involves planning and selecting the proper `path' to follow in accomplishing her/his research project, the development of a high quality product, and the reporting of the progress to the management. The researcher may be involved in several activities, such as courseware development, publishing, and tool development.
- *Software Engineer*: The software engineer guides the development of tools throughout the software lifecycle, and also provides technical services when necessary. S/he reports to the management, too.

- *Task Finisher*: S/he makes sure that the group fulfils its commitments and monitors group progress and communicates emerging problems to the group [70].
- *Author*: The author is involved in the publishing activity (book or CD-ROM production). In the courseware development the author acts as a subject matter expert. S/he has adequate knowledge of the subject matter, the objectives and the audience.
- *Clerical Worker*: S/he is responsible for the secretarial and administrative tasks, being closely associated with the roles related to management and external relations activities.

In addition to the above roles, the role of facilitator has been identified to be vital for the organization, as the person that handles effectively the diversity of human and technological interactions. A description of the role of facilitator is presented below.

4.3.4 Facilitator

The group facilitator has both the skill and the experience in understanding and mediating group processes and help in the diagnosis and treatment of any problems that arise [47]. The facilitator's role is significant and powerful, and the relationship between facilitator and the group is unique. The art and study of facilitating computer-supported group-interactions is still in its infancy. There may exist various combinations of task allocation between a facilitator and a groupware system [87, 24]. It requires particular skills to facilitate the diversity of human and technological interactions.

In this study the Plan-Facilitator was the person in charge of controlling the successful operation of the Plan-Document. Among the main tasks of a facilitator is to monitor the activities of the group and these consist of the relationships between the cooperators, the amount of participation of each one in fulfilling a common task, the time they spend for a particular job, their mood, etc. Then, the unit of work and the process activities should be recorded for analysis. At the analysis stage, potential problems are identified (such as conflicts among the collaborators, miscommunication, timetabling and deadlines) and should be resolved by the intervention of the facilitator. Thus, the facilitator needs to keep open communication channels with all the participants [29].

Furthermore, the members of the organization might feel the need to use the system for their own benefit, since the facilitator would be keeping track of their progress and be willing to provide advice and guidance when they reach a deadlock or face other constraints in performing their task. The Plan-Facilitator, by controlling the use of the tool (Plan-Document), could help in managing and resolving any conflict within the group, commonly arising from different responsibilities, perception and knowledge among the collaborators. The facilitator might be able to suggest a strategy or an action plan directed towards common commitment of the participants in a project. Thus, the facilitator might support extensive use of the system by the group members as an individual or collaborative planning tool. Also, the users of the system might anticipate critical deadlines for the completion of a task and priority conflicts.

4.3.5 Messages

Messages constitute the intermediate products of each activity. There are two types of messages: (i) messages specific to each activity and to the tasks associated with each activity (see example below) and (ii) general messages related to the management of the organization. In the current model of the organization the following types of general messages have been identified:

- Email messages, used especially for task assignment or notification of a problem, or even to satisfy the needs of collaborative parties;
- MUCH reports, obtained from and sent to the system for the projects' progress monitoring;
- Face-to-Face requests, among the various members of the organization; and
- Broadcasting notifications, especially advantageous for fault reporting, annotating, and real-time interaction services.

MUCH reports and Broadcasting notifications are integrated in the Organizational Manual or Plan-Document.

4.3.6 Functions

System functions include: Two types of functions have been identified: (i) system functions have been designed to coordinate the use of the Plan-Document when role instances perform an activity and (ii) role functions which are available to specific roles. Role functions are determined in the Plan-Document and they are valid for the corresponding tasks. A brief description of the above function follows.

System functions include:
- Create node: the user of the system creates a new node and links it to the Plan-Document.
- Save node: save the contents of an updated node.
- Delete node: remove node from the Plan-Document.
- Modify node: Modify the position of a node within the Plan-Document
- Rename node: change the name of a node
- Make annotation: generate a comment attached to a node in the Plan-Document.
- Send broadcasting message: send a general message to other users of the Plan-Document.
- Store broadcasting message: save a general message in a private repository within the Plan-Document.

Role functions include:
- Review: check a plan, document, or report to validate it.
- Delegate: delegate a person assuming a particular role to do something.
- Create: create a plan, document, or report.

4.3.7 Rules

Two types of rules have been determined: global and local rules. In order to solve or ease the problems of higher communication loads and modification difficulties, global rules may go to a partial mediator which only controls rules that involve activity participants on a global scale, while local rules are particularly associated with specific participants, so that participants themselves can invoke appropriate local rules without consulting the global mediator. This kind of re-allocation of rules can also emphasize the profile of a role which a specific participant plays.

4.3.8 Organizational Manual - Plan Document

Organizations, as referred previously, establish an information process, integrate structures and processes in a single and flexible management system, and evaluate their activities and performance. It is common for organizations to construct either short-term (such as monthly, annual, etc. plans) time-related goals for their productivity or much longer objectives, for their growth [32]. In that sense, the structure of Plan-Document was designed to contain longer-term, reference material in the first part (Vision and objectives), followed by short-term activities at the level of the sub-groups Management, External Relations, Tools, Courseware, and Publications.

The individual group members have nodes within their groups. Each set of group nodes reflects the needs of the particular group. In the case of ToolGroup, for example, information from the longer-term part (Vision and Objectives) has been summarized in the group node (Short, Medium and Long Term Plans) for convenience, making the group document more complete and relatively self-contained. Longer-term visions have been mapped through objectives to weekly activities. The development of weekly activities is reported at an individual level, then a consolidated report is prepared for the group. Management thereby have access both to an overview of the sub-group's activities and details at the individual level.

Individuals maintain a record of their goals and accomplishments and also have access to the wider planning horizons and activities of other groups, so that they can appreciate the relevance of their efforts in the wider group context. Management is also kept informed through access to the current activities and this assists with managerial planning and decision-making. Management group meetings continue to be held to exchange news and views, which informally reinforces the more formal Plan-Document's function. In the case of ToolGroup, the Document has actually reduced the frequency of meetings.

It is among the facilitator's tasks to ensure that the document is kept up to date and to perform regular archiving of older material. This role has been given to a top manager who was mainly responsible for the administrative aspects of the work in the organization. Each group in the organization has its own manager instead of a separate facilitator. As it usually happens when most control is required, there is only one facilitator for all the groups together [31].

Another tradition in management adhered concerns management meetings. Although the application of the Plan-Document made most of the low level management meetings no longer necessary, the weekly top management meetings were retained. The participants at these meetings were: the facilitator, the leader of the organization, and the top managers. The groupware tool was "behind the scenes" at face-to-face meetings, being used exclusively by the facilitator to help him obtain an overall idea and guide the discussion.

The managerial support that the Plan-Document has the potential to offer involves the maintenance of detailed record of the activities of each and every group on a weekly basis, and also the availability of general information about the nature of each and every project. The messaging tool that the system offers could help the Plan-Facilitator to contact the members of the groups that face problems (when face-to-face or audio communication is not possible, i.e. in asynchronous modes), thus managing and resolving conflicts in an easier way. In addition, the overall information about the group activities the system can provide might help in better preparation of the agenda for the meetings, the creation of new options for discussion and allow for better structures for the decision-making to be suggested by the Plan-Facilitator.

5 Implementation - The MUCH System

MUCH has been developed primarily as a collaborative hypermedia authoring system. It is based on semantic net principles of nodes and links with a fold/unfold outline browser and navigation facility. Nodes may be created and read by all group members (subject to access permission - all members have been granted full read/write permission). Concurrent editing is automatically controlled by means of locking and releasing nodes.

Since MUCH has a hypertext structure and stores information in discrete nodes, it allows for group memory management [46]. Each group member can retrieve information in a flexible way according to her/his preference using various indexing structures available. The two most common structures used are outlines generated under the traversal-based and the lexicon-based methods [74]. Figures 2 and 3 illustrate the MUCH tool in action.

Figure 2 shows the basic MUCH system. Each line in the hierarchical browser on the left represents a node, which points to a file. A node which is preceded by an asterisk (*) is the root of a sub-tree, with one or more sub-nodes, which can themselves be roots of further sub-trees. Clicking on a root node with the mouse (right button) causes the corresponding tree of the root node to unfold to the next level. This effect is illustrated in Figure 3. In Figure 3, the MUCH System node has been expanded and then also several of its sub-nodes, including MUCHA User Guide, Author's Guide, and System Administration. These sub-nodes have themselves been expanded further, as is seen with Publishing Utilities. A node that has no asterisk is a leaf node (has no further sub-nodes of its own). The depth of the expansion is shown by the extent of the left indentation. The contents of a node are inspected by clicking the left mouse button. In Figure 3, the node that has been selected for reading is Menu Functions and it is highlighted in bold print to indicate that it has been selected.

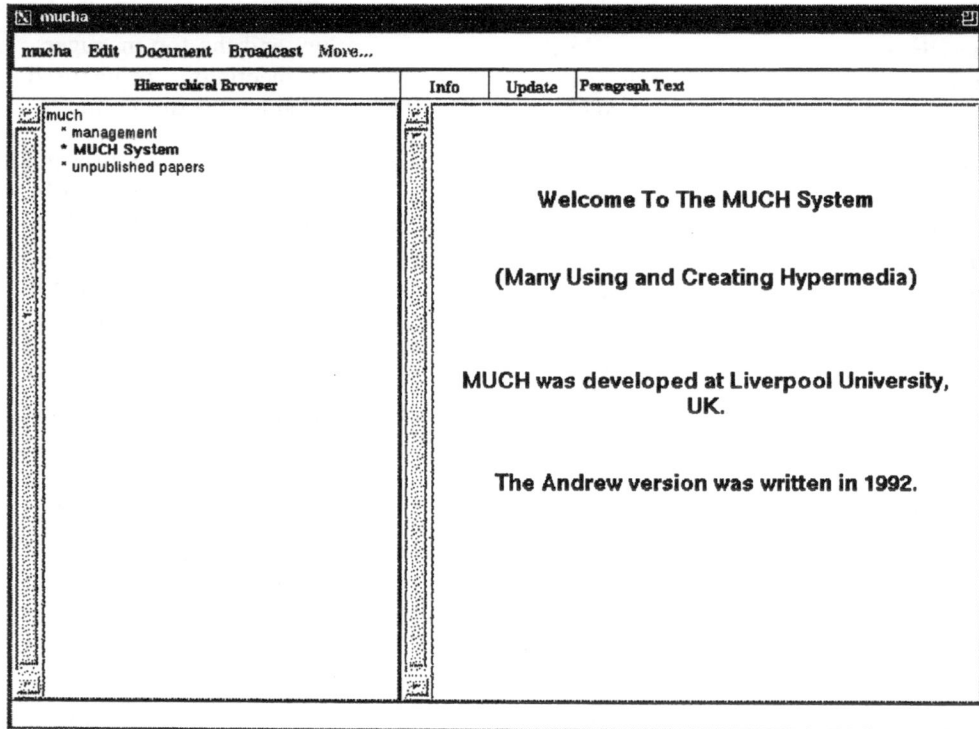

Figure 2: MUCH. The screen illustrates the top level of the outline, containing three nodes; management, MUCH System and unpublished papers.

One of the most powerful features of MUCH in a group setting is its ability to act as a central noticeboard and focal point for the group members. Although this method of communication is generally passive, in the sense of relying on individuals' self-motivation to keep up to date with the latest information , it is supplemented by the availability of a broadcast facility, which can be used in case there is a pressing requirement to notify one or more members to look at some node(s) or request an action to be performed. The broadcast message interrupts whatever task an on-line user is engaged in and requires the user to `dismiss' the message, confirming that it has been read. This treatment ensures that a message is given top priority and should be responded to quickly as compared with, for example, email, where a user might choose to leave the mailbox unread for some time. If a user is not logged on, the message is stored and re-presented at the start of the next log-in session (the broadcaster can determine who is logged on at the time of broadcasting). Broadcasting thus permits real-time communication [18].

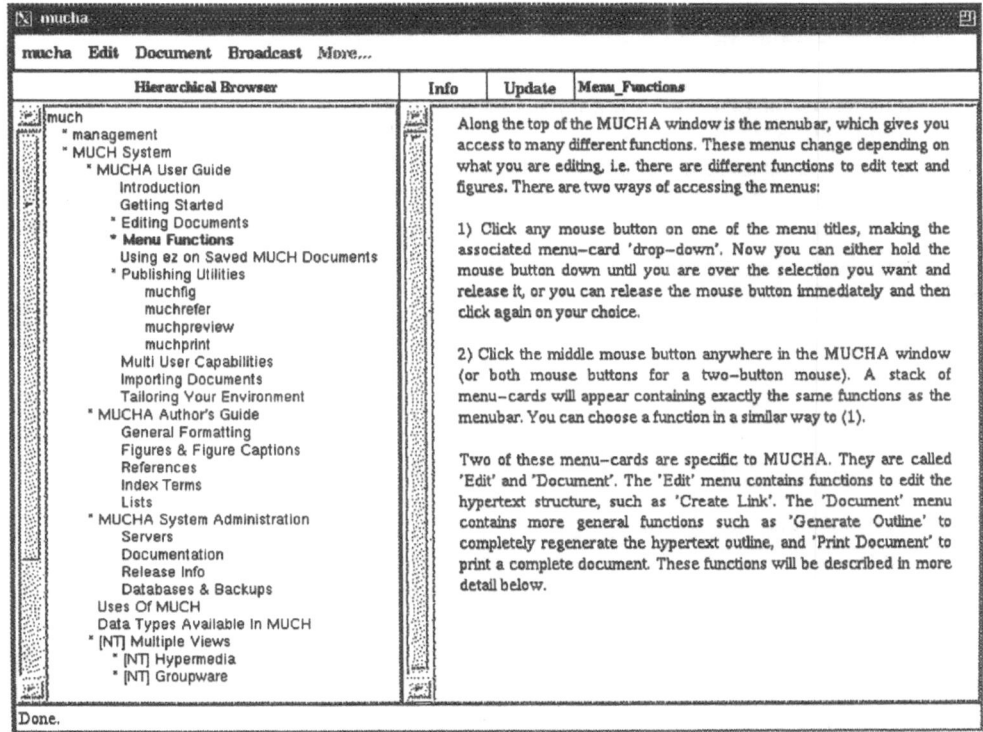

Figure 3: MUCH. The outline has been expanded to a depth of several levels.

The electronic noticeboard can provide a useful managerial function by disseminating information. Also, since members may annotate nodes, MUCH can be used for issue-based discussion. With the noticeboard, the group has a mechanism whereby members can be aware of current activities and kept up to date on their status. Additionally, they are encouraged to contribute to and participate in the direction and evolution of group activities, since the noticeboard serves also as a suggestion box for posting ideas. The disadvantage of noticeboards in general, however, is that on the one hand they can easily become cluttered and out of date. On the other hand, the opposite extreme of lack of information must be avoided, as it might give users the impression that there are no newsworthy items to report.

6 Conclusion

This chapter focused in the study of groupware support for management. Groupware is considered to be a promising collaborative technology capable of supporting managerial work in organizations by reducing the uncertainty and equivocality constraints in the information processing. The case study described involved the application of a groupware tool in coordinating group work in an R&D organization. The organizational structure was modeled according to a object-based system and it was mapped in the design of the groupware tool. This tool supported communication, planning, decision-making and other management-related needs, by both acting as an organizational memory and a focal point for the coordination of the organizational activities. The use of the system aimed at increasing the awareness of the members of the organization of their roles, the collectives goals and expectations, and the progress of the various projects. Moreover, it was the concept of facilitator of the system that was introduced as the person with the skill of handling efficiently the diversity of human and technological interactions that occur in such environments.

References

[1] Ahuja, S. R., J. R. Ensor, and D. N. Horn, The Rapport Multimedia Conferencing System, *Proceedings of the Conference on Office Information Systems,* R. Allen (ed.), ACM Press, Palo Alto, California, March 23-25, 1988, pp. 1-8.

[2] Bannon, L. and K. Schmidt, CSCW: Four Characters in Search of a Context, *Proceedings of the First European Conference on Computer-Supported Cooperative Work,* Gatwick, England, September 13-15, 1989, pp. 358-372.

[3] Bedau, H., Ethical Aspects of Group Decision Making, *Group Decision Making,* Sage Publications, W. C. Swap and Associates, Beverly Hills, 1984, pp. 115-150.

[4] Bedeian, A. G., *Management,* Dryden P., Publ., Chicago, 1989, 2nd ed.

[5] Benford, S., Requirements of Activity Management, *Proceedings of the First European Conference on Computer Supported Cooperative Work,* Gatwick, UK, September 13-15, 1989. pp. 276-286.

[6] Bittner, E., The Concept of Organization, *Social Research,* 32, 1965, pp. 239-255.

[7] Blanning, R. W., D. R. King, J. R. Marsden, and A. C. Seror, Intelligent Models of Human Organizations: The State of the Art, *Journal of Organizational Computing,* 2 (2), 1992, pp. 123-130.

[8] Broome, B. J., and M. Chen, Guidelines for Computer-Assisted Group Problem Solving: Meeting the Challenges of Complex Issues, *Small Group Research,* 23 (2), May 1992, pp. 216-236.

[9] Cashman, P. M., and D. Stroll, Developing the Management Systems of the 1990s: The Role of Collaborative Work, *Technological Support for Work Group Collaboration,* M. H. Olson (ed.), Lawrence Erlbaum Associates, Publishers, Hillsdale, N.J., 1989, pp. 129-146.

[10] Chen, M., Y. I. Liou, and E. S. Weber, Developing Intelligent Organizations: A Context-Based Approach to Individual and Organizational Effectiveness, *Journal of Organizational Computing,* 2 (2), 1992, pp. 181-202.

[11] Ciborra, C., and M. Olson, Encountering Electronic Work Groups, *Proceedings of the Conference on Computer-Supported Cooperative Work,* Portland, Oregon, September 26-29, 1988, pp. 94-101.

[12] Conklin, J. and M. Begeman, gIBIS: A Hypertext Tool for Team Design Deliberation, *Proceedings Hypertext '87,* 1987, pp. 247-252.

[13] Conklin, J., Hypertext: An Introduction and Survey, *Computer,* 20 (9), September 1987, pp. 17-41.

[14] Conrath, D. W., Communications Environment and Its Relationship to Organizational Structure, *Management Science,* 20 (4), December 1973, Part II, pp. 586-603.

[15] Corcoran, E., She Incites Revolutions with Chips and Networks, Lynn Conway; Collaboration Technology or Groupware, *IEEE Spectrum,* 24 , December 1987, pp. 46-51.

[16] Daft, R. L. and R. H. Lengel, Organizational Information Requirements, Media Richness, and Structural Design, *Management Science,* 32 (5), May 1986, pp. 554-571.

[17] Daft, R. L., R. H. Lengel, and L. K. Trevino, Message Equivocality, Media Selection, and Manager Performance: Implications for Information Systems, *MIS Quarterly,* 11 (3), September 1987, pp. 355-366.

[18] Deakin, A., P. I. Gouma, and R. Rada, The Plan-Facilitator and the Plan Document: A New Aspect of Computer Supported Management, to appear in *Journal of Intelligent Systems,* a Special Issue on Computer Supported Cooperative Work: Towards an Integration of the Social and Technical, December 1993.

[19] De Cindio, F., G. De Michelis, C. Simone, R. Vassallo, and A. Zanaboni, CHAOS as a Coordination Technology, *Proceedings of the Conference on Computer-Supported Cooperative Work (CSCW '86),* ACM Press, Austin, Texas, December 3-5, 1986, pp. 325-342.

[20] de Jong, P., A Framework for the Development of Distributed Organizations, *Journal of Organizational Computing,* 2 (1), 1992, pp. 77-94.

[21] Dennis, A. R., J. F. George, L. M. Jessup, J. F. Nunamaker Jr., and D. R. Vogel, Information Technology to Support Electronic Meetings, *MIS Quarterly,* 12 (4),

December 1988, pp. 591-619.

[22] DeSanctis, G. and R. B. Gallupe, A Foundation for the Study of Group Decision Support Systems, *Management Science,* 33 (5), May 1987, pp. 589-609.

[23] Dhar, V. and M. H. Olson, Assumptions Underlying Systems that Support Work Group Collaboration, *Technological Support for Work Group Collaboration,* M. H. Olson (ed.), Lawrence Erlbaum Associates, 1989, pp. 33-50.

[24] Dubs, S. and S. C. Hayne, Distributed Facilitation: A Concept whose Time has Come?, *Proceedings of the Conference of Computer-Supported Cooperative Work (CSCW '92),* Toronto, Canada, October 31 - November 4, 1992, pp. 314-321.

[25] Ellis, C. A., S. J. Gibbs, and G. L. Rein, Groupware: Some Issues and Experiences, *Communications of the ACM,* 34 (1), January 1991, pp. 39-58.

[26] Engelbart, D. and H. Lehtman, Working Together, *BYTE,* December 1988.

[27] Eveland, J. D. and T. K. Bikson, Work Group Structures and Computer Support: A Field Experiment, *Proceedings of the Conference on Computer-Supported Cooperative Work (CSCW '88),* ACM Press, Portland, Oregon, September 26-28, 1988, pp. 324-343.

[28] Finholt, T., L. Sproull, and S. Kiesler, Communication and Performance in Ad Hoc Task Groups, *Intellectual Teamwork: Social Foundations of Cooperative Work,* J. Galegher, R. E. Kraut, and C. Egido (eds.), Lawrence Erlbaum Associates, Hilsdale, New Jersey, 1990, pp. 291-326.

[29] Finlay, P. and C. Marples, A Review of Group Decision Support Systems, *OR Insight,* 4 (4), October -December 1991.

[30] Flores, F., M. Graves, B. Hartfield, and T. Winograd, Computer Systems and the Design of Organizational Interaction, *ACM Transactions on Office Information Systems,* 6, 1988, pp. 153-157.

[31] Friedman, P. G., Upstream Facilitation: A Proactive Approach to Managing Problem-Solving Groups, *Management Communication Quarterly,* 3 (1), August 1989, pp. 33-50.

[32] Gersick, C. J. G., Time and Transition in Work Teams: Toward a New Model of Group Development, *Academy of Management Journal,* 31 (1), 1988, pp. 9-41.

[33] Gibbs, S. J., LIZA: An Extensible Groupware Toolkit, *Proceedings of the ACM SIGCHI Conference on Human Factors in Computing Systems,* Austin, Texas, April 30 - May 4, ACM Press, New York, 1989, pp. 29-35.

[34] Gouma, P. I., A. Deakin, and R.Rada, Group Management and the MUCH System: An Evaluation of the Plan-Document, *Technical Report,* University of Liverpool, January 1993.

[35] Greenbaum, J. S., *In the Name of Efficiency: Management Theory and Shopfloor Practice in Data-Processing Work,* Temple University Press, Philadelphia, 1979.

[36] Greenberg, S. (ed.), *Computer-Supported Collaborative Work and Groupware,* Academic Press, London, 1991.

[37] Greenberg, S., Personalizable Groupware: Accommodating Individual Roles and Group Differences, *Proceedings of the Second European Conference on Computer-Supported Cooperative Work (EC-CSCW '91),* L. Bannon, M. Robinson, and K. Schmidt (eds.), Kluwer, Amsterdam, September 24-27, 1991, pp. 17-31.

[38] Grief, I. (ed.), *Computer-Supported Cooperative Work: A Book of Readings,* Morgan Kaufmann, San Mateo, California, 1988.

[39] Gurbaxani, V. and E. Shi, Computers and Coordination in Manufacturing, *Journal of Organizational Computing,* 2 (1), 1992, pp. 27-46.

[40] Hayes, F., The Groupware Dilemma, *Unix World,* 9 (2), February 1992, pp. 46-50.

[41] Heath, C. and P. Luff, Disembodied Conduct: Communication through Video in a Multi-Media Environment, *Proceedings of the ACM SIGCHI Conference on Human Factors in Computing Systems, CHI '91,* New Orleans, Luisiana, April 28-May 2, 1991, pp. 99-103.

[42] Hsu, J. and T. Lockwood, Collaborative Computing: Computer-Aided Teamwork will Change your Office Culture Forever, *Byte,* 18 (3), March 1993, pp. 113-120.

[43] Isenberg, D. J., How Senior Managers Think, *Harvard Business Review,* November-December 1984, pp. 81-90.

[44] Islei, G., G. Lockett, B. Cox, and M. Stratford, A Decision Support System Using Judgmental Modeling: A Case of R&D in the Pharmaceutical Industry", *IEEE Transactions on Engineering Management,* 38 (3), August 1991, pp. 202-209.

[45] Jirotka, M., N. Gilbert, and P. Luff, On the Social Organisation of Organisations, *Computer Supported Collaborative Work - An International Journal,* 1 (1-2), 1992, pp. 95-118.

[46] Johansen, R., *GROUPWARE: Computer Support for Business Teams,* The Free Press, 1988.

[47] Keltner, J. S., Facilitation: Catalyst for Group Problem Solving, *Management Communication Quarterly,* 3 (1), August 1989, pp. 8-32.

[48] Kling, R., Cooperation, Coordination and Control in Computer-Supported Work", *Communications of the ACM,* 34 (12), December 1991, pp. 83-88.

[49] Knight, D. O. and M. L. Wall, Using Group Technology for Improving Communication and Coordination among Teams of Workers in Manufacturing Cells, *Industrial Engineering,* 21 (1), 1989, pp. 28-34.

[50] Kotter, J. P., What Effective General Managers Really Do, *Harvard Business Review,* November-December 1982, pp. 156-167.

[51] Kraemer, K. L. and J. L. King, Computer-Based Systems for Cooperative Work and Group Decision Making, *ACM Computing Surveys,* 20 (2), June 1988, pp. 115-146.

[52] Krasner, H., J. McInroy, and D. B. Walz, Groupware Research and Technology Issues with Application to Software Process Management, *IEEE Transactions On Systems, Man, and Cybernetics,* 21 (4), July-August 1991, pp. 704-712.

[53] Lockwood, G. and J. Fielden, *Planning and Management in Universities: A study of British Universities,* CHATTO & WINDUS, Sussex University Press, 1973.

[54] Lotus Development Corporation, *A Quick Tour of Notes,* Cambridge, Massachusetts: Lotus Development Corporation, 1991.

[55] Mackay, W., More than just a Communication System: Diversity in the Use of Electronic Mail, *Proceedings of the Conference on Computer-Supported Cooperative Work (CSCW '88),* Portland, Oregon, September 26-29,1988, pp. 344-353.

[56] Malone, T. W., K. R. Grant, F. A. Turbak, S. B. Brobst, and M. D. Cohen, Intelligent Information Sharing-Systems, *Communications of the ACM,* 30 (5), May 1987, pp. 390-402.

[57] Malone, T. W. and K. Crowston, What Is Coordination Theory and how Can It Help Design Cooperative Work Systems, *Proceedings of the third conference on Computer-Supported Cooperative Work (CSCW '90),* Los Angeles, California, October 8-10, ACM Press, New York, 1990, pp. 357-370.

[58] Martinko, M. J. and W. L. Gardner, Structured Observation of Managerial Work: A Replication and Synthesis, *Journal of Management Studies,* 27 (3), May 1990, pp. 329-357.

[59] McGrath, J. E., Small Group Research, *American Behavioral Scientist,* 21 (5), May-June 1978, pp. 651-674.

[60] Mintzberg, H., *The Nature of Managerial Work,* Harper & Row, Publishers, 1973.

[61] Mintzberg, H., Managerial Work: Analysis from Observation, *Management Science,* 18 (2), October 1971, pp. 97-110.

[62] Michailidis, A., R. Rada, and W. Wang, Matching Roles and Technology for Collaborative Work: An Empirical Assessment, *Wirtschaftsinformatik,* 35 (2), 1993, pp. 138-148.

[63] Motiwalla, L. and J. F. Nunamaker Jr., Mail-Man: A Knowledge-Based MAIL Assistant for Managers, *Journal of Organizational Computing,* 2 (2), 1992, pp. 131-154.

[64] Novick, D. G. and J. Walpole, Enhancing the Efficiency of Multiparty Interaction through Computer Mediation, *Interacting with Computers,* 2 (2), 1990, pp. 229-246.

[65] Nunamaker, J. F., A. R. Dennis, J. S. Valacich, D. R. Vogel, and J. F. George, Electronic Meeting Systems to Support Group Work, *Communications of the ACM,* 34 (7), July 1991, pp. 40-61.

[66] Opper, S. and H. Fersko-Weiss, *Technology for Teams: Enhancing productivity in Networked Organizations,* Van Nostrand Reinholt, 1992.

[67] Panko, R. R., Managerial Communication Patterns, *Journal of Organizational Computing,* 2 (1), 1992, pp. 95-122.

[68] Perin, C., Electronic Social Fields in Bureaucracies, *Communications of the ACM,* 34 (12), December 1991, pp. 75-82.

[69] Phillipakis, A. and M. Goul, Concepts and Models of Group Membership in Computer-Supported Knowledge and Decision Tasks, *Journal of Organizational Computing,* 2 (3-4), 1992, pp. 243-262.

[70] Platt, S., R. Peipe, and J. Smyth, *Teams: A Game to Develop Group Skills,* Gower, 1988.

[71] Ploszajski, G., M. G. Singh, and K. S. Hindi, An Overview of Some Computer-Aided Production Management Issues, *Information and Decision Technologies,* 18, 1993, pp. 405-413.

[72] Poole, M. S., An Information Task Approach to Organizational Communication,

Academic Management Review, 30, 1978, pp. 493-504.

[73] Rada, R., *Hypertext: From Text to Expertext,* McGraw-Hill, 1991.

[74] Rada, R. and G. You, Balanced Outlines and Hypertext, *Journal of Documentation,* 48 (1), March 1992, pp. 20-44.

[75] Rice, R. E. and D. E. Shook, Relationships of Job Categories and Organizational Levels to Use of Communication Channels, Including Electronic Mail: A Meta-Analysis and Extension, *Journal of Management Studies,* 27 (2), March 1990, pp. 195-229.

[76] Rittel, H. and M. Webber, Dilemmas in a General Theory of Planning, *Policy Sciences,* 4, 1973.

[77] Rodden, T., A Survey of CSCW Systems, *Inteacting with Computers,* 3 (3), 1991, pp. 319-353.

[78] Rodden, T., Technological Support for Cooperation, *CSCW in Practice: An Introduction and Case Studies,* D. Diaper and C. Sanger (eds.), Springer-Verlag, 1993, pp. 1-22.

[79] Root, W. R., Design of a Multi-Media Vehicle for Social Browsing, *Proceedings of the Conference on Computer-Supported Cooperative Work (CSCW '88),* ACM Press, Portland, Oregon, September 26-28, 1988, pp. 25-38.

[80] Schmidt, K., Riding a Tiger or Computer Supported Collaborative Work", *Proceedings of the Second European Conference on Computer-Supported Cooperative Work (ECCSCW '91),* L. Bannon, M. Robinson, and K. Schmidt (eds.), Kluwer, Amsterdam, September 24-27, 1991, pp. 1-16.

[81] Searle, J. R., *Speech Acts: An Essay in the Philosophy of Language,* Cambridge University Press, 1969.

[82] Singh, B., Invited Talk on Coordination Systems at the *Organizational Computing Conference,* November 13-14, 1989, Austin, Texas.

[83] Smith, H. T., P. A. Hennessy, and G. Lunt, The Activity Model Environment: An Object Oriented Framework for Describing Organisational Communication, *Proceedings of the First European Conference on Computer Supported Cooperative Work (ECSCW '89),* Gatwick, U.K, 1989, pp. 160-172.

[84] Smith, C. L., *Integrating Groupware into a Professional Services Firm: Findings from an Electronic Survey,* Unpublished M.Sc. Thesis, Massachusetts Institute of Technology, 1992.

755

[85] Sproull, L. and S. Kiesler, *Connections: New Ways of Working in the Networked Organization,* MIT Press, 1991.

[86] Valacich, J. S., A. R. Dennis, and J. F. Nunamaker Jr., Electronic Meeting Support: The GroupSystems Concept, *International Journal of Man Machine Studies,* 34 (2), 1991, pp. 262.

[87] Viller, S., *Computer Support for Group Facilitators: An Investigation,* Unpublished M.Sc. thesis, University of Manchester, 1990.

[88] Viller, S., The Group Facilitator: A CSCW Perspective, *Proceedings of the Second European Conference on Computer-Supported Cooperative Work (ECSCW '91),* L. Bannon, M. Robinson, and K. Schmidt (eds.), Kluwer, Amsterdam, September 24-27, 1991, pp. 81-95.

[89] Vroom, V. H., *Methods of Organizational Research,* University of Pittsburgh Press, 1968.

[90] Watabe, K., S. Sakata, K. Maeno, H. Fukuoka, and T. Ohmori, Distributed Multiparty Desktop Conferencing System: Mermaid, *Proceedings of the Conference on Computer Supported Cooperative Work (CSCW '90),* ACM Press, Los Angeles, California, 1990, pp. 27-38.

[91] Winograd, T., Where the Action Is, *BYTE,* December 1988, pp. 256-259.

[92] Woo, C. C. and M. K. Chang, An Approach to Facilitate the Automation of Semistructured and Recurring Negotiations in Organizations, *Journal of Organizational Computing,* 2 (1), 1992, pp. 47-76.

INDEX

758